T0180953

Lecture Notes in Computer Science   13112

More information about this subseries at https://link.springer.com/bookseries/7409

Michal Feldman · Hu Fu ·
Inbal Talgam-Cohen (Eds.)

Web and Internet Economics

17th International Conference, WINE 2021
Potsdam, Germany, December 14–17, 2021
Proceedings

 Springer

Editors
Michal Feldman 🅳
Tel Aviv University
Tel Aviv, Israel

Inbal Talgam-Cohen 🅳
Technion – Israel Institute of Technology
Haifa, Israel

Hu Fu
Shanghai University of Finance
and Economics
Shanghai, China

ISSN 0302-9743   ISSN 1611-3349   (electronic)
Lecture Notes in Computer Science
ISBN 978-3-030-94675-3   ISBN 978-3-030-94676-0   (eBook)
https://doi.org/10.1007/978-3-030-94676-0

LNCS Sublibrary: SL3 – Information Systems and Applications, incl. Internet/Web, and HCI

This Springer imprint is published by the registered company Springer Nature Switzerland AG
The registered company address is: Gewerbestrasse 11, 6330 Cham, Switzerland

Preface

This volume contains all regular papers and abstracts presented at the 17th Conference on Web and Internet Economics (WINE 2021). WINE 2021 was held as an Internet event during December 14–17, 2021, organized at the Hasso Plattner Institute, Potsdam, Germany.

Over the last 17 years, the WINE conference series has become a leading interdisciplinary forum for the exchange of ideas and scientific progress across continents on incentives and computation arising in diverse areas, such as theoretical computer science, artificial intelligence, economics, operations research, and applied mathematics. WINE 2021 built on the success of previous editions of WINE (named the Workshop on Internet and Network Economics until 2013) which were held annually from 2005 to 2020.

We tried for the first time having a Senior Program Committee, which was composed of 27 researchers from the field; the Program Committee had 63 researchers. The committees reviewed 136 submissions and decided to accept 41 papers. Each paper had at least three reviews, with additional reviews solicited as needed. We are very grateful to all members of the Senior Program Committee and the Program Committee for their insightful reviews and discussions. We thank EasyChair for providing a virtual platform to organize the review process. We also thank Springer for publishing the proceedings and offering support for Best Paper and Best Student Paper Awards.

In addition to the contributed talks, the program included four invited talks by leading researchers in the field: Eli Ben Sasson (Technion – Israel Institute of Technology), Colin Camerer (California Institute of Technology, USA), Annie Liang (Northwestern University, USA), and Jenn Wortman Vaughan (Microsoft Research, USA).

Also for the first time we invited Spotlights Beyond WINE talks; the nomination and selection of these talks was adjudicated by Costis Daskalakis, Fuhito Kojima, Ruta Mehta, and Jamie Morgenstern.

Our special thanks go to the general chairs, Ágnes Cseh and Pascal Lenzner, and the local organization team, as well as the poster chair Gagan Aggarwal and the global outreach chair Francisco Marmolejo.

November 2021

Michal Feldman
Hu Fu
Inbal Talgam-Cohen

Organization

General Chairs

Ágnes Cseh   Hasso Plattner Institute, Germany
Pascal Lenzner   Hasso Plattner Institute, Germany

Program Committee Chairs

Michal Feldman   Tel-Aviv University, Israel
Hu Fu   Shanghai University of Finance and Economics, China
Inbal Talgam-Cohen   Technion – Israel Institute of Technology, Israel

Steering Committee

Xiaotie Deng   Peking University, China
Paul Goldberg   University of Oxford, UK
Christos Papadimitriou   Columbia University, USA
Paul Spirakis   University of Liverpool, UK
Rakesh Vohra   University of Pennsylvania, USA
Andrew Yao   Tsinghua University, China
Yinyu Ye   Stanford University, USA

Senior Program Committee

Itai Ashlagi   Stanford University, USA
Omar Besbes   Columbia University, USA
Liad Blumrosen   Hebrew University, Israel
Ioannis Caragiannis   Aarhus University, Denmark
Jose Correa   Universidad de Chile, Chile
Paul Duetting   Google Research, Switzerland
Amos Fiat   Tel-Aviv University, Israel
Nick Gravin   Shanghai University of Finance and Economics, China
Amy R. Greenwald   Brown University, USA
Thomas Kesselheim   University of Bonn, Germany
Scott Kominers   Harvard University, USA
Elias Koutsoupias   University of Oxford, UK
Ron Lavi   Technion – Israel Institute of Technology, Israel,
                                 and University of Bath, UK
Stefano Leonardi   Sapienza Università di Roma, Italy
Tracy Xiao Liu   Tsinghua University, China
Brendan Lucier   Microsoft Research, USA
Vahideh Manshadi   Yale University, USA

Thanh Nguyen	Purdue University, USA
Noam Nisan	Hebrew University, Israel
Malesh Pai	Rice University, USA
Renato Paes Leme	Google Research, USA
Ariel Procaccia	Harvard University, USA
Daniela Saban	Stanford University, USA
Rann Smorodinsky	Technion – Israel Institute of Technology, Israel
Éva Tardos	Cornell University, USA
Adrian Vetta	McGill University, Canada
Gabriel Weintraub	Stanford University, USA

Program Committee

Saeed Alaei	Google Research, USA
Amine Allouah	Columbia University, USA
Eric Balkanski	Columbia University, USA
Hedyeh Beyhaghi	Toyota Technological Institute at Chicago, USA
Kshipra Bhalwalkar	Google Research, USA
Arpita Biswas	Harvard University, USA
Yang Cai	Yale University, USA
Andrea Celli	Facebook Core Data Science, UK
Yu Cheng	Microsoft Research, USA
Avi Cohen	Tel-Aviv University, Israel
Yuan Deng	Google Research, USA
Tomer Ezra	Tel-Aviv University, Israel
Yiding Feng	Northwestern University, USA
Aris Filos-Ratsikas	University of Liverpool, UK
Rupert Freeman	University of Virginia, USA
Federico Fusco	Sapienza Università di Roma, Italy
Nikhil Garg	Cornell Tech, USA
Ganesh Ghalme	Technion – Israel Institute of Technology, Israel
Yiannis Giannakopoulos	Friedrich-Alexander-Universität Erlangen-Nürnberg (FAU), Germany
Vasilis Gkatzelis	Drexel University, USA
Swati Gupta	Georgia Tech, USA
Nima Haghpanah	Pennsylvania State University, USA
Xin Huang	Technion – Israel Institute of Technology, Israel
Shahin Jabbari	Harvard University, USA
Anson Kahng	Carnegie Mellon University, USA
Max Klimm	Technische Universität Berlin, Germany
Kostas Kollias	Google Research, USA
Yuqing Kong	Peking University, China
Moran Koren	Tel-Aviv University, Israel
Annamaria Kovaks	University of Frankfurt, Germany
Ron Kupfer	Hebrew University, Israel
Irene Lo	Stanford University, USA

Jiaqi Lu	Columbia University, USA
Will Ma	Columbia University, USA
Jieming Mao	Google Research, USA
Simon Mauras	Research Institute on the Foundations of Computer Science, France
Divyarthi Mohan	Princeton University, USA
Ioannis (Yannis) Panageas	University of California, Irvine, USA
Chara Podimata	Harvard University, USA
Emmanouil Pountourakis	Drexel University, USA
Alexandros Psomas	Purdue University, USA
Qi Qi	Hong Kong University of Science and Technology, China
Manish Raghavan	Cornell University, USA
Nidhi Rathi	Aarhus University, Denmark
Rebecca Reiffenhauser	Sapienza Università di Roma, Italy
Fedor Sandomirskiy	California Institute of Technology, USA
Jon Schneider	Google Research, USA
Okke Schrijvers	Facebook, USA
Shreyas Sekar	University of Toronto, Canada
Ran Shorrer	Pennsylvania State University, USA
Sujoy Sikdar	Binghamton University, USA
Sahil Singla	Princeton University and Institute for Advanced Study, USA
Balu Sivan	Google Research, USA
Piotr Skowron	University of Warsaw, Poland
Warut Suksompong	National University of Singapore, Singapore
Zhihao Tang	Shanghai University of Finance and Economics, China
Laura Vargas-Koch	RWTH Aachen University, Germany
Ellen Vitercik	Carnegie Mellon University, USA
Zihe Wang	Renmin University, China
Haifeng Xu	University of Virginia, USA
Fang-Yi Yu	Harvard University, USA
Manolis Zampetakis	University of California, Berkeley, USA
Mingfei Zhao	Yale University, USA

Additional Reviewers

Jerry Anunrojwong
Omer Ben-Porat
Ben Berger
Martin Bullinger
Linda Cai
George Christodoulou
Riccardo Colini Baldeschi
Natalie Collina

Thirumulanathan D.
Argyrios Deligkas
Antoine Desir
Maya Dotan
Katharina Eickhoff
Hadi Elzayn
Meryem Essaidi
Alireza Farhadi

Gabriele Farina
Zhe Feng
Giannis Fikioris
Yotam Gafni
Jugal Garg
Ira Globus-Harris
Denizalp Goktas
Minbiao Han
Gregory Herschlag
Cyrus Hettle
Kazuyuki Higashi
Martin Hoefer
Alexandros Hollender
Chien-Chung Huang
Ayumi Igarashi
Shweta Jain
Michał Jaworski
Yaonan Jin
Christopher Jung
Fivos Kalogiannis
Martin Knaack
Anand Krishna
Pooja Kulkarni
Rucha Kulkarni
Omer Lev
Bo Li
Weian Li
Jason Cheuk Nam Liang
Shuze Liu
Ali Makhdoumi
Azarakhsh Malekian
Aghaheybat Mammadov
Claire Mathieu
Stephen McAleer
Kaleigh Mentzer
Julian Mestre
Hassan Mortagy
Omar Mouchtaki
Shivika Narang
Eric Neyman

Argyris Oikonomou
William Parker
Neel Patel
Dominik Peters
Grzegorz Pierczyński
Tristan Pollner
Sharon Qian
Frederick Qiu
Mayank Ratan Bhardwaj
Niklas Rieken
Scott Rodilitz
Jad Salem
Daniel Schoepflin
Ariel Schvartzman
Seyed Ali Shameli
Ravi Sojitra
Gogulapati Sreedurga
Maximilian Stahlberg
Samuel Taggart
Yifeng Teng
Ece Teoman
Clayton Thomas
Robin Vacus
Carmine Ventre
Alexandros Voudouris
Jonathan Wagner
Chang Wang
Hongao Wang
Kangning Wang
Philipp Warode
Matt Weinberg
Andre Wibisono
Jibang Wu
Xiaowei Wu
Mobin Yahyazadeh
Hanrui Zhang
Qiankun Zhang
Yuhao Zhang
Shuran Zheng

Contents

Matching, Markets and Equilibria

Learning, Fairness, Privacy and Behavioral Models

Social Choice and Cryptocurrencies

Abstracts

Mechanism Design and Pricing

Two-Way Greedy: Algorithms
for Imperfect Rationality

Diodato Ferraioli[1], Paolo Penna[2], and Carmine Ventre[3(✉)]

[1] Università di Salerno, Via Giovanni Paolo II, 132, 84084 Fisciano, SA, Italy
`dferraioli@unisa.it`
[2] ETH Zürich, Rämistrasse 101, 8092 Zürich, Switzerland
`paolo.penna@inf.ethz.ch`
[3] King's College London, Strand, London WC2R 2LS, UK
`carmine.ventre@kcl.ac.uk`

Abstract. The realization that selfish interests need to be accounted for in the design of algorithms has produced many interesting and valuable contributions in computer science under the general umbrella of algorithmic mechanism design. Novel algorithmic properties and paradigms have been identified and studied in the literature. Our work stems from the observation that selfishness is different from rationality; agents will attempt to strategize whenever they perceive it to be convenient according to their imperfect rationality. Recent work in economics [18] has focused on a particular notion of imperfect rationality, namely absence of contingent reasoning skills, and defined obvious strategyproofness (OSP) as a way to deal with the selfishness of these agents. Essentially, this definition states that to care for the incentives of these agents, we need not only pay attention about the relationship between input and output, but also about the way the algorithm is run. However, it is not clear to date what algorithmic approaches ought to be used for OSP. In this paper, we rather surprisingly show that, for binary allocation problems, OSP is fully captured by a natural combination of two well-known and extensively studied algorithmic techniques: forward and reverse greedy. We call two-way greedy this underdeveloped algorithmic design paradigm.

Our main technical contribution establishes the connection between OSP and two-way greedy. We build upon the recently introduced cycle monotonicity technique for OSP [9]. By means of novel structural properties of cycles and queries of OSP mechanisms, we fully characterize these mechanisms in terms of extremal implementations. These are protocols that ask each agent to consistently separate one extreme of their domain at the current history from the rest. Through the natural connection with the greedy paradigm, we are able to import a host of known approximation bounds to OSP and strengthen the strategic properties of this family of algorithms. Finally, we begin exploring the full power of two-way greedy (and, in turns, OSP) in the context of set systems.

Diodato Ferraioli is supported by GNCS-INdAM and the Italian MIUR PRIN 2017 Project ALGADIMAR "Algorithms, Games, and Digital Markets". Carmine Ventre acknowledges funding from the UKRI Trustworthy Autonomous Systems Hub (EP/V00784X/1).

M. Feldman et al. (Eds.): WINE 2021, LNCS 13112, pp. 3–21, 2022.
https://doi.org/10.1007/978-3-030-94676-0_1

1   Introduction

An established line of work in computer science recognizes the important role played by self interests. If ignored, these self interests can misguide the algorithm or protocol at hand and lead to suboptimal outcomes. Mechanism design has emerged as the framework of reference to deal with this selfishness. Mechanisms are protocols that interact with the selfish agents involved in the computation; the information elicited through this interaction is used to choose a certain outcome (via an algorithm). The goal of a mechanism is that of reconciling the potentially contradictory aims of agents with that of the designer (i.e., optimize a certain objective function). The agents attach a utility (typically defined as quasi-linear function of the *transfers* defined by the mechanism and the agent's *type* – i.e., cost or valuation – for the solution) to each outcome and are therefore incentivized to force the output of an outcome that maximizes their utility (rather than maximizing the objective function). The quality of a mechanism is assessed against how well it can approximate the objective function whilst giving the right incentives to the agents.

In this context, one seeks to design *strategyproof (SP) mechanisms*—these guarantee that agents will not strategize as it will be in their best interest to adhere to the rules set by the mechanism—and aims to understand what is the best possible approximation that can be computed for the setting of interest. For example, it is known how for *utilitarian* problems (roughly speaking, those whose objective function is the sum of all the agents' types) it is possible to simultaneously achieve optimality and strategyproofness, whilst some non-utilitarian objective (such as, min-max) cannot be approximated too well (irrespectively of computational considerations), see, e.g., [21]. These results can be proved purely from an algorithmic perspective – that ignores incentives and selfishness – in that it is known how strategyproofness is equivalent to a certain *monotonicity property* of the algorithm used by the mechanism to compute the outcome. This monotonicity relates the outcomes of two instances, connected by SP constraints, and limits what the algorithm can do on them. For example, if an agent is part of the solution computed on instance I and becomes "better" (e.g., faster) in instance I' then the algorithm must select the agent also in the solution returned for instance I', all other things unchanged.

Recent research in mechanism design has highlighted how cognitive limitations of the agents might mean that SP is too weak a desideratum for mechanisms. Even for the simplest setting of one-item auction, there is experimental evidence that people strategize against the sealed-bid implementation of second-price auction, whilst ascending-price auction seems easier to grasp [1,14]. The concept of *obvious strategyproofness* (OSP) has been defined in [18] to capture this particular form of imperfect rationality, which is shown to be equivalent to the absence of contingent reasoning skills. Intuitively, for an agent it is obvious to understand if a strategy is better than another in that the worst possible outcome for the former is better than the best one for the latter.

Can we, similarly to SP, derive bounds on the quality of OSP mechanisms that are oblivious to strategic considerations?

There are two obstacles to getting a fully algorithmic approach to OSP mechanisms due to their structure. Whereas SP mechanisms are pairs comprised of an algorithm and transfer (a.k.a., payment) function, in OSP we have a third component (the so-called *implementation tree*) which encapsulates the execution details (e.g., sealed bid vs ascending price) of the mechanisms and the obviousness of the strategic constraints. (For OSP, in fact, the implementation details matter and the classical Revelation Principle does not hold [18].) A technique, known as cycle monotonicity (CMON), allows to express the existence of SP payments for an algorithm in terms of the weight of the cycles in a suitably defined graph. Specifically, it is known that it is sufficient to look at cycles of length two for practically all optimization problems of interest [22]—this yields the aforementioned property of monotone algorithms. Recent work [8,9] extends CMON to OSP and allows to focus only on algorithms and implementation trees. Whilst this has allowed some progress towards settling our main question in the context of single-parameter agents, some unsatisfactory limitations are still present. Firstly, handling two interconnected objects, namely algorithm and implementation tree, simultaneously is hard to work with: e.g., novel ad-hoc techniques (dubbed CMON two-ways in [9]) had to be developed to prove lower bounds. Secondly, the CMON extension to OSP is shown to require the study of cycles of any length, thus implying that the "monotonicity" of the combination algorithm/implementation tree needs to hold amongst an arbitrary number of instances, as opposed to two as in the case of SP. Thirdly, the mechanisms constructed in [8,9] only work for three-value domains since they rely on the simpler two-instance monotonicity (referred to as monotonicity henceforth).

Our Contributions. The technical challenge left open by previous work was to relate monotonicity to many-instance monotonicity. In this paper, we solve this challenge by providing a characterization of OSP mechanisms for binary allocation problems (for which the outcome for each agent is either to be selected or not). This enables us to show that the shape of the implementation tree is essentially fixed and answer the question above in the positive. It turns out that the exact algorithmic structure of OSP mechanisms is intimately linked with a (slight generalization of a) well known textbook paradigm:

OSP can be achieved if and only if the algorithm is two-way greedy.

What does it mean for an algorithm to be two-way greedy? The literature in computer science and approximation algorithms has extensively explored what we call *forward greedy*. These are algorithms that use a (possibly adaptive) (in-)priority function and incrementally build up a solution by adding therein the agent with the highest priority, if this preserves feasibility. It is known that if the priority rule is monotone in each agent's type then this leads to a SP direct-revelation mechanism (see, e.g., [17]). What we show here is that the strategyproofness guarantee is actually much stronger and can deal with imperfect rationality. This is achieved with a simple implementation of forward greedy that sweeps through each agent's domain from the best possible type to the worst. Another relevant approach known in the literature is Deferred Acceptance auctions (DAAs) or reverse greedy algorithms [7,12]. These use a (possibly adaptive) (out-)priority function and build a feasible solution by incrementally throwing out the agents whose type is not good enough with respect the current priority (i.e., whose cost (valuation) is higher (lower) than the out-priority) until a feasible solution is found. It is already known that DAAs are OSP [20] but not the extent to which focusing on them would be detrimental to finding out the real limitations of OSP mechanisms. Two-way greedy algorithms *combine in- and out-priorities*; each agent faces either a greedy in- or out-priority; in the former case, they are included in the solution if feasibility is preserved while in the latter they are excluded from it if the current solution is not yet feasible. The direction faced can depend on which agents have been included in or thrown out from the eventual solution at that point of the execution; in this sense, these are particular adaptive priority algorithms. For a formal definition, please see Sect. 4 and Algorithm 3.

Two-way greedy algorithms stem from our characterization of OSP mechanisms in terms of "extremal implementation trees"; roughly speaking, in these mechanisms we always query each agents about *(the same) extreme* of their domain at the current history. To prove this characterization, we first give a couple of structural properties of OSP mechanisms. We specifically show (i) when a query can be made to guarantee OSP; and, (ii) how a mechanism that is monotone but not many-instance monotone looks like. We use the former property to show that, given an OSP mechanism, we can modify the structure of its implementation tree to make it extreme whilst guaranteeing that the many-instance monotonicity is preserved (i.e., the structure (ii) is not possible). We also show that extremal mechanisms are monotone and that structure (ii) can never arise, thus proving the sufficient condition of our characterization. One caveat about these extremal mechanisms and two-way greedy is necessary. This has to do with a technical exception to the rule of never interleaving top queries (asking for the maximum of the current domain) with bottom queries (asking for the minimum) to an agent. An OSP mechanism can in fact interleave those when, at the current history, an agent becomes *revealable*, that is, the threshold separating winning bids from losing ones, becomes known. In other words, this is a point in which the outcome for this agent (but not necessarily the entire solution) is determined for all but one of her types. OSP mechanisms can at this point use any query ordering to find out what the type of the agent is; this does not affect the incentives of the agents. Accordingly, in a two-way greedy algorithm an agent can face changes of priority direction (e.g., from in- to out-priority) in these circumstances.

To the best of our knowledge this is one of the first known cases of a relationship between strategic properties and an algorithmic paradigm, as opposed to a property about the solution output by the algorithm. Two possible interpretations of this connection can be given. On a conceptual level, the fact that the

Revelation Principle does not hold true for OSP means that we care about the implementation details and thus the right algorithmic nature has to be paradigmatic rather than being only about the final output. On a technical level, given that monotonicity (i.e., two-cycles) is not sufficient for OSP, the need to study many-instance monotonicity (i.e., any cycle) requires to go beyond an output property and look for the way in which the algorithm computes *any* solution.

Our OSP characterization is related to the Personal Clock Auctions (PCAs) in [18]. Roughly speaking, Li proves that for binary allocation problems, each agent faces either an ascending-price auction (where there is an increasing transfer going rate to be included in the solution) or a descending-price auction (where there is a decreasing transfer going rate to be excluded from the solution). There are conceptual and technical differences between our characterization and Li's. His focus is on characterizing the auction format (i.e., social choice function, payments and implementation tree) whereas ours concentrates on algorithms. Studying the approximation guarantee of PCAs requires to disentangle these three components. We defer to the full version of this paper a concrete example taken from the corrigendum of [18] that shows how the technical definition of PCA (contrarily to its intuitive and informal description) does not allow for a simple algorithmic characterization in terms of greedy. It turns out that PCAs require to reason about strategies over the extensive-form implementation as opposed to type profiles – this makes them unsuitable and underspecified from the algorithmic point of view. From the technical perspective, Li requires *continuous domains* whereas we assume that these are *finite*, mainly because of the inherent limitations of the OSP CMON technique [9]. For one, our setup is arguably more interesting for OSP, as it is notoriously harder to understand how to execute extensive-form games in the continuous case. Secondly, our proof technique cannot rely on the existence of a unique threshold (there are two threshold values in discrete domains, i.e., extreme winning and losing reports do not "meet" in the limit) unlike [18]. Importantly, our results allow for a more workable notion of OSP mechanisms for binary allocation problems; our approach and terminology are closer to computer science and algorithms and give a specific recipe to reason about design and analysis of these mechanisms.

We give a host of bounds on the approximation guarantee of OSP mechanisms by relying on our characterization and the known approximation guarantees of forward greedy algorithms, cf. Table 1 below. The strategic equivalence of forward and reverse greedy is one of the most far reaching consequences of our results, given (i) the rich literature on the approximation of forward greedy, and (ii) the misconception about the apparent weaknesses of accepting, rather than rejecting, auctions [20] (see Sect. 4). We expect our work to spawn further research about OSP, having fully extracted the algorithmic nature of these mechanisms. The power and limitations of OSP can now be fully explored, in the context of binary allocation problems. We present some initial bounds on the quality of these algorithms/mechanisms (see Sect. 4). Notably, we close the gap for the approximation guarantee of OSP mechanisms for the knapsack auctions studied in [7]. We show that the logarithmic upper bound provided therein is basically tight, not just for reverse greedy (as shown in [7]) but for the whole class.

Since our main objective is that of establishing the power of OSP mechanisms, in terms of algorithmic tools and their approximation, we do not primarily focus on their computational complexity. Consequently, our lower bounds are unconditional. We discuss this aspect and more opportunities for further research in the conclusions (see Sect. 5). The proofs missing due to lack of space are deferred to the full version of this paper.

2   Preliminaries and Notation

We define a set N of n *selfish agents* and a set of feasible *outcomes* $\mathcal{S}$. Each agent i has a *type* $t_i \in D_i$, where D_i is the *domain* of i. The type t_i is assumed to be *private knowledge* of agent i. We let $t_i(X) \in \mathbb{R}$ denote the *cost* of agent i with type t_i for the outcome $X \in \mathcal{S}$. When costs are negative, it means that the agent has a profit from the solution, called *valuation*.

A *mechanism* has to select an outcome $X \in \mathcal{S}$. For this reason, the mechanism interacts with agents. Specifically, agent i takes *actions* (e.g., saying yes/no) that may depend on her presumed type $b_i \in D_i$ (e.g., saying yes could "signal" that the presumed type has some properties that b_i enjoys). To stress this we say that agent i takes *actions compatible with (or according to) b_i* Note that the presumed type b_i can be different from the real type t_i. For a mechanism M, we let $M(\mathbf{b})$ denote the outcome returned by M when agents take actions according to their presumed types $\mathbf{b} = (b_1, \ldots, b_n)$ (i.e., each agent i takes actions compatible with the corresponding b_i). This outcome is given by a pair (f, p), where $f = f(\mathbf{b}) = (f_1(\mathbf{b}), \ldots, f_n(\mathbf{b}))$ (termed *social choice function* or, simply, *algorithm*) maps the actions taken by the agents according to $\mathbf{b}$ to a feasible solution in $\mathcal{S}$, and $p = p(\mathbf{b}) = (p_1(\mathbf{b}), \ldots, p_n(\mathbf{b})) \in \mathbb{R}^n$ maps the actions taken by the agents according to $\mathbf{b}$ to *payments*. Note that payments need not be positive.

Each selfish agent i is equipped with a *quasi-linear utility function* $u_i \colon D_i \times \mathcal{S} \to \mathbb{R}$: for $t_i \in D_i$ and for an outcome $X \in \mathcal{S}$ returned by a mechanism M, $u_i(t_i, X)$ is the utility that agent i has for the implementation of outcome X when her type is t_i, i.e., $u_i(t_i, M(b_i, \mathbf{b}_{-i})) = p_i(b_i, \mathbf{b}_{-i}) - t_i(f(b_i, \mathbf{b}_{-i}))$. In this work we will focus on *single-parameter* settings, that is, the case in which the private information of each bidder i is a single real number t_i and $t_i(X)$ can be expressed as $t_i w_i(X)$ for some publicly known function w_i. To simplify the notation, we will write $t_i f_i(\mathbf{b})$ when we want to express the cost of a single-parameter agent i of type t_i for the output of social choice function f on input the actions corresponding to a bid vector $\mathbf{b}$. In particular, we will consider *binary allocation problems*, where $f_i(\mathbf{b}) \in \{0, 1\}$, i.e., each agent either belongs to the returned solution ($f_i(\mathbf{b}) = 1$) or not ($f_i(\mathbf{b}) = 0$). A class of binary allocation problems of interest are *set systems* $(E, \mathcal{F})$, where E is a set of elements and $\mathcal{F} \subseteq 2^E$ is a family of feasible subsets of E. Each element $i \in E$ is controlled by a selfish agent, that is, the cost for including i is known only to agent i and is equal to some value t_i. The social choice function f must choose a feasible subset of elements X in $\mathcal{F}$ that minimizes $\sum_{i=1}^{n} b_i(X)$. When the t_i's are non-negative (non-positive, respectively) then this objective is called social cost minimization (social welfare maximization, respectively).

Extensive-Form Mechanisms and Obvious Strategyproofness. We now introduce the concept of implementation tree and we formally define (deterministic) obviously strategy-proof mechanisms. Our definition is built on [19] rather than the original definition in [18]. Specifically, our notion of implementation tree is equivalent to the concept of round-table mechanisms in [19], and our definition of OSP is equivalent to the concept of SP-implementation through a round table mechanism, that is proved to be equivalent to the original definition.

Let us first formally model how a mechanism works. An *extensive-form mechanism* M is a triple $(f, p, \mathcal{T})$ where, as above, the pair (f, p) determines the outcome of the mechanism, and $\mathcal{T}$ is a tree, called *implementation tree*, s.t.:

- Every leaf ℓ of the tree is labeled with a possible outcome of the mechanism $(X(\ell), p(\ell))$, where $X(\ell) \in \mathcal{S}$ and $p(\ell) \in \mathbb{R}$;
- Each node u in the implementation tree $\mathcal{T}$ defines the following:
  - An agent $i = i(u)$ to whom the mechanism makes some query. Each possible answer to this query leads to a different child of u.
  - A subdomain $D^{(u)} = (D_i^{(u)}, D_{-i}^{(u)})$ containing all types that are *compatible* with u, i.e., with all the answers to the queries from the root down to node u. Specifically, the query at node u defines a partition of the current domain of i, $D_i^{(u)}$ into $k \geq 2$ subdomains, one for each of the k children of node u. Thus, the domain of each of these children will have as the domain of i, the subdomain of $D_i^{(u)}$ corresponding to a different answer of i at u, and an unchanged domain for the other agents.

Observe that, according to the definition above, for every profile $\mathbf{b}$ there is only one leaf $\ell = \ell(\mathbf{b})$ such that $\mathbf{b}$ belongs to $D^{(\ell)}$. Similarly, to each leaf ℓ there is at least a profile $\mathbf{b}$ that belongs to $D^{(\ell)}$. For this reason, we say that $M(\mathbf{b}) = (X(\ell), p(\ell))$. Two profiles $\mathbf{b}, \mathbf{b}'$ are said to *diverge* at a node u of $\mathcal{T}$ if this node has two children v, v' such that $\mathbf{b} \in D^{(v)}$, whereas $\mathbf{b}' \in D^{(v')}$. For every such node u, we say that $i(u)$ is the *divergent agent* at u.

Definition 1 (OSP mechanisms). *An extensive-form mechanism M is obviously strategy-proof (OSP) if for every agent i with real type t_i, for every vertex u such that $i = i(u)$, for every $\mathbf{b}_{-i}, \mathbf{b}'_{-i}$ (with $\mathbf{b}'_{-i}$ not necessarily different from $\mathbf{b}_{-i}$), and for every $b_i \in D_i$, with $b_i \neq t_i$, s.t. $(t_i, \mathbf{b}_{-i})$ and $(b_i, \mathbf{b}'_{-i})$ are compatible with u, but diverge at u, it holds that $u_i(t_i, M(t_i, \mathbf{b}_{-i})) \geq u_i(t_i, M(b_i, \mathbf{b}'_{-i}))$.*

Roughly speaking, OSP requires that, at each time step agent i is asked to take a decision that depends on her type, the worst utility that she can get if she behaves according to her true type is at least the best utility she can get by behaving differently. We stress that our definition does not restrict the alternative behavior to be consistent with a fixed type. Each leaf of the tree rooted in u, denoted $\mathcal{T}_u$, corresponds to a profile $\mathbf{b} = (b_i, \mathbf{b}'_{-i})$ compatible with u: then, our definition implies that the utility of i in the leaves where she plays truthfully is at least as much as the utility in every other leaf of $\mathcal{T}_u$.

Cycle-Monotonicity Characterizes OSP Mechanisms. We next describe the main tools in [9] showing that OSP can be characterized by the absence of negative-weight cycles in a suitable weighted graph over the possible strategy profiles. For ease of exposition, we will focus on non-negative costs but the results hold no matter the sign. We consider a mechanism M with implementation tree $\mathcal{T}$ for a social choice function f, and define the following concepts:

- **Separating Node:** A node u in the implementation tree $\mathcal{T}$ is $(\mathbf{a}, \mathbf{b})$-separating for agent $i = i(u)$ if $\mathbf{a}$ and $\mathbf{b}$ are compatible with u (that is, $\mathbf{a}, \mathbf{b} \in D^{(u)}$), and the two types a_i and b_i belong to two different subdomains of the children of u (thus implying $a_i \neq b_i$).
- **OSP-graph:** For every agent i, we define a directed weighted graph $\mathcal{O}_i^{\mathcal{T}}$ having a node for each profile in $D = \times_i D_i$. The graph contains edge $(\mathbf{a}, \mathbf{b})$ if and only if $\mathcal{T}$ has some node u which is $(\mathbf{a}, \mathbf{b})$-separating for $i = i(u)$, and the weight of this edge is $w(\mathbf{a}, \mathbf{b}) = a_i(f_i(\mathbf{b}) - f_i(\mathbf{a}))$. Throughout the paper, we will denote with $\mathbf{a} \rightarrow \mathbf{b}$ an edge $(\mathbf{a}, \mathbf{b}) \in \mathcal{O}_i^{\mathcal{T}}$, and with $\mathbf{a} \rightsquigarrow \mathbf{b}$ a path among these two profiles in $\mathcal{O}_i^{\mathcal{T}}$.
- **OSP Cycle Monotonicity (OSP CMON):** We say that the OSP cycle monotonicity (OSP CMON) holds if, for all i, the graph $\mathcal{O}_i^{\mathcal{T}}$ does not contain negative-weight cycles. Moreover, we say that the OSP two-cycle monotonicity (OSP 2CMON) holds if the same is true when considering cycles of length two only, i.e., cycles with only two edges.

Theorem 1 ([9]). *A mechanism with implementation tree $\mathcal{T}$ for a social function f is OSP on finite domains if and only if OSP CMON holds.*

3   A Characterization of OSP Mechanisms

Given the theorem above, we henceforth assume that the agents have finite domains. In this section we present our characterization of OSP mechanisms for binary allocation problems. We begin by setting the scene and then give three useful structural properties of OSP mechanisms.

Observation 1 (Basic Properties of the OSP-graph). *The weight of an edge $(\mathbf{a}, \mathbf{b})$ is non-zero for: $f_i(\mathbf{a}) = 0$ and $f_i(\mathbf{b}) = 1$, in which case the weight is $w(\mathbf{a}, \mathbf{b}) = +a_i$ (**positive-weight edge**); or, $f_i(\mathbf{a}) = 1$ and $f_i(\mathbf{b}) = 0$, in which case the weight is $w(\mathbf{a}, \mathbf{b}) = -a_i$ (**negative-weight edge**). If OSP 2CMON holds, then (i) for every positive-weight edge as above, we have $b_i < a_i$ and, by symmetry, (ii) for every negative-weight edge as above, we have $a_i < b_i$. Note that these inequalities are strict since $a_i \neq b_i$ for every edge as above.*

Definition 2 (0-always, 1-always, unclear). *For sel $\in \{0, 1\}$, indicating whether i is selected, a generic type $t_i \in D_i^{(u)}$, and a node u of the implementation tree, we define: (i) t_i is sel-always if $f_i(\mathbf{t}) = $ sel for all $\mathbf{t}_{-i} \in D_{-i}^{(u)}$; (ii) t_i is* sel-*sometime if* $f_i(\mathbf{t}) = $ sel *for some* $\mathbf{t}_{-i} \in D_{-i}^{(u)}$. *Moreover,* t_i *is unclear if it is neither 0-always nor 1-always, i.e., both 0-sometime and 1-sometime.*

Structure of the Implementation Tree. We first observe that, w.l.o.g., we can always assume that each node u in the implementation tree has two children.

Observation 2. *For any OSP mechanism* $M = (f, p, \mathcal{T})$ *where* $\mathcal{T}$ *is not a binary tree, there is an OSP mechanism* $M' = (f, p, \mathcal{T}')$ *where* $\mathcal{T}'$ *is a binary tree.*

The mechanism then partitions $D_i^{(u)}$ into two subdomains $L^{(u)}$ and $R^{(u)}$. In general, the two parts $L^{(u)}$ and $R^{(u)}$ can be *any* partition of the subdomain $D_i^{(u)}$, and they are not necessarily ordered. For example, if $D_i^{(u)} = \{1, 2, 5, 6, 7\}$, a query "is your type even?" results in $L^{(u)} = \{1, 5, 7\}$ and $R^{(u)} = \{2, 6\}$, with these two subdomains being "incomparable". However, we will see that in an OSP mechanism these sets $L^{(u)}$ and $R^{(u)}$ share a special structure.

Structure of Admissible Queries. We begin with a simple observation about the structure of the parts implied by a simple application of OSP 2CMON.

Observation 3. *Assume OSP 2CMON holds. At every node* u *where the subdomain is separated into two parts* L *and* R *the following holds. Every 0-sometime type* l *in one side (say* L*) implies that all types in the other side that are bigger* $(r \in R$ *with* $r > l)$ *must be 0-always. Similarly, every 1-sometime type* l *in one side implies that all types in the other side that are smaller must be 1-always.*

Next lemma characterizes the admissible queries in OSP mechanisms.

Lemma 1 (Admissible Queries). *Let* M *be an OSP mechanism with implementation tree* $\mathcal{T}$ *and let* u *be a node of* $\mathcal{T}$ *where the query separates the current subdomain into* L *and* R*. Then one of the following conditions must hold:*

1. *At least one of the two parts, some* $P \in \{L, R\}$*, is homogeneous meaning that either all* $p \in P$ *are 0-always or all* $p \in P$ *are 1-always.*
2. *Agent* i *is revealable at node* u*, meaning that* $D_i^{(u)}$ *has the following structure:*

$$D_i^{(u)} = \{\underbrace{d_1 < d_2 < \cdots < d_a}_{1\text{-}always} < d_* < \underbrace{d_1' < d_2' < \cdots d_b'}_{0\text{-}always}\} \tag{1}$$

where each subset of 1-always and 0-always types may be empty, and d_* *may or may not be 1-always or 0-always. Moreover, the two parts* $P \in \{L, R\}$ *must have the following structure:*

$$P = \{\underbrace{p_1 < \cdots < p_a}_{1\text{-}always} < p_* < \underbrace{p_1' < \cdots < p_b'}_{0\text{-}always}\} \tag{2}$$

with at most one type in $D_i^{(u)}$ *being unclear (neither 0-always nor 1-always).*

Structure of Negative-Weight Cycles. A crucial step is to provide a simple *local-to-global* characterization of OSP mechanisms which (essentially) involves only four type profiles. The following theorem states that, if there is a negative-weight cycle with *more than two* edges but no length-two negative cycle (OSP CMON is violated but OSP 2CMON holds), then there exists a cycle with a rather *special structure*, and this special structure is fully specified by only *four* profiles (the cycle itself may involve several profiles though).

Theorem 2 (Four-Profile Characterization). *Let M be a mechanism with implementation tree T and social choice function f that is OSP 2CMON but not OSP CMON. Then, every negative-weight cycle C in some OSP graph $\mathcal{O}_i^T$ is of the following form: $C = \mathbf{b}^{(2)} \rightarrow \mathbf{b}^{(1)} \rightsquigarrow \mathbf{b}^{(3)} \rightarrow \mathbf{b}^{(4)} \rightsquigarrow \mathbf{b}^{(2)}$ where these four profiles satisfy (i) $b_i^{(1)} < b_i^{(2)} < b_i^{(3)} < b_i^{(4)}$, (ii) $f_i(\mathbf{b}^{(1)}) = f_i(\mathbf{b}^{(3)}) = 1$, and (iii) $f_i(\mathbf{b}^{(2)}) = f_i(\mathbf{b}^{(4)}) = 0$. Moreover, there is no edge between $\mathbf{b}^{(2)}$ and $\mathbf{b}^{(3)}$ in $\mathcal{O}_i^T$.*

This characterization will enable us to provide a simple local transformation of the queries of an OSP mechanisms where we use only top or bottom queries.

Definition 3 (Top and Bottom Queries). *Let $i = i(u)$ for a node u of the implementation tree. If the query at u partitions $D_i^{(u)}$ into $\{\min D_i^{(u)}\}$ and $D_i^{(u)} \setminus \{\min D_i^{(u)}\}$ then we call the query at u a bottom query. A top query at u, instead, separates the maximum of $D_i^{(u)}$ from the rest.*

3.1 OSP is Equivalent to Weak Interleaving

In this section, we show that without loss of generality, we can focus on OSP mechanisms where each agent is asked only top queries or only bottom queries, except when her type becomes revealable (Condition 2. in Lemma 1). In that sense, these mechanisms interleave top and bottom queries for an agent only in a "weak" form.[1] Specifically, let us begin by providing the following definition.

Definition 4 (Extremal, No Interleaving, Weak Interleaving). *A mechanism is extremal if every query is a bottom query or a top query (both types of queries may be used for the same agent).*

An extremal mechanism makes no interleaving queries if each agent is consistently asked only top queries or only bottom queries at each history where she is divergent (some agents may be asked top queries only, and other agents bottom queries only).

A weak interleaving mechanism satisfies the condition that, if top queries and bottom queries are interleaved for some agent i at some node u in the implementation tree, then agent i is revealable at u in the sense of Condition 2. in Lemma 1.

---

[1] It may appear that an alternative formalization of the interleaving between in- and out-priorities could be a query where the type is fully revealed; this would not work as there is still one type for which the outcome is undetermined.

Theorem 3. *For each binary outcome problem, an OSP mechanism exists if and only if an extremal mechanism with weak interleaving exists.*

The Proof. We start with the necessary condition and prove the following chain of implications: *OSP mechanism* $\Rightarrow$ *OSP extremal mechanism* $\Rightarrow$ *OSP weak interleaving mechanism*. Intuitively, given the structure of admissible queries in Lemma 1, we show below that we can locally replace every query with a homogeneous part (Condition 1.) by a "homogeneous" sequence of only top queries or only bottom queries. Moreover, these queries can also be used when the agent becomes revealable (Condition 2.). To this aim, we use the four-profile characterization of negative-weight cycles in Theorem 2.

Theorem 4. *Any OSP mechanism $M = (f, p, \mathcal{T})$ can be transformed into an equivalent extremal OSP mechanism $M' = (f, p, \mathcal{T}')$.*

Proof Sketch. Since the mechanism M is OSP, then OSP CMON must hold (Theorem 1). Let u be a node of $\mathcal{T}$ where M is not extreme, and let $i = i(u)$ be the corresponding divergent agent at u. Assume that all previous queries of i in the path from the root to u are extremal (if not, we can apply the argument to the first query that is not extremal and reiterate).

We locally modify $\mathcal{T}$ in order to make a suitable sequence of bottom queries and top queries about a certain subset Q of types, before we make any further query in the two subtrees of u. It can be shown that this local modification does not affect OSP CMON of agents different from i and it preserves OSP 2CMON for all agents. The proof uses the following main steps and key observations:

1. If OSP CMON is no longer true for $\mathcal{T}'$, then there exists a negative-weight cycle C' which was not present in $\mathcal{O}_i^{\mathcal{T}}$ and therefore must use some added edges $\mathbf{a}^{(1)} \rightarrow \mathbf{a}^{(2)}$ that were not present in $\mathcal{O}_i^{\mathcal{T}}$ and that have been added to $\mathcal{O}_i^{\mathcal{T}'}$ because of the new queries for types in Q;
2. The negative-weight cycle C' must be of the form specified by Theorem 2; in particular, the edge $\mathbf{b}^{(2)} \rightarrow \mathbf{b}^{(3)}$ does not exist in $\mathcal{O}_i^{\mathcal{T}'}$ (which helps to determine properties of the four profiles characterizing C');
3. We use the following *bypass argument* to conclude that in the original graph there is a negative-weight cycle, thus contradicting OSP CMON of the original mechanism. Specifically, for every added edge $\mathbf{a}^{(1)} \rightarrow \mathbf{a}^{(2)}$, the original graph $\mathcal{O}_i^{\mathcal{T}}$ contains a bypass path $\mathbf{a}^{(1)} \rightarrow \mathbf{b}^{(bp)} \rightarrow \mathbf{a}^{(2)}$ such that $w(\mathbf{a}^{(1)} \rightarrow \mathbf{b}^{(bp)} \rightarrow \mathbf{a}^{(2)}) \leq w(\mathbf{a}^{(1)} \rightarrow \mathbf{a}^{(2)})$. By replacing every added edge in C' with the corresponding bypass path, we get a cycle C with negative weight, $w(C) \leq w(C') < 0$. □

We are now ready to show that one can think of weak interleaving OSP mechanisms without loss of generality.

Theorem 5. *Let M be an OSP extremal mechanism with implementation tree $\mathcal{T}$. For any node $u \in \mathcal{T}$, if agent $i = i(u)$ is not revealable at node u, then M has no interleaving for agent i at u.*

The proof of Theorem 3 is completed below with the sufficiency.

Theorem 6. *An extremal mechanism with weak interleaving is OSP.*

4 Two-Way Greedy Algorithms

We reiterate that our characterization does not require costs to be positive; so it holds true for negative costs, that is, valuations. We will then now talk simply about type when we refer to the agents' private information, be it costs or valuations. We will also sometimes use the terminology of in-query (out-query, respectively) to denote a bottom (top) query for costs and a top (bottom) query for valuations.

Definition 5 ((Anti-)Monotone Functions). *We say that a function is monotone (antimonotone, resp.) in the type if it is decreasing (increasing, resp.) in the cost, and increasing (decreasing, resp.) in the valuation.*

In this section, we translate our characterization into algorithmic insights on OSP mechanisms and show a connection between their format and a certain family of adaptive priority algorithms that they use. As a by-product, we show the existence of a host of new mechanisms. We are able to provide the first set of upper bounds on the approximation guarantee of OSP mechanisms, that are independent from domain size (as in [8,9]) or assumptions on the designer's power to catch and punish lies (as in [10,11]), see Table 1. The table also contains the new bounds we can prove on the approximation of OSP mechanisms, by leveraging our algorithmic characterization (i.e., Theorems 7–10).

Table 1. Bounds on the approximation guarantee of OSP mechanisms. (The result for CAs has been observed in [6]. The mechanisms for matroids also follows from Corollary 2.)

Problem	Bound
Known Single-Minded Combinatorial Auctions (CAs)	$\sqrt{m}$ ([17] + Corollary 1)
MST (& weighted matroids)	1 ([15] + Corollary 1)
Max Weighted Matching	2 ([2] + Corollary 1)
p-systems‡	p ([13] + Corollary 1)
Weighted Vertex Cover	2 ([5] + Corollary 1)
Shortest Path	∞ (Theorem 7)
Restricted Knapsack Auctions	$\Omega(\sqrt{n})$ (Theorem 8)
Asymmetric Restricted Knapsack Auctions (3 values)	$\sqrt{n-1}$ (Theorem 9)
Knapsack Auctions	$\Omega(\sqrt{\ln n})$ (Theorem 10)

‡A p-system is a downward-closed set system $(E, \mathcal{F})$ where there are at most p circuits, that is, minimal subsets of E not belonging to $\mathcal{F}$ [13].

Immediate-Acceptance Auctions and Forward Greedy. Let us begin by discussing forward greedy.

Definition 6 (Forward Greedy). *A forward greedy algorithm uses functions* $g_i^{(in)} : \mathbb{R} \times \mathbb{R}^k \to \mathbb{R}$, $k \leq n-1$, *to rank the bids of the players and builds a solution by iteratively adding the agent with highest rank if that preserves feasibility (cf. Algorithm 1). A forward greedy algorithm is monotone if each* $g_i^{(in)}$ *is monotone in* i*'s private type (i.e., its first argument).*

Algorithm 1: Forward greedy algorithm

---

1 Let $b_1, \ldots, b_n$ the input bids
2 $\mathcal{P} \leftarrow \mathcal{F}$ ($\mathcal{P}$ is the set of all feasible solutions)
3 $\mathcal{I} \leftarrow \emptyset$ ($\mathcal{I}$ is the set of infeasible agents, i.e., agents that cannot be included in the final solution)
4 $\mathcal{A} \leftarrow N$ ($\mathcal{A}$ is the set of active or infeasible agents)
5 **while** $|\mathcal{P}| > 1$ **do**
6   Let $i = \arg\max_{k \in \mathcal{A} \setminus \mathcal{I}} g_k^{(in)}(b_k, \mathbf{b}_{N \setminus \mathcal{A}})$
7   **if** *there are solutions* S *in* $\mathcal{P}$ *such that* $i \in S$ **then**
8   Drop from $\mathcal{P}$ all the solutions S such that $i \notin S$ (if any)
9          $\mathcal{A} \leftarrow \mathcal{A} \setminus \{i\}$
10   **else**
11          $\mathcal{I} \leftarrow \mathcal{I} \cup \{i\}$

12 Return the only solution in $\mathcal{P}$

A few observations are in order for Algorithm 1. Firstly, it is not too hard to see that it will always return a solution (i.e., there will eventually be a unique feasible solution in $\mathcal{P}$). To see this consider the case in which (at least) two solutions S, S' are in $\mathcal{P}$. Then there must exist an agent j such that $j \in S \Delta S'$, Δ denoting the symmetric difference between sets. This means that $j \in \mathcal{A} \setminus \mathcal{I}$ and the forward greedy algorithm will decide in the next steps whether j is part of the solution or not (and consequently whether to keep S or S'). Secondly, we stress how forward greedy algorithms belong to the family of adaptive priority algorithms [4]. At Line 6, Algorithm 1 (potentially) updates the priority as a function of the bids of those bidders who have left the auction (i.e., that are not active anymore). A peculiarity of the algorithm (to do with its OSP implementation) is that the adaptivity does not depend on bidders who despite their high priority cannot be part of the eventual solution (i.e., the agents we add to $\mathcal{I}$).[2] This distinction would not be necessary if the problem at hand were *upward closed* (i.e., if a solution S is feasible then any $S' \supset S$ would be in $\mathcal{F}$ too). Thirdly, a notable subclass of forward greedy algorithms are fixed-priority algorithms, where Line 6 and the while loop are swapped (and priority functions only depend on the agents' types). (These algorithms do not need to keep record

---

[2] Note that a syntactically (but not semantically) alternative definition of forward greedy algorithms could do without $\mathcal{I}$ by requiring an extra property on the priority functions (i.e., adaptively floor all the priorities of infeasible players).

of $\mathcal{I}$ either.) This algorithmic paradigm has been applied to different optimization problems (e.g., Kruskal's Minimum Spanning Tree (MST) algorithm [15]). We define *Immediate-Acceptance Auctions (IAAs)* as mechanisms using only in-queries.

Corollary 1. *An IAA that uses algorithm f is OSP if and only if f is monotone forward greedy.*

We note how all fixed-priority algorithms are OSP but not all OSP mechanism must use a fixed-priority algorithm. In fact, for a fixed-priority algorithm the sufficiency proof alone would go through; the second parameter of the priority functions is only needed for the opposite direction.

Algorithm 2: Reverse greedy algorithm

---

1 Let $b_1, \ldots, b_n$ the input bids
2 $\mathcal{P} \leftarrow \mathcal{F}$ ($\mathcal{P}$ is the set of all feasible solutions)
3 $\mathcal{I} \leftarrow \emptyset$ ($\mathcal{I}$ is the set of in agents, i.e., those that cannot be dropped)
4 $\mathcal{A} \leftarrow N$ ($\mathcal{A}$ is the set of active or in agents)
5 **while** $|\mathcal{P}| > 1$ **do**
6   Let $i = \arg\max_{k \in \mathcal{A} \setminus \mathcal{I}} g_k^{(out)}(b_k, \mathbf{b}_{N \setminus \mathcal{A}})$
7   **if** *there are solutions S in $\mathcal{P}$ such that $i \notin S$* **then**
8   Drop from $\mathcal{P}$ all the solutions S such that $i \in S$ (if any)
9          $\mathcal{A} \leftarrow \mathcal{A} \setminus \{i\}$
10   **else**
11         $\mathcal{I} \leftarrow \mathcal{I} \cup \{i\}$
12 Return the only solution in $\mathcal{P}$

It is important to compare the result above with [20, Footnote 15], where it is observed how the strategic properties of forward and reverse greedy algorithms are different. The remark applies to auctions where these algorithms are augmented by the so-called threshold payment scheme. Our result shows that there exist alternative OSP payment schemes for this important algorithmic design paradigm.

Deferred-Acceptance Auctions and Reverse Greedy. Another algorithmic approach that can be used to obtain OSP mechanisms is reverse greedy (see Algorithm 2), that is, having an out-priority function that is antimonotone with each agent's type and drops agents accordingly. In its auction format, this is known as Deferred-Acceptance Auctions (DAAs), see, e.g., [20].

Definition 7 (Reverse Greedy). *A reverse greedy algorithm uses functions $g_i^{(out)} : \mathbb{R} \times \mathbb{R}^k \to \mathbb{R}$, $k \leq n - 1$, to rank the bids of the players and iteratively excludes the player with highest rank (if feasible) until only one solution is left (cf. Algorithm 2). A reverse greedy algorithm is antimonotone if each $g_i^{(out)}$ is antimonotone in i's private type (i.e., its first argument).*

Similarly to the case of forward greedy, the algorithm need not use $\mathcal{I}$ for *downward-closed* problems (i.e., every subset of a feasible solution is feasible as well). Incidentally, this is the way it is discussed in [7].

Corollary 2 ([20]). *A DAA using algorithm f is OSP if and only if f is anti-monotone reverse greedy.*

Algorithm 3: Two-way greedy algorithm

1   Let $b_1, \ldots, b_n$ the input bids
2   $\mathcal{P} \leftarrow \mathcal{F}$ ($\mathcal{P}$ is the set of all feasible solutions)
3   $\mathcal{I} \leftarrow \emptyset$ ($\mathcal{I}$ is the set of infeasible and in agents)
4   $\mathcal{A} \leftarrow N$ ($\mathcal{A}$ is the set of all agents that are active or infeasible or in)
5   **while** $|\mathcal{P}| > 1$ **do**
6   Let $i^{(in)} = \arg\max_{k \in \mathcal{A} \setminus \mathcal{I}} g_k^{(in)}(b_k, \mathbf{b}_{N \setminus \mathcal{A}})$
7   Let $i^{(out)} = \arg\max_{k \in \mathcal{A} \setminus \mathcal{I}} g_k^{(out)}(b_k, \mathbf{b}_{N \setminus \mathcal{A}})$
8   Let i be the agent corresponding to
     $\max \left\{ g_{i^{(in)}}^{(in)}(b_{i^{(in)}}, \mathbf{b}_{N \setminus \mathcal{A}}), g_{i^{(out)}}^{(out)}(b_{i^{(out)}}, \mathbf{b}_{N \setminus \mathcal{A}}) \right\}$
9   **if** $i = i^{(in)} \wedge$ *there are solutions S in $\mathcal{P}$ such that $i \in S$* **then**
10   Drop from $\mathcal{P}$ all the solutions S such that $i \notin S$ (if any)
11       $\mathcal{A} \leftarrow \mathcal{A} \setminus \{i\}$
12   **else if** $i = i^{(out)} \wedge$ *there are solutions S in $\mathcal{P}$ such that $i \notin S$* **then**
13   Drop from $\mathcal{P}$ all the solutions S such that $i \in S$ (if any)
14       $\mathcal{A} \leftarrow \mathcal{A} \setminus \{i\}$
15   **else**
16       $\mathcal{I} \leftarrow \mathcal{I} \cup \{i\}$
17   Return the only solution in $\mathcal{P}$

It is convenient to discuss the relative power of forward and reverse greedy algorithms. We now know from Corollaries 1 and 2 that they are strategically equivalent. Algorithmically, however, there are some differences. There are algorithms and problems, such as, Kruskal algorithm for MST, where we can take the reverse version of forward greedy (e.g., for MST, start with the entire edge set, go through the edges from the most expensive to the cheaper, and remove an edge whenever it does not disconnect the graph) without any consequence to the approximation guarantee. For the minimum spanning tree problem (and, more generally, for finding the minimum-weight basis of a matroid), the reverse greedy algorithm is just as optimal as the forward one. In general (and even for, e.g., bipartite matching), the reverse version of a forward greedy algorithm with good approximation guarantee can be bad [7].

OSP Mechanisms and Two-Way Greedy. We show that the right algorithmic technique for OSP is a suitable combination of forward and reverse greedy.

Definition 8 (Two-way Greedy). *A two-way greedy algorithm uses functions* $g_i^{(in)} : \mathbb{R} \times \mathbb{R}^k \to \mathbb{R}$ *and* $g_i^{(out)} : \mathbb{R} \times \mathbb{R}^k \to \mathbb{R}$, $k \leq n-1$, *for each agent* i, *to rank the bids, and iteratively and greedily includes (if highest priority is defined by* $g^{(in)}$*) or excludes (if highest priority is defined by* $g^{(out)}$*) whenever possible the player with the highest rank until one feasible solution is left (cf. Algorithm 3). A two-way greedy is all-monotone if each* $g_i^{(in)}$ *is monotone in* i*'s private type and* $g_i^{(out)}$ *is antimonotone in* i*'s private type (i.e., their first argument).*

To fully capture the strategic properties of two-way greedy algorithms, we need to define one more property for which we need some background definitions. Consider the total increasing[3] ordering φ of the $2\prod_i |D_i|$ functions[4] $\{g_i^{(\star)}(b, \mathbf{b})\}_{i \in N, b \in D_i, \mathbf{b} \in D_{-i}, \star \in \{in, out\}}$ used by a two-way greedy algorithm. Given $\varphi_\ell = g_i^{(\star)}(b_i, \mathbf{b}_{N \backslash \mathcal{A}_\ell})$, the ℓ-th entry of φ, we let $D_j^{\prec}(\ell)$ ($D_j^{\succ}(\ell)$, respectively) denote the set of types $b \in D_j$ such that $g_j^{(\star)}(b, \mathbf{b}_{N \backslash \mathcal{A}}) > \varphi_\ell$ ($g_j^{(\star)}(b, \mathbf{b}_{N \backslash \mathcal{A}}) < \varphi_\ell$, respectively) with $\mathcal{A} \supseteq \mathcal{A}_\ell$ ($\mathcal{A} \subseteq \mathcal{A}_\ell$, respectively). Moreover, we add b_i (the bid defining φ_ℓ) to $D_i^{\prec}(\ell)$. In words, once in the ordering we reach the ℓ-th entry for a certain agent type and a given "history" (i.e., $\mathbf{b}_{N \backslash \mathcal{A}_\ell}$) with $D_j^{\prec}(\ell)$ we denote all the types that the algorithm has already explored for agent j at this point (that is, for a compatible prior history $\mathbf{b}_{N \backslash \mathcal{A}_\ell}$ with $\mathcal{A} \supseteq \mathcal{A}_\ell$). Similarly, $D_j^{\succ}(\ell)$ denotes those types in D_j that are yet to be considered from this history onwards. Finally, for $d \in \{in, out\}$ we let $\overline{d}$ be a shorthand for the other direction.

Definition 9 (Interleaving Algorithm). *We say that a two-way greedy algorithm is interleaving if for each* i *and* $\mathcal{A} \subseteq N$ *the following occurs. For each* $\varphi_\ell = g_i^{(d)}(b, \mathbf{b}_{N \backslash \mathcal{A}})$ *such that for some* $\mathcal{A}' \subseteq \mathcal{A}$ *it holds* $g_i^{(\overline{d})}(b', \mathbf{b}_{N \backslash \mathcal{A}'}) = \varphi_{\ell'}$ *with* $b' \in D_i^{\succ}(\ell)$ *(and then* $\ell' > \ell$*) we have* $g_i^{(\star)}(x, \mathbf{b}_{N \backslash \mathcal{A}'}) > g_j^{(\star)}(y, \mathbf{b}_{N \backslash \mathcal{A}'})$ *for each* $y \in D_j^{\succ}(\ell')$ *and for all (but at most one)* x *in* $D_i^{\succ}(\ell')$ *(* $\star \in \{in, out\}$*).*

The definition above captures in algorithmic terms the weak interleaving property of extensive-form implementations. Whenever there is a change of direction (from d to $\overline{d}$) for a certain agent i and two compatible histories (cf. condition $\mathcal{A}' \subseteq \mathcal{A}$) then it must be the case that i is revealable and all the other unexplored types (but at most one) *must* be explored next.

Example 1 (Interleaving Algorithm). Consider a setting with three agents, called x, y and z. The *valuation* domain is the same for all the agents and has maximum $t_{\max}$ and minimum $t_{\min}$. Consider the two-way greedy algorithm with the following ordering φ: $g_x^{(in)}(t_{\max}) > g_y^{(in)}(t_{\max}) > g_z^{(out)}(t_{\min}) > g_y^{(out)}(t_{\min}) > \ldots$ (where the second argument is omitted since it is $\emptyset$). Let us focus on $\varphi_2 =$

---

[3] For notational simplicity, we here assume that there are not ties between the priority functions.

[4] The algorithm must not necessarily have a definition for the priority functions for all the combinations of type/history as some might never get explored. In this case, we set all the undefined entries to sufficiently small (tie-less, for simplicity) values.

$g_y^{(in)}(t_{\max})$. Here $D_x^{\prec}(2)$ and $D_y^{\prec}(2)$ is $\{t_{\max}\}$ (as it has been already considered for both agents) whilst $D_x^{\succ}(2)$ and $D_y^{\succ}(2)$ is equal to the original domain but $t_{\max}$. For z, instead, $D_z^{\prec}(2) = \emptyset$ and $D_z^{\succ}(2)$ is still the original domain. At φ_4 there is a change of direction for agent y. The two-way greedy is interleaving if the domain has only three types (since there are no constraints on the in/out priority for third type in the domain of y). For larger domains, instead, we need to look at the next entries of φ to ascertain whether the algorithm is interleaving or not.

Corollary 3. *A mechanism using algorithm f is OSP if and only if f is an all-monotone interleaving two-way greedy algorithm.*

Given the corollary above, we will henceforth simply say two-way greedy (algorithm) and avoid stating the properties of all-monotonicity and interleaving. We next analyse the approximation guarantee of two-way greedy algorithms, whilst a comparison with forward/reverse greedy is deferred to the full version.

Approximation Guarantee of Two-Way Greedy Algorithms. We next prove that the approximation ratio of two-way greedy algorithms is unbounded in set systems where we want to output the solution with minimum social cost *for some* (minimal) structure of $\mathcal{F}$. Examples include the shortest path problem. For such problems, a two-way greedy algorithm must commit immediately to a solution after the first decision in either Line 10 or 13.

Theorem 7 (Social Cost). *For any $\rho \geq 1$, there exists a set system such that no two-way greedy algorithm returns a ρ-approximation to the optimal social cost, even if there are only two feasible solutions and four agents.*

We now consider the case where the players have valuations and we are interested in maximizing the social welfare. We call this setup where no further assumption on the structure of the feasible solutions in $\mathcal{F}$ can be made, a *restricted knapsack auction* problem.

Theorem 8 (Social Welfare). *There exists a set system for which any two-way greedy algorithm has approximation $\Omega(\sqrt{n})$ to the optimal social welfare, even if there are only two feasible solutions.*

Interestingly the result above uses a so-called *asymmetric instance* [7] where one solution is a singleton and the other is comprised of all the remaining bidders. We now prove that the analysis above is tight at least for three-value domains.

Theorem 9. *There is a $\sqrt{n-1}$-approximate forward greedy algorithm for the asymmetric restricted knapsack auctions, when bidders have a three-value domain $\{t_{\min}, t_{\mathrm{med}}, t_{\max}\}$.*

We now turn our attention to downward-closed set systems for social welfare maximization. This is a generalization of the setting studied in [7], called *knapsack auctions*: There are n bidders and m copies of one item; each bidder has a private valuation v_i to receive at least s_i copies of the item, s_i being public knowledge. A solution is feasible if the sum of items allocated to bidders is at most m. The objective is social welfare maximization. The authors of [7] give a $O(\ln m)$-approximate DAA/reverse greedy algorithm and prove a lower bound of $\ln^\tau m$, for a positive constant τ, *limited to* DAA/reverse greedy. We next show that the upper bound is basically tight for the entire class of OSP mechanisms.

Theorem 10 (Social Welfare Downward-Closed Set Systems). *There is a downward-closed set system for which every two-way greedy algorithm has approximation $\Omega(\sqrt{\ln n})$ to the optimal social welfare.*

5 Conclusions

OSP has attracted lots of interest in computer science and economics, see, e.g., [6,8–11,16,19]. Our work can facilitate the study of this notion of incentive-compatibility for imperfectly rational agents. Just as the characterization of DAAs in terms of reverse greedy [20] has given the first extrinsic reason to study the power and limitations of these algorithms [7,12], we believe that our characterization of OSP in terms of two-way greedy will lead to a better understanding of this algorithmic paradigm. In this work, we only began to investigate their power and much more is left to be done. For example, different optimization problems and objective functions could be considered. Moreover, whilst in general OSP mechanisms do not compose sequentially, see, e.g., [3], we could study under what conditions two-way greedy algorithms compose.

References

1. Ausubel, L.M.: An efficient ascending-bid auction for multiple objects. AER **94**(5), 1452–1475 (2004)
2. Avis, D.: A survey of heuristics for the weighted matching problem. Networks **13**(4), 475–493 (1983)
3. Bade, S., Gonczarowski, Y.: Gibbard-satterthwaite success stories and obvious strategyproofness. In: EC, p. 565 (2017)
4. Borodin, A., Nielsen, M.N., Rackoff, C.: (Incremental) priority algorithms. Algorithmica **37**, 295–326 (2003)
5. Clarkson, K.L.: A modification of the greedy algorithm for vertex cover. Inf. Process. Lett. **16**(1), 23–25 (1983)
6. de Keijzer, B., Kyropoulou, M., Ventre, C.: Obviously strategyproof single-minded combinatorial auctions. In: ICALP, pp. 71:1–71:17 (2020)
7. Dütting, P., Gkatzelis, V., Roughgarden, T.: The performance of deferred-acceptance auctions. Math. Oper. Res. **42**(4), 897–914 (2017)
8. Ferraioli, D., Meier, A., Penna, P., Ventre, C.: Automated optimal OSP mechanisms for set systems. In: Caragiannis, I., Mirrokni, V., Nikolova, E. (eds.) WINE 2019. LNCS, vol. 11920, pp. 171–185. Springer, Cham (2019). https://doi.org/10.1007/978-3-030-35389-6_13
9. Ferraioli, D., Meier, A., Penna, P., Ventre, C.: Obviously strategyproof mechanisms for machine scheduling. In: ESA, pp. 46:1–46:15 (2019)

10. Ferraioli, D., Ventre, C.: Probabilistic verification for obviously strategyproof mechanisms. In: IJCAI, pp. 240–246 (2018)
11. Ferraioli, D., Ventre, C.: Approximation guarantee of OSP mechanisms: the case of machine scheduling and facility location. Algorithmica **83**(2), 695–725 (2021)
12. Gkatzelis, V., Markakis, E., Roughgarden, T.: Deferred-acceptance auctions for multiple levels of service. In: EC (2017)
13. Hausmann, D., Korte, B., Jenkyns, T.A.: Worst case analysis of greedy type algorithms for independence systems. In: Padberg, M.W. (ed.) Combinatorial Optimization, pp. 120–131. Springer, Heidelberg (1980). https://doi.org/10.1007/BFb0120891
14. Kagel, J.H., Harstad, R.M., Levin, D.: Information impact and allocation rules in auctions with affiliated private values: a laboratory study. Econometrica **55**(6), 1275–1304 (1987)
15. Kruskal, J.B.: On the shortest spanning subtree of a graph and the traveling salesman problem. Proc. Am. Math. Soc. **7**(1), 48–50 (1956)
16. Kyropoulou, M., Ventre, C.: Obviously strategyproof mechanisms without money for scheduling. In: AAMAS, pp. 1574–1581 (2019)
17. Lehmann, D., O'Callaghan, L., Shoham, Y.: Truth revelation in approximately efficient combinatorial auctions. J. ACM **49**(5), 577–602 (2002)
18. Li, S.: Obviously strategy-proof mechanisms. AER **107**(11), 3257–87 (2017)
19. Mackenzie, A.: A revelation principle for obviously strategy-proof implementation. Games Econ. Behav. **124**, 512–533 (2018)
20. Milgrom, P., Segal, I.: Clock auctions and radio spectrum reallocation. J. Polit. Econ. **128**(1), 1–31 (2020)
21. Nisan, N., Roughgarden, T., Tardos, E., Vazirani, V. (eds.) Algorithmic Game Theory. Cambridge University Press, Cambridge (2007)
22. Saks, M., Yu, L.: Weak monotonicity suffices for truthfulness on convex domains. In: EC (2005)

Bayesian Persuasion in Sequential Trials

Shih-Tang Su$^{(\boxtimes)}$, Vijay G. Subramanian, and Grant Schoenebeck

University of Michigan, Ann Arbor, USA
{shihtang,vgsubram,schoeneb}@umich.edu

Abstract. We consider a Bayesian persuasion problem where the sender tries to persuade the receiver to take a particular action via a sequence of signals. This we model by considering multi-phase trials with different experiments conducted based on the outcomes of prior experiments. In contrast to most of the literature, we consider the problem with constraints on signals imposed on the sender. This we achieve by fixing some of the experiments in an exogenous manner; these are called determined experiments. This modeling helps us understand real-world situations where this occurs: e.g., multi-phase drug trials where the FDA determines some of the experiments, start-up acquisition by big firms where late-stage assessments are determined by the potential acquirer, multi-round job interviews where the candidates signal initially by presenting their qualifications but the rest of the screening procedures are determined by the interviewer. The non-determined experiments (signals) in the multi-phase trial are to be chosen by the sender in order to persuade the receiver best. With a binary state of the world, we start by deriving the optimal signaling policy in the only non-trivial configuration of a two-phase trial with binary-outcome experiments. We then generalize to multi-phase trials with binary-outcome experiments where the determined experiments can be placed at arbitrary nodes in the trial tree. Here we present a dynamic programming algorithm to derive the optimal signaling policy that uses the two-phase trial solution's structural insights. We also contrast the optimal signaling policy structure with classical Bayesian persuasion strategies to highlight the impact of the signaling constraints on the sender.

Keywords: Information design · Bayesian persuasion · Signaling games

1 Introduction

Information design studies how informed agents (senders) persuade uninformed agents (receivers) to take specific actions by influencing the uninformed agents'

S.-T. Su—Supported in part by NSF grant ECCS 2038416 and MCubed 3.0.

V. G. Subramanian—Supported in part by NSF grants ECCS 2038416, CCF 2008130, and CNS 1955777.

G. Schoenebeck—Supported in part by NSF grants CCF 2007256 and CAREER 1452915.

M. Feldman et al. (Eds.): WINE 2021, LNCS 13112, pp. 22–40, 2022.
https://doi.org/10.1007/978-3-030-94676-0_2

beliefs via information disclosure in a game. The canonical Kamenica-Gentzkow model [16] is one where the sender can commit to an information disclosure policy (signaling strategy) before learning the true state. Once the state is realized, a corresponding (randomized) signal is sent to the receiver. Then, the receiver takes an action, which results in payoffs for both the sender and the receiver. Senders in information design problems only need to manipulate the receivers' beliefs with properly chosen signals. The manipulated beliefs will create the right incentives for the receiver to spontaneously take specific actions that benefit the sender (in expectation). In (classical) mechanism design, however, the story is different: the designer is unaware of the agents' private information, and the agents communicate their private information to the designer, who then has to provide incentives via (monetary) transfers or other means. The flexibility afforded by information design that allows the sender to benefit from information disclosure without implementing utility-transfer mechanisms has led to greater applicability of the methodology: various models and theories can be found in survey papers such as [3] and [15].

Our work is motivated by many real-world problems where persuasion schemes are applicable, but the sender is constrained in the choice of signals available for information design. Specifically, we are interested in problems that are naturally modeled via multi-phase trials where the interim outcomes determine the subsequent experiments. Further, we insist that some of the experiments are given in an exogenous manner. This feature imposes restrictions on the sender's signaling space, and without it, we would have a classical Bayesian persuasion problem with an enlarged signal space. Our goal is to study the impact of such constraints on the optimal signaling scheme, and in particular, to contrast it with the optimal signaling schemes in classical Bayesian persuasion.

The following motivating example describes a possible real-world scenario.

Example 1 (Motivating example - Acquiring funds from a venture capital firm).
We consider a scenario where a start-up is seeking funds from a venture capital firm. The process for this will typically involve multiple rounds of negotiation and evaluation: some of these will be demonstrations of the start-up's core business idea, and the others will be assessments by the venture capital firm following their own screening procedures. The start-up will have to follow the venture capital firm's screening procedures but chooses its product demonstrations. Based on

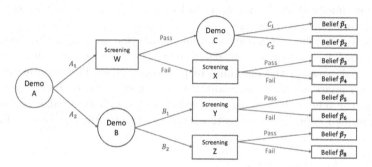

Fig. 1. Example of a negotiation process – a startup vs. a venture capital firm.

these stipulations, the start-up needs to design its demonstrations to maximize its chance of getting funded.

In the example above, the start-up (sender) has to generate an information disclosure scheme to get the desired funds from the venture capital firm (receiver). Then the screening procedures set by the venture capital firm are analogous to our determined experiments, and the demonstrations carried out by the start-up are the (sender) designed experiments. For example, in Fig. 1 we present one plausible interaction where the start-up company designs demonstrations A, B, and C (circles in the figure) and the venture capital firm has predetermined screening examinations W, X, Y, and Z (rectangles in the figure). Whereas we have illustrated this example via a balanced tree, if we have an unbalanced tree owing to the receiver deciding in the middle, we can modify it to a balanced tree by adding the required number of dummy stages.

The sender's reduced flexibility on her signaling strategies under some predetermined experiments with arbitrary positions and informativeness differentiates our work from the growing literature on dynamic information design. Our model considers a problem with the following features: a static state space, a sequential information disclosure environment, and a signaling space restricted by some exogenous constraints whose harshness may depend on the proposed singling schemes. Models with a static state space, an unrestricted signal space but a variety of sequential information disclosure environments have been studied to capture features in different real-world problems: e.g., with multiple senders [12,19], with costly communication [14,22], allowing for sequential decision making [10], or with partial commitment [1,22]. Models with dynamic states and sequential information disclosure environments are usually studied under an informed sender with the knowledge of dynamically changing state(s); a variety of works in this category lie in state change detection [9,11] or routing games [21]. Although several works [7,13,18] also consider constrained signaling schemes, these works either consider the signal space to be smaller than the action space [7,13] or consider a noisy signaling environment [18]. Models with exogenous information [4,5,17], can be viewed as sequential information disclosure problems with exogenous determined experiments placed in determined phases. The sequential information disclosure in our model, which actually enlarges the signal space, makes our work different from above works. To keep the focus of the paper on sequential trials, we discuss the broader literature on constrained senders, algorithmic information design, and works related to the receiver's experiment design[1] in our online version [23].

According to the motivating example illustrated in Fig. 1, the persuasion problem considers a sequence of experiments where experiments further along in the tree depend on the outcomes of previous phases. The experiment to be run in each phase is either exogenously determined or chosen by the sender. In the game, the sender chooses designed experiments with knowledge of the prior, the determined experiments, and the receiver's utility function, but before the

---

[1] See Section 5 in [23] for details.

state of the world is realized. After the sender commits to the experiments (i.e., the signaling strategy), the state of the world is realized, and a specific sequence of experiments is conducted based on the realization of the underlying random variables. The receiver then takes an action depending on the entire sequence of outcomes. The prior, the sender's and receiver's utility functions, the determined experiments, and the designed experiments (after the sender finalizes them) are assumed to be common knowledge. We study this problem for binary states of the world, first for two-phase binary-outcome trials, and then generalized to multi-phase binary-outcome trials. We then generalize to non-binary experiments (still with an underlying binary state space). In the online version [23] we add games with an additional stage where the receiver moves before the sender to decide some or all of the determined experiments, perhaps with some constraints.

Contributions: The main contributions of this work are:

1. To the best of our knowledge, within the multi-phase Bayesian persuasion framework, we are the first to study the design of sender's signaling schemes with exogenously determined experiments in arbitrary positions. Our results highlight the difference between "exogenously determined experiments" and "exogenously given information" in the dynamic information design, where the former gives greater flexibility and allows for more heterogeneity.
2. We explicitly solve the optimal signaling scheme in two-phase trials. Moreover, using structural insights gained from two-phase trials, we present a dynamic programming algorithm to derive the optimal signaling in general multi-phase trials via backward iteration.
3. We analyze the impact of constraints on the sender via the determined experiments by contrasting the performance with the classical Bayesian persuasion setting and when using classical Bayesian persuasion optimal signaling schemes when the sender is constrained. As a part of this, we provide a sufficient condition for when a sequential trial is equivalent to classical Bayesian persuasion with a potentially enlarged signal space.

2 Problem Formulation

There are two agents, a sender (Alice) and a receiver (Bob), participating in the game. We assume binary states of the world, $\Theta = \{\theta_1, \theta_2\}$, with a prior belief $p := \mathbb{P}(\theta_1)$ known to both agents. The receiver has to take an action $\Phi \in \{\phi_1, \phi_2\}$ which can be thought of as a prediction of the true state. We assume that the receiver's utility is given by $u_r(\phi_i, \theta_j) = 1_{\{i=j\}}$ for all $i, j \in \{1, 2\}$. To preclude discussions on trivial cases and to simplify the analysis, we assume that the sender always prefers the action ϕ_1, and her utility is assumed to be $u_s(\phi_1, \theta_i) = 1$ and $u_s(\phi_2, \theta_i) = 0$ for all $i \in \{1, 2\}$.

Before the receiver takes his action, a trial consisting of multiple phases will be run, and the outcome in each phase will be revealed to him. In each phase, one experiment will be conducted, which is chosen according to the outcomes in earlier phases. Hence, the experiment outcomes in earlier phases not only affect the interim belief but also influence the possible (sequence of) experiments that will be conducted afterward. In the most sender-friendly setup where the sender can choose any experiment in each phase without any constraints, the problem is equivalent to the classical Bayesian persuasion problem with an enlarged signal space. However, when some experiments are pre-determined conditional on a set of outcomes, the sender must take these constraints into account to design her optimal signaling structure.

To present our results on the influence of multiple phases on the sender's signaling strategy, we start with a model of two-phase trials with binary-outcome experiments in the rest of this section. We then analyze the optimal signaling strategy of this model in Sect. 3. After that, we will introduce the general model of multiple-phase trials with binary-outcome experiments and propose a systematic approach to analyze the optimal signaling structure in Sect. 4.

2.1   Model of Binary-Outcome Experiments in Two-Phase Trials

There are two phases in the trial: phase I and phase II. Unlike in the classical Bayesian persuasion problem, our goal is for the sender to not have the ability to choose the experiments to be conducted in both phases of the trial. We will start by assuming that the sender can choose any binary-outcome experiment in phase I, but both the phase-II experiments (corresponding to the possible outcomes in phase I) are determined. Formally, in phase I, there is a binary-outcome experiment with two possible outcomes $\omega_1 \in \Omega_1 = \{\omega_A, \omega_B\}$ and each outcome corresponds to a determined binary-outcome experiment, E_A or E_B, which will be conducted in phase II, respectively. The sender can design the experiment in phase I via choosing a probability pair $(p_1, p_2) \in [0, 1]^2$, where $p_i = \mathbb{P}(\omega_A | \theta_i)$. Once the probability pair (p_1, p_2) is chosen, the interim belief of the true state $\mathbb{P}(\theta_1 | \omega_1)$ can be calculated (while respecting the prior) as follows:

$$\mathbb{P}(\theta_1 | \omega_A) = \frac{pp_1}{pp_1 + (1-p)p_2} \text{ and } \mathbb{P}(\theta_1 | \omega_B) = \frac{p(1-p_1)}{p(1-p_1) + (1-p)(1-p_2)}. \quad (1)$$

On the other hand, the phase-II experiments are given in an exogenous manner beyond the sender's control. In phase II, one of the binary-outcome experiments, $E \in \{E_A, E_B\}$ will be conducted according to the outcome, ω_A or ω_B of the phase-I experiment. If ω_A is realized, then experiment E_A will be conducted in phase II; if ω_B is realized, the experiment E_B will be conducted in phase II. Similarly, we can denote the possible outcomes $\omega_2 \in \Omega_{2X} = \{\omega_{XP}, \omega_{XF}\}$ when the experiment E_X is conducted, where notation P, F can be interpreted as passing or failing the experiment. Likewise, the phase-II experiments can be represented by two probability pairs $E_1 = (q_{A1}, q_{A2})$ and $E_2 = (q_{B1}, q_{B2})$, where q_{Xi} denotes the probability that the outcome ω_{XP} is realized conditional on the experiment E_X and the state θ_i, i.e., $q_{Xi} = \mathbb{P}(\omega_{XP} | \theta_i, E_X)$.

In real-world problems, regulations, physical constraints, and natural limits are usually known to both the sender and the receiver before the game starts. Hence, we assume that the possible experiments E_1, E_2 that will be conducted in phase II are common knowledge. Given the pairs $(q_{A1}, q_{A2}), (q_{B1}, q_{B2})$, the sender's objective is to maximize her expected utility by manipulating the posterior belief (of state θ_1) in each possible outcome of phase II. However, since the phase II experiments are predetermined, the sender can only indirectly manipulate the posterior belief by designing the probability pair (p_1, p_2) of the phase-I experiment. As the sender prefers the action ϕ_1 irrespective of the true state, her objective is to select an optimal probability pair (p_1, p_2) to maximize the total probability that the receiver is willing to take action ϕ_1. Recalling the receiver's utility function discussed above, the receiver's objective is to maximize the probability of the scenarios where the action index matches the state index. Thus, the receiver will take action ϕ_1 if the posterior belief $\mathbb{P}(\theta_i|\omega) \geq \frac{1}{2}$ and take action ϕ_2 otherwise. After taking the receiver's objective into the account, the sender's optimization problem can be formulated as below:

$$\max_{p_1, p_2} \sum_{\omega_2 \in \{\omega_{AP}, \omega_{AF}, \omega_{BP}, \omega_{BF}\}} \mathbb{P}(\phi_1, \omega_2) \tag{2}$$

$$\text{s.t.} \left(\mathbb{P}(\theta_1|\omega_{AY}, q_{A1}, q_{A2}, p_1, p_2) - \tfrac{1}{2}\right)\left(\mathbb{P}(\phi_1, \omega_{AY}) - \tfrac{1}{2}\right) \geq 0 \; \forall \, Y \in \{P, F\},$$

$$\left(\mathbb{P}(\theta_1|\omega_{BY}, q_{B1}, q_{B2}, p_1, p_2) - \tfrac{1}{2}\right)\left(\mathbb{P}(\phi_1, \omega_{BY}) - \tfrac{1}{2}\right) \geq 0 \; \forall \, Y \in \{P, F\},$$

$$\mathbb{P}(\omega_{AP}) = pp_1 q_{A1} + (1-p)p_2 q_{A2},$$

$$\mathbb{P}(\omega_{BP}) = p(1-p_1)q_{B1} + (1-p)(1-p_2)q_{B2},$$

$$\mathbb{P}(\omega_{AP}) + \mathbb{P}(\omega_{AF}) = pp_1 + (1-p)p_2,$$

$$\mathbb{P}(\omega_{BP}) + \mathbb{P}(\omega_{BF}) = p(1-p_1) + (1-p)(1-p_2),$$

$$\mathbb{P}(\phi_1, \omega_2) \in [0, 1] \; \forall \omega_2 \in \{\omega_{AP}, \omega_{AF}, \omega_{BP}, \omega_{BF}\}, \qquad p_1, p_2 \in [0, 1].$$

In the sender's optimization problem (2), the first two inequalities are constraints of incentive-compatibility (IC) that preclude the receiver's deviation. The IC constraints can be satisfied when both terms in the brackets are positive or negative. That is to say; the sender can only persuade the receiver to take action ϕ_1/ϕ_2 when the posterior belief (of θ_1) is above/below 0.5. While we have written the IC constraints in a nonlinear form for compact presentation, in reality they're linear constraints. The next four equations are constraints that make the sender's commitment (signaling strategy) Bayes plausible[2]. Hence, there are 4 IC constraints and 4 Bayes-plausible constraints in the optimization problem for a two-phase trial. However, in an N-phase trial, both the number of IC constraints and the number of Bayes-plausible constraints will expand to 2^N each. Although the linear programming (LP) approach can solve this optimization problem, solving this LP problem in large Bayesian persuasion problems can be computationally hard [8]. Hence, instead of solving this optimization problem via an LP, we aim to leverage structural insights discovered in the problem to derive the sender's optimal signaling structure.

---

[2] A commitment is Bayes-plausible [16] if the expected posterior probability of each state equals its prior probability, i.e., $\sum_{\omega \in \Omega} \mathbb{P}(\omega)\mathbb{P}(\theta_i|\omega) = \mathbb{P}(\theta_i)$.

We end this section by emphasizing that this model is the only non-trivial two-phase trial configuration when determined and designed experiments coexist. In other configurations such that some of the phase-II experiments can be designed by sender, the model can be reduced to a corresponding single-phase trial in the sense that the single-phase trial will yield the same payoffs for both sender and receiver when they play optimally. (Note that the reduced model may have a different prior if the experiment in phase-I is determined).

3   Binary-Outcome Experiments in Two-Phase Trials

In this section the sender's optimization problem presented in (2) Sect. 2.1, is solved starting with the simplest non-trivial case. There are only two phases in the trial studied here, and from this we will develop more insight into how different types of experiments (determined versus sender-designed) influence the optimal signaling strategy of the sender. To be more specific, we will analyze how two determined experiments (in phase II) and one sender-designed experiment (in phase I) will impact the sender's optimal signaling strategy. Before we present the general case, we discuss a subset class of two-phase trials that are similar to single-phase trials. In this class of two-phase trials, in one of the phase-II experiments, called a *trivial* experiment, the outcome distribution is independent of the true state. Trivial experiments [2], also called (Blackwell) non-informative experiments in some literature, are frequently used as benchmarks to compare the agents' expected utility change under different signaling schemes/mechanisms, e.g., [20–22]. This two-phase model with a trivial experiment tries to capture real-world problems with one actual (and costly) experiment, e.g., clinical trials, venture capital investments, or space missions. Since the experiment is costly, a screening procedure is provided to decide whether it is worth conducting the experiment. We will then analyze the optimal signaling strategy in the general scenario, where both experiments in phase II are *non-trivial*.

3.1   Experiments with Screenings

We start by analyzing the sender's optimal strategy (signaling structure) in a simple scenario where there is one *non-trivial* experiment conducted in phase II. The sender's authority on choosing the probability pair (p_1, p_2) controls the screening process. To avoid any ambiguity, we first define what a *trivial* experiment is.

Definition 1. *An experiment E is trivial if the distribution of its outcomes $\triangle_E$ is independent of the state of the world: $\triangle_E = \triangle_{E|\theta_i}$ for all $\theta_i \in \Theta$.*

When a trivial experiment (in phase II) is conducted, the posterior belief of the state stays the same as the interim belief derived in (1). When there exists a trivial experiment in the two phase-II trial options, then Lemma 1 states that the sender and the receiver's expected utility under the optimal signaling strategy is the same as in the (single-phase) classical Bayesian persuasion problem.

Lemma 1. *When the state space is binary, both sender and receiver's expected utilities are the same in the following two Bayesian persuasion schemes under each scheme's optimal signaling strategy:*

1. *Bayesian persuasion in a single-phase trial,*
2. *Bayesian persuasion in a two-phase trial with a sender-designed phase-I experiment and a trivial experiment in phase II.*

In the single-trial classical Bayesian persuasion setting, the optimal signaling strategy only mixes the two possible states in one outcome (e.g., when the prosecutor claims the suspect is guilty). On the other outcome, the sender reveals the true state with probability one (e.g., when the prosecutor says the suspect is innocent). When there is a trivial experiment in phase II, the other experiment (supposing that it will be conducted at outcome ω_B) will be rendered defunct by the sender's choice of experiments in phase I. This phenomenon occurs because the sender can always choose to reveal the true state when the non-trivial experiment is to be conducted, i.e., by setting $\mathbb{P}(\theta_1|E_B) = 1$ or $\mathbb{P}(\theta_2|E_B) = 1$; and the classical Bayesian persuasion strategy can be replicated. In essence, having a trivial experiment in the phase-II trial does not constrain the sender.

3.2   Assumptions and Induced Strategies

Next we detail the optimal signaling strategy in our two-phase trial setting with general binary-outcome experiments. To aid in the presentation and to avoid repetition, we make two assumptions without loss of generality and introduce several explanatory concepts before the analysis.

Lemma 2. *We can make the following two assumptions WLOG.*

1. *The probability of passing a phase-II experiment under θ_1 is greater than or equal to the probability under θ_2, i.e., $q_{A1} \geq q_{A2}$ and $q_{B1} \geq q_{B2}$.*
2. *When the true state is θ_1, the experiment conducted when outcome ω_A occurs is more informative[3] than the experiment conducted when outcome ω_B occurs, i.e., $q_{A1} \geq q_{B1}$.*

The sender's strategy consists of the following: choice of phase-I experiment parameters (p_1, p_2) and the persuasion strategies in phase-II for each outcome of the phase-I experiment. To understand better the choices available to the sender and her reasoning in determining her best strategy, we will study the possible persuasion strategies in phase-II; these will be called induced strategies to distinguish them from the entire strategy. Given the assumptions above on phase-II experiments, it'll turn out we can directly rule out one class of induced strategies from the sender's consideration. The other set of induced strategies will need careful assessment that we present next.

---

[3] In terms of the Blackwell informativeness from [6].

Claim. When the inequalities of the assumption $q_{A1} \geq q_{A2}$ and $q_{B1} \geq q_{B2}$ in Lemma 2 are strict, for any phase-II experiment $E_X \in \{E_A, E_B\}$), taking action ϕ_1 when E_X fails but taking action ϕ_2 when E_X passes is not an incentive compatible strategy for the receiver for any interim belief $\mathbb{P}(\theta_1|E_X) \in [0,1]$.

The above claim can be verified by comparing the posterior belief $\mathbb{P}(\theta_1|\omega)$ of each possible outcome $\omega \in \{\omega_{XP}, \omega_{XF}\}$ and the receiver's corresponding best response. Therefore, upon the outcome of a phase-I experiment being revealed (to be either ω_A or ω_B), the sender only has three different classes of *"induced strategies"* by which to persuade the receiver in phase II:

(α_X) Suggest action ϕ_1 only when the phase-II experiment outcome is a pass;
(β_X) Suggest action ϕ_1 no matter the result of phase-II experiment; and
(γ_X) Suggest action ϕ_2 irrespective of the result of phase-II experiment, which is equivalent to not persuading the receiver to take the sender-preferred action.

Given these three classes of induced strategies and the freedom to choose different induced strategies based on the phase-I experiment's outcome, the sender can use any combination of these 3^2 choices to form a set of strategies $\mathcal{S}$. To simplify the representation, we use (c_A, d_B), $c, d \in \{\alpha, \beta, \gamma\}$ to represent a *"type of strategy"* of the sender. Note that to specify a strategy S within the set of strategies, i.e., $S \in \mathcal{S}$, the probability pair (p_1, p_2) has to be determined first. Before we analyze the different strategies, we discuss the relationship between the given phase-II experiments, induced strategies, and the incentive-compatibility requirements from the sender's side (to avoid profitable deviations by the receiver). To avoid ambiguity, hereafter, when we mention incentive compatibility/incentive compatible requirements/IC strategies, we mean the condition/requirements/strategies of a sender's commitment satisfying the following statement: for every possible realized signal under this commitment, the receiver taking the sender-suggested action is incentive-compatible.

3.3 Constraints Given by Phase-II Experiments

By her choice of the experiment in phase I, the sender decides how to split the prior into the interim beliefs for the two experiments available in phase-II. The resulting interim-beliefs then lead to certain induced strategies at stage-II being applicable, i.e., incentive compatible (for the receiver). In other words, the probability pair (p_1, p_2) must make each (applied) induced strategy yield the maximum utility for the receiver. These requirements constrain the sender's choice of (p_1, p_2), and the sender needs to account for the (reduced) choice while deciding the split of the prior. Table 1 summarizes the impact in terms of the parameters of the phase-I experiment via primary requirements on (p_1, p_2) driven by the incentive compatibility while using each class of induced strategies. Hereafter, when we use IC requirements without additional specification, we mean primary IC requirements. From the entries in the table, it is clear that the phase-II experiments (indirectly) limit the sender's strategy selection where this limitation arises due the receiver's IC requirements for each induced strategy (when that induced strategy is used).

With this in mind, the sender's experiment design in phase I, essentially, is to select between different combinations of these induced strategies such that each induced strategy satisfies the constraint listed in Table 1. Hence, we next seek to understand how these IC constraints collectively determine the sender's strategy selection. To answer this, we first discuss the relationship between induced strategies, IC requirements, and the sender's expected utility.

Table 1. IC requirements of the sender's commitment based on the induced strategy

Induced strategy	Primary IC requirement	Induced strategy	Primary IC requirement
α_A	$p_1 \geq \frac{1-p}{p}\frac{q_{A2}}{q_{A1}}p_2$	α_B	$1 - p_1 \geq \frac{1-p}{p}\frac{q_{B2}}{q_{B1}}(1-p_2)$
β_A	$p_1 \geq \frac{1-p}{p}\frac{1-q_{A2}}{1-q_{A1}}p_2$	β_B	$1 - p_1 \geq \frac{1-p}{p}\frac{1-q_{B2}}{1-q_{B1}}(1-p_2)$
γ_A	$p_1 \leq \frac{1-p}{p}\frac{q_{A2}}{q_{A1}}p_2$	γ_B	$1 - p_1 \leq \frac{1-p}{p}\frac{q_{B2}}{q_{B1}}(1-p_2)$

From the sender's perspective, each induced strategy and its corresponding signals provide a path to persuade (or dissuade) the receiver to take action ϕ_1. Since the sender's objective is to maximize the probability that action ϕ_1 is taken, she would like to use the "most efficient"[4] pair of induced strategies to persuade the receiver[5]. To better understand the "efficiency" of induced strategies, we evaluate each induced strategy under a given phase-II experiment E_X:

- α_X **strategy:** To persuade receiver to take action ϕ_1 via this induced strategy, the sender needs to ensure that $\mathbb{P}(\theta_1|\omega_{XP}) \geq \frac{1}{2}$. Hence, the interim belief $\mathbb{P}(\theta_1|E_X)$ must satisfy $\mathbb{P}(\theta_1|E_X) \in \left[\frac{q_{X2}}{q_{X1}+q_{X2}}, \frac{1-q_{X2}}{2-q_{X1}-q_{X2}}\right]$, otherwise a commitment using α_X induced strategy will never be incentive-compatible. From the sender's perspective, the most efficient strategy to persuade the receiver using α_X induced strategy is to design the phase-I experiment such that $\mathbb{P}(\theta_1|E_X) = \frac{q_{X2}}{q_{X1}+q_{X2}}$. At this interim belief, the sender experiences a relative expected utility $2q_{X1}$ (with respect to the prior). When $\mathbb{P}(\theta_1|E_X) \in \left(\frac{q_{X2}}{q_{X1}+q_{X2}}, \frac{1-q_{X2}}{2-q_{X1}-q_{X2}}\right)$, the sender's marginal expected utility when the interim belief increases is $q_{X1} - q_{X2}$.

- β_X **strategy:** To persuade receiver to take action ϕ_1 with this induced strategy the sender needs to ensure that both inequalities $\mathbb{P}(\theta_1|\omega_{XP}) \geq \frac{1}{2}$ and $\mathbb{P}(\theta_1|\omega_{XF}) \geq \frac{1}{2}$ hold. Given the assumption in Lemma 2, namely $q_{X1} \geq q_{X2}$, the only constraint that can be tight is $\mathbb{P}(\theta_1|\omega_{XF}) \geq \frac{1}{2}$. Hence, IC commitments using β_X induced strategy exist only when the interim belief $\mathbb{P}(\theta_1|E_X) \geq \frac{1-q_{X2}}{2-q_{X1}-q_{X2}}$. From the sender's perspective, the most efficient

---

[4] The efficiency of a strategy is defined as $\frac{\mathbb{P}(\phi_1|\text{interim belief, the induced strategy used})}{\mathbb{P}(\theta_1|\text{interim belief, the induced strategy used})}$.

[5] When the prior falls in the region where the optimal signaling strategy is non-trivial.

strategy to persuade the receiver using a β_X induced strategy is to design the phase-I experiment such that $\mathbb{P}(\theta_1|E_X) = \frac{1-q_{X2}}{2-q_{X1}-q_{X2}}$ with the resulting relative expected utility $1 + \frac{1-q_{X1}}{1-q_{X2}}$. Unlike an α_X induced strategy where the sender still gets a positive utility gain when the interim belief increases, for a β_X induced strategy the sender's marginal expected utility gain when the interim belief increases is 0 when $\mathbb{P}(\theta_1|E_X) > \frac{1-q_{X2}}{2-q_{X1}-q_{X2}}$.

- γ_X **strategy:** Since the sender suggests the receiver to take action ϕ_2 in this strategy, the sender's expected utility is 0 when using this induced strategy.

According to the discussion above, it is clear that the sender will not use the set of strategies corresponding to (γ_A, γ_B) unless the prior $p = 0$. Besides, we know that different induced strategies provide different relative expected utility to the sender. When induced strategies are used in the most efficient manner, the relative expected utility under a α_X induced strategy is at most $2q_{X1}$, and the average expected utility under a β_X induced strategy is at most $1 + \frac{1-q_{X1}}{1-q_{X2}}$.

Since these two values capture the best scenario that the sender can achieve by tailoring the interim belief under the given experiment, we define this pair of ratios, $(2q_{X1}, 1 + \frac{1-q_{X1}}{1-q_{X2}})$ as a function of (the given) experiment, denoted by $PerP(E_X)$; this pair is called the *persuasion potential*.

Definition 2. *Given an experiment $E_X = (q_{X1}, q_{X2})$, the persuasion potential of this experiment, $PerP(E_X)$, is the pair $\left(2q_{X1}, 1 + \frac{1-q_{X1}}{1-q_{X2}}\right)$.*

To provide some insights on the importance and use of the persuasion potential we preview Corollary 1. Corollary 1 states that the sender only uses induced strategies in the most efficient manner, i.e., $\mathbb{P}(\theta_1|E_X) = \frac{q_{X2}}{q_{X1}+q_{X2}}$ when an induced strategy α_X is used and $\mathbb{P}(\theta_1|E_X) = \frac{1-q_{X2}}{2-q_{X1}-q_{X2}}$ when an induced strategy β_X is used. Thus, the persuasion potential can simplify the sender's search for the optimal signaling strategy. When a particular induced strategy is used in the most efficient manner described in the above parameter, the interim belief is now determined. Therefore, the sender does not need to search for the optimal signaling strategy from the whole set of IC strategies but only needs to search from a small number of strategies that generate the particular interim beliefs.

3.4   Persuasion Ratio and the Optimal Signaling Structure

Since the sender wants to maximize the total probability of action ϕ_1, she needs to compare different sets of strategies formed by different pairs of induced strategies. To compare each set of strategies, we introduce the persuasion ratio of a set of strategies for a given value of the prior.

Definition 3. *Given a set of incentive-compatible strategies $\mathcal{S}$, e.g., $\mathcal{S} = (c_A, d_B)$ with $c, d \in \{\alpha, \beta, \gamma\}$ which satisfying IC requirements, the persuasion ratio of the set of strategies $\mathcal{S}$ is the maximum total probability of action ϕ_1 is taken (under a strategy within the set) divided by the prior p, $PR(\mathcal{S}, p) = \max_{\mathcal{S} \in \mathcal{S}} \frac{\mathbb{P}(\phi_1|\mathcal{S})}{p}$.*

Careful readers may notice that if we multiply the persuasion ratio with the prior, the value will be the (maximum) expected utility the sender can achieve from the given set of strategies. Since the sender's expected utility is monotone increasing in the prior p regardless of which set of strategies the sender adopts, the persuasion ratio under a given prior can be viewed as the relative utility gain this set of strategies can offer to the sender. Hence, given a specific prior, if a set of strategies has a higher persuasion ratio with respect to another set of strategies, the sender should use a strategy in the former instead of the latter.

According to this discussion, we can draw a persuasion ratio curve for each set of strategies as the prior is varied in $[0, 1]$. Abusing notation, we represent the persuasion ratio curve by $PR(\mathcal{S})$. It may appear that an optimization needs to be carried out for each value of the prior. However, structural insights presented in the following two lemmas considerably simplify the analysis. Properties presented in Lemma 3 narrow down the space where the sender needs to search for the optimal signaling strategy. This allows us to depict persuasion ratio curves $PR(S)$ for some basic strategies. On top of that, Lemma 4 provides a systematic approach to derive persuasion ratio curves for all types of strategies.

Lemma 3. *Given a type of strategy $\mathcal{S}$ and a prior p, there exists a (sender's) optimal strategy $S \in \mathcal{S}$ which satisfies one of the following two conditions:*

1. *At least one IC requirement of the constituent induced strategies is tight;*
2. *There is a signal that will be sent with probability 1 under S.*

Before discussing the simplifications that Lemma 3 yields in terms of the key properties for solving the problem, we give an intuitive outline of the proof of Lemma 3. When the IC requirements of the two induced strategies are not tight and both signals are sent with non-zero probability in a strategy S, the sender can increase her expected utility by slightly raising the probability of the signal with a higher persuasion ratio (and adjust the probability of the other signal to respect the prior) to form a strategy S_+. The sender can keep doing this 'slight' modification of her strategies until either one of the IC conditions is satisfied or the signal is sent with probability one.

Given this lemma, the persuasion ratio curve of the following types of strategies: $(\alpha_A, \gamma_B), (\beta_A, \gamma_B), (\gamma_A, \alpha_B), (\gamma_A, \beta_B)$ can be determined immediately since the IC requirement can never be tight for the γ class induced strategy. For the remaining four types of strategies: $(c_A, d_B), c, d \in \{\alpha, \beta\}$, the following lemma aids in solving for the strategy meeting the persuasion ratio without a point-wise calculation. As a preview of result of Lemma 5, once we derive the persuasion ratio curve for each type of strategies via Lemma 4, we can immediate identify the optimal signaling strategy by overlaying those curves in one figure.

Lemma 4. *Given the persuasion ratio curves of the types of strategies (c_A, γ_B) and (γ_A, d_B), denoted by $PR((c_A, \gamma_B), p)$ and $PR((\gamma_A, d_B), p)$, respectively, the persuasion ratio curve of the set of incentive compatible strategies (c_A, d_B), denoted by $PR((c_A, d_B), p)$, is the generalized concave hull[6] of the functions*

---

[6] If $PR((c_A, \gamma_B), \cdot)$ and $PR((\gamma_A, d_B), \cdot)$ were the same function $f(\cdot)$, then this would be its convex hull.

$PR((c_A, \gamma_B), p)$ and $PR((\gamma_A, d_B), p)$ when $\mathcal{S}_{(c_A, d_B)}(p) \neq \emptyset$:

$$PR((c_A, d_B), p) = \max_{\substack{x, u, v \in [0,1], \\ xu + (1-x)v = p}} xPR((c_A, \gamma_B), u) + (1-x)PR((\gamma_A, d_B), v)), \quad (3)$$

where $\mathcal{S}_{(c_A, d_B)}(p)$ is the set of IC strategies of (c_A, d_B) at p.

The proof of Lemma 4 uses the structure of the sender's expected utility when no γ induced strategy is used in the types of strategies employed. The sender's expected utility function under (c_A, d_B) at prior p can be represented as a linear combination of her utility function under (c_A, γ_B) at prior u and her utility function under (γ_A, d_B) at prior v, then the optimization problem of solving the optimal phase-I experiment parameters (p_1, p_2) can be transformed to the maximization problem in the statement of Lemma 4. As we have discussed the means to determine each type of strategy's persuasion ratio curve, Lemma 5 illustrates the persuasion ratio curve of the optimal signaling strategy.

Lemma 5. *The persuasion ratio curve of the optimal signaling strategy $PR^*(p)$ is the upper envelope of the different types of strategies' persuasion curves. Further, the optimal signaling strategy (under a given prior) is the strategy that reaches the frontier of the persuasion ratio curve (at that prior).*

Since a higher persuasion ratio indicates a higher (sender's) expected utility for every given prior, the sender will choose the upper envelope of the persuasion curves of the different types of strategies. Because the set $\mathcal{S}_{(c_A, d_B)}(p)$ could be empty for some prior values with a corresponding persuasion ratio $PR(c_A, d_B), p) = 0$, the main effort in proving Lemma 5 is to show the existence of an incentive-compatible commitment on the frontier of the persuasion ratio curve at every possible prior. Finally, once the persuasion ratio curve of the optimal signaling strategy is determined, we can immediately infer an optimal signaling strategy S^* under a specific prior.

For a two-phase trial or to solve the last two phases of a trial with more than two phases studied in Sect. 4, the following corollary can further simplify the sender's optimization procedure.

Corollary 1. *Let $\Pi^*(p)$ represents the optimal signaling strategy at prior p. If $\mathbb{P}(\phi_1 | \Pi^*(p)) < 1$, then the following two statements are true:*

1. *When α_X is used in $\Pi^*(p)$, the interim belief is $\mathbb{P}_{\Pi^*(p)}(\theta_1 | E_X) = \frac{q_{X2}}{q_{x1} + q_{X2}}$.*
2. *When β_X is used in the optimal signaling strategy $\Pi^*(p)$, the interim belief is $\mathbb{P}_{\Pi^*(p)}(\theta_1 | E_X) = \frac{1 - q_{X2}}{2 - q_{x1} - q_{X2}}$.*

With Corollary 1, the Eq. (3) in Lemma 4 reduces to a linear equation. Hence, the comparison in Lemma 5 and the computation in Lemma 4 can be reduced to a comparison of the (unique) corresponding IC strategies (if one exists) under the interim belief listed in Corollary 1 for different types of strategies.

3.5 Comparison with Classical Bayesian Persuasion Strategies

Given the optimal signaling strategy derived in Lemma 5, one natural follow-up question is the quantification of the sender's utility improvement obtained by adopting the optimal signaling strategy in comparison to using strategies structurally similar to the optimal strategies in classical Bayesian persuasion for a binary state of the world. Owing the page limit, we directly define a class of strategies structurally similar to the classical Bayesian persuasion strategy below and provide the justification in our online version [23].

Definition 4. *With binary states of the world, a (binary-state) Bayesian persuasion (BBP) strategy is a strategy that "mixes two possible states in one signal and reveals the true state on the other signal".*

Given the model defined in Sect. 2.1, a BBP strategy is forced to use at least one γ_X induced strategy[7]. Given a fixed type of strategy, e.g., (α_A, γ_B), an optimal BBP strategy using this type of strategy can be solved by the concavification approach after the calculation of the sender's expected utility under interim beliefs. After solving the optimal BBP strategy of a given strategy type via concavification respectively, the optimal BBP strategy is the strategy in the set of $\{(\alpha_A, \gamma_B), (\beta_A, \gamma_B), (\gamma_A, \alpha_B), (\gamma_A, \beta_B)\}$ which yields the highest expected utility for the sender. Figure 2 plots the sender's expected utility for the optimal signaling strategy and the optimal BBP strategy under a given pair of phase-II experiments: $(q_{A1}, q_{A2}) = (0.8, 0.2), (q_{B1}, q_{B2}) = (0.7, 0.3)$. The blue line in Fig. 2 is the benchmark of the sender's maximum expected utility in a single-phase scenario where the sender chooses the experiment. Note this would also be the optimal performance if one of phase-II trials were changed to a trivial experiment. As we can see, the sender's expected utility is lowered owing to the determined phase-II experiments. For low-priors, the optimal signaling strategy derived in Sect. 3.4 and the optimal BBP strategy give the sender the same expected utility[8]. However, as the prior increases, a utility gap between

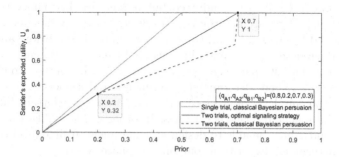

Fig. 2. Sender's utility under different problem settings and strategies (Color figure online)

---

[7] Because using either α_X, β_X requires a mixture of two possible states in one signal.

[8] Because the optimal signaling strategy in the low-prior region is (α_A, γ_B) here.

the optimal signaling strategy derived in Sect. 3.4 and the optimal BBP strategy appears and then increases until the receiver will take ϕ_1 with probability
one. The utility gap starts when the optimal signaling strategy uses strategies
(α_A, β_B) or (α_B, β_A) which are not considered in BBP strategies.

4 Binary-Outcome Experiments in Multi-phase Trials

This section generalizes the structural results in Sect. 3 to multi-phase trials.
First, we generalize the model in Sect. 2.1 to multi-phase trials and then propose
a dynamic programming algorithm to solve for the optimal signaling strategy.
The state for the dynamic program will be the interim belief on the state of the
world that results at any node in the extensive-form delineation of the problem.
As the belief at each level is determined based on the actions in earlier stages (if
any), in the backward iteration procedure, we will determine the optimal choice
of experiments by the sender (if there is a choice) for any possible interim belief.
In this dynamic programming, there is only a terminal reward that arises from
the receiver's action based on the outcome of the final trial and based on the
receiver's resulting posterior beliefs.

4.1   Model of Binary-Outcome Experiments in Multi-phase Trials

There are N phases in a trial where one binary-outcome experiment will be
conducted in each phase. However, as in the two-phase-trial settings, the specific experiment conducted in each phase is determined by the earlier phases'
outcomes. Therefore, we can model an N-phase trial by a height-N binary tree
where each leaf node represents an outcome revealed to the receiver, and each
non-leaf node represents an experiment. With this binary tree, experiment $E_{i,j}$
represents the jth experiment to be conducted at level i. When j is odd, an experiment $E_{i,j}$ will be conducted only if the experiment $E_{i-1,(j+1)/2}$ is conducted
and passed. Similarly, when j is even, an experiment $E_{i,j}$ will be conducted only
if the experiment $E_{i-1,j/2}$ is conducted but it fails. In this binary tree, some
experiments, e.g., $E_{i,j}$, are determined. However, some experiments, e.g., $E_{k,l}$,
can be designed by the sender; all the parameters are chosen before any experiment is conducted and are common knowledge. In such experiments, the sender
can choose a probability pair, e.g., $(p_{kl1}, p_{kl2}) \in [0,1]^2$. In contrast to the model
defined in Sect. 2.1, here determined experiments and sender-designed experiments can be at any level of the tree with the placement arbitrary but carried
out before the sender receives her signals. In other words, unlike the model in
Sect. 2.1, a sender may be able to design an experiment at final level (phase N)
owing to a determined experiment outcome at phase $N-1$.

In this model, the prior, the experiments that are determined (their location
on the tree and their parameters), and the sender-designed experiments (their
location on the tree) are common knowledge. The sender has to design all her
experiments simultaneously and before the game starts (when the state is realized); the designed experiments' parameters are then revealed to the receiver

(again before the game starts). Given the experiments designed by the sender and the realized outcome of a sequence of experiments, the receiver will take an action to guess the true state of the world. For simplicity of analysis, we keep the sender and the receiver's utilities the same as in Sect. 2. Then, the sender's objective is to jointly design the set of experiments that she has the flexibility to choose to maximize her expected utility, which is nothing but the probability of the receiver taking action ϕ_1. Before proceeding, we point out that this model can be easily generalized to unbalanced binary trees straightforwardly by adding dummy nodes with determined trivial experiments defined in Sect. 3.1 to construct an equivalent balanced binary tree.

4.2   Determined Versus Sender-Designed Experiments

Given the model, the sender can manipulate the phase-K interim belief only when designing an experiment at phase $K - 1$. If an experiment at phase $K - 1$ is determined, then the phase-K interim belief is a function of interim belief at phase $K - 1$. Therefore, figuring out how these two types of experiments, determined and sender-designed experiments, will influence every given phase's interim belief is the key to solving for the optimal signaling strategy. We start by noting that if the posterior belief at a leaf node is given, then the receiver's action is determined - he will take the action with the highest posterior probability unless there is a tie, in which case he is indifferent and will follow the sender's recommendation. Therefore, we can use backward iteration and the principle of optimality to determine the optimal signaling. We start by considering the last phase's experiments when the sender can design them.

Experiments at Phase N. Recall the result we have discussed in Sect. 3.2, a determined experiment in the last phase (phase-II in Sect. 3.2) limits the sender's strategy choice to one of three induced strategies. Besides, the best scenario that the sender can achieve via using these induced strategies (without violating the IC requirement) is captured by the persuasion potential of the determined experiment. However, when there is a sender-designed experiment at phase N and the interim belief[9] $\tilde{p} \leq \frac{1}{2}$, the sender can always design an experiment which makes two states equally likely when this experiment passes and reveal the less-preferred state (by sender) when it fails. If we cast this sender-designed experiment in terms of a determined experiment, the sender-designed experiment will have a persuasion potential $(2,2)$[10]. Thus, no matter the type of experiment at phase N, we can capture the sender's optimal set of induced strategies via a persuasion potential.

---

[9] See the corresponding footnote in [23] for the discussion of the interim belief $\tilde{p} > \frac{1}{2}$.
[10] See the corresponding footnote in [23] for the detailed derivation.

ALGORITHM 1: Dynamic programming approach for multi-phase trials

---

Input: The set of determined experiments $\mathbf{E_D}$, the binary tree structure

Output: The optimal persuasion ratio curve

1. For each experiment at phase N, $E_{N,i}$ $i \in \{1, ..., 2^N\}$, solve its persuasion potential $Prep(E_{N,i})$

2. For each experiment at phase $N-1$, $E_{N-1,i}$ $i \in \{1, ..., 2^{N-1}\}$, find the optimal persuasion ratio curve using $(Prep(E_{N,2i-1}), Prep(E_{N,2i}))$.

3. K=N-2

4. **while** $K > 0$ **do**

> For each experiment at phase K, $E_{K,i}$ $i \in \{1, ..., 2^K\}$, find the optimal persuasion ratio curve using equation (4) or Claim 2
>
> K=K-1

end

5. Return the optimal persuasion ratio curve at phase 1

Experiments in Phase $N-1$. In the second-last phase, results in Sect. 3.4 describe a sender-designed experiment's role in the optimal signaling strategy: *pick the strategy on the frontier of all persuasion-ratio curves.* However, if the experiment is determined in the second-last phase, an additional constraint on the interim belief between the second-last phase and the last phase is enforced. That is to say, the set of (feasible) strategies will shrink. Fortunately, after enforcing the constraints, the process of searching for the optimal signaling strategy under a determined experiment is the same as the sender-designed experiment, i.e., pick the strategy in the frontier of all persuasion ratio curves. Therefore, at each possible branch of phase $N-1$, we can plot an optimal persuasion ratio curve capturing the sender's optimal signaling strategy at phase $N-1$ and phase N.

Experiments in Earlier Phases. Now we consider experiments in earlier phases. When we have a determined experiment in phase-K, e.g., $E_{k,i} = (q_{Ki1}, q_{Ki2})$, and we have solved the optimal persuasion ratio curves of its succeeding phase $((K+1))$, i.e., $PR^*_{K+1,2i-1}(p)$ and $PR^*_{K+1,2i}(p)$, then the optimal persuasion ratio curve at this determined phase-K experiment $E_{k,i}$ is just a linear combination of $PR^*_{K+1,2i-1}(p)$ and $PR^*_{K+1,2i}(p)$ can be written as follows:

$$PR^*_{K,i}(p) = (pq_{Ki1} + (1-p)q_{Ki2})PR^*_{K+1,2i-1}\left(\frac{pq_{Ki1}}{pq_{Ki1}+(1-p)q_{Ki2}}\right) + (p(1-q_{Ki1})$$

$$+ (1-p)(1-q_{Ki2}))PR^*_{K+1,2i}\left(\frac{p(1-q_{Ki1})}{p(1-q_{Ki1})+(1-p)(1-q_{Ki2})}\right) \qquad (4)$$

For a sender-designed experiment at phase K, e.g., $E_{K,j}$, if we have already solved the optimal persuasion ratio curves of its succeeding phase-$(K+1)$ experiments $E_{K+1,2j-1}$ and $E_{K+1,2j}$, the sender's best design at $E_{K,j}$ is to find a linear combination of $PR^*_{K+1,2j-1}(p)$ and $PR^*_{K+1,2j}(p)$ which yield the highest persuasion ratio for every phase-K interim belief p. Since the persuasion ratio curve is monotone decreasing in the belief, the optimal persuasion ratio curve can be constructed similar to Lemma 4 as shown in Corollary 2.

Corollary 2. *Given two persuasion ratio curves at phase $K+1$, $PR^*_{K+1,2j-1}(p)$ and $PR^*_{K+1,2j}(p)$, the optimized persuasion ratio curve $PR^*_{K,j}(p)$ at phase K is the maximum convex combination of $PR^*_{K+1,2j-1}(p)$ and $PR^*_{K+1,2j}(p)$, i.e.,*

$$PR^*_{K,j}(p) = \max_{x,u,v\in[0,1], xu+(1-x)v=p} xPR^*_{K+1,2j-1}(u) + (1-x)PR^*_{K+1,2j}(v) \quad (5)$$

Non-binary Outcome Experiments. When the experiments have non-binary outcomes, the same approach derived above works with an increased number of phases (if complexity is not an issue).

For general non-binary experiments, see the proof of Lemma 6 in [23] for a detailed construction from non-binary to binary experiments.

Lemma 6. *Given a non-binary experiment $E = \{q_{1,1}, ..., q_{1,n}; q_{2,1}, ..., q_{2,n}\}$, we can replace it by $\lceil \log_2 n \rceil$ levels of binary outcome experiments.*

4.3   Multi-phase Model and Classical Bayesian Persuasion

At the end of this section, we mention a class of special multi-phase trials where the sender's expected utility under the optimal signaling strategy is equivalent to utility obtained from a single-phase Bayesian persuasion model. Inspired by the two-phase example with a trivial experiment in Lemma 1, the sender can implement a signaling strategy similar to single-phase Bayesian persuasion when there exists a trivial experiment in the last phase and she can design experiments in earlier phases. When the sender can design all earlier phases, she can voluntarily reduce the signal space in effect via designing the experiment $E_{i,j}$ to be $E_{i,j} = (1,1)$ or $E_{i,j} = (0,0)$, i.e., a non-informative experiment. By doing this, the sender can reduce the multi-phase trial to an equivalent two-phase trial model and then a straightforward extension of Lemma 1 will hold when there exists a trivial experiment in the last phase. The following Lemma 7 further generalizes the class of multi-phase models where the sender has the same expected utility as a single-phase Bayesian persuasion problem with a necessary pruning process defined in Definition 5. Owing to the page limit, explanations and some preliminary analysis about the robustness of signaling strategies under small perturbations from trials satisfying Lemma 7 is only available in our online version [23].

Definition 5. *Given an N-phase trial model M, a pruned N-phase trial model $Prun(M)$ is a model which recursively replaces every subtree of M by a revealing experiment $E_\theta = (1,0)$ if the subtree satisfies the following condition, starting from the leaves: the (sub)root of this subtree has a trivial (determined) experiment E_X with at least one of its succeeding experiments non-trivial (but determined).*

Note that the pruned tree will potentially be unbalanced.

Lemma 7. *Given an N-phase trial M with binary-outcome experiments, if there exists a pruned N-phase trial model $Prun(M)$ such that the following two conditions hold, then the sender's expected utility is given by an equivalent single-phase Bayesian persuasion model.*

1. *For every non-trivial determined experiment, its sibling is either a trivial or a sender-designed experiment.*
2. *There exists a least one sender-designed experiment in each (from root to leaf) experiment sequence of $Prun(M)$.*

References

1. Au, P.H.: Dynamic information disclosure. Rand J. Econ. **46**(4), 791–823 (2015)
2. Basu, D.: Statistical information and likelihood [with discussion]. Sankhyā: Indian J. Stat. Series A 1–71 (1975)
3. Bergemann, D., Morris, S.: Information design: a unified perspective. J. Econ. Lit. **57**(1), 44–95 (2019)
4. Bizzotto, J., Rüdiger, J., Vigier, A.: Dynamic persuasion with outside information. Am. Econ. J. Microecon. **13**(1), 179–94 (2021)
5. Bizzotto, J., Vigier, A.: Can a better informed listener be easier to persuade? Econ. Theor. **72**(3), 705–721 (2020). https://doi.org/10.1007/s00199-020-01321-w
6. Blackwell, D.: Equivalent comparisons of experiments. Ann. Math. Statist. **24**, 265–272 (1953)
7. Dughmi, S., Kempe, D., Qiang, R.: Persuasion with limited communication. In: Proceedings of the 2016 ACM Conference on Economics and Computation, pp. 663–680 (2016)
8. Dughmi, S., Xu, H.: Algorithmic Bayesian persuasion. SIAM J. Comput. **50**(3), 68–97 (2019)
9. Ely, J.C.: Beeps. Am. Econ. Rev. **107**(1), 31–53 (2017)
10. Ely, J.C., Szydlowski, M.: Moving the goalposts. J. Polit. Econ. **128**(2), 468–506 (2020)
11. Farhadi, F., Teneketzis, D.: Dynamic information design: a simple problem on optimal sequential information disclosure. Available at SSRN 3554960 (2020)
12. Forges, F., Koessler, F.: Long persuasion games. J. Econ. Theory **143**(1), 1–35 (2008)
13. Gradwohl, R., Hahn, N., Hoefer, M., Smorodinsky, R.: Algorithms for persuasion with limited communication. In: Proceedings of the 2021 ACM-SIAM Symposium on Discrete Algorithms (SODA), pp. 637–652. SIAM (2021)
14. Honryo, T.: Dynamic persuasion. J. Econ. Theory **178**, 36–58 (2018)
15. Kamenica, E.: Bayesian persuasion and information design. Ann. Rev. Econ. **11**, 249–272 (2019)
16. Kamenica, E., Gentzkow, M.: Bayesian persuasion. Am. Econ. Rev. **101**(6), 2590–2615 (2011)
17. Kolotilin, A., Mylovanov, T., Zapechelnyuk, A., Li, M.: Persuasion of a privately informed receiver. Econometrica **85**(6), 1949–1964 (2017)
18. Le Treust, M., Tomala, T.: Persuasion with limited communication capacity. J. Econ. Theory **184**, 104940 (2019)
19. Li, F., Norman, P.: On Bayesian persuasion with multiple senders. Econ. Lett. **170**, 66–70 (2018)
20. Li, J., Zhou, J.: Blackwell's informativeness ranking with uncertainty-averse preferences. Games Econom. Behav. **96**, 18–29 (2016)
21. Meigs, E., Parise, F., Ozdaglar, A., Acemoglu, D.: Optimal dynamic information provision in traffic routing. arXiv preprint arXiv:2001.03232 (2020)
22. Nguyen, A., Tan, T.Y.: Bayesian persuasion with costly messages. Available at SSRN 3298275 (2019)
23. Su, S.T., Subramanian, V., Schoenebeck, G.: Bayesian persuasion in sequential trials. arXiv preprint arXiv:2110.09594 (2021)

The Optimality of Upgrade Pricing

Dirk Bergemann[1], Alessandro Bonatti[2], Andreas Haupt[3]([✉]),
and Alex Smolin[4]

[1] Department of Economics, Yale University, New Haven, USA
`dirk.bergemann@yale.edu`
[2] Sloan School of Management, Massachusetts Institute of Technology,
Cambridge, USA
`bonatti@mit.edu`
[3] Institute for Data, Systems, and Society, Massachusetts Institute of Technology,
Cambridge, USA
`haupt@mit.edu`
[4] Toulouse School of Economics, University of Toulouse Capitole, Toulouse, France

Abstract. We consider a multiproduct monopoly pricing model. We provide sufficient conditions under which the optimal mechanism can be implemented via upgrade pricing—a menu of product bundles that are nested in the strong set order. Our approach exploits duality methods to identify conditions on the distribution of consumer types under which (a) each product is purchased by the same set of buyers as under separate monopoly pricing (though the transfers can be different), and (b) these sets are nested.

We exhibit two distinct sets of sufficient conditions. The first set of conditions weakens the monotonicity requirement of types and virtual values but maintains a *regularity* assumption, i.e., that the product-by-product revenue curves are single-peaked. The second set of conditions establishes the optimality of upgrade pricing for type spaces with *monotone marginal rates of substitution (MRS)*—the relative preference ratios for any two products are monotone across types. The monotone MRS condition allows us to relax the earlier regularity assumption.

Under both sets of conditions, we fully characterize the product bundles and prices that form the optimal upgrade pricing menu. Finally, we show that, if the consumer's types are monotone, the seller can equivalently post a vector of single-item prices: upgrade pricing and separate pricing are equivalent.

Keywords: Revenue maximization · Mechanism design · Strong duality · Upgrade pricing

1 Introduction

1.1 Motivation and Results

Pricing multiple goods with market power is a canonical problem in the theory of mechanism design. It is also a challenge of growing importance and complexity

© Springer Nature Switzerland AG 2022
M. Feldman et al. (Eds.): WINE 2021, LNCS 13112, pp. 41–58, 2022.
https://doi.org/10.1007/978-3-030-94676-0_3

for online retailers and service providers, such as Amazon and Netflix. Both in theory and in practice, designing the optimal *mixed bundling* mechanism, (i.e., pricing every subset of products) becomes exceedingly complex in the presence of a large number of goods.

A natural question is then whether simpler pricing schemes are optimal under suitable demand conditions. A simple, commonly used mechanism consists of *upgrade pricing*, whereby the available options are ranked by set inclusion, i.e., some goods are only available as add-ons, Ellison (2005). For example, many online streaming services use a tiered subscription model, whereby users can pay to upgrade to a "premium package"—a subscription with a larger selection of the provider's content relative to the "basic package", Philips (2017).

In this paper, we obtain sufficient conditions under which upgrade pricing maximizes the seller's revenue. Our approach consists of first identifying conditions under which the consumer's types can be ordered in terms of their absolute or relative willingness to pay for the seller's goods, and then ranking the goods themselves by the profitability of selling them to larger sets of consumer types. Our sufficient conditions not only establish the optimality of *some* upgrade pricing menu: they also show that the optimal bundles are deterministic, and they reveal the order in which they are ranked in the menu. That is, we identify all the nested bundles that appear in the seller's menu, and the profit-maximizing price for each one.

Our results consist of two distinct sets of conditions. The first set of conditions (Theorem 1) illustrates the essence upgrade pricing optimality in what we label as "regular" settings. While these conditions are reminiscent of regularity in one dimension, they are in fact weaker than the monotonicity of the buyer's multidimensional types and of the (item by item) Myersonian virtual values. What we require is for the consumer's types to be ranked in such a way that the virtual values for each item are negative over an initial and positive over a final segment. Furthermore, we require any consumer with a positive virtual value for an item to also have a larger value for that item, relative to any type with a negative virtual value. At the optimal prices, the lowest type buying each good is indifferent between buying it and not buying it. Finally, the sets of types buying each item are nested under the *weak* monotonicity property, which implies the optimal allocation can be implemented via upgrade pricing.

The second set of conditions (Theorem 2) describes our best attempt at extending our approach to non-regular distribution of types. In order to further weaken the regularity requirement, we restrict attention to type spaces for which the relative preference ratios for any two goods are monotone across types. An example of ordered relative preferences is if higher types have a stronger preference for good 2 over good 1. We refer to such a condition as "monotone marginal rates of substitution" (monotone MRS).

The intuition for our two results can be grasped by considering the demand functions for each good separately. Under monotonicity and monotone MRS, the optimal monopoly prices for each of the goods are ranked. In the special case where the Myersonian virtual values for our ordered types

$$\phi_i^k = \theta_i^k - \frac{1 - F_i}{f_i}\left(\theta_{i+1}^k - \theta_i^k\right)$$

are also monotone for each item k, the first set of conditions applies.

When virtual values are not monotone, however, they can cross zero more than once. In that case, the result still holds, but the proof requires the right ironing procedure. Our ironing procedure relaxes the standard approach of Myerson (1981) and the literature up to Haghpanah and Hartline (2020). Specifically, we do not iron with the goal of monotone virtual values, which corresponds to a concave revenue curve. Rather we iron towards single-crossing virtual values which leads to a *quasi*concave revenue curve. We then use the structure implied by monotone MRS to derive a dual certificate of optimality.

Under either set of conditions, each good is purchased by the same set of buyers that would buy it if that were the seller's *only* product. We further show (Theorem 3) that, if the consumer's types are (not weakly) monotone, the seller can equivalently post the vector of single-item monopoly prices—i.e., bundling is redundant. For example, in the case of two goods sold separately, monotone type spaces mean that no consumer type will buy good 2 without also buying good 1. More generally, the seller benefits from restricting the set of bundles the consumer can purchase through a proper menu of options with the upgrade property. However, examples also show that implementability through separate pricing is neither necessary nor sufficient for the optimality of upgrade pricing.

1.2 Related Literature

First and foremost, our paper contributes to the economics literature on product bundling. The profitability of mixed bundling relative to separate pricing was first examined by Adams and Yellen (1976), and further generalized by McAfee et al. (1989). More recently, a number of contributions have studied the optimal selling mechanisms in the case of two or three goods, and derived conditions for the optimality of pure bundling (see, for example, Manelli and Vincent (2006) and Pavlov (2011)). Daskalakis et al. (2017) use duality methods to characterize the solution of the multiproduct monopolist's problem, and show how the optimal mechanism may involve a continuum of lotteries over items. Bikhchandani and Mishra (2020) derive conditions under which the optimal mechanism is deterministic when the buyer's utility is not necessarily additive. Finally, Ghili (2021) establishes conditions for the optimality of pure bundling when buyers' values are interdependent. Relative to all these papers, we focus on a specific class of simple mechanisms, which includes pure bundling as a special case.

Hart and Nisan (2017) and Babaioff et al. (2014) also study the properties of simpler schemes. The former derives a lower bound on the revenue obtained from separate item pricing. The latter obtains an upper bound on the revenue of the optimal mechanism, relative to the better of pure bundling and separate pricing.

In the context of nonlinear pricing, Wilson (1993) suggested a "demand profile" approach that determines the price of each incremental unit by treating it as separate market. This approach is particularly attractive in settings where there is a natural ordering over the items. This in particular is the case when there is a homogeneous good that is offered in various quantities, such as in energy markets for electricity or water. This approach naturally generates a sequence of upgrade prices. The demand profile approach, and in particular the incremental pricing rule implied by it, does not always yield an optimal mechanism as consumers may wish to obtain earlier units in order to obtain the later units. Thus, a contribution of the current paper is to determine when upgrade pricing is exactly optimal and then to find the upgrade prices as solutions to the global revenue maximization problem rather than the incremental item problem. Other papers make assumptions that make sure that a demand profile-type approach yields an optimal mechanism. In Johnson and Myatt (2003), buyers have unit demand and sellers offer different varieties of a single good. The approach in their paper is to assume a quality ranking on the varieties and to solve for the upgrade prices—the additional payments required to buy a better variety. The survey of the nonlinear pricing literature by Armstrong (2016) covers related approaches that optimize upgrades separately.

Our formulation of the dual problem follows Cai et al. (2016), who present a general duality approach to Bayesian mechanism design. Cai et al. (2016) formulate virtual valuations in terms of dual variables, state the weak and the strong duality results, and use them to establish lower bounds for relative performance of simple mechanisms. An important contribution by Haghpanah and Hartline (2020) exploits the duality machinery to provide sufficient conditions for the exact optimality of a specific, simple mechanism—pure bundling—consisting of offering a maximal bundle at a posted price. Under their sufficient conditions, the dual variables can be recovered from a single-dimensional problem in which the seller is restricted to bundle all items together.

We follow the approach of Haghpanah and Hartline (2020) by leveraging the duality approach to provide sufficient conditions for the optimality of a particular class of mechanisms. Haghpanah and Hartline (2020) gave a characterization of the optimality of the grand bundle, we provide a characterization for upgrade pricing. As upgrade pricing allows multiple bundles to be present in the menu, we cannot assign the dual variables by solving a one-dimensional problem. Instead, we develop a novel ironing algorithm that generates these variables by ironing different item's revenue curves for different types. Under our sufficient conditions, the so-constructed virtual surplus is maximized by an element-wise monotone allocation that can be implemented by upgrade pricing; by complementary slackness, this certifies the optimality of upgrade pricing. Because pure bundling is one instance of upgrade pricing, our conditions differ from those of Haghpanah and Hartline (2020).

Our ironing differs from existing ironing approaches using duality and tackles a more general problem. In comparison to Haghpanah and Hartline (2020), we prove optimality for mechanisms with menu size surpassing two. Fiat et al. (2016) studies a two-parameter model, and uses an ironing approach that leads from the revenue curves to their concave closure. Devanur et al. (2020) generalizes

Fiat et al. (2016) to more general orders on the second parameter. Our approach tackles optimality for an arbitrary finite number of items and varies the ironing procedure. On a technical level, our ironing procedure yields quasi-concave ironed revenue curves, whereas the ironed revenue curves in Haghpanah and Hartline (2020); Fiat et al. (2016); Devanur et al. (2020) are concave.

Our results also feed into a literature specifying optimal finite mechanisms for multi-dimensional types. (Daskalakis et al. 2017, section 7) for example characterizes the optimal mechanisms for the two-good monopolist problem if the optimal mechanism has a particular structure. While Daskalakis et al. (2017) requires that the region of the type space that is not allocated any item is not adjacent to all regions getting specific constant allocations, upgrade pricing mechanisms consistently break this requirement.

1.3   Structure of the Paper

The model is introduced in Sect. 2. The first set of sufficient condition is presented in Sect. 3. In Sect. 4, we present our results for monotone MRS type spaces. In Sect. 5, we discuss the relationship between separate pricing and upgrade pricing. We conclude in Sect. 6.

2   Model

We consider a standard multiple-good monopoly setting. There is a single seller of $d \geq 1$ goods and a single buyer. The seller's marginal costs of production are normalized to zero. The buyer's utility function is additive across goods. We refer to the vector of marginal utilities $\theta_i \in \mathbb{R}^d$ as the buyer's type. Therefore, the utility of buyer type θ_i from the consumption vector $q \in [0,1]^d$ is given by

$$U(\theta_i, q) = \sum_{k=1}^{d} \theta_i^k q^k.$$

We also adopt the shorthand notation $\langle \theta_i, q \rangle := \sum_{k=1}^{d} \theta_i^k q^k$. As a convention, we denote types by subscripts and items by superscripts. The buyer's utility is quasi-linear in transfers and her outside option is also normalized to zero.

The buyer knows her type. From the seller's perspective, the buyer's type is distributed over a finite set $\Theta \subseteq \mathbb{R}_+^d$, with $|\Theta| = n$, according to the distribution $f \in \Delta(\Theta)$. For any positive integer n, we adopt the convention that $[n] := \{1, 2, \ldots, n\}$, and we index types by $i \in [n]$. We let $f_i := f(\theta_i)$ and denote the cumulative distribution sequence by $F_i = \sum_{j=1}^{i} f_j$, $i \in [n]$.

The seller aims to maximize revenue. By the revelation principle, we can focus on direct mechanisms $(q, t) = (q_i, t_i)_{i \in \{0\} \cup [n]}$. These mechanisms can be interpreted as menus with $n+1$ items so that item i delivers consumption vector q_i at price t_i and item $(q_0, t_0) := (0, 0)$ captures the buyer's outside option.

We call a menu *upgrade pricing* if $\{q_0, q_1, \ldots, q_n\}$ can be ordered in the component-wise partial order on $\mathbb{R}^d$ given by $q \leq q' \Leftrightarrow \forall k \in [d]: q^k \leq (q')^k$. Our main goal is to provide conditions under which upgrade pricing maximizes the seller's revenue among all direct mechanisms.

3 Optimal Mechanisms for Regular Distributions

We will make prominent use of the (partial) Lagrangian duality-based certificate of optimality used by Cai et al. (2016). We state the underlying duality result to fix notation.

3.1 Duality

In what follows, we will associate with λ_{ji} the Lagrange multiplier of the incentive compatibility constraint of type θ_j deviating to type θ_i, $j \in [n], i \in \{0\} \cup [n]$:

$$\langle q_j, \theta_j \rangle - t_j \geq \langle q_i, \theta_j \rangle - t_i.$$

We note that the incentive constraints corresponding to λ_{j0}, $j \in [n]$ are type j's individual rationality constraints. As a main tool in our analysis, we define the multi-dimensional *virtual values* associated with Lagrange multipliers $\lambda \in \mathbb{R}^n \times \mathbb{R}^{n+1}$ as

$$\phi_i^\lambda := \theta_i - \frac{1}{f_i} \sum_{j=1}^n \lambda_{ji}(\theta_j - \theta_i). \tag{1}$$

Lemma 1. *A mechanism* $(q_i, t_i)_{i \in \{0\} \cup [n]}$ *maximizes revenue if and only if there exist multipliers* λ_{ji}, $j \in [n]$, $i \in \{0\} \cup [n]$ *such that*

1. $\lambda_{ji} \geq 0$ *(Non-Negativity)*
2. $(q_i)_{i \in [n]}$ *optimizes* $\max_{(q_i)_{i \in [n]} \in [0,1]^n} \sum_{i=1}^n f_i \langle q_i \cdot \phi_i^\lambda \rangle$ *(Virtual Welfare Maximization)*
3. $f_i = \sum_{j=0}^n \lambda_{ij} - \sum_{j=1}^n \lambda_{ji}$ *for all* $i \in [n]$ *(Feasibility of Flow)*
4. $\lambda_{ji}(\langle q_j, \theta_j \rangle - t_j - \langle q_i, \theta_j \rangle - t_i) = 0$ *for all* $j \in [n], i \in \{0\} \cup [n]$ *(Complementary Slackness)*
5. *There are transfers* t *such that* (q, t) *is incentive compatible and individually rational (Implementability)*

We call the dual variables λ_{ji}, $j \in [n]$, $i \in [n] \cup \{0\}$ *flows* from type j to type i whenever they are non-negative and satisfy Lemma 1 item 3. This name is inspired by flow conservation constraints from the maximum flow and minimum cost flow problem in discrete mathematics (Korte and Vygen 2011).

The proof of this lemma is contained in the full version of this paper Bergemann et al. (2021a).

3.2 A Sufficient Condition for Regular Distributions

Our first set of sufficient conditions for upgrade pricing optimality consists of a weak monotonicity condition and a regularity condition.

We call a type distribution F *weakly monotone* with cutoffs $i^1, i^2, \ldots, i^d \in [n]$ if for any $i, j \in [n]$ and $k \in \{1, 2, \ldots, d\}$,

$$i \leq i^k \leq j \implies \theta_i^k \leq \theta_j^k.$$

Note that weak monotonicity is strictly weaker than monotonicity: for each item, only order comparisons with respect to a cutoff type need to hold, whereas types above or below the cutoff can be arbitrarily ordered.

Similarly, a type distribution F is *regular* with respect to cutoffs $i^1, i^2, \ldots, i^d \in [n]$ if for any $i, j \in [n]$ and $k \in \{1, 2, \ldots, d\}$,

$$i \leq i^k \leq j \implies \phi_i^k \leq 0 \leq \phi_j^k, \tag{2}$$

where ϕ_i denotes the *initial d-dimensional virtual values*

$$\phi_i := \theta_i - \frac{1 - F_i}{f_i}(\theta_{i+1} - \theta_i). \tag{3}$$

The initial d-dimensional virtual values can be seen as multi-dimensional versions of the virtual values in Myerson (1981).

We say that a type distribution F is *compatibly* weakly monotone and regular if it is both weakly monotone and regular with respect to the same set of cutoffs. When such cutoffs i^k exist, they are essentially unique except between contiguous types of vanishing virtual value ϕ_i^k and monotone types θ_i^k, $i \in [n]$, $k \in [d]$. Subfigure 1a illustrates a type distribution with this property.

Our regularity condition can be equivalently stated in terms of the *pseudo-revenues*

$$R_i^k := (1 - F_{i-1})\theta_i^k. \tag{4}$$

Subfigure 1c depicts pseudo-revenues. We call (4) *pseudo*-revenue because, without an assumption that the values are monotone with respect to the component-wise partial order, the pseudo-revenue does not correspond to the revenue from sales of item k at a posted price of θ_i^k. In particular, because we have

$$\begin{aligned}
\frac{R_i^k - R_{i+1}^k}{f_i} &= \frac{(1 - F_{i-1})\theta_i^k - (1 - F_i)\theta_{i+1}^k}{f_i} \\
&= \frac{f_i\theta_i^k - (1 - F_i)(\theta_{i+1}^k - \theta_i^k)}{f_i} = \theta_i^k - \frac{1 - F_i}{f_i}(\theta_{i+1}^k - \theta_i^k) = \phi_i^k, \quad (5)
\end{aligned}$$

imposing regularity with respect to the cutoffs i^k is equivalent to requiring that R_i^k is single-peaked with peak i^k. While pseudo-revenues do not have immediate economic meaning, they are an important technical tool, in particular for our analysis of non-regular distributions in Sect. 4.

Theorem 1. *If the type distribution F is compatibly weakly monotone and regular with respect to cutoffs $(i^k)_{k \in [d]}$, then upgrade pricing is optimal. In particular, the following mechanism is optimal:*

$$q_i^k := \begin{cases} 1 & i \geq i^k \\ 0 & else. \end{cases}, \quad i \in [n], k \in [d]. \tag{6}$$

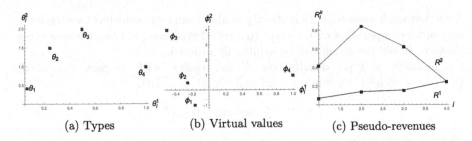

(a) Types   (b) Virtual values   (c) Pseudo-revenues

Fig. 1. Types, virtual values and pseudo-revenues for type space $\Theta = \{(9/128,$ $27/64), (1/4, 3/2), (1/2, 2), (1, 1)\}$ and type distribution $f = (7/16, 3/16, 1/8, 1/4)$. The optimal mechanism sells good 2 at a price of 1 and good 1 as an upgrade, also at a price of 1. All types except θ_1 buy good 2, and only type θ_4 buys good 1.

Proof of Theorem 1. Define the dual variables

$$\hat{\lambda}_{ji} = \begin{cases} 1 - F_i & \text{if } j = i+1 \\ 0 & \text{else.} \end{cases} \tag{7}$$

Observe that, by definition, $\hat{\lambda}$ induces the initial virtual values, $\phi_i = \phi_i^{\hat{\lambda}}$.

We check the properties of Lemma 1. Virtual welfare maximization, Condition 2, follows from

$$q_i^k = 1 \overset{(6)}{\Longleftrightarrow} R_i^k \geq R_{i+1}^k \overset{(5)}{\Longleftrightarrow} \phi_i^k \geq 0.$$

For flow preservation, Condition 3, observe that

$$\sum_{j=1}^{n} \hat{\lambda}_{ij} - \sum_{j=0}^{n} \hat{\lambda}_{ji} = 1 - F_{i-1} - (1 - F_i) = f_i.$$

The mechanism is implementable, Condition 5, by assumption of compatible weak monotonicity and regularity.

Finally, we need to check that complementary slackness (Condition 4) holds. Observe that $\hat{\lambda}_{ij} > 0$ implies $j = i - 1$. Hence, all types must be indifferent between their allocation and payment and the allocation and payment of the next lower type. If the next lower type has the same allocation and payment, this is clearly satisfied. Otherwise, this is the first type buying an upgrade. If this type were not indifferent between buying it and not buying it, the price of the upgrade could be raised, and the revenue increased, without affecting other types' incentives. Thus, this type must be indifferent between their allocation (and payment) and the next lower type's allocation. □

Our assumptions of regularity and weak monotonicity relax the monotonicity of types and Myersonian virtual values by allowing for permutations above and below the monopoly price. These assumptions nonetheless require that the set of types that buy each object remains an upper selection, and conversely the set of types that do not buy remains a lower selection. The intuition for why this works is similar to the idea that the monopoly price does not depend on the valuations of types that are not marginally buying, just as long as they do not become marginal buyers.

These assumptions depend on the fixed order of types we have introduced in the model. Thus, if there exists an order that satisfies these assumptions, upgrade pricing is optimal. Furthermore, multiple orders of types might satisfy the theorem's conditions for a given type distribution F. In this case, the theorem can be used to certify optimality of mechanism (6), based on the different orders. As optimality of a mechanism for a distribution F does not depend on the order on types, the revenue of (6) must be the same for all orders with which the conditions of Theorem 1 are satisfied.

Our next set of conditions imposes similar requirements, strengthened appropriately to allow for non-regular type distributions, which require ironing.

4 Optimal Mechanisms for Non-regular Distributions

We now establish the optimality of an upgrade pricing mechanism in settings without regularity. The weaker sufficient conditions will replace the regularity condition and will allow for ironing to be part of the optimal mechanism. The new sufficient conditions will serve to allow us to perform the ironing procedure item-by-item, and limit the interaction of constraints across items. We say that a type space Θ has *monotone marginal rates of substitution* if

$$1 \leq i \leq j \leq n \text{ and } 1 \leq k \leq l \leq d \implies \frac{\theta_i^l}{\theta_i^k} \leq \frac{\theta_j^l}{\theta_j^k}.$$

for any $i, j \in [n]$, $l, k \in [d]$.

Recall that *pseudo-revenue* is given by $R_i^k = (1 - F_i)\theta_i^k$.

We call a scalar sequence $(R_i)_{i \in [n]}$, *quasi-concave* if there is a cutoff $i' \in [n]$ such that $i' \leq i \leq j$ or $j \leq i \leq i'$ implies $R_i \geq R_j$. We call the point-wise smallest quasi-concave sequence that point-wise dominates $(R_i)_{i \in [n]}$ its *quasi-concave closure* and denote it by $(\overline{R}_i)_{i \in [n]}$.

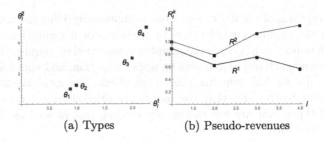

(a) Types   (b) Pseudo-revenues

Fig. 2. Type space and pseudo-revenues for type space $\Theta = \{(^{57}/_{64}, 1), (1, ^5/_4),$ $(2, 3), (^9/_4, 5)\}$ and type distribution $f = (^3/_8, ^1/_4, ^1/_8, ^1/_4)$. The optimal mechanism sells good 1 at a price of $^{57}/_{64}$, and good 2 as an upgrade at a price of 5. All types buy good 1, and only type θ_4 buys good 2.

We will make regular use of the sequence $(\overline{R}_i^k)_{i \in [n]}$, the quasi-concave closure of the pseudo-revenue for item k.

To allow for our construction of a dual certificate of optimality, we need additional assumptions. These will be formulated in terms of *candidate ironing intervals*. For a pseudo-revenue R, we call a set of contiguous types $I \subseteq [n]$ with

$$\overline{R}_i^k \neq R_i^k \tag{8}$$

for all $i \in I$ such that there is no superset of contiguous types $I' \supseteq I$ such that (8) holds for all $i \in I'$, a *candidate ironing interval for item k*. (With slight abuse of language, we refer to discrete sets of contiguous types as *intervals*.) Every item k may have several candidate ironing intervals, and every type can be contained in a candidate ironing interval for different items.

We relax the regularity assumption on pseudo-revenues R_i^k. Instead of assuming regularity, i.e. R_i^k to be single-peaked with peak i^k, we assume two properties that are in combination weaker than regularity. We call a type distribution F *mostly regular* if for some cutoffs $i^k \in \arg\max_{i \in [n]} R_i^k$ and any i such that $i^k < i \leq i^{k+1}$, the following hold:

1. (No partial overlap) If I is a candidate ironing interval of item k and J is a candidate ironing interval of item $k+1$, then either $I \cap J = \emptyset$ and there is $i \in [n]$ such that $I < i < J$ or $J < i < I$, or one of I, J is a subset of the other excluding its endpoints.
2. (No ironing on neighboring maxima) For any ironing candidate interval I of item k, $i^k, i^{k+1} \notin I$.
3. (Not too shuffled) For any candidate ironing interval $I \subseteq \{i^k + 1, i^k + 2, \ldots, i^{k+1} - 1\}$ and $i \in I$,

$$\theta_{\min I}^{k+1} \leq \theta_i^{k+1} \qquad\qquad \theta_{\max I}^k \leq \theta_{\max I+1}^k$$

Finally, we call a distribution *compatibly* weakly monotone and mostly regular if it is weakly monotone and mostly regular with respect to the same cutoffs i^k, $k \in [d]$.

Note that monotone MRS by itself is not a restrictive assumption. For example, in two dimensions, every type set can be ordered in order of monotone MRS. In combination with compatible weak monotonicity and mostly regularity, this assumption becomes stronger.

Subfigure 2a shows the type space of a compatibly weakly monotone and mostly regular type distribution, and Subfig. 2b its pseudo-revenues.

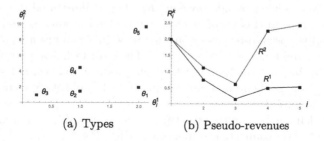

(a) Types   (b) Pseudo-revenues

Fig. 3. Failure of no overlap: $\{2,3\}$ is a candidate ironing interval for item 2, $\{3,4\}$ is a candidate ironing interval for item 1.

Conversely, Fig. 3 shows an instance of a distribution over a monotone MRS type space that is *not* mostly regular. In particular, this example fails the first condition, because it involves overlapping candidate ironing intervals. We will use our assumptions to construct dual variables $(\lambda_{ij})_{i,j\in[n]}$ by ironing pseudo-revenues for each item. In our proof that there is an optimal mechanism with an upgrade pricing allocation, we will use monotone MRS to show that for each type, ironing is only needed for two items, the lowest item in the MRS order that the type bought, and the highest item in the MRS order that she didn't buy. We will use the first two conditions of mostly regularity to show that from these two items, we can select a single item to iron at a time, while not changing the other item's virtual values in a way that will break virtual welfare maximization of the allocation. As in Theorem 1, weak monotonicity ensures implementability of an upgrade pricing allocation, i.e., the existence of a price vector $(t_i)_{i\in[n]}$ such that the mechanism (q,t) is incentive compatible and individually rational. To allow for our ironing procedure to work, we also need a mild requirement on the monotonicity of types beyond weak monotonicity. While weak monotonicity was a requirement that could be formulated item-by-item, this requirement links the type order of neighboring items.

Theorem 2. *Let Θ have monotone marginal rates of substitution. If the type distribution F is compatibly weakly monotone and mostly regular with respect to cutoffs $(i^k)_{k\in[d]}$, then upgrade pricing is optimal. In particular, the following mechanism is optimal:*

$$q_i^k := \begin{cases} 1 & i \geq i^k \\ 0 & else, \end{cases} \quad i \in [n], k \in [d]. \tag{9}$$

Note that the allocation (9) is the allocation that arises from separate monopoly pricing.[1]

To prove Theorem 2, we will construct a sequence of flows λ_i from $i = n$ down to $i = 1$, starting with $\hat{\lambda}$, the initial flow that induces Myersonian multi-dimensional virtual values. Given a definition of pseudo-revenue implied by a flow, our Ironing Algorithm will, for each type i and for at least one item k, iron to match induced pseudo-revenue with the quasi-concave closure of multi-dimensional Myersonian pseudo-revenue, $\overline{R}_i^k$. This is illustrated in Fig. 4.

The main steps in this proof are to show that the ironing is well-defined in that such implied pseudo-revenue is attainable with a non-negative and feasible flow (Lemma 6 and Lemma 4, respectively). The most technical part of the proof consists of showing that the Ironing Algorithm produces dual variables that maximize virtual welfare (Lemma 3 and Lemma 7 (a)), and satisfy complementary slackness (Lemma 7 (b)).

Our first lemma is a main structural tool to link different items' virtual values and is tightly connected to monotone MRS. For $k \in [d]$, $i \in [n]$, and flow λ, denote the normalized virtual value by

$$\nu_i^{\lambda,k} := \frac{\phi_i^{k,\lambda}}{\theta_i^k}.$$

The property that we will use repeatedly is that $\nu_i^{\lambda,k}$ has the same sign as $\phi_i^{\lambda,k}$. We call a flow *downward* if $\lambda_{ji} > 0$ for $i,j \in [n]$ implies that $j > i$.

Lemma 2. *Let Θ have monotone MRS. For any non-negative downward flow λ, $\nu_i^{\lambda,k} \geq \nu_i^{\lambda,l}$ for any $1 \leq k \leq l \leq d$ and $i \in [n]$.*

Proof. It follows from definitions and monotone marginal rates of substitution that

$$\frac{\phi_i^{\lambda,k}}{\theta_i^k} = \frac{\theta_i^k - \frac{1}{f_i}\sum_{j=1}^n \lambda_{ji}(\theta_j^k - \theta_i^k)}{\theta_i^k} = 1 + \frac{1}{f_i}\sum_{j=i}^n \lambda_{ji} - \frac{1}{f_i}\sum_{j=i}^n \lambda_{ji}\frac{\theta_j^k}{\theta_i^k}$$

$$\geq 1 + \frac{1}{f_i}\sum_{j=i}^n \lambda_{ji} - \frac{1}{f_i}\sum_{j=i}^n \lambda_{ji}\frac{\theta_j^l}{\theta_i^l} = \frac{\theta_i^l - \frac{1}{f_i}\sum_{j=1}^n \lambda_{ji}(\theta_j^l - \theta_i^l)}{\theta_i^l} = \frac{\phi_i^{\lambda,l}}{\theta_i^l}.$$

□

The next Lemma shows that virtual welfare maximization reduces to virtual welfare maximization for the neighboring items, i.e., the last item that a type buys and the first item that a type does not buy—with respect to the MRS order.

---

[1] In Sect. 5, we further explore the relationship between upgrade pricing and separate pricing, by showing conditions under which the allocation (9) can be implemented by a vector of single-item prices.

Lemma 3. *Assume Θ has monotone MRS and mostly regular and that there exists a non-negative downward flow λ such that for any $i \in [n]$ such that $i^k \leq i \leq i^{k+1}$, we have $\phi_i^{\lambda,k} \geq 0$ and $\phi_i^{\lambda,k+1} \leq 0$. Then, the allocation in (9) maximizes virtual welfare.*

Proof. Fix $i \in [n]$ such that $i^k \leq i \leq i^{k+1}$. Note that as $\phi_i^{\lambda,k}$ and $\nu_i^{\lambda,k}$ are positive multiples of each other, Lemma 2 implies the implications

$$\phi_i^{\lambda,k+1} \leq 0 \implies \phi_i^{\lambda,l} \leq 0, \qquad l \geq k+1$$
$$\phi_i^{\lambda,k} \geq 0 \implies \phi_i^{\lambda,l} \geq 0, \qquad l < k.$$

Therefore, the assumption implies that $\phi_i^{\lambda,l} \leq 0$ for any $l > k$ and $\phi_i^{\lambda,l} \geq 0$ for any $l \leq k$, which ensures virtual welfare maximization of (9).   □

For $k = 0$ and $k = d$ this Lemma reduces virtual welfare maximization for all items, and ironing for all items, to virtual welfare maximization for the first resp. last item. Finding a flow that maximizes virtual welfare reduces to ironing the (one-dimensional) virtual values ϕ_i^1 and ϕ_i^d. For types $i \leq i^1$ and $i \geq i^d$, we can hence use techniques from one-dimensional ironing and iron the pseudo-revenue to its concave closure in a discrete variant of Myerson (1981)'s procedure. From now, our discussion therefore focuses on $k \in [d-1]$ and $i \in [i^k + 1, i^{k+1}]$, i.e. types where an ironing that ensures virtual welfare maximization for both item k and item $k+1$ is needed.

The following algorithm will make use of $\hat{\lambda}$ as defined in (7), the *initial flow* and of a generalization of the pseudo-revenue. The pseudo-revenue associated to a flow λ, $R_i^{\lambda,k}$ is

$$R_i^{\lambda,k} = \sum_{j=i}^{n} f_j \phi_j^{\lambda,k}.$$

This generalization is intuitive, as virtual values are, as in (5), slopes of pseudo-revenues

$$\frac{R_i^{\lambda,k} - R_{i+1}^{\lambda,k}}{f_i} = \frac{\sum_{j=i}^{n} f_j \phi_j^{\lambda,k} - \sum_{j=i+1}^{n} f_j \phi_j^{\lambda,k}}{f_i} = \phi_i^{\lambda,k}. \qquad (10)$$

Our algorithm will adjust a flow by raising one point in a revenue sequence at a time, from right to left. We will prove that this will yield slopes of revenue sequences—i.e. virtual values—which have the correct sign for virtual welfare maximization of (9). This is non-trivial, as pseudo-revenues for different items might not move in the same direction when dual variables are changed.

$$\lambda \leftarrow \hat{\lambda};$$

for $i = n$ **to** 1 **do**

Let $\gamma_i \in [0,1]$ be maximal such that for

$$\lambda'_{ji} \leftarrow \gamma_i \lambda_{ji}, \quad \forall j : n > j > i$$
$$\lambda'_{j(i-1)} \leftarrow \lambda_{j(i-1)} + (1 - \gamma_i)\lambda_{ji}, \quad \forall j : n > j > i \qquad (11)$$
$$\lambda'_{i(i-1)} \leftarrow \lambda_{i(i-1)} - (1 - \gamma_i) \sum_{i'=i}^{n} \lambda_{i'i},$$

$R_i^{\lambda',\kappa(i)} = \overline{R}_i^{\kappa(i)}$ holds;

$\lambda \leftarrow \lambda';$

Return λ';

Algorithm: Ironing, parameterized by an ironing mapping $\kappa \colon [n] \to [d]$

The flow (11) was used earlier in Haghpanah and Hartline (2020). An important difference is that Haghpanah and Hartline (2020) choose γ_i to iron the revenue sequence of the grand bundle to the concave closure of pseudo-revenue. Instead, we iron to the *quasi*-concave closure of (their equivalent of) pseudo-revenue of an item $\kappa(i)$. The parameter γ_i can be found as solution to a system of linear equations. We show that a solution $\gamma_i \in [0,1]$ exists in Lemma 6.

We first observe that the Ironing Algorithm outputs a flow which is non-negative and feasible. The proofs of the next two statements are in the full version of this paper Bergemann et al. (2021a).

Lemma 4. *The output of the Ironing Algorithm is a flow, i.e. non-negative and satisfies flow preservation, Lemma 1 Item 3.*

Next observe that in the Ironing Algorithm, iteration i changes the revenue (for any item k) only for type i. Hence, our ironing algorithm raises pseudo-revenue for one type at a time.

Lemma 5. *For any iteration i, $R_j^{\lambda',k} = R_j^{\lambda,k}$ for any $j \neq i$. In particular, $\phi_j^{\lambda',k} = \phi_j^{\lambda,k}$ for $j \notin \{i-1, i\}$.*

Before showing that γ_i in the algorithm always exists, we define the ironing function $\kappa(i)$.

By no ironing on neighboring maxima, each candidate ironing interval I must be contained in an interval $\{i^k, i^k+1, \ldots, i^{k+1}\}$. By this condition, in addition to no partial overlap, for each type i, there is a unique inclusion maximal candidate among the candidate ironing intervals for items k and $k+1$. We let $\kappa(i)$ denote the item this interval is a candidate ironing interval for. If i is not part of any ironing interval, we set $\kappa(i)$ arbitrarily in $\{k, k+1\}$. We call $\kappa(i)$ the *ironed item for type i* and piece-wise constant intervals of κ *ironing intervals*.

Lemma 6. *Assume that F is mostly regular. Then, for each $i \in [n]$, γ_i such that $R_i^{\lambda_i(\gamma_i),\kappa(i)} = \overline{R}_i^{\kappa(i)}$ exists. In particular, the Ironing Algorithm is well-defined.*

Proof. We prove this statement by induction from $i = n$ down to 1. Let $i \in [n]$ and assume that $R_{i+1}^{\lambda,\kappa(i)} = \overline{R}_{i+1}^{\lambda,\kappa(i)}$. If i is not part of an ironing interval, then by definition of ironing intervals and Lemma 5, $R_i^{\lambda,k} = \overline{R}_i^{\lambda,k}$, and the induction step is trivial by choosing $\gamma_i = 1$, yielding $R_i^{\lambda'(1),k} = \overline{R}_i^{\lambda'(1),k}$. Otherwise, i is in an ironing interval. Let $\kappa(i) = k$. By no partial overlap, if $i + 1$ is part of an ironing interval, it must be part of the same ironing interval, in particular must have been ironed for item k. Hence, by the induction hypothesis, $R_{i+1}^{\lambda,k} = \overline{R}_{i+1}^{\lambda,k}$.

Denote

$$\overline{\phi}_i^k = \frac{\overline{R}_i^k - \overline{R}_{i+1}^k}{f_i}$$

the slope of the quasi-concave closure of pseudo-revenue of item k at type i. By definition of the quasi-concave closure, the slope of the revenue curve must be non-positive,

$$\overline{\phi}_i^k \leq 0.$$

As all types are non-negative, we get that

$$\overline{\phi}_i^k \leq 0 \leq \theta_i^k = \phi_i^{\lambda'(0),k}. \tag{12}$$

Again by Lemma 5, $\overline{R}_{i+1}^k = R_{i+1}^{\lambda'(0),k}$. Therefore

$$\overline{R}_i^k = f_i \overline{\phi}_i^k + \overline{R}_{i+1}^k = f_i \overline{\phi}_i^k + R_{i+1}^{\lambda'(0),k}$$
$$\leq f_i \phi_i^{\lambda'(0),k} + R_{i+1}^{\lambda'(0),k} = R_i^{\lambda_i(0),k}.$$

In particular, $\overline{R}_i^k \leq R_i^{\lambda_i(0),k}$.

Also, by Lemma 5 and the definition of the quasi-concave closure, $R_i^{\lambda'(1),k} = R_i^{\lambda,k} \leq \overline{R}_i^k$. As $\gamma \mapsto R_i^{\lambda'(\gamma),k}$ is a continuous function, the existence of the desired $\gamma \in [0,1]$ follows from the Intermediate Value Theorem. □

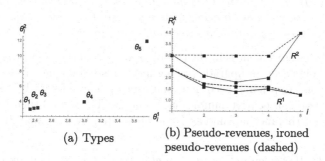

(a) Types

(b) Pseudo-revenues, ironed pseudo-revenues (dashed)

Fig. 4. Ironing of virtual values and corresponding pseudo-revenues.

The last lemma before the proof of Theorem 2 shows that the output of the algorithm satisfies complementary slackness and the condition of Lemma 3, which is sufficient for virtual welfare maximization.

Lemma 7. *Assume that Θ is has monotone MRS, and that F is mostly regular. Then, q maximizes virtual welfare and satisfies the requirements of Lemma 3 with respect to λ', the output of the Ironing Algorithm.*

The proof of this statement is in the full version of this paper Bergemann et al. (2021a). Having this result, we are ready to finish the proof of Theorem 2.

Proof of Theorem 2. Implementability follows from weak monotonicity and the definition of the optimal mechanism, (9). Non-negativity and feasibility of flow are properties of the Ironing Algorithm shown in Lemma 4. Virtual welfare maximization and complementary slackness have been shown in Lemma 7.   □

5   Upgrade Pricing and Separate Pricing

In both Theorem 1 and Theorem 2, we established the optimality of an upgrade pricing mechanism that yields the same allocation as separate (item by item) monopoly pricing, though not necessarily the same transfers. We will show in this section that, under monotonicity with respect to the component-wise partial order, separate pricing and upgrades become equivalent—upgrade pricing is redundant.

We say that the type space Θ is *monotone* if $\theta_i^k \leq \theta_j^k$ for any $i < j \in [n]$ and $k \in [d]$.

We call a mechanism separate pricing if a type separately chooses whether to buy each item k at a price p_k. Formally, a mechanism satisfies separate pricing if it can be written as:

$$q_i^k = \begin{cases} 1 & \theta_i^k \geq p_k \\ 0 & \text{else,} \end{cases} \qquad t_i = \sum_{k=1}^{d} p_k \mathbb{1}_{q_i^k=1}.$$

Theorem 3. *If the type space Θ is monotone, then the outcome of any upgrade pricing mechanism can be implemented via separate pricing, and conversely. When the type space is not monotone, neither implication needs to hold.*

The proof of this statement is in the complete version of the paper Bergemann et al. (2021a). Whenever an upgrade pricing mechanism implements the allocation of optimal separate pricing, each marginal type $\underline{\theta}_k$ is indifferent by construction between the two consecutive bundles b_{k-1} and b_k. Theorem 3 then implies that the outcome of this mechanism can be implemented by the separate monopoly prices.

Corollary 1. *If Θ is monotone, q is an allocation of an optimal upgrade pricing mechanism, and q is the allocation of separate monopoly pricing, then separate monopoly pricing is optimal.*

Adding a monotonicity condition to both of our main theorems, Theorem 1 and Theorem 2, we hence obtain two sets of sufficient conditions under which separate monopoly pricing is optimal.

Corollary 2. *If Θ is monotone and F is regular, separate monopoly pricing is optimal.*

Corollary 3. *If Θ is monotone and has a monotone marginal rates of substitution, and F is mostly regular, then separate monopoly pricing is optimal.*

6 Conclusion

It is a common practice for a seller to offer bundles of products or services that are ordered in a way that more expensive bundles contain all items from less expensive bundles as well as some extra items. In this paper, we provide sufficient conditions under which such "upgrade pricing" schemes are exactly optimal for a monopolist seller.

There are several ways in which the current analysis could be extended. First, our conditions could be relaxed to account for richer type spaces and type distributions, such as a continuum of types in the d-dimensional space. One natural extension can be obtained immediately: assume that a type distribution can be split into several type cohorts, in fact quantized type space, such that each type cohort satisfies the conditions of our theorems. Our results imply that the optimal mechanisms in each respective cohort are upgrade pricing. In this respect, Bergemann et al. (2021b) show that in nonlinear pricing problems, the revenue of the continuous type space is generally well approximated by a finite quantized type space.

Second, our sufficient conditions for the optimality of upgrade pricing may be complemented by necessary conditions. In doing so, one may want to distinguish between conditions on type distributions and type spaces. For example, one may ask which type spaces guarantee that upgrade pricing is optimal irrespective of the type distribution.

Finally, throughout the paper we highlight the interplay between optimality of different pricing schemes: bundling, upgrade pricing, and separate sales. It would be instructive to provide a more complete characterization of the cases in which one of these schemes strictly outperforms another.

Acknowledgements. We thank Mark Armstrong and the seminar audience at MIT for helpful comments. Bergemann and Bonatti acknowledge financial support through NSF SES 1948336. Smolin acknowledges funding from the French National Research Agency (ANR) under the Investments for the Future (Investissements d'Avenir) program (grant ANR-17-EURE-0010).

References

Adams, W.J., Yellen, J.L.: Commodity bundling and the burden of monopoly. Quart. J. Econ. **90**(3), 475–498 (1976)

Armstrong, M.: Nonlinear pricing. Ann. Rev. Econ. **8**, 583–614 (2016)

Babaioff, M., Immorlica, N., Lucier, B., Weinberg, S.M.: A simple and approximately optimal mechanism for an additive buyer. In: 2014 IEEE 55th Annual Symposium on Foundations of Computer Science, pp. 21–30 (2014)

Bergemann, D., Bonatti, A., Haupt, A., Smolin, A.: The optimality of upgrade pricing. arXiv:2107.10323 (2021a)

Bergemann, D., Yeh, E., Zhang, J.: Nonlinear pricing with finite information. Games Econom. Behav. **130**, 62–84 (2021b)

Bikhchandani, S., Mishra, D.: Selling two identical objects. arXiv:2009.11545 (2020)

Cai, Y., Devanur, N.R., Weinberg, S.M.: A duality based unified approach to Bayesian mechanism design. In: Proceedings of the Annual ACM Symposium on Theory of Computing (2016)

Daskalakis, C., Deckelbaum, A., Tzamos, C.: Strong duality for a multiple-good monopolist. Econometrica **85**(3), 735–767 (2017)

Devanur, N.R., Goldner, K., Saxena, R.R., Schvartzman, A., Weinberg, S.M.: Optimal mechanism design for single-minded agents. In: Proceedings of the 21st ACM Conference on Economics and Computation, pp. 193–256 (2020)

Ellison, G.: A model of add-on pricing. Quart. J. Econ. **120**(2), 585–637 (2005)

Fiat, A., Goldner, K., Karlin, A.R., Koutsoupias, E.: The fedex problem. In: Proceedings of the 2016 ACM Conference on Economics and Computation, pp. 21–22 (2016)

Ghili, S.: A characterization for optimal bundling of products with interdependent values. Technical report, Yale University (2021)

Haghpanah, N., Hartline, J.: When is pure bundling optimal? Rev. Econ. Stud. **88**(3), 1127–1156 (2020)

Hart, S., Nisan, N.: Approximate revenue maximization with multiple items. J. Econ. Theory **172**, 313–347 (2017)

Johnson, J.P., Myatt, D.P.: Multiproduct quality competition: fighting brands and product line pruning. Am. Econ. Rev. **93**(3), 748–774 (2003)

Korte, B.H., Vygen, J.: Combinatorial Optimization. Springer, Heidelberg (2011). https://doi.org/10.1007/978-3-642-24488-9

Manelli, A.M., Vincent, D.R.: Bundling as an optimal selling mechanism for a multiple-good monopolist. J. Econ. Theory **127**(1), 1–35 (2006)

McAfee, R.P., McMillan, J., Whinston, M.D.: Multiproduct monopoly, commodity bundling, and correlation of values. Quart. J. Econ. **104**(2), 371–383 (1989)

Myerson, R.: Optimal auction design. Math. Oper. Res. **6**, 58–73 (1981)

Pavlov, G.: Optimal mechanism for selling two goods. BE J. Theor. Econ. **11**(1), 1–33 (2011)

Philips, J.: Don't look now, but the great unbundling has spun into reverse. New York Times (2017)

Wilson, R.: Nonlinear Pricing. Oxford University Press, Oxford (1993)

On Symmetries in Multi-dimensional Mechanism Design

Meryem Essaidi$^{(\boxtimes)}$ and S. Matthew Weinberg

Princeton University, Princeton, NJ 08540, USA
{messaidi,smweinberg}@cs.princeton.edu
https://www.cs.princeton.edu/~messaidi
https://www.cs.princeton.edu/~smattw

Abstract. We consider a revenue-maximizing seller with multiple items for sale to a single population of buyers. Our main result shows that for a single population of additive buyers with independent (but not necessarily identically distributed) item values, bundling all items together achieves a constant-factor approximation to the revenue-optimal item-symmetric mechanism.

We further motivate this direction via fairness in ad auctions. In ad auction domains the items correspond to views from particular demographics, and recent works have therefore identified a novel *fairness constraint*: equally-qualified users from different demographics should be shown the same desired ad at equal rates. Prior work abstracts this to the following fairness guarantee: if an advertiser places an identical bid on two users, those two users should view the ad with the same probability [27,34]. We first propose a relaxation of this guarantee from worst-case to Bayesian settings, which circumvents strong impossibility results from these works, and then study this guarantee through the lens of symmetries, as any item-symmetric auction is also fair (by this definition). Observe that in this domain, bundling all items together corresponds to concealing all demographic data [23].

Keywords: Symmetry · Fairness · Multi-dimensional ad auctions

1 Introduction

Ad auctions are a significant source of revenue for numerous firms, causing their theoretical study to be a mainstay in both the Economics and Computer Science communities. Classical works typically design and analyze auctions that optimize the participants' collective utility (i.e. the sum of bidders' values for items they receive, also called the *welfare*) [20,30,49], or perhaps just the auctioneer's utility (i.e., her *revenue*) [42]. Recently, the ubiquity of ad auctions in domains where fairness constraints are first-order concerns has motivated a new desideratum for consideration: the *items'* utility for the outcome selected.

While it makes little sense to consider the utility of an apple or orange, recall that the items in ad auction domains are in fact users. That is, when an advertiser

© Springer Nature Switzerland AG 2022
M. Feldman et al. (Eds.): WINE 2021, LNCS 13112, pp. 59–75, 2022.
https://doi.org/10.1007/978-3-030-94676-0_4

(bidder) wins an item (impression on a particular user), that item (the user) also enjoys some utility.

In practice, these utilities are hard to quantify (even moreso than typical values for an item), and this side of the market is typically not monetized. As a result, there are no 'bids' or 'utilities' of the items/users to consider. However, some examples of high-utility ads include those for desirable jobs, low-interest loans, etc. and are subject to anti-discrimination laws. Specifically, it is considered unfair for users who are equally qualified for jobs/loans/etc. to view protected ads at different rates. Therefore, recent works have proposed considering the utility of users (items) through the lens of *fairness* [27,34]. That is, these works propose to still consider the utility of the auctioneer and bidders in the classical sense, but to additionally ensure that the outcomes are fair to the users/items.

In practice, a non-discriminatory advertiser might submit identical bids for equally-qualified users of different demographics. But [27] observes that this *fair behavior* is insufficient to achieve a *fair outcome*. Indeed, protected ads are bidding against non-protected ads (e.g. Men's Shoes, Maternity Clothes, etc.) which legally place discriminatory bids. If discriminatory advertisers place higher bids on demographic A than B, then the price of impressions for demographic A will be higher, and then *even a non-discriminatory ad will be displayed in a discriminatory manner*.

To have a simple example in mind (taken from [27]), consider the case that the auctioneer runs a second-price auction on each of two items. This auction format is ostensibly fair: there is nothing in its description that seems to bias it against any item or bidder. But consider when Bidder One (a protected ad) submits a bid of 1 for both items, and Bidder Two (a non-protected ad) submits a bid of 2 for item one and 0 for item two. Then Bidder One wins item two, and Bidder Two wins item one. As a result, the demographic corresponding to item one views no protected ads, while the demographic corresponding to item two views protected ads with probability one. That is, despite the fact that the protected advertiser bids in a non-discriminatory manner, and that the auction is ostensibly fair, the result is an unfair outcome (assuming that users prefer to see the protected ad).

While the above example is clearly stylized, this phenomenon is not just of theoretical concern. Indeed, automated systems are constantly making decisions that affect our daily lives. These systems rely on advanced algorithms and big data to make decisions which, in theory, have the potential to be better informed and more equitable. However, studies have shown that they may instead internalize and perpetuate societal biases [1,2,19,35,38,44]. Efforts on mitigating bias have ranged from examinations of the data [8,48,50], to machine learning algorithmic contributions [7,46,52], to theoretical analyses [14,27,34]. In our domain of study, works indeed find that impressions for female users are more expensive than impressions for male users, and that female users are less likely to see ads for high paying jobs [25] and STEM jobs [37], even when advertisers for such jobs are unbiased [3].

[27,34] posit that it is the job of the auction designer to guarantee fair outcomes, and propose formal fairness definitions motivated by individual fairness [26]. Requiring an auction to satisfy these constraints of course limits the auctioneer's ability to optimize its revenue and/or the bidders' welfare, so these works study the tradeoff between fairness and optimality. Our work follows this same paradigm, but deviates from prior work in a few fundamental ways, highlighted below.[1]

Bayesian vs. Worst-Case. [14,27,34] consider worst-case definitions of fairness. For example, they might demand that for all possible bids of non-protected advertisers, if a protected advertiser submits an identical bid for two demographics, those demographics view the protected ad with (almost) equal probability. These definitions arise naturally from the fairness literature upon which they build, but unfortunately also lead to strong impossibility results.

Ad auctions, however, are executed millions of times daily, and auctioneers have quite extensive Bayesian priors. Indeed, revenue optimization is typically studied in Bayesian settings, where the designer seeks to maximize their expected revenue. *We propose to also consider a Bayesian, rather than worst-case, notion of fairness.* Indeed, unfairness is undesirable exactly when it is systemic, and Bayesian notions are best suited to capture systemic phenomena. A formal definition appears in Sect. 2, and a discussion appears in Section A.1 of the long version. By considering a Bayesian notion of fairness, we're able to circumvent the impossibility results proved in [34].

Revenue vs. Welfare. [14,27,34] consider auctions that attempt to maximize welfare (more specifically: they attempt to maximize the "declared welfare," but don't assume that the declared bids correspond to the bidders' actual values). In the absence of fairness constraints, such auctions are extremely well-understood, and are particularly simple in the settings considered (e.g. if the bidders are additive[2], welfare is maximized by awarding the item to the highest bidder).

However, if an auctioneer truly wishes to optimize their expected revenue, it is well-understood that revenue-optimal auctions are *significantly* more complex [17,21,22,32,41,43,47]. *We propose to consider an auctioneer who wishes to optimize their expected revenue in a multi-dimensional Bayesian setting,* rather than one who wishes to optimize the declared welfare. Our model is the standard setup for multi-dimensional mechanism design (formally defined in Sect. 2): the auctioneer has multiple items for sale, and bidders' values for the items are drawn independently.

Direct vs. Indirect Competition. [14,27,34] consider advertisers who directly compete to display their ads to a limited supply of users. In such settings, even a benevolent platform must fail to show some ads to some users. While our model is rich enough to capture this setting, our analysis isolates a different source of competition.

---

[1] The below paragraphs distinguish our model from [14,27,34]. We discuss other related works such as [11] in Sect. 1.2.

[2] A valuation function $v(\cdot)$ is additive if for all sets S of items, $v(S) = \sum_{j \in S} v(\{j\})$.

Specifically, even when there is unlimited supply, a revenue-maximizing designer may choose not to show every ad to every user. Indeed, if they cannot offer different prices to different advertisers, they may achieve greater revenue by setting a high price that excludes some advertisers from purchasing. In this sense, advertisers indirectly compete with each other: one advertiser's bids affect the impressions sold to another due to the fact that the seller wishes to optimize their revenue, rather than due to limited supply. *We study the seller's revenue objective (rather than the limited supply of users) as a driving source of unfairness.*

Connection to Symmetries. We adopt (a Bayesian version of) the individual fairness notion proposed in [27]: an auction is fair if whenever a bidder submits an identical bid for two items, they receive those items with the same probability (in expectation over other bidders' bids). We further observe that this notion is implied by the stronger definition of *item-symmetric* [24]. Specifically, an auction is item-symmetric if whenever a bidder swaps their bids for two items, this swaps the probabilities with which they receive those items (in expectation over other bidders' bids).

Item-symmetric auctions have been studied in the multi-dimensional mechanism design literature for their own sake as a tool to optimize revenue in a computationally-efficient manner [24,36], but we use them here as a tool to guarantee fair outcomes. Specifically, *we target the design of item-symmetric auctions.*

1.1   Results and Technical Highlights

The previous paragraphs motivate our modeling decisions within the multi-dimensional mechanism design domain: we consider a single seller with m items for sale, and the buyers' values are drawn from a distribution known to the seller (Bayesian vs. Worst-Case). The seller's goal is to design a truthful auction that optimizes their expected revenue (Revenue vs. Welfare). We focus on the case where there is unlimited supply, which can alternatively be represented by a single population of potential bidders (Direct vs. Indirect Competition).[3]

From here, we wish to study item-symmetric auctions, and do so through the lens of simplicity vs. optimality [12,13,31,33]: *is there a simple, approximately-optimal item-symmetric auction?* For example, one particularly simple auction is to *bundle the items together* (that is, pick a price p and allow the buyer to receive all items for price p, or no items). It is also easy to see that this auction is item-symmetric (and therefore fair). Indeed, in the language of ad auctions, it corresponds to an auction which does not use personalized data at all, and chooses to display an ad to whatever user shows up independently of their demographics [23].

---

[3] To quickly see why the unlimited supply setting is equivalent to the single bidder setting: Because there are no supply constraints, it is feasible to pick any single-bidder mechanism and just use it for every bidder.

Another particularly simple auction is to *sell the items separately* (that is, for each item j, pick a price p_j, and allow the buyer to pick any set of items S to purchase at price $\sum_{j \in S} p_j$). Auctions of this format, while simple, are not necessarily item-symmetric (nor fair): if $p_1 > p_2$, then an advertiser could submit an identical bid of p_2 for both items yet receive only item two. A proper subclass of such auctions, which is item-symmetric (and therefore fair), is to *sell the items separately and symmetrically* (that is, set a single price p, and allow the buyer to pick any set of items and pay p per item). In the language of ad auctions, selling separately and symmetrically corresponds to setting a price of p to display an ad (independently of any data), but letting the advertisers choose any subsets of demographics to display their ads.

Our main result is that bundling the items together achieves a constant-factor approximation to the revenue-optimal item-symmetric mechanism, and that this factor can be improved by considering the better of bundling together and selling separately and symmetrically.

Main Result (See Theorems 1 and 2): For a single additive buyer, and any number of independent items, bundling together achieves a $O(1)$-approximation to the revenue-optimal item-symmetric mechanism. The maximum between bundling together and selling separately and symmetrically improves this constant.

We also provide several auxiliary results in Appendix F of the long version, which study the relationships between simple mechanisms such as bundling together, selling separately, and selling separately and symmetrically.

Brief Technical Highlight. Almost all prior work on simple vs. optimal multi-dimensional mechanism design consider selling separately as a simple mechanism, and are perfectly content to argue that the maximum between selling separately and some other simple mechanism achieves a constant-factor approximation [5,6,9,10,16,28,29,45,51]. These works proceed by proving elegant upper bounds on the optimal achievable revenue, and breaking these bounds into terms which can be approximated by simple mechanisms. In particular, there is usually a term that corresponds to "revenue achieved when a bidder has an unusually high value for some item" (e.g., "the tail" in [4,5,16,39,45], SINGLE in [6,9,10,28,29]), and this term is easily approximated by the revenue of selling separately (typically, this term is also the most straight-forward to approximate).

In our setting however, selling separately is not a symmetric mechanism, and in fact could be up to a factor of $\Omega(\#\text{items})$ better than the optimal symmetric mechanism (See Example 3 in Appendix C.2 of the long version). Therefore we need to target an upper bound that in some cases *is even tighter than the revenue achieved by selling separately*.

At a very high level, prior bounds leverage the fact that "the auctioneer cannot both extract revenue $\approx v$ when the buyer has a high value v for item j *and* revenue $\approx 2v$ when the buyer has an even higher value $2v$ for item j, because the buyer with value $2v$ can always lie and pretend that their value is v instead." Our bound instead must leverage the fact that "the auctioneer cannot both extract revenue $\approx v$ when the buyer has a high value v for *some item j* and also revenue $\approx 2v$ when the buyer has an even higher value $2v$ for *some other item ℓ.*" This is because if the auctioneer extracts revenue $\approx v$ when the buyer has value v for item j, they must also extract revenue $\approx v$ when the buyer has value v for item ℓ (by item-symmetry), and then the buyer with value $2v$ for item ℓ can always pretend that their value is instead v. The main technical challenge is figuring out a way to leverage this intuition into a concrete bound. Once we find the right approach, the complete proof is fairly clean, and is able to leverage existing machinery in the simple vs. optimal literature for multi-dimensional mechanism design.

1.2   Related Work

Multi-dimensional Mechanism Design. At a technical level, the most closely related field to our work is that of simple vs. optimal multi-dimensional mechanism design. For example, our proof outline is reminiscent of [5,39], and we provide an alternative proof outline reminiscent of [9]. Note also that our main result (that bundling together is a constant-factor approximation of the optimal revenue from any symmetric auction) implies one result from [39] (that bundling together is a constant-factor approximation of the optimal revenue when all items are i.i.d.).[4] [23] were the first to explicitly note the connection between the sale of an "uncertain item" (e.g. an impression to a user whose demographic is known only to the designer) and the classic multi-dimensional mechanism design setting.

The simple vs. optimal agenda within multi-dimensional mechanism design was initiated in seminal works [12,13,31] (and the simple vs. optimal agenda in general was initiated in [33]), and there is now a vast body of works building on these techniques [5,6,9,10,15,16,18,28,29,39,40,45,51]. At a technical level, our work bears some similarity to these works, because we rely on the same fundamental building blocks. The technical novelty in our work derives from consideration of symmetries. Prior work has also considered item-symmetric auctions, but only by finding the *optimal* item-symmetric auction computationally efficiently [24], or using the optimal item-symmetric auction to approximate the asymmetric optimum [36]. In contrast, our work aims to approximate the optimal item-symmetric auction with something simpler.

Individual Fairness in Auction Design. To our knowledge, [27] were the first to consider fairness in auction design from a theoretical perspective. Their work provides fairness definitions based on individual fairness and motivating examples demonstrating that unfair outcomes can arise from ostensibly fair auctions and non-discriminatory behavior of auctioneers. Follow-up works such as [14,34] proceed in similar models. These works provide strong impossibility results in the worst-case, but also provide matching positive results (and improve the guarantees of these positive results under restrictions on the otherwise worst-case input). Section 1 discusses extensively several ways in which our work contributes to this line of work, along with the technical differences.

---

[4] This follows as the optimal auction when all items are i.i.d. is in fact item-symmetric.

The other related work in this direction is [11], who also consider the theoretical design of fair auctions from the perspective of a revenue-maximizing seller. The biggest difference between their work and ours is that they essentially consider a single-dimensional setting (that is, they seek to optimize the Myersonian virtual value of the winning bidder for each item), but place fairness constraints across auctions for different items. They formulate a linear program in their setting and optimize their problem (exactly) computationally efficiently. Put another way, their work exclusively considers auctions which "sell items separately" from a revenue perspective, but with cross-item constraints concerning fairness.

One simple way to compare our work to these is that we focus on a simple notion of fairness, but in the sophisticated multi-dimensional mechanism design setting, whereas these prior works consider more sophisticated/quantitative notions of fairness but in simpler auction settings (either welfare-maximization or single-dimensional revenue-maximization).

Empirical Studies of Fairness in Auction Design. Empirical studies on the rate at which ads are displayed to different demographics motivate the line of work to which we contribute [3,25,37]. For example, [25] finds that ads for high-paying jobs are shown to more men than women, and [37] draws the same conclusion for STEM jobs. The empirical studies in [3] support the conjecture that this is due to "spillover effects" caused by higher competition for female views.

2   Preliminaries

Our main results consider a single seller (the advertising platform) with m items for sale (each item corresponds to an impression for a different demographic of user) to a single buyer (representing the population of advertisers, who do not directly compete for limited supply). However, we instantiate our model with $n \geq 1$ buyers (directly competing advertisers), in order to best compare with prior work.

Each buyer i has a value v_{ij} for each item j, and has value $\sum_{j \in S} v_{ij}$ for set S (that is, the bidders are additive). Each v_{ij} is drawn independently from a distribution D_{ij}, and we define $D_i := \times_j D_{ij}$ to represent the ith population of advertisers,[5] and a particular $\vec{v}_i$ represents a particular advertiser. We denote by $D := \times_i D_i$ as the entire population, and $\vec{v}$ as a particular profile of advertisers. When there is just a single bidder, we abuse notation and let $D := D_1$, and $\vec{v} := \vec{v}_1$. For discrete distributions, we let $f^D(\vec{v}) := \Pr_{\vec{w} \leftarrow D}[\vec{w} = \vec{v}]$. For a single

---

[5] For example, it could be that $D_i = D_{i'}$ for all i, i', and each bidder is drawn from the same population. This represents settings where the platform cannot price-discriminate based on properties of the advertiser. It could also be that $D_i \neq D_{i'}$. In such settings, perhaps D_i is the population of 'big' advertisers, and $D_{i'}$ is the population of 'small' advertisers, and the platform knows from which population each individual advertiser is drawn.

variable discrete distribution D, we let $F^D(\cdot)$ denote the CDF, so $F^D(v) := \sum_{w \in [0,v]} f^D(w)$.

The seller's goal is to design a truthful mechanism that maximizes their expected revenue. Specifically, a mechanism consists of a mapping from valuation profiles $\vec{v}$ to ex-post allocation probabilities $x_{ij}(\vec{v})$ for all bidders i and items j, and an ex-post price $p_i(\vec{v})$ for all bidders i. This denotes the probability that bidder i gets item j when the full vector of bids is $\vec{v}$, and the price that bidder i pays (respectively). The interim allocation rule is a mapping from valuation vector $\vec{v}_i$ to the interim allocation probability $\pi_{ij}(\vec{v}_i) := \mathbb{E}_{\vec{v}_{-i} \leftarrow D_{-i}}[x_{ij}(\vec{v}_i; \vec{v}_{-i})]$. The interim price paid is $q_i(\vec{v}_i) := \mathbb{E}_{\vec{v}_{-i} \leftarrow D_{-i}}[p_i(\vec{v}_i; \vec{v}_{-i})]$.[6] These quantities denote the probability that bidder i receives item j and the price bidder i pays (respectively), conditioned on reporting $\vec{v}_i$ *and in expectation over the remaining bids being drawn from* D_{-i}. If we wish to emphasize that these terms come from a specific mechanism M, we will write $x_{ij}^M(\cdot)$, $p_i^M(\cdot)$, etc. We say that a mechanism is truthful if it is Bayesian individually rational (BIR) and Bayesian incentive compatible (BIC). That is:

$$\sum_j v_{ij} \cdot \pi_{ij}(\vec{v}_i) - q_i(\vec{v}_i) \geq 0, \quad \forall i, \ \vec{v}_i. \tag{BIR}$$

$$\sum_j v_{ij} \cdot \pi_{ij}(\vec{v}_i) - q_i(\vec{v}_i) \geq \sum_j v_{ij} \cdot \pi_{ij}(\vec{v}_i') - q_i(\vec{v}_i'), \forall i, \ \vec{v}_i, \vec{v}'. \tag{BIC}$$

The seller's goal is to find, over all truthful mechanisms, the one maximizing her expected revenue (which can be written either as $\mathbb{E}_{\vec{v} \leftarrow D}[\sum_i p_i(\vec{v})]$ or $\sum_i \mathbb{E}_{\vec{v}_i \leftarrow D_i}[q_i(\vec{v}_i)]$).

Fairness and Symmetries. Motivated by the discussion in Sect. 1, we define an auction to be *fair* if whenever an advertiser places the same bid for an impression for two different demographics, those two demographics view the ad with the same probability. After mapping from advertiser to buyer, and demographics to items, this yields the following two definitions, depending on whether we seek a guarantee ex-post or in the interim.

Definition 1 (Fair Auction). *An auction is* ex-post fair *with respect to bidder i if for all valuation profiles $\vec{v}$, and items j, k:*

$$v_{ij} = v_{ik} \ \Rightarrow \ x_{ij}(\vec{v}) = x_{ik}(\vec{v}) \ \forall i.$$

An auction is interim fair *if for all bidders i, valuation vectors $\vec{v}_i$, and items j, k:*

$$v_{ij} = v_{ik} \Rightarrow \pi_{ij}(\vec{v}_i) = \pi_{ik}(\vec{v}_i).$$

Intuitively, an auction is ex-post fair with respect to bidder i if no matter the bids of the other bidders (advertisers), when bidder i places an identical bid for two items (views from particular demographics), they receive those items with

---

[6] We use the standard notation $\vec{v}_{-i}$ to refer to the vector of bids excluding bidder i, and D_{-i} to refer to the distribution over valuation profiles, excluding bidder i.

the same probability (those demographics view the ad with the same probability). An auction is interim fair with respect to bidder i if when bidder i places an identical bid for two items, they receive those items with the same probability in expectation over the other bidders' bids (assuming they are drawn from D_{-i}).

Both definitions are implied by the following stronger definitions (respectively), which require that the auction be invariant under relabeling of items. Below, the notation $\sigma(\vec{v})$ refers to a vector satisfying $(\sigma(\vec{v}))_{i\sigma(j)} = \vec{v}_{ij}$ for all i,j (that is, the items/demographics have been relabeled according to σ).

Definition 2 (Symmetric Auction). *An auction is ex-post symmetric with respect to bidder i if for all permutations σ on $[m]$, valuation vectors $\vec{v}_i$, and partial valuation profiles $\vec{v}_{-i}$:*

$$\sigma(\vec{x}_i(\vec{v}_i; \vec{v}_{-i})) = \vec{x}_i(\sigma(\vec{v}_i), \vec{v}_{-i}).$$

An auction is interim symmetric if for all permutations σ, bidders i, and valuation vectors $\vec{v}_i$:

$$\sigma(\vec{\pi}_i(\vec{v}_i)) = \vec{\pi}_i(\sigma(\vec{v}_i)).$$

Intuitively, an auction is symmetric if permuting a valuation vector by σ permutes the allocation vector by σ as well. We briefly observe that symmetry implies fairness.

Observation 1. *If an auction is ex-post (respectively, interim) symmetric, it is also ex-post (respectively, interim) fair.*

Proof. Let $v_{ij} = v_{ik}$, and consider the permutation σ which swaps j and k. Then $\sigma(\vec{v}_i) = \vec{v}_i$. Symmetry therefore implies[7] that $x_{ij}(\vec{v}_i; \vec{v}_{-i}) = x_{i\sigma(j)}(\sigma(\vec{v}_i); \vec{v}_{-i}) = x_{ik}(\vec{v}_i; \vec{v}_{-i})$. This completes the proof for ex-post.

Similarly by symmetry: $\pi_{ij}(\vec{v}_i) = \pi_{i\sigma(j)}(\sigma(\vec{v}_i)) = \pi_{ik}(\vec{v}_i)$. This completes the proof for interim.

A Stronger Fairness Guarantee via Symmetry. The fairness guarantees above (and those in prior work) demand that equally-valued users are shown an ad with the same probability. A stronger fairness guarantee might instead demand that if demographic i is *valued higher* by an advertiser than demographic j, then users from demographic i are shown that ad at least as often as those from demographic j. We term this property *strong monotonicity in fairness*, defined below (after mapping from advertiser to buyer, and demographics to items).

Definition 3 (Strong Monotonicity in Fairness). *An auction satisfies ex-post strong monotonicity in fairness with respect to bidder i, if for all valuation profiles $\vec{v}$, and items j, k:*

$$v_{ij} \geq v_{ik} \;\Rightarrow\; x_{ij}(\vec{v}) \geq x_{ik}(\vec{v}) \; \forall i.$$

---

[7] To see this, recall that $\sigma(\vec{x}(\vec{v}_i; \vec{v}_{-i}))$ is a vector that puts $x_{ij}(\vec{v}_i; \vec{v}_{-i})$ in the $i, \sigma(j)$ coordinate.

An auction is interim strong monotonicity in fairness *if for all bidders i, valuation vectors $\vec{v}_i$, and items j, k:*

$$v_{ij} \geq v_{ik} \Rightarrow \pi_{ij}(\vec{v}_i) \geq \pi_{ik}(\vec{v}_i).$$

Note that ex-post (resp. interim) fairness does not imply ex-post (resp. interim) strong monotonicity in fairness. Interestingly, however, [24] has already previously studied interim strong monotonicity in fairness (under the name strong monotonicity), and shown that it is implied by interim symmetry! That is, while previously studied notions of fairness alone do not imply this stronger fairness notion, symmetry does. Below we briefly repeat their observation (and it's short proof, for completeness).

Observation 2 ([24]). *If an auction is interim symmetric and Bayesian Incentive Compatible, then it satisfies interim strong monotonicity in fairness.*

Proof. By Observation 1, any auction that is interim symmetric is also interim fair. This means that if $v_{ij} = v_{ik}$, then $\pi_{ij}(\vec{v}_i) \geq \pi_{ik}(\vec{v}_i)$, so the conditions for interim strong monotonicity in fairness hold whenever $v_{ij} = v_{ik}$, and we need only consider the case when $v_{ij} > v_{ik}$.

Assume for contradiction that $v_{ij} > v_{ik}$ but $\pi_{ij}(\vec{v}) < \pi_{ik}(\vec{v})$. Advertiser i could lie and swap v_{ij} and v_{ik}. By symmetry, this swaps $\pi_{ij}(\vec{v})$ and $\pi_{ik}(\vec{v})$, and strictly increases the advertiser's interim expected value (by $(v_{ij} - v_{ik}) \cdot (\pi_{ik}(\vec{v}) - \pi_{ij}(\vec{v}))$). The auctioneer still charges advertiser i the same interim expected price (also by symmetry), giving them strictly more expected utility, and contradicting that the auction is Bayesian Incentive Compatible.

We briefly note that ex-post symmetry and ex-post incentive compatibility also imply ex-post strong monotonicity in fairness, and the proof outline is identical (but we will not formally state/prove this, as we did not formally define ex-post incentive compatibility).

Observation 3. *If an auction satisfies ex-post (respectively, interim) strong monotonicity in fairness, then it is also ex-post (respectively, interim) fair.*

Proof. Let $v_{ij} = v_{ik}$, then $v_{ij} \geq v_{ik}$ and $v_{ij} \leq v_{ik}$. By ex-post strong monotonicity in fairness, we get $x_{ij}(\vec{v}) \geq x_{ik}(\vec{v})$ and $x_{ij}(\vec{v}) \leq x_{ik}(\vec{v})$. Therefore, $x_{ij}(\vec{v}) = x_{ik}(\vec{v})$.

Similarly, by interim strong monotonicity in fairness, we get $\pi_{ij}(\vec{v}) \geq \pi_{ik}(\vec{v})$ and $\pi_{ij}(\vec{v}) \leq \pi_{ik}(\vec{v})$. Therefore, $\pi_{ij}(\vec{v}) = \pi_{ik}(\vec{v})$. This completes the proof for ex-post.

Selling Separately and Bundling Together. For a single bidder distribution D, we use the following notation:

- $\text{REV}_M(D)$: the revenue of a particular mechanism M for distribution D.
- $\text{REV}(D)$: the optimal revenue achieved by any truthful mechanism for D (formally, this is: $\sup_{M,\text{M is truthful}}\{\text{REV}_M(D)\}$). Observe that mechanisms achieving $\text{REV}(D)$ are not necessarily fair nor symmetric.

- SYMREV(D): the optimal revenue achieved by any truthful and interim symmetric mechanism for D. By definition, the mechanism giving SYMREV(D) is symmetric.
- SREV(D): the optimal revenue achieved by *selling separately* to a bidder drawn from D. That is, the seller sets a price $p_j := \arg\max_p \{p \cdot \Pr_{v \leftarrow D_j}[v \geq p]\}$ on item j, and the buyer purchases all items for which $v_j \geq p_j$. Observe that such mechanisms are not necessarily fair nor symmetric. In the context of our running example, this corresponds to the platform setting a different price to display an ad to each demographic, and allowing each advertiser to choose which demographic views to purchase.
- SSREV(D): the optimal revenue achieved by *symmetrically selling separately* to a bidder drawn from D. That is, the seller sets the same price $p := \arg\max_q \{q \cdot \sum_j \Pr_{v \leftarrow D_j}[v \geq q]\}$ on all items, and the buyer purchases all items for which $v_j \geq p$. Observe that such mechanisms are both fair and symmetric. In the context of our running example, this corresponds to the platform setting the same price to display an ad to each demographic, and allowing each advertiser to choose which demographic views to purchase.
- BREV(D): the optimal revenue achieved by *bundling together*. That is, the seller sets a price $p := \arg\max_q \{q \cdot \Pr_{\vec{v} \leftarrow D}[\sum_j v_j \geq q]\}$ on the grand bundle of all items, and the buyer either purchases all items at total price p (when $\sum_j v_j \geq p$), or nothing. Observe that such mechanisms are both fair and symmetric. In the context of our running example, this corresponds to the platform ignoring all demographic information, and allowing advertisers to show their ads to all users or none.

Mapping Between ad Auctions and Multi-dimensional Mechanism Design. We briefly repeat the connection between ad auctions and the classical multi-item auction setup formally identified by [23, Theorem 2]. An item j in the classic setting corresponds to a demographic j in the ad auction domain. Moreover, awarding the buyer item j with probability x_j corresponds to showing a user with type j their ad with probability x_j. Therefore, if advertisers have a value of v_j per click from demographic j, and demographic j represents a d_j fraction of the population, then the advertiser's value for an allocation $\vec{x}$ is $\sum_j v_j \cdot d_j x_j$. Observe that when each demographic represents the same fraction of the population (e.g. male/female) that the advertiser's valuation is simply additive. Therefore, our main results on item-symmetric mechanisms with an additive buyer directly have bite in the ad auction domain when each demographic is equally likely.[8]

We also remind the reader that [23, Theorem 2] observes that bundling items together in the classic setting corresponds to concealing demographic data in the ad auction setting. Similarly, selling separately in the classic setting corresponds to setting a price p_j to display an ad to demographic j, and letting advertisers

---

[8] We also note that it is an interesting open direction to extend our main results from an additive bidder to a 'scaled additive' bidder so that this connection holds even for non-uniform demographic distributions.

choose which subset of demographics to target. Selling separately and symmetrically further enforces that $p_i = p_j$ for all i, j.

Now that our model is formally defined, we revisit the discussion of Sect. 1 with concrete examples in Appendix A of the long version. In the next section, we present our main results.

3   Main Result: BRev is a Constant-Factor Approximation to SymRev

In this section, we prove our main result: BREV is a constant factor approximation to SYMREV. Recall that in our setting, BREV corresponds to the optimal revenue achieved by a mechanism which ignores demographic data entirely.

Theorem 1. *Let D be any additive single-bidder distribution over any number of independent items. Then $204\text{BREV}(D) \geq \text{SYMREV}(D)$.*

We prove Theorem 1 in two steps. The first step is the main step, and proves a theorem reminiscent of the main result of [5], establishing that either $\text{BREV}(D)$ or $\text{SSREV}(D)$ is a constant-factor approximation to $\text{SYMREV}(D)$ (Theorem 2). The second step argues that in fact $\text{BREV}(D)$ is a constant factor approximation to $\text{SSREV}(D)$ (Proposition 1).

Theorem 2. *Let D be any additive single-bidder distribution over any number of independent items. Then $24\text{BREV}(D) + 20\text{SSREV}(D) \geq \text{SYMREV}(D)$.*

Proposition 1. *Let D be any additive single-bidder distribution over any number of independent items. Then $9\text{BREV}(D) \geq \text{SSREV}(D)$.*

We defer the proof of Theorem 1 to Appendix B of the long version. In Appendix B.1 we show how to upper bound $\text{SYMREV}(D)$ with $\text{REV}(D')$ for a modified distribution D'. To prove Theorem 2, we will first provide a modified distribution D', show that its revenue is close to that of D, and then design a flow for D'. In Appendix B.2 we upper bound $\text{REV}(D')$ with $24\text{BREV}(D) + 20\text{SSREV}(D)$, and we provide a proof based on tools used in [31]. In Sect. 3.1 we prove Proposition 1.

In Appendix C of the long version, we provide an alternative proof based on the [9] duality framework (for the case when the distribution of the bidder's maximum value for the items is regular). In Appendix C.2 we also overview a naive attempt at applying their framework (using their "canonical flow"), which helps provide intuition for the need to go through D'.

3.1   Comparing BRev to SSRev

In this section, we prove Proposition 1: BREV is a constant factor approximation to SSREV. Recall that in our setting, BREV corresponds to the optimal revenue achieved by a mechanism which ignores demographic data entirely, while SSREV corresponds to the optimal revenue achieved by a mechanism which sets the same price to display an ad to each demographic, and allows each advertiser to choose which demographic views to purchase.

To prove Proposition 1, consider any mechanism that sets price p on each item separately, and let $q_j(p)$ denote the probability that the bidder purchases item j (that is, that $v_j \geq p$), and let $q(p) := \sum_j q_j(p)$ denote the expected number of items purchased at price p (and therefore, $\mathrm{SSREV}(D) := \sup_p\{p \cdot q(p)\}$). We will show that there is always a price p' for the grand bundle that collects a constant fraction of $p \cdot q(p)$.

We will first consider the case where $q(p) \leq 8$ (that is, at most 8 items are sold at price p in expectation). Unsurprisingly, in this case it suffices to set a price $p' := p$ on the grand bundle.

Lemma 1. *Let $q(p) \leq 8$. Then selling the grand bundle at price p generates expected revenue at least $p \cdot q(p)/9$.*

Proof. The proof follows from straight-forward calculations:

$$\Pr_{\vec{v} \leftarrow D}\left[\sum_j v_j \geq p\right] \geq \Pr_{\vec{v} \leftarrow D}[\max_j\{v_j\} \geq p] = 1 - \Pr_{\vec{v} \leftarrow D}[\forall j, v_j < p]$$

$$= 1 - \prod_j \Pr_{v_j \leftarrow D_j}[v_j < p] = 1 - \prod_j (1 - q_j(p))$$

$$\geq 1 - \prod_i e^{-q_j(p)} = 1 - e^{-q(p)}$$

$$\geq q(p)/9.$$

Above, the first line holds since the distribution for $\sum_j v_j$ stochastically dominates $\max_j v_j$. The second line follows as values are independent. The third line holds as $e^{-q_j(p)} \geq 1 - q_j(p)$, and the last line holds for all values $q(p) \leq 8$. In particular, this means that we can set price p on the grand bundle, and it will sell with probability at least $q(p)$, completing the proof.

This completes our analysis of the first case. We now consider the case where $q(p) > 8$, and we set the grand bundle price to be $p' = q(p)p/2$. We will also use the notation $\sigma^2(p)$ to denote the variance of the random variable $\sum_j \mathbb{I}(v_j \geq p)$. We quickly observe a bound on $\sigma^2(p)$, which follows as all values are independent.

Observation 4. $\sigma^2(p) = \sum_j q_j(p)(1 - q_j(p)) \leq \sum_j q_j(p) = q(p)$.

Lemma 2. *Let $q(p) > 8$. Then selling the grand bundle at price $p \cdot q(p)/2$ generates expected revenue at least $p \cdot q(p)/9$.*

Proof. Observe that certainly $\sum_j v_j \geq p \cdot q(p)/2$ when there are at least $q(p)/2$ items with value greater than p. We lower bound the probability of this event using Chebyshev's inequality:

$$\Pr_{\vec{v}\leftarrow D}[\sum_j v_j \geq p \cdot q(p)/2] \geq \Pr_{\vec{v}\leftarrow D}[\sum_j \mathbb{I}(v_j \geq p) \geq \frac{1}{2}x(p)]$$

$$\geq \Pr_{\vec{v}\leftarrow D}[|\sum_i \mathbb{I}(v_j \geq p) - q(p)| \leq \frac{1}{2}x(p)]$$

$$\geq 1 - \frac{\sigma^2(p)}{q(p)^2/4}$$

$$\geq 1 - 4/q(p) \geq 1/2$$

Proposition 1 now follows from Lemma 1 and Lemma 2.

While not related to our main result, we also explore the relationship between SSREV and BREV in the other direction, and include a proof in Appendix E of the long version. The outline is similar to a related claim in [5].

Theorem 3 (SSRev is a log approximation of BRev). *For any distribution D for a single additive buyer and m not necessarily independent items,* $\mathrm{BREV}(D) \leq 5\log(m)\mathrm{SSREV}(D)$.

In Appendix D of the long version, we analyze several examples demonstrating the relationship between BREV and SSREV.

4   Conclusions

Motivated by recent works which consider fairness constraints in welfare-maximizing or single-dimensional auctions [11,14,27,34], we introduce fairness considerations in multi-dimensional mechanism design. We study interim (rather than worst-case) notions of fairness, and use this to motivate the study of simple item-symmetric auctions. Our main technical result is that the simple auction which ignores demographic information entirely is a constant-factor approximation to the optimal item-symmetric auction.

References

1. Google translate's gender problem (and bing translate's, and systran's) (2013). https://www.fastcompany.com/3010223/google-translates-gender-problem-and-bing-translates-and-systrans
2. Google photos labeled black people 'gorillas'. USA Today (2015). https://www.usatoday.com/story/tech/2015/07/01/google-apologizes-after-photos-identify-black-people-as-gorillas/29567465/
3. Ali, M., Sapiezynski, P., Bogen, M., Korolova, A., Mislove, A., Rieke, A.: Discrimination through optimization: how Facebook's ad delivery can lead to biased outcomes. Proc. ACM Hum. Comput. Interact. **3**(CSCW), 199:1–199:30 (2019). https://doi.org/10.1145/3359301
4. Babaioff, M., Gonczarowski, Y.A., Nisan, N.: The menu-size complexity of revenue approximation. In: Proceedings of the 49th Annual ACM SIGACT Symposium on Theory of Computing, STOC 2017, Montreal, QC, Canada, 19–23 June 2017, pp. 869–877 (2017). https://doi.org/10.1145/3055399.3055426. http://doi.acm.org/10.1145/3055399.3055426

5. Babaioff, M., Immorlica, N., Lucier, B., Weinberg, S.M.: A simple and approximately optimal mechanism for an additive buyer. In: 55th IEEE Annual Symposium on Foundations of Computer Science, FOCS 2014, Philadelphia, PA, USA, 18–21 October 2014, pp. 21–30 (2014). https://doi.org/10.1109/FOCS.2014.11
6. Beyhaghi, H., Weinberg, S.M.: Optimal (and benchmark-optimal) competition complexity for additive buyers over independent items. In: Proceedings of the 51st ACM Symposium on Theory of Computing Conference (STOC) (2019)
7. Bolukbasi, T., Chang, K.W., Zou, J.Y., Saligrama, V., Kalai, A.T.: Man is to computer programmer as woman is to homemaker? Debiasing word embeddings. In: Advances in Neural Information Processing Systems, pp. 4349–4357 (2016)
8. Buolamwini, J., Gebru, T.: Gender shades: intersectional accuracy disparities in commercial gender classification. In: ACM Conference on Fairness, Accountability, Transparency (FAccT) (2018)
9. Cai, Y., Devanur, N., Weinberg, S.M.: A duality based unified approach to Bayesian mechanism design. In: Proceedings of the 48th ACM Conference on Theory of Computation (STOC) (2016)
10. Cai, Y., Zhao, M.: Simple mechanisms for subadditive buyers via duality. In: Proceedings of the 49th Annual ACM SIGACT Symposium on Theory of Computing, STOC 2017, Montreal, QC, Canada, 19–23 June 2017, pp. 170–183 (2017). https://doi.org/10.1145/3055399.3055465. http://doi.acm.org/10.1145/3055399.3055465
11. Celis, L.E., Mehrotra, A., Vishnoi, N.K.: Toward controlling discrimination in online ad auctions. In: Chaudhuri, K., Salakhutdinov, R. (eds.) Proceedings of the 36th International Conference on Machine Learning, ICML 2019, Long Beach, California, USA, 9–15 June 2019. Proceedings of Machine Learning Research, vol. 97, pp. 4456–4465. PMLR (2019). http://proceedings.mlr.press/v97/mehrotra19a.html
12. Chawla, S., Hartline, J.D., Kleinberg, R.D.: Algorithmic pricing via virtual valuations. In: The 8th ACM Conference on Electronic Commerce (EC) (2007)
13. Chawla, S., Hartline, J.D., Malec, D.L., Sivan, B.: Multi-parameter mechanism design and sequential posted pricing. In: The 42nd ACM Symposium on Theory of Computing (STOC) (2010)
14. Chawla, S., Jagadeesan, M.: Fairness in ad auctions through inverse proportionality. CoRR abs/2003.13966 (2020). https://arxiv.org/abs/2003.13966
15. Chawla, S., Malec, D.L., Sivan, B.: The power of randomness in Bayesian optimal mechanism design. In: The 11th ACM Conference on Electronic Commerce (EC) (2010)
16. Chawla, S., Miller, J.B.: Mechanism design for subadditive agents via an ex ante relaxation. In: Proceedings of the 2016 ACM Conference on Economics and Computation, EC 2016, Maastricht, The Netherlands, 24–28 July 2016, pp. 579–596 (2016). https://doi.org/10.1145/2940716.2940756. http://doi.acm.org/10.1145/2940716.2940756
17. Chen, X., Diakonikolas, I., Orfanou, A., Paparas, D., Sun, X., Yannakakis, M.: On the complexity of optimal lottery pricing and randomized mechanisms. In: IEEE 56th Annual Symposium on Foundations of Computer Science, FOCS 2015, Berkeley, CA, USA, 17–20 October 2015, pp. 1464–1479 (2015). https://doi.org/10.1109/FOCS.2015.93
18. Cheng, Yu., Gravin, N., Munagala, K., Wang, K.: A simple mechanism for a budget-constrained buyer. In: Christodoulou, G., Harks, T. (eds.) WINE 2018. LNCS, vol. 11316, pp. 96–110. Springer, Cham (2018). https://doi.org/10.1007/978-3-030-04612-5_7

19. Chouldechova, A.: Fair prediction with disparate impact: A study of bias in recidivism prediction instruments. arXiv preprint arXiv:1703.00056 (2017)
20. Clarke, E.H.: Multipart Pricing of Public Goods. Public Choice **11**(1), 17–33 (1971)
21. Daskalakis, C., Deckelbaum, A., Tzamos, C.: The complexity of optimal mechanism design. In: The 25th ACM-SIAM Symposium on Discrete Algorithms (SODA) (2014)
22. Daskalakis, C., Deckelbaum, A., Tzamos, C.: Strong duality for a multiple-good monopolist. Econometrica **85**(3), 735–767 (2017)
23. Daskalakis, C., Papadimitriou, C.H., Tzamos, C.: Does information revelation improve revenue? In: Conitzer, V., Bergemann, D., Chen, Y. (eds.) Proceedings of the 2016 ACM Conference on Economics and Computation, EC 2016, Maastricht, The Netherlands, 24–28 July 2016, pp. 233–250. ACM (2016). https://doi.org/10.1145/2940716.2940789
24. Daskalakis, C., Weinberg, S.M.: Symmetries and optimal multi-dimensional mechanism design. In: Proceedings of the 13th ACM Conference on Electronic Commerce, EC 2012, Valencia, Spain, 4–8 June 2012, pp. 370–387 (2012). https://doi.org/10.1145/2229012.2229042
25. Datta, A., Tschantz, M.C., Datta, A.: Automated experiments on ad privacy settings. Proc. Priv. Enhancing Technol. **2015**(1), 92–112 (2015). https://doi.org/10.1515/popets-2015-0007
26. Dwork, C., Hardt, M., Pitassi, T., Reingold, O., Zemel, R.S.: Fairness through awareness. In: Goldwasser, S. (ed.) Innovations in Theoretical Computer Science 2012, Cambridge, MA, USA, 8–10 January 2012, pp. 214–226. ACM (2012). https://doi.org/10.1145/2090236.2090255
27. Dwork, C., Ilvento, C.: Fairness under composition. In: Blum, A. (ed.) 10th Innovations in Theoretical Computer Science Conference, ITCS 2019, San Diego, California, USA, 10–12 January 2019. LIPIcs, vol. 124, pp. 33:1–33:20. Schloss Dagstuhl - Leibniz-Zentrum für Informatik (2019). https://doi.org/10.4230/LIPIcs.ITCS.2019.33
28. Eden, A., Feldman, M., Friedler, O., Talgam-Cohen, I., Weinberg, S.M.: The competition complexity of auctions: a bulow-klemperer result for multi-dimensional bidders. In: Proceedings of the 2017 ACM Conference on Economics and Computation, EC 2017, Cambridge, MA, USA, 26–30 June 2017, p. 343 (2017). https://doi.org/10.1145/3033274.3085115. http://doi.acm.org/10.1145/3033274.3085115
29. Eden, A., Feldman, M., Friedler, O., Talgam-Cohen, I., Weinberg, S.M.: A simple and approximately optimal mechanism for a buyer with complements: abstract. In: Proceedings of the 2017 ACM Conference on Economics and Computation, EC 2017, Cambridge, MA, USA, 26–30 June 2017, p. 323 (2017). https://doi.org/10.1145/3033274.3085116. http://doi.acm.org/10.1145/3033274.3085116
30. Groves, T.: Incentives in teams. Econometrica **41**(4), 617–631 (1973)
31. Hart, S., Nisan, N.: Approximate revenue maximization with multiple items. J. Economic Theory **172**, 313–347 (2017). https://doi.org/10.1016/j.jet.2017.09.001
32. Hart, S., Reny, P.J.: Maximizing revenue with multiple goods: nonmonotonicity and other observations. Theor. Econ. **10**(3), 893–922 (2015)
33. Hartline, J.D., Roughgarden, T.: Simple versus optimal mechanisms. In: ACM Conference on Electronic Commerce, pp. 225–234 (2009)
34. Ilvento, C., Jagadeesan, M., Chawla, S.: Multi-category fairness in sponsored search auctions. In: Hildebrandt, M., Castillo, C., Celis, E., Ruggieri, S., Taylor, L., Zanfir-Fortuna, G. (eds.) FAT* 2020: Conference on Fairness, Accountability, and Transparency, Barcelona, Spain, 27–30 January 2020, pp. 348–358. ACM (2020). https://doi.org/10.1145/3351095.3372848

35. Kay, M., Matuszek, C., Munson, S.A.: Unequal representation and gender stereo-types in image search results for occupations. In: Proceedings of the 33rd Annual ACM Conference on Human Factors in Computing Systems, pp. 3819–3828. ACM (2015)

36. Kothari, P., Mohan, D., Schvartzman, A., Singla, S., Weinberg, S.M.: Approxima-tion schemes for a buyer with independent items via symmetries. In: The 60th Annual IEEE Symposium on Foundations of Computer Science (FOCS) (2019)

37. Lambrecht, A., Tucker, C.: Algorithmic bias? An empirical study of apparent gender-based discrimination in the display of STEM career ads. Manag. Sci. **65**(7), 2966–2981 (2019). https://doi.org/10.1287/mnsc.2018.3093

38. Levin, S.: A beauty contest was judged by AI and the robots didn't like dark skin, September 2016

39. Li, X., Yao, A.C.C.: On revenue maximization for selling multiple independently distributed items. Proc. Natl. Acad. Sci. **110**(28), 11232–11237 (2013)

40. Liu, S., Psomas, C.: On the competition complexity of dynamic mechanism design. In: Proceedings of the Twenty-Ninth Annual ACM-SIAM Symposium on Discrete Algorithms, SODA 2018, New Orleans, LA, USA, 7–10 January 2018, pp. 2008–2025 (2018). https://doi.org/10.1137/1.9781611975031.131

41. Manelli, A.M., Vincent, D.R.: Multidimensional mechanism design: revenue maxi-mization and the multiple-good monopoly. J. Econ. Theory **137**(1), 153–185 (2007)

42. Myerson, R.B.: Optimal auction design. Math. Oper. Res. **6**(1), 58–73 (1981)

43. Pavlov, G.: Optimal mechanism for selling two goods. B.E. J. Theor. Econ. **11**(3) (2011)

44. Plaugic, L.: Faceapp's creator apologizes for the app's lightening 'hot' filter, April 2017

45. Rubinstein, A., Weinberg, S.M.: Simple mechanisms for a subadditive buyer and applications to revenue monotonicity. In: Proceedings of the Sixteenth ACM Con-ference on Economics and Computation, EC 2015, Portland, OR, USA, 15–19 June 2015, pp. 377–394 (2015). https://doi.org/10.1145/2764468.2764510

46. Ryu, H., Adam, H., Mitchell, M.: Inclusivefacenet: improving face attribute detec-tion with race and gender diversity. In: Proceedings of FAT/ML (2018)

47. Thanassoulis, J.: Haggling over substitutes. J. Econ. Theory **117**, 217–245 (2004)

48. Torralba, A., Efros, A.A.: Unbiased look at dataset bias. In: IEEE Conference on Computer Vision and Pattern Recognition (CVPR) (2011)

49. Vickrey, W.: Counterspeculations, auctions, and competitive sealed tenders. J. Finance **16**(1), 8–37 (1961)

50. Wang, A., Narayanan, A., Russakovsky, O.: Vibe: a tool for measuring and mit-igating bias in image datasets. CoRR abs/2004.07999 (2020). https://arxiv.org/abs/2004.07999

51. Yao, A.C.C.: An n-to-1 bidder reduction for multi-item auctions and its applica-tions. In: The Twenty-Sixth Annual ACM-SIAM Symposium on Discrete Algo-rithms (SODA) (2015)

52. Zemel, R., Wu, Y., Swersky, K., Pitassi, T., Dwork, C.: Learning fair representa-tions. In: Proceedings of the 30th International Conference on Machine Learning (ICML 2013), pp. 325–333 (2013)

Welfare-Preserving ε-BIC to BIC Transformation with Negligible Revenue Loss

Vincent Conitzer[1], Zhe Feng[2(✉)], David C. Parkes[3], and Eric Sodomka[4]

[1] Duke University, 2080 Duke University Road, Durham, NC 27708, USA
conitzer@cs.duke.edu
[2] Google Research, 1600 Amphitheatre Parkway, Mountain View, CA 94043, USA
zhef@google.com
[3] Harvard John A. Paulson School of Engineering and Applied Sciences,
150 Western Avenue, Allston, MA 02134, USA
parkes@eecs.harvard.edu
[4] Facebook Research, 1 Rathbone Square, London W1T 1FB, UK
sodomka@fb.com

Abstract. In this paper, we provide a transform from an ε-BIC mechanism into an exactly BIC mechanism without any loss of social welfare and with additive and negligible revenue loss. This is the first ε-BIC to BIC transformation that preserves welfare and provides negligible revenue loss. The revenue loss bound is tight given the requirement to maintain social welfare. Previous ε-BIC to BIC transformations preserve social welfare but have no revenue guarantee [4], or suffer welfare loss while incurring a revenue loss with both a multiplicative and an additive term, e.g., [9,14,28]. The revenue loss achieved by our transformation is incomparable to these earlier approaches and can be significantly less. Our approach is different from the previous replica-surrogate matching methods and we directly make use of a directed and weighted type graph (induced by the types' regret), one for each agent. The transformation runs a *fractional rotation step* and a *payment reducing step* iteratively to make the mechanism Bayesian incentive compatible. We also analyze ε-expected ex-post IC (ε-EEIC) mechanisms [18]. We provide a welfare-preserving transformation in this setting with the same revenue loss guarantee for uniform type distributions and give an impossibility result for non-uniform distributions. We apply the transform to linear-programming based and machine-learning based methods of automated mechanism design.

Keywords: BIC transformation · Automated mechanism design · Approximately IC mechanism

Part of the work was done when Zhe Feng was a PhD student at Harvard University, where he was supported by a Google PhD fellowship. See the full version of this paper for the complete proofs and Appendix at https://arxiv.org/pdf/2007.09579.pdf.

© Springer Nature Switzerland AG 2022
M. Feldman et al. (Eds.): WINE 2021, LNCS 13112, pp. 76–94, 2022.
https://doi.org/10.1007/978-3-030-94676-0_5

1   Introduction

Optimal mechanism design is very challenging in multi-dimensional settings such as those for selling multiple items, such as those that arise in the sale of wireless spectrum licenses or the allocation of advertisements to slots in internet advertising. Recognizing this challenge, there is considerable interest in adopting algorithmic approaches to address these problems of economic design. These include polynomial-time black-box reductions from multi-dimensional revenue maximization to the algorithmic problem for virtual welfare optimization, e.g., [6–8], and the application of methods from linear programming [12,13] and machine learning [16,18,19] to automated mechanism design.

Moreover, it is common in practical settings that it is important consider both social welfare (efficiency) and revenue. For example, national governments that use auctions to sell wireless spectrum licenses care both about the efficiency of the allocation as this promotes valuable use as well as the revenue that flows from auctions into the budget. In regard to online advertising, there are various works that explore this trade-off between welfare and revenue. Display advertising has focused on yield optimization (i.e., maximizing a combination of revenue and the quality of ads shown) [3], and work in sponsored search auctions has considered a squashing parameter that trades off efficiency and revenue [25]. At the same time, there is a surprisingly small theoretical literature that considers both welfare and revenue properties together (e.g., [15]).

At the same time, the use of computational methods for economic design often comes with a limitation, which is that the output mechanism may only be approximately incentive compatible (IC); e.g., the black-box reductions are approximately IC when the algorithmic problems are solved in polynomial time, the LP approach works on a discretized space to reduce computational cost but thereby achieves a mechanism that is only approximately IC in the full space, and the machine learning approaches train a mechanism over finite training data and achieve approximate IC on the full type distribution. While it has been debated as to whether approximate incentive compatibility may suffice, e.g., [1,11,26], this does add an additional layer of unpredictability to the performance of a designed mechanism. First, the fact that an agent can gain only a small amount from deviating does not preclude strategic behavior—perhaps the agent can easily identify a useful deviation, for example through repeated interactions, that reliably provides increased profit. This can be a problem when strategic responses lead to an unraveling of the desired economic properties of the mechanism (we provide such an example in this paper). The possibility of strategic reports by participants has additional consequences as well, for example making it more challenging for a designer to confidently measure ex-post welfare after outcomes are realized.

For the above reasons, there is considerable interest in methods to transform an ε-Bayesian incentive compatible (ε-BIC) mechanism to an exactly BIC mechanism [9,14,28], or an ε-expected ex-post IC (ε-EEIC) mechanisms [16,18] into an exactly BIC mechanism. The main question we want to answer in this paper is:

Given an ε-BIC/ε-EEIC mechanism, is there an exact BIC mechanism that maintains social welfare and achieves negligible revenue loss compared with the original mechanism under truthful reports? If so, can we find the transformed, BIC mechanism efficiently?

In this paper, we provide the first ε-BIC to BIC transform that is welfare-preserving while also ensuring only negligible revenue loss relative to the baseline mechanism. This simultaneous attention to the properties of both welfare and revenue is of practical importance. An immediate corollary of our main result is the well known result from economic theory, namely that efficient allocations can be implemented in an incentive-compatible way. For example, the transform can be applied to a first-price, sealed-bid auction to achieve an efficient and BIC auction.

Our approach is different from the previous replica-surrogate matching methods and we directly make use of a directed and weighted type graph (induced by the types' regret), one for each agent. The transformation runs a *fractional rotation step* and a *payment reducing step* iteratively to make the mechanism Bayesian incentive compatible.

The transform also satisfies another appealing property, which is that of *allocation-invariance*. The transformed mechanism maintains the same distribution on outcomes, allocations for example, as the baseline mechanism (focusing here on the non-monetary part of the output of the mechanism).[1] This property is useful in many scenarios. Consider, for example, a principal such as Amazon that is running a market and also incurs a resource cost for different outcomes (e.g. warehouse storage cost). With this allocation-invariance property, then not only is the welfare the same (or better) and the revenue loss negligible, but the resource cost (averaged over iterations of the mechanism) of the principal is preserved by the transform.

1.1 Model and Notation

We consider a general mechanism design setting with a set of n agents $N = \{1, \ldots, n\}$. Each agent i has a private type t_i. We denote the entire type profile as $t = (t_1, \ldots, t_n)$, which is drawn from a joint distribution $\mathcal{F}$. Let $\mathcal{F}_i$ be the marginal distribution of agent i and $\mathcal{T}_i$ be the support of $\mathcal{F}_i$. Let t_{-i} be the joint type profile of the other agents, $\mathcal{F}_{-i}$ be the associated marginal type distribution. Let $\mathcal{T} = \mathcal{T}_1 \times \cdots \times \mathcal{T}_n$ and $\mathcal{T}_{-i}$ be the support of $\mathcal{F}$ and $\mathcal{F}_{-i}$, respectively. In this setting, there is a set of feasible *outcomes* denoted by $\mathcal{O}$, typically an allocation of items to agents. Later in the paper, we sometimes also use "outcome" to refer to the output of the mechanism, namely the allocation together with the payments, when this is clear from the context.

We focus on the discrete type setting, i.e., $\mathcal{T}_i$ is a finite set containing m_i possible types, i.e., $|\mathcal{T}_i| = m_i$. Let $t_i^{(j)}$ denote the jth possible type of agent i,

---

[1] This allocation-invariance is *ex ante*, i.e., it is with respect to the prior distribution over types.

where $j \in [m_i]$. For all i and t_i, $v_i : (t_i, o) \rightarrow \mathbb{R}_{\geq 0}$ is a valuation that maps a type t_i and outcome o to a non-negative real number. A *direct revelation* mechanism $\mathcal{M} = (x, p)$ is a pair of *allocation rule* $x_i : \mathcal{T} \rightarrow \Delta(\mathcal{O})$, possibly randomized, and *expected payment rule* $p_i : \mathcal{T} \rightarrow \mathbb{R}_{\geq 0}$. We slightly abuse notation, and also use v_i to define the expected value of bidder i for mechanism $\mathcal{M}$, with the expectation taken with respect to the randomization used by the mechanism, that is

$$\forall i, \hat{t} \in \mathcal{T}, v_i(t_i, x(\hat{t})) = \mathbb{E}_{o \sim x(\hat{t})}[v_i(t_i, o)], \tag{1}$$

for true type t_i and reported type profile $\hat{t}$. When the reported types are $\hat{t} = (\hat{t}_1, \dots, \hat{t}_n)$, the output of mechanism $\mathcal{M}$ for agent i is denoted as $\mathcal{M}_i(\hat{t}) = (x_i(\hat{t}), p_i(\hat{t}))$. We define the *utility* of agent i with true type t_i and a reported type $\hat{t}_i$ given the reported type profile $\hat{t}_{-i}$ of other agents as a quasilinear function,

$$u_i(t_i, \mathcal{M}(\hat{t})) = v_i(t_i, x(\hat{t})) - p_i(\hat{t}). \tag{2}$$

For a multi-agent setting, it will be useful to also define the *interim rules*.

Definition 1 (Interim Rules of a Mechanism). *For a mechanism $\mathcal{M}$ with allocation rule x and payment rule p, the interim allocation rule X and payment rule P are defined as,* $\forall i, t_i \in \mathcal{T}_i, X_i(t_i) = \mathbb{E}_{t_{-i} \in \mathcal{F}_{-i}}[x_i(t_i; t_{-i})], P_i(t_i) = \mathbb{E}_{t_{-i} \in \mathcal{F}_{-i}}[p_i(t_i; t_{-i})].$

In this paper, we assume we have oracle access to the interim quantities of mechanism $\mathcal{M}$.

Assumption 1 (Oracle Access to Interim Quantities). *For any mechanism $\mathcal{M}$, given any type profile $t = (t_1, \dots, t_n)$, we receive the interim allocation rule $X_i(t_i)$ and payments $P_i(t_i)$, for all i, t_i.*

We define the menu of a mechanism $\mathcal{M}$ in the following way.

Definition 2 (Menu). *For a mechanism $\mathcal{M}$, the menu of bidder i is the set $\{\mathcal{M}_i(t)\}_{t \in \mathcal{T}}$. The menu size of agent i is denoted as $|\mathcal{M}_i|$.*

In mechanism design, there is a focus on designing *incentive compatible* mechanisms, so that truthful reporting of types is an equilibrium. This is without loss of generality by the revelation principle.

It has also been useful to work with approximate-IC mechanisms, and these have been studied in various papers, e.g. [2,9,10,14,16,18–20,24,28].

In this paper, we focus on two definitions of approximate incentive compatibility, ε-BIC and ε-expected ex post incentive compatible (ε-EEIC).

Definition 3 (ε-BIC Mechanism). *A mechanism $\mathcal{M}$ is ε-BIC iff for all i, t_i,*

$$\mathbb{E}_{t_{-i} \sim \mathcal{F}_{-i}}[u_i(t_i, \mathcal{M}(t))] \geq \max_{\hat{t}_i \in \mathcal{T}_i} \mathbb{E}_{t_{-i} \sim \mathcal{F}_{-i}}[u_i(t_i, \mathcal{M}(\hat{t}_i; t_{-i}))] - \varepsilon$$

Definition 4 (ε-expected ex post IC (ε-EEIC) Mechanism) [18]. *A mechanism $\mathcal{M}$ is ε-EEIC if and only if for all i,*

$$\mathbb{E}_t \left[\max_{\hat{t}_i \in \mathcal{T}_i} u_i(t_i, \mathcal{M}(\hat{t}_i; t_{-i})) - u_i(t_i, \mathcal{M}(t)) \right] \leq \epsilon$$

A mechanism $\mathcal{M}$ is ε-EEIC iff no agent can gain more than ε ex post regret, in expectation over all type profiles $t \in \mathcal{T}$ (where ex post regret is the amount by which an agent's utility can be improved by misreporting to some $\hat{t}_i$ given knowledge of t, instead of reporting its true type t_i). A 0-EEIC mechanism is essentially DSIC.[2]

We can also consider an interim version of ε-EEIC, termed as ε-expected interim IC (ε-EIIC), which is defined as

$$\mathbb{E}_{t_i \sim \mathcal{F}_i} \left[\mathbb{E}_{t_{-i} \sim \mathcal{F}_{-i}} \left[u_i(t_i, \mathcal{M}(t_i; t_{-i})) \right] \right] \geq \mathbb{E}_{t_i \sim \mathcal{F}_i} \left[\max_{t'_i \in \mathcal{T}_i} \mathbb{E}_{t_{-i} \sim \mathcal{F}_{-i}} \left[u_i(t_i, \mathcal{M}(t'_i; t_{-i})) \right] \right] - \varepsilon$$

All our results for ε-EEIC to BIC transformation hold for ε-EIIC mechanism. Indeed, we prove that any ε-EEIC mechanism is also ε-EIIC in Lemma 1 in Appendix.

Another important property of mechanism design is *individual rationality* (IR), and we define two standard versions of IR (ex-post/interim IR) in Appendix. The transformation that we provide from ε-BIC/ε-EEIC to BIC preserves individual rationality: if the original mechanism is interim IR then the mechanism achieved after transformation is interim IR, and if the original mechanism is ex-post IR then the mechanism achieved after transformation is ex-post IR.

For a mechanism $\mathcal{M}$, let $R^{\mathcal{M}}(\mathcal{F})$ and $W^{\mathcal{M}}(\mathcal{F})$ represent the expected revenue and social welfare, respectively, when agent types are sampled from $\mathcal{F}$ and they play $\mathcal{M}$ *truthfully*[3]. This definition applies equally to an IC or non-IC mechanism.

Definition 5 (Expected Social Welfare and Revenue). *For a mechanism $\mathcal{M} = (x, p)$ with agents' types drawn from distribution $\mathcal{F}$, the expected revenue for truthful reports is $R^{\mathcal{M}}(\mathcal{F}) = \mathbb{E}_{t \sim \mathcal{F}}[\sum_{i=1}^n p_i(t)]$, and the expected social welfare for truthful reports is $W^{\mathcal{M}}(\mathcal{F}) = \mathbb{E}_{t \sim \mathcal{F}}[\sum_{i=1}^n v_i(t_i, x(t))]$.*

In this work, we focus on the following transformation.

Definition 6 (Welfare-preserving Transformation with Negligible Revenue Loss). *Given an ε-BIC/ε-EEIC mechanism $\mathcal{M}$ over type distribution $\mathcal{F}$, a welfare-preserving transform that provides negligible revenue loss outputs a mechanism $\mathcal{M}'$ such that, $W^{\mathcal{M}'}(\mathcal{F}) \geq W^{\mathcal{M}}(\mathcal{F})$ and $R^{\mathcal{M}'}(\mathcal{F}) \geq R^{\mathcal{M}}(\mathcal{F}) - r(\varepsilon)$, where $r(\varepsilon) \to 0$ as $\varepsilon \to 0$.*

---

[2] For discrete type settings, 0-EEIC is exactly DSIC. For the continuous type case, a 0-EEIC mechanism is DSIC up to zero measure events.

[3] In this paper, we consider the revenue and welfare performance of the untruthful mechanisms with truthful reports, which is commonly used in the literature. It is an interesting future direction to consider the performance of untruthful mechanisms under equilibrium reporting.

1.2   Previous ε-BIC to BIC Transformations

There are existing algorithms for transforming any ε-BIC mechanism to an exactly BIC mechanism with only negligible revenue loss [9,14,28]. The central tools and reductions in these papers build upon the method of *replica-surrogate matching* [4,21,22]. Here we briefly introduce replica-surrogate matching and its application to an ε-BIC to BIC transformation.

Replica-Surrogate Matching. For each agent i, construct a bipartite graph $G_i = (\mathcal{R}_i \cup \mathcal{S}_i, E)$. The nodes in $\mathcal{R}_i$ are called *replicas*, which are types sampled i.i.d. from the type distribution of agent i, $\mathcal{F}_i$. The nodes in $\mathcal{S}_i$ are called *surrogates*, and also sampled from $\mathcal{F}_i$. In particular, the true type t_i is added in $\mathcal{R}_i$. There is an edge between each replica and each surrogate. The weight of the edge between a replica $r_i^{(j)}$ and a surrogate $s_i^{(k)}$ is induced by the mechanism, and defined as

$$w_i(r_i^{(j)}, s^{(k)}) = E_{t_{-i} \in \mathcal{F}_{-i}} \left[v_i(r_i^{(j)}, x(s_i^{(k)}, t_{-i})) \right] - (1 - \eta) \cdot \mathbb{E}_{t_{-i} \in \mathcal{F}_{-i}} \left[p_i(s_i^{(k)}, t_{-i}) \right]. \quad (3)$$

The replica-surrogate matching computes the maximum weight matching in G_i.

ε-BIC to BIC Transformation by Replica-Surrogate Matching [14]. We briefly describe this transformation, deferring the details to Appendix. Given a mechanism $\mathcal{M} = (x, p)$, this transformation constructs a bipartite graph between replicas (include the true type t_i) and surrogates, as described above. The approach then runs VCG matching to compute the maximum weighted matching for this bipartite graph, and charges each agent its VCG payment. For unmatched replicas in the VCG matching, the method randomly matches a surrogate. Let $\mathcal{M}' = (x, (1 - \eta)p)$ be the modified mechanism. If the true type t_i is matched to a surrogate s_i, then agent i uses s_i to compete in $\mathcal{M}'$. The outcome of $\mathcal{M}'$ is $x(s)$, given matched surrogate profile s, and the payment of agent i (matched in VCG matching) is $(1 - \eta)p_i(s)$ plus the VCG payment from the VCG matching, where η is the parameter in replica-surrogate matching. If t_i is not matched in the VCG matching, the agent gets nothing and pays zero.

This replica-surrogate matching transform does not preserve welfare. Indeed, the replica-surrogate matching transformation must suffer welfare loss in some cases.[4] Turning to revenue, the revenue loss of the replica-surrogate matching mechanism relative to the orginal mechanism $\mathcal{M}$ is guaranteed to be at most $\eta \text{Rev}(\mathcal{M}) + \mathcal{O}\left(\frac{n\varepsilon}{\eta}\right)$ [14,28], and has both a multiplicative and an additive term. Cai et al. [9] propose a polynomial time algorithm for performing this transform with only sample access to the type distribution and query access to the

---

[4] The previous ε-BIC to BIC transformations [9,14,28] don't state the welfare loss guarantee clearly. Consider Example 1 shown in Sect. 1.3, the original ε-BIC mechanism already maximizes welfare and the optimal allocation is unique, any unmatched type in replica-surrogate matching creates a welfare loss. Particularly, the welfare loss is *unbounded* when (inappropriately) choosing $\eta < \frac{\varepsilon}{\sqrt{m-1}}$ in replica-surrogate matching.

original ε-BIC mechanism. The transform extends replica-surrogate matching and *Bernoulli factory* techniques proposed by [17] to handle negative weights in the bipartite graph and provides the same revenue property as the previous work [14, 28], without preserving social welfare.[5] In this work, we assume oracle access to the interim quantities of the original ε-BIC mechanism, following the model of [4, 14, 21, 22, 28]. How to generalize the proposed transform to the setting that only has sample access to the type distribution and runs in polynomial time will be an interesting future work.

The black-box reduction of [4] focuses on preserving welfare only. Indeed, it can be regarded as a special case of this replica-surrogate matching method, where the weight of the bipartite graph only depends on the valuations and not the prices ($\eta = 1$ in Eq. (3)), and the replicas and surrogates are both $\mathcal{T}_i$ (there is no sampling for replicas and surrogates). For this reason, the transform described in [4] can preserve social welfare but may provide arbitrarily bad revenue (see Example 1).

1.3   Our Contributions

We first state the main result of the paper, which provides a welfare-preserving transform from approximate BIC to exact BIC with negligible revenue loss. This result holds for the general mechanism design setting with $n \geq 1$ agents and independent private types and is not restricted to allocation problems.

Main Result 1 (Theorem 6). *With $n \geq 1$ agents and independent private types, and an ε-BIC and IR mechanism $\mathcal{M}$ that achieves W expected social welfare and R expected revenue given truthful reports, there exists a BIC and IR mechanism $\mathcal{M}'$ that achieves at least W social welfare and $R - \sum_{i=1}^{n} |\mathcal{T}_i|\varepsilon$ revenue. The transformation is (ex ante) allocation-invariant. Given an oracle access to the interim quantities of $\mathcal{M}$, the running time of the transformation from $\mathcal{M}$ to $\mathcal{M}'$ is at most* $\mathtt{poly}(\sum_i |\mathcal{T}_i|)$.

The transformation works directly on the type graph of each agent, and it is this that allows us to maintain social welfare— indeed, we may even improve social welfare in our transformation. In contrast, the transformation from [4] can incur unbounded revenue loss (see Example 1, in which it loses all revenue) and existing approaches [9, 10, 14, 28] with negligible revenue loss can lose social welfare (see Example 1).

---

[5] Dughmi et al. [17] propose a general transformation from any black-box algorithm $\mathcal{A}$ to a BIC mechanism that only incurs negligible loss of welfare, with only polynomial number queries to $\mathcal{A}$, by using Bernoulli factory techniques. This approach has no guarantee on the revenue loss. Cai et al. [9] generalize Bernoulli factory techniques in the replica-surrogate matching to transform any ε-BIC mechanism to a BIC mechanism that only incurs negligible loss of revenue, with polynomial number queries to the original ε-BIC mechanism and polynomial number samples from the type distribution.

Choosing $\eta = \sqrt{\varepsilon}$, the revenue loss of existing transforms [9,10,14,28] is at most $\sqrt{\varepsilon}\text{Rev}(\mathcal{M}) + O(n\sqrt{\varepsilon})$, with both a multiplicative and an additive-loss in revenue, while our revenue loss is additive. In the case that the original revenue, $\text{Rev}(\mathcal{M})$, is order-wise smaller than the number of types, i.e., $\text{Rev}(\mathcal{M}) = o(\sum_i |T_i|)$, the existing transforms provide a better revenue bound (at some cost of welfare loss). But when the revenue is relatively larger than the number of types, i.e., $\text{Rev}(\mathcal{M}) = \Omega(\sum_i |T_i|)$, our transformation can achieve strictly better revenue than these earlier approaches while also preserving welfare.

Before describing our techniques we illustrate these properties through a single agent, two outcome example in Example 1. We show that even for the case that $\text{Rev}(M) = o(\sum_i |T_i|)$, our transformation can strictly outperform existing transforms w.r.t revenue loss.

Example 1. Consider a single agent with m types, $\mathcal{T} = \{t^{(1)}, \cdots, t^{(m)}\}$, where the type distribution is uniform. Suppose there are two outcomes, the agent with type $t^{(j)}(j = 1, \ldots, m-1)$ values outcome 1 at 1 and values outcome 2 at 0. The agent with type $t^{(m)}$ values outcome 1 at $1 + \varepsilon$ and outcome 2 at $\sqrt{m}$. The mechanism $\mathcal{M}$ we consider is: if the agent reports type $t^{(j)}, j \in [m-1]$, $\mathcal{M}$ gives outcome 1 to the agent with a price of 1, and if the agent reports type $t^{(m)}$, $\mathcal{M}$ gives outcome 2 to the agent with a price of $\sqrt{m}$. $\mathcal{M}$ is ε-BIC, because the agent with type $t^{(m)}$ has a regret ε. The expected revenue achieved by $\mathcal{M}$ is $1 + \frac{\sqrt{m}-1}{m}$. In addition, $\mathcal{M}$ maximizes social welfare, $1 + \frac{\sqrt{m}-1}{m}$.

Our transformation decreases the payment of type $t^{(m)}$ by ε for a loss of $\frac{\varepsilon}{m}$ revenue and preserves the social welfare.

The transformation by [4] preserves the social welfare, however, the VCG payment (envy-free prices) is 0 for each type. Therefore, the approach proposed in [4] loses all revenue.

Moreover, the approaches that make use of replica-surrogate matching [9, 14,28, e.g.] lose at least $\frac{\varepsilon}{m} + \frac{\varepsilon}{\sqrt{m}-1}$ revenue, which is about $(\sqrt{m}+1)$ times larger than the revenue loss of our transformation. We argue this claim by a case analysis,

- If $\eta \geq \frac{\varepsilon}{\sqrt{m}-1}$, the VCG matching is the identical matching and the VCG payment is 0 for each type. In total, the agent loses at least $\eta \cdot \frac{\sqrt{m}+m-1}{m} \geq \frac{\varepsilon}{m} + \frac{\varepsilon}{\sqrt{m}-1}$ expected revenue.

- If $\eta < \frac{\varepsilon}{\sqrt{m}-1}$, the agent with type $t^{(m)}$ will be assigned outcome 1 ($t^{(m)}$ is matched to some $t^{(j)}, j \in [m-1]$, in VCG matching) and the VCG payment is η. Thus, type $t^{(m)}$ loses at least $\sqrt{m} - (1 - \eta) - \eta = \sqrt{m} - 1$ revenue. For any type $t^{(j)}, j \in [m-1]$, if $t^{(j)}$ is matched in VCG matching, the VCG payment is 0, since it will be matched to another type $t^{(k)}, k \in [m-1]$. Each type $t^{(j)}, j \in [m-1]$ loses at least η revenue. Overall the agent loses at least $\frac{\sqrt{m}-1}{m}$ expected revenue. In addition, since the type $t^{(m)}$ is assigned outcome 1, we lose at least $\frac{\sqrt{m}-1-\varepsilon}{m}$ expected social welfare.

Moreover, there is a chance that a type is not matched, in which case the social welfare is reduced.

Our transformation satisfies also satisfies an appealing allocation-invariance property (see Definition 7). Given an ε-BIC mechanism $\mathcal{M} = (x, p)$, the transform outputs a BIC mechanism $\mathcal{M}' = (x', p')$ that satisfies $\sum_{t \in \mathcal{T}} f(t) x'(t) = \sum_{t \in \mathcal{T}} f(t) x(t)$. As noted above, this property would be of interest, for example, to a principal who is operating the logistics for provisioning goods sold through the mechanism. Because of allocation-invariance, the principal knows that the distribution on goods sold is unchanged as a result of the transform and thus logistical aspects in regard to inventory storage are unchanged. The previous transformations [4,9,14,28] don't satisfy this allocation-invariance property.

We also support ε-expected ex-post IC (ε-EEIC), which is motivated by work on the use of machine learning to achieve approximate IC mechanisms in multi-dimensional settings [16,18,19]. In comparison with ε-BIC, the ε-EEIC metric only guarantees at most ε ex-post gain in expectation over type profiles, with no interim guarantee for any particular type. It is incomparable in strength with ϵ-BIC because ε-EEIC also strengthens ε-BIC in working with ex-post regret rather than interim regret. Our second main result shows how to transform an ε-EEIC mechanism to a BIC mechanism. For this, we need the additional assumption of a uniform type distribution and prove that this is necessary to achieve a transform with suitable properties.

Main Result 2 (Informal Theorems 5 and 6). *For $n \geq 1$ agents with independent uniform type distribution, our ε-BIC to BIC transformation can be applied to an ε-EEIC mechanism and all results in Main Result 1 hold here. For a non-uniform type distribution, we show an impossibility result for an ε-EEIC to BIC, welfare-preserving transformation with only negligible revenue loss, even for the single agent case.*

Moreover, we also argue that our revenue loss bounds are tight given the requirement to maintain social welfare. This holds for both ε-BIC mechanisms and ε-EEIC mechanisms.

Main Result 3 (Informal Theorems 2 and 7). *There exists an ε-BIC/ε-EEIC and IR mechanism for $n \geq 1$ agents with independent uniform type distribution, for which any welfare-preserving transformation must suffer $\Omega(\sum_i |\mathcal{T}_i| \varepsilon)$ revenue loss.*

We also apply the transform to automated mechanism design in Sect. 5, considering both a linear-programming and machine learning framework and looking to maximize a linear combination of expected revenue and social welfare, i.e., $\mu_\lambda(\mathcal{M}, \mathcal{F}) = (1-\lambda) R^{\mathcal{M}}(\mathcal{F}) + \lambda W^{\mathcal{M}}(\mathcal{F})$, for some $\lambda \in [0,1]$ and type distribution $\mathcal{F}$. We summarize the result of this application.

Main Result 4 (Informal Theorems 10 and 11). *For n agents with independent type distribution $\times_{i=1}^{n} \mathcal{F}_i$ on $\mathcal{T} = \mathcal{T}_1 \times \cdots \times \mathcal{T}_n$ and an α-approximation LP algorithm* ALG *to output an ε-BIC (ε-EEIC) and IR mechanism $\mathcal{M}$ on*

$\mathcal{F}$ with $\mu_\lambda(\mathcal{M}, \mathcal{F}) \geq \alpha\text{OPT}$, there exists a BIC and IR mechanism $\mathcal{M}'$, s.t., $\mu_\lambda(\mathcal{M}', \mathcal{F}) \geq \alpha\text{OPT} - (1 - \lambda)\sum_{i=1}^{n}|\mathcal{T}_i|\varepsilon$. Given oracle access to the interim quantities of $\mathcal{M}$, the running time to output the mechanism $\mathcal{M}'$ is at most $\text{poly}(\sum_{i=1}|\mathcal{T}_i|, rt_{\text{ALG}}(x))$, where $rt_{\text{ALG}}(\cdot)$ is the running time of ALG and x is the bit complexity of the input. Similar results hold for a machine-learning based approach, in a PAC learning manner.

Compared with the previous transformations that are able to achieve negligible revenue loss [9,14,28], our transformation achieves a better blended objective of welfare and revenue when λ is close to 1 since we preserve welfare of the original mechanism after transformation.

1.4   Our Techniques

Instead of constructing a bipartite replica-surrogate graph, our transformation makes use of a directed, weighted type graph, one for each agent. For simplicity of exposition, we can consider a single agent with a uniform type distribution.

Given an ε-BIC mechanism, $\mathcal{M}$, we construct a graph $G = (\mathcal{T}, E)$, where each node represents a possible type of the agent and there is an edge from node $t^{(j)}$ to $t^{(k)}$ if the output of the mechanism for type $t^{(k)}$ is weakly preferred by the agent for true type $t^{(j)}$ in $\mathcal{M}$, i.e. $u(t^{(j)}, \mathcal{M}(t^{(k)})) \geq u(t^{(j)}, \mathcal{M}(t^{(j)}))$. The weight w_{jk} of edge $(t^{(j)}, t^{(k)})$ is defined as the *regret* of type $t^{(j)}$ by not misreporting $t^{(k)}$, i.e.,

$$w_{jk} = u(t^{(j)}, \mathcal{M}(t^{(k)})) - u(t^{(j)}, \mathcal{M}^\varepsilon(t^{(j)})). \tag{4}$$

The transformation method then iterates over the following two steps, constructing a transformed mechanism from the original mechanism. We briefly introduce the two steps here and defer to Appendix for detailed description.

Step 1. If there is a cycle $\mathcal{C}$ in the type graph with at least one positive-weight edge, then all types in this cycle weakly prefer their descendant in the cycle and one or more strictly prefers their descendant. In this case, we *"rotate"* the outcome and payment of types against the direction of the cycle, to let each type receive a weakly better outcome compared with its current outcome. We repeat Step 1 until all cycles in the type graph are removed.

Step 2. We pick a source node, if any, with a positive-weight outgoing edge (and thus regret for truthful reporting). We decrease the payment made by this source node, as well as decreasing the payment made by each one of its ancestors (note the lack of cycles at this point) by the same amount, until we create a new edge in the type graph with weight zero, such that the modification to payments is about to increase regret for some type. If at any point we create a cycle, we move to Step 1. Otherwise, we repeat Step 2 until there are no source nodes with positive-weight, outgoing edges.

The algorithm works on the type graph induced by the original, approximately IC mechanism, $\mathcal{M}$, and directly modifies the mechanism for each type, to make the mechanism IC. This allows the transformation to preserve welfare and provides negligible revenue loss. Step 2 has no effect on welfare, since it only changes (interim) payment for each type. Step 1 is designed to remove cycles created in Step 2 so that we can run Step 2, while preserving welfare simultaneously. Both steps reduce the total weight of the type graph, which is equivalent to reducing the regret in the mechanism to make it IC. We show the transform in Fig. 1.

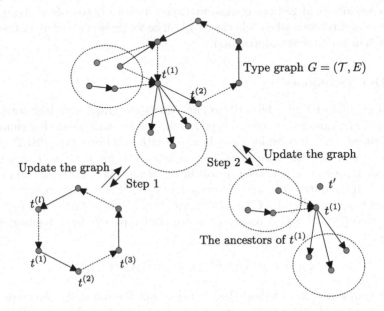

Fig. 1. Visualization of the transformation for a single agent with a uniform type distribution: we start from a type graph $G(\mathcal{T}, E)$, where each edge $(t^{(1)}, t^{(2)})$ represents the agent weakly prefers the allocation and payment of type $t^{(2)}$ rather than his true type $t^{(1)}$. The weight of each edge is denoted in Eq. (4). In the graph, we use solid lines to represent the positive-weight edges, and dashed lines to represent zero-weight edges. We first find a shortest cycle, and rotate the allocation and payment along the cycle and update the graph (Step 1). We keep doing Step 1 to remove all cycles. Then we pick a source node $t^{(1)}$, and decrease the payment of type $t^{(1)}$ and all the ancestors of $t^{(1)}$ until we reduce the weight of one outgoing edge from $t^{(1)}$ to zero or we create a new zero-weight edge from t' to $t^{(1)}$ or one of the ancestors of $t^{(1)}$ (Step 2).

For a single agent with non-uniform type distribution, we handle the unbalanced density probability of each type by redefining the type graph, where the weight of the edge in type graph is weighted by the product of the probability of the two nodes that are incident to an edge. We propose a new Step 1 by introducing *fractional rotation*, such that for each cycle in the type graph, we rotate the allocation and payment with a fraction for any type $t^{(j)}$ in the cycle. By carefully choosing the fraction for each type in the cycle, we can argue that our transformation preserves welfare and provides negligible revenue loss.

The multi-agent setting reduces to the single-agent case, building a type graph for each agent induced by the interim rules (see Appendix for the construction of this type graph). With oracle access to the interim quantities of the original mechanism, we build the type graph of each agent i in $\texttt{poly}(|\mathcal{T}_i|)$ time. We then apply the transform for each type graph of agent i, induced by the interim rules.

This is analogous to a replica-surrogate matching approach, which also defines the weights between replicas and surrogates by interim rules and runs the replica-surrogate matching for the reported type of each agent. Replica-surrogate matching uses this sampling technique to make the distribution of reported types of each agent equal to the distribution of the true type. In comparison, Steps 1 and 2 of our transform leave the type distribution unchanged, so that the transform attains this property for free. Then we can apply our transformation for each type graph separately. The new challenge in our transformation is feasibility, i.e., establishing consistency of the agent-wise rotations to interim quantities. We show the transformation for each type graph guarantees feasibility by appeal to Border's lemma [5]. Our transformation can also be directly applied to an ε-EEIC mechanism in the case that each agent has an independent uniform type distribution.[6]

2 Warm-Up: Single Agent with Uniform Type Distribution

In this section, we consider the case of a single agent and a uniformly distributed type distribution $\mathcal{F}$, i.e. $\forall j \in [m], f(t^{(j)}) = \frac{1}{m}$. Even for this simple case, the proof is non-trivial. Moreover, the technique for this simple case can be extended to handle more intricate cases. The main result in this section is Theorem 1, which makes use of a constructive proof to modify a ε-EEIC/ε-BIC mechanism to a BIC mechanism. An interesting observation is that ε-EEIC may only provide $m\varepsilon$-BIC for a uniform type distribution, which indicates that transforming ε-EEIC may incur a worse revenue loss bound. However, Theorem 1 shows we can achieve the same revenue loss bound for both ε-BIC and ε-EEIC.

Theorem 1. *Consider a single agent, with m types $\mathcal{T} = \{t^{(1)}, t^{(2)}, \cdots, t^{(m)}\}$, and a uniform type distribution $\mathcal{F}$. Given an ϵ-EEIC/ε-BIC and IR mechanism $\mathcal{M}$, which achieves W expected social welfare and R expected revenue, there exists a BIC and IR mechanism $\mathcal{M}'$ that achieves at least W expected social welfare and $R - m\varepsilon$ revenue. Given an oracle access to $\mathcal{M}$, the running time of the transformation from $\mathcal{M}$ to $\mathcal{M}'$ is at most $\texttt{poly}(|\mathcal{T}|)$.*

---

[6] This need to transform an infeasible, IC mechanism into a feasible and IC mechanism also arises in [27], who use a method from [23] to correct for feasibility violations that result from statistical machine learning while preserving strategy-proofness.

2.1 Lower Bound on Revenue Loss

In the transformation shown in Fig. 2 in Appendix, the revenue loss is bounded by $m\varepsilon$. This revenue loss bound is tight up to a constant factor while insisting on maintaining social welfare.

Theorem 2. *There exists an ε-BIC (ε-EEIC) and IR mechanism $\mathcal{M}$ for a single agent with uniform type distribution for which any ε-BIC and IR to BIC and IR transformation (without loss of social welfare) must suffer at least $\Omega(m\varepsilon)$ revenue loss.*

2.2 Tighter Bound of Revenue Loss for Settings with Finite Menus

In some settings, the total number of possible types of an agent may be very large and yet the menu size can remain relatively small. In particular, suppose that a mechanism $\mathcal{M}$ has a small number of outputs, i.e., $|\mathcal{M}| = C$ and $C \ll m$, where m is the number of types and C is the menu size. Given this, we can provide a tighter bound on revenue loss for this setting in the following theorem. The complete proof is deferred to Appendix.

Theorem 3. *Consider a single agent with m types $\mathcal{T} = \{t^{(1)}, t^{(2)}, \cdots, t^{(m)}\}$, sampled from a uniform type distribution $\mathcal{F}$. Given an ε-BIC mechanism $\mathcal{M}$ with C different menus ($C \ll m$) that achieves S expected social welfare and R revenue, there exists an BIC mechanism $\mathcal{M}'$ that achieves at least S social welfare and $R - C\varepsilon$ revenue.*

3 Single Agent with General Type Distribution

In this section, we consider a setting with a single agent that has a non-uniform type distribution. A naive idea is that we can "divide" a type with a larger probability to several copies of the same type, each with equal probability, and then apply our proof of Theorem 1 to get a BIC mechanism. However, this would result in a weak bound on the revenue loss, since we would divide the m types into multiple, small pieces. This section is divided into two parts. First we show our transformation for an ε-BIC mechanism in this setting. Second, we show an impossibility result for an ε-EEIC mechanism, that is, without loss of welfare, no transformation can achieve negligible revenue loss.

3.1 ε-BIC to BIC Transformation

We propose a novel approach for a construction for the case of a single agent with a non-uniform type distribution. The proof is built upon Theorem 1, however, there is a technical difficulty to directly apply the same approach for this non-uniform type distribution case. Since each type has a different probability, we cannot rotate the allocation and payment in the same way as in Step 1 in the proof of Theorem 1.

We instead redefine the type graph $G = (\mathcal{T}, E)$, where the weight of the edge is now weighted by the product of the probability of the two nodes that are incident to an edge. We also modify the original rotation step shown in Fig. 3 in Appendix: for each cycle in the type graph, we rotate the allocation and payment with the fraction of $\frac{f(t^{(k)})}{f(t^{(j)})}$ for any type $t^{(j)}$ in the cycle, where $f(t^{(k)})$ is the smallest type probability of the types in the cycle. This step is termed as "fractional rotation step." We summarize the results in Theorem 4 and show the proof in Appendix.

Theorem 4. *Consider a single agent with m types, $\mathcal{T} = \{t^{(1)}, t^{(2)}, \cdots, t^{(m)}\}$ drawn from a general type distribution $\mathcal{F}$. Given an ε-BIC and IR mechanim $\mathcal{M}$ that achieves W expected social welfare and R expected revenue, there exists a BIC and IR mechanism $\mathcal{M}'$ that achieves at least W social welfare and $R - m\varepsilon$ revenue.*

Allocation-Invariant Transformation. In addition to the welfare and revenue guarantee achieved by this transformation, the transform has another desired property, as defined below.

Definition 7 (Allocation-invariance property). *Two mechanisms $\mathcal{M} = (x, p)$ and $\mathcal{M}' = (x', p')$ are (ex ante) allocation-invariant if and only if $\sum_{t \in \mathcal{T}} f(t)x(t) = \sum_{t \in \mathcal{T}} f(t)x'(t)$.*

For the single agent setting with a general type distribution, the transform only changes the allocation rules in Step 1. Since we use the *fractional rotation* in Step 1, the quantity $\sum_{t \in \mathcal{T}} f(t)x(t)$ is maintained after each Step 1. Then, it is straightforward to show that the transform satisfies this allocation-invariance property.[7]

3.2 Impossibility Result for ε-EEIC Transformation

As mentioned above, given any ε-BIC for a single agent with a general type distribution, we can transform to an exactly BIC mechanism with no loss of welfare and negligible loss of revenue. However, the same claim doesn't hold for ε-EEIC. Theorem 5 shows that no transformation can achieve negligible revenue loss while insisting on welfare preservation. The proof is provided in Appendix.

Theorem 5. *There exists a single agent with a non-uniform type distribution, and an ε-EEIC and IR mechanism, for which there is no IC transformation that preserves social welfare and IR and achieves negligible revenue loss.*

---

[7] By contrast, the previous transformations [9,14,28] cannot preserve the distribution of the allocation, even for the single agent and uniform type distribution case.

4   Multiple Agents with Independent Private Types

First, we state our positive result for a setting with multiple agents and independent, private types (Theorem 6). We assume each agent i's type t_i is independently drawn from $\mathcal{F}_i$ ($\mathcal{F}_i$ can be non-uniform). Then $\mathcal{F}$ is a product distribution that can be denoted as $\times_{i=1}^n \mathcal{F}_i$. The complete proof of the following theorem is shown in Appendix.

Theorem 6. *With n agents and independent private types, and an ε-BIC and IR mechanism $\mathcal{M}$ that achieves W expected social welfare and R expected revenue, there exists a BIC and IR mechanism $\mathcal{M}'$ that achieves at least W social welfare and $R - \sum_{i=1}^n |\mathcal{T}_i| \varepsilon$ revenue. The same result holds for an ε-EEIC mechanism with multiple agents, in the case that each agent has an independent uniform type distribution. Given an oracle access to the interim quantities of $\mathcal{M}$, the running time of the transformation from $\mathcal{M}$ to $\mathcal{M}'$ is at most* $\mathtt{poly}(\sum_i |\mathcal{T}_i|)$.

Allocation-Invariant Transformation. The transformation for multiple agents with independent private types is also allocation-invariant. To prove this, we can observe for $\mathcal{M} = (x, p)$ that

$$\sum_{t \in \mathcal{T}} f(t)x(t) = \sum_{t_i \in \mathcal{T}_i} f_i(t_i) \cdot \mathbb{E}_{t_{-i} \sim \mathcal{F}_{-i}}[x(t_i, t_{-i})] = \sum_{t_i \in \mathcal{T}_i} f_i(t_i)X_i(t_i).$$

Then, we have $\sum_{t \in \mathcal{T}} f(t)x'(t) = \sum_{t \in \mathcal{T}} f(t)x(t)$ for the transformed mechanism $\mathcal{M}' = (x', p')$, by Eq. (8) in the proof of Theorem 6 in Appendix.

Lower Bound on Revenue Loss. Similarly to single agent case, we can also prove a lower bound of revenue loss of any welfare-preserving transformation for multiple agents with independent private types. We summarize this result in Theorem 7, and show the proof in Appendix.

Theorem 7. *For any number $n \geq 1$ of agents with independent uniform type distribution, there exists an ε-BIC/ε-EEIC and IR mechanism, for which any welfare-preserving transformation must suffer at least* $\Omega(\sum_i |\mathcal{T}_i| \varepsilon)$ *revenue loss.*

4.1   Impossibility Results

In our main positive result (Theorem 6), we assume independent private types and the target of transformation is BIC mechanism. These two assumptions are near-tight. The proofs are deferred to Appendix.

Theorem 8 (Failure of interdependent type). *There exists an ε-BIC mechanism $\mathcal{M}$ w.r.t an interdependent type distribution $\mathcal{F}$ (see Definition 9 in Appendix), such that no BIC mechanism over $\mathcal{F}$ can achieve negligible revenue loss compared with $\mathcal{M}$.*

Theorem 8 provides a counterexample to show that if we allow for interdependent types, where the value of one agent depends on the type of another, there is no way to construct a BIC mechanism without negligible revenue loss

compared with the original ε-BIC mechanism even if we remove the requirement of welfare preservation. This leaves an open question is whether there is a
counterexample for an ε-BIC transform for correlated, private types.

Theorem 9 (Failure of DSIC target). *There exists an ε-BIC mechanism
$\mathcal{M}$ defined on a type distribution $\mathcal{F}$, such that no DSIC mechanism over $\mathcal{F}$ can
achieve negligible revenue loss compared with $\mathcal{M}$.*

Theorem 9 gives an impossibility result for the setting that we start from
an ε-BIC mechanism We leave open the question as to whether it is possible
to transform an ε-EEIC mechanism to a DSIC mechanism with zero loss of
social welfare and negligible loss of revenue, for multiple agents with independent
uniform type distribution.

5   Application to Automated Mechanism Design

In this section, we apply the transform to linear-programming based and
machine-learning based approaches to automated mechanism design (AMD) [12],
where the mechanism is automatically created for the setting and objective at
hand.

We state the main results for the following, blended design objective of revenue and welfare, for a given $\lambda \in [0,1]$ and type distribution $\mathcal{F}$,

$$\mu_\lambda(\mathcal{M}, \mathcal{F}) = (1 - \lambda) R^{\mathcal{M}}(\mathcal{F}) + \lambda W^{\mathcal{M}}(\mathcal{F}). \tag{5}$$

Let $\texttt{OPT} = \max_{\mathcal{M}:\mathcal{M} \text{ is BIC and IR}} \mu_\lambda(\mathcal{M}, \mathcal{F})$ be the optimal objective achieved
by a BIC and IR mechanism defined on $\mathcal{F}$. We consider two different AMD
approaches, an LP-based approach and a machine-learning based approach.

LP-based AMD. As explained in more details in Appendix, an LP-based approach to BIC mechanism design introduces a decision variable for each outcome
and each type profile. In practice, the type space of each agent may be exponential in the number of items for multi-item auctions, and the number of type
profiles is exponential in the number of agents. To address this challenge, it is
necessary to discretize $\mathcal{T}_i$ to a coarser space $\mathcal{T}_i^+$, $(|\mathcal{T}_i^+| \ll |\mathcal{T}_i|)$ and construct the
coupled type distribution $\mathcal{F}_i^+$. (e.g., by rounding down to the nearest points in
$\mathcal{T}_i^+$, that is, the mass of each point in $\mathcal{T}_i$ is associated with the nearest point
in $\mathcal{T}_i^+$.) Then we can apply an LP-based AMD approach for type distribution
$\mathcal{F}^+ = (\mathcal{F}_1^+, \cdots, \mathcal{F}_n^+)$. Even though the LP returns an mechanism defined only
on $\mathcal{T}^+$, the mechanism $\mathcal{M}$ can be defined on $\mathcal{T}$, by the same coupling technique.
For example, given any type profile $t \in \mathcal{T}$, there is a coupled $t^+ \in \mathcal{T}^+$, and
the mechanism $\mathcal{M}$ takes t^+ as the input. This coupling technique makes the
mechanism only approximately IC. Suppose, in particular, that we have an α-
approximation LP algorithm that outputs an ε-BIC and IR mechanism $\mathcal{M}$ over
$\mathcal{F}$, such that $\mu_\lambda(\mathcal{M}, \mathcal{F}) \geq \alpha\texttt{OPT}$. By an application of the transform to $\mathcal{M}$, we
have the following theorem.

Theorem 10 (LP-based AMD). *For n agents with independent type distribution $\times_{i=1}^{n} \mathcal{F}_i$, and an LP-based AMD approach for coarsened distribution $\mathcal{F}^+$ on coarsened type space $\mathcal{T}^+$ that gives an ε-BIC and IR mechanism $\mathcal{M}$ on $\mathcal{F}$, with $(1 - \lambda)R + \lambda W \geq \alpha \text{OPT}$, for some $\lambda \in [0,1]$, and some $\alpha \in (0,1)$, then there exists a BIC and IR mechanism $\mathcal{M}'$ such that $\mu_\lambda(\mathcal{M}', \mathcal{F}) \geq \alpha \text{OPT} - (1 - \lambda) \sum_{i=1}^{n} |\mathcal{T}_i| \varepsilon$. Given oracle access to the interim quantities of $\mathcal{M}$ on $\mathcal{F}$ and an α-approximation LP solver with running time $rt_{LP}(x)$, where x is the bit complexity of the input, the running time to output the mechanism $\mathcal{M}'$ is at most $\texttt{poly}(\sum_i |\mathcal{T}_i|, rt_{LP}(\texttt{poly}(\sum_i |\mathcal{T}_i^+|, \frac{1}{\varepsilon})))$.*

Machine-Learning Based AMD. RegretNet uses an artificial neural network to learn approximately-incentive compatible auctions for multi-dimensional mechanism design [16]. See Appendix for more details of the application of RegretNet to a setting in which the design goal is a blend of revenue and welfare. RegretNet outputs an ε-EEIC mechanism. Suppose that RegretNet is used in a setting with an independent, uniform type distribution $\mathcal{F}$. To train RegretNet, we randomly draw S samples from $\mathcal{F}$ to form a training data $\mathcal{S}$ and train the model on $\mathcal{S}$. Let $\mathcal{H}$ be the function space modeled by RegretNet and suppose a PAC-learner that outputs an ε-EEIC mechanism $\mathcal{M} \in \mathcal{H}$ on $\mathcal{F}$, such that $\mu_\lambda(\mathcal{M}, \mathcal{F}) \geq \sup_{\hat{\mathcal{M}} \in \mathcal{H}} \mu_\lambda(\hat{\mathcal{M}}, \mathcal{F}) - \varepsilon$ holds with probability at least $1 - \delta$, by observing $S = S(\varepsilon, \delta)$ i.i.d samples from $\mathcal{F}$. By an application of the transform to $\mathcal{M}$, we have the following theorem.

Theorem 11 (RegretNet AMD). *For n agents with independent uniform type distribution $\times_{i=1}^{n} \mathcal{F}_i$ over $\mathcal{T} = (\mathcal{T}_1, \cdots, \mathcal{T}_n)$, and RegretNet to generate an ε-EEIC and IR mechanism $\mathcal{M}$ on $\mathcal{F}$ with $\mu_\lambda(\mathcal{M}, \mathcal{F}) \geq \sup_{\hat{\mathcal{M}} \in \mathcal{H}} \mu_\lambda(\hat{\mathcal{M}}, \mathcal{F}) - \varepsilon$ holds with probability at least $1 - \delta$, for some $\lambda \in [0,1]$, trained on $S = S(\varepsilon, \delta)$ i.i.d samples from $\mathcal{F}$, where $\mathcal{H}$ is the function class modeled by RegretNet, then there exists a BIC and IR mechanism $\mathcal{M}'$, with probability at least $1 - \delta$, such that $\mu_\lambda(\mathcal{M}', \mathcal{F}) \geq \sup_{\hat{\mathcal{M}} \in \mathcal{H}} \mu_\lambda(\hat{\mathcal{M}}, \mathcal{F}) - (1 - \lambda) \sum_{i=1}^{n} |\mathcal{T}_i| \varepsilon - \varepsilon$. Given oracle access to the interim quantities of $\mathcal{M}$ on $\mathcal{F}$ and a PAC-learner with running time $rt_{Net}(x)$, where x is the bit complexity of the input, the running time to output the mechanism $\mathcal{M}'$ is at most $\texttt{poly}(\sum_i |\mathcal{T}_i|, \varepsilon, rt_{Net}(\texttt{poly}(S, \frac{1}{\varepsilon})))$.*

References

1. Azevedo, E.M., Budish, E.: Strategy-proofness in the Large. Rev. Econ. Stud. **86**(1), 81–116 (2019)
2. Balcan, M., Sandholm, T., Vitercik, E.: Estimating approximate incentive compatibility. In: Proceedings of the 2019 ACM Conference on Economics and Computation, EC 2019, 867p (2019)
3. Balseiro, S.R., Feldman, J., Mirrokni, V., Muthukrishnan, S.: Yield optimization of display advertising with ad exchange. Manage. Sci. **60**(12), 2886–2907 (2014)
4. Bei, X., Huang, Z.: Bayesian incentive compatibility via fractional assignments. In: Proceedings of the 2011 Annual ACM-SIAM Symposium on Discrete Algorithms, pp. 720–733 (2011)

5. Border, K.C.: Implementation of reduced form auctions: a geometric approach. Econometrica **59**(4), 1175–1187 (1991)
6. Cai, Y., Daskalakis, C., Weinberg, M.S.: Optimal multi-dimensional mechanism design: reducing revenue to welfare maximization. In: Proceedings of the 53rd IEEE Symposium on Foundations of Computer Science, pp. 130–139 (2012)
7. Cai, Y., Daskalakis, C., Weinberg, S.M.: An algorithmic characterization of multi-dimensional mechanisms. In: Proceedings of the 44th ACM Symposium on Theory of Computing (2012)
8. Cai, Y., Daskalakis, C., Weinberg, S.M.: Understanding incentives: mechanism design becomes algorithm design. In: Proceedings of the 54th IEEE Symposium on Foundations of Computer Science, pp. 618–627 (2013)
9. Cai, Y., Oikonomou, A., Velegkas, G., Zhao, M.: An efficient ε-BIC to BIC transformation and its application to black-box reduction in revenue maximization. In: Proceedings of the 32nd Annual ACM-SIAM Symposium on Discrete Algorithms (SODA) (2021)
10. Cai, Y., Zhao, M.: Simple mechanisms for subadditive buyers via duality. In: Proceedings of the 49th ACM Symposium on Theory of Computing, pp. 170–183 (2017)
11. Carroll, G.: When are local incentive constraints sufficient? Econometrica **80**(2), 661–686 (2012)
12. Conitzer, V., Sandholm, T.: Complexity of mechanism design. In: Proceedings of the 18th Conference on Uncertainty in Artificial Intelligence, pp. 103–110 (2002)
13. Conitzer, V., Sandholm, T.: Self-interested automated mechanism design and implications for optimal combinatorial auctions. In: Proceedings of the 5th ACM Conference on Electronic Commerce, pp. 132–141 (2004)
14. Daskalakis, C., Weinberg, S.M.: Symmetries and optimal multi-dimensional mechanism design. In: Proceedings of the 13th ACM Conference on Electronic Commerce, pp. 370–387 (2012)
15. Diakonikolas, I., Papadimitriou, C., Pierrakos, G., Singer, Y.: Efficiency-revenue trade-offs in auctions. In: Czumaj, A., Mehlhorn, K., Pitts, A., Wattenhofer, R. (eds.) ICALP 2012. LNCS, vol. 7392, pp. 488–499. Springer, Heidelberg (2012). https://doi.org/10.1007/978-3-642-31585-5_44
16. Duetting, P., Feng, Z., Narasimhan, H., Parkes, D., Ravindranath, S.S.: Optimal auctions through deep learning. In: Chaudhuri, K., Salakhutdinov, R. (eds.) Proceedings of the 36th International Conference on Machine Learning. Proceedings of Machine Learning Research, vol. 97. PMLR, Long Beach, California, USA, 9–15 June 2019
17. Dughmi, S., Hartline, J.D., Kleinberg, R., Niazadeh, R.: Bernoulli factories and black-box reductions in mechanism design. In: Proceedings of the 49th Annual ACM SIGACT Symposium on Theory of Computing (2017)
18. Dütting, P., Fischer, F., Jirapinyo, P., Lai, J., Lubin, B., Parkes, D.C.: Payment rules through discriminant-based classifiers. ACM Trans. Econ. Comput. **3**(1), 5 (2014)
19. Feng, Z., Narasimhan, H., Parkes, D.C.: Deep learning for revenue-optimal auctions with budgets. In: Proceedings of the 17th International Conference on Autonomous Agents and Multiagent Systems, pp. 354–362 (2018)
20. Feng, Z., Schrijvers, O., Sodomka, E.: Online learning for measuring incentive compatibility in ad auctions. In: The World Wide Web Conference, WWW 2019, pp. 2729–2735 (2019)

21. Hartline, J.D., Kleinberg, R., Malekian, A.: Bayesian incentive compatibility via matchings. In: Proceedings of the Twenty-Second Annual ACM-SIAM Symposium on Discrete Algorithms (2011)
22. Hartline, J.D., Lucier, B.: Bayesian algorithmic mechanism design. In: Proceedings of the Forty-Second ACM Symposium on Theory of Computing (2010)
23. Hashimoto, T.: The generalized random priority mechanism with budgets. J. Econ. Theory **177**, 708–733 (2018)
24. Lahaie, S., Medina, A.M., Sivan, B., Vassilvitskii, S.: Testing incentive compatibility in display ad auctions. In: Proceedings of the 27th International World Wide Web Conference (WWW) (2018)
25. Lahaie, S., Pennock, D.M.: Revenue analysis of a family of ranking rules for keyword auctions. In: Proceedings of the 8th ACM Conference on Electronic Commerce (EC), pp. 50–56 (2007)
26. Lubin, B., Parkes, D.C.: Approximate strategyproofness. Curr. Sci. **103**(9), 1021–1032 (2012)
27. Narasimhan, H., Parkes, D.C.: A general statistical framework for designing strategy-proof assignment mechanisms. In: Proceedings of the Conference on Uncertainty in Artificial Intelligence (2016)
28. Rubinstein, A., Weinberg, M.S.: Simple mechanisms for a subadditive buyer and applications to revenue monotonicity. ACM Trans. Econ. Comput. **6**, 1–25 (2018)

Strategyproof Facility Location
in Perturbation Stable Instances

Dimitris Fotakis[✉] and Panagiotis Patsilinakos

School of Electrical and Computer Engineering, National Technical
University of Athens, 15780 Athens, Greece
fotakis@cs.ntua.gr, patsilinak@corelab.ntua.gr

Abstract. We study the approximability of k-Facility Location games
on the real line by strategyproof mechanisms without payments. To circumvent impossibility results for $k \geq 3$, we focus on γ-*(perturbation)*
stable instances, where the optimal agent clustering is not affected by
moving any subset of consecutive agent locations closer to each other by a
factor at most $\gamma \geq 1$. We show that the optimal solution is strategyproof
in $(2 + \sqrt{3})$-stable instances, if it does not include any singleton clusters,
and that allocating the facility to the agent next to the rightmost one
in each optimal cluster is strategyproof and $(n - 2)/2$-approximate for
5-stable instances (even if singleton clusters are present), where n is the
number of agents. On the negative side, we show that for any $k \geq 3$ and
any $\delta > 0$, deterministic anonymous strategyproof mechanisms suffer an
unbounded approximation ratio in $(\sqrt{2} - \delta)$-stable instances. Moreover,
we prove that allocating the facility to a random agent of each optimal
cluster is strategyproof and 2-approximate in 5-stable instances.

1 Introduction

We consider k-*Facility Location games*, where $k \geq 2$ facilities are placed on
the real line based on the preferences of n strategic agents. Such problems are
motivated by natural scenarios in Social Choice, where a local authority plans
to build a fixed number of public facilities in an area (see e.g., [36]). The choice
of the locations is based on the preferences of local people, or *agents*. Each agent
reports her ideal location, and the local authority applies a (deterministic or
randomized) *mechanism* that maps the agents' preferences to k facility locations.

The agents evaluate the mechanism's outcome according to their *connection
cost*, i.e., the distance of their ideal location to the nearest facility. The agents
seek to minimize their connection cost and may misreport their ideal locations in
an attempt of manipulating the mechanism. Therefore, the mechanism should be

---

This work was supported by the Hellenic Foundation for Research and Innovation
(H.F.R.I.) under the "First Call for H.F.R.I. Research Projects to support Faculty
members and Researchers and the procurement of high-cost research equipment grant",
project BALSAM, HFRI-FM17-1424. The full version of this work is available at
https://arxiv.org/abs/2107.11977.

M. Feldman et al. (Eds.): WINE 2021, LNCS 13112, pp. 95–112, 2022.
https://doi.org/10.1007/978-3-030-94676-0_6

strategyproof, i.e., it should ensure that no agent can benefit from misreporting her location, or even *group strategyproof*, i.e., resistant to coalitional manipulations. The local authority's objective is to minimize the *social cost*, namely the sum of agent connections costs. In addition to allocating the facilities in a incentive compatible way, which is formalized by (group) strategyproofness, the mechanism should result in a socially desirable outcome, which is quantified by the mechanism's approximation ratio to the optimal social cost.

Since Procaccia and Tennenholtz [37] initiated the research agenda of *approximate mechanism design without money*, k-Facility Location has served as the benchmark problem in the area and its approximability by deterministic or randomized strategyproof mechanisms has been studied extensively in virtually all variants and generalizations. For instance, previous work has studied multiple facilities on the line (e.g., [25,26,34]), in general metric spaces [24,33]), different objectives (e.g., social cost, maximum cost, the L_2 norm of agent connection costs [22,26,37]), restricted metric spaces more general than the line (cycle, plane, trees, see e.g., [2,16,29,35]), facilities that serve different purposes (e.g., [31,32,42]), and different notions of private information about the agent preferences that should be declared to the mechanism (e.g., [15,20]).

The basic question of approximating the optimal social cost by strategyproof mechanisms for k-Facility Location on the line is relatively well-understood. For $k = 1$, placing the facility at the median location is optimal and group strategyproof. For $k = 2$ facilities, the best possible approximation ratio is $n-2$ and is achieved by a natural group strategyproof mechanism that places the facilities at the leftmost and rightmost locations [25,37]. Yet, for $k \geq 3$ facilities, there do not exist any deterministic anonymous[1] strategyproof mechanisms with a bounded (in terms of n and k) approximation ratio [25]. On the positive side, there is a randomized anonymous group strategyproof mechanism with an approximation ratio of n [26] (see also Sect. 1.1 for a selective list of more references).

Perturbation Stability in k-Facility Location Games. Our work aims to circumvent the strong impossibility result of [25] and is motivated by the recent success on the design of polynomial-time exact algorithms for perturbation stable clustering instances (see e.g., [3,9–11,38]). An instance of a clustering problem, like k-Facility Location (a.k.a. k-median), is γ-*perturbation stable* (or simply, γ-*stable*), for some $\gamma \geq 1$, if the optimal clustering is not affected by scaling down any subset of the entries of the distance matrix by a factor at most γ. Perturbation stability was introduced by Bilu and Linial [12] and Awasthi, Blum and Sheffet [7] (and has motivated a significant volume of followup work since then, see e.g., [3,9,11,38] and the references therein) in an attempt to obtain a theoretical understanding of the superior practical performance of relatively simple clustering algorithms for well known NP-hard clustering problems. Intuitively, the optimal clusters of a γ-stable instance are somehow well separated, and thus, relatively easy to identify (see also the main properties of stable instances in Sect. 3). As a result, natural extensions of simple algorithms, like single-linkage,

---

[1] A mechanism is *anonymous* if its outcome depends only on the agent locations, not on their identities.

can recover the optimal clustering in polynomial time, provided that $\gamma \geq 2$ [3], and standard approaches, like dynamic programming (resp. local search), work in almost linear time for $\gamma > 2 + \sqrt{3}$ (resp. $\gamma > 5$) [1].

In this work, we investigate whether restricting our attention to stable instances allows for improved strategyproof mechanisms with bounded (and ideally, constant) approximation guarantees for k-Facility Location on the line, with $k \geq 2$. We note that the impossibility results of [25] crucially depend on the fact that the clustering (and the subsequent facility placement) produced by any deterministic mechanism with a bounded approximation ratio must be sensitive to location misreports by certain agents (see also Sect. 6). Hence, it is very natural to investigate whether the restriction to γ-stable instances allows for some nontrivial approximation guarantees by deterministic or randomized strategyproof mechanisms for k-Facility Location on the line.

To study the question above, we adapt to the real line the stricter[2] notion of γ-metric stability [3], where it is also required that the distances form a metric after the γ-perturbation. In our notion of *linear* γ-*stability*, the instances should retain their linear structure after a γ-perturbation. Hence, a γ-perturbation of a linear k-Facility Location instance is obtained by moving any subset of pairs of consecutive agent locations closer to each other by a factor at most $\gamma \geq 1$. We say that a k-Facility Location instance is γ-stable, if the original instance and any γ-perturbation of it admit the same unique optimal clustering. Interestingly, for γ sufficiently large, γ-stable instances of k-Facility Location have additional structure that one can exploit towards the design of strategyproof mechanisms with good approximation guarantees (see also Sect. 3).

From a conceptual viewpoint, our work is motivated by a reasoning very similar to that discussed in [13] and summarized in *"clustering is hard only when it doesn't matter"* by Roughgarden [40]. In a nutshell, we expect that when k public facilities (such as schools, libraries, hospitals, representatives) are to be allocated to some communities (e.g., cities, villages or neighborhoods, as represented by the locations of agents on the real line) the communities are already well formed, relatively easy to identify and difficult to radically reshape by small distance perturbations or agent location misreports. Moreover, in natural practical applications of k-Facility Location games, agents tend to misreport "locally" (i.e., they tend to declare a different ideal location in their neighborhood, trying to manipulate the location of the local facility), which usually does not affect the cluster formation. In practice, this happens because the agents do not have enough knowledge about locations in other neighborhoods, and because "large non-local" misreports are usually easy to identify by combining publicly avail-

---

[2] The notion of γ-metric stability is "stricter" than standard γ-stability in the sense that the former excludes some perturbations allowed by the latter. Hence, the class of γ-metric stable instances includes the class of γ-stable instances. More generally, the stricter a notion of stability is, the larger the class of instances qualified as stable, and the more general the positive results that one gets. Similarly, for any $\gamma' > \gamma \geq 1$, the class of γ(-metric) stable instances includes the class of γ'(-metric) instances. Hence, a smaller value of γ makes a positive result stronger and more general.

able information about the agents (e.g., occupation, address, habits, lifestyle). Hence, we believe that the class of γ-stable instances, especially for relatively small values of γ, provides a reasonably accurate abstraction of k-Facility Location instances that a mechanism is more likely to deal with in practice. We feel that our work takes a small first step towards justifying that (not only clustering but also) strategyproof facility location is hard only when it doesn't matter.

Contributions and Techniques. Our conceptual contribution is that we initiate the study of efficient (wrt. their approximation ratio for the social cost) strategyproof mechanisms for the large and natural class of γ-stable instances of k-Facility Location on the line. Our technical contribution is that we show the existence of deterministic (resp. randomized) strategyproof mechanisms with a bounded (resp. constant) approximation ratio for 5-stable instances and any number of facilities. Moreover, we show that the optimal solution is strategyproof for $(2 + \sqrt{3})$-stable instances, if the optimal clustering does not include any singleton clusters (which is likely to be the case in virtually all practical applications). To provide evidence that restriction to stable instances does not make the problem trivial, we strengthen the impossibility result of [25]. Specifically, we show that for any $k \geq 3$ and any $\delta > 0$, there do not exist any deterministic anonymous strategyproof mechanisms for k-Facility Location on $(\sqrt{2} - \delta)$-stable instances with bounded (in terms of n and k) approximation ratio.

At the conceptual level, we interpret the stability assumption as a prior on the class of true instances. Namely, we assume that the mechanism has only to deal with γ-stable true instances, a restriction motivated by (and fully consistent with) how the stability assumption is used in the literature on efficient algorithms for stable clustering (see e.g., [3,9,11,12], where the algorithms are analyzed for stable instances only). More specifically, our mechanisms expect as input a declared instance such that in the optimal clustering, the distance between any two consecutive clusters is at least $\frac{(\gamma-1)^2}{2\gamma}$ times larger than the diameters of the two clusters (a.k.a. *cluster-separation* property, see Lemma 1). This condition is necessary (but not sufficient) for γ-stability and can be easily checked. If the declared instance does not satisfy the cluster-separation property, our mechanisms do not allocate any facilities. Otherwise, our mechanisms allocate k facilities (even if the instance is not stable). We prove that for all γ-stable true instances (with the exact stability factor γ depending on the mechanism), if agents can only deviate so that the declared instance satisfies the cluster-separation property (and does not have singleton clusters, for the optimal mechanism), our mechanisms are strategyproof and achieve the desired approximation guarantee. Hence, if we restrict ourselves to γ-stable true instances and to agent deviations that do not obviously violate γ-stability, our mechanisms should only deal with γ-stable declared instances, due to strategyproofness. On the other hand, if non-stable true instances may occur, the mechanisms cannot distinguish between a stable true instance and a declared instance, which appears to be stable, but is obtained from a non-stable instance through location misreports.

The restriction that the agents of a γ-stable instance are only allowed to deviate so that the declared instance satisfies the cluster-separation property conceptually resembles the notion of *local verification* (see e.g., [4,6,14,27,28]), where the set of each agent's allowable deviations is restricted to a so-called *correspondence set*, which typically depends on the agent's true type, but not on the types of the other agents. Instead of restricting the correspondence set of each individual agent independently, we impose a structural condition on the entire declared instance, which restricts the set of the agents' allowable deviations, but in an observable sense. Hence, we can actually implement our notion of verification, by checking some simple properties of the declared instance, instead of just assuming that any deviation outside an agent's correspondence set will be caught and penalized (which is the standard approach in mechanisms with local verification [4,14], but see e.g., [6,24] for noticeable exceptions).

On the technical side, we start, in Sect. 3, with useful properties of stable instances of k-Facility Location. We show (i) the *cluster-separation* property (Lemma 1), i.e., that in any γ-stable instance, the distance between any two consecutive clusters is at least $\frac{(\gamma-1)^2}{2\gamma}$ times larger than their diameters; and (ii) the so-called *no direct improvement from singleton deviations* property (Lemma 2), i.e., that in any 3-stable instance, no agent who deviates to a location, which becomes a singleton cluster in the optimal clustering of the resulting instance, can improve her connection cost through the facility of that singleton cluster.

In Sect. 4, we show that for $(2+\sqrt{3})$-stable instances whose optimal clustering does not include any singleton clusters, the optimal solution is strategyproof (Theorem 1). For the analysis, we observe that a misreport cannot be profitable for an agent, unless it results in a different optimal clustering. The key step is to show that for $(2 + \sqrt{3})$-stable instances without singleton clusters, a profitable misreport cannot change the optimal clustering, unless the instance obtained from the misreport violates the cluster-separation property. To the best of our knowledge, the idea of penalizing (and thus, essentially forbidding) a whole class of potentially profitable misreports by identifying how they affect a key structural property of the original instance, which becomes possible due to our restriction to stable instances, has not been used before in the design of strategyproof mechanisms for k-Facility Location.

We should also motivate our restriction to stable instances without singleton clusters in their optimal clustering. So, let us consider the rightmost agent x_j of an optimal cluster C_i in a γ-stable instance $\vec{x}$. No matter the stability factor γ, it is possible that x_j performs a so-called *singleton deviation*. Namely, x_j deviates to a remote location x' (potentially very far away from any location in $\vec{x}$), which becomes a singleton cluster in the optimal clustering of the resulting instance. Such a singleton deviation might cause cluster C_i to merge with (possibly part of the next) cluster C_{i+1}, which in turn, might bring the median of the new cluster much closer to x_j. It is not hard to see that if we stick to the optimal solution, where the facilities are located at the median of each optimal cluster, there are γ-stable instances[3], with arbitrarily large $\gamma \geq 1$, where some agents

---

[3] E.g., let $k = 2$ and consider the $\Theta(\gamma)$-stable instance $(0, 1 - \varepsilon, 1, 6\gamma, 6\gamma + \varepsilon, 6\gamma + 1, 6\gamma + 1 + \varepsilon, 6\gamma + 2)$, for any $\gamma \geq 1$. Then, the agent at location 6γ can decrease its connection cost (from 1) to ε by deviating to location $(6\gamma)^2$.

can deviate to a remote location and gain, by becoming singleton clusters, while maintaining the desirable stability factor of the declared instance.

To deal with singleton deviations, we place the facility either at a location close to an extreme one, as we do in Sect. 5 with the ALMOSTRIGHTMOST mechanism, or at a random location, as we do in Sect. 7 with the RANDOM mechanism. More specifically, in Sect. 5, we show that the ALMOSTRIGHTMOST mechanism, which places the facility of any non-singleton optimal cluster at the location of the second rightmost agent, is strategyproof for 5-stable instances (even if their optimal clustering includes singleton clusters) and achieves an approximation ratio at most $(n - 2)/2$ (Theorem 2). Moreover, in Sect. 7, we show that the RANDOM mechanism, which places the facility of any optimal cluster at a location chosen uniformly at random, is strategyproof for 5-stable instances and achieves an approximation ratio of 2 (Theorem 4).

To obtain a deeper understanding of the challenges behind the design of strategyproof mechanisms for stable instances of k-Facility Location on the line, we strengthen the impossibility result of [25, Theorem 3.7] so that it applies to γ-stable instances with $\gamma < \sqrt{2}$ (Sect. 6). Through a careful analysis of the image sets of deterministic strategyproof mechanisms, we show that for any $k \geq 3$, any $\delta > 0$, and any $\rho \geq 1$, there do not exist any ρ-approximate deterministic anonymous strategyproof mechanisms for $(\sqrt{2}-\delta)$-stable instances (Theorem 3). The proof of Theorem 3 requires additional ideas and extreme care (and some novelty) in the agent deviations, so as to only consider stable instances, compared against the proof of [25, Theorem 3.7]. Interestingly, singleton deviations play a crucial role in the proof of Theorem 3.

1.1   Other Related Work

Previous work has shown that deterministic strategyproof mechanisms can only achieve a bounded approximation ratio for k-Facility Location on the line, only if we have at most 2 facilities [25,37]. Notably, stable (called *well-separated* in [25]) instances with $n = k+1$ agents play a key role in the proof of inapproximability of k-Facility Location by deterministic anonymous strategyproof mechanisms [25, Theorem 3.7]. On the other hand, randomized mechanisms are known to achieve a better approximation ratio for $k = 2$ facilities [34], a constant approximation ratio if we have $k \geq 2$ facilities and only $n = k + 1$ agents [18,26], and an approximation ratio of n for any $k \geq 3$ [26]. Fotakis and Tzamos [24] considered winner-imposing randomized mechanisms that achieve an approximation ratio of $4k$ for k-Facility Location in general metric spaces. In fact, the approximation ratio can be improved to $\Theta(\ln k)$, using the analysis of [5].

For the objective of maximum agent cost, Alon et al. [2] almost completely characterized the approximation ratios achievable by randomized and deterministic strategyproof mechanisms for 1-Facility Location in general metrics and rings. Fotakis and Tzamos [26] presented a 2-approximate randomized group strategyproof mechanism for k-Facility Location on the line and the maximum cost objective. For 1-Facility Location on the line and the objective of minimizing the sum of squares of the agent connection costs, Feldman and Wilf [22] proved that the best approximation ratio is 1.5 for randomized and 2 for deterministic mechanisms. Golomb and Tzamos [30] presented tight (resp. almost tight) additive approximation guarantees for locating a single (resp. multiple) facilities on the line and the objectives of the maximum cost and the social cost.

Regarding the application of perturbation stability, we follow the approach of *beyond worst-case analysis* [38], where researchers seek a theoretical understanding of the superior practical performance of certain algorithms by formally analyzing them on practically relevant instances. The beyond worst-case approach is not anything new for Algorithmic Mechanism Design. *Bayesian* analysis is standard in revenue maximization when we allocate private goods (see e.g., [39]) and has led to many strong and elegant results for social welfare maximization in combinatorial auctions by truthful posted price mechanisms (see e.g., [17,21]). However, in this work, instead of assuming (similar to Bayesian analysis) that the mechanism designer has a relatively accurate knowledge of the distribution of agent locations on the line (and use e.g., an appropriately optimized percentile mechanism [43]), we employ a deterministic restriction on the class of instances (namely, perturbation stability), and investigate if deterministic (resp. randomized) strategyproof mechanisms with a bounded (resp. constant) approximation ratio are possible for locating any number $k \geq 2$ facilities on such instances. To the best of our knowledge, the only previous work where the notion of perturbation stability is applied to Algorithmic Mechanism Design (to combinatorial auctions, in particular) is [23] (but see also [8,19] where the similar in spirit assumption of endowed valuations was applied to combinatorial markets).

2   Notation, Definitions and Preliminaries

We let $[n] = \{1, \ldots, n\}$. For any $x, y \in \mathbb{R}$, we let $d(x, y) = |x - y|$ be the distance of locations x and y on the real line. For a tuple $\vec{x} = (x_1, \ldots, x_n) \in \mathbb{R}^n$, we let $\vec{x}_{-i}$ denote the tuple $\vec{x}$ without coordinate x_i. For a non-empty set S of indices, we let $\vec{x}_S = (x_i)_{i \in S}$ and $\vec{x}_{-S} = (x_i)_{i \notin S}$. We write $(\vec{x}_{-i}, a)$ to denote the tuple $\vec{x}$ with a in place of x_i, $(\vec{x}_{-\{i,j\}}, a, b)$ to denote the tuple $\vec{x}$ with a in place of x_i and b in place of x_j, and so on. For random variable X, $\mathbb{E}(X)$ is the expectation of X. For an event E in a sample space, $\Pr(E)$ is the probability that E occurs.

Instances. We consider k-Facility Location with $k \geq 2$ facilities and $n \geq k + 1$ agents on the real line. We let $N = \{1, \ldots, n\}$ be the set of agents. Each agent $i \in N$ resides at a location $x_i \in \mathbb{R}$, which is i's private information. We usually refer to a locations profile $\vec{x} = (x_1, \ldots, x_n) \in \mathbb{R}^n$, $x_1 \leq \cdots \leq x_n$, as an *instance*. By slightly abusing the notation, we use x_i to refer both to the agent i's location and sometimes to the agent i (i.e., the strategic entity) herself.

Mechanisms. A *deterministic mechanism* M for k-Facility Location maps an instance $\vec{x}$ to a k-tuple $(c_1, \ldots, c_k) \in \mathbb{R}^k$, $c_1 \leq \cdots \leq c_k$, of facility locations. We let $M(\vec{x})$ denote the outcome of M in instance $\vec{x}$, and let $M_j(\vec{x})$ denote c_j, i.e., the j-th smallest coordinate in $M(\vec{x})$. We write $c \in M(\vec{x})$ to denote that $M(\vec{x})$ places a facility at location c. A *randomized mechanism* M maps an instance $\vec{x}$ to a probability distribution over k-tuples $(c_1, \ldots, c_k) \in \mathbb{R}^k$.

Connection Cost and Social Cost. Given a k-tuple $\vec{c} = (c_1, \ldots, c_k)$, $c_1 \leq \cdots$
$\leq c_k$, of facility locations, the connection cost of agent i wrt. $\vec{c}$, denoted
$d(x_i, \vec{c})$, is $d(x_i, \vec{c}) = \min_{1 \leq j \leq k} |x_i - y_j|$. Given a deterministic mechanism M
and an instance $\vec{x}$, $d(x_i, M(\vec{x}))$ denotes i's connection cost wrt. the outcome
of $M(\vec{x})$. If M is a randomized mechanism, i' expected connection cost is
$\mathbb{E}_{\vec{c} \sim M(\vec{x})}(d(x_i, \vec{c}))$. The *social cost* of a deterministic mechanism M for instance $\vec{x}$
is $cost(\vec{x}, M(\vec{x})) = \sum_{i=1}^{n} d(x_i, M(\vec{x}))$. The social cost of a facilities profile $\vec{c} \in \mathbb{R}^k$
is $cost(\vec{x}, \vec{c}) = \sum_{i=1}^{n} d(x_i, \vec{c})$. The *expected social cost* of a randomized mecha-
nism M $\vec{x}$ is $cost(\vec{x}, M(\vec{x})) = \sum_{i=1}^{n} \mathbb{E}_{\vec{c} \sim M(\vec{x})}(d(x_i, \vec{c}))$. The *optimal social cost* for
an instance $\vec{x}$ is $cost^*(\vec{x}) = \min_{\vec{c} \in \mathbb{R}^k} \sum_{i=1}^{n} d(x_i, \vec{c})$. For k-Facility Location, the
optimal social cost can be found in $O(kn \log n)$ time by dynamic programming.

Approximation Ratio. A mechanism M has an approximation ratio of $\rho \geq 1$,
if for any instance $\vec{x}$, $cost(\vec{x}, M(\vec{x})) \leq \rho \, cost^*(\vec{x})$. We say that the approximation
ratio ρ of M is *bounded*, if ρ is bounded from above either by a constant or by
a (computable) function of n and k.

Strategyproofness. A deterministic mechanism M is *strategyproof*, if no agent
can benefit by misreporting her location. I.e., M is strategyproof, if for all loca-
tion profiles $\vec{x}$, any agent i, and all locations y, $d(x_i, M(\vec{x})) \leq d(x_i, M((\vec{x}_{-i}, y)))$.
A randomized mechanism M is strategyproof (in expectation), if for all location
profiles $\vec{x}$, any agent i, and all $y \in \mathbb{R}$, $\mathbb{E}_{\vec{c} \sim M(\vec{x})}(d(x_i, \vec{c})) \leq \mathbb{E}_{\vec{c} \sim M((\vec{x}_{-i}, y))}(d(x_i, \vec{c}))$.

Clusterings. A *clustering* (or k-clustering, if k is not clear from the context) of
an instance $\vec{x}$ is any partitioning $\vec{C} = (C_1, \ldots, C_k)$ of $\vec{x}$ into k sets of consecutive
agent locations. We index clusters from left to right. I.e., $C_1 = \{x_1, \ldots, x_{|C_1|}\}$,
$C_2 = \{x_{|C_1|+1}, \ldots, x_{|C_1|+|C_2|}\}$, and so on. We refer to a cluster C_i with $|C_i| = 1$
as a *singleton* cluster. We sometimes use $(\vec{x}, \vec{C})$ to highlight that we consider $\vec{C}$
as a clustering of instance $\vec{x}$.

Two clusters C and C' are identical, denoted $C = C'$, if they include the
exact same locations. Two clusterings $\vec{C} = (C_1, \ldots, C_k)$ and $\vec{Y} = (Y_1, \ldots, Y_k)$
of an instance $\vec{x}$ are the same, if $C_i = Y_i$, for all $i \in [k]$. Abusing the notation,
we say that a clustering $\vec{C}$ of an instance $\vec{x}$ is identical to a clustering $\vec{Y}$ of a
γ-perturbation $\vec{x}'$ of $\vec{x}$ (see also Definition 1), if $|C_i| = |Y_i|$, for all $i \in [k]$.

We let $x_{i,l}$ and $x_{i,r}$ denote the leftmost and the rightmost agent of each
cluster C_i. Then, $x_{i-1,r} < x_{i,l} \leq x_{i,r} < x_{i+1,l}$, for all $i \in \{2, \ldots, k-1\}$. We
extend this notation to refer to other agents by their relative location in each
cluster. Namely, $x_{i,l+1}$ (resp. $x_{i,r-1}$) is the second agent from the left (resp.
right) of cluster C_i. The *diameter* of a cluster C_i is $D(C_i) = d(x_{i,l}, x_{i,r})$. The
distance of clusters C_i and C_j is $d(C_i, C_j) = \min_{x \in C_i, y \in C_j} \{d(x, y)\}$.

A k-facilities profile $\vec{c} = (c_1, \ldots, c_k)$ forms a clustering $\vec{C} = (C_1, \ldots, C_k)$ of
an instance $\vec{x}$ by assigning each agent/location x_j to the cluster C_i with facility
c_i closest to x_j. The *optimal clustering* of an instance $\vec{x}$ is the clustering of $\vec{x}$
induced by the facility locations profile with minimum social cost.

The social cost of a clustering $\vec{C}$ induced by a k-facilities profile $\vec{c}$ on $\vec{x}$ is
simply $cost(\vec{x}, \vec{c})$. We sometimes refer to the social cost $cost(\vec{x}, \vec{C})$ of a clustering
$\vec{C}$ for $\vec{x}$, without any explicit reference to the corresponding facilities profile.

3   Perturbation Stability: Definition and Properties

Next, we introduce the notion of γ-(linear) stability and prove some useful properties of γ-stable instances of k-Facility Location.

Definition 1 (γ-Pertrubation and γ-Stability). *Let $\vec{x} = (x_1, \ldots, x_n)$ be a locations profile. A locations profile $\vec{x}' = (x'_1, \ldots, x'_n)$ is a γ-perturbation of $\vec{x}$, for some $\gamma \geq 1$, if $x'_1 = x_1$ and for every $i \in [n-1]$, $d(x_i, x_{i+1})/\gamma \leq d(x'_i, x'_{i+1}) \leq d(x_i, x_{i+1})$. A k-Facility Location instance $\vec{x}$ is γ-perturbation stable (or simply, γ-stable), if $\vec{x}$ has a unique optimal clustering $(C_1, \ldots, C_k)$ and every γ-perturbation $\vec{x}'$ of $\vec{x}$ has the same unique optimal clustering $(C_1, \ldots, C_k)$.*

Our notion of linear perturbation stability naturally adapts the notion of metric perturbation stability [3, Definition 2.5] to the line. We note, the class of γ-stable linear instances, according to Definition 1, is at least as large as the class of metric γ-stable linear instances, according to [3, Definition 2.5].

Similarly to [3, Theorem 3.1] (see also [40, Lemma 7.1] and [7, Corollary 2.3]), we can show that for all $\gamma \geq 1$, every γ-stable instance $\vec{x}$, which admits an optimal clustering $C_1, \ldots, C_k$ with optimal centers $c_1, \ldots, c_k$, satisfies the following γ-center proximity property: For all cluster pairs C_i and C_j, with $i \neq j$, and all locations $x \in C_i$, $d(x, c_j) > \gamma d(x, c_i)$. We regularly use the following consequence of γ-center proximity (see [40, Lemma 7.2]).

Proposition 1. *Let $\gamma \geq 2$ and let $\vec{x}$ be any γ-stable instance, with unique optimal clustering $C_1, \ldots, C_k$ and optimal centers $c_1, \ldots, c_k$. Then, for all clusters C_i and C_j, with $i \neq j$, and all locations $x \in C_i$ and $y \in C_j$, $d(x, y) > (\gamma - 1)d(x, c_i)$.*

We next show that for γ large enough, the optimal clusters of a γ-stable instance are well-separated, in the sense that the distance of two consecutive clusters is larger than their diameters.

Lemma 1 (Cluster-Separation Property). *For any γ-stable instance on the line with optimal clustering $(C_1, \ldots, C_k)$ and all clusters C_i and C_j, with $i \neq j$, $d(C_i, C_j) > \frac{(\gamma-1)^2}{2\gamma} \max\{D(C_i), D(C_j)\}$.*

The cluster-separation property of Lemma 1 is proven in [1] as a consequence of γ-cluster proximity. Setting $\gamma \geq 2 + \sqrt{3}$, we get that:

Corollary 1. *Let $\gamma \geq 2 + \sqrt{3}$ and let $\vec{x}$ be any γ-stable instance with unique optimal clustering $(C_1, \ldots, C_k)$. Then, for all clusters C_i and C_j, with $i \neq j$, $d(C_i, C_j) > \max\{D(C_i), D(C_j)\}$.*

The following is an immediate consequence of the cluster-separation property.

Observation 1 *Let $\vec{x}$ be a k-Facility Location with a clustering $\vec{C} = (C_1, \ldots, C_k)$ such that for any two clusters C_i and C_j, $\max\{D(C_i), D(C_j)\} < d(C_i, C_j)$. Then, if in the optimal clustering of $\vec{x}$, there is a facility at the location of some $x \in C_i$, no agent in C_i is served by a facility at $x_j \notin C_i$.*

Next, we establish the so-called *no direct improvement from singleton deviations* property, used to show the strategyproofness of the ALMOSTRIGHTMOST and RANDOM mechanisms.

Lemma 2. *Let $\vec{x}$ be a γ-stable instance with $\gamma \geq 3$ and optimal clustering $\vec{C} = (C_1, ..., C_k)$ and cluster centers $(c_1, ..., c_k)$, and let an agent $x_i \in C_i \setminus \{c_i\}$ and a location x' such that x' is a singleton cluster in the optimal clustering of the resulting instance $(\vec{x}_{-i}, x')$. Then, $d(x_i, x') > d(x_i, c_i)$.*

The following shows that for 5-stable instances $\vec{x}$, an agent cannot form a singleton cluster, unless she deviates by a distance larger than the diameter of her cluster in $\vec{x}$'s optimal clustering.

Lemma 3. *Let $\vec{x}$ be any γ-stable instance with $\gamma \geq 5$ and optimal clustering $\vec{C} = (C_1, ..., C_k)$. Let $x_i \in C_i \setminus \{c_i\}$ be any agent and x' any location such that x' is a singleton cluster in the optimal clustering of instance $\vec{x}' = (\vec{x}_{-i}, x')$, where x_i has deviated to x'. Then, $d(x', x_i) > D(C_i)$.*

4 Optimal is Strategyproof for $(2 + \sqrt{3})$-Stable Instances

We next show that allocating the facilities optimally is strategyproof for $(2+\sqrt{3})$-stable instances of k-Facility Location, if the optimal clustering does not include any singleton clusters. More specifically, in this section, we analyze Mechanism 1.

Since placing the facility at the median in a single cluster is strategyproof, a deviation can be profitable only if it results in a k-clustering different from the optimal clustering $(C_1, \ldots, C_k)$ of $\vec{x}$. For γ sufficiently large, γ-stability implies that the optimal clusters are well identified, so that any attempt to alter the optimal clustering (without introducing singleton clusters and without violating the cluster separation property) results in an increased cost for the deviating agent. We should highlight that Mechanism 1 may also "serve" non-stable instances that satisfy the cluster separation property. We next prove that OPTIMAL is strategyproof if the true instance is $(2 + \sqrt{3})$-stable and its optimal clustering does not include any singleton clusters:

Theorem 1. *The OPTIMAL mechanism applied to $(2 + \sqrt{3})$-stable instances of k-Facility Location without singleton clusters in their optimal clustering is strategyproof and minimizes the social cost.*

Mechanism 1: OPTIMAL

    Result: An allocation of k facilities
    Input: A k-Facility Location instance $\vec{x}$.
1 Compute the optimal clustering $(C_1, \ldots, C_k)$ of $\vec{x}$.
2 Let c_i be the left median point of each cluster C_i.
3 **if** $\left(\exists i \in [k] \text{ with } |C_i| = 1\right)$ *or* $\left(\exists i \in [k-1] \text{ with } \right.$
    $\max\{D(C_i), D(C_{i+1})\} \geq d(C_i, C_{i+1})\right)$ **then**
    | **Output:** "FACILITIES ARE NOT ALLOCATED".
4 **else**
    | **Output:** The k-facility allocation $(c_1, \ldots, c_k)$.

Proof. We let $cost(X, \vec{C}) = \sum_{x \in X} d(x_j, \vec{C})$ denote the cost of a set of agents X in a clustering $\vec{C} = (C_1, \ldots, C_k)$ of an instance $\vec{x}$. Moreover, $cost(\vec{y}, \vec{C})$ denotes the cost of instance $\vec{y}$ in clustering $\vec{C}$ with the same centers as in $\vec{C}$ for $\vec{x}$.

Since optimality is given, we only need to establish strategyproofness. We show the following: Let $\vec{x}$ be any $(2+\sqrt{3})$-stable k-Facility Location instance with optimal clustering $\vec{C} = (C_1, \ldots, C_k)$. For any agent i and any location y, let $\vec{Y}$ be the optimal clustering of instance $\vec{y} = (\vec{x}_{-i}, y)$ resulting from i's deviation from x_i to y. Then, if y is not a singleton cluster in $(\vec{y}, \vec{Y})$, either $d(x_i, \vec{C}) < d(x_i, \vec{Y})$, or there is an $i \in [k-1]$ for which $\max\{D(Y_i), D(Y_{i+1})\} \geq d(Y_i, Y_{i+1})$.

So, we let $x_i \in C_i$ deviate to a location y, resulting in $\vec{y} = (\vec{x}_{-i}, y)$ with optimal clustering $\vec{Y}$. Since y is not a singleton cluster, it is clustered with agents belonging in one or two clusters of $\vec{C}$, say either in cluster C_j or in clusters C_{j-1} and C_j. By optimally of $\vec{C}$ and $\vec{Y}$, the number of facilities serving $C_{j-1} \cup C_j \cup \{y\}$ in $(\vec{y}, \vec{Y})$ is no less than the number of facilities serving $C_{j-1} \cup C_j$ in $(\vec{x}, \vec{C})$. Hence, there is at least one facility in either C_{j-1} or C_j.

Wlog., suppose that a facility is allocated to an agent in C_j in $(\vec{y}, \vec{Y})$. By Corollary 1 and Observation 1, no agent in C_j is served by a facility in $\vec{x} \setminus C_j$ in $\vec{Y}$. Thus we get the following cases:

Case 1: *y is not allocated a facility in $\vec{Y}$:* This can happen in one of two ways:

> **Case 1a** y is clustered together with some agents from cluster C_j and no facility placed in C_j serves agents in $\vec{x} \setminus C_j$ in $\vec{Y}$.
>
> **Case 1b:** y is clustered together with some agents from a cluster C_j and at least one of the facilities placed in C_j serve agents in $\vec{x} \setminus C_j$ in $\vec{Y}$.

Case 2: *y is allocated a facility in $\vec{Y}$.* This can happen in one of two ways:

> **Case 2a:** y only serves agents that belong in C_j (by optimality, y must be the median location of the new cluster, which implies that either $y < x_{i,l}$ and y only serves $x_{i,l}$ or $x_{j,l} \leq y \leq x_{j,r}$).
>
> **Case 2b:** In $\vec{Y}$, y serves agents that belong in both C_{j-1} and C_j.

We next show that $cost(\vec{y}, \vec{C}) < cost(\vec{y}, \vec{Y})$. Hence, OPTIMAL would also select $\vec{C}$ for $\vec{y}$, rendering x_i's deviation to y non-profitable. Specifically,

$$cost(\vec{y}, \vec{C}) < cost(\vec{y}, \vec{Y}) \Leftrightarrow$$
$$cost(\vec{x}, \vec{C}) + d(y, \vec{C}) - d(x_i, \vec{C}) < cost(\vec{x}, \vec{Y}) + d(y, \vec{Y}) - d(x_i, \vec{Y}) \Leftrightarrow$$
$$d(y, \vec{C}) - d(y, \vec{Y}) < cost(\vec{x}, \vec{Y}) - cost(\vec{x}, \vec{C}) + d(x_i, \vec{C}) - d(x_i, \vec{Y})$$

Since x_i gains by deviating to y, $d(x_i, \vec{C}) - d(x_i, \vec{Y}) > 0$. So, it suffices to show:

$$d(y, \vec{C}) - d(y, \vec{Y}) \leq cost(\vec{x}, \vec{Y}) - cost(\vec{x}, \vec{C})$$
$$= cost(C_j, \vec{Y}) - cost(C_j, \vec{C}) + cost(\vec{x} \setminus C_j, \vec{Y}) - cost(\vec{x} \setminus C_j, \vec{C}) \quad (1)$$

We start with Case 1a and Case 2a, i.e., the cases where $\vec{Y}$ allocates facilities to agents of C_j (between $x_{j,l}$ and $x_{j,r}$) serving only agents in C_j. Note that in

Case 2a, y can also be located outside C_j and serve only $x_{i,l}$. We treat this case as Case 1a, since it is equivalent to placing the facility on $x_{i,l}$ serving y.

Then, (1) holds if $\vec{Y}$ allocates a single facility to agents in $C_j \cup \{y\}$, because the facility is allocated to the median of $C_j \cup \{y\}$, hence $d(y, \vec{C}) - d(y, \vec{Y}) = cost(C_j, \vec{Y}) - cost(C_j, \vec{C})$, while $cost(\vec{x} \setminus C_j, \vec{Y}) - cost(\vec{x} \setminus C_j, \vec{C}) \geq 0$, since $\vec{C}$ is optimal for $\vec{x}$. So, we focus on the most interesting case where the agents in $C_j \cup \{y\}$ are allocated at least two facilities. We observe that (1) follows from:

$$d(y, \vec{C}) - d(y, \vec{Y}) \leq \tfrac{1}{\gamma}\left(cost(\vec{x} \setminus C_j, \vec{Y}) - cost(\vec{x} \setminus C_j, \vec{C})\right) \tag{2}$$

$$cost(C_j, \vec{C}) - cost(C_j, \vec{Y}) \leq \left(1 - \tfrac{1}{\gamma}\right)\left(cost(\vec{x} \setminus C_j, \vec{Y}) - cost(\vec{x} \setminus C_j, \vec{C})\right) \tag{3}$$

To establish (2) and (3), we first consider the valid γ-perturbation of instance $\vec{x}$ where all distances between consecutive agent pairs to the left of C_j (i.e. agents $\{x_1, x_2, \ldots, x_{j-1,r}\}$) and between consecutive agent pairs to the right of C_j (i.e. agents $\{x_{j+1,l}, \ldots, x_{k,r}\}$) are scaled down by γ. By stability, the clustering $\vec{C}$ remains the unique optimal clustering for the perturbed instance $\vec{x}'$. Moreover, since agents in $\vec{x} \setminus C_j$ are not served by a facility in C_j in $\vec{C}$ and $\vec{Y}$, and since all distances outside C_j are scaled down by γ, while all distances within C_j remain the same, the cost of the clusterings $\vec{C}$ and $\vec{Y}$ for the perturbed instance $\vec{x}'$ is $cost(C_j, \vec{C}) + cost(\vec{x} \setminus C_j, \vec{C})/\gamma$ and $cost(C_j, \vec{Y}) + cost(\vec{x} \setminus C_j, \vec{Y})/\gamma$, respectively. Using $cost(\vec{x}', \vec{C}) < cost(\vec{x}', \vec{Y})$ and $\gamma \geq 2$, we obtain:

$$cost(C_j, \vec{C}) - cost(C_j, \vec{Y}) < \tfrac{1}{\gamma}\left(cost(\vec{x} \setminus C_j, \vec{Y}) - cost(\vec{x} \setminus C_j, \vec{C})\right) \tag{4}$$

$$\leq \left(1 - \tfrac{1}{\gamma}\right)\left(cost(\vec{x} \setminus C_j, \vec{Y}) - cost(\vec{x} \setminus C_j, \vec{C})\right) \tag{5}$$

Moreover, if $C_j \cup \{y\}$ is served by at least two facilities in $\vec{Y}$, the facility serving y (and some agents of C_j) is placed at the median location of $\vec{Y}$'s cluster that contains y. Wlog., we assume that y lies on the left of the median of C_j. Then, the decrease in the cost of y due to the additional facility in $\vec{Y}$ is equal to the decrease in the cost of $x_{i,l}$ in $\vec{Y}$, which bounds from below the total decrease in the cost of C_j due to the additional facility in $\vec{Y}$. Hence,

$$d(y, \vec{C}) - d(y, \vec{Y}) \leq cost(C_j, \vec{C}) - cost(C_j, \vec{Y}) \tag{6}$$

We conclude these cases, by observing that (2) follows from (6) and (4).

Finally, we study Case 1b and Case 2b, i.e., the cases where some agents of C_j are clustered with agents of $\vec{x} \setminus C_j$ in $\vec{Y}$. Let C'_{j1} and C'_{j2} denote the clusters of $(\vec{y}, \vec{Y})$ including all agents of C_j (i.e., $C_j \subseteq C'_{j1} \cup C'_{j2}$). By hypothesis, at least one of C'_{j1} and C'_{j2} contains an agent $z \in \vec{x} \setminus C_j$. Suppose this cluster is C'_{j1}. Then, $D(C'_{j1}) > D(C_j)$, since by Corollary 1, for any $\gamma \geq (2 + \sqrt{3})$, the distance of any agent z outside C_j to the nearest agent in C_j is larger than C_j's diameter. But since both C'_{j1} and C'_{j2} contain agents of C_j, we have that $d(C'_{j1}, C'_{j2}) < D(C_j)$. Thus, $D(C'_{j1}) > d(C'_{j1}, C'_{j2})$, violating the cluster-separation property. Hence instance $\vec{y}$ is not γ-stable and Mechanism 1 allocates no facilities.   □

5   Dealing with Singleton Deviations Deterministically

Next, we present a deterministic strategyproof mechanism for 5-stable instances whose optimal clustering may include singleton clusters. To make singleton cluster deviations non profitable, cluster merging has to be discouraged by the facility allocation rule. So, we allocate facilities near the edge of each optimal cluster, ending up with $\Theta(n)$-approximation and a requirement for larger stability, in order to achieve strategyproofness. Specifically, we now need to ensure that no agent can become a singleton cluster close enough to her true location. Also, we need to easily ensure that no agent can gain by being allocated a facility as a member a neighboring cluster, which is achieved by allocating on *near* to edge agents of clusters. Finally, since agents can now gain by splitting their (true) optimal cluster, we need to ensure that such deviations are either non-profitable or violate the cluster-separation property. Formalizing, we can prove the following:

Theorem 2. ALMOSTRIGHTMOST *(Mechanism 2) is strategyproof for 5-stable instances of k-Facility Location and achieves an approximation ratio of $(n-2)/2$.*

Mechanism 2: ALMOSTRIGHTMOST

Result: An allocation of k facilities
Input: A k-Facility Location instance $\vec{x}$.
1 Find the optimal clustering $\vec{C} = (C_1, \ldots, C_k)$ of $\vec{x}$.
2 **if** *there are two consecutive clusters C_i and C_{i+1} with*
$\quad \max\{D(C_i), D(C_{i+1})\} \geq d(C_i, C_{i+1})$ **then**
$\quad$| **Output:** "FACILITIES ARE NOT ALLOCATED".
3 **for** $i \in \{1, \ldots, k\}$ **do**
4 $\quad$ **if** $|C_i| > 1$ **then**
5 $\quad\quad$| Allocate a facility to the second rightmost agent of C_i, i.e., $c_i \leftarrow x_{i,r-1}$.
6 $\quad$ **else**
7 $\quad\quad$| Allocate a facility to the single agent location of C_i: $c_i \leftarrow x_{i,l}$.
8 $\quad$ **end**
9 **end**
$\quad$ **Output:** The k-facility allocation $\vec{c} = (c_1, \ldots, c_k)$.

---

6   Inapproximability by Deterministic Mechanisms

We next extend the impossibility result of [25, Theorem 3.7] to $\sqrt{2}$-stable instances of k-Facility Location on the line, with $k \geq 3$. We start with some basic facts about strategyproof mechanisms and by adapting the technical machinery of well-separating instances from [25, Sect. 2.2] to stable instances.

Image Sets and Holes. Given a mechanism M, the *image set* $I_i(\vec{x}_{-i})$ of an agent i wrt. instance $\vec{x}_{-i}$ is the set of facility locations i can obtain by varying her reported location. Formally, $I_i(\vec{x}_{-i}) = \{a \in \mathbb{R} : \exists y \in \mathbb{R} \text{ with } M(\vec{x}_{-i}, y) = a\}$.

If M is strategyproof, any image set $I_i(\vec{x}_{-i})$ is a collection of closed intervals (see e.g., [41, p. 249]). Moreover, M places a facility at the location in $I_i(\vec{x}_{-i})$ nearest to the declared location of agent i. Formally, for any agent i, all instances $\vec{x}$, and all locations y, $d(y, M(\vec{x}_{-i}, y)) = \inf_{a \in I_i(\vec{x}_{-i})}\{d(y, a)\}$.

Some care is due, because we consider mechanisms that need to be strategyproof only for γ-stable instances $(\vec{x}_{-i}, y)$. The image set of such a mechanism M is well defined (possibly by assuming that all facilities are placed to essentially $+\infty$), whenever $(\vec{x}_{-i}, y)$ is not γ-stable. Moreover, the requirement that M places a facility at the location in $I_i(\vec{x}_{-i})$ nearest to the declared location y of agent i holds only if the resulting instance $(\vec{x}_{-i}, y)$ is stable. We should underline that all instances considered in the proof of Theorem 3 are stable (and the same holds for the proofs of the propositions adapted from [25, Sect. 2.2]).

Any (open) interval in the complement of an image set $I \equiv I_i(\vec{x}_{-i})$ is called a *hole* of I. Given a $y \notin I$, we let $l_y = \sup_{a \in I}\{a < y\}$ and $r_y = \inf_{a \in I}\{a > y\}$ be the locations in I nearest to y on the left and on the right, respectively. Since I is a collection of closed intervals, l_y and r_y are well-defined and satisfy $l_y < y < r_y$. For convenience, given a $y \notin I$, we refer to the interval (l_y, r_y) as a y-hole in I.

Well-Separated Instances. Let M be a deterministic strategyproof mechanism with a bounded approximation $\rho \geq 1$ for k-Facility Location. An instance $\vec{x}$ is $(x_1|\cdots|x_{k-1}|x_k, x_{k+1})$-*well-separated* if $x_1 < \cdots < x_k < x_{k+1}$ and $\rho d(x_{k+1}, x_k) < \min_{i \in \{2,\ldots,k\}}\{d(x_{i-1}, x_i)\}$. We call x_k and x_{k+1} the *isolated pair* of the well-separated instance $\vec{x}$. Hence, given a ρ-approximate mechanism M, a well-separated instance includes a pair of nearby agents at distance to each other less than $1/\rho$ times the distance between any other pair of consecutive agents. Therefore, any ρ-approximate mechanism serves the two nearby agents by the same facility and serve each of the remaining "isolated" agents by a different facility. We remark that well-separated instances are also ρ-stable.

We can adapt some useful properties of well-separated instances from [25, Sect. 2.2] so that they hold for $\sqrt{2}$-stable instances.

Lemma 4 (Proposition 2.2, [25]). *Let M be any deterministic strategyproof mechanism with approximation ratio $\rho \geq 1$. For any $(x_1|\cdots|x_{k-1}|x_k, x_{k+1})$-well-separated instance $\vec{x}$, $M_k(\vec{x}) \in [x_k, x_{k+1}]$.*

Lemma 5 (Proposition 2.4, [25]). *Let M be any deterministic strategyproof mechanism with approximation ratio $\rho \geq 1$, and let $\vec{x}$ be a $(x_1|\cdots|x_{k-1}|x_k, x_{k+1})$-well-separated instance with $M_k(\vec{x}) = x_{k+1}$. Then, for every $(x_1|\ldots|x_{k-1}|x_k', x_{k+1}')$-well-separated instance $\vec{x}'$ with $x_{k+1}' \leq x_{k+1}$, $M_k(\vec{x}') = x_{k+1}'$.*

The Proof of the Impossibility Result. For the following, we build on the proof of [25, Theorem 3.7]. However, we need some additional ideas and to be more careful with agent deviations, since we can only rely on $\sqrt{2}$-stable instances.

Theorem 3. *For every $k \geq 3$ and any $\delta > 0$, any deterministic anonymous strategyproof mechanism for $(\sqrt{2} - \delta)$-stable instances of k-Facility Location on the real line with $n \geq k + 1$ agents has an unbounded approximation ratio.*

Proof. We only consider the case where $k = 3$ and $n = 4$ (the proof applies to any $k \geq 3$ and $n \geq k + 1$). To reach a contradiction, let M be any deterministic anonymous strategyproof mechanism for $(\sqrt{2} - \delta)$-stable instances of 3-Facility Location with $n = 4$ agents and with an approximation ratio of $\rho \geq 1$.

We consider a $(x_1|x_2|x_3, x_4)$-well-separated instance $\vec{x}$. For a large enough $\lambda \gg \rho$ and a very large (practically infinite) $B \gg 6\rho\lambda$, we let $\vec{x} = (0, \lambda, 6B + \lambda, 6B + \lambda + \varepsilon)$, for some small enough $\varepsilon > 0$ ($\varepsilon \ll \lambda/\rho$). By choosing λ and ε appropriately, the instance $\vec{x}$ becomes γ-stable, for $\gamma \gg \sqrt{2}$.

By Lemma 4, $M_3(\vec{x}) \in [x_3, x_4]$. Wlog., we assume that $M_3(\vec{x}) \neq x_3$ (the case where $M_3(\vec{x}) \neq x_4$ is fully symmetric). Then, by moving agent 4 to $M_3(\vec{x})$, which results in a well-separated instance and, by strategyproofness, requires that M keeps a facility there, we can assume wlog. that $M_3(\vec{x}) = x_4$.

Since $\vec{x}$ is well-separated and M is ρ-approximate, both x_3 and x_4 are served by the facility at x_4. Hence, there is a x_3-hole $h = (l, r)$ in the image set $I_3(\vec{x}_{-3})$. Since $M(\vec{x})$ places a facility at x_4 and not in x_3, the right endpoint r of h lies between x_3 and x_4, i.e. $r \in (x_3, x_4]$. Moreover, since M is ρ-approximate and strategyproof for $(\sqrt{2} - \delta)$-stable instances, agent 3 should be served by a facility at distance at most $\rho\lambda$ to her, if she is located at $4B$. Hence, the left endpoint of the hole h is $l > 3B$. We distinguish two cases based on the distance of the left endpoint l of h to x_4.

Case 1: $x_4 - l > \sqrt{2}\lambda$. We consider the instance $\vec{y} = (\vec{x}_{-3}, a)$, where $a > l$ is arbitrarily close to l (i.e., $a \gtrsim l$) so that $d(a, x_4) = \sqrt{2}\lambda$. Since $d(x_1, x_2) = \lambda$, $d(x_2, a)$ is quite large, and $d(a, x_4) = \sqrt{2}\lambda$, the instance $\vec{y}$ is $(\sqrt{2} - \delta)$-stable, for any $\delta > 0$. By strategyproofness, $M(\vec{y})$ places a facility at l, since $l \in I_3(\vec{x}_{-3})$.

Now, we consider the instance $\vec{y}' = (\vec{y}_{-4}, l)$. Since we can choose $a > l$ so that $d(l, a) \ll \lambda$, the instance $\vec{y}'$ is $(x_1|x_2|l, a)$-well-separated and $(\sqrt{2} - \delta)$-stable. Hence, by strategyproofness, $M(\vec{y}')$ keeps a facility at l, because $l \in I_4(\vec{y}_{-4})$.

Then, by Lemma 5, $y_4' = a \in M(\vec{y}')$, because for the $(x_1|x_2|x_3, x_4)$-well-separated instance $\vec{x}$, $M_3(\vec{x}) = x_4$, and $\vec{y}'$ is a $(x_1|x_2|l, a)$-well-separated instance with $y_4' \leq x_4$. Since both $l, a \in M(\vec{y}')$, either agents 1 and 2 are served by the same facility of $M(\vec{y}')$ or agent 2 is served by the facility at l. In both cases, the social cost of $M(\vec{y}')$ becomes arbitrarily larger than $a - l$, which is the optimal social cost of the 3-Facility Location instance $\vec{y}'$.

Case 2: $x_4 - l \leq \sqrt{2}\lambda$. This case is similar to Case 1. $\square$

7   A Randomized $O(1)$-Approximate Mechanism

We show that a simple randomized mechanism is strategyproof for 5-stable instances, deals with singleton clusters and achieves approximation ratio of 2.

The intuition is that AlmostRightmost can be easily transformed to a randomized mechanism, using the same key properties to guarantee strategyproofness, but achieving a 2-approximation. Specifically, Random (see also Mechanism 3) again finds the optimal clusters, but then places a facility at the location of an agent selected uniformly at random from each optimal cluster. We use the cluster-separation property, as a necessary condition for stability of the optimal clustering. The stability properties required to guarantee strategyproofness are very similar to those required by AlmostRightmost, because the set of possible profitable deviations is very similar for AlmostRightmost and Random. Finally, we note that the cluster-separation property step of RANDOM (step 2) now makes use that due to Lemma 1, it must be that $1.6 \cdot \max\{D(C_i), D(C_{i+1})\} < d(C_i, C_{i+1})$ for 5-stable instances.

Mechanism 3: RANDOM

    Result: An allocation of k facilities
    Input: A k-Facility Location instance $\vec{x}$.
 1 Find the optimal clustering $\vec{C} = (C_1, \ldots, C_k)$ of $\vec{x}$.
 2 **if** *there are two consecutive clusters C_i and C_{i+1} with*
       $1.6 \cdot \max\{D(C_i), D(C_{i+1})\} \geq d(C_i, C_{i+1})$ **then**
     | **Output:** "FACILITIES ARE NOT ALLOCATED".
 3 **for** $i \in \{1, \ldots, k\}$ **do**
 4 | Allocate the facility to an agent c_i selected uniformly at random from the
       | agents of cluster C_i
 5 **end**
    Output: The k-facility allocation $\vec{c} = (c_1, \ldots, c_k)$.

Theorem 4. RANDOM *(Mechanism 3) is strategyproof and achieves an approximation ratio of 2 for 5-stable instances of k-Facility Location on the line.*

Proof (Sketch.). We present here only the main steps. The approximation guarantee is straightforward. We need to cover the key deviation cases. In particular, we show that a deviating agent $x_i \in C_j$ cannot gain in the following cases:

Case 1: x_i deviates and becomes a member of another cluster;
Case 2: x_i deviates and becomes a self-serving center;
Case 3: x_i deviates and causes C_j either to merge or to split.

    The most interesting case is Case 1: $x_i \in C_j$ deviates to y and is clustered together with agents from a different cluster of $\vec{C}$, in order to gain, without splitting C_j. We show that all such deviations will cause the condition of step 2 to be satisfied, hence becoming non-profitable.   □

References

1. Agarwal, P., Chang, H., Munagala, K., Taylor, E., Welzl, E.: Clustering under perturbation stability in near-linear time. In: Proceedings of the 40th IARCS Conference on Foundations of Software Technology and Theoretical Computer Science (FSTTCS 2020). LIPIcs, vol. 182, pp. 8:1–8:16 (2020)
2. Alon, N., Feldman, M., Procaccia, A., Tennenholtz, M.: Strategyproof approximation of the minimax on networks. Math. Oper. Res. **35**(3), 513–526 (2010)
3. Angelidakis, H., Makarychev, K., Makarychev, Y.: Algorithms for stable and perturbation-resilient problems. In: Proceedings of the 49th ACM Symposium on Theory of Computing (STOC 2017), pp. 438–451 (2017)
4. Archer, A., Kleinberg, R.: Truthful germs are contagious: a local-to-global characterization of truthfulness. In: Proceedings of the 9th ACM Conference on Electronic Commerce (EC 2008), pp. 21–30 (2008)
5. Arthur, D., Vassilvitskii, S.: k-means++: the advantages of careful seeding. In: Proceedings of the 18th ACM-SIAM Symposium on Discrete Algorithms (SODA 2007), pp. 1027–1035. SIAM (2007)

6. Auletta, V., Prisco, R.D., Penna, P., Persiano, G.: The power of verification for one-parameter agents. J. Comput. Syst. Sci. **75**, 190–211 (2009)

7. Awasthi, P., Blum, A., Sheffet, O.: Center-based clustering under perturbation stability. Inf. Process. Lett. **112**(1–2), 49–54 (2012)

8. Babaioff, M., Dobzinski, S., Oren, S.: Combinatorial auctions with endowment effect. In: Proceedings of the 2018 ACM Conference on Economics and Computation (EC 2018), pp. 73–90 (2018)

9. Balcan, M., Haghtalab, N., White, C.: k-center clustering under perturbation resilience. In: Proceedings of the 43rd International Colloquium on Automata, Languages and Programming (ICALP 2016). LIPIcs, vol. 55, pp. 68:1–68:14 (2016)

10. Balcan, M.F., Blum, A., Gupta, A.: Clustering under approximation stability. J. ACM **60**(2), 1–34 (2013)

11. Balcan, M., Liang, Y.: Clustering under perturbation resilience. SIAM J. Comput. **45**(1), 102–155 (2016)

12. Bilu, Y., Linial, N.: Are stable instances easy? In: Proceedings of the 1st Symposium on Innovations in Computer Science (ICS 2010), pp. 332–341. Tsinghua University Press (2010)

13. Bilu, Y., Daniely, A., Linial, N., Saks, M.E.: On the practically interesting instances of MAXCUT. In: Portier, N., Wilke, T. (eds.) Proceedings of the 30th Symposium on Theoretical Aspects of Computer Science (STACS 2013). LIPIcs, vol. 20, pp. 526–537. Schloss Dagstuhl - Leibniz-Zentrum für Informatik (2013)

14. Caragiannis, I., Elkind, E., Szegedy, M., Yu, L.: Mechanism design: from partial to probabilistic verification. In: Proceedings of the 13th ACM Conference on Electronic Commerce (EC 2012), pp. 266–283 (2012)

15. Chen, Z., Fong, K.C., Li, M., Wang, K., Yuan, H., Zhang, Y.: Facility location games with optional preference. Theoret. Comput. Sci. **847**, 185–197 (2020)

16. Dokow, E., Feldman, M., Meir, R., Nehama, I.: Mechanism design on discrete lines and cycles. In: Proceedings of the 13th ACM Conference on Electronic Commerce (EC 2012), pp. 423–440 (2012)

17. Düetting, P., Feldman, M., Kesselheim, T., Lucier, B.: Prophet inequalities made easy: stochastic optimization by pricing non-stochastic inputs. In: Proceedings of the 58th Symposium on Foundations of Computer Science (FOCS 2017), pp. 540–551 (2017)

18. Escoffier, B., Gourvès, L., Kim Thang, N., Pascual, F., Spanjaard, O.: Strategyproof mechanisms for facility location games with many facilities. In: Brafman, R.I., Roberts, F.S., Tsoukiàs, A. (eds.) ADT 2011. LNCS (LNAI), vol. 6992, pp. 67–81. Springer, Heidelberg (2011). https://doi.org/10.1007/978-3-642-24873-3_6

19. Ezra, T., Feldman, M., Friedler, O.: A general framework for endowment effects in combinatorial markets. In: Proceedings of the 2020 ACM Conference on Economics and Computation (EC 2020) (2020)

20. Feigenbaum, I., Li, M., Sethuraman, J., Wang, F., Zou, S.: Strategic facility location problems with linear single-dipped and single-peaked preferences. Auton. Agent. Multi-agent Syst. **34**(2), 49 (2020)

21. Feldman, M., Gravin, N., Lucier, B.: Combinatorial auctions via posted prices. In: Proceedings of the 26th ACM-SIAM Symposium on Discrete Algorithms, pp. 123–135 (2014)

22. Feldman, M., Wilf, Y.: Randomized strategyproof mechanisms for Facility Location and the mini-sum-of-squares objective. CoRR abs 1108.1762 (2011)

23. Ezra, T., Feldman, M., Friedler, O.: A general framework for endowment effects in combinatorial markets. In: Proceedings of the 2020 ACM Conference on Economics and Computation (EC 2020) (2020)

24. Fotakis, D., Tzamos, C.: Winner-imposing strategyproof mechanisms for multiple facility location games. Theoret. Comput. Sci. **472**, 90–103 (2013)
25. Fotakis, D., Tzamos, C.: On the power of deterministic mechanisms for facility location games. ACM Trans. Econ. Comput. **2**(4), 15:1–15:37 (2014)
26. Fotakis, D., Tzamos, C.: Strategyproof facility location for concave cost functions. Algorithmica **76**(1), 143–167 (2016)
27. Fotakis, D., Tzamos, C., Zampetakis, M.: Mechanism design with selective verification. In: Proceedings of the 2016 ACM Conference on Economics and Computation (EC 2016), pp. 771–788. ACM (2016)
28. Fotakis, D., Zampetakis, E.: Truthfulness flooded domains and the power of verification for mechanism design. ACM Trans. Econ. Comput. **3**(4), 20:1–20:29 (2015)
29. Goel, S., Hann-Caruthers, W.: Coordinate-wise median: not bad, not bad, pretty good. CoRR abs/2007.00903 (2020). https://arxiv.org/abs/2007.00903
30. Golomb, I., Tzamos, C.: Truthful facility location with additive errors. CoRR abs/1701.00529 (2017). http://arxiv.org/abs/1701.00529
31. Kyropoulou, M., Ventre, C., Zhang, X.: Mechanism design for constrained heterogeneous facility location. In: Fotakis, D., Markakis, E. (eds.) SAGT 2019. LNCS, vol. 11801, pp. 63–76. Springer, Cham (2019). https://doi.org/10.1007/978-3-030-30473-7_5
32. Li, M., Lu, P., Yao, Y., Zhang, J.: Strategyproof mechanism for two heterogeneous facilities with constant approximation ratio. In: Proceedings of the 29th International Joint Conference on Artificial Intelligence (IJCAI 2020), pp. 238–245 (2020)
33. Lu, P., Sun, X., Wang, Y., Zhu, Z.: Asymptotically optimal strategy-proof mechanisms for two-facility games. In: Proceedings of the 11th ACM Conference on Electronic Commerce (EC 2010), pp. 315–324 (2010)
34. Lu, P., Wang, Y., Zhou, Y.: Tighter bounds for facility games. In: Leonardi, S. (ed.) WINE 2009. LNCS, vol. 5929, pp. 137–148. Springer, Heidelberg (2009). https://doi.org/10.1007/978-3-642-10841-9_14
35. Meir, R.: Strategyproof facility location for three agents on a circle. In: Fotakis, D., Markakis, E. (eds.) SAGT 2019. LNCS, vol. 11801, pp. 18–33. Springer, Cham (2019). https://doi.org/10.1007/978-3-030-30473-7_2
36. Miyagawa, E.: Locating libraries on a street. Soc. Choice Welfare **18**, 527–541 (2001)
37. Procaccia, A., Tennenholtz, M.: Approximate mechanism design without money. In: Proceedings of the 10th ACM Conference on Electronic Commerce (EC 2009), pp. 177–186 (2009)
38. Roughgarden, T.: Beyond the Worst-Case Analysis of Algorithms. Cambridge University Press, Cambridge (2020)
39. Roughgarden, T., Talgam-Cohen, I.: Approximately Optimal Mechanism Design. CoRR (2018). http://arxiv.org/abs/1812.11896
40. Roughgarden, T.: Lecture 6: Perturbation-stable clustering. CS264: Beyond Worst-Case Analysis (2017). http://timroughgarden.org/w17/l/l6.pdf
41. Schummer, J., Vohra, R.: Mechanism design without money. Algorithmic Game Theory **10**, 243–299 (2007)
42. Serafino, P., Ventre, C.: Heterogeneous facility location without money. Theoret. Comput. Sci. **636**, 27–46 (2016)
43. Sui, X., Boutilier, C., Sandholm, T.: Analysis and optimization of multi-dimensional percentile mechanisms. In: Proceedings of the 23rd International Joint Conference on Artificial Intelligence (IJCAI 2013), pp. 367–374. IJCAI/AAAI (2013)

Contract Design for Afforestation Programs

Wanyi Dai Li[1]([✉]), Nicole Immorlica[2], and Brendan Lucier[2]

[1] Stanford University, Stanford, CA, USA
wanyili@stanford.edu
[2] Microsoft Research, Cambridge, MA, USA
{nicimm,brlucier}@microsoft.com

Abstract. Trees on farms provide environmental benefits to society and improve agricultural productivity for farmers. We study incentive schemes for afforestation on farms through the lens of contract theory, designing conditional cash transfer schemes that encourage farmers to sustain tree growth. We capture the tree growth process as a Markov chain whose evolution is affected by the agent's (farmer's) choice of costly effort. The principal has imperfect information about the agent's costs and chosen effort, and wants to find the minimal payments that maximize long-run tree survival. We derive the form of optimal contract structure and show how to calculate optimal payments in polynomial time. Notably, even when costs are time-invariant, the optimal contract can involve time-varying payments that are typically higher in earlier periods and may end early. We surveyed farmers partnered with an afforestation program in Uganda to collect data on tree maintenance costs and we derive the optimal payment contract for the reported costs.

1 Introduction

The UN's Sustainable Development Goal #15 challenges society to sustainably manage forests, combat desertification, halt and reverse land degradation, and halt biodiversity loss. Meanwhile, governments and corporations around the world have set ambitious carbon removal and reduction goals. Microsoft, for instance, has pledged to be carbon negative by 2030 [21]. Afforestation and reforestation, which attempt to reverse land degradation and sustain biodiversity as well as capturing carbon, have recently gained attention as a clear next step towards achieving these intertwined goals [11]. One opportunity for afforestation is to grow trees on agricultural land owned by farmers. Mature indigenous trees on farms not only improve biodiversity and carbon storage capacity, but also deliver robust water and soil quality which improves the long-term health of the farm. However, smallholder farmers in developing countries often either do not grow

The authors want to thank Microsoft AI for Earth and the World Agroforestry Center (ICRAF). This research is partly funded by NSF EAGER CCF-1841550: Algorithmic Approaches for Developing Markets, Global Challenges in Economics and Computation 2020 grant, and Stanford University King Center on Global Development graduate student research grant. This research project including the survey conducted in Uganda has been reviewed by the Stanford Institutional Review Board and has been determined to be under the exempt category (eProtocol number 53157).

© Springer Nature Switzerland AG 2022
M. Feldman et al. (Eds.): WINE 2021, LNCS 13112, pp. 113–130, 2022.
https://doi.org/10.1007/978-3-030-94676-0_7

trees or abandon them before reaching maturity. While our survey results indicate farmers value trees on their farms, the slow and risky nature of the growing process often prohibits them from having trees present. This is exacerbated by farmers' cash constraint and limited labor capacity. This raises a large set of questions regarding how best to help farmers overcome barriers and how to operationalize afforestation programs, giving rise to a growing literature on payments for environmental goods [6,7,15].

A typical afforestation program offers a many-year-long contract which is a payment schedule conditional on tree survival. Such programs have benefited from recent advancements in machine learning and AI to process remote sensing data (satellite imagery), which have made it possible to monitor land use change cheaply [3,8,16,19]. Participants get paid a percentage of the total payment at each monitoring round but the exact payment might be adjusted downwards if not all the trees have survived. For example, in [14], the author conducted a randomized controlled trial in Malawi where farmers were asked to grow trees over a period of 3 years. They were paid in equal installments after 6 months, then, 1, 2 and 3 years adjusted by the number of survived trees. The author observed that farmers had private information regarding their likelihood of following through to the end of the 3 year period, and that some farmers dropped out of the contract over time even though they all initially agreed to take up the task. Motivated by this experiment, we ask: if afforestation programs are designed for the long term and at a large scale, can the program designer optimize over the contract space to minimize abandonment of this slow and costly task?

Our research is a first attempt to create an analytical framework for afforestation incentive schemes using contract theory. We set up a principal-agent model, where the principal (referred to as she) can be an NGO, a local government, or any buyer of ecosystem services, and the agent (referred to as he) represents a smallholder farmer. The principal contracts with the agent to procure tree-growing services on his farm. The goal of the principal is to maximize the success of the program as measured by density of mature trees grown and maintained by the agent, and to minimize the payments needed to incentivize this growth. There is a large literature on canonical contract design [9,20], dynamic contract design [4] and, more recently, robust contract design [2,5]. A closely related work with ours is [18] which is also a dynamic contracting problem where both adverse selection and moral hazard may be present; further, there is two-sided limited liability. The author shows that it is sufficient to consider stationary contracts when searching for optimal dynamic contracts. This work contributes to a growing literature on payments for environmental goods [6,7,15].

A key feature of our model is that environmental benefits depend on the state of the world and are not just a stochastic outcome of the agent's action. The dynamics of tree growth follow a Markov Chain where the state space is the tree age (or, similarly, tree height or canopy size). Natural risks and the agent's efforts both influence the steady state distribution of this Markov chain. Further, the effort choices of the agent are influenced by the natural risks, the payments from the principal, and also their private type (e.g., whether the farmer has a "green-thumb" or how well-suited their land is to growing trees) which determines the cost of their effort. Agent types are multi-dimensional, with costs that may vary as a function of the maturity of their trees. As our model has a temporal aspect, we assume that agents are forward-looking and time-discounting, and seek to maximize total time-discounted lifetime utility.

The principal can observe the state of the tree, but does not know an agent's type nor effort. The population of agents is assumed to have been pre-selected to participate in the afforestation program, so the principal's task is to construct a contract such that each agent will *take up* the contract and *follow through* in the long run. The principal is assumed to be more patient than the agent, since agents are individual farmers and the principal is a large centralized program that cares about long-run afforestation outcomes. Thus, among contracts that incentivize the population of farmers hired by the principal to follow through by exerting effort each round, the principal prefers those with lower average payments in the steady-state of the resulting Markov process.

We show that the optimal (i.e., minimum-cost) payment structure can be computed in time polynomial in the type space and has clear economic interpretations. For example, if tree maintenance costs are constant over time then the optimal payment structure can be computed in linear time. We show payments are deceasing with time in this case. This is caused by two forces. As the principal is more patient than the agents, she prefers to front-load payments. However, she is constrained by how much she can do so as she needs to satisfy the incentive constraints (to prevent agents from dropping out and re-entering just to collect the payments). We conclude with some discussion of practical considerations for real-world implementation.

To further illustrate the application of our model, we run our algorithm on data collected from a survey of farmers participating in an afforestation program in Uganda managed by the World Agroforestry Center (ICRAF). Our survey included self-reported estimates of costs to grow and maintain trees at different levels of maturity. We find that the optimal payment contract under these estimates has a very natural structure, where payments are initially high, quickly fall to a regular baseline payment that is maintained for the majority of the program, then fall to zero as the tree approaches maturity. We find that this general structure is quite robust to perturbations of the agent costs and model parameters, suggesting a natural form of contract with practical applicability.

Outline. We set up our model in Sect. 2, then show how to compute the optimal contract in Sect. 3. In Sect. 4 we provide additional theoretical insight into the structure of the optimal contract for the special case of time-invariant costs. Then in Sect. 5 we apply our algorithm to real-world survey data and discuss the structure of the resulting contracts.

2 Model

State Space. We use a Markov chain with finite state space $\mathcal{S} = \{0, 1, \ldots, M\}$ to model the state of a tree. M is the number of periods (years) that a tree takes to mature. State 0 represents no tree, states 1 to $M - 1$ represents the growing process and the final state M represents maturity. At time step $t = 0$, the agent starts in state $s = 0$. We assume $M \geq 2$, so that there is at least one intermediate state before reaching maturity. We assume the principal has a monitoring technology available, so the state is publicly observable.

Agent Actions and Type. In every period, the agent chooses a binary action a from set $\mathcal{A} = \{0, 1\}$ where $a = 0$ means no effort and $a = 1$ means exerting effort. The action is *unobservable* to the principal. Action $a = 0$ is costless but action $a = 1$ has a non-negative cost c_s when taken in state $s, \forall s \in \{0, \ldots, M-1\}$ and cost 0 at maturity state M.[1] The profile of costs $\mathbf{c} = (c_s)_{s<M}$ are private information known to the agent, and form the agent's type. The type $\mathbf{c}$ is drawn from a distribution F with support $[0, \bar{c}]^M$. The principal has the knowledge of the distribution F but not the agent's individual type.[2] Note that we will tend to write $\mathbf{c}_{-s}$ to mean all elements of $\mathbf{c}$ except for the one indexed by s, so that $(\mathbf{c}_{-s}, c_s')$ means profile $\mathbf{c}$ with entry s replaced with c_s', and similarly for other profiles and vectors.

We say that the agents satisfy *uniform costs* if, for each type $\mathbf{c} \in \text{SUPP}(F)$, we have $c_s = c_{s'}$ for each s and s'. In a slight abuse of notation we will tend to think of a type in the uniform costs model as a real number $c \geq 0$ such that $c_s = c$ for all s. As another special case, we say that the agents satisfy *fixed relative costs* if there exists a sequence of values $(f_s)_{s<M}$ such that, for each $\mathbf{c} \in \text{SUPP}(F)$, there exists a constant $t \geq 0$ such that $c_s = t \cdot f_s$ for each $s < M$. That is, all agents agree on the relative cost to care for trees in different stages, but differ in their absolute costs. Note that uniform costs is a special case of fixed relative costs, in which $f_s = 1$ for all s.

Exogenous Shocks. In every period, a tree might die due to some natural risk out of the farmer's control. We model this risk as a probabilistic exogenous shock – in each period where the tree is in state s, a shock does *not* occur with probability q_s. For most of the paper we assume constant shock probability $q_s = q$, but our results extend to state-dependent shocks (such as younger trees being more vulnerable to natural risk).

Transition Probability. The transition probabilities depend on the agent's behavior and exogenous shocks. If, at time t, the state is $s_t \in \{0, 1, \ldots, M\}$, then $s_{t+1} = \min\{s_t + 1, M\}$ (i.e., the tree grows or stays at maturity if already mature) if and only if the agent chooses $a = 1$ and no exogenous shock occurs. Otherwise, $s_{t+1} = 0$ (i.e., the tree is lost) if the agent chooses $a = 0$ or if the shock occurs.

Value at Maturity. A tree in state s delivers value $v_s \geq 0$ to an agent. We will assume that $v_s = 0$ for $s \neq M$, meaning that the agent has no value for an immature tree.[3] Indeed, we can think of "maturity" as representing the stage of growth at which the tree provides intrinsic value to the agent, rather than a biologically terminal state, so that the tree's value v_M is enjoyed in each round where the state is M.

Principal's Payment. In each round the principal can transfer a payment to the agent. Payments from the principal can depend on the evolution of the state but not the agent's action. Informally, we would like the payment to depend on the current state, normalized so that the payment in state 0 is 0. This is without loss, as we focus on Markovian

---

[1] Our survey results suggest that the cost is positive but small in the final state, and less than the value of a mature tree. Our model can accommodate this by additively shifting the cost and value in state M. Our chosen normalization is for notational convenience.

[2] In our construction the principal will only make use of knowledge about the support of F, but not the distribution itself.

[3] Though we assume these values are 0, we still introduce the notation for technical convenience when describing utilities.

strategies. One slight subtlety is that because state M transitions to itself, we would like to distinguish between payment made when first transitioning from state $M - 1$ to M, and the payment made when transitioning from M to itself. (Indeed, if the value v_M is high enough, one might hope to avoid making transfers to maintain a mature tree.)

With this in mind, we describe a contract as a profile of payments $\mathbf{p} = \{p_s\}_{s \in S} \in \mathbb{R}_+^{M+1}$. For each $s < M$, p_s is the payment transferred upon reaching state $s + 1$ from state s. We emphasize the indexing: for example, p_0 is the payment made when the agent reaches state 1. (This choice is for notational convenience when comparing payments with costs, as we describe below.) This leaves p_M, which is the payment transferred in state M when the previous state was also M (i.e., the payment the agent receives for keeping a mature tree alive). Note that no transfer occurs in state 0.

Summary: Timing. To summarize the model up to this point, we describe the timing within each round. Round t begins in some state $s_t \in S$. The agent then chooses an action $a_t \in \{0, 1\}$, and pays cost c_{s_t} if $a_t = 1$. We then resolve the exogenous shock to determine whether the tree survives. If the tree does survive, the agent receives payoff v_{s_t}. The state then transitions to s_{t+1}, determined by the exogenous shock and action a_t. Finally, the principal observes the new state and provides payment p_{s_t}. This ends round t, and the game proceeds to round $t + 1$.

Agent Utility. The agent is risk neutral and his utility is linear in the payment. In what follows, we assume the action choice of the agent is Markovian, i.e., does not depend on the history. This assumption is without loss in Markovian games. Suppose round t begins in state s. If the agent chooses action $a_t = 0$, then his stage utility for round t will be 0 regardless of the state s. This is because he exerts no costly effort and his payment will necessarily be 0. If he chooses action $a_t = 1$, his stage utility will be $p_s + v_s - c_s$ if the tree survives and $-c_s$ otherwise. Taken together, his expected stage utility (over the realization of the exogenous shock) is $q_s(p_s + v_s) - c_s$. Note that the stage utility depends only on the action taken and the current state s. We will therefore write $u_s(a)$ for the stage utility of taking action a in state s. A Markovian strategy is then described by an action a_s to take in each state s. We focus on deterministic strategies without loss of generality, and we assume that the agent always chooses to exert effort (the option preferred by the principal) when the agent is indifferent.

The agent discounts the future with discounting rate $\delta < 1$. Assuming the agent employs strategy $\mathbf{a} = (a_s)$, his continuation utility beginning in state s_0 is $\sum_{t=1}^{\infty} \delta^t u_{s_t}(a_{s_t})$ where s_t is a random variable denoting the state after t transitions, following strategy $\mathbf{a}$, beginning from state s_0. Thus the total expected utility of such an agent, beginning in state s_0, is

$$\mathbb{E}\left[\sum_{t=0}^{\infty} \delta^t u_{s_t}(a_{s_t})\right]. \tag{1}$$

Given a choice of contract $\mathbf{p} = \{p_s\}$, we can therefore write $\mathbf{a}^{\mathbf{p}}(\mathbf{c})$ for the utility-optimizing strategy (i.e., choice of action in each state) for an agent of type $\mathbf{c}$ given contract $\mathbf{p}$. That is, the agent will choose to take action $a_s^{\mathbf{p}}(\mathbf{c})$ in state s. Given any fixed strategy $\mathbf{a} = (a_s)$, we will write $\mathcal{D}^{\mathbf{a}}$ for the steady-state distribution of the Markov process that results from the agent applying strategy a. It will also be convenient to write $\mathcal{D}^1$ for the steady-state distribution for the strategy that chooses action 1 (exert effort) every round.

2.1 The Principal's Problem

The principal's objective is to find the least-costly contract such that all agent types $\mathbf{c} \in \text{SUPP}(F)$ (i.e., all agents in the population of farmers hired to grow trees) exert effort each round. That is, the objective function is

$$\min_{\mathbf{p}} \mathbb{E}_{s \sim \mathcal{D}^1} [p_s] \tag{2}$$

$$\text{s.t. } a_s^{\mathbf{P}}(\mathbf{c}) = 1 \quad \text{for all } \mathbf{c} \in \text{SUPP}(F) \text{ and } s \in \mathcal{S}.$$

where the expectation is over the steady state distribution $\mathcal{D}^1$ of the Markov chain conditional on all agents choosing to exert effort every period, and the choice of $\mathbf{p} = \{p_s\}$ is subject to the constraint that $a_s^{\mathbf{P}}(\mathbf{c}) = 1$ for all $\mathbf{c}$ and s. Our focus on the steady-state is motivated by the fact that, in practice, the principal will be interacting with many agents who have many trees, in which case the steady-state of the process is a proxy for the aggregate outcome. The principal focuses on contracts that induce its contracted agents to always exert effort and hence successfully grow trees. Given a choice of $\mathbf{p}$, the principal's payoff is determined by the agent's utility-maximizing choice of actions, which is endogenous to this payment and induces the steady state of the Markov chain. There is an implicit expectation over $\mathbf{c} \in F$ in the objective (2), but it can be omitted because of the constraint that the contract induces effort from each type, which means the steady-state (and hence long-run payments) are independent of the agent's type. We also assume the principal faces a limited liability constraint, which in this setting means that all payments are non-negative.

The minimization problem in (2) is restricted to the set of contracts for which the utility-maximizing choice of action is to exert effort each round. In Sects. 3.1 and 3.2 we show how to encode these constraints with respect to the model primitives.

3 Computing the Optimal Contract

3.1 The Agent's Perspective

The agent can choose between two actions in every state, leading to 2^{M+1} potential strategies. But we note that due to the structure of the Markov process it suffices to consider the following restricted set of stationary strategies.

Definition 1 (Agent strategies). *The set of strategies, denoted as $\phi \in \{0, \dots, M, \infty\}$, correspond to choosing to exert effort only up to a certain state. Explicitly, in strategy $\phi \in \{1, \dots, M\}$ the agent chooses $a_s = 1$ in states $s < \phi$ and chooses $a_s = 0$ in states $s \geq \phi$. Strategy $\phi = 0$ corresponds to choosing $a = 0$ for all states (not participating) and ϕ_∞ corresponds to choosing $a = 1$ in every state.*

To see why it is without loss to consider this restricted class of strategies, note that if $a_s = 0$ for some s then the Markov process will never reach beyond state s. The agent's payoffs and the state evolution therefore depend only on the longest prefix of states for which the agent always exerts effort.

Expected Costs and Payments. For each strategy ϕ, the expected cost, evaluated at some state s, is the sum of the current period cost and the discounted expected future cost of choosing ϕ. We denote this as $\mathbb{E}C_s^\phi(c)$. Since strategy ϕ only ever reachest states $s \le \phi$, we define $\mathbb{E}C_s^\phi(c)$ only for $s \le \phi$. Recall here that $\mathbf{c} = (c_s)_{s<M}$ denotes the type of the agent, which we think of as being fixed. The expectation is therefore only with respect to the evolution of the Markov chain given strategy ϕ. We can compute this cost for each $\phi < \infty$ by solving a set of $\phi + 1$ linear equations following the Markov chain dynamics. These linear equations are as follows:

$$\mathbb{E}C_s^\phi(\mathbf{c}) = c_s + \delta q \mathbb{E}C_{s+1}^\phi(\mathbf{c}) + \delta(1-q)\mathbb{E}C_0^\phi(\mathbf{c}), \forall s \in \{0, \ldots, \phi - 1\} \quad (3)$$

$$\mathbb{E}C_\phi^\phi(\mathbf{c}) = \delta \mathbb{E}C_0^\phi(\mathbf{c}) \quad (4)$$

The first term in Eq. (3) represents the cost of effort in state s. The second two terms are weighted by the agents' discount factor δ. The second term captures the future discounted cost given no exogenous shock; the third captures the case with an exogenous shock. Equation (4) uses the fact that, in round ϕ, strategy ϕ does not exert effort and therefore suffers no costs and transitions to state 0 with probability 1. For notional convenience, we define $\mathbb{E}C^\phi(c) \equiv \mathbb{E}C_0^\phi(c)$ to be the expected total (future discounted) cost for the agent c in equilibrium, as evaluated at *state 0*.

Solving the set of inequalities described by (3) and (4) yields

$$\mathbb{E}C^\phi(\mathbf{c}) = \frac{1}{Z^\phi}\sum_{s=0}^{\phi-1}(\delta q)^s c_s, \quad Z^\phi = 1 - \delta(\delta q)^\phi - \delta(1-q)\cdot\frac{1-(\delta q)^\phi}{1-\delta q} \quad (5)$$

where Z^ϕ is a fixed normalizing constant independent of the type $\mathbf{c}$.

We can similarly calculate the expected total cost of strategy $\phi = \infty$, as evaluated at state 0. The only difference is that (4) is replaced with $\mathbb{E}C_M^\infty(\mathbf{c}) = \delta q \mathbb{E}C_M^\infty(\mathbf{c}) + \delta(1-q)\mathbb{E}C_0^\infty(\mathbf{c})$ to account for the possibility of remaining in the mature state in the absence of an exogenous shock. Solving for $\mathbb{E}C_M^\infty(\mathbf{c})$ then yields

$$\mathbb{E}C^\infty(\mathbf{c}) = \frac{1}{Z^\infty}\sum_{s=0}^{M-1}(\delta q)^s c_s, \quad Z^\infty = 1 - \delta(1-q)\cdot\frac{1}{1-\delta q} \quad (6)$$

Similarly, we can calculate the agent's expected total payments from a given payment plan and from having mature trees. The expected payments of strategy ϕ given payment plan $\mathbf{p} = (p_s)_{s<M}$ is denoted $\mathbb{E}B^\phi(\mathbf{p})$. Note that $\mathbb{E}B^0 = 0$ as agents are not paid unless they enroll and plant a seedling. For $\phi < \infty$, we then have

$$\mathbb{E}B_s^\phi(\mathbf{p}) = qp_s + \delta q \mathbb{E}B_{s+1}^\phi(\mathbf{p}) + \delta(1-q)\mathbb{E}B_0^\phi(\mathbf{p}), \forall s \in \{0, \ldots, \phi - 1\} \quad (7)$$

$$\mathbb{E}B_\phi^\phi(\mathbf{p}) = \delta \mathbb{E}B_0^\phi(\mathbf{p}) \quad (8)$$

Note that the only distinction with (3) and (4) is the additional factor of q that multiplies the stage payments; this accounts for the fact that the agent is only paid in the event that an exogenous shock does not prevent the growth of the tree. Solving for $\mathbb{E}C^\phi(c)$, and modifying for $\phi = \infty$ in the same way as costs, yields

$$\mathbb{E}B^\phi(\mathbf{p}) = \frac{1}{Z^\phi} \cdot q \cdot \sum_{s=0}^{\phi-1} (\delta q)^s p_s \text{ for all } \phi < \infty, \qquad \mathbb{E}B^\infty(\mathbf{p}) = \frac{1}{Z^\infty} \cdot q \cdot \sum_{s=0}^{M-1} (\delta q)^s p_s \tag{9}$$

where Z^ϕ and Z^∞ are the same normalizing constants as in (5) and (6).

Finally, we denote the total long-run expected value of a tree when the agent chooses strategy ϕ, evaluated at stage 0, to be $\mathbb{E}v_A^\phi$. Note that $\mathbb{E}v_A^\phi = 0$ for all strategies except for $\phi = \infty$. We can solve for $\mathbb{E}v_A^\infty$ in a manner similar to cost and payments to yield

$$\mathbb{E}v_A^\infty = \frac{1}{Z^\infty} \cdot \frac{(\delta q)^M}{1 - \delta q} \cdot v_M. \tag{10}$$

Given a payment plan $\mathbf{p} = (p_s)_{s<M}$, an agent of type $\mathbf{c}$ entering the program will choose a strategy $\phi \in \{0, \dots, M, \infty\}$ that maximizes his expected utility evaluated at stage 0. We will write this utility as $\mathbb{E}U^\phi(\mathbf{c}, \mathbf{p}) = \mathbb{E}B^\phi(\mathbf{p}) - \mathbb{E}C^\phi(\mathbf{c}) + \mathbb{E}v_A^\phi$.

3.2 Encoding the Principal's Problem

Recall that the principle's goal is to find the least-costly contract such that for each $\mathbf{c} \in \text{Supp}(F)$, an agent with cost $\mathbf{c}$ will choose $\phi = \infty$. We will show how to compute this optimal payment schedule.[4]

In order for the agent with cost $\mathbf{c}$ to choose $\phi = \infty$, we must satisfy the incentive compatibility (IC) constraint that no agent would gain by deviating to a different strategy from any stage. That is, we require that $\mathbb{E}U_s^\phi \leq \mathbb{E}U_s^\infty$ for all ϕ and s. Because agents are Bayesian and hence time-consistent, it suffices to restrict attention to agent decisions at stage $s = 0$, leading to the following M IC constraints: $\forall \phi \in 1, \dots, M$ and $\mathbf{c} \in \text{Supp}(F)$,

$$\mathbb{E}B^\phi(\mathbf{p}) - \mathbb{E}C^\phi(\mathbf{c}) \leq \mathbb{E}B^\infty(\mathbf{p}) - \mathbb{E}C^\infty(\mathbf{c}) + \mathbb{E}v_A^\infty. \tag{11}$$

We also require the individual rationality constraint (IR) that $\mathbb{E}U^\infty \geq 0$ for each agent $\mathbf{c} \in \text{Supp}(F)$, which can be written as

$$\mathbb{E}B^\infty(\mathbf{p}) - \mathbb{E}C^\infty(\mathbf{c}) + \mathbb{E}v_A^\infty \geq 0. \tag{12}$$

The principle's problem can therefore be expressed as minimizing the total long-run expected average payment, (2), subject to constraints (11) and (12) for each type $\mathbf{c} \in \text{Supp}(F)$.

3.3 How Much to Pay

We now construct a contract that minimizes the expected total payments required to have a given population $\mathcal{C} \subseteq \text{Supp}(F)$ of agent types always complete the tree-growing process. The principal's problem can then be solved by taking $\mathcal{C} = \text{Supp}(F)$. Our

---

[4] Note that this assumes that such a contract exists, which we have not yet shown.

ALGORITHM 1: Algorithm FINDPRICES

---

1 Initialize $p_s \leftarrow 0$ for all $s \in \{0, 1, \ldots, M-1\}$;
2 $X \leftarrow \max_{\mathbf{c} \in C} \mathbb{E}C^{\infty}(\mathbf{c}) - \mathbb{E}v_A^{\infty}$;
3 **for** $s = 0, 1, \ldots, M-1$ **do**
4 $\quad$ $p_s^{IC} \leftarrow$
$\quad$ $\max \{ p_s' \geq 0 : B^{s+1}(\mathbf{p}_{-s}, p_s') \leq \mathbb{E}C^{s+1}(\mathbf{c}) + X - \mathbb{E}C^{\infty}(\mathbf{c}) + \mathbb{E}v_A^{\infty} \; \forall \mathbf{c} \in C \}$;
5 $\quad$ $p_s^{IR} \leftarrow \min \{ p_s' \geq 0 : B^{\infty}(\mathbf{p}_{-s}, p_s') = X \}$;
6 $\quad$ $p_s \leftarrow \min \{ p_s^{IC}, p_s^{IR} \}$;
7 $\quad$ **if** $p_s^{IR} \leq p_s^{IC}$ **then break**
8 **end**
9 $p_M \leftarrow \min \{ p_M' \geq 0 : B^{\infty}(\mathbf{p}_{-M}, p_M') = X \}$;
10 **return** $(p_0, p_1, \ldots, p_M)$

---

algorithm is listed in pseudocode as Algorithm 1. Roughly speaking, our algorithm first determines the minimum value of $\mathbb{E}B^{\infty}(\mathbf{p})$ that would satisfy the IR constraint (12) for all agent types. Then, starting with all payments set to 0, it raises prices greedily, making earlier payments as high as possible subject to IC constraints (11). It does so until it constructs a schedule of payments $\mathbf{p}$ such that the target value $B^{\infty}(\mathbf{p})$ is reached, at which point the algorithm terminates.

In more detail, we first define $X = \max_{\mathbf{c} \in C} \mathbb{E}C^{\infty}(\mathbf{c}) - \mathbb{E}v_A^{\infty}$. This is the maximum long-run disutility of any agent for exerting effort every period, which determines our IR constraint. Our algorithm will compute payments such that the total lifetime benefit from payments when exerting effort every round, $\mathbb{E}B^{\infty}(\mathbf{p})$, is precisely equal to X, matching the IR constraint (12). We then calculate payments iteratively. We initially set $p_s = 0$ for all s. Then, for each $s = 0, 1, \ldots$ in sequence, we calculate the largest payment p_s that satisfies the IC constraints: each agent type $\mathbf{c} \in C$ would prefer strategy $\phi = \infty$ to strategy $\phi = s$, given all previous payments. This largest payment is p_s^{IC}. We also calculate the smallest payment p_s^{IR} that would set the total lifetime benefit from payments equal to the target level X; this is p_s^{IR}.[5] If $p_s^{IC} < p_s^{IR}$, then the IC constraints bind: we set p_s equal to p_s^{IC} and move on to setting the payment for round $s+1$. If $p_s^{IR} \leq p_s^{IC}$ then we have met the IR constraint while satisfying all IC constraints, so we set p_s equal to p_s^{IR} and terminate. Note that in this case all payments for subsequent rounds are set to 0. Finally, if the IR constraint is still not satisfied when we exit the loop, we set payment p_M high enough so that $B^{\infty}(p_0, \ldots, p_{M-1}, p_M)$ is equal to our target X. Note that this can only occur if we run through all iterations of the for loop without triggering the break condition; otherwise we will have $p_M = 0$.

We now prove that the resulting payments satisfy all IC and IR constraints, and moreover form the minimal-cost contract that does so. We first show that the payments are valid, in that they satisfy all IC and IR constraints and hence every agent type will choose to exert effort in every state.

Theorem 1. *Given any subpopulation* $C \subseteq$ SUPP(F), *Algorithm* FINDPRICES *computes a payment schedule* $\mathbf{p}$ *such that* $\mathbf{a}_s^{\mathrm{P}}(\mathbf{c}) = 1$ *for all* $\mathbf{c} \in C$.

---

[5] Note that the value of p_s^{IR} that sets $B^{\infty}(\mathbf{p}_{-s}, p_s') = X - \mathbb{E}v_A^{\infty}$ will be generically unique.

The proof is omitted due to space constraints. The most technical part of the argument is showing that even if the algorithm exits the main loop on some iteration s, the IC and IR constraints will still hold for strategies $s' > s$. Intuitively, since payments beyond round s are set to 0, the constraints for any strategy $s' > s$ achieving non-negative utility will be dominated by the constraints for strategy s.

We next show that the prices returned by Algorithm FINDPRICES are in fact optimal, in the sense that they minimize expected long-run payments by the principle among all contracts that satisfy the IC and IR constraints.

Theorem 2. *Given any subpopulation $C \subseteq \text{SUPP}(F)$, Algorithm FINDPRICES computes the least-cost payment schedule such that each $c \in C$ will choose to exert effort in every period.*

The proof of Theorem 2 proceeds in two steps. We first show that, given our choice to have the IR constraint bind with equality for some agent type, our method of constructing prices greedily (by frontloading prices as much as possible given the IC constraints) is optimal. The second step is to show that this choice to have the IR constraint bind with equality is optimal. That is, it is optimal to choose a $\mathbf{p}$ that minimizes $B^\infty(\mathbf{p})$ subject to the IR constraint. Note that this second step is not immediate! It might seem intuitive that we wouldn't want to "waste money" by giving agents more total payment than necessary to get them to enroll in the program. However, a higher total payment relaxes the IC constraints, which enables higher payments in earlier rounds where money goes further due to time-discounting. Our proof shows that the higher early payments will not make up for the loss due to increasing $B^\infty(\mathbf{p})$. The details are omitted due to space constraints.

3.4 Computation

We note that Algorithm 1 runs in time $O(M \times |\mathcal{C}|)$, where $|\mathcal{C}|$ is the size of the type space, as we iterate over all IC constraints (one for each type in $\mathcal{C}$) for each strategy $s = 0, 1, \ldots, M - 1$. When F is represented as a discrete list of types along with probabilities, this method runs in polynomial time. More generally, if F is described implicitly as a continuous distribution, one could proceed by discretizing the space of possible types. Assuming an upper bound $\bar{c}$ on the maximum possible cost within a single round, one can discretize costs up to a small error $\epsilon > 0$ resulting in $(\bar{c}/\epsilon)^M$ possible types, at an additive loss of ϵ in the resulting efficiency and incentive properties.

This general reduction results in an algorithm with runtime $O(M(\bar{c}/\epsilon)^M)$, which may be plausible when M (the number of rounds of growth) is small. For larger M, one might be able to reduce the number of potential types under some structural assumptions on agent costs. For example, recall under the assumption of fixed relative costs, the cost for agent i to care for a tree in stage s is $c_s^i = t_i \cdot f_s$ where $t_i \in [0, \bar{c}]$. Under such a model, one could discretize the space of multipliers t_i to multiples of $\epsilon > 0$, resulting in a total runtime of $O(M \cdot \bar{c}/\epsilon)$, again at an additive loss of ϵ in the resulting efficiency and incentive properties.

3.5   Discussion: Payment to Maintain a Mature Tree

In the optimal payment schedule, the final state payment p_M rewards transitions from M to M. In an afforestation program, it's intuitive to compensate a participant for tree growing efforts which involve non-zero payments in $p_0, \ldots, p_{M-1}$. Once a tree matures, it is no longer costly for the agent to keep the tree alive. A positive payment p_M may appear to be unnecessary but in the optimal solution, the principal may have to keep paying the agent even after a tree reaches M so that the agent does not cut it down and reenter the program.[6] However, the principal does not need to pay the agent a positive price to keep a mature tree if the value of a mature tree to the agent is large.

Corollary 1. *For every F and $C \subseteq \text{SUPP}(F)$, there is a sufficient large value of v_M such that, in the optimal payment schedule $\mathbf{p}$ where all $\mathbf{c} \in C$ exert effort every round, we have $p_M = 0$.*

Although v_M is exogenous in our model, in reality it can be partially affected by design. By working with local stakeholders, the principal can offer tree seedling options preferred by farmers and educate them the benefit of agroforestry.[7]

4   Special Case: Uniform Costs

In the special case that cost vectors are uniform across time for each agent, a type is described by a real number $c \geq 0$ such that $c_s = c$ for each round s. In this case we can further analyze and interpret the structure of the optimal contract that incentivizes all cost types in a sub-interval $[c_l, c_h]$ to grow and maintain mature trees. As we will show, the early payments are dictated by the incentive constraints of the high-cost type whereas the late payments are dictated by the incentive constraints of the low-cost types. In particular, it is important to curtail late-stage payments to prevent low-cost types from dropping out as soon as a tree reaches maturity and re-entering the program to accrue cash benefits.

4.1   The Agent's Perspective – Uniform Costs

When costs are uniform, the constraints in (5) that define an agent's costs for strategy ϕ immediately imply that costs scale linearly with c. That is,

$$\mathbb{E}C_0^\phi(c) = K_\phi c, \forall \phi \in \{1, \ldots, M\}, \tag{13}$$

where the coefficients K_ϕ depend on q, δ, and ϕ. We can also define $K_0 \equiv 0$. We are then able to observe that

$$0 = K_0 < K_1 < \cdots < K_M, \quad K_M > K_\infty > 0. \tag{14}$$

---

[6] In reality, a reentering decision may look like planting more trees. In our model, we consider an agent who have reached his land capacity. Thus, the agent only reenter through cutting down existing trees, claiming to have lost them through an exogenous shock, and planting new ones.

[7] See, for example, the "Trees on Farms for Biodiveristy" project at the World Agroforestry Center: https://treesonfarmsforbiodiversity.com/about-trees-on-farms/.

In other words, strategies $\phi \leq M$ get more costly as ϕ increases, since the agent spends more time exerting effort in expectation. The exception is ϕ_∞, which is less costly than strategy $\phi = M$ because the agent can keep the tree mature without any cost.

4.2 Who Drops Out and When

One might expect that if a contract causes the agent with a high cost to follow through (i.e., choose strategy ∞), one with a lower cost would follow through as well. However, this is not always the case: as we next show, the agent with lower costs might be incentivized to drop out of the program "late" (i.e., just before the tree reaches maturity) then reenter with a new tree in order to collect more payments. This is because an agent with lower costs obtains higher net utility from early payments, and may find this utility from collecting cash rewards more attractive than maintaining the tree at maturity.

We now make this more formal. Suppose that $\mathbb{E}B^\infty(\mathbf{p})$ is fixed. Denote $\mathbb{E}\bar{B}^\phi(c)$ to be the minimal expected payments that type c needs to receive in order to weakly prefer strategy ϕ to ∞ given $\mathbb{E}B^\infty(\mathbf{p})$. By definition, $\mathbb{E}\bar{B}^\phi(c) \equiv \mathbb{E}C^\phi(c) - \mathbb{E}C^\infty(c) + \mathbb{E}B^\infty(\mathbf{p}) + \mathbb{E}v_A^\infty$. The term $\mathbb{E}\bar{B}^\phi(c)$ measures how much expected payment from strategy ϕ an agent requires in order to cause deviation from the principal's desired strategy, ∞. The following lemma demonstrates that high-cost types are prone to choosing early drop-out strategies and low-cost types are prone to choosing late drop-out strategies.

Lemma 1. *For any pair of $c_l < c_h$ in the support of $F(c)$, there exists a state $\hat{s} \in \{0,\ldots,M\}$ such that $\forall \phi < \hat{s}, c_h = \arg\min_{c\in[c_l,c_h]} \mathbb{E}\bar{B}^\phi(c)$ and $\forall \phi \geq \hat{s}, c_l = \arg\min_{c\in[c_l,c_h]} \mathbb{E}\bar{B}^\phi(c)$.*

Proof. From the definition of $\mathbb{E}\bar{B}^\phi(c)$ and Equation (14), we can write $\mathbb{E}\bar{B}^\phi(c) = (K_\phi - K_\infty)c + \kappa$ for some constant κ. As $K_0 < K_\infty < K_M$ and K_ϕ is increasing for $0 \leq \phi \leq M$, there must be some intermediate state $\hat{s}$ such that $K_\phi - K_\infty < 0$ for all $\phi < \hat{s}$ and $K_\phi - K_\infty \geq 0$ for all $\phi \geq \hat{s}$. When $K_\phi - K_\infty < 0$, the minimum of $\mathbb{E}\bar{B}^\phi(c)$ is achieved by c_h proving the first claim. Likewise, when $K_\phi - K_\infty \geq 0$ the minimum of $\mathbb{E}\bar{B}^\phi(c)$ is achieved by c_l proving the second claim.

4.3 Structure of Optimal Payments Under Uniform Costs

We now discuss the structure of the contract that minimizes the expected total payments required to have a given subpopulation always complete the tree-growing process. This contract is described in Sect. 3.3, but we can describe it more explicitly in the case of uniform costs. Our first observation is that if the agent with costs c_l and c_h choose $\phi = \infty$, then any agent with cost $c \in [c_l, c_h]$ will choose it as well. It therefore suffices to consider subpopulations corresponding to cost intervals. Theorem 3 describes the optimal payments for a given cost interval.

Theorem 3. *In the case of uniform costs, given any subpopulation $[c_l, c_h] \subseteq \text{SUPP}(F)$, the least cost contract is the payment schedule $\mathbf{p}^*$, that solves the following set of $M+1$ equations. For all $s \in \{1,\ldots,\hat{s}-1\}$ (where $\hat{s}$ is as defined in Lemma 1):*

$$\mathbb{E}B^s(\mathbf{p}) = \mathbb{E}C^s(c_h), \tag{15}$$

Otherwise, for $s \geq \hat{s}$, either

$$\mathbb{EB}^s(\mathbf{p}) = \mathbb{EC}^s(c_l) + \mathbb{EB}^\infty(\mathbf{p}) + \mathbb{E}v_A^\infty - \mathbb{EC}^\infty(c_l), \tag{16}$$

or $p_s = 0$. Finally,

$$\mathbb{EC}^\infty(c_h) = \mathbb{EB}^\infty(\mathbf{p}) + \mathbb{E}v_A^\infty. \tag{17}$$

The proof follows directly by interpreting the construction from Sect. 3.3 in the special case where agent types are totally ordered. In this case, the value of X in Algorithm 1 will always be determined by the highest type agent, c_h. Then, as in Sect. 3.3, it is optimal for the principal to greedily maximize payments in earlier states as long as the constraints in Eq. (11) are satisfied. As in Theorem 2, this is due to the difference in discounting factors between the agent and the principal. Second, we identify a small set of binding constraints. We reduce the number of constraints from $(|c|+1) \times (M+1)$ to $M+1$ by utilizing the analysis in Sect. 3.1 and Lemma 1. In particular, the IC constraint will bind always for the highest agent type c_h up until some iteration $\hat{s}$, after which the IC constraint will bind for the lowest agent type c_l.[8]

Discussion: Binding Constraints. It's intuitive to see that in the optimal solution, the IR constraint for agent c_h binds (Eq. (17)). The agent c_h is indifferent between $\phi = \infty$ and not participating. It's sufficient to have IR satisfied for c_h because IR will also be satisfied for all lower types. Regarding IC constraints, although the principal prefers to shift payments to earlier states, there is a limit on how much shifting is possible. As we increases early stage payments $(p_s, \forall s \in \{0, \ldots, \hat{s} - 2\})$, the IC constraint of the agent c_h will be violated first; this causes c_h to choose early drop-out strategies (Eq. (15)). Similarly, late-stage payments $(p_s, \forall s \in \{\hat{s} - 1, \ldots, M\})$ cannot exceed the minimal payments that keeps low-type c_l indifferent between dropping out in intermediate stages and strategy $\phi = \infty$ (Eq. (16)). The threshold state $\hat{s}$ comes from Lemma 1.

Take-Up and Follow-Through Behavior. What happens if we target a subpopulation $[c_l, c_h]$ that is a strict subset of SUPP(F)? Given the optimal payment schedule, all intermediate types in $[c_l, c_h]$ choose $\phi = \infty$, any types higher than c_h choose ϕ_0, and the types lower than c_l choose ϕ_M. Even though the principal only intends to have $c \in [c_l, c_h]$ to take up and follow through the contract, she cannot prevent lower types $c < c_l$ from taking up but not following through. This is consistent with the self-selection observations made in [12]. Further, the low cost agent ($c < c_l$) chooses to drop out before the tree reaches maturity even in the absence of exogenous shocks. If the contract is not properly designed, then this intentional drop out behavior will be exacerbated. In [13], the authors argue that exogenous shocks cause participants to not follow through, thus lowering program cost-efficiency. While this is consistent with our model, we further contribute to this discussion by showing that another possible reason for drop-out is that front-loaded payments (even the optimal payments) can incentivize some agent type to join the program to collect early payouts but then abandon trees. For

---

[8] $|c|$ can be thought as the size of the discretization of the agent type space between c_l and c_h; there are M IC constraints and 1 IR constraint for each type; there are $M + 1$ number of limited liability constraints.

this reason, a principal must be careful when allowing open enrollment to an afforestation program, as agents with costs outside the anticipated range may reduce program
effectiveness due to incentivized drop-outs.

5   Application: Optimal Payments for Real-World Cost Data

We partnered with the World Agroforestry Center (ICRAF) to collect survey data from a
collection of farmers enrolled in an afforestation program in Uganda. The survey asked
for demographic information and information about the quantity and type of trees on
the farmer's land. Most relevant to this work, the survey asked about estimated costs to
care for trees at different stages of growth. This was described in the language of hours
per week: farmers were asked to estimate care time for trees that are 0–3 months old,
4–6 months, 6–12, 12–24, and 24–48. We also asked farmers to indicate an effective
value (in money) for time spent on informal labor, both with respect to how much they
are typically paid for informal work and how much they pay others. We received 408
responses from farmers enrolled in the ICRAF program. After excluding responses with
invalid or missing cost data, we were left with 382 entries.

We use the responses from the survey to estimate an instance of our model. There
are 12 rounds, each corresponding to a 3-month period. Our default estimate of time
preferences is a discount rate of 0.35 per year, corresponding to $\delta = (0.35)^{1/4} = 0.77$
in our model. This default choice reflects the observation that poorer individuals, like
many in our survey response pool, tend to discount the future more heavily than wealthy
families [17].[9] We estimate $q = 0.5^{1/4} = 0.84$ (probability of tree survival given effort)
using a recent study that estimates a 50% survival rate of trees in farm afforestation
programs, conditional on the agent exerting effort.[10]

To estimate the pool of agent types $\mathcal{C}$, we fit the raw survey response data to a
parameterized model of fixed relative costs. In this model, each agent i has a type $c^i = (c_s^i)$ of the form $c_s^i = t_i \cdot f_s$ where $t_i \geq 0$ and $f_s = (e^{\alpha s + \beta} + \gamma)$. Here we think of
$t_i \geq 0$ as an agent-specific type, and the parameterized function $f_s = (e^{\alpha s + \beta} + \gamma)$ as
modeling the declining cost of maintaining trees as they grow.[11] We then fit the data to a
choice of parameters (α, β, γ) and agent-specific types t_i. The best-fit parameterization
has an R^2 of 0.87. We omitted outliers in which the factor t_i deviates from the median
by a factor greater than 10, leaving 371 non-outlier responses. The resulting profile of
relative costs, along with the distribution of scaling factors, are presented in Fig. 1.

Finally, to estimate v_M, we consider the reported cost to maintain a tree for those
respondents who report keeping trees on their property without further compensation.

---

[9] As noted in many studies ([1,17,22]), attempts to empirically estimate annual discount factors
for particular populations (including poor urban and farming populations) have been inconsis-
tent, leading to estimates ranging from near-zero to 0.75. We chose an intermediate value of
0.35 as our baseline, and include a robustness check that compares different choices for δ.

[10] Ideally we would obtain tree survival estimates more specifically tailored to the ICRAF environment, but the afforestation program is still in its infancy and does not yet have that data
available. We therefore compare the outcomes with many different values of q to check robustness to this choice of parameter.

[11] The particular form of this model was chosen post-hoc after noting that most reported costs
tended to decline exponentially to a baseline minimal effort level.

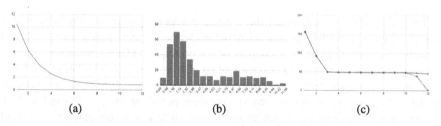

Fig. 1. Surveyed tree maintenance costs and resulting optimal payments. Reported costs were fit to a parametric model of the form $c_s^i = t_i \cdot f_s$, where s denotes time in quarterly periods. (a) The fit curve of relative costs f_s, with round s on the x-axis. (b) A histogram showing the distribution of multipliers t_i in the surveyed population. (c) The optimal schedule of quarterly payments for this collection of costs, with s along the x-axis, calculated using higher (blue, circle) and lower (orange, triangle) estimates of tree value. (Color figure online)

To obtain a conservative estimate of v_M, we set it equal to the largest reported cost among non-outliers to maintain trees at 24–48 months. The motivation is that such respondents are intrinsically motivated to at least maintain near-mature trees at this level of effort. We also consider a less-conservative estimate in which we use reported costs to maintain trees at $0 - 3$ months as a bound on tree value. The idea being that, given a hypothetical guarantee that such a tree would survive and immediately mature, it would be worthwhile to plant and care for it.

Given these estimates, we implemented our algorithm for computing optimal payments. See Fig. 1(c). As expected, we find that the optimal contract front-loads payments toward the earlier part of the care period, and reduces payments over time. It was common for the optimal contract to then settle into a stable quarterly payment, set at a level that disincentivizes departure from the program. In the last few periods, the payments can decline as they are replaced by the intrinsic value of maintaining a mature tree. When the value of a tree is estimated to be very low (conservative estimation method), we find that it is necessary to maintain positive payments even for mature trees. This is because of the presence of high-cost individuals in the dataset who necessitate very large early payments; such payments encourage low-cost types to drop out and re-enter the program unless encouraged to keep their trees in perpetuity. When the value of a tree is estimated higher using the less conservative estimation method, the optimal contract can reduce payments to 0 for near-mature and mature trees.

5.1   Restricting the Pool of Agents

The optimal payment contract for the cost reports we collected involves a very large up-front payment. Looking at the distribution of agent value multipliers in Fig. 1(b), one might guess that this is due to the impact of a small minority of high-cost respondents who must be incentivized to participate. Given such a candidate pool, the principal may wish to structure payments that do not incentivize participation from the highest types, and instead target a smaller subpopulation that includes all agents with multipliers in a range $[0, \theta]$. This results in fewer trees due to lower enrollment, but potentially at a much lower payment per unit of mature trees maintained.

To explore this, we calculated the optimal payment contract for the subset of n lowest-cost farmers (with respect to our fit parameterized model), for each $n \leq 371$, and calculated the long-run effective payment of the principal under the resulting contract. As these contracts incentivize full exertion of effort from the target population, the long-run quantity of mature trees scales linearly with the chosen n. What will change non-linearly is the resulting aggregate long-run payment per agent. We plot this in Fig. 2(a). As one can see, the long-run payment *per farmer* increases moderately up to around $n = 250$, then increase more quickly up to $n = 350$, at which point it increases sharply. This suggests that it is indeed a tail of high-cost individuals who have an outsized effect on the budget necessary to maintain all trees at maturity. In Fig. 2(b) we also plot out the optimal contract for different choices of n. As we exclude more of the high-cost agents, we find that the resulting savings are the result of multiple effects: the high initial payment and baseline stable level of quarterly payments are both reduced, and the point at which payments decrease from the stable level to 0 occurs earlier.

(a)   (b)

Fig. 2. The impact of excluding high-cost agents on total payment and optimal contract structure. (a) The average long-run payment at steady-state *per agent* for the contract that targets the lowest-cost n respondents. (b) The optimal contract that targets the n lowest-cost respondents.

5.2 Robustness Checks

To check our choice to fit reported data to a parameterized model with fixed relative costs, we also computed optimal payments for the raw reported costs. See Fig. 3(a) for a comparison, plotted in the scenario with a higher estimate of tree value. We find that the overall structure of the optimal contract is unchanged, though payments for the raw costs are consistently higher. One possible explanation for this difference is a floor effect, where even respondents with very low costs tended to report at least one hour per week of maintenance time, which distorts the relative costs for agents at the low end of the cost curve. In contrast, our parameterized model extrapolates values below 1 h for those with low reported costs, leading to lower costs on average.

To check our choices for δ and q, we also explored the effect of changing these parameters on the resulting payments. This again results in a similar pattern of optimal payment; see Fig. 3. When farmers are more patient, the intrinsic value of a tree is higher at intermediate rounds and hence payments can be reduced more quickly. Also, when tree survival rates increase, there is less inherent risk and hence lower total cost, and hence payments can be uniformly decreased.

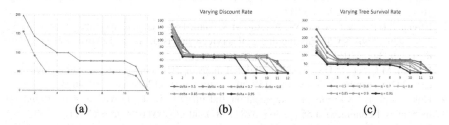

Fig. 3. Robustness checks. (a) Comparing the optimal payments for raw reported costs (red) versus fit data (blue). (b) Optimal payments under different choices of δ. (c) Optimal payments under different choices of q. (Color figure online)

6 Limitations and Extensions

In our model, the agent's cost type is the sole source of heterogeneity. Other potential sources of heterogeneity to consider include the agent's discounting rate and risk preference [10, 14]. We assume that the principal has the knowledge of the support of possible agent costs throughout the tree growing stages. This further advocates the need for robustness considerations. The fact that the optimal contract depends only on the support of the type distribution, and not the distribution itself, provides some measure of robustness, but it would be useful to also explore the impact of perturbations to the support.

One might also consider a model with stochastic agent costs, where in every period the agent cost is redrawn from some distribution. The stochastic agent cost model aims to answer how payments can be designed to alleviate dropout due to stochastic shocks, including income shocks, which might be modeled as a sudden increase in effective tree-maintenance cost.

References

1. Andersen, S., Harrison, G.W., Lau, M.I., Rutström, E.E.: Eliciting risk and time preferences. Econometrica **76**(3), 583–618 (2008)
2. Carroll, G.: Robustness and linear contracts. Am. Econ. Rev. **105**(2), 536–563 (2015). https://doi.org/10.1257/aer.20131159
3. Dao, D., et al.: GainForest: scaling climate finance for forest conservation using interpretable machine learning on satellite imagery. In: ICML Climate Change Workshop at 36th International Conference on Machine Learning, pp. 5–7 (2019)
4. DeMarzo, P.M., Sannikov, Y.: Optimal security design and dynamic capital structure in a continuous-time agency model. J. Finance **61**(6), 2681–2724 (2006). https://doi.org/10.1111/j.1540-6261.2006.01002.x

5. Dütting, P., Roughgarden, T., Talgam-Cohen, I.: Simple versus optimal contracts. In: ACM EC 2019 - Proceedings of 2019 ACM Conference on Economics and Computation, pp. 369–387 (2019). https://doi.org/10.1145/3328526.3329591

6. Engel, S., Pagiola, S., Wunder, S.: Designing payments for environmental services in theory and practice: an overview of the issues. Ecol. Econ. **65**(4), 663–674 (2008). https://doi.org/10.1016/j.ecolecon.2008.03.011

7. Fenichel, E.P., Adamowicz, W., Ashton, M.S., Hall, J.S.: Incentive systems for forest-based ecosystem services with missing financial service markets. J. Assoc. Environ. Resour. Econ. **6**(2), 319–347 (2019)

8. Hansen, M.C., et al.: High-resolution global maps of 21st-century forest cover change. Science **342**(November), 850–854 (2013). https://doi.org/10.1126/science.1244693

9. Holmstrom, B., Milgrom, P.: Aggregation and linearity in the provision of intertemporal incentives. Econometrica **55**(2), 303–328 (1987). http://www.jstor.org/stable/1913238

10. Ihli, H.J., Chiputwa, B., Musshoff, O.: Do changing probabilities or payoffs in lottery-choice experiments affect risk preference outcomes? Evidence from rural Uganda. J. Agric. Resour. Econ. **41**(2), 324–345 (2016)

11. IPCC: Climate Change and Land: an IPCC special report on climate change, desertification, land degradation, sustainable land management, food security, and greenhouse gas fluxes in terrestrial ecosystems (2019). https://www.ipcc.ch/srccl/

12. Jack, B.K., Jayachandran, S.: Self-selection into payments for ecosystem services programs. Proc. Natl. Acad. Sci. **116**, 201802868 (2018). https://doi.org/10.1073/pnas.1802868115, http://www.pnas.org/lookup/doi/10.1073/pnas.1802868115

13. Jack, B.K., Oliva, P., Severen, C., Walker, E., Bel, S.: Technology adoption under uncertainty: take up and subsequent investment in Zambia. Natl. Bur. Econ. Res. **21**, 52 (2015). https://doi.org/10.3386/w21414

14. Jack, K.K.: Private information and the allocation of land use subsidies in Malawi. Am. Econ. J. Appl. Econ. **5**(3), 113–135 (2013). https://doi.org/10.1257/app.5.3.113

15. Jayachandran, S., Laat, J.D., Lambin, E.F., Stanton, C.Y., Audy, R., Thomas, N.E.: Cash for carbon: a randomized trial of payments for ecosystem services to reduce deforestation. Science **357**(6348), 267–273 (2017). FS-1035

16. Jean, N., Wang, S., Samar, A., Azzari, G., Lobell, D., Ermon, S.: Tile2Vec: unsupervised representation learning for spatially distributed data. Proc. AAAI Conf. Artif. Intell. **33**, 3967–3974 (2019). https://doi.org/10.1609/aaai.v33i01.33013967

17. Lawrance, E.C.: Poverty and the rate of time preference: evidence from panel data. J. Polit. Econ. **99**(1), 54–77 (1991)

18. Levin, J.: Relational incentive contracts. Am. Econ. Rev. **93**(3), 835–857 (2003)

19. Lütjens, B., Liebenwein, L., Kramer, K.: Machine learning-based estimation of forest carbon stocks to increase transparency of forest preservation efforts. In: NeurIPS Workshop on Tackling Climate Change with AI (2019). http://arxiv.org/abs/1912.07850

20. Ross, S.: The economic theory of agency?: the principal's problem linked references are available on JSTOR for this article: the economic theory of agency: the principal's problem. Am. Econ. Rev. **63**(2), 134–139 (1973)

21. Smith, B.: Microsoft will be carbon negative by 2030, July 2020. https://blogs.microsoft.com/blog/2020/01/16/microsoft-will-be-carbon-negative-by-2030/

22. Tanaka, T., Camerer, C.F., Nguyen, Q.: Risk and time preferences: linking experimental and household survey data from Vietnam. Am. Econ. Rev. **100**(1), 557–71 (2010)

Relaxing the Independence Assumption in Sequential Posted Pricing, Prophet Inequality, and Random Bipartite Matching

Ioannis Caragiannis[1]([✉])[ID], Nick Gravin[2][ID], Pinyan Lu[2], and Zihe Wang[3][ID]

[1] Department of Computer Science, Aarhus University, Aarhus, Denmark
iannis@cs.au.dk
[2] ITCS, Shanghai University of Finance and Economics, Shanghai, China
{nikolai,lu.piyan}@mail.shufe.edu.cn
[3] Gaoling School of Artificial Intelligence, Renmin University of China,
Beijing, China
wang.zihe@ruc.edu.cn

Abstract. We reexamine three classical settings of *optimization under uncertainty*, which have been extensively studied in the past, assuming that the several random events involved are mutually independent. Here, we assume that such events are only pair-wise independent; this gives rise to a much richer space of instances. Our aim has been to explore whether positive results are possible even under the more general assumptions. We show that this is indeed the case.

Indicatively, we show that, when applied to pair-wise independent distributions of buyer values, sequential posted pricing mechanisms get at least $\frac{1}{1.299}$ of the revenue they get from mutually independent distributions with the same marginals. We also adapt the well-known prophet inequality to pair-wise independent distributions of prize values to get a 1/3-approximation using a non-standard uniform threshold strategy. Finally, in a stochastic model of generating random bipartite graphs with pair-wise independence on the edges, we show that the expected size of the maximum matching is large but considerably smaller than in Erdős-Renyi random graph models where edges are selected independently. Our techniques include a technical lemma that might find applications in other interesting settings involving pair-wise independence.

Keywords: Posted pricing · Auctions · Prophet inequality · Revenue maximization · Bipartite matching

1 Introduction

Optimization in environments with uncertainty has received much attention in several research areas. It plays a central role in modern EconCS research (see, e.g., [7,10] for early surveys in bayesian mechanism design) and is also pervasive,

© Springer Nature Switzerland AG 2022
M. Feldman et al. (Eds.): WINE 2021, LNCS 13112, pp. 131–148, 2022.
https://doi.org/10.1007/978-3-030-94676-0_8

more broadly, in TCS (the rich theory of random graphs [3] is an example). Uncertainty manifests itself in many different ways as the following three settings indicate:

Sequential Posted Pricing: A seller has a single item to sell to n buyers. Each buyer has a random value v_i for the item distributed as $v_i \sim F_i$. The seller knows distributions $\{F_i\}_{i=1}^n$ and approaches buyers one by one in an arbitrary or fixed order. She offers the item to buyer i at a price p_i, which i takes when $v_i \geq p_i$ and pays p_i to the seller. The goal is to find a pricing scheme that maximizes the seller's revenue in expectation.

Optimal Stopping: A gambler plays a series of n games, each game $i \in [n]$ has a prize v_i distributed according to distribution F_i. The order of the games and the distribution of the prize values are known in advance to the gambler. Once the prize v_i for game i is realized, the gambler must decide whether to keep this prize and abandon the remaining games, or to discard this prize and continue playing. The gambler wants to maximize the expected reward.

Random Graph Models: Well-known models for the generation of random graphs assume a fixed set of nodes and produce each edge e between a pair of distinct nodes with a marginal probability p_e. Several graph parameters (e.g., the size of the maximum matching) have important meaning in areas like brain science, networking, or social sciences, and bounding these parameters is the subject of much research in many fields.

A simplifying assumption in most studies of the above settings is that the marginal distributions are mutually independent, i.e., the joint distribution is a product distribution. Under such an assumption, it is well-known that sequential posted pricing yields approximately-optimal revenue in single-parameter settings and generalizes nicely to multi-parameter environments [6]. Also, the optimal stopping strategy for the gambler can be computed by backward induction. A celebrated result, known as the *prophet inequality* [15], suggests that a simple threshold strategy can give an expected reward to the gambler that is at least 50% of the reward that could be achieved by a very powerful prophet, who has access to the maximum realized prize value [10,14]. Finally, in the random graph model where edges among pairs of n fixed nodes are drawn independently with the same probability p, a value of $p = \Omega(\ln n/n)$ is sufficient so that a hamiltonian cycle and, hence, a perfect matching exists, with high probability (see [3] for related results in random graphs). Unfortunately, such results (i.e., tight bounds or good approximations to revenue, gambler reward, or size of the maximum matching) do not hold for arbitrary joint distributions.

On the other hand, a recent line of work [5,9] on the monopoly problem for an additive buyer has proposed an alternative *correlation-robust framework* to study general distributions from the robust optimization perspective (see also [1,2]). In this framework, the algorithm designer knows only marginal distributions $\{F_i\}_{i=1}^n$ of each piece of the input and is given no information about correlation across different pieces in the joint distribution. The evaluation of the algorithm's performance is then taken in the worst-case, over the uncertainty of the problem, i.e., over all possible joint distributions with the specified set of marginal distributions $\{F_i\}_{i=1}^n$. The fact that the joint distribution is not explicitly given has an important practical advantage. Indeed, the representation and sampling complexity for learning correlated multidimensional distributions is exponential in the dimension n. In this respect, learning and operating with information about separate marginals is a much simpler task that does not suffer from the curse of dimensionality. However, the correlation-robust framework does not allow to incorporate any extra information about the distribution beyond the marginals. For example in the monopoly problem, there is no obstacle for the seller to acquire additional information, say, about dependencies between pairs of items by doing more extensive market research. On the other hand, the expected performance guarantees in the correlation robust framework are rather pessimistic compared to the mutually independent case and the worst-case joint distribution of the input often admits strong dependencies between its parts. The latter is not something we usually expect in practice, where it is more likely to see rather weak dependencies and significant variability between any given pair of input's components.

Our goal in this work is to model and study such situations with potential (weak) dependencies between input components. A straightforward approach would be to extend the correlation robust framework as follows: (i) specify the set of marginal distributions for any pair of input components $\{F_{i,j}\}_{i,j\in[n]}$ (e.g., the joint distribution of values for each pair of items $(v_i, v_j) \sim F_{i,j}$) and (ii) evaluate the expectation in the worst-case over all feasible joint distributions that agree with $\{F_{i,j}\}_{i,j\in[n]}$. Unfortunately, not all such pair-wise distributions (even consistent with singleton marginals $\{F_i\}_{i=1}^n$) would admit a feasible joint distribution[1]. Even when there exists a feasible joint distribution the set of pair-wise marginal distributions can sometimes uniquely identify the joint distribution. Moreover, the extra information does not necessarily help when we compare to the worst case distribution π^* for $\{F_i\}_{i=1}^n$. Indeed, one can take the worst-case joint distribution π^* and write down the restriction of π^* to $\{\pi_{i,j}^*\}_{i,j\in[n]}$. Then, the performance for this set of pair-wise marginals would not be better than that for π^*.

To avoid these complications we consider an important special case where pair-wise marginal distributions $\{F_{i,j}\}_{i,j\in[n]}$ are all independent, i.e., $F_{i,j} = F_i \times F_j$ for all $i, j \in [n]$. In other words, we assume that joint distribution is *pair-wise independent*. At first glance, pair-wise independence might appear rather similar to the standard mutually independence assumption. However, there are some important differences which we discuss below. We will also highlight the importance of this robust optimization approach assuming a pair-wise independent joint distribution.

1. *(Statistics vs. probability model)*. The idealistic model with mutually independent distributions is a probability model that is not easy to verify in the proper statistical sense. Indeed, the joint distribution has exponential dependency on the number of components, so it would take a super-polynomial number of samples to confirm that the input distribution is close in total

---

[1] For example, if value of item B is always equal to the values of items A and C, then items A and C cannot be independent.

variation distance to the specific product distribution. On the other hand, the pair-wise independence condition is a statistical condition that can be checked in practice with only polynomially many samples.

2. *(Robustness to weak dependencies of data).* In practice, a multi-dimensional data distribution usually exhibits some form of mutual dependency that might be noticeable at the level of pair-wise marginal distributions. However, these dependencies are often weak and it is still reasonable to approximate each marginal pair distribution $F_{i,j}$ as the product distribution $F_i \times F_j$. In other words, we might want to allow small approximation errors in our model (using a pair-wise independent distribution) to the real joint distribution. The mutually independent model is a specific distribution, and as such can be too far from the most likely distribution matching the data. On the other hand, a robust guarantee for any pair-wise independent distribution is still meaningful even if the pair-wise marginals of the true input distribution are slightly perturbed compared to $F_i \times F_j$.

3. *(Large class of distributions).* To understand the size of the class of pair-wise independent distributions, let us consider the case of finite discrete supports, i.e., each marginal distribution F_i has finite support of size $|F_i|$. In this case the dimension of the simplex of feasible joint distributions is $|F_1| \cdot \ldots \cdot |F_n|$ and the mutually independent distribution is a single point. On the other hand, there are not more than $\sum_{i<j} |F_i| \cdot |F_j|$ linear constraints in the description of a pair-wise independent distribution. The product distribution is pair-wise independent and has positive probability (i.e., the inequality $\mathbf{Pr}\,[\mathbf{v}] \geq 0$ is not tight) for any point $\mathbf{v}$ in the support. Hence, the dimension of the pair-wise independent distribution is at least $\prod_{i=1}^{n} |F_i| - \sum_{i<j} |F_i| \cdot |F_j|$.

1.1   Our Results and Techniques

In this paper we study the three settings we discussed in the beginning of this section. We show that *any* sequential posted pricing mechanism with a given set of prices $\{p_i\}_{i=1}^{n}$ has an expected revenue that is at most 1.299 times larger in the case of mutually independent distributions of buyer values compared to the case of pair-wise independent distributions with the same marginals. Our result only requires that prices $\{p_i\}_{i=1}^{n}$ are offered in Pareto-optimal order, i.e., from higher to lower prices. The main tool we exploit to prove this result is a lemma that is conceptually similar to Lovász Local Lemma (LLL; see [13]). Recall that LLL bounds the probability that none among a series of events happens in terms of the marginal probabilities of these events, provided that they have a certain structure of dependencies. Our LLL-type statement bounds the probability that none among a series of pair-wise independent events happen in terms of their marginal probabilities. We believe that this lemma will find application beyond the scope of the current paper. We give an example that shows that our lemma is essentially tight; this implies that our bound on the revenue of sequential posted pricing is tight as well.

We also present variations of the prophet inequality when prizes have pair-wise independent values. A non-standard uniform threshold strategy yields the following guarantee. The worst expected reward of the gambler among all pair-wise independent prize value distributions with given marginals is at least 1/3 of the best expected prophet's reward over all pair-wise independent distributions with the same marginals. Again, we exploit an alternative expression of our local lemma. Interestingly, we show that uniform threshold strategies cannot yield a guarantee better than 40%, in contrast to the 50% guarantee of the classical prophet inequality [15] (see also [10,14]) for mutually independent distributions. A non-uniform threshold strategy (exploiting ideas from [6]) is shown to break this barrier and achieve a 41.4% guarantee, at least for continuous pair-wise independent distributions. It is slightly more complicated though and requires additional information on the joint distribution besides the marginals.

Notice that the prophet inequality bounds are not universal like the ones for sequential posted pricing. Specifically, we show that there exists a uniform threshold strategy that achieves constant approximation to the prophet's reward for a mutually independent prize value distribution but achieves very low expected reward for a pair-wise independent distribution with the same marginals. Our results indicate that sequential posted pricing and optimal stopping are two economic settings where positive results are possible by relaxing independence to pair-wise independence. We demonstrate that such results are not possible for second price auctions. Broadening the class of economic problems that are "friendly" to the pair-wise independence assumption is an important direction for future research.

Finally, we consider a stochastic model for generating random bipartite graphs with n nodes in each side of the bipartition, so that each edge exists with some (non-necessarily uniform) probability. We assume that the expected degree of any node is Δ. When edges exist in the graph mutually independently, folklore results (e.g., see [3]) suggest that a perfect matching exists almost certainly, provided that $\Delta = \Omega(\ln n)$. Furthermore, we can show that the expected size of the maximum matching is $n - n \cdot O(\exp(-\Delta))$. In contrast, in the case of pair-wise independence (on the existence of edges), the lower bound we can show is $n - n/\sqrt{\Delta}$, which leaves open the possibility of non-existence of perfect matchings for all interesting range of values for parameter Δ. Our proof is based on a second-moment argument and exploits the fact that the maximum matching has the same size with the minimum vertex cover in a bipartite graph. We also present a non-trivial pair-wise independent distribution over bipartite graphs that shows that our bound is essentially tight. These results indicate that a revision of classical results on random graphs under the pair-wise independence lens might reveal a very interesting new picture.

1.2   Roapmap

The rest of the paper is structured as follows. We present our local-lemma-type statement in Sect. 2. Sequential posted pricing is studied in Sect. 3. Our results for prophet inequalities are presented in Sect. 4. Section 5 is devoted to proving our bounds for random bipartite matchings.

2   A Local-Lemma-Type Probability Statement

We begin by proving an LLL-style probability statement for pair-wise independent distributions. The lemma will be particularly useful in Sects. 3 and 4 but we believe that it will find applications in other settings as well.

Let $\{E_i\}_{i=1}^n$ be a set of random events with $\mathbf{Pr}\,[E_i] = q_i$. We are interested in the probability that at least one of the random events happens. If these events are mutually independent, this probability is exactly

$$\mathbf{Pr}\left[\bigvee_{i=1}^n E_i\right] = 1 - \prod_{i=1}^n (1 - q_i).$$

We want to lower bound this probability when the events are only known to be pair-wise independent. In Lovász local lemma (LLL), these random events are either mutually independent or worst-case correlated, and LLL gives a low bound on the probability that none of the events happens. In a sense, the two lemmas both relax the independence assumption to a "local" assumption but in different directions. LLL models situations where the dependencies are only happening locally: an event is mutually independent with all other events except its neighbors. In our lemma, we model the setting where independence is only guaranteed locally: any pair of events are independent with each other but not necessarily globally. We prove that, in any pair-wise independent distribution, the probability is at least a constant fraction of the probability in the mutually independent setting.

Lemma 1. *Let $\{E_i\}_{i=1}^n$ be a set of random events. Let $\mathbf{F}_{ind}$ and $\mathbf{F}_\pi$ be a mutually independent and a pair-wise independent distribution over these events, respectively, with $\underset{\mathbf{F}_{ind}}{\mathbf{Pr}}\,[E_i] = \underset{\mathbf{F}_\pi}{\mathbf{Pr}}\,[E_i] = q_i$. Then,*

$$\underset{\mathbf{F}_\pi}{\mathbf{Pr}}\left[\bigvee_{i=1}^n E_i\right] \geq \frac{\sum_{i=1}^n q_i}{1 + \sum_{i=1}^n q_i}$$

and

$$\underset{\mathbf{F}_\pi}{\mathbf{Pr}}\left[\bigvee_{i=1}^n E_i\right] \geq \frac{1}{1.299}\,\underset{\mathbf{F}_{ind}}{\mathbf{Pr}}\left[\bigvee_{i=1}^n E_i\right] = \frac{1}{1.299}\left(1 - \prod_{i=1}^n (1 - q_i)\right).$$

Proof. We prove only the first inequality here; the proof of the second one is omitted. We denote by X_i the indicator random variable for event E_i and define the random variable $X = \sum_{i=1}^n X_i$. Then $\bigvee_{i=1}^n E_i$ (the event we are interested in) is the random event $\{X > 0\}$. By definition, we have $\mathbf{E}\,[X_i] = q_i$ and thus $\mathbf{E}\,[X] = \sum_{i=1}^n q_i$.

Since the random variables $\{X_i\}_{i=1}^n$ are pair-wise independent, we have

$$\mathbf{Var}\,[X] = \sum_{i=1}^n \mathbf{Var}\,[X_i] + 2\sum_{i<j} \mathrm{Cov}(X_i, X_j) = \sum_{i=1}^n \mathbf{Var}\,[X_i] = \sum_{i=1}^n q_i(1 - q_i).$$

Let $f_k = \Pr_{X \sim \mathbf{F}_\pi} [X = k]$ for all $i \in [n]$. Using this notation and applying Cauchy-Schwartz's inequality, we have

$$\Pr_{X \sim \mathbf{F}_\pi} [X > 0] = \sum_{k=1}^{n} f_k \geq \frac{\left(\sum_{k=1}^{n} k \cdot f_k\right)^2}{\sum_{k=1}^{n} k^2 \cdot f_k} = \frac{\mathbf{E}[X]^2}{\mathbf{E}[X^2]}$$

$$= \frac{\mathbf{E}[X]^2}{\mathbf{Var}[X] + \mathbf{E}[X]^2} = \frac{(\sum_{i=1}^{n} q_i)^2}{\sum_{i=1}^{n} q_i(1 - q_i) + (\sum_{i=1}^{n} q_i)^2}.$$

This immediately gives us $\Pr_{\mathbf{F}_\pi} [\bigvee_{i=1}^{n} E_i] = \Pr_{X \sim \mathbf{F}_\pi} [X > 0] \geq \frac{(\sum_{i=1}^{n} q_i)^2}{\sum_{i=1}^{n} q_i + (\sum_{i=1}^{n} q_i)^2} = \frac{\sum_{i=1}^{n} q_i}{1 + \sum_{i=1}^{n} q_i}.$ $\qquad \square$

The upper bound of 1.299 in the statement of Lemma 1 is almost tight. Here, we give an example where the gap between $\Pr_{\mathbf{F}_{ind}} [\bigvee_{i=1}^{n} E_i]$ and $\Pr_{\mathbf{F}_\pi} [\bigvee_{i=1}^{n} E_i]$ is at least 1.296. In the example, $q_i = q = 2/(n-1)$ for all $i \in [n]$. For the distribution $\mathbf{F}_{ind}$, we have $\Pr_{\mathbf{F}_{ind}} [\bigvee_{i=1}^{n} E_i] = 1 - (1-q)^n$, which approaches $1 - e^{-2}$ as n goes to infinity.

Now consider the following probability distribution. With probability $\frac{2n}{3(n-1)}$ a set of exactly three events among $\{E_i\}_{i=1}^{n}$ happen. These three events are chosen uniformly at random among the $\binom{n}{3}$ possible choices. With the remaining probability of $1 - \frac{2n}{3(n-1)}$ no event happens. In this distribution, the probability that at least one event happens approaches $\frac{2}{3}$ as n goes to infinity. The ratio between the two probabilities approaches $(1 - e^{-2})/(\frac{2}{3}) = 1.29699$.

It remains to show that this distribution is pair-wise independent. Indeed, for any $i \neq j$, we have:

$$\Pr[E_i \wedge E_j] = \frac{2n}{3(n-1)} \cdot \frac{\binom{n-2}{1}}{\binom{n}{3}} = \frac{4}{(n-1)^2} = q^2,$$

$$\Pr[E_i \wedge \overline{E_j}] = \Pr[\overline{E_i} \wedge E_j] = \frac{2n}{3(n-1)} \cdot \frac{\binom{n-2}{2}}{\binom{n}{3}} = \frac{2(n-3)}{(n-1)^2} = q(1-q), \text{ and}$$

$$\Pr[\overline{E_i} \wedge \overline{E_j}] = 1 - q^2 - 2q(1-q) = (1-q)^2,$$

as desired.

3 Sequential Posted Pricing

In this section, we consider the setting with a seller, who aims to sell a single item to n potential buyers. Buyer $i \in [n]$ has value v_i distributed according to distribution F_i. The seller uses a *sequential posted pricing* mechanism. He considers the buyers one by one according to their index; when considering buyer i, the seller offers her the item at price p_i. We consider Pareto-efficient pricing schemes that satisfy $p_1 \geq p_2 \geq ... \geq p_n$. In this way, the seller does not risk to loose revenue to low price buyers, who will have their chance only after the item is offered to other buyers at a higher price. Our aim is to analyze sequential posted pricing mechanisms assuming that the distributions F_i are pair-wise independent and compare their revenue to the revenue they would have when applied to mutually independent valuations with the same marginals.

Let us consider the simple case where a uniform price p is offered to all buyers. In this case, we can directly use Lemma 1 to conclude that the expected revenue over any pair-wise independent distribution is at least $\frac{1}{1.299}$ of the revenue of a corresponding mutually independent distribution. Indeed, denote by $E_i = \{v_i \geq p\}$ the event that buyer i accepts the price. As the price is the same for all buyers, the probability guarantee immediately translates into the revenue guarantee. Hence, the revenue of the mechanism when the events E_i are pair-wise independent is at least $\frac{1}{1.299}$ times the revenue of the mechanism when these events are mutually independent but have the same probabilities.

For the general case where prices can be different, we also need to pay attention to who gets the item. To do so, we apply Lemma 1 n times, each time considering the first k buyers for $k = 1, 2, \cdots, n$. Let $\lambda = 1.299$ be the approximation guarantee from Lemma 1. Let X_i be the random variable indicating whether the value of buyer i is at least p_i, and let q_i be the probability that $X_i = 1$. Using the second inequality from Lemma 1, we have

$$\lambda \cdot \mathbf{Pr}\left[\sum_{i=1}^{k} X_i > 0\right] \geq 1 - \prod_{i=1}^{k}(1 - q_i).$$

We multiply this inequality by the price difference $p_k - p_{k+1}$ to get

$$\lambda(p_k - p_{k+1})\mathbf{Pr}\left[\sum_{i=1}^{k} X_i > 0\right] \geq (p_k - p_{k+1})\left(1 - \prod_{i=1}^{k}(1 - q_i)\right),$$

where $p_{n+1} = 0$. After summing these inequalities for $k \in \{1, \ldots, n\}$, we get

$$\lambda \cdot \sum_{k=1}^{n} p_k \cdot \mathbf{Pr}\left[X_k = 1, \forall i < k, X_i = 0\right] \geq \sum_{k=1}^{n} p_k q_k \prod_{i=1}^{k-1}(1 - q_i) \qquad (1)$$

Observe that the LHS of equation (1) is equal to λ times the revenue generated by the sequential posted pricing mechanism, while the RHS of (1) is equal to the revenue the mechanism would have if the valuations of buyers were mutually independent. We summarize this observation in the following statement.

Theorem 1. *A posted pricing mechanism under a pairwise independent distribution of buyer valuations achieves at least $1/1.299$ fraction of the revenue under a mutually independent distribution with the same marginals.*

Our lower bound from Sect. 2 directly translates to a posted pricing instance with uniform prices. Thus, a sequential posted pricing mechanism for mutually independent distributions can generate revenue at least 1.296 time more than for pair-wise independent distributions.

We note that such robust properties do not necessary hold in other mechanism design settings. In particular, in a second price auction, the revenue gap in the cases of pair-wise independent and mutually independent buyer valuations can be huge. The reason is that in the mutually independent case, the second largest bid is high with large probability, while in the pairwise independent case, this probability can be very small.

First, consider the following setting. There are n i.i.d. buyers. Each buyer has value 1 with probability $\frac{1}{n-1}$ and value 0 otherwise. The revenue of the second price auction is then the probability that at least two buyers have value 1, i.e.,

$$1 - \left(1 - \frac{1}{n-1}\right)^n - \frac{n}{n-1}\left(1 - \frac{1}{n-1}\right)^{n-1} = 1 - 2\left(1 - \frac{1}{n-1}\right)^{n-1} \geq 1 - 2/e.$$

We now construct a pairwise independent distribution in which the value of each buyer is 1 with probability $\frac{1}{n-1}$ and 0 otherwise, and under which the generated revenue is very small. In this distribution, there are two kinds of valuation profiles. The first profile appears with probability $\frac{1}{(n-1)^2}$ and all buyers have value 1. In the second profile, one buyer, selected uniformly at random, has value 1 and the rest of the buyers have value 0. The second price auction has expected revenue of $\frac{1}{(n-1)^2}$ as it gets a revenue of 1 only on the first profile. One can easily verify that the probability distribution is indeed pair-wise independent.

4   Prophet Inequality

Sequential posted pricing is closely related to the *prophet inequality* from optimal stopping theory [15]. In this scenario a gambler plays sequentially a series of n games. Each game $i \in [n]$ has prize v_i distributed according to distribution F_i. The order of the games and the distribution of the prize values are known in advance to the gambler. Once the prize v_i for game i is realized, the gambler must decide whether to keep this prize and abandon the remaining games, or to discard this prize and continue playing. A prophet in this setting knows the realization of all prizes in advance and therefore can stop at the right moment and take the highest prize.

4.1   A Uniform Threshold Policy

It is well-known that the gambler can achieve a 2-approximation of the optimal prize by following a simple *uniform threshold strategy*, which is given by a single threshold $\widehat{v}$ and requires the gambler to accept the first prize i with $v_i \geq \widehat{v}$. The standard assumption in the prophet inequality literature is that the prize distributions $\{F_i\}_{i=1}^n$ for different games are mutually independent.[2] Here, we only assume that the distributions $\{F_i\}_{i=1}^n$ are pair-wise independent.

---

[2] We note that prophet inequalities for classes of prize distributions with limited correlation have been studied before. The survey of Hill and Kertz [11] discusses early related results in stopping theory while the papers [4,8,12] are representative of recent related work by the EconCS community. However, such results are rarely based on the use of simple threshold strategies as the one we use here.

Theorem 2. *For any set of marginal prize distributions $\{F_i\}_{i=1}^n$, there exists a threshold $\hat{v}$ such that the expected reward of the uniform threshold strategy for any pair-wise independent joint distribution is at least $1/3$ of the expected value of the maximum prize[3].*

Proof. Let REF denote the expected reward of the prophet, i.e., the expected maximum prize $\mathbf{E}\left[\max_i v_i\right]$, and APX denote the expected reward of the gambler with a uniform threshold strategy. We denote by $\hat{v}$ the uniform threshold (to be defined later). Let $x = \sum_{i=1}^n \mathbf{Pr}\left[v_i \geq \hat{v}\right]$.

We use the same upper bound for REF as in the standard exposition of the prophet inequality (e.g., see [10,14]).

$$\mathsf{REF} = \mathop{\mathbf{E}}_{\mathbf{v}\sim\mathbf{F}}\left[\max_i v_i\right] \leq \hat{v} + \mathop{\mathbf{E}}_{\mathbf{v}\sim\mathbf{F}}\left[\max_i(v_i - \hat{v})^+\right] \leq \hat{v} + \sum_i \mathop{\mathbf{E}}_{v_i\sim F_i}\left[(v_i - \hat{v})^+\right].$$

We note that this upper bound holds for any joint distribution with the given marginal distributions (not necessary pair-wise independent). In this bound the RHS depends only on the marginal distributions.

We will split the gambler's reward APX into two parts: (i) the first part, APX_1, is the guaranteed contribution of $\hat{v}$ if some reward is taken and (ii) the second part, APX_2, is the extra contribution of $v_i - \hat{v}$ when i is chosen. To bound APX_1, we use Lemma 1:

$$\mathsf{APX}_1 = \mathbf{Pr}_\pi\left[\max_i v_i \geq \hat{v}\right]\cdot\hat{v} \geq \frac{\sum_{i=1}^n \mathbf{Pr}\left[v_i \geq \hat{v}\right]}{1+\sum_{i=1}^n \mathbf{Pr}\left[v_i \geq \hat{v}\right]}\cdot\hat{v} = \frac{x}{1+x}\cdot\hat{v}, \quad (2)$$

where $x \stackrel{\text{def}}{=} \sum_{i=1}^n \mathbf{Pr}\left[v_i \geq \hat{v}\right]$. In general, the notation $\mathbf{Pr}_\pi\left[\cdot\right]$ is used when pair-wise independent prizes are considered.

To bound APX_2, we define the event $\mathcal{E}_{i,v} \stackrel{\text{def}}{=} \{\mathbf{v}\,|v_i = v;\ \forall j\neq i, v_j < \hat{v}\}$ for every $v \geq \hat{v}$, i.e., the reward in game i is $v_i = v$ while all the remaining prizes are below the threshold. The crucial property of any joint pair-wise independent distribution π is that $\mathbf{Pr}_\pi\left[\mathcal{E}_{i,v}\right] \geq (1-x)\mathbf{Pr}_{F_i}\left[v_i = v\right]$ which we show below. Indeed, by definition

$$\mathbf{Pr}_\pi\left[\mathcal{E}_{i,v}\right] = \mathbf{Pr}_{F_i}\left[v_i = v\right]\cdot\mathbf{Pr}_\pi\left[\bigcap_{j\neq i}[v_j < \hat{v}]\,\middle|\,v_i = v\right].$$

By the union bound, we have

$$\mathop{\mathbf{Pr}}_\pi\left[\bigcap_{j\neq i}[v_j < \hat{v}]\,\middle|\,v_i = v\right] \geq 1 - \sum_{j\neq i}\mathop{\mathbf{Pr}}_\pi\left[v_j \geq \hat{v}|v_i = v\right].$$

---

[3] We remark that, while the expectation for the threshold strategy is taken in the *worst case* over any pair-wise independent distribution, the expectation for the prophet is taken in the *best case* over any distribution with the given marginals.

Due to pair-wise independence, $\mathbf{Pr}_\pi\left[v_j \geq \widehat{v}|v_i = v\right] = \mathbf{Pr}_{F_j}\left[v_j \geq \widehat{v}\right]$. By definition of x, we know that $\sum_{j \neq i} \mathbf{Pr}\left[v_j \geq \widehat{v}\right] \leq x$. Hence,

$$\mathbf{Pr}\left[\mathcal{E}_{i,v}\right] \geq \mathbf{Pr}\left[v_i = v\right] \cdot \left(1 - \sum_{j \neq i} \mathbf{Pr}\left[v_j \geq \widehat{v}\right]\right) \geq (1-x)\mathbf{Pr}\left[v_i = v\right].$$

When $\mathcal{E}_{i,v_i}$ happens, we get the additional contribution of $v_i - \widehat{v}$. As all random events $\{\mathcal{E}_{i,v_i}\}_{i \in [n], v_i \geq \widehat{v}}$ are disjoint, we have

$$\mathsf{APX}_2 \geq (1-x) \sum_{i=1}^{n} \int_{v_i \geq \widehat{v}} (v_i - \widehat{v})\, \mathrm{d}F_i(v_i) = (1-x) \sum_{i=1}^{n} \mathbf{E}\left[(v_i - \widehat{v})^+\right]. \qquad (3)$$

Since $\sum_{i=1}^{n} \mathbf{E}\left[(v_i - \widehat{v})^+\right]$ is a continuous function of $\widehat{v}$ decreasing to zero when $\widehat{v} \to \infty$, we can choose[4] the threshold $\widehat{v}$ so that

$$\widehat{v} = 2 \cdot \sum_{i=1}^{n} \mathbf{E}\left[(v_i - \widehat{v})^+\right].$$

Then, by the definition of REF, we have

$$\widehat{v} \geq \frac{2}{3} \cdot \mathsf{REF}, \quad \text{and} \quad \sum_{i=1}^{n} \mathbf{E}\left[(v_i - \widehat{v})^+\right] \geq \frac{1}{3} \cdot \mathsf{REF}.$$

If $x \geq 1$, then the lower bound (3) is trivial and we only use (2) to get

$$\mathsf{APX} \geq \mathsf{APX}_1 \geq \frac{1}{2} \cdot \widehat{v} \geq \frac{1}{3} \cdot \mathsf{REF}.$$

Otherwise, if $0 \leq x \leq 1$, we combine (2) and (3) to get

$$\mathsf{APX} = \mathsf{APX}_1 + \mathsf{APX}_2 \geq \left(\frac{2}{3}\frac{x}{1+x} + \frac{1}{3}(1-x)\right) \mathsf{REF}$$

$$\geq \left(\frac{2}{3}\frac{x}{2} + \frac{1}{3}(1-x)\right) \mathsf{REF} = \frac{1}{3} \cdot \mathsf{REF}.$$

This completes the proof.   □

---

[4] If the distributions were continuous, we could choose the threshold $\widehat{v}$ so that $x = \frac{\sqrt{5}-1}{2}$. For this value of x we have $\frac{x}{1+x} = 1 - x = \frac{3-\sqrt{5}}{2} = 0.382$. Then, we could get a lower bound on APX by combining (2) and (3) as follows:

$$\mathsf{APX} = \mathsf{APX}_1 + \mathsf{APX}_2 \geq 0.382 \left(\widehat{v} + \sum_{i=1}^{n} \mathbf{E}\left[(v_i - \widehat{v})^+\right]\right) \geq 0.382 \cdot \mathsf{REF}.$$

In our proof, we assume that distributions can be discontinuous and, thus, we may not be able to set $\widehat{v}$ to get a particular value of x.

We now present limitations of uniform threshold strategies for pair-wise independent distributions. These limitations come in contrast with the case of mutually independent distributions, where some of such policies can give at least 50% of the prophet's value as reward.

Theorem 3. *Uniform threshold strategies cannot guarantee more than 40% of prophet's value for some pair-wise independent distributions of prize values.*

Proof. In the proof, we use the following distribution. There are $n + 2$ items. Item 1 has a deterministic value of 1. Item $n + 2$ has value n with probability $\frac{1}{n}$ and value 0 otherwise. The values of items $2, ..., n + 1$ have identical marginal distributions: value 2 with probability $\frac{1}{n}$ and 0 with probability $\frac{n-1}{n}$. The joint distribution π is constructed as follows. When item $n + 2$ has value n:

- With probability $\frac{1}{n} - \frac{1}{n^2}$, all items $2, ..., n + 1$ have value 0.
- With probability $1 - \frac{1}{n}$, exactly one item among $2, ..., n + 1$ has value 2 and the remaining items have value 0. The high-value item is selected uniformly at random among the items $2, ..., n + 1$.
- With probability $\frac{1}{n^2}$, items $2, ..., n + 1$ have all values 2.

When item $n + 2$ has value 0,

- With probability $\frac{n-1}{2n}$, items $2, ..., n + 1$ have all value 0.
- With probability $\frac{1}{n}$, exactly one item (selected uniformly at random) among $2, ..., n + 1$ has value 2 and the remaining items have value 0.
- With probability $\frac{n-1}{2n}$, exactly two items among $2, ..., n + 1$ have value 2 and the remaining items have value 0. The two items are chosen uniformly at random among the $\binom{n}{2}$ possible pairs.

It is straightforward to verify that this is a pair-wise independent distribution. The expected value of the prophet is

$$n \cdot \frac{1}{n} + \left(1 - \frac{1}{n}\right) \cdot \left(1 \cdot \frac{n-1}{2n} + 2 \cdot \frac{n+1}{2n}\right) = \frac{5}{2} - \frac{1}{n} - \frac{1}{2n^2}.$$

Using a threshold that is smaller than 1, the reward of the gambler is (deterministically) 1. Using a threshold higher than 2, her expected reward is $n \cdot \frac{1}{n} = 1$, too. Now, assume that a threshold in $(1, 2]$ is used. Then, the expected reward of the gambler is

$$\frac{1}{n}\left(n\left(\frac{1}{n} - \frac{1}{n^2}\right) + 2\left(1 - \frac{1}{n} + \frac{1}{n^2}\right)\right) + \left(1 - \frac{1}{n}\right)\left(2 \cdot \frac{n+1}{2n}\right) = 1 + \frac{3}{n} - \frac{4}{n^2} + \frac{2}{n^3}.$$

Hence, as n approaches infinity, the reward of the prophet approaches $5/2$, while the reward of the gambler approaches 1 when using any uniform threshold strategy. The theorem follows. □

We have proved that there exists a threshold such that prophet inequality still holds with a slightly worse constant. However, unlike the case of sequential posted pricing where a constant gap holds for any choice of prices, this is not true for all choices of thresholds in the prophet inequality setting. We give an example where a certain threshold strategy achieves constant fraction of the maximum welfare in the mutually independent case, but it gets almost zero fraction in the pairwise independent case.

Our example has four items. The values of the first three items are 0 and 1, equally likely; the fourth item has large deterministic value $V > 1$. Assuming mutually independent values, the expected gain of the gambler when she uses a uniform threshold strategy with the threshold 1 is $\frac{7+V}{8}$. Now, consider the following pair-wise independent distribution, in which the gambler always gets a value of 1 when she uses 1 as a uniform threshold. The first three items have values $(1, 1, 1), (0, 1, 0), (1, 0, 0), (0, 0, 1)$ with equal probabilities. Our claim follows for large values of V.

4.2   Non-uniform Threshold Strategies

We now demonstrate that non-uniform threshold strategies can be more powerful than uniform ones. We adapt a technique from [6]. The gambler uses different thresholds $\tau_1, \tau_2, ..., \tau_n$ and n coins where the probability of the i-th coin toss to be heads is q_i. At step i, if the award has not been given before and the prize v_i exceeds the thresholds τ_i, the gambler tosses the i-th coin and gets the prize if it comes heads. In our analysis, we assume that the prize values follow continuous distributions.

Theorem 4. *For any set of continuous marginal prize distributions $\{F_i\}_{i=1}^{n}$, there exist thresholds $(\tau_i)_{i \in [n]}$ and probabilities $(q_i)_{i \in [n]}$ so that the expected reward of the gambler's strategy is at least $\sqrt{2} - 1 \approx 41.4\%$ of the expected value of the maximum prize.*

Proof. For $i = 1, 2, ..., n$, let $p_i = \mathbf{Pr}\left[v_i \geq v_j, \forall j \in [n]\right]$ and define τ_i to be such that $\mathbf{Pr}\left[v_i \geq \tau_i\right] = p_i$. Then, $\mathbf{E}\left[v_i \cdot \mathbb{1}\{v_i \geq v_j, \forall j \in [n]\}\right] \leq \mathbf{E}\left[v_i \cdot \mathbb{1}\{v_i \geq \tau_i\}\right]$ and

$$\mathbf{E}\left[\max_i v_i\right] \leq \sum_i \mathbf{E}\left[v_i \cdot \mathbb{1}\left\{v_i \geq \tau_i\right\}\right]. \tag{4}$$

For $i = 1, 2, ..., n$, let R_i be the event that no award has been given at steps $1, 2, ..., i-1$, P_i the event that $v_i \geq \tau_i$ (i.e., $\mathbf{Pr}\left[P_i\right] = p_i$), and Q_i the event that the random coin toss at step i comes heads (i.e., $\mathbf{Pr}\left[Q_i\right] = q_i$). For $i \geq 2$, we have

$$\mathbf{Pr}\left[R_i | v_i = v\right] = \mathbf{Pr}\left[R_{i-1} \wedge \overline{P_{i-1} \wedge Q_{i-1}} \mid v_i = v\right]$$

$$\geq \mathbf{Pr}\left[R_{i-1} \mid v_i = v\right] + \mathbf{Pr}\left[\overline{P_{i-1} \wedge Q_{i-1}} \mid v_i = v\right] - 1$$

$$\geq \mathbf{Pr}\left[R_{i-1} \mid v_i = v\right] - p_{i-1}q_{i-1}. \tag{5}$$

The first inequality uses the property $\mathbf{Pr}\left[A \wedge B\right] = \mathbf{Pr}\left[A\right] + \mathbf{Pr}\left[B\right] - \mathbf{Pr}\left[A \vee B\right] \geq \mathbf{Pr}\left[A\right] + \mathbf{Pr}\left[B\right] - 1$. The second inequality follows since the events

P_{i-1} and Q_{i-1} and $\{v_i = v\}$ are independent. Summing the inequalities (5) for $i = 2, ..., n$ together with the obvious fact $\mathbf{Pr}\left[R_1|v_i = v\right] = 1$, we get

$$\mathbf{Pr}\left[R_i|v_i = v\right] \geq 1 - \sum_{j<i} p_j q_j. \tag{6}$$

The expected award APX of the gambler is

$$\begin{aligned}
\mathsf{APX} &= \sum_{i=1}^{n} \int_{\tau_i}^{\infty} v\mathbf{Pr}\left[R_i|v_i = v\right] q_i \, dF_i(v_i) \\
&\geq \sum_{i=1}^{n} q_i(1 - \sum_{j<i} p_j q_j) \int_{\tau_i}^{\infty} v \, dF_i(v_i) \\
&\geq \min_i \left\{ q_i \left(1 - \sum_{j<i} p_j q_j \right) \right\} \cdot \sum_i \mathbf{E}\left[v_i \, \mathbb{1} \left\{ v_i \geq \tau_i \right\}\right] \\
&\geq \min_i \left\{ q_i \left(1 - \sum_{j<i} p_j q_j \right) \right\} \cdot \mathbf{E}\left[\max_i v_i\right]. \tag{7}
\end{aligned}$$

The first inequality follows by (6). The second one is obvious and the third one follows by (4).

We will now define the q_i's appropriately so that $\min_i \left\{ q_i \left(1 - \sum_{j<i} p_j q_j \right) \right\} \geq \sqrt{2} - 1$. The theorem will then follow by (7). Let $\alpha = \frac{\sqrt{2}-1}{2}$ and $\beta = (1 + \sqrt{2})^2$ and define the function $g : [0, 1] \to \mathbb{R}_{\geq 0}$ with $g(x) = \sqrt{\frac{\alpha}{\beta - x}}$. It can be verified by tedious calculations that

$$g(x) \left(1 - \int_0^x g(t) dt \right) = 2\alpha = \sqrt{2} - 1$$

for every $x \in [0, 1]$. Now, let $q_i = g\left(\sum_{j<i} p_j\right)$ and observe that $\sum_{j<i} p_j q_j \leq \int_0^{\sum_{j<i} p_j} g(t) dt$ (as $g(t)$ is a decreasing function and the integral in the right hand side is larger than its Riemann sum for the partition into the intervals of lengths $(p_j)_{j<i}$). Hence, for every $i \in [n]$, we have

$$q_i \left(1 - \sum_{j<i} p_j q_j \right) \geq g\left(\sum_{j<i} p_j\right)\left(1 - \int_0^{\sum_{j<i} p_j} g(t) dt \right) = \sqrt{2} - 1$$

as desired. $\square$

5   Matchings in Random Bipartite Graphs

In this section, we consider a stochastic graph model for bipartite graphs, extending the classical Erdős-Renyi model. In particular, the stochastic model

$\mathbf{G}(L, R, \{p_e\}_{e \in L \times R})$ is a distribution over bipartite graphs $G = (L, R, E)$ with $E \subseteq R \times L$, such that the marginal probability of each edge $e \in R \times L$ to appear in G is equal to $\mathbf{Pr}_{G \sim \mathbf{G}}[e \in E(G)] = p_e$. We are interested in the case of stochastically Δ-regular n-vertex models, which generate bipartite graphs with $|L| = |R| = n$ and average degree Δ, i.e.,

$$\mathop{\mathbf{E}}_{G \sim \mathbf{G}}[\deg(u)] = \sum_{e:\{u\} \times R} p_e = \Delta \quad \text{and} \quad \mathop{\mathbf{E}}_{G \sim \mathbf{G}}[\deg(v)] = \sum_{e:L \times \{v\}} p_e = \Delta$$

for every vertex $u \in L$ and $v \in R$, respectively.

Note that there might be many Δ-regular n-vertex models with fixed marginal probabilities. The most well studied case is the adaptation of the Erdős-Renyi model $\mathbf{G}_{\mathrm{ind}}$, where the events $e \in E(G)$ are mutually independent for $e \in L \times R$ with $p_e = p$ for all $e \in L \times R$. Here, we focus on models $\mathbf{G}_\pi$ where these events are pair-wise independent. Our aim is to prove bounds on the expected size $\mu(G)$ of the maximum matching of graph $G \sim \mathbf{G}_\pi$. It is well-known that for the model $\mathbf{G}_{\mathrm{ind}}$, the expected size of the maximum matching is $n - O(\exp(-\Delta))$ and, hence, perfect matchings exist with high probability when Δ becomes (super)logarithmic. Such results are not possible in the more general pair-wise independent case; still, the expected size of the maximum matching is quite large.

Theorem 5. *Let $\mathbf{G}_\pi$ be a stochastic Δ-regular n-vertex model with marginals $\{p_e\}_{e \in L \times R}$ such that the events $\{e \in E(G)\}_{e \in L \times R}$ for $G \sim \mathbf{G}_\pi$ are pair-wise independent. Then, the expected size of the maximum matching of a randomly generated graph $G \sim \mathbf{G}_\pi$ is at least $\mathbf{E}_{G \sim \mathbf{G}_\pi}[\mu(G)] \geq n - n/\sqrt{\Delta}$.*

Proof. The main idea of the proof is to look at the aggregate distribution of the vertex degrees of the whole graph G. On the one hand, the pairwise independence condition allows us to calculate precisely the expectation and variance of the degree of any particular vertex. On the other hand, the non-existence of a large matching $\mu(G)$ in a realized graph $G \sim \mathbf{G}_\pi$ implies a large deviation of degrees of many vertices from their mean Δ. This allows us to get the desired bound on the following random variable $f(G)$, where $G \sim \mathbf{G}_\pi$.

$$f(G) \stackrel{\text{def}}{=} \sum_{v \in L \cup R} (d_v - \Delta)^2, \quad \text{where } d_v \text{ is the degree of each vertex } v \text{ in } G.$$

In any graph $G \sim \mathbf{G}_\pi$, let $E_v \stackrel{\text{def}}{=} \{e \in E(G) | e \text{ is incident to } v\}$ for each $v \in L \cup R$. The variance of a vertex degree d_v is equal to

$$\mathop{\mathbf{E}}_{G \sim \mathbf{G}_\pi}\left[(d_v - \Delta)^2\right] = \mathop{\mathbf{Var}}_{G \sim \mathbf{G}_\pi}[d_v] = \mathop{\mathbf{Var}}_{G \sim \mathbf{G}_\pi}\left[\sum_{e \in E_v} \mathbb{1}\{e \in E(G)\}\right]$$

$$= \sum_{e \in E_v} \mathop{\mathbf{Var}}_{G \sim \mathbf{G}_\pi}[\mathbb{1}\{e \in E(G)\}] = \sum_{e \in E_v} p_e \cdot (1 - p_e)$$

$$= \sum_{e \in E_v} p_e - \sum_{e \in E_v} p_e^2 \leq \Delta,$$

where the first equality is due to the definition of variance and the fact that
$\mathbf{E}[d_v] = \Delta$, the third equality is due to the property of variance and the fact
that random variables $\mathbb{1}\{e \in E(G)\}$ are pairwise independent for all $e \in E_v$,
and the fourth equality follows since $e \in E(G)$ is a Bernoulli random variable.
Therefore,

$$
\mathop{\mathbf{E}}_{G \sim \mathbf{G}_\pi} [f(G)] = \mathop{\mathbf{E}}_{G \sim \mathbf{G}_\pi} \left[\sum_{v \in L \cup R} (d_v - \Delta)^2 \right]
$$
$$
= \sum_{v \in L \cup R} \mathop{\mathbf{E}}_{G \sim \mathbf{G}_\pi} [(d_v - \Delta)^2] \leq 2 \cdot n \cdot \Delta. \tag{8}
$$

We also observe that if a realized graph $G \sim \mathbf{G}_\pi$ has a small maximum matching
$\mu(G) < n$, then many vertex degrees in G must significantly deviate from Δ. To
this end, we first establish the following lemma (its proof is omitted).

Lemma 2. *Let d_v be the degree in G of each vertex $v \in L \cup R$. Then, $\forall \delta \geq 0$*

$$
\sum_{v \in L \cup R} (d_v - \delta)^2 \geq \frac{2\delta^2(n - \mu(G))^2}{n}.
$$

We can now combine Lemma 2 for $\delta = \Delta$ with (8) to get

$$
2n\Delta \geq \mathop{\mathbf{E}}_{G \sim \mathbf{G}_\pi} [f(G)] = \mathop{\mathbf{E}}_{G \sim \mathbf{G}_\pi} \left[\sum_{v \in L \cup R} (d_v - \Delta)^2 \right]
$$
$$
\geq \mathop{\mathbf{E}}_{G \sim \mathbf{G}_\pi} \left[\frac{2\Delta^2(n - \mu(G))^2}{n} \right] \geq \frac{2\Delta^2}{n} \mathop{\mathbf{E}}_{G \sim \mathbf{G}_\pi} [n - \mu(G)]^2
$$

Thus, $\mathop{\mathbf{E}}_{G \sim \mathbf{G}_\pi} [n - \mu(G)] \leq \frac{n}{\sqrt{\Delta}}$ and the theorem follows. $\qquad\square$

5.1 A Tight Upper Bound

We now show that our bound in Theorem 5 is tight for a wide range of values
of parameter Δ (compared to n). We do so using the following stochastic model
$\mathbf{G}_\pi$. In our construction we assume that $n - \Delta = \Omega(n)$ and $\Delta \geq 2$ is an integer[5].

1. With probability $1 - \alpha$ (where α is a parameter which we will specify later),
   we select uniformly at random a Δ-regular bipartite graph with $|L| = |R| = n$ vertices. Denote by $\mathbf{D}_{reg}$ the uniform probability distribution over these
   graphs.
2. With the remaining probability α, we select uniformly at random a subset
   $A \subset L$ of size $|A| = \frac{n}{2}\left(1 - \frac{1}{c\sqrt{\Delta}}\right)$, where c is the closest to 1 number such
   that $|A|$ is an integer. Note that c would get arbitrary close to 1 as n goes

---

[5] Our construction can be extended to cover the case of non-integer Δ with some
minor adjustments.

to infinity. Similarly, we select uniformly at random a subset $B \subset R$ of size $|B| = \frac{n}{2}\left(1 + \frac{1}{c\sqrt{\Delta}}\right) = n - |A|$. Next, we describe two distributions $\mathbf{D}(x_1)$ and $\mathbf{D}(x_2)$, each parametrized by a selection probability. In each distribution, we draw edges between the sets A and B and between the sets $L \backslash A$ and $R \backslash B$ i.i.d. with probability x_1 in $\mathbf{D}(x_1)$ and with probability x_2 in $\mathbf{D}(x_2)$. In particular:
(a) with probability 0.5, we generate a bipartite graph $G \sim \mathbf{D}(x_1)$;
(b) with probability 0.5, we generate a bipartite graph $G \sim \mathbf{D}(x_2)$.
We choose x_1 and x_2 so that the expected degree of the graph G drawn from the mixture of $\mathbf{D}(x_1)$ and $\mathbf{D}(x_2)$ is exactly Δ. In particular, we set $x_1 \stackrel{\text{def}}{=} x(1 - \delta)$ and $x_2 \stackrel{\text{def}}{=} x(1 + \delta)$, where $x \stackrel{\text{def}}{=} \frac{\Delta \cdot n}{2 \cdot |A| \cdot |B|}$ and $\delta^2 \stackrel{\text{def}}{=} \frac{1}{(n-1)c^2\Delta - n}$.

We choose probability

$$\alpha \stackrel{\text{def}}{=} \frac{(n - \Delta)\left((n-1)c^2\Delta - n\right)}{n(n-1)c^2\Delta - n^2 + \Delta n^2 - 2(n-1)c^2\Delta^2}. \tag{9}$$

Theorem 6. *The model* $\mathbf{G}_\pi$ *is pairwise independent over the set of edges, has probability* $p = \frac{\Delta}{n}$ *for every edge to be realised, and generates bipartite graphs with a maximum matching of expected size* $\mathbf{E}_{G \sim \mathbf{G}_\pi}\left[\mu(G)\right] \leq n\left(1 - \Omega\left(\frac{1}{\sqrt{\Delta}}\right)\right)$ *as long as* $n - \Delta = \Omega(n)$ *and* $\Delta \geq 2$.

The formal proof of Theorem 6 is omitted due to lack of space.

Acknowledgements. We thank Yuan Zhou for helpful discussions on the expected size of maximum matchings in random bipartite graphs. This work was partially supported by Science and Technology Innovation 2030 - "New Generation of Artificial Intelligence" Major Project No. 2018AAA0100903; COST Action 16228 "European Network for Game Theory"; Innovation Program of Shanghai Municipal Education Commission; Program for Innovative Research Team of Shanghai University of Finance and Economics (IRTSHUFE) and the Fundamental Research Funds for the Central Universities; National Natural Science Foundation of China (Grant No. 61806121); Beijing Outstanding Young Scientist Program (No. BJJWZYJH012019100020098) Intelligent Social Governance Platform; Major Innovation & Planning Interdisciplinary Platform for the "Double-First Class" Initiative, Renmin University of China.

References

1. Babaioff, M., Feldman, M., Gonczarowski, Y.A., Lucier, B., Talgam-Cohen, I.: Escaping cannibalization? Correlation-robust pricing for a unit-demand buyer. In: Proceedings of the 21st ACM Conference on Economics and Computation (EC), p. 191 (2020)
2. Bei, X., Gravin, N., Lu, P., Tang, Z.G.: Correlation-robust analysis of single item auction. In: Proceedings of the 30th Annual ACM-SIAM Symposium on Discrete Algorithms (SODA), pp. 193–208 (2019)
3. Bollobás, B.: Random Graphs, 2nd edn. Cambridge University Press, Cambridge (2001)

 4. Cai, Y., Oikonomou, A.: On simple mechanisms for dependent items. In: Proceedings of the 22nd ACM Conference on Economics and Computation (EC), pp. 242–262 (2021)
 5. Carroll, G.: Robustness and separation in multidimensional screening. Econometrica **85**(2), 453–488 (2017)
 6. Chawla, S., Hartline, J.D., Malec, D.L., Sivan, B.: Multi-parameter mechanism design and sequential posted pricing. In: Proceedings of the 42nd Annual ACM Symposium on Theory of Computing (STOC), pp. 311–320 (2010)
 7. Chawla, S., Sivan, B.: Bayesian algorithmic mechanism design. SIGecom Exchanges **13**(1), 5–49 (2014)
 8. Correa, J.R., Dütting, P., Fischer, F.A., Schewior, K., Ziliotto, B.: Unknown I.I.D. prophets: better bounds, streaming algorithms, and a new impossibility (extended abstract). In: Proceedings of the 12th Innovations in Theoretical Computer Science Conference (ITCS), pp. 86:1–86:1 (2021)
 9. Gravin, N., Lu, P.: Separation in correlation-robust monopolist problem with budget. In: Proceedings of the 29th Annual ACM-SIAM Symposium on Discrete Algorithms, (SODA), pp. 2069–2080 (2018)
10. Hartline, J.D.: Mechanism Design and Approximation. Graduate Textbook (in preparation) (2020)
11. Hill, T., Kertz, R.: A survey of prophet inequalities in optimal stopping theory. Contemp. Math. **125**, 191–207 (1992)
12. Immorlica, N., Singla, S., Waggoner, B.: Prophet inequalities with linear correlations and augmentations. In: Proceedings of the 21st ACM Conference on Economics and Computation (EC), pp. 159–185 (2020)
13. Motwani, R., Raghavan, P.: Randomized Algorithms. Cambridge University Press, Cambridge (1995)
14. Roughgarden, T.: Twenty Lectures on Algorithmic Game Theory. Cambridge University Press, Cambridge (2016)
15. Samuel-Cahn, E.: Comparison of threshold stop rules and maximum for independent nonnegative random variables. Ann. Probab. **12**, 1213–1216 (1984)

Allocating Indivisible Goods to Strategic Agents: Pure Nash Equilibria and Fairness

Georgios Amanatidis[1]([✉]), Georgios Birmpas[2], Federico Fusco[2], Philip Lazos[2,3], Stefano Leonardi[2], and Rebecca Reiffenhäuser[2]

[1] Department of Mathematical Sciences, University of Essex, Colchester, UK
`georgios.amanatidis@essex.ac.uk`
[2] Department of Computer, Control, and Management Engineering "Antonio Ruberti", Sapienza University of Rome, Rome, Italy
`{birbas,fuscof,lazos,leonardi,rebeccar}@diag.uniroma1.it`
[3] IOHK, London, UK
`philip.lazos@iohk.io`

Abstract. We consider the problem of fairly allocating a set of indivisible goods to a set of *strategic* agents with additive valuation functions. We assume no monetary transfers and, therefore, a *mechanism* in our setting is an algorithm that takes as input the reported—rather than the true—values of the agents. Our main goal is to explore whether there exist mechanisms that have pure Nash equilibria for every instance and, at the same time, provide fairness guarantees for the allocations that correspond to these equilibria. We focus on two relaxations of envy-freeness, namely *envy-freeness up to one good* (EF1), and *envy-freeness up to any good* (EFX), and we positively answer the above question. In particular, we study two algorithms that are known to produce such allocations in the non-strategic setting: Round-Robin (EF1 allocations for any number of agents) and a cut and choose algorithm of Plaut and Roughgarden [35] (EFX allocations for two agents). For Round-Robin we show that all of its pure Nash equilibria induce allocations that are EF1 with respect to the underlying true values, while for the algorithm of Plaut and Roughgarden we show that the corresponding allocations not only are EFX but also satisfy *maximin share fairness*, something that is not true for this algorithm in the non-strategic setting! Further, we show that a weaker version of the latter result holds for any mechanism for two agents that always has pure Nash equilibria which all induce EFX allocations.

1 Introduction

Fair division refers to the problem of distributing a set of resources among a set of agents in such a way that everyone is "happy" with the overall allocation.

This work was supported by the ERC Advanced Grant 788893 AMDROMA "Algorithmic and Mechanism Design Research in Online Markets", the MIUR PRIN project ALGADIMAR "Algorithms, Games, and Digital Markets", and the NWO Veni project No. VI.Veni.192.153.

M. Feldman et al. (Eds.): WINE 2021, LNCS 13112, pp. 149–166, 2022.
https://doi.org/10.1007/978-3-030-94676-0_9

Capturing this "happiness" can be elusive, as it may be determined by complicated underlying social dynamics; however, two well-motivated (and mathematically conducive) interpretations are those of *envy-freeness* [22,23,38] and *proportionality* [37]. When an allocation is envy-free, each agent values the set of resources that she receives at least as much as the set of any other agent, while when an allocation is proportional, each agent receives at least $1/n$ of her total value for all the goods, assuming there are n agents. Since the first mathematically formal treatment of fair division by Banach, Knaster and Steinhaus [37], the multifaceted questions that arise for the different variants of the problem have been studied in a diverse group of fields, including mathematics, economics, and political science. As many of these questions are inherently algorithmic, fair division questions, especially the ones related to the existence, computation, and approximation of different fairness notions have been very actively studied during the last two decades (see, e.g., [11,32,36] for surveys of recent results).

In the standard discrete fair division setting we study here, the resources are indivisible goods and the agents have additive valuation functions over them. Typically, there is the additional assumption that all the goods need to be allocated. This discrete setting poses a significant conceptual challenge, as the classic notions of fairness originally introduced for *divisible* goods, such as envy-freeness and proportionality, are impossible to satisfy. The example that illustrates this situation needs only two agents and just one positively valued good. Whoever does not receive the good will not consider the result to be either envy-free or proportional. However, this should not necessarily be considered an *unfair* outcome, as it is done out of necessity, not malice: the only other (deterministic) option would be to deprive both agents of the good. To define what is *fair* in this context, a number of weaker fairness notions have been proposed. Among the most prevalent of those are *envy-freeness up to one good* (EF1), *envy-freeness up to any good* (EFX), and *maximin share fairness* (MMS). The notions of EF1 and EFX were introduced by Lipton et al. [31], Budish [15], and Gouvrès et al. [27], Caragiannis et al. [17] respectively, and they can be seen as additive relaxations of envy-freeness. Both of them are based on the following rationale: an agent may envy another agent but only by the value of the most (for EF1), or the least (for EFX) desirable good in the other agent's bundle. It is straightforward that EF1 is weaker than EFX, and indeed this is reflected to the known results for the two notions. The concept of the *maximin share* of an agent was introduced by Budish [15] as a relaxation to the proportionality benchmark. The corresponding fairness notion, *maximin share fairness*, requires that each agent receives the maximum value that this agent would obtain if she was allowed to partition the goods into n bundles and then take the worst of them (see Definitions 3 and 4).

From an algorithmic point of view, there are many results regarding the existence and the computation of these notions (see Related Work). Here, however, we are interested in exploring the problem from a game theoretic perspective. We assume that the agents are *strategic*, which means that it is possible for an agent to intentionally misreport her values for (some of) the goods in order to end up with a bundle of higher total value. We see this as a very natural direction, as it captures what may happen in practice in real-life scenarios where fair division solutions can be applied, e.g., in a divorce settlement. It should be noted here that, in accordance to the existing literature on truthful allocation mechanisms [1,3,16,21,28,33,34], we assume there are *no monetary transfers*. Therefore, a *mechanism* in our setting is just an algorithm that takes as input the, possibly misreported, values that the agents declare. The existence of *truthful* mechanisms, i.e., mechanisms where no agent has an incentive to lie, was studied in the same setting by Amanatidis et al. [1] who showed that, even for two agents, truthfulness and fairness are incompatible by providing impossibility results for several fairness notions. So, the next natural question to ask is:

Are there non-truthful mechanisms whose equilibria define fair *allocations?*

Our main quest is to investigate whether there exist mechanisms that have *pure Nash equilibria* for every instance and each allocation corresponding to an equilibrium provides fairness guarantees with respect to the *true* valuation functions of the agents. The stability notion of a pure Nash equilibrium describes a state where each agent plays a deterministic strategy (namely, reports her value for each good) and no agent can gain more value by deviating to a different strategy.

Our Contribution. To the best of our knowledge, our work is the first to consider the above question. Our results are mostly positive, as we show that the class of mechanisms that are implementable in polynomial time, have pure Nash equilibria for every instance, and provide some fairness guarantee at these equilibria is non-empty. Specifically, in Sect. 3, we study a mechanism adaptation of the Round-Robin algorithm which is known to produce EF1 allocations in the non-strategic setting [17] and, under some mild assumptions which we show that can be lifted, always has pure Nash equilibria [7]. Further, in Sect. 4, we consider the stronger notion of EFX. We focus on the case of two agents and study a mechanism adaptation of the algorithm of Plaut and Roughgarden [35], Mod-Cut&Choose, which is known to always produce EFX allocations in the non-strategic setting. Our main results can be summarized as follows:

- Round-Robin always has pure Nash equilibria (see the full version [2]) and these induce allocations that are EF1 with respect to the underlying true values (Theorem 3). That is, Round-Robin retains its fairness properties at its equilibria, even when the input is given by strategic agents! To show this, we rely on a novel recursive construction of "nicely structured" bid profiles.
- For two agents, Mod-Cut&Choose always has pure Nash equilibria and these induce allocations that are EFX and MMS with respect to the underlying true values (Theorem 5). It should be noted that in the non-strategic setting the allocations returned by Mod-Cut&Choose are not necessarily MMS!
- We generalize a weaker version of the latter. All mechanisms that have pure Nash equilibria for every instance with two agents and these equilibria induce allocations that are always EFX provide stronger MMS guarantees in these allocations than generic EFX allocations do (Theorems 6 and 7).

Related Work. The non-strategic version of the problem of fairly allocating goods to additive agents has been studied extensively. We provide a summary of indicative results mostly for the notions that we consider. In particular, EF1 allocations always exist and can be computed in polynomial time [17,31,32]. For the stronger notion of EFX, the picture is not that clear. It is known that such allocations always exist for 2 or 3 agents [17,18,27], and in the former case they can be efficiently computed using Mod-Cut&Choose [35]. The existence of complete EFX allocations for 4 or more agents remains one of the most intriguing open problems in fair division. Finally, MMS allocations always exist for only 2 agents, although computing them is NP-hard [39], but for 3 or more agents existence is not guaranteed [30]. However, there are algorithms that run in polynomial time and produce constant factor approximation guarantees [5,9,24,26,30], with $\frac{3}{4} + \frac{1}{12n}$ being the current state of the art [25].

The works of Caragiannis et al. [16] and Amanatidis et al. [1,3] are very relevant to ours in the sense that they all studied the exact same strategic discrete fair division setting. As we mentioned earlier, though, their focus was different as they were only interested in truthful mechanisms. Amanatidis et al. [1] provided strong impossibility results in this direction: for instances with two agents, no truthful mechanism can consistently produce EF1 (and thus EFX) allocations when there are more than 4 goods, while the best possible approximation with respect to MMS declines linearly with the number of goods.

Aziz et al. [7] studied the existence of pure Nash equilibria of Round-Robin and showed that when no agent values any two goods equally, there always exists a pure Nash equilibrium. In addition, they provided a linear time algorithm that computes the preference rankings (i.e., the orderings of the goods that correspond to the reported values) that leads to this equilibrium, thus giving a constructive solution. Aziz et al. [6] showed that computing best responses for Round-Robin, and for *sequential mechanisms* more generally, is NP-hard, fixing an error in the work of Bouveret and Lang [12] on the same topic.

We conclude by pointing out that in contrast to the case of indivisible goods, the problem of fairly allocating a set divisible goods to a set of strategic agents has been repeatedly studied. For some indicative papers in this line of work, we refer the reader to [10,13,14,19,20].

2   Preliminaries

We consider the problem of allocating a set of indivisible goods to a set of agents in a fair manner under the presence of incentives. For $a \in \mathbb{N}$ we use $[a]$ to denote the set $\{1, 2, \ldots, a\}$. An instance to our problem is an ordered triple $(N, M, \mathbf{v})$, where $N = [n]$ is a set of n agents, $M = \{g_1, \ldots, g_m\}$ is a set of m goods, and $\mathbf{v} = (v_1, \ldots, v_n)$ is a vector of the agents' additive valuation functions. In particular, each agent i has a non-negative value $v_i(\{g\})$ (or simply $v_i(g)$) for each good $g \in M$, and for every $S, T \subseteq M$ with $S \cap T = \emptyset$ we have $v_i(S \cup T) = v_i(S) + v_i(T)$. Equivalently, the value of an agent is simply the sum of the values of the goods that she got. We assume there is no free disposal, which means that all the goods must be allocated. Thus, an allocation $(A_1, \ldots, A_n)$, where A_i is the *bundle* of agent i, is a partition of M. It is often useful to refer to the order of preference an agent has over the goods. We say that a valuation function v_i *induces a preference ranking* $\succeq_i$ if $g \succeq_i g' \Leftrightarrow v_i(g) \geq v_i(g')$ for all $g, g' \in M$. We use $\succ_i$ if the corresponding preference ranking is *strict*, i.e., when $g \succeq_i g' \wedge g' \succeq_i g \Rightarrow g = g'$, for all $g, g' \in M$.

2.1 Fairness Notions

There is a significant number of different notions one can use to determine which allocations are "fair". The most prominent such notions are *envy-freeness* (EF) [22,23,38] and *proportionality* (PROP) [37], and, in the discrete setting we study here, their relaxations, namely *envy-freeness up to one good* (EF1) [15], *envy-freeness up to any good* (EFX) [17], and *maximin share fairness* (MMS) [15]. Particularly for additive valuation functions, we have that EF $\Rightarrow$ EFX $\Rightarrow$ EF1 and EF $\Rightarrow$ PROP $\Rightarrow$ MMS, where $X \Rightarrow Y$ means that any allocation that satisfies fairness criterion X always satisfies fairness criterion Y as well.

Definition 1. *An allocation $(A_1, \ldots, A_n)$ is*

- *envy-free (EF), if for every $i, j \in N$, $v_i(A_i) \geq v_i(A_j)$.*
- *envy-free up to one good (EF1), if for every pair of agents $i, j \in N$, with $A_j \neq \emptyset$, there exists a good $g \in A_j$, such that $v_i(A_i) \geq v_i(A_j \setminus \{g\})$.*
- *envy-free up to any good (EFX), if for every pair $i, j \in N$, with $A_j \neq \emptyset$ and every good $g \in A_j$ with $v_i(g) > 0$, it holds that $v_i(A_i) \geq v_i(A_j \setminus \{g\})$.*

While these notions rely on comparisons among the agents, proportionality focuses on everyone receiving at least a $1/n$ fraction of the total value.

Definition 2. *An allocation $(A_1, \ldots, A_n)$ is proportional (PROP), if for every $i \in N$, $v_i(A_i) \geq v_i(M)/n$.*

In the same direction, but adjusted for indivisible goods, a number of fairness notions have been based on the notion of *maximin shares* [15]. Imagine that agent i is asked to partition the goods into n bundles, under the condition that she will receive the worst bundle among those. If the resources were divisible, then she would clearly split everything evenly into n bundles of value $v_i(M)/n$ each, thus capturing the benchmark required for proportionality. However, now that the goods are indivisible, agent i would like to create a partition maximizing the minimum value of a bundle. This value is her maximin share.

Definition 3. *Given a subset $S \subseteq M$ of goods, the n-maximin share of agent i with respect to S is $\mu_i(n, S) = \max_{\mathcal{A} \in \Pi_n(S)} \min_{A_j \in \mathcal{A}} v_i(A_j)$, where $\Pi_n(S)$ is the set of all partitions of S into n bundles.*

From the definition and the preceding discussion, we have that $n \cdot \mu_i(n, S) \leq v_i(S)$. When $S = M$, we call $\mu_i(n, M)$ the *maximin share* of agent i and denote it by μ_i as long as it is clear what n and M are.

Definition 4. *An allocation* $\mathcal{A} = (A_1, \ldots, A_n)$ *is called an* α-*maximin share fair* (α-*MMS*) *allocation if* $v_i(A_i) \geq \alpha \cdot \mu_i$, *for every* $i \in N$. *When* $\alpha = 1$ *we just say that* $\mathcal{A}$ *is an MMS allocation.*

Besides MMS, there exist other fairness criteria based on maximin shares, like *pairwise maximin share fairness* (PMMS) [17] and *groupwise maximin share fairness* (GMMS) [8]. While we are not going into more details about them, we note that PMMS $\Rightarrow$ EFX [17] and that for $n = 2$, MMS, PMMS, and GMMS coincide. In particular, we need the following result of Caragiannis et al. [17].

Theorem 1 (Follows from Theorem 4.6 of [17]). *For* $n = 2$, *any MMS allocation is also an EFX allocation.*

One can also consider how the approximate versions of EF1, EFX and MMS relate to each other (see [4]). Here we need the following result about the worst case MMS guarantee of an EFX allocation for the case of two agents.

Theorem 2 (Follows from Proposition 3.3 of [4]). *For* $n = 2$, *any EFX allocation is also a* $\frac{2}{3}$-*MMS allocation. This is tight, i.e., for every* $\delta > 0$ *there exists an EFX allocation that is not a* $\left(\frac{2}{3} + \delta\right)$-*MMS allocation, for any* $m \geq 4$.

2.2 Mechanisms and Equilibria

We are interested in *mechanisms* that produce allocations with fairness guarantees. In our setting there are *no payments*, so an allocation mechanism $\mathcal{M}$ is an algorithm that takes its input from the agents and allocates all the goods. We use this distinction in terminology to highlight that this reported input may differ from the actual valuation functions. In particular, we assume that each agent i reports a *bid vector* $\boldsymbol{b}_i = (b_{i1}, \ldots, b_{im})$, where $b_{ij} \geq 0$ is the value agent i claims to have for good $g_j \in M$. A mechanism $\mathcal{M}$ takes as input a *bid profile* $\mathbf{b} = (\boldsymbol{b}_1, \ldots, \boldsymbol{b}_n)$ of bid vectors and outputs an allocation $\mathcal{M}(\mathbf{b})$. In our setting we assume that the agents are *strategic*, i.e., an agent may misreport her true values if this results to a better allocation from her point of view. Hence, in general, $\boldsymbol{b}_i \neq (v_i(g_1), \ldots, v_i(g_m))$. While $\boldsymbol{b}_i$ is defined as a vector, for a generic good $h \in M$ it is often convenient to use the function notation $\boldsymbol{b}_i(h)$ to denote the bid value $b_{i\ell}$, where ℓ is such that $h = g_\ell$; extending this we may write $\boldsymbol{b}_i(S)$ for $\sum_{h \in S} \boldsymbol{b}_i(h)$. We say that a bid vector $\boldsymbol{b}_i$ induces a preference ranking $\succeq_i$ if $g \succeq_i g' \Leftrightarrow \boldsymbol{b}_i(g) \geq \boldsymbol{b}_i(g')$ for all $g, g' \in M$, and use $\succ_i$ for strict rankings.

We focus on the fairness guarantees of the pure equilibria of the mechanisms we study. Given a profile $\mathbf{b} = (\boldsymbol{b}_1, \ldots, \boldsymbol{b}_n)$, we write $\mathbf{b}_{-i}$ to denote $(\boldsymbol{b}_1, \ldots, \boldsymbol{b}_{i-1}, \boldsymbol{b}_{i+1}, \ldots, \boldsymbol{b}_n)$ and, given a bid vector $\boldsymbol{b}_i'$, we use $(\boldsymbol{b}_i', \mathbf{b}_{-i})$ to denote the profile $(\boldsymbol{b}_1, \ldots, \boldsymbol{b}_{i-1}, \boldsymbol{b}_i', \boldsymbol{b}_{i+1}, \ldots, \boldsymbol{b}_n)$. For the next definition we abuse the notation slightly: given an allocation $\mathcal{A} = (A_1, \ldots, A_n)$, we write $v_i(\mathcal{A})$ instead of $v_i(A_i)$.

Definition 5. *Let* $\mathcal{M}$ *be an allocation mechanism and consider a profile* $\mathbf{b} = (\boldsymbol{b}_1, \ldots, \boldsymbol{b}_n)$. *We say that* $\boldsymbol{b}_i$ *is a* best response *to* $\mathbf{b}_{-i}$ *if for every* $\boldsymbol{b}_i' \in \mathbb{R}^m_{\geq 0}$, *we have* $v_i(\mathcal{M}(\boldsymbol{b}_i', \mathbf{b}_{-i})) \leq v_i(\mathcal{M}(\mathbf{b}))$. *The profile* $\mathbf{b}$ *is a* pure Nash equilibrium *(PNE) if, for each* $i \in N$, $\boldsymbol{b}_i$ *is a best response to* $\mathbf{b}_{-i}$.

When **b** is a PNE and $\mathcal{M}(\mathbf{b})$ has a fairness guarantee, e.g., $\mathcal{M}(\mathbf{b})$ is EF1, we attribute the same guarantee to the profile itself, i.e., we say that **b** is EF1.

Remark 1. The mechanisms we consider run in polynomial time. However there are computational complexity questions that go beyond the mechanisms themselves. For instance, how does an agent compute a best response or how do all the agents reach an equilibrium? We do not study such questions here and we only focus on the fairness properties of PNE. It should be noted, however, that such problems are typically hard. For instance, computing a best response for Round-Robin is NP-hard in general [6] (although for fixed n it is not [40]), and we show that the same is true for Mod-Cut&Choose (Proposition 1).

Remark 2. An easy observation on the main question of this work is that *any* PNE of *any* α-approximation mechanism for computing MMS allocations is an α-MMS allocation. Indeed, this is true, not only for MMS but for any fairness notion that depends on agents achieving specific value benchmarks that depend on their own valuation function, e.g., it is also true for PROP. While this is definitely interesting to note, nothing is known on the existence of PNE of any constant factor approximation algorithm for computing MMS allocations in the literature. Clearly, an existence result for any such algorithm [5,9,24–26,30] would imply an analogue of Theorem 3 for approximate MMS.

3   Fairness of Nash Equilibria of Round-Robin

In this section we focus on one of the simplest and most well-studied allocation algorithms, Round-Robin, a draft algorithm where the agents take turns and in each turn the active agent receives her most preferred available (i.e., unallocated) good. Below we state Round-Robin as a mechanism (Mechanism 1) that takes as input a bid profile rather than the valuation functions of the agents. In its full generality, Round-Robin should also take a permutation N as an input to determine the priority of the agents. Here, for the sake of presentation, we assume that the agents in each *round* (lines 3–6) are always considered according to their "name", i.e., agent 1 is considered first, agent 2 s, and so on. This is without loss of generality, as it only requires renaming the agents accordingly.

While it is long known that truth-telling is generally not a PNE in sequential allocation mechanisms (a special case of which is Round-Robin) [29], we present here a minimal example that illustrates the mechanics of manipulation. Let $N = \{1, 2\}$ and $M = \{a, b, c\}$ with the valuation functions being as shown in the table on the left. The circles show the allocation returned by Round-Robin when the agents bid their true values, while the superscripts indicate in which order were the goods assigned. Given that agent 2 is not particularly interested to good a, agent 1 can manipulate the mechanism into giving her $\{a, b\}$ instead $\{a, c\}$ by claiming that these are her top goods as in the table on the right.

$$
\begin{array}{c c c c}
 & a & b & c \\
v_1 : & \textcircled{6}^1 & 5 & \textcircled{4}^3 \\
v_2 : & 4 & \textcircled{6}^2 & 5
\end{array}
\qquad\qquad
\begin{array}{c c c c}
 & a & b & c \\
b_1 : & \textcircled{5}^3 & \textcircled{6}^1 & 4 \\
v_2 : & 4 & 6 & \textcircled{5}^2
\end{array}
$$

Mechanism 1. Round-Robin($b_1, \ldots, b_n$)

1: $S = M$; $(A_1, \ldots, A_n) = (\emptyset, \ldots, \emptyset)$; $k = \lceil m/n \rceil$
2: **for** $r = 1, \ldots, k$ **do**   // Each value of r determines the corresponding *round*.
3:      **for** $i = 1, \ldots, n$ **do**
4:          $g = \arg\max_{h \in S} b_i(h)$ // Break ties lexicographically (hence we use "=").
5:          $A_i = A_i \cup \{g\}$ // Current agent receives her "favorite" available good.
6:          $S = S \setminus \{g\}$ // The good is no longer available.
7: **return** $\mathcal{A} = (A_1, \ldots, A_n)$

Thus, bidding according to v_1, v_2 is not a PNE. The example is minimal, in the sense that with just 1 agent or less than 3 goods truth-telling is a PNE of Round-Robin almost trivially.

Before moving to the main technical part of this section, we discuss some assumptions that again are without loss of generality. As we have mentioned in the Introduction, it is known that, as an algorithm, Round-Robin outputs EF1 allocations when all agents have additive valuation functions [17,32]. Also Round-Robin as a mechanism is known to have PNE for any instance where *no agent values two goods exactly the same*, and at least some such equilibria (namely, the ones consistent with the so-called *bluff profile*) are easy to compute [7]. From a technical point of view, this assumption that all the valuation functions induce strict preference rankings is convenient, as it greatly reduces the number of corner cases one has to deal with. However, as we show in the full version of this paper [2], this assumption is benign and the result of Aziz et al. [7] extends to general additive valuation functions. On a different but related note, we assume, for the remainder of this section, that all the bid vectors induce strict preference rankings (but not necessarily consistent with the preference rankings induced by the corresponding valuation functions). This is without loss of generality, because even if a bid vector contains some bids that are equal to each other, a strict preference ranking is imposed by the lexicographic tie-breaking of the mechanism itself. So, formally, when we abuse the notation and write $g \succ_i h$ we mean that either $b_i(g) > b_i(h)$, or $b_i(g) = b_i(h)$ and g has a lower index than h in the standard naming of goods as $g_1, g_2, \ldots, g_m$.

We next state the main result of our work. Despite its proof being rather involved, the intuition behind it is simple. On one hand, whenever an agent bids truthfully, she sees the resulting allocation as being EF1. On the other hand, no matter what an agent bids, we show it is possible to "replace" her with an imaginary version of herself who does not affect the allocation, and not only bids truthfully, but she considers the bundles of the allocation to be as valuable as the original agent thought they were. The rather elaborate formal argument relies on the recursive construction of auxiliary valuation functions and bids, and on the fact that small changes in a single preference ranking minimally change the "history" of available goods during the execution of the mechanism.

Theorem 3. *For any fair division instance $\mathcal{I} = (N, M, \mathbf{v})$, every PNE of the Round-Robin mechanism is EF1 with respect to the valuation functions $v_1, \ldots, v_n$.*

As we will see shortly, proving Theorem 3 reduces to showing that the agent who "picks first" in the Round-Robin mechanism views the final allocation as envy-free, as long as she bids a best response to other agents' bids. While Theorem 4 sounds very much like the standard statement about the first agent in the algorithmic setting, its proof relies on a technical lemma that carefully builds a "nice" instance which is equivalent, in some sense, to the original. Recall that we have assumed that the agents' priority is indicated by their indices.

Theorem 4. *For any fair division instance $\mathcal{I} = (N, M, \mathbf{v})$, if the reported bid vector $\boldsymbol{b}_1$ of agent 1 is a best response to the (fixed) bid vectors $\boldsymbol{b}_2, \ldots, \boldsymbol{b}_n$ of all other players, then agent 1 does not envy (with respect to v_1) any bundle in the allocation outputted by Round-Robin$(\boldsymbol{b}_1, \ldots, \boldsymbol{b}_n)$.*

Note that since we are interested in PNE, it is always the case that each agent's bid is a best response to other agents' bids. As mentioned above, Theorem 4 is essentially a corollary to Lemma 1. The lemma shows the existence of an alternative version of agent 1 who is truthful, her presence does not affect the original allocation, and, as long as the allocation is the same, she shares the same values with the original agent 1. While its proof is rather involved, the high level idea is that we recursively construct a sequence of bids and valuation functions, each pair of which preserves the original allocation and the view of agent 1 for it, while being closer to being truthful. To achieve this we occasionally move value between the goods originally allocated to agent 1 and update the bid accordingly.

Lemma 1. *Suppose that the valuation function v_1 induces a strict preference ranking on the goods, i.e., for any $g, h \in M$, $v_1(g) = v_1(h) \Rightarrow g = h$. Let $\mathbf{b} = (\boldsymbol{b}_1, \boldsymbol{b}_2, \ldots, \boldsymbol{b}_n)$ be such that $\boldsymbol{b}_1$ is a best response of agent 1 to $\mathbf{b}_{-i} = (\boldsymbol{b}_2, \ldots, \boldsymbol{b}_n)$. Then there exists a valuation function v_1^* with the following properties:*

- *If $\boldsymbol{b}_1^* = (v_1^*(g_1), \ldots, v_1^*(g_m))$, i.e., $\boldsymbol{b}_1^*$ is the truthful bid for v_1^*, then Round-Robin$(\mathbf{b})$ and Round-Robin$(\boldsymbol{b}_1^*, \mathbf{b}_{-1})$ return the same allocation $(A_1, \ldots, A_n)$.*
- *$v_1^*(A_1) = v_1(A_1)$.*
- *For every good $g \subseteq M \setminus A_1$, it holds that $v_1^*(g) = v_1(g)$.*

Due to space constraints, we defer the proof of the lemma to the full version of our paper [2] and move on to the proofs of Theorems 3 and 4.

Proof of Theorem 4. Consider an arbitrary instance $\mathcal{I} = (N, M, \mathbf{v})$ and assume that the input of Round-Robin is $\mathbf{b} = (\boldsymbol{b}_1, \boldsymbol{b}_2, \ldots, \boldsymbol{b}_n)$, where $\boldsymbol{b}_1$ is a best response of agent 1 to $\mathbf{b}_{-i} = (\boldsymbol{b}_2, \ldots, \boldsymbol{b}_n)$ according to her valuation function v_1. Let $(A_1, \ldots, A_n)$ be the output of Round-Robin$(\mathbf{b})$. In order to apply Lemma 1, we need v_1 to induce a strict preference ranking over the goods. For the sake of presentation, we assume here that this is indeed the case, and we treat the general case formally in the full version [2]. So, we now consider the hypothetical scenario implied by Lemma 1 in this case: keeping agents 2 through n fixed, suppose that the valuation function of agent 1 is the function v_1^* given by the lemma, and her bid $\boldsymbol{b}_1^*$ is the truthful bid for v_1^*. The first part of Lemma 1 guarantees that the output of Round-Robin$(\boldsymbol{b}_1^*, \mathbf{b}_{-i})$ remains $(A_1, \ldots, A_n)$.

It is known that, no matter what others bid, if the agent with the highest priority (here agent 1 with v_1^*) reports her true values to Round-Robin, the resulting allocation is EF from her perspective (see, e.g., the proof of Theorem 12.2 in [32]). In our hypothetical scenario this is the case for agent 1 and it translates into having $v_1^*(A_1) \geq v_1^*(A_i)$ for all $i \in N$. Then the second and third parts of the lemma imply that $v_1(A_1) \geq v_1(A_i)$ for all $i \in N$, i.e., agent 1 does not envy any bundle in the original instance.   □

Given Theorem 4, the proof of Theorem 3 is of similar flavor to the proof on Round-Robin producing EF1 allocations in the non-strategic setting [32].

Proof of Theorem 3. Let $\mathbf{b} = (\boldsymbol{b}_1, \ldots, \boldsymbol{b}_n)$ be a PNE of the Round-Robin mechanism for the instance $\mathcal{I}$. By Theorem 4, it is clear that the allocation returned by Round-Robin($\mathbf{b}$) is EF, and hence EF1, from the point of view of agent 1. We fix an agent ℓ, where $\ell \geq 2$. For $i \in [\ell-1]$, let h_i be the good that agent i *claims to be* her favourite among the goods that are available when it is her turn in the first round, i.e., $h_i = \arg \max_{h \in M \setminus \{h_1 \ldots, h_{i-1}\}} b_i(h)$. Right before agent ℓ is first assigned a good, all goods in $H = \{h_1, \ldots, h_{\ell-1}\}$ have already been allocated. We consider the instance $\mathcal{I}' = (N', M', \mathbf{v}')$ in which all goods in H are missing. That is, $N' = N$, $M' = M \setminus H$, and $\mathbf{v}' = (v_1', \ldots, v_n')$ where $v_i' = v_i|_{M'}$, for $i \in [n]$, is the restriction of the function v_i on M'. Similarly define $\boldsymbol{b}_i' = \boldsymbol{b}_i|_{M'}$, for $i \in [n]$, the restrictions of the bids to the available goods, and $\mathbf{b}' = (\boldsymbol{b}_1', \ldots, \boldsymbol{b}_n')$. Finally, we consider the version of Round-Robin, call it Round-Robin$_\ell$, that starts with agent ℓ and then follows the indices in increasing order.

We claim that for Round-Robin$_\ell$ the bid $\boldsymbol{b}_\ell'$ is a best response for agent ℓ assuming that the restricted bid vectors of all the other agents are fixed. To see this, notice that for any $\boldsymbol{c}_\ell = (c_{\ell 1}, c_{\ell 2}, \ldots, c_{\ell m})$, the bundles given to agent ℓ by Round-Robin($\boldsymbol{c}_\ell, \mathbf{b}_{-\ell}$) and Round-Robin$_\ell(\boldsymbol{c}_\ell|_{M'}, \mathbf{b}_{-\ell}')$ are the same! In fact, the execution of Round-Robin$_\ell(\boldsymbol{c}_\ell|_{M'}, \mathbf{b}_{-\ell}')$ is identical to the execution of Round-Robin($\boldsymbol{c}_\ell, \mathbf{b}_{-\ell}$) from its ℓth step onward. So, if $\boldsymbol{b}_\ell'$ was not a best response in the restricted instance, then there would be a profitable deviation for agent ℓ, say $\boldsymbol{b}_\ell^*$, so that ℓ would prefer her bundle in Round-Robin$_\ell(\boldsymbol{b}_\ell^*, \mathbf{b}_{-\ell}')$ to her bundle in Round-Robin$_\ell(\mathbf{b}')$. This would imply that any extension of $\boldsymbol{b}_\ell^*$ to a bid vector for all goods in M (by arbitrarily assigning numbers to goods in H) would be a profitable deviation for agent ℓ in the profile $\mathbf{b}$ for Round-Robin, contradicting the fact that $\mathbf{b}$ is a PNE.

Now we may apply Theorem 4 for Round-Robin$_\ell$ (where agent ℓ plays the role of agent 1 of the theorem's statement) for instance $\mathcal{I}'$ and bid profile $\mathbf{b}'$. The theorem implies that agent ℓ does not envy any bundle in the allocation $(A_1, \ldots, A_n)$ outputted by Round-Robin$_\ell(\mathbf{b}')$, i.e., $v_\ell'(A_\ell) \geq v_\ell'(A_i)$, for all $i \in [n]$. Using the observation made above about the execution of Round-Robin$_\ell(\mathbf{b}')$ being identical to the execution of Round-Robin($\mathbf{b}$) after $\ell - 1$ goods have been allocated, we have that Round-Robin($\mathbf{b}$) returns the allocation $(A_1 \cup \{h_1\}, \ldots, A_{\ell-1} \cup \{h_{\ell-1}\}, A_\ell, \ldots, A_n)$. So, for any $i < \ell$ we have $v_\ell(A_\ell) = v_\ell'(A_\ell) \geq v_\ell'(A_i) = v_\ell(A_i) = v_\ell(A_i \cup \{h_i\}) - v_\ell(h_i)$, while for $i > \ell$ we simply have $v_\ell(A_\ell) = v_\ell'(A_\ell) \geq v_\ell'(A_i) = v_\ell(A_i)$. Thus, the allocation returned by Round-Robin($\mathbf{b}$) is EF1 from the point of view of agent ℓ.   □

4   Towards EFX Equilibria: The Case of Two Agents

As we saw, Round-Robin has PNE for every instance, and the corresponding allocations are always EF1. The natural next question is *can we have a similar guarantee for a stronger fairness notion?* In particular, we want to explore whether an analogous result is possible when we consider envy-freeness up to *any* good. When the agents are not strategic, it is known that EFX allocations exist when we have at most 3 agents [17,18]. It should be noted that for the case of 3 agents no polynomial time algorithm is known, and it is unclear whether the constructive procedure of Chaudhury et al. [18] has any PNE. For $n \geq 4$, the existence of EFX allocations remains a major open problem. Therefore, we turn our attention to the case of two agents.

4.1   A Mechanism with EFX Nash Equilibria

A polynomial-time algorithm that outputs EFX allocations when we have two agents is given by Plaut and Roughgarden [35]. This is a *cut and choose* algorithm where the cut (lines 3–5) is produced using a variant of the *envy-cycle-elimination* algorithm of Lipton et al. [31] on two copies of agent 1, and then agent 2 "chooses" the best bundle among the two (line 6). We state it as mechanism Mod-Cut&Choose below (recall the notation $b_i(S)$ for $\sum_{h \in S} b_i(h)$). We should point out that this mechanism is not truthful, since there is no truthful mechanism for two agents that produces EF1 (or EFX for that matter) allocations for more than four goods [1]. Interestingly, we show that although not truthful, Mod-Cut&Choose always has at least one PNE for any instance, while all its equilibria are MMS and, by Theorem 1, EFX.

Mechanism 2. Mod-Cut&Choose(b_1, b_2)

1: $(E_1, E_2) = (\emptyset, \emptyset)$
2: $(h_1, h_2, \ldots, h_m)$ is M sorted in decreasing order with respect to v_1
                                                      // Break ties lexicographically.
3: **for** $i = 1, \ldots, m$ **do**
4:        $j = \arg \min_{k \in [2]} b_1(E_k)$ // Identify worst bundle w.r.t. b_1; ties in favor of E_1.
5:        $E_j = E_j \cup \{h_i\}$ // Add the next good to that bundle.
6: $\ell = \arg \max_{k \in [2]} b_2(E_k)$ // Identify best bundle w.r.t. b_2; ties in favor of E_1.
7: **return** $\mathcal{A} = (M \setminus E_\ell, E_\ell)$ // Give this to agent 2 and the other bundle to agent 1.

By Theorem 2, there is no reason to expect that the mechanism would guarantee more than $2\mu_i/3$ to each agent. Indeed, seen as an algorithm, it does not always produce MMS allocations unless P = NP! To see this, first notice that when the agents are identical, that would mean that we can run the algorithm to exactly find their maximin share. This is equivalent to having an oracle for the classic PARTITION problem. As the algorithm's running time is polynomial, that would imply that P = NP. The same simple argument shows that it is NP-hard to compute a best response bid vector for agent 1.

Proposition 1. *Computing a best response b_1 of agent 1 for Mod-Cut&Choose, given b_2, is NP-hard.*

We begin with the following lemma on the "cut" part of Mod-Cut&Choose, stating that agent 1 may create any desirable partition of the goods (up to the ordering of the two sets). This is a necessary component of the proof of the main result of this section. Its proof is deferred to the full version [2].

Lemma 2. *Let (X_1, X_2) be a partition of M. Agent 1, by bidding accordingly, can force Mod-Cut&Choose to construct E_1, E_2 in lines 3–5, such that $\{E_1, E_2\} = \{X_1, X_2\}$.*

In particular, agent 1 can force the mechanism to construct E_1, E_2, such that $\min\{v_1(E_1), v_1(E_2)\} = \mu_1$. Such a pair (E_1, E_2) is called a μ_1-*partition*. At least one μ_1-partition exists, by the definition of μ_1.

Corollary 1. *Agent 1 can force Mod-Cut&Choose to construct a μ_1-partition in lines 3–5.*

We can now proceed to the main theorem of this section on the existence and fairness properties of the PNE of Mod-Cut&Choose.

Theorem 5. *For any instance $\mathcal{I} = (\{1, 2\}, M, \mathbf{v})$, the Mod-Cut&Choose mechanism has at least one PNE. Moreover, every PNE of the mechanism is MMS and EFX with respect to the valuation functions v_1, v_2.*

Proof. Given a partition $\mathcal{X} = (X_1, X_2)$ we are going to slightly abuse the notation—as we do in our pseudocode—and consider $\arg\min_{X \in \mathcal{X}} v_2(X)$ to be a single set in $\mathcal{X}$ rather than a subset of $\{X_1, X_2\}$. To do so, we assume that ties are broken in favor of the highest indexed set (here X_2) and tie-breaking is applied by the arg min operator.

We will define a profile (b_1, b_2) and show that it is a PNE. First, let $b_2 = (v_2(g_1), v_2(g_2), \ldots, v_2(g_m))$ be the truthful bid of agent 2. Next b_1 is the bid vector (as defined within the proof of Lemma 2) that results in Mod-Cut&Choose constructing a partition in

$$\arg\max_{\mathcal{X} \in \Pi_2(M)} v_1\left(\arg\max_{X \in \mathcal{X}} v_2(X)\right).$$

To see that there exists such b_1, notice that the set $\Pi_2(M)$ of all possible partitions is finite, and, by Lemma 2, every possible partition can be produced by Mod-Cut&Choose given the appropriate bid vector of agent 1. So, agent 1 forces the partition that maximizes, according to v_1, the value of the least desirable bundle according to v_2. Now it is easy to see that given the bidding strategy of agent 2, i.e., playing truthfully, there is no deviation for agent 1 that is profitable (by definition). Moreover, agent 2 gets the best of the two bundles according to her valuation function (regardless of the partition, truth telling is a dominant strategy for her), thus there is no profitable deviation for her either. Therefore, (b_1, b_2) is a PNE for $\mathcal{I}$.

Regarding the second part of the statement, suppose for a contradiction that there is a PNE **b**, where an agent i does not achieve her μ_i in the allocation returned by Mod-Cut&Choose(**b**). If this agent is agent 1, then according to Corollary 1, there is a bid vector b'_1 she can report, so that the algorithm will produce a μ_1- partition. By deviating to b'_1, regardless of the set given to agent 2, agent 1 will end up with a bundle she values at least μ_1. As this would be a strict improvement over what she currently gets, it would contradict the fact that **b** is a PNE. So, it must be the case where agent 2 gets a bundle she values strictly less than μ_2. Notice that, regardless of the partition which Mod-Cut&Choose to constructs in lines 3–5, by declaring her truthful bid, agent 2 gets a bundle of value at least $v_2(M)/2$. By Definition 3, it is immediate to see that this value is at least μ_2, i.e., deviating to her truthful bid is a strict improvement over what she currently gets by Mod-Cut&Choose(**b**), which is a contradiction.

It remains to show that the allocation returned by Mod-Cut&Choose(**b**) is also EFX. However, since here $n = 2$, this directly follows from Theorem 1.   □

4.2   The Enhanced Fairness of EFX Nash Equilibria

As it was discussed before Proposition 1, it is surprising that the EFX equilibria of Mod-Cut&Choose impose stronger fairness guarantees compared to generic EFX allocations or even EFX allocations produced by Mod-Cut&Choose itself in the non-strategic setting. In this section we explore whether something similar holds for *every* mechanism with EFX equilibria. Specifically, we consider the (obviously non-empty) class of mechanisms that have PNE for every instance and these equilibria always lead to EFX allocations. Our goal is to determine if these allocations have better fairness guarantees (with respect to the underlying true valuation functions) than EFX allocations in general. To this end, we start by examining instances of two agents and 4 goods and we prove that for every mechanism of this class, all allocations at a PNE are MMS allocations. The reason we start from this restricted set of instances is that it already provides a clear separation with the non-strategic setting. Recall from Theorem 2 that there are instances with just 4 goods where an EFX allocation may not be a $\left(\frac{2}{3} + \delta\right)$-MMS allocation, for any $\delta > 0$.

Theorem 6. *Let $\mathcal{M}$ be a mechanism that has PNE for any instance $(\{1, 2\}, M, (v_1, v_2))$ with $|M| = 4$, and all these equilibria lead to EFX allocations with respect to v_1, v_2. Then each such EFX allocation is also an MMS allocation.*

The proof of Theorem 6 (which is deferred to the full version [2]) relies on extensive case analysis, where in each case assuming that the allocation is EFX but not MMS eventually contradicts the fact that the current profile is a PNE. When we consider instances with 5 or more goods, this approach is not fruitful anymore. The reason for that is not solely the increased number of cases one has to handle, but rather the fact that now some of the cases do not seem to lead to a contradiction at all.

Although we suspect that the theorem is no longer true for more than 4 goods, we are able prove a somewhat weaker property that still separates the EFX allocations in PNE from generic EFX allocations in the non-strategic setting. In particular, for general mechanisms that have PNE for every instance and these equilibria are always EFX, we show that the corresponding allocations always guarantee an approximation to MMS that is *strictly better* than 2/3.

Theorem 7. *Let $\mathcal{M}$ be a mechanism that has PNE for any instance $(\{1,2\}, M, (v_1, v_2))$, and all these equilibria lead to EFX allocations with respect to v_1, v_2. Then each such EFX allocation is also an α-MMS allocation for some $\alpha > 2/3$.*

Proof. Suppose for a contradiction that this is not the case. This means that there exists such a mechanism $\mathcal{M}$ and an instance $(\{1,2\}, M, (v_1, v_2))$, for which there is a PNE $\mathbf{b} = (\mathbf{b}_1, \mathbf{b}_2)$ that results in an EFX allocation $\mathcal{A} = (A_1, A_2)$, where $v_i(A_i) \leq 2\mu_i/3$ for at least one $i \in [2]$. Without loss of generality, assume $v_1(A_1) \leq 2\mu_1/3$ and notice that this means that $v_1(A_1) = 2\mu_i/3$, as $v_1(A_1)$ cannot be smaller than $2\mu_i/3$, by Theorem 2. This implies that $v_1(A_2) \geq 4\mu_i/3$, since $v_1(M) \geq 2\mu_1$ by Definition 3.

Initially, we will restrict the number of the goods with positive value (according to v_1) in A_2. Let $S \subseteq A_2$ be the set of such goods, i.e., $S = \{g \in A_2 \mid v_1(g) > 0\}$. Let $|S| = k$ and notice that k cannot be 0 or 1 since otherwise $v_1(A_1) \geq \mu_1$. Finally, let $x \in \arg\min_{g \in S} v_1(g)$ be a minimum valued good for agent 1 in S. We have

$$\frac{2}{3}\mu_1 = v_1(A_1) \geq v_1(S \setminus \{x\}) \geq v_1(S) - \frac{v_1(S)}{k} = \frac{(k-1)}{k} v_1(A_2) \geq \frac{(k-1)}{k} \frac{4}{3}\mu_1,$$

where the first inequality follows from (A_1, A_2) being EFX. Given our observation that $k \leq 2$, the above implies that $k = 2$. Name h_1 and h_2 the goods of S, and observe that if $v_1(A_2) = v_1(\{h_1, h_2\}) > 4\mu_1/3$, then (A_1, A_2) cannot be EFX from the perspective of agent 1. Thus, we get that $v_1(A_2) = 4\mu_1/3$, which in conjunction with EFX implies $v_1(h_1) = v_1(h_2) = 2\mu_1/3$.

Next we argue that A_1 contains at least 2 goods that have positive value for agent 1. Indeed, if all the goods in A_1 had zero value, then we would have $v_1(A_1) = 0 < 2\mu_1/3$ as A_2 contains two positively valued goods, while if there was just one positively valued good in A_1, this would imply that only three goods have positive value for agent 1, and each one of them has value $2\mu_1/3$. The latter would make the existence of a μ_1-partition impossible, which is a contradiction. So, since there are at least two positively valued goods in A_1 for agent 1, we arbitrarily choose two of them, and we name them h_3 and h_4. We arbitrarily name the remaining goods $h_5, h_6, \ldots, h_m$.

Consider now a different valuation instance $\mathbf{v}^* = (v^*, v^*)$ where the agents have identical values over the goods. The valuation function is defined as

$$v^*(h_j) = \begin{cases} 1.2 & j = 1 \\ 1 & j \in \{2, 3\} \\ \varepsilon & j \in \{4, \ldots, m\} \end{cases}$$

where $\varepsilon > 0$ and $(m-3) \cdot \varepsilon < 0.2$. It is easy to see that for this valuation instance there are only two EFX allocations, namely, $\mathcal{X} = (\{h_1, h_4, \ldots, h_m\}, \{h_2, h_3\})$, and its symmetric $\mathcal{Y} = (\{h_2, h_3\}, \{h_1, h_4, \ldots, h_m\})$. According to our assumption, there must be a bidding vector $\mathbf{b}^* = (\boldsymbol{b}_1^*, \boldsymbol{b}_2^*)$ that is a PNE of $\mathcal{M}$ for the instance $(\{1, 2\}, M, \mathbf{v}^*)$, and since all PNE of $\mathcal{M}$ are also EFX, $\mathcal{M}(\mathbf{b}^*)$ must output one of $\mathcal{X}$ and $\mathcal{Y}$. Moreover, observe that the value agent 2 receives (with respect to V^*) in these allocations is 2 and $1.2 + (m-3)\varepsilon < 1.4$ respectively.

For now assume that $\boldsymbol{b}_1 \neq \boldsymbol{b}_1^*$ and $\boldsymbol{b}_2 \neq \boldsymbol{b}_2^*$. We will show that, in this case, running $\mathcal{M}$ with input $\mathbf{b}' = (\boldsymbol{b}_1^*, \boldsymbol{b}_2)$ results to agent 2 receiving a bundle of value strictly better than 2 according to v^*. This contradicts the fact that $\mathbf{b}^* = (\boldsymbol{b}_1^*, \boldsymbol{b}_2^*)$ is a PNE under $\mathbf{v}^* = (v^*, v^*)$. Recall that $\mathbf{b} = (\boldsymbol{b}_1, \boldsymbol{b}_2)$ is a PNE under $\mathbf{v} = (v_1, v_2)$, that $v_1(h_1) = v_1(h_2) = v_1(A_1) = 2\mu_1/3$, and that $v_1(h_3), v_1(h_4)$ are strictly positive. So, let us examine what each agent may get if agent 1 deviates from $\mathbf{b}$ to $\mathbf{b}' = (\boldsymbol{b}_1^*, \boldsymbol{b}_2)$:

- In case the bundle of agent 1 contains good h_1, it cannot contain any good from $\{h_2, h_3, h_4\}$; otherwise $\mathbf{b} = (\boldsymbol{b}_1, \boldsymbol{b}_2)$ would not be a PNE under $\mathbf{v} = (v_1, v_2)$. Thus, $\{h_2, h_3, h_4\}$ is part of the bundle of agent 2.
- In case the bundle of agent 1 contains good h_2, it cannot contain any good from $\{h_1, h_3, h_4\}$; otherwise $\mathbf{b} = (\boldsymbol{b}_1, \boldsymbol{b}_2)$ would not be a PNE under $\mathbf{v} = (v_1, v_2)$. Thus, $\{h_1, h_3, h_4\}$ is part the bundle of agent 2.
- In case the bundle of agent 1 does not contain any of h_1 and h_2, then it is possible for her to get any subset $T \subseteq \{h_3, h_4, \ldots, h_m\}$. However, $\{h_1, h_2\}$ is part the bundle of agent 2.

Thus, in the allocation returned by $\mathcal{M}(\mathbf{b}')$, agent 2 gets a bundle that contains $\{h_2, h_3, h_4\}$ or $\{h_1, h_3, h_4\}$ or $\{h_1, h_2\}$. Consider the value of these sets according to v^*:

$$v^*(\{h_2, h_3, h_4\}) = 2 + \varepsilon, \quad v^*(\{h_1, h_3, h_4\}) = 2.2 + \varepsilon, \quad v^*(\{h_1, h_2\}) = 2.2.$$

That is, in every single case the value agent 2 derives under $\mathbf{v}^* = (v^*, v^*)$ when the profile $\mathbf{b}' = (\boldsymbol{b}_1^*, \boldsymbol{b}_2)$ is played is strictly better than 2. However, 2 is the maximum possible value that agent 2 could derive under $\mathbf{v}^*$ when the profile $\mathbf{b}^*$ is played. This contradicts the fact that $\mathbf{b}^*$ is a PNE under $\mathbf{v}^*$, as $\boldsymbol{b}_2$ is a profitable deviation for agent 2.

The remaining corner cases are straightforward to deal with. To begin with, it is not possible to have $\boldsymbol{b}_1 = \boldsymbol{b}_1^*$ and $\boldsymbol{b}_2 = \boldsymbol{b}_2^*$, as $\mathcal{X} \neq \mathcal{A}$ and $\mathcal{Y} \neq \mathcal{A}$.

Next, assume that $\boldsymbol{b}_1 = \boldsymbol{b}_1^*$ and $\boldsymbol{b}_2 \neq \boldsymbol{b}_2^*$. This directly contradicts the fact that $\mathbf{b}^*$ is a PNE under $\mathbf{v}^* = (v^*, v^*)$. To see this, starting from $\mathbf{b}^*$ let agent 2 deviate to $\boldsymbol{b}_2$. She then gets A_2 which contains h_1, h_2 and has value for her $v^*(A_2) \geq 2.2 > 2$.

Finally, assume that $\boldsymbol{b}_1 \neq \boldsymbol{b}_1^*$ and $\boldsymbol{b}_2 = \boldsymbol{b}_2^*$. This directly contradicts the fact that $\mathbf{b}$ is a PNE under $\mathbf{v} = (v_1, v_2)$. To see this, starting from $\mathbf{b}$ let agent 1 deviate to $\boldsymbol{b}_1^*$. She either gets $\{h_1, h_4, \ldots, h_m\}$ of value at least $v_1(h_1) + v_1(h_4) > 2\mu_1/3 = v_1(A_1)$ or she gets $\{h_2, h_3\}$ of value $v_1(h_2) + v_1(h_3) > 2\mu_1/3 = v_1(A_1)$.

Since every possible case leads to a contradiction, we conclude that every allocation that corresponds to a PNE of a mechanism in the class of interest, guarantees to each agent i value that is strictly better than $2\mu_i/3$, for $i \in [2]$. □

5   Discussion

In this work we studied the problem of fair allocating a set of indivisible goods, to a set of strategic agents. Somewhat surprising—given the existing strong impossibilities for truthful mechanisms—our results are mostly positive. In particular, we showed that there exist mechanisms that have PNE for every instance, and at the same time the allocations that correspond to PNE have strong fairness guarantees with respect to the true valuation functions.

We believe that there are several interesting directions for future work that follow our research agenda. For instance, it would be interesting to explore how algorithms that compute EF1 allocations for richer valuation function domains (e.g., the Envy-Cycle-Elimination algorithm [31]) behave in the strategic setting we study in this work. Here the question is twofold. On one hand, it is unclear whether such algorithms have PNE for every valuation instance, while on the other, it would be important to determine if they maintain their fairness properties at their equilibria or not. The existence of PNE for algorithms that compute approximate MMS allocation is on a similar direction and, as we mentioned in Sect. 2, in this case we get the MMS guarantee on the equilibria for free.

Theorems 6 and 7 leave an open question on the MMS guarantee that the equilibria of mechanisms that always have PNE and these are EFX. Although we suspect that the corresponding allocations are not always MMS, such a result would immediately imply that for every such mechanism which runs in polynomial time, finding a best response of an agent is a computationally hard problem. Going beyond the case of two agents here seems to be a highly nontrivial problem as it is not very plausible that the current state of the art for the non-strategic setting could be analysed under incentives.

Finally, while we did not really focus on complexity questions, it is clear that computing best responses is generally hard. However, when they are not, for instance when the number of agents in Round-Robin is fixed [40], we would like to know if best response dynamics always converge to a PNE or there might be cyclic behavior (as it happens with better response dynamics [7]).

References

1. Amanatidis, G., Birmpas, G., Christodoulou, G., Markakis, E.: Truthful allocation mechanisms without payments: characterization and implications on fairness. In: Proceedings of the 2017 ACM Conference on Economics and Computation, EC 2017, pp. 545–562. ACM (2017)
2. Amanatidis, G., Birmpas, G., Fusco, F., Lazos, P., Leonardi, S., Reiffenhäuser, R.: Allocating indivisible goods to strategic agents: pure Nash equilibria and fairness. CoRR abs/2109.08644 (2021). https://arxiv.org/abs/2109.08644
3. Amanatidis, G., Birmpas, G., Markakis, E.: On truthful mechanisms for maximin share allocations. In: Proceedings of the 25th International Joint Conference on Artificial Intelligence, IJCAI 2016, pp. 31–37. IJCAI/AAAI Press (2016)
4. Amanatidis, G., Birmpas, G., Markakis, E.: Comparing approximate relaxations of envy-freeness. In: Proceedings of the 27th International Joint Conference on Artificial Intelligence, IJCAI 2018, pp. 42–48 (2018). ijcai.org

5. Amanatidis, G., Markakis, E., Nikzad, A., Saberi, A.: Approximation algorithms for computing maximin share allocations. ACM Trans. Algorithms **13**(4), 52:1–52:28 (2017)
6. Aziz, H., Bouveret, S., Lang, J., Mackenzie, S.: Complexity of manipulating sequential allocation. In: Proceedings of the 31st AAAI Conference on Artificial Intelligence, AAAI 2017, pp. 328–334. AAAI Press (2017)
7. Aziz, H., Goldberg, P., Walsh, T.: Equilibria in sequential allocation. In: Rothe, J. (ed.) ADT 2017. LNCS (LNAI), vol. 10576, pp. 270–283. Springer, Cham (2017). https://doi.org/10.1007/978-3-319-67504-6_19
8. Barman, S., Biswas, A., Murthy, S.K.K., Narahari, Y.: Groupwise maximin fair allocation of indivisible goods. In: Proceedings of the 32nd AAAI Conference on Artificial Intelligence, AAAI 2018, pp. 917–924. AAAI Press (2018)
9. Barman, S., Krishnamurthy, S.K.: Approximation algorithms for maximin fair division. ACM Trans. Econ. Comput. **8**(1), 5:1–5:28 (2020)
10. Bei, X., Chen, N., Huzhang, G., Tao, B., Wu, J.: Cake cutting: envy and truth. In: Proceedings of the 26th International Joint Conference on Artificial Intelligence, IJCAI 2017, pp. 3625–3631 (2017). ijcai.org
11. Bouveret, S., Chevaleyre, Y., Maudet, N.: Fair allocation of indivisible goods. In: Handbook of Computational Social Choice, pp. 284–310. Cambridge University Press (2016)
12. Bouveret, S., Lang, J.: Manipulating picking sequences. In: Proceedings of the 21st European Conference on Artificial Intelligence - ECAI 2014, vol. 263, pp. 141–146. IOS Press (2014)
13. Brânzei, S., Caragiannis, I., Kurokawa, D., Procaccia, A.D.: An algorithmic framework for strategic fair division. In: Proceedings of the 30th AAAI Conference on Artificial Intelligence AAAI 2016, pp. 418–424. AAAI Press (2016)
14. Brânzei, S., Gkatzelis, V., Mehta, R.: Nash social welfare approximation for strategic agents. In: Proceedings of the 2017 ACM Conference on Economics and Computation, EC 2017, pp. 611–628. ACM (2017)
15. Budish, E.: The combinatorial assignment problem: approximate competitive equilibrium from equal incomes. J. Polit. Econ. **119**(6), 1061–1103 (2011)
16. Caragiannis, I., Kaklamanis, C., Kanellopoulos, P., Kyropoulou, M.: On low-envy truthful allocations. In: Rossi, F., Tsoukias, A. (eds.) ADT 2009. LNCS (LNAI), vol. 5783, pp. 111–119. Springer, Heidelberg (2009). https://doi.org/10.1007/978-3-642-04428-1_10
17. Caragiannis, I., Kurokawa, D., Moulin, H., Procaccia, A.D., Shah, N., Wang, J.: The unreasonable fairness of maximum Nash welfare. ACM Trans. Econ. Comput. **7**(3), 12:1–12:32 (2019)
18. Chaudhury, B.R., Garg, J., Mehlhorn, K.: EFX exists for three agents. In: Proceedings of the 2020 ACM Conference on Economics and Computation, EC 2020, pp. 1–19. ACM (2020)
19. Chen, Y., Lai, J.K., Parkes, D.C., Procaccia, A.D.: Truth, justice, and cake cutting. Games Econ. Behav. **77**(1), 284–297 (2013)
20. Cole, R., Gkatzelis, V., Goel, G.: Mechanism design for fair division: allocating divisible items without payments. In: Proceedings of the 14th ACM Conference on Electronic Commerce, EC 2013, pp. 251–268. ACM (2013)
21. Ehlers, L., Klaus, B.: Coalitional strategy-proof and resource-monotonic solutions for multiple assignment problems. Soc. Choice Welf. **21**(2), 265–280 (2003)
22. Foley, D.K.: Resource allocation and the public sector. Yale Economics Essays **7**, 45–98 (1967)

23. Gamow, G., Stern, M.: Puzzle-Math. Viking press, New York (1958)
24. Garg, J., McGlaughlin, P., Taki, S.: Approximating maximin share allocations. In: Proceedings of the 2nd Symposium on Simplicity in Algorithms, SOSA@SODA 2019. OASICS, vol. 69, pp. 20:1–20:11. Schloss Dagstuhl - Leibniz-Zentrum für Informatik (2019)
25. Garg, J., Taki, S.: An improved approximation algorithm for maximin shares. In: Proceedings of the 2020 ACM Conference on Economics and Computation, EC 2020, pp. 379–380. ACM (2020)
26. Ghodsi, M., Hajiaghayi, M.T., Seddighin, M., Seddighin, S., Yami, H.: Fair allocation of indivisible goods: improvements and generalizations. Math. Oper. Res. 46(3), 1038–1053 (2021)
27. Gourvès, L., Monnot, J., Tlilane, L.: Near fairness in matroids. In: Proceedings of the 21st European Conference on Artificial Intelligence - ECAI 2014, vol. 263, pp. 393–398. IOS Press (2014)
28. Klaus, B., Miyagawa, E.: Strategy-proofness, solidarity, and consistency for multiple assignment problems. Int. J. Game Theory 30(3), 421–435 (2002)
29. Kohler, D.A., Chandrasekaran, R.: A class of sequential games. Oper. Res. 19(2), 270–277 (1971)
30. Kurokawa, D., Procaccia, A.D., Wang, J.: Fair enough: guaranteeing approximate maximin shares. J. ACM 65(2), 8:1–8:27 (2018)
31. Lipton, R.J., Markakis, E., Mossel, E., Saberi, A.: On approximately fair allocations of indivisible goods. In: Proceedings of the 5th ACM Conference on Electronic Commerce, EC 2004, pp. 125–131. ACM (2004)
32. Markakis, E.: Approximation algorithms and hardness results for fair division with indivisible goods. In: Trends in Computational Social Choice, Chap. 12. AI Access (2017)
33. Papai, S.: Strategyproof multiple assignment using quotas. Rev. Econ. Design 5(1), 91–105 (2000)
34. Papai, S.: Strategyproof and nonbossy multiple assignments. J. Public Econ. Theory 3(3), 257–271 (2001)
35. Plaut, B., Roughgarden, T.: Almost envy-freeness with general valuations. SIAM J. Discret. Math. 34(2), 1039–1068 (2020)
36. Procaccia, A.D.: Cake cutting algorithms. In: Handbook of Computational Social Choice, pp. 311–330. Cambridge University Press (2016)
37. Steinhaus, H.: Sur la division pragmatique. Econometrica 17(Supplement), 315–319 (1949)
38. Varian, H.R.: Equity, envy and efficiency. J. Econ. Theory 9, 63–91 (1974)
39. Woeginger, G.J.: A polynomial-time approximation scheme for maximizing the minimum machine completion time. Oper. Res. Lett. 20(4), 149–154 (1997)
40. Xiao, M., Ling, J.: Algorithms for manipulating sequential allocation. In: Proceedings of the 34th AAAI Conference on Artificial Intelligence, AAAI 2020, pp. 2302–2309. AAAI Press (2020)

On the Benefits of Being Constrained
When Receiving Signals

Shih-Tang Su[1]([⊠]), David Kempe[2], and Vijay G. Subramanian[1]

[1] Electrical Engineering and Computer Science, University of Michigan,
Ann Arbor, USA
shihtang@umich.edu
[2] Department of Computer Science, University of Southern California,
Los Angeles, USA

Abstract. We study a Bayesian persuasion setting in which the receiver is trying to match the (binary) state of the world. The sender's utility is partially aligned with the receiver's, in that conditioned on the receiver's action, the sender derives higher utility when the state of the world matches the action.

Our focus is on whether in such a setting, being constrained helps a receiver. Intuitively, if the receiver can only take the sender's preferred action with smaller probability, the sender might have to reveal more information, so that the receiver can take the action more specifically when the sender prefers it. We show that with a binary state of the world, this intuition indeed carries through: under very mild nondegeneracy conditions, a more constrained receiver will always obtain (weakly) higher utility than a less constrained one. Unfortunately, without additional assumptions, the result does not hold when there are more than two states in the world, which we show with an explicit example.

Keywords: Bayesian persuasion · Information design · Signaling games

1 Introduction

In this paper, we study situations akin to the following stylized dialog, which will likely be familiar to anyone who has ever served on hiring committees:

ALICE: I see that you wrote strong recommendation letters for your Ph.D. graduates Carol and Dan. Can you compare them for us?
BOB: They are both great! Carol made groundbreaking contributions to . . .; Dan made groundbreaking contributions to . . . .
ALICE: Which of the two would you say is stronger?

S.-T. Su—Supported in part by NSF grant ECCS 2038416 and MCubed 3.0.
V. G. Subramanian—Supported in part by NSF grants ECCS 2038416, CCF 2008130, and CNS 1955777.

© Springer Nature Switzerland AG 2022
M. Feldman et al. (Eds.): WINE 2021, LNCS 13112, pp. 167–185, 2022.
https://doi.org/10.1007/978-3-030-94676-0_10

BOB: They are hard to compare. You really need to interview both of them!
ALICE: We can only invite one of them for an interview.
BOB: I guess Carol is a bit stronger.

What happened in this example? Alice and Bob were involved in a signaling setting, in which Bob had an informational advantage. Bob's goal was to get as many of his students interviews as possible, while Alice wanted to only invite the strong students. While Bob knew which of his students were strong (or how strong), Alice had to rely on information from Bob. As is standard in signaling settings, Bob could use this fact to improve his own utility. In this sense, the example initially was virtually identical to the standard "judge/prosecutor" example in the seminal paper of Kamenica and Gentzkow [21].

However, a change happened along the way. When Alice revealed that she was constrained in her actions (to one interview at most), this changed the utility that Bob could obtain from his previous strategy. For example, if he had insisted on not ranking the students, Alice might have flipped a coin. Implicitly, while Bob wanted *both* of his students to obtain interviews, when forced to choose, he knew he would obtain higher utility from the stronger of his students being interviewed. In this sense, his utility function was "partially aligned" with Alice's; this partial alignment, coupled with Alice's constraint, resulted in Alice obtaining more information, and thus higher utility.

The main goal of the present paper is to investigate to what extent the behavior illustrated informally in the dialog above arises in a standard model of Bayesian persuasion. Specifically, if the utilities of the sender and receiver are "partially aligned," will it always benefit a receiver to be more constrained in how she can choose her actions?

1.1 The Model: An Overview

Our model—described fully in Sect. 2—is based on the standard Bayesian persuasion model of Kamenica and Gentzkow [21]. For our main result, we assume that the state space is binary: $\Theta = \{\theta_1, \theta_2\}$. These states could correspond to a student being bad/good in our introductory example, a defendant being innocent/guilty in the example of Kamenica and Gentzkow [21], or a stock about to go up or down. The sender and receiver share a common prior Γ for the distribution of the state θ. In addition, the sender will observe the actual state θ, but only after committing to a signaling scheme (also called information structure).

A signaling scheme is a (typically randomized) mapping $\phi : \Theta \to \Sigma$. The receiver observes the (typically random) signal $\sigma = \phi(\theta)$; based on this observation, she takes an action $a \in A$. Here, we assume that—like the state space—the action space is binary, i.e., $A = \{a_1, a_2\}$. Based on the true state of the world and the action taken by the receiver, both the sender and receiver derive utilities $U_S(\theta, a)$, $U_R(\theta, a)$. The receiver will choose her action (upon observing σ) to maximize her own expected utility; the sender, knowing that the receiver is rational, will commit to a signaling scheme to maximize his expected utility under rational receiver behavior.

Motivated by many practical applications, we assume that the receiver prefers to match the state of the world, in the sense that $U_R(\theta_1, a_1) \geq U_R(\theta_1, a_2)$ and $U_R(\theta_2, a_2) \geq U_R(\theta_2, a_1)$. For instance, in our introductory example, Alice prefers to interview strong candidates and to not interview weak ones; in the judge-prosecutor example, the judge prefers convicting exactly the guilty defendants; and an investor prefers to buy stocks that will go up and sell stocks that will go down. Our assumption about the "partial alignment" of the sender and receiver utilities is formalized as an *action-matching* preference of the sender, stated as follows: $U_S(\theta_1, a_1) \geq U_S(\theta_2, a_1)$ and $U_S(\theta_2, a_2) \geq U_S(\theta_1, a_2)$. That is, if a candidate is being interviewed, Bob prefers it to be a strong candidate over a weak one (but may still prefer a weak candidate being interviewed over a strong/weak candidate *not* being interviewed); similarly, if a prosecutor sees a defendant convicted, he would prefer the defendant to be guilty (but may still prefer an innocent defendant being convicted over going free); similarly, an investment platform may want to entice a client to buy stock, but conditioned on the client buying stock, the platform may prefer for the stock to go up.

In addition to the assumption of partial alignment, our main addition to the standard Bayesian persuasion model is to consider constraints on the receiver's actions. Specifically, we assume that there are (lower and upper) bounds $\underline{b}$, $\overline{b}$ on the probability with which the receiver is allowed to take action[1] a_1. Such a constraint corresponds to a department only being willing to interview at most 10% of their applicants, a judge having a quota for how many defendants (at most) to convict, or a conference having an upper bound on its number/fraction of accepted papers. Such a constraint creates dependencies between the receiver's actions under different received signals, and may force her to randomize between actions, contrary to the standard Bayesian persuasion setting in which the receiver may deterministically choose any utility-maximizing action conditioned on the observed signal σ. To see this, consider a prior under which a candidate is strong with probability $\frac{1}{3}$, and the receiver obtains utility 1 from interviewing a strong candidate and -1 from interviewing a weak candidate (and 0 from not interviewing). If the sender reveals no information, the receiver would prefer to interview no candidates, but a lower-bound constraint may force her to do so, in which case she would randomize the decision to interview the smallest total number of candidates. We write $\pi : \Sigma \to A$ for the receiver's (typically randomized) mapping from signals to actions. Note that the constraint applies across all sources of randomness (the state of the world, the sender's randomization, and the receiver's randomization), so it is required that $\underline{b} \leq \mathbb{P}_{\Gamma, \phi, \pi}[\pi(\phi(\theta)) = a_1] \leq \overline{b}$.

To avoid trivialities, we assume that $\mathbb{P}_\Gamma[\theta = \theta_1] \in [\underline{b}, \overline{b}]$, that is, if the sender revealed the state of the world perfectly, the receiver would be allowed to match it. We say that a receiver with constraints $(\underline{b}', \overline{b}')$ is *more constrained* than one with constraints $(\underline{b}, \overline{b})$ iff $\underline{b}' \geq \underline{b}$ and $\overline{b}' \leq \overline{b}$.

---

[1] This implies constraints of $1 - \overline{b}$, $1 - \underline{b}$ on the probability of taking action a_2. A more general model and its specialization to binary actions is discussed in Sect. 2.3.

1.2 Our Results

Our main result is that when the state of the world is binary, a receiver is always (weakly) better off when more constrained. We state this result here informally, and revisit it more formally (and prove it) in Sect. 3.

Theorem 1 (Main Theorem (stated informally)). *Consider a Bayesian persuasion setting in which the state and action spaces are binary, the receiver is trying to match the state of the world, and the sender is action-matching. Then, a more constrained receiver always obtains (weakly) higher utility than a less constrained one.*

Unfortunately, this insight does not extend to more fine-grained states of the world: even for a ternary state of the world, there are examples with partially aligned sender and receiver in which a more constrained receiver is strictly worse off. We discuss such an example in depth in Sect. 4. It is possible to obtain some positive results recovering versions of Theorem 1 by imposing additional constraints on the sender's and receiver's utility functions. However, many of these constraints are strong, and have only limited applicability to real-world settings. We discuss some of these approaches in Sect. 5—whether there are less restrictive conditions recovering Theorem 1 for more states of the world is an interesting direction for future work.

1.3 Related Work

In general, information design as an area is concerned with situations in which a better-informed sender or information designer can influence the behavior of other agents via the provision of information. The literature generally studies problems in which the underlying game between the agents is given and fixed, but where the sender can influence the outcome by an appropriate choice of information to be disclosed. The core difference between Bayesian persuasion [3,5,6,20,21] and other standard paradigms that study information transmission (such as cheap talk [11], verifiable messages [17,27] or signaling games [33]) is the commitment power of the sender. In Bayesian persuasion models, the sender moves first and commits to a (typically randomized) mapping from states of the world to signals. Subsequently, the sender observes the state of the world and applies the mapping. Based on the mapping and the observed signal, the rational recipients (called receivers) choose actions.

The study of Bayesian persuasion was initiated in the seminal work of Kamenica and Gentzkow [21] and Rayo and Segal [31]. In their work, the sender can *commit to sending any distribution of messages* before (accurately) observing the state of the world; the receiver, on the other hand, only has knowledge of the prior. The full commitment setting allows for an equivalence to an alternate model where the sender publicly chooses the amount of information (regarding the state of the world) he will privately observe and then (strategically) decides how much of this information to share with the receiver via verifiable messages. Follow-up work of Bergemann and Morris [3,5] established a useful and important equivalence between the set of outcomes achievable via information design and Bayes correlated equilibria. Since these seminal works, there has been a large body of work on Bayesian persuasion with theoretical developments as well as a multitude of applications. To keep our discussion focused, for the broader literature, we refer the reader to survey articles [6,20].

The literature closest to our work studies information design with a constrained sender: the constraints arise through a diverse set of assumptions. The work in [29,30] shows that pooling equilibria result if the receiver either prefers lower complexity (for a certification process) or performs a validation of the sender's signal; this holds whether the signals of the sender are exogenously constrained or not. A growing body of work considers constraints on the sender that arise either due to communication costs for signaling [10,16,19,28], capacity limitations for signaling [13,25], the sender's signal serving multiple purposes (such as convincing a third party to take a payoff-relevant action) [7], or costs to the receiver for acquiring additional information [26]. The contributions are then to characterize either the applicability of the concavification approach [21], the optimal signaling structure, or the conditions for the optimality of certain signaling structures. In [22], constraints on the sender arise from the receiver having access to some publicly available information. Within this context, Kolotilin [22] studies comparative statics on the sender's utility based on the quality of the sender's information or the public information. There is also a burgeoning literature on constraints on the sender arising from a privately informed receiver (e.g., [8,9,12,18,24]). The main contributions in this line of research are to characterize the optimal signaling structure with a key aspect being the fact that the sender constructs a different signal for each receiver type.

Based on the discussion above, clearly there is significant literature studying a constrained sender's optimal signaling scheme and utility. However, work that studies constraints on the *receiver*, or their impact on the receiver's utility, is extremely limited. To the best of our knowledge, [2] is the only work to analyze a constrained receiver problem. The authors impose *ex ante* and *ex post* constraints on the receiver's posterior beliefs, characterize the dimensionality of the optimal signaling structure and develop low-complexity approximate welfare maximizing algorithms. In our work, we have two important differences: first, we impose constraints on the receiver's *actions* as opposed to posterior beliefs; and second, we explore when these constraints result in increased utility for the receiver.

2   Problem Formulation

Our model is based on the standard Bayesian persuasion model [21]. Two players, a sender and a receiver, interact in a signaling game. The sender can observe the state of the world, while the receiver can take an action. The sender can convey information about the state of the world to the sender. Both players receive utility as a function of both the state of the world and the action chosen by the receiver. Since their utility functions typically do not align, the sender will be strategic in the information he reveals to the receiver.

2.1   State of the World, Actions, and Utilities

The (random) state of the world θ is drawn from a state space Θ. For our main result, we assume that the state space is binary ($\Theta = \{\theta_1, \theta_2\}$); however, we define the model in more generality. The sender and receiver share a common-knowledge prior distribution $\Gamma \in \Delta(\Theta)$ for θ. When the state space is binary, this prior is fully characterized by $p = \mathbb{P}_\Gamma[\theta = \theta_1]$.

Only the receiver can take an action $a \in A$. Again, for our main result, we assume that the action space is binary: $A = \{a_1, a_2\}$. Both the sender's and the receiver's utilities are functions of the true state θ and the action taken; they are captured by the functions $U_S : \Theta \times A \to \mathbb{R}$ and $U_R : \Theta \times A \to \mathbb{R}$. As discussed in Sect. 1.1, we assume that the receiver tries to match the state of the world with her action.

Definition 1 (State-Matching Receiver). *We say that the receiver's utility function is* state-matching *if it satisfies the following: for all i, j, k with $i \leq j \leq k$ or $i \geq j \geq k$, we have that*

$$U_R(\theta_i, a_j) \geq U_R(\theta_i, a_k). \tag{1}$$

When the state of the world is binary, the condition simplifies to:

$$U_R(\theta_1, a_1) \geq U_R(\theta_1, a_2) \text{ and } U_R(\theta_2, a_2) \geq U_R(\theta_2, a_1). \tag{2}$$

In words, a state-matching receiver always prefers an action closer to the true state; however, the definition does not enforce any comparisons between choosing an action that is too high vs. too low compared to the true state.

The key notion for our analysis is a partial alignment of the sender's utility with the receiver's. This is captured by the fact that the sender, given any fixed action, would prefer states closer to the action, expressed in Definition 2:

Definition 2 (Action-Matching Sender). *We say that the sender's utility function is* action-matching *if it satisfies the following: for all i, j, k with $i \leq j \leq k$ or $i \geq j \geq k$, we have that*

$$U_S(\theta_j, a_i) \geq U_S(\theta_k, a_i). \tag{3}$$

When the state of the world is binary, the condition simplifies to:

$$U_S(\theta_1, a_1) \geq U_S(\theta_2, a_1) \text{ and } U_S(\theta_2, a_2) \geq U_S(\theta_1, a_2). \tag{4}$$

In words, an action-matching sender always prefers a state of the world closer to the action chosen by the receiver; again, we do not enforce any comparisons between states that are higher vs. lower than the chosen action.

Notice the difference between Inequalities (3) and (4) vs. (1) and (2): (1) and (2) compare the receiver's utilities when the state of the world is fixed and the action is changed, while (3) and (4) compare the sender's utilities when the action is fixed and the state of the world is changed. That is, given that the receiver takes a particular action, the sender derives higher utility when that action more closely matches the state of the world than when it does not. Again, a justification for this assumption is discussed in Sect. 1.1.

2.2   Signaling Schemes

Before the receiver takes her action, the sender can send a signal σ to reveal (partial) information about the state of the world. More precisely, prior to observing the state of the world θ, the sender commits to a signaling scheme ϕ, which is a mapping $\phi : \Theta \to \Delta(\Sigma)$. For our purposes ϕ is conveniently characterized by the probability with which each signal is sent conditional on the state. We write $\phi_{i,j} = \mathbb{P}[\phi(\theta) = \sigma_j \mid \theta = \theta_i] \in [0,1]$ for the probability that signal σ_j is sent conditional on the state of the world being θ_i. We write $\overline{\phi}_j = \sum_i \mathbb{P}_\Gamma[\theta = \theta_i] \cdot \phi_{i,j}$ for the probability of sending the signal σ_j.

The receiver is Bayes-rational, and her objective is to maximize her expected utility after observing the signal. The expected utility derived from action a when observing σ_j can be written as

$$\overline{U}_R(\sigma_j, a) = \sum_{\theta_i \in \Theta} \mathbb{P}[\theta = \theta_i \mid \phi(\theta) = \sigma_j] \cdot U_R(\theta_i, a) = \sum_{\theta_i \in \Theta} \frac{\mathbb{P}[\theta = \theta_i] \cdot \phi_{i,j}}{\overline{\phi}_j} \cdot U_R(\theta_i, a).$$

Thus, barring other constraints (which we will introduce below), the receiver chooses an action a in $\mathrm{argmax}_a \overline{U}_R(\sigma_j, a)$. Following most of the literature in the field of information design, we assume that the receiver breaks ties in favor of an action most preferred by the sender. The following very useful alternative view has been observed in the prior literature (see, e.g., [4]): instead of sending abstract signals, the sender can without loss of generality send the receiver a recommended action a_j. The sender must ensure that ϕ is such that the receiver will always voluntarily follow the recommendation. In other words, the recommended action a_j must always be in $\mathrm{argmax}_a \overline{U}_R(\sigma_j, a)$. This constraint ensures *ex-post incentive compatibility (EPIC)* of the signaling scheme, and is often referred to as an *obedience constraint*.

We write $\pi : \Sigma \to \Delta(A)$ for the receiver's (possibly randomized) best-response function. In the setting described so far, there is actually no need for the receiver to randomize, and she can always choose any arbitrary deterministic $\pi(\sigma_j) \in \mathrm{arg\,max}_a \overline{U}_R(\sigma_j, a)$. However, as we will see in Sect. 2.3, the situation changes when the receiver is constrained. For a receiver strategy π, we write $\pi_{i,j} = \mathbb{P}[\pi(\sigma_j) = a_i]$ for the probability that the receiver, upon observing signal σ_j, chooses action a_i.

The sender's objective is to design a signaling strategy which maximizes his expected utility in the subgame perfect equilibrium. That is, he chooses ϕ so as to maximize his expected utility (under all sources of randomness)

$$\mathbb{E}_{\theta \sim \Gamma, \sigma \sim \phi(\theta), a \sim \pi^{(\phi)}(\sigma)} \left[U_S(\theta, a) \right],$$

assuming a best response $\pi^{(\phi)}$ from the receiver.

2.3   Constrained Receiver

Our main conceptual departure from prior work is that we consider constraints on the receiver, restricting the probability with which actions can be chosen. In a general setting, such constraints are lower and upper bounds on the probability of taking each action, i.e., $\underline{b}_a$ and $\overline{b}_a$ for each a. Formally, we require that for each action a_i, the combination of the sender's signaling scheme ϕ and the receiver's response π satisfy

$$\underline{b}_{a_i} \leq \sum_j \overline{\phi}_j \cdot \pi_{i,j} \leq \overline{b}_{a_i}. \tag{5}$$

The constraints are common knowledge among the sender and receiver. When the state space is binary, the constraints can be simplified: they are fully characterized by the lower and upper bounds $\underline{b} = \max(\underline{b}_{a_1}, 1 - \overline{b}_{a_2}), \overline{b} = \min(\overline{b}_{a_2}, 1 - \underline{b}_{a_1})$ for the probability with which the receiver can choose action a_1.

The focus of our work is on whether being (more) constrained helps the receiver, by forcing an action-matching sender to disclose "more" information. Without any further assumptions, this is trivially false. For example, suppose that the state of the world is known to be θ_1 with probability 1, and both the sender and the receiver obtain utility 1 when the receiver chooses action a_1, and 0 otherwise. If the constraint specified that a_1 must be taken with probability 0, and a_2 with probability 1, then of course, the receiver (and the sender) would be worse off. In order to allow us to clearly articulate the question of whether a constrained receiver obtains more information, we require that perfect state matching would always be feasible for the receiver, if the true state were revealed:

Definition 3 (Implementable and Feasible Constraints). *Consider constraints $\langle \underline{b}_{a_i}, \overline{b}_{a_i} \rangle$ for all $a_i \in A$. We say that the constraints are* implementable *iff $\sum_i \underline{b}_{a_i} \leq 1 \leq \sum_i \overline{b}_{a_i}$.*

The constraints are feasible *iff $\underline{b}_{a_i} \leq \mathbb{P}_\Gamma[\theta = \theta_i] \leq \overline{b}_{a_i}$ for all i.*

For the special case of a binary state space, a constraint $\langle \underline{b}, \overline{b} \rangle$ is feasible iff $\underline{b} \leq p \leq \overline{b}$.

Notice that when constraints are not implementable, there is no strategy for the receiver to satisfy all constraints. When constraints are feasible, then with full information, perfect state matching can be implemented by the receiver.

We say that the constraints $\langle \underline{b}_{a_i}, \overline{b}_{a_i} \rangle$ are *more binding* (or the receiver is more constrained by them) than $\langle \underline{b}'_{a_i}, \overline{b}'_{a_i} \rangle$ if and only if $\underline{b}'_{a_i} \leq \underline{b}_{a_i}$ and $\overline{b}_{a_i} \leq \overline{b}'_{a_i}$ for all i. When the state space is binary, the condition simplifies: the constraint $\langle \underline{b}, \overline{b} \rangle$ is more binding than $\langle \underline{b}', \overline{b}' \rangle$ if and only if $\underline{b}' \leq \underline{b}$ and $\overline{b} \leq \overline{b}'$.

We note that the presence of a constraint may force the receiver to randomize between actions, even possibly actions that are not optimal. For a simple example, suppose that the state of the world is binary and determined by a fair coin flip, and the receiver obtains utility 2 from matching state θ_2, 1 from matching state θ_1, and 0 for not matching the state. If the sender reveals no information, then a receiver constrained by—say—$\underline{b} = \overline{b} = \frac{1}{2}$, would have to flip a fair coin to decide which action to choose, even though the optimal strategy would be to always choose a_2.

While the receiver's best response π may in general (have to) be randomized, we show that there is always an optimal signaling strategy for the sender such that the receiver will play a deterministic strategy π. Notice that the following proposition does not even require feasibility in the sense that the prior distribution satisfies the constraints: it merely requires that the constraints allow for the existence of *any* signaling scheme and corresponding receiver strategy.

Proposition 1. *Assume that $|\Sigma| \geq |A|$, and let $\langle \underline{b}_{a_i}, \overline{b}_{a_i} \rangle$ (for all i) be implementable constraints on the receiver. Then, for any signaling scheme $\hat{\phi}$, there exists another signaling scheme ϕ under which the sender has at least the same utility as under $\hat{\phi}$, and such that the receiver's best response $\pi^{(\phi)}$ is deterministic. In particular, there is a sender-optimal strategy under which the receiver responds deterministically.*

Proof. We will give an explicit construction of such a strategy. Let $\hat{\phi}$ be any signaling scheme. Let $\pi^{(\hat{\phi})}$ be the receiver's (randomized) best response. Recall that $\pi_{i,j}$ is the probability with which the receiver plays a_i when receiving the signal σ_j. We will first construct an intermediate signaling scheme ϕ', and from it the final signaling scheme ϕ.

As a first step, the signaling scheme maps to an expanded space $\Sigma' = \Sigma \times A$. When observing the state θ_k, the sender sends the signal (σ_j, a_i) with probability $\phi_{k,j} \cdot \pi_{i,j}$. In other words, the sender performs exactly the randomization that the receiver would perform, and makes the corresponding recommendation to the receiver. Conditioned on the signal σ_j, the signal's second component a_i reveals no information about the state of the world. Therefore, because the distribution of a_i is exactly the distribution that $\pi^{(\hat{\phi})}(\sigma_j)$ uses, it is a best response for the receiver (and satisfies the constraints) to deterministically[2] follow the sender's "recommendation" a_i when receiving the signal (σ_j, a_i).

Then, following the standard approach for reducing the size of the signal space, we "compress" all signals under which the receiver chooses the same action into one signal. That is, under the final signaling scheme ϕ, whenever the sender was going to send (σ_j, a_i) for any j under ϕ', the sender simply sends a_i. Because it is a best response for the receiver to deterministically choose a_i for all received (σ_j, a_i), it is still a best response to follow the recommendation a_i.

Thus, we have constructed a signaling scheme ϕ such that the receiver plays a deterministic best response, and the number of signals employed by the sender is at most $|A|$.

Finally, to prove the existence of a sender-optimal signaling scheme with deterministic receiver response, let $\hat{\phi}$ be any sender-optimal signaling scheme. The existence of a signaling scheme, and thus a sender-optimal one, follows because the constraints are implementable by assumption. Then, applying the previous argument to $\hat{\phi}$ gives the desired optimal signaling scheme with deterministic receiver responses.   □

---

[2] Note that it is optimal for the receiver to follow the recommendation due to the overall constraints. In isolation, the receiver may be better off deviating for some signals—however, doing so would violate a constraint, or come at the expense of having to choose an even more suboptimal action under another signal.

In general, most of the literature on Bayesian persuasion assumes that the signal space is at least as large as the action space (which is enough to obtain sender-optimal strategies, and find them via an LP [23] when EPIC holds). Hence, we make the same assumption that $|\Sigma| \geq |A|$ in Proposition 1.

Henceforth, we will restrict attention to signaling schemes with deterministic best response functions π without loss of optimality. However, the sender still has to ensure that following the deterministic recommendation is incentive compatible for the receiver. Since the receiver is constrained, her space of deviations is only to best-response functions satisfying the constraints. This is captured by the following definition:

Definition 4. *Let $\phi : \Theta \to \Sigma$ be a direct signaling scheme for the sender, i.e., making action recommendations and assuming $\Sigma = A$. Let Π be the set of all randomized mappings $\pi : \Sigma \to A$ (characterized by $\pi_{i,j}$) satisfying the following inequalities for all actions a_j:*

$$\underline{b}_{a_j} \leq \sum_i \overline{\phi}_i \cdot \pi_{i,j} \leq \overline{b}_{a_j}.$$

Then, ϕ is ex ante incentive compatible *iff for all feasible response functions $\pi \in \Pi$,*

$$\sum_i \overline{\phi}_i \cdot \overline{U}_R(\sigma_i, a_i) \geq \sum_i \sum_j \overline{\phi}_i \cdot \pi_{i,j} \cdot \overline{U}_R(\sigma_i, a_j).$$

Note that the presence of constraints forces us to deviate from the standard EPIC requirement in the literature. Definition 4 bears similarity to definitions in [2,9,12], where ex ante constraints are considered.

3   Our Main Result

In this section, we present the main result of this paper.

Theorem 2. *Consider a Bayesian persuasion setting in which the state and action spaces are binary. The receiver is state-matching, and the sender is action-matching. Let $\langle \underline{b}, \overline{b} \rangle$ and $\langle \underline{b}', \overline{b}' \rangle$ be two feasible constraints such that $\langle \underline{b}, \overline{b} \rangle$ is more binding than $\langle \underline{b}', \overline{b}' \rangle$, and let Φ, Φ' be the set of all sender-optimal signaling schemes under these constraints, respectively.*

Let $\phi \in \mathrm{argmax}_{\phi \in \Phi} \, \mathbb{E}_{\theta \sim \Gamma, \sigma \sim \phi(\theta), a \sim \pi^{(\phi)}(\sigma)} \left[U_R(\theta, a) \right]$ maximize the receiver's utility over Φ, and $\phi' \in \mathrm{argmax}_{\phi' \in \Phi'} \, \mathbb{E}_{\theta \sim \Gamma, \sigma \sim \phi'(\theta), a \sim \pi^{(\phi')}(\sigma)} \left[U_R(\theta, a) \right]$ maximize the receiver's utility over Φ'. Then the receiver is no worse off under ϕ than under ϕ', i.e., $\mathbb{E}_{\theta \sim \Gamma, \sigma \sim \phi(\theta), a \sim \pi^{(\phi)}(\sigma)} \left[U_R(\theta, a) \right] \geq \mathbb{E}_{\theta \sim \Gamma, \sigma \sim \phi'(\theta), a \sim \pi^{(\phi')}(\sigma)} \left[U_R(\theta, a) \right]$.

Proof. At a high level, the intuition for the proof is as follows. Based on the discussion in Sect. 2.3, the constraints on the receiver actually translate into constraints on the sender in the optimization problem. Because the sender's signaling schemes are more constrained, he has to reveal more information. However, this intuition is not complete—after all, the constraints may entice the sender to reveal *less* information. Furthermore, as we see in Sect. 4, when the state space is not binary, a more constrained receiver may be worse off.

Let ϕ, ϕ' be as defined in the statement of the theorem, and let $\phi_{i,j}$, $\phi'_{i,j}$ be their corresponding conditional probabilities of sending the signal σ_j in state θ_i. By Proposition 1, w.l.o.g., under the sender-optimal strategies ϕ, ϕ', the sender recommends an action to the receiver, and the receiver deterministically follows the recommendation. That is, the signal σ_i can be associated with the action a_i for $i = 1, 2$. Our proof is based on distinguishing four cases, depending on the sender's utility:

1. $U_S(\theta_1, a_1) \geq U_S(\theta_1, a_2)$ and $U_S(\theta_2, a_2) \geq U_S(\theta_2, a_1)$
   In this case, for every state, the sender prefers the same action as the receiver. Since the sender's and the receiver's preferences are fully aligned, the sender's optimal strategy is to fully reveal the state of the world. Since the constraints are feasible, the receiver can perfectly match the state of the world under both constraints, and hence obtains the same utility under both constraints.
2. $U_S(\theta_1, a_1) \geq U_S(\theta_1, a_2)$ and $U_S(\theta_2, a_2) \leq U_S(\theta_2, a_1)$
   In this case, the sender always prefers action a_1. Since the sender is action-matching, $U_S(\theta_1, a_1) \geq U_S(\theta_2, a_1)$ and $U_S(\theta_2, a_2) \geq U_S(\theta_1, a_2)$. Combining these inequalities, we obtain that the sender's utility function satisfies the following total order:

$$U_S(\theta_1, a_1) \geq U_S(\theta_2, a_1) \geq U_S(\theta_2, a_2) \geq U_S(\theta_1, a_2).$$

   This implies that

$$U_S(\theta_1, a_1) - U_S(\theta_1, a_2) \geq U_S(\theta_2, a_1) - U_S(\theta_2, a_2). \tag{6}$$

   We now show that $\phi_{1,2} = 0$. An identical proof also shows that $\phi'_{1,2} = 0$. We distinguish two cases:

   (a) If $\phi_{1,2} > 0$ and $\phi_{2,1} > 0$, then the sender could move some probability mass $\epsilon > 0$ from recommending a_2 under θ_1 to recommending a_1, and in return move the same amount from recommending a_1 under θ_2 to recommending a_2. Because the receiver is state-matching, she will still follow the sender's recommendation, and the total probability with which each action is played stays unchanged, so the strategy is still feasible. By Eq. (6), the sender's utility (weakly) increases. By choosing ϵ as large as possible, we arrive at the claim or at the following case.
   (b) If $\phi_{1,2} > 0$ and $\phi_{2,1} = 0$, then $\overline{\phi}_{a_1} = p \cdot \phi_{1,1} < p \leq \overline{b}$. Therefore, it is feasible for the sender to always send the signal σ_1 when the state is θ_1 (i.e., decrease $\phi_{1,2}$ to 0 and increase $\phi_{1,1}$ by the same amount). Again, because the receiver is state-matching, she will still follow the sender's recommendation, and the sender is weakly better off because $U_S(\theta_1, a_1) \geq U_S(\theta_1, a_2)$.
   Because $U_S(\theta_2, a_1) \geq U_S(\theta_2, a_2)$ and $\phi_{1,1} = 1$ (as proved above), the sender will also send σ_1 with as much probability as possible when the state is θ_2, subject to not violating the receiver's incentive to play a_1 and not exceeding

the upper bound $\bar{b}$ (or $\bar{b}'$). In other words, the sender maximizes $\phi_{1,2}$ subject to $\mathbb{E}_{\theta \sim \Gamma, \sigma \sim \phi(\theta)}[U_R(\theta, a_1)|\sigma_1] \geq \mathbb{E}_{\theta \sim \Gamma, \sigma \sim \phi(\theta)}[U_R(\theta, a_2)|\sigma_1]$ and $\bar{b} \geq \bar{\phi}_{a_1}$ (or $\bar{b}' \geq \bar{\phi}_{a_1}$). Using $\phi_{1,1} = 1$, the incentive constraint is equivalently expressed as $\phi_{2,1} \leq \frac{p \cdot (U_R(\theta_1, a_1) - U_R(\theta_1, a_2))}{(1-p) \cdot (U_R(\theta_2, a_2) - U_R(\theta_2, a_1))}$. Since this inequality is independent of the bound and $\bar{b}'$ is more restricted than $\bar{b}$, the receiver is weakly better off under the constraint $\bar{b}$ than under $\bar{b}'$.

3. $U_S(\theta_1, a_1) \leq U_S(\theta_1, a_2)$ and $U_S(\theta_2, a_2) \geq U_S(\theta_2, a_1)$

   This case is symmetric to the previous one. Here, the roles of a_1 and a_2 (and θ_1 and θ_2) are reversed, and the important constraint becomes the lower bound $\underline{b}$ (and $\underline{b}'$) rather than the upper bound $\bar{b}$.

4. $U_S(\theta_1, a_1) \leq U_S(\theta_1, a_2)$ and $U_S(\theta_2, a_2) \leq U_S(\theta_2, a_1)$

   In this case, the fact that the sender is action-matching together with the assumed inequalities implies that

$$U_S(\theta_2, a_2) \overset{AM}{\geq} U_S(\theta_1, a_2) \geq U_S(\theta_1, a_1) \overset{AM}{\geq} U_S(\theta_2, a_1) \geq U_S(\theta_2, a_2).$$

Thus, the sender's utility is the same, regardless of the state and action. As a result, the sender is indifferent between all signaling schemes. In particular, fully revealing the state is an optimal strategy for the sender for any constraint; clearly, this would be best for the receiver.

Thus, for all four cases, the receiver will be no worse off under the more binding constraint. □

3.1   Necessity of Partial Alignment

Our main Theorem 2 assumes that the sender is partially aligned with the receiver (in addition to the state space being binary). One may ask whether the partial alignment is necessary, or whether a more constrained receiver is *always* better off with binary state and action spaces. Here, we show that the assumption is necessary, by giving a 2×2 example under which the receiver is worse off when more constrained.

The sender's and receiver's utility functions are given in Table 1. Here, $0 < \epsilon \ll 1$. The prior distribution over states is $p = \frac{1}{4}$.

Table 1. Sender's and Receiver's Utility in the example without partial alignment

	θ_1	θ_2
a_1	2	3
a_2	1	0

	θ_1	θ_2
a_1	1	ϵ
a_2	0	1

(a) Sender's Utility   (b) Receiver's Utility

The sender prefers action a_1 in both states, and the receiver is state-matching. Notice that the sender is not action-matching: when the receiver plays a_1, the sender prefers the state θ_2 over θ_1. We write σ_i for the sender's signal suggesting action a_i, $i \in \{1,2\}$.

We will compare the receiver's expected utilities in the following two settings:

1. There are (effectively) no constraints, i.e., $\bar{b}_{a_1} = 1, \underline{b}_{a_1} = 0, \bar{b}_{a_2} = 1, \underline{b}_{a_2} = 0$.
2. The constraint profile binds the sender-preferred action to at most its prior probability, i.e., $\bar{b}'_{a_1} = \frac{1}{4}, \underline{b}'_{a_1} = 0, \bar{b}'_{a_2} = 1, \underline{b}'_{a_2} = 0$.

The first setting is the classical Bayesian persuasion problem: the sender's optimal signaling strategy can be obtained by the concavification approach presented in [21], and is the following: Send σ_1 with $\phi_{1,1} = 1$ and $\phi_{2,1} = \frac{1}{3}$; send σ_2 with $\phi_{1,2} = 0$ and $\phi_{2,2} = \frac{2}{3}$. Given this commitment, the receiver's expected utility is $\frac{1+\epsilon}{2}$ when receiving σ_1 (because θ_1 and θ_2 are equally likely to occur), and her expected utility is 1 when receiving σ_2. Thus, the receiver's overall expected utility is $\frac{3+\epsilon}{4}$.

In the second setting, the sender cannot send the signal σ_1 as frequently as in the unconstrained case. When the sender is forced to reduce $\mathbb{P}[\sigma_1]$, he prefers to reduce the probability $\phi_{1,1}$ instead of $\phi_{1,2}$. This is because $U_S(\theta_2, a_1) - U_S(\theta_2, a_2) > U_S(\theta_1, a_1) - U_S(\theta_1, a_2)$. However, reducing $\phi_{1,1}$ solely may cause the signal σ_1 to not be persuasive any more, when the posterior belief violates the incentive constraint. Hence, the sender's optimal signaling strategy requires him to maximize the total probability of σ_1, under the constraint that the receiver is still willing to take action a_1 under σ_1. Thus, the sender's optimal signaling scheme is the following: Send σ_1 with $\phi_{1,1} = \frac{1}{2}$ and $\phi_{2,1} = \frac{1}{6}$; send σ_2 with $\phi_{1,2} = \frac{1}{2}$ and $\phi_{2,2} = \frac{5}{6}$.

Against this signaling scheme, the receiver's best response to σ_1 is taking action a_1, with an expected utility of $\frac{1+\epsilon}{2}$. Her best response to σ_2 is taking action a_2, with an expected utility of $\frac{5}{6}$. Hence, the receiver's expected utility is $\frac{6+\epsilon}{8}$ under the constraints $\langle \underline{b}', \bar{b}' \rangle$.

In summary, the receiver's expected utility of $\frac{3+\epsilon}{4}$ in the first setting is higher than her utility of $\frac{6+\epsilon}{8}$ in the second setting. Thus, we have exhibited an example where a more constrained receiver is worse off than a less constrained one.

4   Failure of the Main Result with Larger State Spaces

Unfortunately, contrary to the case of binary state and action spaces, when the state and action spaces are larger, a state-matching receiver and action-matching sender (and feasible constraints) are not enough to ensure that the receiver is always better off when more constrained. Consider the utilities given in Table 2. There are three states in the world, and correspondingly three actions. The prior over the states is uniform.

Table 2. An example where a constrained receiver is worse off

	θ_1	θ_2	θ_3
a_1	10	10	0
a_2	0	2	2
a_3	0	0	1

	θ_1	θ_2	θ_3
a_1	4	2	0
a_2	0	3	1
a_3	0	1	3

	θ_1	θ_2	θ_3
σ_1	1	$\frac{1}{2}$	0
σ_2	0	$\frac{1}{2}$	$\frac{1}{2}$
σ_3	0	0	$\frac{1}{2}$

(a) Sender's Utility   (b) Receiver's Utility   (c) Sender-optimal signaling scheme when $\bar{b}_{a_1} = \frac{1}{2}$

Notice that the receiver is state-matching, and the sender is action-matching.

Unconstrained Receiver. First, consider an unconstrained receiver. The sender's optimal signaling scheme ϕ is to recommend action a_1 whenever the state of the world is θ_1 or θ_2, and recommend action a_3 otherwise.

To verify that the receiver follows the recommendation, one simply compares the utility from the alternative actions: when the sender recommends a_1, following the recommendation gives the receiver expected utility $\frac{1}{2} \cdot 4 + \frac{1}{2} \cdot 2 = 3$, while a_2 would give utility $\frac{1}{2} \cdot 0 + \frac{1}{2} \cdot 3 = \frac{3}{2}$, and a_3 would give $\frac{1}{2} \cdot 0 + \frac{1}{2} \cdot = \frac{1}{2}$. For the recommendation of a_3, the receiver gets to match the state deterministically, so following the recommendation is optimal. Because the signaling scheme is even ex post incentive compatible for the receiver, it is most definitely ex ante incentive compatible.

To see that this signaling scheme is optimal for the sender, first observe that for states θ_1 and θ_2, the sender obtains the maximum possible utility of 10 over all actions. For state θ_3, the sender would prefer the receiver to play action a_2. However, the only way to get the sender to play a_2 is to mix at least one unit of probability of θ_2 per unit of probability of θ_3. While this increases the sender's utility for the unit of probability from θ_3 from 1 to 2, it decreases his utility for the unit of probability from θ_2 from 10 (since the receiver played a_1) to 2. Thus, the given signaling scheme is sender-optimal.

Under this signaling scheme, the receiver's expected utility can be calculated as $\frac{2}{3} \cdot (\frac{1}{2} \cdot 4 + \frac{1}{2} \cdot 2) + \frac{1}{3} \cdot 3 = 3$.

Adding a Non-trivial Constraint. Now, consider a receiver constrained by an upper bound $\bar{b}_{a_1} = \frac{1}{2}$. Table 2c shows the sender-optimal signaling scheme. Here, the entries show the conditional probability $\phi_{i,j}$ of recommending action a_j (i.e., sending signal σ_j) when the state is θ_i.

First, notice that action a_1 is recommended with probability $\frac{1}{2}$, so the constraint is satisfied. Second, the receiver will follow the sender's recommendation, as can be checked by comparing her utility from each of the three actions conditioned on any signal. (In the case of receiving σ_2, she is indifferent between a_2 and a_3—recall that we assume tie breaking in favor of the sender.) Again, the given signaling scheme is even ex post incentive compatible, so in particular, it is also ex ante incentive compatible.

To see that the signaling scheme is optimal for the sender, first notice that he induces action a_1 (under states θ_1 or θ_2) with the maximum probability of $\frac{1}{2}$.

Also, notice that using all of the probability from θ_1 to induce a_1 is optimal for the sender, because under θ_1, if any action other than a_1 is played, the sender's utility is 0. Because $\frac{1}{6}$ unit of probability from θ_2 yields a recommendation of a_1, at most $\frac{1}{6}$ can yield a recommendation of a_2, which gives the next-highest utility for the sender. And because the receiver will choose a_2 only when the conditional probability of θ_2 is at least as large as that of θ_3, action a_2 is induced with the maximum possible probability of $\frac{1}{3}$. Inducing any other actions for any of the states would yield the sender utility 0. Hence, the given signaling scheme is optimal for the sender.

Under this signaling scheme, the receiver's expected utility is $\frac{1}{2} \cdot (\frac{2}{3} \cdot 4 + \frac{1}{3} \cdot 2) + \frac{1}{3}(\frac{1}{2} \cdot 3 + \frac{1}{2} \cdot 1) + \frac{1}{6} \cdot 3 = \frac{17}{6}$.

Thus, the constrained receiver's utility of $\frac{17}{6}$ is lower than the unconstrained receiver's of 3.

5 Discussion

We showed that a state-matching receiver, facing an action-matching sender under a binary state space, obtains weakly higher utility when more constrained. We believe that such behavior is in fact observed in the real world: for example, recommenders tend to be more careful in whom they nominate for particularly selective awards or positions.

5.1 Larger State/Action Spaces

As we discussed in Sect. 4, our results do not carry over to larger state spaces. Indeed, even for state spaces with three states, in which the receiver tries to minimize the distance between the action and the state of the world, there are counter-examples under which a constrained receiver is worse off.

While the result does not hold in full generality with three (or more) states, by imposing additional conditions, a positive result can be recovered:

Proposition 2. *Assume that the state space has size* $|\Theta| = 3$, *and that the receiver is state-matching and the sender is action-matching. In addition, assume that the following two conditions are satisfied.*

1. *The sender has a monotone[3] preference over actions across all states, i.e.,* $U_S(\theta_i, a_1) \geq U_S(\theta_i, a_2) \geq U_S(\theta_i, a_3)$ *for all* i.
2. *For every state* i, *the receiver is worse off choosing an action* $j < i$ *that is too low compared to choosing an action* $k > i$ *that is too high[4]: that is,* $U_R(\theta_i, a_j) \leq U_R(\theta_i, a_k)$ *for all* $j < i < k$.

Then, a more constrained receiver is never worse off than a less constrained one.

---

[3] The result holds symmetrically if the order is reversed.
[4] Notice that in the case $|\Theta| = 3$, this constraint only applies to $i = 2, j = 1, k = 3$. We phrase it more generally to set the stage for a further generalization below.

The additional assumptions on the sender side capture a stronger version of the utility relationship of the interesting cases in the proof of Theorem 2. They are motivated in many of our cases: for instance, a letter writer may want to obtain the highest possible honor (or salary) for a student, or a prosecutor may want to maximize the sentence of a defendant.

The additional assumption on the receiver side would capture a cautious department or judge, who would prefer to err on the side of not inviting weak candidates (or giving awards to undeserving candidates), or giving the defendant a sentence that is too low rather than ever giving too high of a sentence.

While Proposition 2 shows that with enough assumptions, a positive result can be recovered, we believe that the assumptions are still rather restrictive, meaning that the proposition is likely of limited interest. The proof involves a long and tedious case distinction, and we therefore do not include it in the paper.

For fully general state spaces (i.e., $n = |\Theta| \geq 3$), we can currently obtain a positive result only by imposing even more assumptions on the utility functions. In addition to the (generalization of) the assumptions from Proposition 2, we can make the following assumptions: (1) Whenever $j < i$, the sender's utility difference between actions $j < j'$ is larger under state θ_i than under state $\theta_{i'}$ for $i' > i$. In other words, when the state of the world is smaller, the sender is more sensitive to changes in the receiver's action. (2) For any fixed state θ_i, the receiver's utility as a function of j (the action) is increasing and *convex* for $j \leq i$, and decreasing and convex for $j \geq i$. By adding these two assumptions, we can again obtain a result that a constrained receiver is always weakly better off than an unconstrained one. While it is possible to construct reasonably natural applications which satisfy these conditions, the conditions are far from covering a broad class of Bayesian persuasion settings. For this reason, we are not including a proof of this result, instead considering the discussion as a point of departure towards identifying less stringent assumptions that may enable positive results.

Whether there is a broad and natural class of Bayesian persuasion instances with more than two states of the world in which the insight "A more constrained receiver is better off" from Theorem 2 carries over is an interesting direction for future research.

5.2   Finding Optimal Signaling Schemes

While the main focus of our work is on the receiver's utility when more constrained, our model also raises an interesting computational question, as briefly discussed in Sect. 2.3. In particular, we do not know whether there is a polynomial-time algorithm which—given the sender's and receiver's utility functions as well as the constraints on the receiver—finds a sender-optimal signaling scheme. Since probability constraints on receivers (quotas) are quite natural in many signaling settings, this constitutes an interesting direction for future work.

The main difficulty in applying standard techniques is that the constraints may force the receiver to play an ex post suboptimal action. The standard LP for the sender's optimization problem [14] maximizes the sender's expected utility subject to the constraint that the receiver is incentivized to play the sender's recommended action. To appreciate the difference, consider a setting in which the state of the world is uniform over $\{\theta_1, \theta_2\}$, and the sender and receiver both obtain utility 1 if the receiver plays action a_1, and 0 otherwise. Without any constraints, the sender need not send any signal, and the receiver would simply play action a_1. But if the receiver is constrained to playing action a_1 with probability exactly $\frac{1}{2}$, then she must randomize, including the (always suboptimal) action a_2 with probability $\frac{1}{2}$. By Proposition 1, the randomization can be pushed to the sender instead, but when the sender recommend action a_2, it will be ex post suboptimal for the receiver to follow the recommendation. Indeed, an LP requiring deterministic ex post obedience from the sender would become infeasible for this setting. Whether the sender's optimization problem can still be cast as a different LP, or solved using other techniques, is an interesting direction.

We remark here that the preceding example does not have a state-matching receiver. If the receiver is state-matching and the constraints are feasible, then full revelation of the state is ex post incentive compatible for the receiver. This implies that the linear program for optimizing the sender's utility over ex post incentive compatible signaling schemes has a feasible solution. However, since the LP is more restricted, it is not at all clear that its optimum solution maximizes the sender's utility when the recommendation does not have to be ex post incentive compatible.

5.3   Receiver's Strategic Behaviors on Constraint Enforcement

We assumed throughout the paper that the receiver's constraints are common knowledge, and that enforcing the constraints is indeed required of the receiver (or in her best interest). Aside from the interview example provided in Sect. 1, such constraints are encountered in real-world scenarios such as a patient's dietary restrictions, the salary cap for a sports team, or the capacity limit of an event or facility.

Given that we showed constraints to be beneficial for the receiver, one may suspect that a receiver could strategically misrepresent how harsh her constraints are, or—along the same lines—claim to be constrained, but not enforce the claimed constraints. This would allow the receiver to obtain more information from a sender. In other words, when constraints are not common knowledge, they become private information of the receiver, which could be strategically manipulated; for instance, in the interview example, Alice could indicate a constraint just to force Bob's hand.

Naturally, allowing strategic manipulation in the model will significantly complicate the problem, either making it a dynamic information design problem [15] with multiple senders [1] or a mechanism design problem with incorporated information design modules [32]. Analyzing a model with private receiver constraints thus constitutes an interesting directions for future work.

References

1. Ambrus, A., Takahashi, S.: Multi-sender cheap talk with restricted state spaces. Theor. Econ. **3**(1), 1–27 (2008)
2. Babichenko, Y., Talgam-Cohen, I., Zabarnyi, K.: Bayesian persuasion under ex ante and ex post constraints. In: Proceedings of the AAAI Conference on Artificial Intelligence, vol. 35, no. 6, pp. 5127–5134 (2021)
3. Bergemann, D., Morris, S.: Robust predictions in games with incomplete information. Econometrica **81**(4), 1251–1308 (2013)
4. Bergemann, D., Morris, S.: The comparison of information structures in games: bayes correlated equilibrium and individual sufficiency. Technical report 1909R, Cowles Foundation for Research in Economics, Yale University (2014)
5. Bergemann, D., Morris, S.: Bayes correlated equilibrium and the comparison of information structures in games. Theor. Econ. **11**(2), 487–522 (2016)
6. Bergemann, D., Morris, S.: Information design: a unified perspective. J. Econ. Lit. **57**(1), 44–95 (2019)
7. Boleslavsky, R., Kim, K.: Bayesian persuasion and moral hazard (2018). Available at SSRN 2913669
8. Candogan, O.: Reduced form information design: persuading a privately informed receiver (2020). Available at SSRN 3533682
9. Candogan, O., Strack, P.: Optimal disclosure of information to a privately informed receiver (2021). Available at SSRN 3773326
10. Carroni, E., Ferrari, L., Pignataro, G.: Does costly persuasion signal quality? (2020). Available at SSRN
11. Crawford, V.P., Sobel, J.: Strategic information transmission. Econometrica **50**(6), 1431–1451 (1982)
12. Doval, L., Skreta, V.: Constrained information design: Toolkit (2018). arXiv preprint arXiv:1811.03588
13. Dughmi, S., Kempe, D., Qiang, R.: Persuasion with limited communication. In: Proceedings 17th ACM Conference on Economics and Computation, pp. 663–680 (2016)
14. Dughmi, S., Xu, H.: Algorithmic Bayesian persuasion. SIAM J. Comput. **50**(3), 68–97 (2019)
15. Farhadi, F., Teneketzis, D.: Dynamic information design: a simple problem on optimal sequential information disclosure. Dyn. Games Appl., 1–42 (2021)
16. Gentzkow, M., Kamenica, E.: Costly persuasion. Am. Econ. Rev. **104**(5), 457–62 (2014)
17. Grossman, S.J.: The informational role of warranties and private disclosure about product quality. J. Law Econ. **24**(3), 461–483 (1981)
18. Guo, Y., Shmaya, E.: The interval structure of optimal disclosure. Econometrica **87**(2), 653–675 (2019)
19. Hedlund, J.: Persuasion with communication costs. Games Econom. Behav. **92**, 28–40 (2015)
20. Kamenica, E.: Bayesian persuasion and information design. Ann. Rev. Economics. **11**, 249–272 (2019)
21. Kamenica, E., Gentzkow, M.: Bayesian persuasion. Am. Econ. Rev. **101**(6), 2590–2615 (2011)
22. Kolotilin, A.: Experimental design to persuade. Games Econom. Behav. **90**, 215–226 (2015)

23. Kolotilin, A.: Optimal information disclosure: a linear programming approach. Theor. Econ. **13**(2), 607–635 (2018)
24. Kolotilin, A., Mylovanov, T., Zapechelnyuk, A., Li, M.: Persuasion of a privately informed receiver. Econometrica **85**(6), 1949–1964 (2017)
25. Le Treust, M., Tomala, T.: Persuasion with limited communication capacity. J. Econ. Theor. **184**, 104940 (2019)
26. Matyskova, L.: Bayesian persuasion with costly information acquisition. In: cERGE-EI Working Paper Series (2018)
27. Milgrom, P.R.: Good news and bad news: representation theorems and applications. Bell J. Econ. **12**(2), 380–391 (1981)
28. Nguyen, A., Tan, T.Y.: Bayesian persuasion with costly messages. J. Econ. Theor. **193**, 105212 (2021)
29. Perez-Richet, E.: Interim Bayesian persuasion: First steps. Am. Econ. Rev. **104**(5), 469–74 (2014)
30. Perez-Richet, E., Prady, D.: Complicating to persuade? (2012, unpublished Manuscript)
31. Rayo, L., Segal, I.: Optimal information disclosure. J. Polit. Econ. **118**(5), 949–987 (2010)
32. Roesler, A.K.: Mechanism design with endogenous information. Technical report, Working Paper, Bonn Graduate School of Economics, Bonn, Germany (2014)
33. Spence, M.: Job market signaling. Q. J. Econ. **87**(3), 355–374 (1973)

Towards a Characterization of Worst Case Equilibria in the Discriminatory Price Auction

Evangelos Markakis[1], Alkmini Sgouritsa[2], and Artem Tsikiridis[1(✉)]

[1] Department of Informatics, Athens University of Economics and Business, Athens, Greece
{markakis,artem}@aueb.gr

[2] Department of Computer Science, University of Liverpool, Liverpool, UK
asgouritsa2@liverpool.ac.uk

Abstract. We study the performance of the discriminatory price auction under the uniform bidding interface, which is one of the popular formats for running multi-unit auctions in practice. We undertake an equilibrium analysis with the goal of characterizing the inefficient mixed equilibria that may arise in such auctions. We consider bidders with capped-additive valuations, which is in line with the bidding format, and we first establish a series of properties that help us understand the sources of inefficiency. Moving on, we then use these results to derive new lower and upper bounds on the Price of Anarchy of mixed equilibria. For the case of two bidders, we arrive at a complete characterization of inefficient equilibria and show an upper bound of 1.1095, which is also tight. For multiple bidders, we show that the Price of Anarchy is strictly worse, improving the best known lower bound for submodular valuations. We further present an improved upper bound of 4/3 for the special case where there exists a "high" demand bidder. Finally, we also study Bayes-Nash equilibria, and exhibit a separation result that had been elusive so far. Namely, already with two bidders, the Price of Anarchy for Bayes-Nash equilibria is strictly worse than that for mixed equilibria. Such separation results are not always true (e.g., the opposite is known for simultaneous second price auctions) and reveal that the Bayesian model here introduces further inefficiency.

1 Introduction

Multi-unit auctions form a popular transaction means for selling multiple units of a single good. They have been in use for a long time, and there are by now several practical implementations across many countries. Some of the most prominent applications involve government sales of treasury securities to investors [6], as well as electricity auctions (for distributing electrical energy) [18]. Apart from governmental use, they are also run in other financial markets, and they are being deployed by various online brokers [16]. In the economics literature, multi-unit

© Springer Nature Switzerland AG 2022
M. Feldman et al. (Eds.): WINE 2021, LNCS 13112, pp. 186–204, 2022.
https://doi.org/10.1007/978-3-030-94676-0_11

auctions have been a subject of study ever since the seminal work of Vickrey [23], and some formats were conceived even earlier, by Friedman [10].

The focus of our work is on the welfare performance of the *discriminatory price auction*, which is also referred to as pay-your-bid auction. In particular, we study the *uniform bidding interface*, which is the format most often employed in practice. Under this format, each bidder submits two parameters, a monetary per-unit bid, along with an upper bound on the number of units desired. Hence, each bidder is essentially asked to declare a *capped-additive* curve (a special case of submodular functions). The auctioneer then allocates the units by satisfying first the demand of the bidder with the highest monetary bid, then moving to the second highest bid, and so on, until there are no units left. As a price, each winning bidder pays his bid multiplied by the number of units received.

It is easy to see that the discriminatory price auction is not a truthful mechanism, and the same holds for other formats used in practice. Consequently, in the more recent years, a series of works have studied the social welfare guarantees that can be obtained at equilibrium. The outcome of these works is quite encouraging for the discriminatory price auction. Namely, pure Nash equilibria are always efficient, whereas for mixed and Bayes-Nash equilibria, the Price of Anarchy is bounded by 1.58 [13] for submodular valuations. These results suggest that simple auction formats can attain desirable guarantees and provide theoretical grounds for the overall success in practice.

Despite these positive findings, there has been no progress on further improving the current Price of Anarchy bounds. The known lower bound of 1.109 by [8] is quite far from the upper bounds derived by the commonly used smoothness-based approaches, [13,22], which however do not seem applicable for producing further improvements. We believe the main difficulty in getting tighter results is that one needs to delve more deeply into the properties of Nash equilibria. But obtaining any form of characterization results for mixed or Bayesian equilibria is a notoriously hard problem. Even with two bidders it is often difficult to describe how the set of equilibria looks like. This is precisely the focus of our work, where we manage to either partially or fully characterize equilibrium profiles towards obtaining improved Price of Anarchy bounds, as we outline below.

1.1   Contribution

Motivated by the previous discussion, in Sect. 3 we initiate an equilibrium analysis for mixed equilibria. We consider bidders with capped-additive valuations, which is a subclass of submodular valuations, and consistent with the bidding format. Our results can be seen as a partial characterization of inefficient mixed equilibria, and our major highlights include both structural properties on the demand profile (see Theorem 3), as well as properties on the distributions of the mixed strategies (see Corollary 2, Theorem 4 and Lemma 7).

In Sect. 4, we use these results to derive new lower and upper bounds on the Price of Anarchy for mixed equilibria. For two bidders, we arrive at a complete characterization of inefficient equilibria and show an upper bound of 1.1095, which is tight.[1] For multiple bidders, we show that the Price of Anarchy is strictly worse, which also improves the best known lower bound for submodular valuations [8]. We further present an improved upper bound of 4/3 for the special case where there exists a "high" demand bidder. We believe these latter instances are representative of the worst-case inefficiency that may arise, and refer to the relevant discussion in Sect. 4.2. To summarize, our results show that in several cases, the Price of Anarchy is even lower than the previous bound of [13] and strengthen the perception that such auctions can work well in practice.

Finally, in Sect. 5, we also study Bayes-Nash equilibria, and we exhibit a separation result that had been elusive so far: already with two bidders, the Price of Anarchy for Bayes-Nash equilibria is strictly worse than for mixed equilibria. Such separation results, though intuitive, do not hold for all auction formats. For example, in simultaneous second price auctions with submodular valuations [7], the known tight bounds for mixed equilibria extend to the Bayesian model via smoothness arguments [19]. This reveals that the Bayesian model in our setting introduces a further source of inefficiency. Note that to obtain this result, we transform the underlying optimization of social welfare at equilibrium to a well-posed variational calculus problem. This technique may be of independent interest and have other applications in mechanism design.

1.2   Related Work

The work of [1] was among the first ones that studied the sources of inefficiency in multi-unit auctions. For the discriminatory price auction, the Price of Anarchy was later studied in [22], and for bidders with submodular valuations, the currently best upper bound of $e/(e-1) \approx 1.58$ has been obtained by [13] (both for mixed and for Bayes-Nash equilibria). These results exploit the smoothness-based techniques, developed by [19,22]. One can also obtain slightly worse upper bounds for subadditive valuations, by using a different methodology, based on [9]. As for lower bounds, the only construction known for submodular valuations is by [8], yielding a bound of at least 1.109. In parallel to these results, there has been a series of works on the inefficiency of many other auction formats, ranging from multi-unit to combinatorial auctions, see among others, [4,5,7,9].

Apart from social welfare guarantees, several other aspects or properties of equilibrium behavior have been studied. Recently in [17], a characterization of equilibria is given for a model where the supply of units can be drawn from a distribution. In the past, several works have focused on revenue equivalence results between the discriminatory price and the uniform price auction, see e.g. [2,20]. On a different direction, comparisons from the perspective of the bidders are carried out in [3].

---

[1] In [8] there is a lower bound of 1.109 that applies to our setting with two bidders and three units. The lower bound we provide here is just slightly better, but most importantly, it is tight and can be seen as a generalization of the instance in [8] to many units.

For a more detailed exposition on multi-unit auctions and their earlier applications, we refer the reader to the books [14] and [15]. For more recent applications, we refer to [6,11,18], for treasury bonds, carbon licence auctions, and electricity auctions, respectively.

2 Notation and Definitions

We consider a discriminatory price multi-unit auction, involving the allocation of k identical units of a single item, to a set $\mathcal{N} = \{1,\ldots,n\}$ of bidders. Each bidder $i \in \mathcal{N}$ has a private value $v_i > 0$, which reflects her value per unit and a private demand $d_i \in \mathbb{Z}_+$ which reflects the maximum number of units bidder i requires. Therefore, if the auction allocates $x_i \leq k$ units to bidder i, her total value will be $\min\{x_i, d_i\} \cdot v_i$. We note that this class of valuations is a subclass of submodular valuations, and includes all additive vectors (when $d_i = k$). We will refer to them as *capped-additive* valuations.

We focus on the following simple format for the discriminatory price auction, which is known as the *uniform bidding* interface. The auctioneer asks each bidder $i \in \mathcal{N}$ to submit a tuple (b_i, q_i), where $b_i \geq 0$, is her monetary bid per unit (not necessarily equal to v_i), and q_i is her demand bid (not necessarily equal to d_i). We denote by $\mathbf{b} = (b_1,\ldots,b_n)$ the monetary bidding vector, and similarly $\mathbf{q}$ will be the declared demand vector. For a bidding profile $(\mathbf{b}, \mathbf{q})$, the auctioneer allocates the units by satisfying first the demand of the bidder with the highest monetary bid, then moving to the second highest bid, and so on, until there are no units left. Hence, all the winners have their reported demand satisfied, except possibly for the one selected last, who may be partially satisfied. Moreover, we assume that in case of ties, a deterministic tie-breaking rule is used, which does not depend on the input bids submitted by the players to the auctioneer (e.g., a fixed ordering of the players suffices).

For every bidding profile $(\mathbf{b}, \mathbf{q})$, we let $x_i(\mathbf{b}, \mathbf{q})$ be the number of units allocated to bidder i, where obviously $x_i(\mathbf{b}, \mathbf{q}) \leq q_i$. In the discriminatory auction, the auctioneer requires each bidder i to pay b_i per allocated unit, hence a total payment of $b_i \cdot x_i(\mathbf{b}, \mathbf{q})$. The utility function of bidder $i \in \mathcal{N}$, given a bidding profile $(\mathbf{b}, \mathbf{q})$, is: $u_i(\mathbf{b}, \mathbf{q}) = \min\{x_i(\mathbf{b}, \mathbf{q}), d_i\}v_i - x_i(\mathbf{b}, \mathbf{q})b_i$.

Viewed as games, these auctions have an infinite pure strategy space, and we also allow bidders to play mixed strategies, which are probability distributions over their set of pure strategies. When each bidder $i \in \mathcal{N}$ uses a mixed strategy G_i, she *independently* draws a bid (b_i, q_i) from G_i. We refer to $\mathbf{G} = \times_{i=1}^n G_i$ as the product distribution of bids. Under mixed strategies, the expected utility of a bidder i is $\mathbb{E}_{(\mathbf{b},\mathbf{q})\sim\mathbf{G}}[u_i(\mathbf{b}, \mathbf{q})]$.

Definition 1. *We say that $\mathbf{G}$ is a mixed Nash equilibrium when for all $i \in \mathcal{N}$, all $b_i' \geq 0$ and all $q_i' \in \mathbb{Z}_+$*

$$\mathbb{E}_{(\mathbf{b},\mathbf{q})\sim\mathbf{G}}[u_i(\mathbf{b}, \mathbf{q})] \geq \mathbb{E}_{(\mathbf{b}_{-i},\mathbf{q}_{-i})\sim\mathbf{G}_{-i}}[u_i((b_i', \mathbf{b}_{-i}), (q_i', \mathbf{q}_{-i}))].$$

We note that in any equilibrium, if a bidder i declares with positive probability a bid that exceeds v_i, she should not be allocated any unit, since such strategies are strictly dominated by bidding the actual value v_i.

Fact 1. *Let* $\mathbf{G}$ *be a mixed Nash equilibrium. The probability that a bidder i is allocated some units, conditioned that she bids higher than v_i, is 0.*

In the sequel, we focus on equilibria, where the monetary bids never exceed the value per unit.

Given a valuation profile $(\mathbf{v}, \mathbf{d})$, we denote by $OPT(\mathbf{v}, \mathbf{d})$ the optimal social welfare (which can be computed very easily by running the allocation algorithm of the auction with the true value and demand vector). We also denote by $SW(\mathbf{G})$ the expected social welfare of a mixed Nash equilibrium $\mathbf{G}$, i.e., equal to $\mathbb{E}_{(\mathbf{b},\mathbf{q})\sim\mathbf{G}}[\sum_i \min\{x_i(\mathbf{b}, \mathbf{q}), d_i\}v_i]$. The Price of Anarchy is the worst-case ratio $\frac{OPT(\mathbf{v},\mathbf{d})}{SW(\mathbf{G})}$, over all valuation profiles $(\mathbf{v}, \mathbf{d})$, and all equilibria $\mathbf{G}$.

We refer to an equilibrium as *inefficient* when its social welfare is strictly less than the optimal.

3   Towards a Characterization of Inefficient Mixed Equilibria

In this section, we derive a series of important properties, that help us understand better how can inefficient equilibria arise. These properties will help us analyze the Price of Anarchy in Sect. 4.

3.1   Mixed Nash Equilibria with Demand Revelation

Our first result is that it suffices to focus on equilibria where bidders truthfully reveal their demand, resulting therefore in a single-parameter strategy space for the bidders (Theorem 1). We further argue that the inefficiency in equilibria appears only when the total demand exceeds k (Lemma 1) and therefore this is what we assume for the rest of the paper.

Theorem 1. *Let* $(\mathbf{v}, \mathbf{d})$ *be a valuation profile, and* $\mathbf{G}$ *be a mixed Nash equilibrium. Then, for every $i \in \mathcal{N}$, and in every pure strategy profile $(b_i, q_i) \sim G_i$, we can replace q_i by d_i so that the resulting distribution remains a mixed Nash equilibrium with the same social welfare.*

Lemma 1. *If $\sum_i d_i \le k$ then the social welfare of any mixed Nash equilibrium is optimal.*

3.2   Existence of Non-empty-handed Bidders

For the rest of the paper we consider only strategy profiles where the bidders' demand bid matches their true demand. The main goal of this subsection is to derive Theorem 3, where we show that in any inefficient mixed equilibrium, there always exists a bidder such that the total demand of the other winners is strictly less than k, meaning that at least one item is allocated to him for sure (with probability one). This is a crucial property for understanding the formation of inefficient mixed equilibria. To proceed, we give first some further notation to be used in this and the following sections.

Further Notation. Given Theorem 1, instead of using distributions on tuples (b_i, q_i), we suppose that each bidder $i \in \mathcal{N}$ independently draws only a monetary bid b_i from a distribution B_i and we refer to $\mathbf{B} = \times_{i=1}^n B_i$ as the product distribution of monetary bids or just bids from now on. For a bidding profile $\mathbf{b}$, the utility of a bidder i will simply be denoted as $u_i(\mathbf{b})$, instead of $u_i(\mathbf{b}, \mathbf{d})$. Definition 1 is also simplified, and we say that $\mathbf{B}$ is an equilibrium if $\mathbb{E}_{\mathbf{b} \sim \mathbf{B}}[u_i(\mathbf{b})] \geq \mathbb{E}_{\mathbf{b}_{-i} \sim \mathbf{B}_{-i}}[u_i((b_i', \mathbf{b}_{-i}))]$, for any i and any $b_i' \geq 0$. Similarly, the social welfare of a mixed Nash equilibrium $\mathbf{B}$ is given by just $SW(\mathbf{B})$.

For a mixed strategy bidding profile $\mathbf{B}$, we denote by $W(\mathbf{B})$ the set of bidders with positive expected utility, i.e., $W(\mathbf{B}) = \{j : \mathbb{E}_{\mathbf{b} \sim \mathbf{B}}[u_j(\mathbf{b})] > 0\}$, and let $\mathbf{B}_W = \times_{i \in W(\mathbf{B})} B_i$. Moreover, the support of a bidder i in $\mathbf{B}$ is the domain of the distribution B_i, that i plays under $\mathbf{B}$, denoted by $Supp(B_i)$. We denote by $\ell(B_i), h(B_i)$ the leftmost and rightmost points, respectively, in the support of bidder i. In particular, if the rightmost part of the domain of B_i is a mass point b or an interval in the form $[a, b]$, then $h(B_i) = b$, and similarly for $\ell(B_i)$. In cases of distributions over intervals, we can safely assume that the domain contains only *closed* intervals, because the endpoints are chosen with zero probability. We further denote by $\ell(\mathbf{B}_W), h(\mathbf{B}_W)$ the leftmost and rightmost points, respectively, of the union of the supports of $W(\mathbf{B})$.

For $i = 1, \ldots, n$ we denote by F_i the CDF of B_i and by f_i their PDF. Moreover, given a profile $\mathbf{b}$, it is often useful in the analysis to consider the vector of bids (thresholds) that a bidder i competes against, denoted by $\beta(\mathbf{b})_{-i} = (\beta_1(\mathbf{b}_{-i}), \ldots, \beta_k(\mathbf{b}_{-i}))$. Here, $\beta_j(\mathbf{b}_{-i})$ is the j-th lowest winning bid of the profile $\mathbf{b}_{-i}$, for $j = 1, \ldots, k$, so that $\beta(b)_{-i}$ describes the winning bids if i didn't participate. This implies that, under profile $\mathbf{b}$, bidder i is allocated $j = 1, \ldots, k-1$ units capped by d_i, when $\beta_j(\mathbf{b}_{-i}) < b_i < \beta_{j+1}(\mathbf{b}_{-i})$ and d_i units, when $\beta_k(\mathbf{b}_{-i}) < b_i$. We note that because we focus on the uniform bidding interface, some consecutive β_j values may coincide and be equal to the bid of the same bidder. When $\mathbf{b}_{-i} \sim \mathbf{B}_{-i}$, for $i = 1, \ldots, n$, we denote the CDF of the random variable $\beta_j(\mathbf{b}_{-i})$ as $\hat{F}_{ij}$, for $j = 1, \ldots, k$. In the next fact, we express the expected allocation of any bidder i for bidding some $\alpha > 0$, in terms of the values $\hat{F}_{ij}(\alpha)$.

Fact 2. *Let $\mathbf{B}_{-i}$ be a product distribution of bids. Then for all $\alpha \geq 0$, where no bidder other than (possibly) i has a mass point,* $\mathbb{E}_{\mathbf{b}_{-i} \sim \mathbf{B}_{-i}}[x_i(\alpha, \mathbf{b}_{-i})] =$

$$\sum_{j=1}^{d_i} \hat{F}_{ij}(\alpha).$$

Given a bidding profile $\mathbf{B}$, for any bidder i we define $\hat{F}_i^{avg}(x) = \frac{\sum_{j=1}^{d_i} \hat{F}_{ij}(x)}{d_i}$, to be the average CDF of the winning bids that bidder i competes against. Note that $\hat{F}_i^{avg}$ is a CDF since it is the average of a number of CDFs.

Remark 1. The $\hat{F}_{ij}$ functions are right continuous, as they are CDFs, and moreover, if the F_i functions have no mass point, the same holds for the $\hat{F}_{ij}$ functions. Additionally, if for any j, the $\hat{F}_{ij}$ functions are continuous, so is $\hat{F}_i^{avg}$, as the average of continuous functions.

We start by ruling out certain scenarios that cannot occur at inefficient equilibria. First, we can safely ignore bidders with zero expected utility, since in any *inefficient* mixed Nash equilibrium they do not receive any units.

Lemma 2. *Any mixed Nash equilibrium $\mathbf{B}$ with at least one bidder with zero expected utility, but positive expected number of allocated units, is efficient.*

Next, we show that to have inefficiency at an equilibrium, there must exist at least two bidders with positive expected utility.

Lemma 3. *Let $(\mathbf{v}, \mathbf{d})$ be a valuation profile and $\mathbf{B}$ be an inefficient mixed Nash equilibrium. Then, $|W(\mathbf{B})| \geq 2$.*

The next warm-up properties involve the expected utility of a bidder under an equilibrium $\mathbf{B}$, conditioned that she bids within a certain interval or at a single point. We start with Fact 3, which is a straightforward implication of the equilibrium definition, and proceed by arguing that no two bidders may bid on the same point with positive probability. Theorem 2 concludes by stating the main property regarding the utility of bidders when bidding in their support.

Fact 3. *Let $\mathbf{B}$ be an equilibrium. For a bidder i, consider a partition of $Supp(B_i)$ (or of a subset of it) into smaller disjoint sub-intervals, say $I_1, \ldots, I_\ell$, such that B_i has a positive probability on each sub-interval (mass points may also be considered as sub-intervals). Then, it should hold that $\mathbb{E}_{\mathbf{b} \sim \mathbf{B}}[u_i(\mathbf{b}) \mid b_i \in I_r] = \mathbb{E}_{\mathbf{b} \sim \mathbf{B}}[u_i(\mathbf{b})]$, for every $r = 1, \ldots, \ell$.*

Based on Fact 3, we can obtain the following point-wise version. Variations of the version below have also appeared in related works, see e.g., [8].

Theorem 2. *Given a mixed Nash equilibrium $\mathbf{B}$, bidder i and $z \in Supp(B_i)$, where no other bidder has a mass point on z, $\mathbb{E}_{\mathbf{b}_{-i} \sim \mathbf{B}_{-i}}[u_i(z, \mathbf{b}_{-i})] = \mathbb{E}_{\mathbf{b} \sim \mathbf{B}}[u_i(\mathbf{b})]$.*

We further give the following observation regarding the existence of mass points on $\ell(\mathbf{B}_W)$.

Observation 1. *In any inefficient mixed Nash equilibrium $\mathbf{B}$, there can be no bidders $i, j \in W(\mathbf{B})$ such that both $Pr[b_i = \ell(\mathbf{B}_W)] > 0$ and $Pr[b_j = \ell(\mathbf{B}_W)] > 0$.*

The main theorem of this section follows, stating the existence of a special bidder, who always receives at least one unit, and is referred to as *non-empty-handed.*

Theorem 3. *Let* $(\mathbf{v}, \mathbf{d})$ *be a valuation profile, and let* $\mathbf{B}$ *be any inefficient mixed Nash equilibrium. Then, there exists a bidder* $i \in W(\mathbf{B})$, *such that*

$$\sum_{j \in W(\mathbf{B}) \setminus \{i\}} d_j \leq k - 1.$$

Proof. On the contrary, suppose that for every $i \in W(\mathbf{B})$, $\sum_{j \in W(\mathbf{B}) \setminus \{i\}} d_j \geq k$. Let i be some bidder with $\ell = \ell(\mathbf{B}_W) \in Supp(B_i)$. We distinguish two cases.

Case 1: There exists an interval in the form $[\ell, \ell + \epsilon]$, on which B_i has a positive probability mass and on which the bidders of $W(\mathbf{B}) \setminus \{i\}$ have a zero mass. We note that we also allow $\epsilon = 0$, i.e., that i has a mass point on ℓ and the other bidders do not. This means that when bidder i bids within $[\ell, \ell + \epsilon]$, all the other bidders from $W(\mathbf{B})$ are above him. Since we assumed that the total demand of $W(\mathbf{B}) \setminus \{i\}$ is at least k, bidder i does not win any units in this case. Since i bids with positive probability in $[\ell, \ell + \epsilon]$, by Fact 3, we have $\mathbb{E}_{\mathbf{b} \sim \mathbf{B}}[u_i(\mathbf{b})] = 0$, which contradicts the fact that $i \in W(\mathbf{b})$.

Case 2: Note that by Observation 1, it cannot happen that both bidder i and at least one bidder $j \in W \setminus \{i\}$, have a mass point on ℓ. Hence, the only remaining case to consider is that any mass point that may exist by the bidders is at some $x > \ell$, and there is also no interval starting from ℓ that is used only by bidder i. Thus, there exists an interval I in the form $I = [\ell, \ell + \epsilon]$ for some small enough $\epsilon > 0$, and a bidder $j \in W(\mathbf{B}) \setminus \{i\}$, such that both B_i and B_j contain I in their support, and have positive probability mass on I without mass points.

By Theorem 2, we obtain that $\mathbb{E}_{\mathbf{b}_{-i} \sim \mathbf{B}_{-i}}[u_i(\ell, \mathbf{b}_{-i})] = \mathbb{E}_{\mathbf{b} \sim \mathbf{B}}[u_i(\mathbf{b})] > 0$. This is a contradiction, because by bidding ℓ, bidder i ranks lower than all other bidders of $W(\mathbf{B})$ with probability one. By our assumption that $\sum_{j \in W(\mathbf{B}) \setminus \{i\}} d_j \geq k$, there are no units left for i when she ranks last among $W(\mathbf{B})$, and therefore, $\mathbb{E}_{\mathbf{b}_{-i} \sim \mathbf{B}_{-i}}[u_i(\ell, \mathbf{b}_{-i})] = 0$. □

The property above already implies the following interesting corollary, that if all bidders have unit demand, any mixed Nash equilibrium is efficient.

Corollary 1. *Let* $(\mathbf{v}, \mathbf{d})$ *be a valuation profile with only unit-demand bidders, i.e.,* $d_i = 1$ *for all* i. *Then any mixed Nash equilibrium* $\mathbf{B}$ *is efficient.*

3.3   The Support and the CDFs of Mixed Nash Equilibria

The existence of a non-empty-handed bidder (Theorem 3) helps us to establish further properties that characterize the structure of inefficient mixed Nash equilibria. These properties (and especially Theorem 4) will be important to establish the inefficiency results that follow. We start with an observation regarding the highest bid of any bidder $i \in W(\mathbf{B})$, which should be strictly less than v_i.

Observation 2. *For any bidder $i \in W(\mathbf{B})$, $h(B_i) < v_i$.*

The next lemma shows that at any equilibrium $\mathbf{B}$, bidders who are not non-empty-handed cannot have higher bids in their support than the support of the non-empty-handed bidders. Moreover, any bidder who is non-empty-handed does not have a reason to use bids that are higher than the maximum bid of all other winning bidders. The reason is that if such differences existed, then there would be incentives to win the same number of units by lowering one's bid. Then, Lemma 5 shows that no bidder will bid alone at any point or interval, and Lemma 6 specifies that no mass points may exist apart from one case.

Lemma 4. *Let $(\mathbf{v}, \mathbf{d})$ be a valuation profile and $\mathbf{B}$ be any inefficient mixed Nash equilibrium. Then, for any non-empty-handed bidder i, it holds that $h(B_i) = h(\mathbf{B}_{W \setminus \{i\}}) = h(\mathbf{B}_W)$.*

Lemma 5. *Let $(\mathbf{v}, \mathbf{d})$ be any valuation profile and $\mathbf{B}$ be any mixed Nash equilibrium. For all $i \in W(\mathbf{B})$, it holds that $Supp(B_i) \subseteq \bigcup_{j \in W(\mathbf{B}) \setminus \{i\}} Supp(B_j)$.*

Lemma 6. *Let $(\mathbf{v}, \mathbf{d})$ be a valuation profile and $\mathbf{B}$ be any inefficient mixed Nash equilibrium.*
1) There exists no bidder $i \in W(\mathbf{B})$ and no point $z \in Supp(B_i) \setminus \{\ell(\mathbf{B}_W)\}$, with $F_i(z) > \lim_{z \to z^-} F_i(z)$, i.e., there are no mass points among the bidders of $W(\mathbf{B})$, except possibly the leftmost endpoint of all bidders' distributions.
2) At most one bidder $i \in W(\mathbf{B})$ may have a mass point on $\ell(\mathbf{B}_W)$, in which case, i is a non-empty-handed bidder.

By combining Theorem 2 and Lemma 6 we get the following Corollary.

Corollary 2. *For any inefficient mixed Nash equilibrium $\mathbf{B}$, the following hold:*
1) For any bidder i and $z \in Supp(B_i) \setminus \{\ell(\mathbf{B}_W)\}$, $\mathbb{E}_{\mathbf{b}_{-i} \sim \mathbf{B}_{-i}}[u_i(z, \mathbf{b}_{-i})] = \mathbb{E}_{\mathbf{b} \sim \mathbf{B}}[u_i(\mathbf{b})]$.
2) If there exists a bidder i with $Pr[b_i = \ell(\mathbf{B}_W)] > 0$, then i is a non-empty-handed bidder and $\mathbb{E}_{\mathbf{b}_{-i} \sim \mathbf{B}_{-i}}[u_i(\ell(\mathbf{B}_W), \mathbf{b}_{-i})] = \mathbb{E}_{\mathbf{b} \sim \mathbf{B}}[u_i(\mathbf{b})]$.
3) If no non-empty-handed bidder exists with mass point on $\ell(\mathbf{B}_W)$, for any bidder i with $\ell(\mathbf{B}_W) \in Supp(B_i)$, $\mathbb{E}_{\mathbf{b}_{-i} \sim \mathbf{B}_{-i}}[u_i(\ell(\mathbf{B}_W), \mathbf{b}_{-i})] = \mathbb{E}_{\mathbf{b} \sim \mathbf{B}}[u_i(\mathbf{b})]$.

Observation 3. *For any inefficient mixed Nash equilibrium $\mathbf{B}$, either there exists a non-empty-handed bidder $i \in W(\mathbf{B})$ with a mass point on $\ell(\mathbf{B}_W)$, or there are at least two non-empty-handed bidders with $\ell(\mathbf{B}_W)$ in their support.*

Given any (inefficient) equilibrium, the next theorem specifies the average CDF of the winning bids that bidder i competes against, i.e., $\hat{F}_i^{avg}$, in i's support.

Theorem 4. *Let $(\mathbf{v}, \mathbf{d})$ be any valuation profile and $\mathbf{B}$ be any inefficient mixed Nash equilibrium. Then, for $i \in W(\mathbf{B})$, the CDF $\hat{F}_i^{avg}$ satisfies*

$$\hat{F}_i^{avg}(z) = \frac{u_i}{d_i(v_i - z)}, \quad \forall z \in Supp(B_i),$$

where $u_i = \mathbb{E}_{\mathbf{b} \sim \mathbf{B}}[u_i(\mathbf{b})] > 0$.

A corollary of Theorem 4 is that the union of the support of the winners is an interval.

Corollary 3. *Let* $(\mathbf{v}, \mathbf{d})$ *be any valuation profile and* $\mathbf{B}$ *be any inefficient mixed Nash equilibrium. Then, for every bidder* $i \in W(\mathbf{B})$, $\bigcup_{j \in W(\mathbf{B}) \setminus \{i\}} Supp(B_j) = [\ell(\mathbf{B}_W), h(\mathbf{B}_W)]$.

The final lemma of this section shows that the rightmost point in the support of $\mathbf{B}$ is a function of the parameters of certain non-empty-handed bidders.

Lemma 7. *Let* $(\mathbf{v}, \mathbf{d})$ *be any valuation profile and* $\mathbf{B}$ *be any inefficient mixed Nash equilibrium. Let* $i \in W(\mathbf{B})$ *be the non-empty-handed bidder such that* $Pr[b_i = \ell(\mathbf{B}_W)] > 0$, *or if no such bidder exists, then let* i *be any non-empty-handed bidder with* $\ell(\mathbf{B}_W)$ *in his support. We have*

$$h(\mathbf{B}_W) = h(B_i) = v_i - \left(k - \sum_{j \in W(\mathbf{B}) \setminus \{i\}} d_j\right) \frac{v_i - \ell(\mathbf{B}_W)}{d_i}.$$

4   Price of Anarchy for Mixed Equilibria

We can now exploit the properties derived so far for mixed equilibria, in order to analyze the inefficiency of the discriminatory price auction. Since we focus on inefficient equilibria, we assume that in any valuation profile considered in this section, there are at least two bidders with a different value per unit.

4.1   The Case of Two Bidders

We pay particular attention to the case of $n = 2$. This is a setting where we can fully characterize in closed form the distributions of the inefficient mixed Nash equilibria, and derive valuable intuitions for the worst-case instances with respect to the Price of Anarchy, that are helpful also for auctions with multiple bidders. The main result of this subsection is the following theorem, showing that the inefficiency is quite limited.

Theorem 5. *For* $k \geq 2$, $n = 2$ *and capped additive valuation profiles, the Price of Anarchy of mixed equilibria is at most* 1.1095, *and this is tight as* k *goes to infinity.*

We postpone the proof of Theorem 5, as we first need to establish some properties regarding the form of inefficient mixed Nash equilibria with two bidders. For $n = 2$, a capped-additive valuation profile can be described as $(\mathbf{v}, \mathbf{d}) = ((v_1, d_1), (v_2, d_2))$. Recall also that it is sufficient to focus our attention only on profiles where $d_1 + d_2 > k$, since otherwise, by Lemma 1 any mixed equilibrium is efficient. We start our analysis by characterizing the support of inefficient mixed Nash equilibria.

Lemma 8. *Let* $(\mathbf{v}, \mathbf{d}) = ((v_1, d_1), (v_2, d_2))$ *be any capped-additive valuation profile of two bidders, and* $\mathbf{B} = (B_1, B_2)$ *be any inefficient mixed Nash equilibrium. Then:*

1. $Supp(B_1) = Supp(B_2) = [\ell(B_1), h(B_1)]$, *and* $\ell(B_1) = 0$.
2. $h(B_1)$ *takes one of the following values*

$$h(B_1) = v_1 \frac{d_1 + d_2 - k}{d_1} \quad or \quad h(B_1) = v_2 \frac{d_1 + d_2 - k}{d_2}.$$

The following theorem specifies the cumulative distribution functions that comprise any inefficient mixed Nash equilibrium, along with a necessary condition for the existence of such equilibria. For a bidder i below, we use the notation v_{-i} and d_{-i} to denote the value and demand of the other bidder.

Theorem 6. *Let* $(\mathbf{v}, \mathbf{d}) = ((v_1, d_1), (v_2, d_2))$ *be a capped-additive valuation profile of two bidders, and* $\mathbf{B} = (B_1, B_2)$ *be any inefficient mixed Nash equilibrium.*

1. *The cumulative distribution function of bidder* i, *for* $i = 1, 2$, *is*

$$F_i(z) = \frac{1}{d_1 + d_2 - k} \left(\frac{d_{-i}(v_{-i} - h(B_i))}{v_{-i} - z} - (k - d_i) \right). \tag{1}$$

2. *Furthermore, for* i *being the non-empty-handed bidder with a mass point at* 0, *or if no such bidder exists, being any non-empty-handed bidder, it holds that* $\frac{v_{-i}}{v_i} \geq \frac{d_{-i}}{d_i}$,

Remark 2. By Lemma 8 and Theorem 6, we can see that there can be at most two inefficient equilibria, depending on how the interval of the support was determined.

We are now ready to prove Theorem 5.

Proof Sketch of Theorem 5. The properties established so far imply a full characterization of instances that have inefficient equilibria. To establish Theorem 5, we will group instances into three appropriate classes and we will solve an appropriately defined optimization problem that approximates the Price of Anarchy for each subclass to arbitrary precision.

Suppose without loss of generality that we are given a value profile $(\mathbf{v}, \mathbf{d}) = ((v_1, d_1), (v_2, d_2))$ of k units, such that $d_1 \geq d_2 > 0$. Let $\bar{d}_1 := \frac{d_1}{k}$ and $\bar{d}_2 = \frac{d_2}{k}$, be the *normalized* demands of the bidders. Essentially, we intend to use $v_1, v_2, \bar{d}_1$ and $\bar{d}_2$ as the variables of the optimization problem mentioned before.

Let $\mathbf{B}$ be any inefficient mixed Nash equilibrium. With a slight abuse of notation we view the term $h(B_i)$ as a function of the valuation profile parameters, as established by Lemma 8, and define the functions $h_i(\mathbf{v}, \bar{\mathbf{d}}) = v_i \frac{\bar{d}_1 + \bar{d}_2 - 1}{\bar{d}_i}$ for $i = 1, 2$. Our goal now is to express the social welfare of $\mathbf{B}$, solely in terms of the value profile parameters, $(\mathbf{v}, \mathbf{d})$ and k, and without dependencies on the underlying equilibrium distributions. To proceed, we define first two auxiliary functions; namely, for $i = 1, 2$, we let $S_i(\mathbf{v}, \bar{\mathbf{d}})$ be equal to:

$$\bar{d}_{-i}(v_{-i} - v_i) \left(1 - \int_0^{h_i(\mathbf{v}, \bar{\mathbf{d}})} \frac{1}{\bar{d}_1 + \bar{d}_2 - 1} \left(\frac{\bar{d}_i(v_i - h_i(\mathbf{v}, \bar{\mathbf{d}}))}{v_i - z} - (1 - \bar{d}_i) \right) \frac{v_{-i} - h_i(\mathbf{v}, \bar{\mathbf{d}})}{(v_{-i} - z)^2} dz \right) + v_i.$$

With these expressions in mind, the following lemma allows us to obtain the social welfare in a form that we can later exploit for producing our upper bound. The lemma follows by Theorem 6, which tells us what the equilibrium CDFs are, in terms of the valuation profile.

Lemma 9. *Let i be a non-empty handed bidder with a mass point at 0. Then, $SW(\mathbf{B}) = kS_i(\mathbf{v}, \bar{\mathbf{d}})$. If no such bidder exists, then either $SW(\mathbf{B}) = kS_1(\mathbf{v}, \bar{\mathbf{d}})$ or $SW(\mathbf{B}) = kS_2(\mathbf{v}, \bar{\mathbf{d}})$.*

To conclude the proof of the upper bound, we solve a sequence of optimization problems as determined by the cases arising in the statement of Lemma 9, and by the ordering of the values v_1, v_2. By solving these problems numerically, we found out that in the worst case instance $v_1 = 1, v_2 \approx 0.526, \bar{d}_1 = 1, \bar{d}_2 \approx 0.357$. It is not hard to convert the variables to the underlying worst case instance, which we present in the next paragraph.

Tight Example. Consider an instance of the discriminatory auction for $k \geq 4$ units and $n = 2$ bidders. Bidder 1 has value $v_1 = 1$ and $d_1 = k$, whereas bidder 2 has a value $v_2 = 0.526$ and $d_2 = \lceil 0.357k \rceil$ units. Let B_1, B_2 be two distributions supported in $[0, \frac{d_2}{k}]$. Note that $v_2 > \frac{d_2}{k}$. In accordance to Equation (1), the cumulative distribution functions of B_1 and B_2 are

$$F_1(z) = \frac{v_2 - \frac{d_2}{k}}{v_2 - z}, \qquad\qquad F_2(z) = \frac{k - d_2}{d_2} \frac{z}{1 - z}.$$

It is easy to verify that $\mathbf{B} = (B_1, B_2)$ is indeed a mixed equilibrium. The optimal allocation is for bidder 1 to obtain all k units and the expected social welfare of $\mathbf{B}$, by Lemma 9, is $SW(\mathbf{B}) = kS_1(\mathbf{v}, \bar{\mathbf{d}})$, since $F_1(0) > 0$. The worst case inefficiency ratio occurs as k grows and is approximately 1.1095. □

4.2 Multiple Bidders

Inspired by the construction in the previous section, we move to instances with more than two bidders and provide first a lower bound on the Price of Anarchy. This bound shows a separation between $n = 2$ and $n > 2$, in the sense that equilibria can be more inefficient with a higher number of bidders. It also improves the best known lower bound of the discriminatory price auction for the class of submodular valuations, which was 1.109, by [8]. The improvement however is rather small.

Theorem 7. *For $n > 2$, and for the class of mixed strategy Nash equilibria, the Price of Anarchy is at least 1.1204.*

The above bound is the best lower bound we have been able to establish, even after some extensive experimentation (driven by the results in the remainder of this section). It is natural to wonder if there is a matching upper bound, which would establish that the Price of Anarchy remains very small even for a large number of bidders. Recall that from [13], we know already a bound of

$e/(e-1) \approx 1.58$. Although we have not managed to settle this question, we will provide an improved upper bound for a special case, for which there is evidence that it captures worst-case scenarios of inefficiency. At the same time, we will be able to characterize the format of such worst case equilibria.

To obtain some intuition, it is instructive to look at the proofs of our two lower bounds, in Theorem 5 and in Theorem 7. One can notice that the main source of inefficiency is the fact that the auctioneer accepts multi-unit demand declarations. When this does not occur, we have already shown in Corollary 1 that mixed Nash equilibria attain optimal welfare. When multi-demand bidders are present, Theorem 5 shows that in the case of two bidders, the most inefficient mixed Nash equilibrium occurs when a participating bidder declares a demand for all the units, whereas the opponent requires a much smaller fraction of the supply. In the proof of Theorem 7 above, we have extended this paradigm for multiple bidders with an arbitrary demand structure, but under the assumption that one of the bidders requires all the units (the additive bidder). Such a setting, of one large-demand bidder facing competition by multiple small-demand bidders has also been discussed in [3]. Furthermore, there exist other auction formats that also needed such a demand profile at their worst case instances, see e.g., [5] for the uniform price auction. To summarize, it seems unlikely that the worst instances involve only bidders with low demand or small variation on their demands.

Given the above, we will analyze the family of instances where there exists an additive bidder (with demand equal to k), and where she also has the highest value per unit. In fact, the latter assumption is needed only for the Price of Anarchy analysis but not for the characterization of the worst-case demand profile and the equilibrium strategies. We strongly believe that this class is representative of the most inefficient mixed Nash equilibria (which is true already for the case of two bidders).

The main result of this section is the following.

Theorem 8. *Consider the class of valuation profiles, where there exists an additive bidder α with the highest value, and an equilibrium $\mathbf{B}$, such that $\alpha \in W(\mathbf{B})$. Then, the Price of Anarchy is at most $4/3$.*

The proof of the theorem is by following a series of steps. The existence of the additive bidder helps in the analysis, because a direct corollary of Theorem 3 is that the additive bidder is the sole non-empty-handed bidder (everyone else faces competition for all the units).

Corollary 4. (by Theorem 3). *Consider a valuation profile $(\mathbf{v}, \mathbf{d})$ with an additive bidder α, that admits an equilibrium $\mathbf{B}$, such that $\alpha \in W(\mathbf{B})$. Then, bidder α is the unique non-empty-handed bidder under $\mathbf{B}$, thus, $\sum_{i \in W(\mathbf{B}) \setminus \{\alpha\}} d_i \leq k-1$.*

To proceed, we ensure that for the instances described by Theorem 8, it suffices to analyze the equilibria where bidder α belongs to $W(\mathbf{B})$, i.e., there cannot exist a more inefficient equilibrium $\mathbf{B}'$ of these instances with $\alpha \notin W(\mathbf{B}')$. This is addressed by the following lemma.

Lemma 10. *Consider a valuation profile, and suppose that it admits two distinct inefficient equilibria,* $\mathbf{B}$ *and* $\mathbf{B}'$. *If* $i \in W(\mathbf{B})$ *is a non-empty-handed bidder in* $\mathbf{B}$, *then* $i \in W(\mathbf{B}')$.

Using Lemma 10 and Corollary 4, from now on, we fix a bidder α and an inefficient equilibrium $\mathbf{B}$, so that α is additive and $\alpha \in W(\mathbf{B})$.

Corollary 4 already gives us an insight about the competition in such an equilibrium $\mathbf{B}$. While bidder α will have to compete against the other bidders of $W(\mathbf{B})$ to win extra units, in addition to those that she is guaranteed to obtain, each bidder in $W(\mathbf{B}) \setminus \{\alpha\}$ only competes against α. Each of them is not guaranteed any units, unless she outbids α (bidder α is the only cause of externality for bidders in $W(\mathbf{B}) \setminus \{\alpha\}$, and anyone bidding lower than α cannot get any units). If bidder α did not exist, the other winners could be automatically granted the demand they are requesting since, in total, it is smaller than k and hence, there is no competition among them.

Observation 4. $\hat{F}_i^{avg}(z) = F_\alpha(z)$, *for every* $i \in W(\mathbf{B}) \setminus \{\alpha\}$, *where* F_α *is the CDF of bidder* α.

We continue with further properties on the support of the mixed strategies.

Lemma 11. *For the equilibrium* $\mathbf{B}$ *under consideration, it is true that:*

1. $Supp(B_\alpha) = [\ell(\mathbf{B}_W), h(\mathbf{B}_W)]$.
2. *For any two bidders* $i, j \in W(\mathbf{B}) \setminus \{\alpha\}$ *such that* $v_i \neq v_j$, *the set* $Supp(B_i) \cap Supp(B_j)$ *is of measure* 0 *(intersection points can occur only at endpoints of intervals).*

Lemma 11 suggests that we can group the bidders according to their values (since only bidders with the same value can overlap in their support). Let $r \leq |W(\mathbf{B}) \setminus \{\alpha\}|$ represent the number of distinct values $v_1, \ldots, v_r$, that bidders in $W(\mathbf{B}) \setminus \{\alpha\}$ have. We can partition the bidders of $W(\mathbf{B}) \setminus \{\alpha\}$ into r groups $W_1(\mathbf{B}), \ldots, W_r(\mathbf{B})$, such that, for $j = 1, \ldots, r$, the bidders in group $W_j(\mathbf{B})$ have value v_j. Similarly, we split the support of the winning bidders $[\ell(\mathbf{B}_W), h(\mathbf{B}_W)]$ into r intervals, i.e., $[\ell(\mathbf{B}_W), h(\mathbf{B}_W)] = \bigcup_{j=1}^r I_j(\mathbf{B})$, where each interval $j \in \{1, \ldots, r\}$ is formed as $I_j(\mathbf{B}) = \bigcup_{i \in W_j(\mathbf{B})} Supp(B_i)$. The following is a direct corollary of Lemma 11.

Corollary 5. *For every* $s, t \in \{1, \ldots, r\}$ *with* $s \neq t$, *the set* $I_s(\mathbf{B}) \cap I_t(\mathbf{B})$ *is of measure* 0.

When all bidders in $W(\mathbf{B}) \setminus \{\alpha\}$ have distinct values there are precisely $|W(\mathbf{B}) \setminus \{\alpha\}|$ intervals, whereas when they all have a common value, they must be bidding on the entire interval $[\ell(W(\mathbf{B})), h(W(\mathbf{B}))]$ (the equilibrium in the 2-bidder case when $d_1 = k$, in Sect. 4.1, is one such example). We sometimes denote as $I_0(\mathbf{B})$ the interval of losing bidders $[0, \ell(\mathbf{B}_W)]$, i.e., for the bidders in $\mathcal{N} \setminus W(\mathbf{B})$. Note that given $\mathbf{B}$, the only criterion for the membership of the support of a bidder i in an interval $I_s(\mathbf{B})$ is their value.

The next step is quite crucial in simplifying the extraction of our upper bound. We show that the worst case demand structure for the bidders in $W(\mathbf{B}) \setminus \{\alpha\}$ is when they all have unit demand.

Theorem 9. *For the value profile* $(\mathbf{v}, \mathbf{d})$ *and the equilibrium* $\mathbf{B}$ *under consideration, there exists another value profile* $(\mathbf{v}', \mathbf{d}')$ *and a product distribution* $\mathbf{B}'$ *such that*

1. $\alpha \in W(\mathbf{B}')$ *is an additive bidder and for every bidder* $i \in W(\mathbf{B}') \setminus \{\alpha\}$, *it holds that* $d_i' = 1$.
2. $\mathbf{B}'$ *is a mixed Nash equilibrium for* $(\mathbf{v}', \mathbf{d}')$.
3. $\frac{OPT(\mathbf{v}, \mathbf{d})}{SW(\mathbf{B})} = \frac{OPT(\mathbf{v}', \mathbf{d}')}{SW(\mathbf{B}')}$.

For the remainder of the section, it suffices to analyze valuation profiles, that possess equilibria where the members of $W(\mathbf{B})$ are either additive or unit-demand. Recall, that due to Corollary 4, there must be a unique additive bidder. Hence, we fix an instance given by a valuation profile $(\mathbf{v}, \mathbf{d})$, so that at the equilibrium $\mathbf{B}$, the set $W(\mathbf{B})$ consists of n unit-demand bidders plus the additive bidder α, i.e., $n = |W(\mathbf{B}) \setminus \{\alpha\}|$. Moreover, due to the following observation we may assume, without loss of generality, that the support of each unit-demand bidder has no overlapping intervals with other bidders from $W(\mathbf{B}) \setminus \{\alpha\}$.

Lemma 12. *Let* $(\mathbf{v}, \mathbf{d})$ *be a value profile, and let* $\mathbf{B}$ *be any mixed Nash equilibrium, such that the members of* $W(\mathbf{B})$ *are all unit-demand bidders aside from one additive bidder. Then, there exists a mixed Nash equilibrium* $\mathbf{B}'$ *with disjoint support intervals such that* $SW(\mathbf{B}) = SW(\mathbf{B}')$.

Therefore, by Corollary 5 and the discussion preceding it, the support of each bidder $i = 1, \ldots, n$ is $[\ell(B_i), h(B_i)]$. Note that due to Lemma 11, the unit-demand bidders must cover the entire interval $[\ell(\mathbf{B}_W), h(\mathbf{B}_W)]$. Hence, for a unit-demand bidder $i = 1, \ldots, n$, it must be that $\ell(B_i) = h(B_{i-1})$, assuming for convenience that $h(B_0) = \ell(\mathbf{B}_W)$.

The next theorem provides a more complete understanding of the support intervals and the distributions of the equilibrium $\mathbf{B}$.

Theorem 10. *For the value profile* $(\mathbf{v}, \mathbf{d})$ *under consideration, the following properties hold:*

1. *For bidder* α, *we have* $h(B_\alpha) = h(B_n) = h(\mathbf{B}_W) = v_\alpha - (k-n)\frac{v_\alpha - \ell(B_\alpha)}{k}$. *Moreover, for every unit-demand bidder* $i = 1, \ldots, n-1$ *it holds that*

$$\ell(B_{i+1}) = h(B_i) = v_\alpha - \frac{(k-n)(v_\alpha - \ell(B_\alpha))}{k - n + i} .$$

2. *The CDF* F_α *of bidder* α, *is a branch function, so that for* $i = 1, \ldots, n$, $F_\alpha(z) = F_\alpha^i(z)$ *for every* $z \in [h(B_{i-1}), h(B_i)]$ *with*

$$F_\alpha^i(z) = \prod_{j=i+1}^{n} \left(\frac{v_j - h(B_j)}{v_j - h(B_{j-1})} \right) \frac{v_i - h(B_i)}{v_i - z} .$$

Before proving our upper bound, we present two additional lemmas. The first is a straightforward inequality, that is a direct consequence of the definition of a mixed equilibrium, and the second is an expression for the social welfare. Both of these are useful for obtaining our final Price of Anarchy upper bound.

Lemma 13. *Consider a value profile* $(\mathbf{v}, \mathbf{d})$*, and any inefficient mixed Nash equilibrium* $\mathbf{B}$*, with* $W(\mathbf{B})$ *consisting only of additive or unit-demand bidders. Then, for* $i = 2, \ldots, n$*,* $m = 1, \ldots, i-1$*, and every* $z \in [h(B_{m-1}), h(B_m)]$*,*

$$\prod_{j=m+1}^{i-1} \frac{v_j - h(B_j)}{v_j - h(B_{j-1})} \leq \frac{v_m - z}{v_m - h(B_m)} \frac{v_i - h(B_{i-1})}{v_i - z}. \tag{2}$$

Lemma 14. *Consider a value profile* $(\mathbf{v}, \mathbf{d})$*, and any inefficient mixed Nash equilibrium* $\mathbf{B}$*, with* $W(\mathbf{B})$ *consisting only of additive or unit-demand bidders. The expected social welfare is*

$$k v_\alpha - (k - n)(v_\alpha - \ell(B_\alpha)) \sum_{i=1}^{n} \prod_{j=i+1}^{n} \left(\frac{v_j - h(B_j)}{v_j - h(B_{j-1})} \right) \int_{h(B_{i-1})}^{h(B_i)} \frac{v_i - h(B_i)}{v_i - z} \frac{v_\alpha - v_i}{(v_a - z)^2} dz.$$

Proof of Theorem 8. For brevity, we denote $\ell(B_a)$ as ℓ and for $j = 1, \ldots, n$, we denote $h(B_j)$ as h_j. Moreover, by assumption $v_a \geq v_n$. To simplify the calculations, we assume that $v_a = 1$ by rescaling all values in the instance.

Given a mixed Nash equilibrium $\mathbf{B}$, we lower bound the expected social welfare $SW(\mathbf{B})$ described in the equation of Lemma 14 as

$$SW(\mathbf{B}) = k - (k - n)(1 - \ell) \sum_{i=1}^{n} \prod_{j=i+1}^{n} \left(\frac{v_j - h_j}{v_j - h_{j-1}} \right) \int_{h_{i-1}}^{h_i} \frac{v_i - h_i}{v_i - z} \frac{1 - v_i}{(1 - z)^2} dz$$

$$= k - (k - n)(1 - \ell) \sum_{i=1}^{n} \prod_{j=i+1}^{n} \left(\frac{v_j - h_j}{v_j - h_{j-1}} \right)$$

$$\left(\int_{h_{i-1}}^{h_i} \frac{v_i - h_i}{v_i - z} \frac{1}{(1 - z)} dz - \int_{h_{i-1}}^{h_i} \frac{v_i - h_i}{(1 - z)^2} dz \right)$$

$$> k - (k - n)(1 - \ell) \sum_{i=1}^{n} \prod_{j=i+1}^{n} \left(\frac{v_j - h_j}{v_j - h_{j-1}} \right) \int_{h_{i-1}}^{h_i} \frac{v_i - h_i}{v_i - z} \frac{1}{(1 - z)} dz$$

$$\geq k - (k - n)(1 - \ell) \int_{\ell}^{h_n} \frac{v_n - h_n}{(v_n - z)(1 - z)} dz$$

$$\geq k - (k - n)(1 - \ell) \int_{\ell}^{h_n} \frac{1 - h_n}{(1 - z)^2} dz \geq k - (k - n)(1 - \ell)$$

$$= k - (k - n)(h_n - \ell) = k - (k - n)\left(\frac{n}{k}(1 - \ell) \right) \geq k - \frac{(k - n)n}{k} \geq \frac{3}{4}k.$$

The first inequality is true since for all bidders $i = 1, \ldots, n$, it holds that $v_i > h_i$ by Observation 2. The second one is an application of the mixed Nash equilibrium property encoded by Eq. (2) of Lemma 13. The next two inequalities occur by observing that the respective functions are increasing in terms of v_n (which, by assumption, we upper bound with $v_n \leq 1$) and ℓ (which we lower bound with $\ell \geq 0$). The last inequality follows by setting $x = \frac{n}{k}$ and minimizing the function $s(x) = 1 - x + x^2$ for $x \in (0,1)$. The theorem follows by observing that the optimal welfare is k, since the additive bidder has the highest value. $\qquad \square$

5   A Separation Between Mixed and Bayesian Cases

In this section we explore the more general solution concept of Bayes Nash equilibrium. We consider the following incomplete information setting. Let (v_i, d_i) be the type of bidder $i \in \mathcal{N}$. We suppose that the private value v_i of a bidder i is drawn independently from a distribution V_i. The second part of bidder i's type is his demand d_i; for the purposes of this section (we only construct a lower bound instance), we assume d_i to be deterministic private information.

Each bidder i is aware of her own value per unit v_i and the product distribution formed by the V_j's, and decides a strategy $(b_i, q_i) \sim G_i(v_i)$ for each value $v_i \sim V_i$. The bidding strategy is in general a mixed strategy. In the special case that bidder i chooses a single bid $(b_i(v_i), q_i)$ for each drawn value v_i, he submits a pure strategy, where q_i is not necessarily d_i.

Definition 2. *Given* $\mathbf{V} = \times_{i=1}^{n} V_i$ *and* $\mathbf{d}$, *a profile* $\mathbf{G}(\mathbf{v})$ *is a Bayes Nash equilibrium if for all* $i \in \mathcal{N}$, v_i *in* V_i's *domain*, $b_i' \geq 0$ *and* $q_i' \in \mathbb{Z}_+$ *it holds that*

$$\mathop{\mathbb{E}}_{\mathbf{v}_{-i} \sim \mathbf{V}_{-i}} \left[\mathop{\mathbb{E}}_{(\mathbf{b},\mathbf{q}) \sim \mathbf{G}(\mathbf{v})} [u_i^{v_i}(\mathbf{b}, \mathbf{q})] \right] \geq$$

$$\mathop{\mathbb{E}}_{\mathbf{v}_{-i} \sim \mathbf{V}_{-i}} \left[\mathop{\mathbb{E}}_{(\mathbf{b}_{-i},\mathbf{q}_{-i}) \sim \mathbf{G}_{-i}(\mathbf{v}_{-i})} [u_i^{v_i}((b_i', q_i'), (\mathbf{b}_{-i}, \mathbf{q}_{-i}))] \right],$$

where $u_i^{v_i}(\cdot)$ *stands for bidder* i's *utility when his value is* v_i.

We can define the Bayesian Price of Anarchy in the same way as before, by comparing against the expected optimal welfare, over the value distributions.

Although in a few other auction formats, the inefficiency does not get worse when one moves to incomplete information games, we exhibit that this is not the case here. We present a lower bound on the Bayesian Price of Anarchy of 1.1204, with two bidders. For mixed equilibria and two bidders, Theorem 5 showed that the Price of Anarchy is at most 1.1095. Although this difference is small, it shows that the Bayesian model is more expressive and can thus create more inefficiency. In particular, we stress that the bound obtained here for two bidders is inspired by the same bound of 1.1204 for mixed equilibria in Theorem 7, where we had to use a large number of bidders.

Theorem 11. *For* $n = 2$, $k \geq 2$, *and capped additive valuation profiles, the Price of Anarchy of Bayes Nash equilibria is at least 1.1204.*

Remark 3. When $k = 1$, there is a lower bound of 1.15 in [12] for the first price auction. However this requires a very large number of bidders. There is a simpler construction with two bidders in [21] but it only yields a lower bound of 1.06.

Acknowledgements. This research was supported by the Hellenic Foundation for Research and Innovation (H.F.R.I.). The first author was supported under the "1st Call for H.F.R.I. Research Projects to support faculty members and researchers and the procurement of high-cost research equipment" grant (Project Number: HFRI-FM17-3512), and the third author was supported under the H.F.R.I. PhD Fellowship grant (Fellowship Number: 289).

References

1. Ausubel, L., Cramton, P.: Demand Reduction and Inefficiency in Multi-Unit Auctions. Technical report, University of Maryland (2002)
2. Ausubel, L., Cramton, P., Pycia, M., Rostek, M., Weretka, M.: Demand reduction and inefficiency in multi-unit auctions. Rev. Econ. Stud. **81**, 1366–1400 (2014)
3. Baisa, B., Burkett, J.: Large multi-unit auctions with a large bidder. J. Econ. Theory **174**, 1–15 (2018)
4. Bhawalkar, K., Roughgarden, T.: Welfare guarantees for combinatorial auctions with item bidding. In: ACM-SIAM Symposium on Discrete Algorithms, SODA 2011, pp. 700–709 (2011)
5. Birmpas, G., Markakis, E., Telelis, O., Tsikiridis, A.: Tight welfare guarantees for pure Nash equilibria of the uniform price auction. Theory Comput. Syst. **63**(7), 1451–1469 (2018). https://doi.org/10.1007/s00224-018-9889-7
6. Brenner, M., Galai, D., Sade, O.: Sovereign debt auctions: uniform or discriminatory? J. Monet. Econ. **56**(2), 267–274 (2009)
7. Christodoulou, G., Kovács, A., Schapira, M.: Bayesian combinatorial auctions. J. ACM **63**(2), 11:1–11:19 (2016)
8. Christodoulou, G., Kovács, A., Sgouritsa, A., Tang, B.: Tight bounds for the price of anarchy of simultaneous first-price auctions. ACM TEAC **4**(2), 9:1–9:33 (2016)
9. Feldman, M., Fu, H., Gravin, N., Lucier, B.: Simultaneous auctions without complements are (almost) efficient. Games Econ. Behav. **123**, 327–341 (2020)
10. Friedman, M.: A Program for Monetary Stability. Fordham University Press, New York (1960)
11. Goldner, K., Immorlica, N., Lucier, B.: Reducing inefficiency in carbon auctions with imperfect competition. In: Innovations in Theoretical Computer Science, ITCS 2020, pp. 15:1–15:21 (2020)
12. Hartline, J., Hoy, D., Taggart, S.: Price of anarchy for auction revenue. In: ACM Conference on Economics and Computation, EC 2014, pp. 693–710 (2014)
13. de Keijzer, B., Markakis, E., Schäfer, G., Telelis, O.: Inefficiency of standard multi-unit auctions. In: Bodlaender, H.L., Italiano, G.F. (eds.) ESA 2013. LNCS, vol. 8125, pp. 385–396. Springer, Heidelberg (2013). https://doi.org/10.1007/978-3-642-40450-4_33
14. Krishna, V.: Auction Theory. Academic Press, San Diego (2002)
15. Milgrom, P.: Putting Auction Theory to Work. Cambridge University Press, Cambridge (2004)
16. Ockenfels, A., Reiley, D.H., Sadrieh, A.: Economics and information systems, chap. 12. Online Auctions, pp. 571–628 (2006)

17. Pycia, M., Woodward, K.: Auctions of homogeneous goods: a case for pay-as-bid. In: ACM Conference on Economics and Computation, EC 2021 (2021)
18. Rio, P.D.: Designing auctions for renewable electricity support. Best practices from around the world. Energy Sustain. Dev. **41**, 1–13 (2017)
19. Roughgarden, T.: The price of anarchy in games of incomplete information. In: ACM Conference on Economics and Computation, EC 2012, pp. 862–879 (2012)
20. Swinkels, J.: Efficiency of large private value auctions. J. Econ. Theory **69**(1), 37–68 (2001)
21. Syrgkanis, V.: Efficiency of mechanisms in complex markets. Ph.D. thesis, Cornell University (2014)
22. Syrgkanis, V., Tardos, E.: Composable and efficient mechanisms. In: ACM Symposium on Theory of Computing, STOC 2013, pp. 211–220 (2013)
23. Vickrey, W.: Counterspeculation, auctions, and competitive sealed tenders. J. Financ. **16**(1), 8–37 (1961)

Matching, Markets and Equilibria

Improved Analysis of RANKING for Online Vertex-Weighted Bipartite Matching in the Random Order Model

Billy Jin$^{(\boxtimes)}$ 🆔 and David P. Williamson 🆔

School of Operations Research and Information Engineering, Cornell University,
Ithaca, NY, USA
{bzj3,davidpwilliamson}@cornell.edu

Abstract. In this paper, we consider the online vertex-weighted bipartite matching problem in the random arrival model. We consider the generalization of the RANKING algorithm for this problem introduced by Huang, Tang, Wu, and Zhang [9], who show that their algorithm has a competitive ratio of 0.6534. We show that assumptions in their analysis can be weakened, allowing us to replace their derivation of a crucial function g on the unit square with a linear program that computes the values of a best possible g under these assumptions on a discretized unit square. We show that the discretization does not incur much error, and show computationally that we can obtain a competitive ratio of 0.6629. To compute the bound over our discretized unit square we use parallelization, and still needed two days of computing on a 64-core machine. Furthermore, by modifying our linear program somewhat, we can show computationally an upper bound on our approach of 0.6688; any further progress beyond this bound will require either further weakening in the assumptions of g or a stronger analysis than that of Huang et al.

Keywords: Bipartite matching · Online algorithms

1 Introduction

In the maximum bipartite matching problem, we are given as input a bipartite graph $G = (U, V, E)$ such that each edge $(u, v) \in E$ has $u \in U$ and $v \in V$. A set $F \subseteq E$ of edges is a *matching* if there is at most one edge of F incident to each vertex $u \in U$ and $v \in V$. The goal is to find a matching of maximum cardinality. This problem has been well-studied and is one of the fundamental problems in combinatorial optimization (see, for example, Schrijver [17, Chapter 16]).

In a classic paper from 1990, Karp, Vazirani, and Vazirani [12] introduce an online version of this problem and the RANKING algorithm for it. In their online version of the problem, the vertices V are known to the algorithm in advance, while the vertices of U are introduced one at a time; we refer to the vertices of

B. Jin—Supported in part by NSERC fellowship PGSD3-532673-2019.

M. Feldman et al. (Eds.): WINE 2021, LNCS 13112, pp. 207–225, 2022.
https://doi.org/10.1007/978-3-030-94676-0_12

V as the *offline* vertices and those of U as the *online* vertices. The algorithm maintains a matching F, initially empty. As each vertex u of U arrives, the edges incident to u are also revealed to the algorithm. Once a vertex u arrives, the algorithm must either choose an edge incident to u to add to F or decide not to add an edge incident to u to the matching F. These choices are irrevocable: no edge incident to u may be added at any later point in time. In the RANKING algorithm, the algorithm initially chooses a random permutation π of the offline vertices V; when a new vertex $u \in U$ arrives, the algorithm adds edge (u, v) to the matching that maximizes $\pi(v)$ over the vertices $v \in V$ that do not have any edge of F already incident (i.e. the *unmatched* vertices of V incident to u), if such a vertex exists, otherwise it leaves u unmatched. Karp, Vazirani, and Vazirani prove that this algorithm achieves a *competitive ratio* of at least $1 - \frac{1}{e}$; that is, the algorithm finds a matching whose expected cardinality is at least $1 - \frac{1}{e}$ times the size of the maximum matching in G. They further show that this ratio is tight; that is, there are instances of the problem such that no online algorithm can achieve a better competitive ratio.

Since this work, there have been many simplifications of the original analysis (e.g. Birnbaum and Mathieu [3]; Devanur, Jain, and Kleinberg [5]), proposed changes in the online model, and extensions to more general matching problems. Of interest to us in this paper are the *random arrival model*, proposed by Goel and Mehta [7], and the maximum *vertex-weighted* online matching problem, introduced by Aggarwal, Goel, Karande, and Mehta [1]. In the random arrival model, the online vertices of U arrive in an order given by a random permutation. Goel and Mehta show that the greedy algorithm attains a competitive ratio of $1 - \frac{1}{e}$ in the random arrival model. Later, Karande, Mehta, and Tripathi [11] and Mahdian and Yan [13] show that the RANKING algorithm has competitive ratio strictly better than $1 - \frac{1}{e}$ in this model, with Mahdian and Yan giving a competitive ratio of 0.696. In the vertex-weighted version of the problem, the offline vertices $v \in V$ have weight $w_v \geq 0$, and the goal is to find a matching F that maximizes the total weight of the matched vertices in V (that is, the vertices in V that have an incident edge in F). Aggarwal et al. show that a generalization of RANKING achieves a $1 - \frac{1}{e}$ competitive ratio for the vertex-weighted version of the problem (with adversarial arrivals). Devanur, Jain, and Kleinberg [5] later interpreted the Aggarwal et al. algorithm as follows. Each offline vertex $v \in V$ draws a value y_v from $[0, 1]$ uniformly at random; when a new vertex $u \in U$ arrives, we add edge (u, v) to matching F for the unmatched v (if any) that maximizes $w_v(1 - g(y_v))$, where $g(y) = e^{y-1}$.

Huang, Tang, Wu, and Zhang [9] studied the combination of these two models, the maximum vertex-weighted online matching problem in the random arrival model. Drawing on the ideas of Devanur et al., they proposed the following further generalization of the RANKING algorithm. In addition to having each offline vertex $v \in V$ draw a value y_v from $[0, 1]$, since the online vertices arrive in random order, they propose having each online vertex $u \in U$ draw a value $y_u \in [0, 1]$ uniformly at random, and have the online vertices arrive in order of nondecreasing y_u. When a new vertex $u \in U$ arrives, we add edge (u, v) to the matching F for the unmatched v (if any) that maximizes $w_v(1 - g(y_v, y_u))$, for a function g with certain properties. Huang et al. assume that $g(x, y) = \frac{1}{2}(h(x) + 1 - h(y))$ for $h : [0, 1] \rightarrow [0, 1]$, and end up choosing $h(x) = \min(1, \frac{1}{2}e^x)$ to achieve a competitive ratio of 0.6534, beating the $1 - \frac{1}{e} \approx 0.632$ competitive ratio achieved by Aggarwal et al. in the adversarial arrival model.

1.1 Our Contributions

We build upon the work of Huang et al. to give a competitive ratio of 0.6629 for the maximum vertex-weighted online matching problem in the random arrival model. We begin by showing that several assumptions Huang et al. make about the form of $g(x, y)$ needed for the analysis of their generalization of RANKING can be relaxed. Instead, we can make several weaker assumptions about the form of $g(x, y)$. These assumptions can be encoded in a linear program that allows us to produce the best possible piecewise-affine function $g : [0, 1]^2 \rightarrow [0, 1]$ under these assumptions for any given discretization of $[0, 1]^2$.

We then need to compute the competitive ratio by finding a point in $[0, 1]^2$ where g reaches a certain minimum of a complicated function of g given by Huang et al. To do this, we show that the error in the competitive ratio achieved by restricting ourselves to finding the minimum in the set of discretized points is linear in the size of the discretization, so we can restrict ourselves to checking just the points in this set if we are willing to tolerate some small error. We note that the checking is easily parallelizable, and we wrote our code to use all the cores of the machine on which it is run. Even so, we still needed two days of a 64-core, 64 GB machine on Amazon's EC2 platform to achieve our competitive ratio of 0.6629.

Because we use a linear program to find the function g, we can also use a slight modification of it to find an upper bound on the best possible competitive ratio obtainable using the Huang et al. analysis with our weakened assumptions on g. We modify the linear program so that any function g with our weakened assumptions is feasible, and modify the objective function so that it gives an upper bound on the ratio obtained via the Huang et al. analysis. Solving the linear program results in an upper bound of 0.6688. Thus any further improvement in the competitive ratio will require either further weakening in the assumptions of g or a stronger analysis than that of Huang et al.

Mahdian and Yan [13] also use linear programming in their paper for online unweighted bipartite matching in the random arrival model; in particular, they use factor-revealing LPs. Our use of LPs is quite different. In the case of [13], the value of the LP gives a bound on the competitive ratio of the algorithm for each size of the graph. Here we use the LP to find a function g used in the analysis of Huang et al. by finding values of the function at points given by a discretization of the unit square, which we then interpolate into a function over the entire square. The competitive ratio is then obtained from this interpolated function as described above.

While in this paper we focus on the *random order* model, another well-studied model for online bipartite matching is the *known IID* model (also referred to in the literature as *stochastic online matching*). In the known IID model, we assume that there is a known distribution over subsets of offline nodes, and each online vertex has its neighbouring set drawn iid from the distribution. This is a strictly stronger assumption than random order; any algorithm that achieves a competitive ratio of α under the random order model will achieve a competitive ratio of at least α under the known IID model [16]. The known IID model was introduced by Feldman et al. [6], who gave a 0.67-competitive algorithm on unweighted graphs under the additional assumption that the expected number of arriving nodes of each type is an integer. The competitive ratio was gradually improved (and the integral rates assumption relaxed in some cases) in a series of works by Bahmani and Kapralov [2], Manshadi et al. [14], Jaillet and Lu [10], Brubach et al. [4], and Huang and Shu [8]. In particular, Huang and Shu [8] recently gave a 0.7009-competitive algorithm for vertex-weighted online bipartite matching under the known iid model. It is worth noting that this does not subsume our result, because known iid is a stronger assumption than random order.

Paper Structure. Our paper is organized as follows. In Sect. 2, we recap the argument of Huang et al. that we will use. In Sect. 3, we introduce the weaker assumptions on the function g that we will use, and prove that the arguments of Huang et al. continue to hold under these weaker assumptions so that we can still use their bound on the competitive ratio under these weaker assumptions. In Sect. 4, we introduce the LP that will define our function g; we show how to define a piecewise-affine function g from the LP solution, and we show that the assumptions we need on g hold for this LP-defined function. In Sect. 5, we provide a bound on the error we incur in the competitive ratio by only checking the Huang et al. bound at discrete points of the unit square. In Sect. 6, we explain the computation that was used to obtain our competitive ratio of 0.6629. Section 7 explains how we modify our linear program to obtain an upper bound on the competitive ratio that is attainable via the Huang et al. analysis with our weakened assumptions on g. We conclude in Sect. 8. For space reasons, many proofs and figures are deferred to the full version of the paper.[1]

2   Background

As stated in the introduction, we assign each offline vertex $v \in V$ a value y_v from $[0, 1]$ chosen uniformly at random, and following Huang et al. we assume that each online vertex $u \in U$ also has a value y_u from $[0, 1]$ chosen uniformly at random, and that the online vertices arrive in nondecreasing order of their y_u value. The variant of the RANKING algorithm for the problem uses a function $g : [0,1]^2 \to [0,1]$ that is increasing in the first argument and decreasing in the

---

[1] The full version of the paper can be accessed at https://arxiv.org/abs/2007.12823.

second. When an online vertex $u \in U$ arrives, it is matched to the unmatched neighbor $v \in V$ that maximizes $w_v(1 - g(y_v, y_u))$.

The analysis of this algorithm by Huang et al. [9] follows that of Devanur, Jain, and Kleinberg [5]. It considers the linear programming relaxation of the vertex-weighted bipartite matching problem and its dual linear program, shown below, with the primal on the left and the dual on the right.

$$
\begin{aligned}
\max \quad & \sum_{(u,v) \in E} w_v x_{uv} & & & \min \quad & \sum_{u \in U} \alpha_u + \sum_{v \in V} \alpha_v \\
\text{s.t.} \quad & \sum_{v:(u,v) \in E} x_{uv} \leq 1 & & \forall u \in U & \text{s.t.} \quad & \alpha_u + \alpha_v \geq w_v & & \forall (u,v) \in E \\
& \sum_{u:(u,v) \in E} x_{uv} \leq 1 & & \forall v \in V & & \alpha_u, \alpha_v \geq 0 & & \forall u \in U, v \in V. \\
& x_{uv} \geq 0 & & \forall (u,v) \in E.
\end{aligned}
$$

The goal of the analysis is to find a set of nonnegative variables α, whose values may depend on the random y values, such that $\sum_{(u,v) \in F} w_v = \sum_{u \in U} \alpha_u + \sum_{v \in V} \alpha_v$ and $E_y[\alpha_u + \alpha_v] \geq \beta \cdot w_v$ for all $(u,v) \in E$. (Here, F is the set of edges in the matching found by the algorithm.) Given the two conditions, it is possible to define a dual solution that is a factor of β away from the total weight of the matched edges, implying a competitive ratio of β. Whenever the algorithm adds a matching edge (u,v) to F, it defines $\alpha_u = w_v \cdot g(y_v, y_u)$ and $\alpha_v = w_v(1 - g(y_v, y_u))$, ensuring that the first condition is met.

The main result of Huang et al. is the following.

Lemma 1 (Lemma 4.1 [9]). *Suppose that $g(x,y) = \frac{1}{2}(h(x) + 1 - h(y))$, for some increasing function $h : [0,1] \to [0,1]$ that satisfies $h'(x) \leq h(x)$. Then for any $u \in U$ and $v \in V$ such that $(u,v) \in E$,*

$$
\frac{1}{w_v} E_y[\alpha_u + \alpha_v] \geq \min_{0 \leq \gamma, \tau \leq 1} f(\gamma, \tau)
$$

for

$$
f(\gamma, \tau) = \left\{ (1-\tau)(1-\gamma) + (1-\tau) \int_0^\gamma g(x,\tau) dx \right.
$$

$$
\left. + \int_0^\tau \min_{\theta \leq \gamma} \left\{ (1 - g(\theta, y)) + \int_0^\theta g(x,y) dx + \int_\theta^\gamma g(x,\tau) dx \right\} dy \right\}.
$$

Thus, the competitive ratio of RANKING is at least $\min_{0 \leq \gamma, \tau \leq 1} f(\gamma, \tau)$.

Huang et al. show that by taking $h(x) = \min(1, \frac{1}{2} e^x)$, they can prove that $f(\gamma, \tau) > 1 - \frac{1}{2} \ln 2 \approx 0.6534$ for all $0 \leq \gamma, \tau \leq 1$, attaining their claimed competitive ratio by the reasoning above.

3 Relaxing Assumptions

Huang et al. assume that $g(x, y) = \frac{1}{2}(1 + h(x) - h(y))$, for some increasing function $h : [0, 1] \to [0, 1]$ that satisfies $h'(x) \le h(x)$. This is a strong assumption and gives several nice properties of g which are useful in the analysis. We relax this assumption and do not constrain g to satisfy this condition. Instead, we replace this condition by several weaker conditions. This allows us to search over a wider class of functions g when trying to maximize the bound in Lemma 1. However, to leverage their result, we must show that the conclusion of Lemma 1 still holds for all g that satisfy these weaker conditions. We prove the following.

Theorem 1. *Let g be a function obeying the following conditions.*

1. *$g(x, y) : [0, 1]^2 \to [0, 1]$ is continuous,*
2. *$g(x, y)$ is increasing in x and decreasing in y,*
3. *$\frac{\partial g(x,y)}{\partial x} \le g(x, y),$[2]*
4. *$\frac{\partial g(x,y)}{\partial y} \ge g(x, y) - 1$, and*
5. *for all x, y, y' with $y' > y$, we have $g(1, y) - g(x, y) \ge g(1, y') - g(x, y')$.*

Then Lemma 4.1 in [9] still holds, and the competitive ratio of the RANKING algorithm is at least

$$
\min_{0 \le \gamma, \tau \le 1} \left\{ (1 - \tau)(1 - \gamma) + (1 - \tau) \int_0^\gamma g(x, \tau) dx \right.
$$
$$
\left. + \int_0^\tau \min_{\theta \le \gamma} \left\{ (1 - g(\theta, y)) + \int_0^\theta g(x, y) dx + \int_\theta^\gamma g(x, \tau) dx \right\} dy \right\}
$$
(1)

Proof. The result in Lemma 4.1 of [9] follows entirely from facts proved in their Lemmas 3.3, 3.4, and 3.5. We show that these lemmas continue to hold given the conditions on g above. These proofs can be found in the full version of the paper.

From now on, we will refer to the five conditions in Theorem 1 as conditions 1–5.

4 LP Formulation

To find a function g that maximizes the bound in Theorem 1, we discretize $[0, 1]^2$ into an $n \times n$ grid for a sufficiently large positive integer n, and write an LP to search for the values of g on this discretized grid.

---

[2] We use notation for partial derivatives, but the result also holds for non-differentiable functions, if we use subgradients, etc. In particular, the result holds for the piecewise-affine functions g we obtain from solving the LP in Sect. 4. To keep the exposition simple, we will continue using partial derivative notation throughout the paper.

In Sect. 4.1, we formulate the conditions 1–5, which are the conditions that any feasible g must satisfy, as constraints in the LP. Next, in Sect. 4.2, we formulate the expression in Theorem 1, which is the bound we are trying to maximize, as an LP objective. Finally, in Sect. 4.3, we will see how to extend the values of g on the discretized $n \times n$ grid, which is what the LP returns, to a function g defined on the entire unit square.

4.1 Formulating the Constraints

In this section, we show how to formulate the conditions 1–5 as constraints in the LP.

Fix a positive integer n and let $x_i = y_i = \frac{i}{n}$, for $i = 0, 1, \ldots, n$. Our LP will have variables $g(x_i, y_j)$, the values of g on the discretized unit square. Next, we encode the conditions 1–5 as constraints of the LP. Below are the conditions, and their corresponding LP constraints:

1. $g(x, y) : [0, 1]^2 \to [0, 1]$ and g is continuous. The corresponding LP constraints are $0 \le g(x_i, y_j) \le 1$, for all $i, j = 0, 1, \ldots, n$. Note that we do not include any constraints to enforce the continuity of g, since the aim of the LP is to determine the value of g at a discretized set of points.
2. $g(x, y)$ is increasing in x and decreasing in y. The corresponding LP constraints are
   - $g(x_i, y_j) \le g(x_k, y_j)$ for all $0 \le i, j, k \le n$ with $i \le k$;
   - $g(x_i, y_j) \ge g(x_i, y_l)$ for all $0 \le i, j, l \le n$ with $j \le l$.
3. $\frac{\partial g(x,y)}{\partial x} \le g(x, y)$. We discretize this constraint to create the following LP constraints:
   - $\frac{g(x_{i+1}, y_j) - g(x_i, y_j)}{x_{i+1} - x_i} \le g(x_i, y_{j+1})$ for all $0 \le i, j \le n - 1$
   - $\frac{g(x_{i+1}, y_n) - g(x_i, y_n)}{x_{i+1} - x_i} \le g(x_i, y_n)$ for all $0 \le i \le n - 1$

Remark 1. It is more natural to encode the constraints as $\frac{g(x_{i+1}, y_j) - g(x_i, y_j)}{x_{i+1} - x_i} \le g(x_i, y_j)$ for all $0 \le i \le n - 1$, $0 \le j \le n$. Since $g(x_i, y_{j+1}) \le g(x_i, y_j)$, our constraints are even stronger. We do this because when we extend g from its discretized values to a function defined on the entire unit square, this slightly stronger version of the constraint will be needed to show that the extended function also satisfies the condition.

4. $\frac{\partial g(x,y)}{\partial y} \ge g(x, y) - 1$. As with the previous constraint, the corresponding LP constraints are
   - $\frac{g(x_i, y_{j+1}) - g(x_i, y_j)}{y_{j+1} - y_j} \ge g(x_{i+1}, y_j) - 1$, for all $0 \le i, j \le n - 1$
   - $\frac{g(x_n, y_{j+1}) - g(x_n, y_j)}{y_{j+1} - y_j} \ge g(x_n, y_j) - 1$, for all $0 \le j \le n - 1$.
5. For all x, y, y' with $y' > y$, $g(1, y) - g(x, y) \ge g(1, y') - g(x, y')$. The corresponding LP constraints are

$$g(x_n, y_j) - g(x_i, y_j) \ge g(x_n, y_l) - g(x_i, y_l) \quad \text{for all } 0 \le i, j \le n \text{ with } l > j.$$

4.2 Formulating the Objective

The expression we are trying to maximize is given in (1). To formulate this approximately as an LP objective, we

1. Approximate the $\min_{0\leq\gamma,\tau\leq1}$ and $\min_{\theta\leq\gamma}$ expressions by minimizing over a finite set of values, and
2. Approximate the integrals by finite sums.

We begin by letting $f(\gamma,\tau)$ be the expression inside the outermost min, so that the bound is equal to $\min_{0\leq\gamma,\tau\leq1} f(\gamma,\tau)$. Since we cannot check all values of γ and τ, we approximate it by $\min_{0\leq i,j\leq n} f(x_i,y_j)$. We write this as a linear objective using the standard trick of introducing a dummy variable t, and maximizing t subject to $t \leq f(x_i,y_j)$ for all $0 \leq i,j \leq n$.

Next, we must write constraints to model $f(x_i,y_j)$. We replace the inner $\min_{\theta\leq x_i}$ by a minimum over the discretized grid: $\min_{\theta\leq x_i}$ becomes $\min_{x_k\leq x_i}$. For each integral that appears in the expression for f, we replace it by a left Riemann sum. For example, the integral $\int_0^{x_i} g(x,y_j)dx$ would be replaced by $\frac{1}{n}\sum_{k=0}^{i-1} g(x_k,y_j)$.

With these approximations, we can approximate $f(x_i,y_j)$ as a linear function $\tilde{f}(x_i,y_j)$ of the $g(x_i,y_j)$ variables:

$$f(x_i,y_j) \approx \tilde{f}(x_i,y_j) = (1-x_i)(1-y_j) + (1-y_j)\cdot\frac{1}{n}\sum_{k=0}^{i-1} g(x_k,y_j)$$

$$+ \frac{1}{n}\sum_{l=0}^{j-1}\min_{k\leq i}\left\{(1-g(x_k,y_l)) + \frac{1}{n}\sum_{d=0}^{k-1}g(x_d,y_l) + \frac{1}{n}\sum_{d=k}^{i-1}g(x_d,y_j)\right\}$$

Hence, to summarize this section and Sect. 4.1, the full linear program we use the compute the values of g on the discretized $n \times n$ grid is as follows:

$$\max\quad t$$
$$\text{s.t.}\quad t \leq \tilde{f}(x_i,y_j)\quad\text{for all }0 \leq i,j \leq n$$

and such that g satisfies the constraints from Sect. 4.1.

4.3 Extending the Discretized Function to the Unit Square

The linear program gives us values of g on any given discretization of $[0,1]^2$, but to use the bound in Theorem 1 we must extend g to be defined on the entire unit square, and show that this extended function satisfies conditions 1–5. To extend g from its values on an $n \times n$ grid to a function defined on the entire unit square, we triangulate the $n \times n$ grid as shown in Fig. 1.

For a point (x,y) on gridpoint, its function value is given by the LP. For any other point (x,y), we define $g(x,y)$ to be a convex combination of the function values on the three vertices of the triangle containing (x,y). More precisely, suppose (x,y) is contained in the triangle with vertices $(a_1,b_1),(a_2,b_2)$, and (a_3, b_3), where the (a_i, b_i) are gridpoints. (See Fig. 2.) Then we define $g(x, y) = \lambda_1 \cdot g(a_1, b_1) + \lambda_2 \cdot g(a_2, b_2) + \lambda_3 \cdot g(a_3, b_3)$, where $\lambda_1, \lambda_2, \lambda_3$ are the unique coefficients that satisfy $\lambda_1, \lambda_2, \lambda_3 \geq 0$, $\lambda_1 + \lambda_2 + \lambda_3 = 1$ and $(x, y) = \lambda_1 \cdot (a_1, b_1) + \lambda_2 \cdot (a_2, b_2) + \lambda_3 \cdot (a_3, b_3)$. Geometrically, the extended function is piecewise affine – it is affine on each triangle.

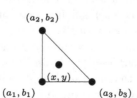

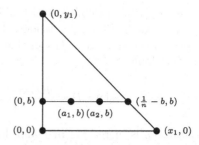

Fig. 1. Triangulating the grid. Here, $n = 4$.

Fig. 2. Extending the function values to a point inside a triangle.

Fig. 3. Illustration of the proof of Condition 3.

We now prove that the extended function satisfies conditions 1–5. We list the conditions below, and prove that the extended function satisfies them:

1. $g(x, y) : [0, 1]^2 \to [0, 1]$ and g is continuous. The extended function takes values in $[0, 1]$ because its values are convex combinations of its values on the discretized grid, which are in $[0, 1]$. It is continuous because it is piecewise affine.

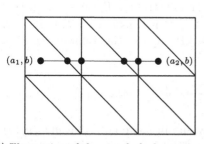

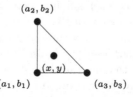

(a) Illustration of the proof of why we may assume (a_1, b) and (a_2, b) are contained in the same triangle.

(b) Illustration of the proof of why $g(a_1, b) \leq g(a_2, b)$ for (a_1, b), (a_2, b) in the same triangle and $a_1 \leq a_2$.

Fig. 4. Figures used in the proof of Condition 2.

2. $g(x, y)$ is increasing in x and decreasing in y. We will show that g is increasing in x; the proof that it is decreasing in y is similar.

Let (a_1, b) and (a_2, b) be points in the unit square, with $a_1 \leq a_2$. We must show that $g(a_1, b) \leq g(a_2, b)$. First, observe that it suffices to show this when the two points are contained in the same triangle. This is because if (a_1, b) and (a_2, b) were contained in different triangles, then the horizontal line segment l from (a_1, b) to (a_2, b) can be divided from left to right into a sequence of segments (say $l_1, \ldots, l_k$), each of which is contained in a single triangle. Then the fact that g is increasing on each smaller segment l_i would imply that g is increasing on l. (For an illustration of this, see Fig. 4a.)

So, we can assume (a_1, b) and (a_2, b) are contained in the same triangle. Without loss of generality, suppose (a_1, b) and (a_2, b) are both contained in the lower-leftmost triangle; that is, the triangle with vertices $(0, 0)$, $(x_1, 0)$, and $(0, y_1)$; the proof for any other triangle is the same. See Fig. 4b.

Note that (a_1, b) and (a_2, b) are both on the line segment from $(0, b)$ to $(\frac{1}{n} - b, b)$. Since g is piecewise affine in any triangle, it follows that $g(a_1, b) = (1 - \lambda_1) \cdot g(0, b) + \lambda_1 \cdot g(\frac{1}{n} - b, b)$, where $0 \leq \lambda_1 \leq 1$ satisfies $\lambda_1 \cdot (\frac{1}{n} - b) = a_1$. Similarly, $g(a_2, b) = (1 - \lambda_2) \cdot g(0, b) + \lambda_2 \cdot g(\frac{1}{n} - b, b)$, where $0 \leq \lambda_2 \leq 1$ satisfies $\lambda_2 \cdot (\frac{1}{n} - b) = a_2$. Now, since $a_1 \leq a_2$, it follows that $\lambda_1 \leq \lambda_2$. Therefore, to show that $g(a_1, b) \leq g(a_2, b)$ it suffices to show that $g(0, b) \leq g(\frac{1}{n} - b, b)$.

To see this, we note that $g(0, b) = (1 - \lambda) \cdot g(0, 0) + \lambda \cdot g(0, \frac{1}{n})$, where $0 \leq \lambda \leq 1$ satisfies $\frac{\lambda}{n} = b$. Similarly, $g(\frac{1}{n} - b, b) = (1 - \lambda) \cdot g(\frac{1}{n}, 0) + \lambda \cdot g(0, \frac{1}{n})$. Since $g(\frac{1}{n}, 0) \geq g(0, 0)$ (this was a constraint in the LP), it follows that $g(0, b) \leq g(\frac{1}{n} - b, b)$, as needed.

3. $\frac{\partial g(x, y)}{\partial x} \leq g(x, y)$. Consider a horizontal line segment l between two adjacent gridpoints, say between (x_i, y_j) and (x_{i+1}, y_j). In the triangulation, l is adjacent to two triangles: one triangle T_1 below it and one triangle T_2 above it. (If $y_i = 0$ or $y_i = 1$, then l is only adjacent to one triangle, but the same argument still goes through.) See Fig. 3 for an illlustration. Because g is piecewise affine in each triangle, it follows that $\frac{\partial g(x, y)}{\partial x}$ is constant on $T_1 \cup T_2$, and is equal to the slope of l. Recall that the LP imposes the following constraint on the slope of l:

$$\text{slope}(l) = \frac{g(x_{i+1}, y_j) - g(x_i, y_j)}{x_{i+1} - x_i} \leq g(x_i, y_{j+1})$$

Because g is increasing in x and decreasing in y, we note that $g(x_i, y_{j+1}) \leq \inf\{g(x, y) : (x, y) \in T_1 \cup T_2\}$. Thus $\frac{\partial g(x, y)}{\partial x} \leq g(x, y)$ holds on $T_1 \cup T_2$. Because any triangle is adjacent to some horizontal line segment in the grid, this argument shows that $\frac{\partial g(x, y)}{\partial x} \leq g(x, y)$ holds for all (x, y) in the unit square, and we are done.

4. $\frac{\partial g(x, y)}{\partial y} \geq g(x, y) - 1$. The proof of this is similar to the proof of the previous condition.

5. For all x, y, y' with $y' > y$, $g(1, y) - g(x, y) \geq g(1, y') - g(x, y')$.

Let $\mathcal{F} = \{0, \frac{1}{n}, \frac{2}{n}, \ldots, 1\}$. If $x, y, y' \in \mathcal{F}$, then the condition holds, because these were constraints imposed by the LP.

Suppose now (x, y) lies in the interior of some triangle T. Fix x and y', and imagine varying y up and down such that (x, y) remains inside T. Let I be the range of values of y such that (x, y) remains inside T. Since g is affine on each triangle, it follows that $\frac{\partial}{\partial y}(g(1, y) - g(x, y))$ is constant for all y in I. Therefore (by moving y in the direction that decreases the LHS of the inequality if necessary), it suffices to prove the inequality in the case (x, y) is on the boundary of a triangle. Similarly, we may assume that (x, y') lies on the boundary of a triangle.

Suppose (x, y) and (x, y') both lie on hypotenuses (see Fig. 5). The case where one or both of the points lie on a base of a triangle is very similar (and easier), so we will omit it here.

Let h_1 and h_2 be the two endpoints of the hypotenuse containing (x, y), with h_1 lower than h_2. Similarly, define h'_1 and h'_2. Let b_1 and b_2 be the two endpoints of the vertical grid segment containing $(1, y)$. Similarly, define b'_1 and b'_2. We will use the fact that the inequality holds for the gridpoints (b_1, h_1, b'_1, h'_1) and the gridpoints (b_2, h_2, b'_2, h'_2) to deduce that it holds for our points.

The inequality on the points (b_1, h_1, b'_1, h'_1) is

$$g(b_1) - g(h_1) \geq g(b'_1) - g(h'_1)$$

The inequality on the points (b_2, h_2, b'_2, h'_2) is

$$g(b_2) - g(h_2) \geq g(b'_2) - g(h'_2)$$

Now let $0 \leq \lambda \leq 1$ be the scalar so that $\lambda b_1 + (1 - \lambda)b_2 = (1, y)$. Observe that we also have $\lambda h_1 + (1 - \lambda)h_2 = (x, y)$, $\lambda b'_1 + (1 - \lambda)b'_2 = (1, y')$, and $\lambda h'_1 + (1 - \lambda)h'_2 = (x, y')$.

Now, multiply the inequality for (b_1, h_1, b'_1, h'_1) by λ, and multiply the inequality for (b_2, h_2, b'_2, h'_2) by $(1 - \lambda)$, then add them together. The result is the inequality

$$g(1, y) - g(x, y) \geq g(1, y') - g(x, y'),$$

which is what we wanted.

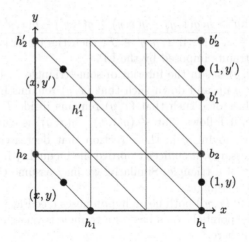

Fig. 5. Illustration of proof of condition 5.

5   Checking the Bound

The linear program gives us function values defined on a discretization of the unit square, which we then extend to a function g defined on the entire unit square via triangulation. It remains now to plug this g into the bound for the competitive ratio given by Theorem 1. We cannot evaluate the bound analytically for the function g returned by the LP; instead, we evaluate it computationally.

For $0 \leq \gamma, \tau \leq 1$, let

$$
f(\gamma, \tau) = (1 - \tau)(1 - \gamma) + (1 - \tau) \int_0^\gamma g(x, \tau) dx
$$

$$
+ \int_0^\tau \min_{\theta \leq \gamma} \left\{ (1 - g(\theta, y)) + \int_0^\theta g(x, y) dx + \int_\theta^\gamma g(x, \tau) dx \right\} dy,
$$

so that, by Theorem 1, the competitive ratio of g is at least $\min_{0 \leq \gamma, \tau \leq 1} f(\gamma, \tau)$.

When we evaluate this bound using a computer, we incur two sources of error:

1. The bound takes a minimum over all (γ, τ) in the unit square. However, using a computer, we can only check a finite number of points (γ, τ).
2. For a fixed (γ, τ), we do not calculate $f(\gamma, \tau)$ exactly. Instead, using a computer, we calculate an approximation $\hat{f}(\gamma, \tau)$, by
   - Approximating the integrals with finite sums, and
   - Replacing the inner minimum over all $\theta \leq \gamma$ by a minimum over a finite set of θ.

In what follows, we will bound the errors above. This proves that the output of the computer program is a valid bound on the competitive ratio. We will show that f is Lipschitz in γ and τ, which implies that checking all values of

(γ, τ) in a sufficiently fine discretization of the unit square is enough to obtain a quantifiable bound on the error.

Before we move on, we remind the reader what it means for a function to be Lipschitz.

Definition 1. *A function* $f : \mathbb{R}^n \rightarrow \mathbb{R}$ *is L-Lipschitz if* $|f(x) - f(y)| \leq L \|x - y\|$ *for all* $x, y \in \mathbb{R}^n$.

It will be convenient for us to work with Lipschitzness in a particular coordinate.

Definition 2. *A function* $f : \mathbb{R}^n \rightarrow \mathbb{R}$ *is L-Lipschitz in its ith coordinate if*

$$|f(x_1, \ldots, x_i, \ldots, x_n) - f(x_1, \ldots, x_i', \ldots, x_n)| \leq L |x_i - x_i'|$$

for all $x_1, \ldots, x_i, x_i', \ldots, x_n \in \mathbb{R}$.

Lemma 2. $f(\gamma, \tau)$ *is 1-Lipschitz in* γ *and 3-Lipschitz in* τ.

The preceding lemma allows us to control the error incurred from checking the bound over all (γ, τ) in a discretization of the unit square instead of the entire unit square. The second source of error is that for a fixed (γ, τ), we evaluate an approximation $\hat{f}(\gamma, \tau)$ to $f(\gamma, \tau)$, because we replace the integrals with discrete sums and the minimization over all $\theta \leq \gamma$ with a minimization over finitely many θ. The following lemma controls the second source of error.

To make notation less cluttered, let

- $p(\gamma, \tau) = (1 - \gamma)(1 - \tau) + (1 - \tau) \int_0^\gamma g(x, \tau)dx$,
- $h(\gamma, \tau, \theta, y) = (1 - g(\theta, y)) + \int_0^\theta g(x, y)dx + \int_\theta^\gamma g(x, \tau)dx$, and
- $q(\gamma, \tau, y) = \min_{\theta \leq \gamma} h(\gamma, \tau, \theta, y)$.

so that $f(\gamma, \tau) = p(\gamma, \tau) + \int_0^\tau q(\gamma, \tau, y)dy$.

Lemma 3. *Fix* $\gamma, \tau \in [0, 1]$, *and let* m *be a positive integer. Let* $\hat{f}(\gamma, \tau)$ *be the approximation to* $f(\gamma, \tau)$ *obtained by:*

- *Replacing the integral* $\int_0^\tau q(\gamma, \tau, y)dy$ *with a trapezoidal sum with subdivision length* $\frac{1}{m}$,
- *Replacing the other three integrals with left Riemann sums with subdivision length* $\frac{1}{m}$, *and*
- *Replacing the minimum over all* $\theta \leq \gamma$ *with a minimum over a discretization with subdivision length* $\frac{1}{m}$.

Then $\hat{f}(\gamma, \tau) \leq f(\gamma, \tau) + \frac{5}{4m}$.

More precisely, $\hat{f}$ *is defined as follows. Define* $x_k = y_k = \frac{k}{m}$ *for* $k = 0, 1, \ldots, m$. *Let* i *and* j *be the integers such that* $x_i \leq \gamma < x_{i+1}$, *and* $y_j \leq \tau < y_{j+1}$. *Then*

$$\hat{f}(\gamma, \tau) = \hat{p}(\gamma, \tau) + \frac{1}{m} \sum_{k=0}^{j-1} \frac{\hat{q}(\gamma, \tau, y_k) + \hat{q}(\gamma, \tau, y_{k+1})}{2}$$

where $\hat{p}$ and $\hat{q}$ are defined to be

$$\hat{p}(\gamma,\tau) = (1-\gamma)(1-\tau) + (1-\tau) \cdot \frac{1}{m} \sum_{k=0}^{i-1} g(x_k,\tau)$$

and

$$\hat{q}(\gamma,\tau,y) = \min_{k \le i+1} \left\{ 1 - g(x_k,y) + \frac{1}{m} \sum_{d=0}^{k-1} g(x_d,y) + \frac{1}{m} \sum_{d=k}^{i-1} g(x,y_{j+1}) \right\}$$

Combining the above two lemmas allows us to quantify the error incurred when we evaluate the bound in Theorem 1 using a computer.

Corollary 1. *Let $\mathcal{F} = \{0, \frac{1}{n}, \frac{2}{n}, \ldots, 1\}^2$ be an $n \times n$ discretization of the unit square. If we minimize over all $(\gamma,\tau) \in \mathcal{F}$, of the function $\hat{f}(\gamma,\tau)$ defined in Lemma 3, then the minimum value satisfies*

$$\min_{(\gamma,\tau)\in\mathcal{F}} \hat{f}(\gamma,\tau) \le \min_{(\gamma,\tau)\in[0,1]^2} f(\gamma,\tau) + \frac{2}{n} + \frac{5}{4m}.$$

Proof. By Lemma 3, we have $\min_{(\gamma,\tau)\in\mathcal{F}} \hat{f}(\gamma,\tau) \le \min_{(\gamma,\tau)\in\mathcal{F}} f(\gamma,\tau) + \frac{5}{4m}$.

Now let $(\gamma^*,\tau^*) = \arg\min_{(\gamma,\tau)\in[0,1]^2} f(\gamma,\tau)$. Let $(\hat{\gamma},\hat{\tau})$ be the closest point to (γ^*,τ^*) in the discretized grid $\mathcal{F}$. Then $|\hat{\gamma} - \gamma^*| \le \frac{1}{2n}$, and $|\hat{\tau} - \tau^*| \le \frac{1}{2n}$. By Lemma 2, we know f is 1-Lipschitz in γ and 3-Lipschitz in τ, which implies that

$$\min_{(\gamma,\tau)\in\mathcal{F}} f(\gamma,\tau) \le f(\hat{\gamma},\hat{\tau}) \le f(\gamma^*,\tau^*) + \frac{1}{2n} + \frac{3}{2n}.$$

Chaining this with the previous displayed inequality, we obtain

$$\min_{(\gamma,\tau)\in\mathcal{F}} \hat{f}(\gamma,\tau) \le f(\gamma^*,\tau^*) + \frac{2}{n} + \frac{5}{4m},$$

as claimed.

6   Computational Results

In this section, we describe the computations that we performed to obtain a competitive ratio of 0.6629. Recall that Theorem 1 states that the competitive ratio of RANKING is bounded below by an expression of the form $\min_{(\gamma,\tau)\in[0,1]^2} f(\gamma,\tau)$, where f depends on the function g that the algorithm uses. To obtain our competitive ratio, we

1. Solve the LP in Sect. 4 for an appropriate discretization of the unit square. (We chose a 50 × 50 discretization here.)
2. Plug the function g obtained from the LP into the bound in Theorem 1.

Note that for the function g obtained from the LP, we can only evaluate the bound in Theorem 1 approximately. This is because g is a piecewise-affine function defined by interpolating its values on a 50×50 grid, so it has no amenable closed form. As described in Sect. 5, we let $\mathcal{F} = \{0, \frac{1}{n}, \ldots, 1\}^2$ for some large enough n, and we evaluate $\min_{(\gamma, \tau) \in \mathcal{F}} \hat{f}(\gamma, \tau)$, where $\hat{f}$ is an approximation to f amenable to computer evaluation. (Again, refer to Sect. 5 for the details.)

Corollary 1 gives us quantifiable bound on the error incurred when we evaluate $\min_{(\gamma, \tau) \in \mathcal{F}} \hat{f}(\gamma, \tau)$ instead of the true bound $\min_{(\gamma, \tau) \in [0,1]^2} f(\gamma, \tau)$. We used a computer to evaluate $\min_{(\gamma, \tau) \in \mathcal{F}} \hat{f}(\gamma, \tau)$ with $n = 2^{14}$ and $m = 2^{10}$, and obtained $\min_{(\gamma, \tau) \in \mathcal{F}} \hat{f}(\gamma, \tau) = 0.66433$. Thus, by Corollary 1, the competitive ratio of the algorithm is at least

$$\min_{(\gamma, \tau) \in \mathcal{F}} \hat{f}(\gamma, \tau) - \frac{2}{n} - \frac{5}{4m} = 0.66298.$$

Computing the bound for the above choice of parameters n and m necessitated the use of clever computation techniques; the naive computation (which simply goes through all $(\gamma, \tau) \in \mathcal{F}$ one by one, evaluating from scratch $\hat{f}(\gamma, \tau)$ for each) is *too slow* for the size of the discretization we required to obtain a good bound. (For a point $(\gamma, \tau) \in \mathcal{F}$, we estimate that evaluating $\hat{f}(\gamma, \tau)$ is roughly a $O(m^3)$ operation. The naive computation, which does this for each of the n^2 points in $\mathcal{F}$, is then a $O(n^2 m^3)$ computation, which is much too slow for the parameters $n = 2^{14}$ and $m = 2^{10}$.) To speed up the computation, we used two techniques: (1) Precomputation of values that are used repeatedly by the code, and (2) parallelization. Even after speeding up the computation using precomputed tables and parallelization, we still needed two days of computing time on a 64-core machine with 64GB of memory.[3] Without either one of these techniques, the computation would not have terminated in a reasonable amount of time. In the remainder of this section, we describe the above techniques in more detail.

Remark 2. The perceptive reader might notice that it would be conceptually simpler to skip the second step given above altogether (i.e. plugging the function g from the LP into the bound of Theorem 1), since the objective of the LP is already an approximation of the bound in Theorem 1. The reason we do not do this is because to be able to prove a good enough bound, we need to evaluate $\min_{(\gamma, \tau) \in \mathcal{F}} \hat{f}(\gamma, \tau)$ for a fine enough discretization. However, solving the LP is prohibitively expensive for large discretizations. To put this into context, we solved the LP on a 50×50 discretization, and used the output of the LP to evaluate $\min_{(\gamma, \tau) \in \mathcal{F}} \hat{f}(\gamma, \tau)$ on an $n \times n$ discretization, where $n = 2^{14}$. From the description of the LP in Sect. 4, it can be seen that for an $n \times n$ discretization, the LP has roughly n^3 variables and n^4 constraints. For $n = 2^{14}$, this would have been too large an LP to solve.

---

[3] We performed this computation on Amazon EC2. We used a compute-optimized `c6g.16xlarge` instance, running the Amazon Linux 2 AMI.

6.1   Precomputing Tables

The computation we are trying to perform is $\min_{(\gamma,\tau)\in\mathcal{F}}\hat{f}(\gamma,\tau)$. Recall from Sect. 5 that $\hat{f}$ is defined as

$$\hat{f}(\gamma,\tau) = \hat{p}(\gamma,\tau) + \frac{1}{m}\sum_{k=0}^{j-1}\frac{\hat{q}(\gamma,\tau,y_k) + \hat{q}(\gamma,\tau,y_{k+1})}{2}$$

where $\hat{p}$ and $\hat{q}$ are defined to be

- $\hat{p}(\gamma,\tau) = (1-\gamma)(1-\tau) + (1-\tau)\cdot\frac{1}{m}\sum_{k=0}^{i-1}g(x_k,\tau)$, and
- $\hat{q}(\gamma,\tau,y) = \min_{k\leq i+1}\left\{1 - g(x_k,y) + \frac{1}{m}\sum_{d=0}^{k-1}g(x_d,y) + \frac{1}{m}\sum_{d=k}^{i-1}g(x_d,y_{j+1})\right\}$,

where in the above expressions, $x_k = y_k = \frac{k}{m}$, and i and j are defined to be the integers such that $x_i \leq \gamma < x_{i+1}$, and $y_j \leq \tau < y_{j+1}$.

The key observation is that for two different points (γ,τ) and (γ',τ'), some parts of the computation of $\hat{f}(\gamma,\tau)$ and $\hat{f}(\gamma',\tau')$ are the same. Thus, we can speed up the code by precomputing these values and storing them in memory, so that they can be fetched instead of being recomputed each time they are needed. We identified two types of values that could be reused, and precomputed a table for each.

A Table to Store the Values of g. From the expression for $\hat{f}$, we see that it involves many evaluations of g. We can precompute these values and store them in an table for future use. Note that we only ever need to evaluate g on points of the form $(x_i, y_j) = (i/n, j/n)$, which results in a $(n+1)\times(n+1)$ table to be stored in memory. For our choice of $n = 2^{14} = 16384$, this resulted in a table of size roughly 6 GB.

A Table to Store the Values of the Inner Minimum. We also precomputed a table to store the values of $\hat{q}(\gamma,\tau,y)$. Note that $\hat{q}(\gamma,\tau,y)$ only depends on x_i, y_j, and y, where $x_i \leq \gamma < x_{i+1}$, and $y_j \leq \tau < y_{j+1}$. That is, the computation of $\hat{q}(\gamma,\tau,y)$ rounds γ and τ to the nearest points on the $\frac{1}{m}$ discretized grid. Thus there are $m+1$ possible values for each of x_i, y_j, and y, which implies that there are $(m+1)^3$ possible values for $\hat{q}(\gamma,\tau,y)$. We precomputed a table to store all of these values in memory. For our choice of $m = 2^{10} = 1024$, this resulted in a table of size roughly 8 GB.

6.2   Parallelization

The other efficiency gain came from parallelizing the code. With the use of precomputed tables, our code runs in two stages. In the first stage, we compute the two tables described above, and in the second, we use the precomputed tables to compute $\min_{(\gamma,\tau)\in\mathcal{F}}\hat{f}(\gamma,\tau)$. Both stages are amenable to parallelization. For the first stage, computing the value of a table entry is independent of computing the value of another table entry, so filling each table can be done in parallel. For the second stage, evaluating $\hat{f}(\gamma,\tau)$ is independent of evaluating $\hat{f}(\gamma',\tau')$ for different pairs (γ,τ) and (γ',τ'), so this can also be done in parallel. We used the `multiprocessing` module in Python to parallelize our code, which we then ran on a 64-core machine on Amazon's EC2. In total, this took about 2 days. We estimate that a non-parallelized version would have taken more than 100 days to run.

7   Limits of Our Method

Our approach cannot obtain a competitive ratio significantly better than 0.6629. Thus, any further progress beyond this bound will require either further weakening in the assumptions of g, or a stronger analysis than that of Huang et al. We can show the following theorem, whose proof is in the full version of the paper.

Theorem 2. *For any g function that satisfies conditions 1–5, the value of the bound in Theorem 1 is at most 0.6688.*

8   Conclusion

Figure 6 compares the contour plot of the function g we used, to the function $g(x,y) = \frac{1}{2}(h(x) + 1 - h(y))$ used by Huang et al. (Here, $h(x) = \min(1, \frac{1}{2}e^x)$.) The plots look qualitatively quite different. One interesting question is to try to extrapolate a simple function g from the LP contour plot, such that when g is plugged into the bound in Theorem 1, it improves upon the competitive ratio in Huang et al. Visually, the LP contour plot suggests trying a piecewise-linear g with two pieces, where one piece only depends on y. However, we can prove that no function in this class can improve upon the competitive ratio in Huang et al.

As we have noted, because of our upper bound, the value of the competitive ratio cannot be improved by much without either weakening the assumptions on g that we use in Theorem 1 or improving the analysis of Huang et al. [9] that we use. Huang et al. give a potentially stronger bound on the competitive ratio in the conclusion of their paper. However, it was unclear to us how to express their stronger bound as a linear program on discrete values of g.

As discussed in Sect. 7 of the full version of the paper, in order to compute the upper bound, we significantly relaxed the points at which the minimums of the function $f(\gamma, \tau)$ are taken, yet this did not change the value of the LP by much. It seems possible that the Huang et al. analysis can be simplified to reflect this fact.

It would also be interesting to derive an improved upper bound for this problem. To our knowledge, the best upper bound is the same as for the unweighted online bipartite matching problem with random arrivals, which is 0.823, and is due to Manshadi, Oveis Gharan, and Saberi [15].

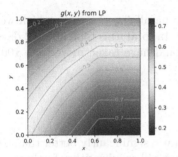
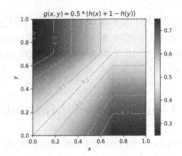

(a) Contour plot of the function values obtained from solving the LP in Section 4. Here, we used a 50 × 50 discretization.

(b) Contour plot of the function $g(x,y) = \frac{1}{2}(h(x) + 1 - h(y))$, where $h(x) = \min(1, \frac{1}{2}e^x)$. This is the function used by Huang et al.

Fig. 6. Side-by-side comparison of the function g we used, versus the function g used by Huang et al.

References

1. Aggarwal, G., Goel, G., Karande, C., Mehta, A.: Online vertex-weighted bipartite matching and single-bid budgeted allocations. In: Proceedings of the Twenty-Second Annual ACM-SIAM Symposium on Discrete Algorithms, SODA 2011, pp. 1253–1264. Society for Industrial and Applied Mathematics (2011)
2. Bahmani, B., Kapralov, M.: Improved bounds for online stochastic matching. In: de Berg, M., Meyer, U. (eds.) ESA 2010. LNCS, vol. 6346, pp. 170–181. Springer, Heidelberg (2010). https://doi.org/10.1007/978-3-642-15775-2_15
3. Birnbaum, B., Mathieu, C.: On-line bipartite matching made simple. SIGACT News **39**, 81–87 (2008)
4. Brubach, B., Sankararaman, K.A., Srinivasan, A., Xu, P.: New algorithms, better bounds, and a novel model for online stochastic matching. CoRR, abs/1606.06395 (2016). http://arxiv.org/abs/1606.06395, arXiv:1606.06395
5. Devanur, N.R., Jain, K., Kleinberg, R.D.: Randomized primal-dual analysis of ranking for online bipartite matching. In: Proceedings of the 24th Annual ACM-SIAM Symposium on Discrete Algorithms, June 2013. https://doi.org/10.1137/1.9781611973105.7
6. Feldman, J., Mehta, A., Mirrokni, V., Muthukrishnan, S.: Online stochastic matching: beating 1−1/e. In: 2009 50th Annual IEEE Symposium on Foundations of Computer Science, pp. 117–126 (2009). https://doi.org/10.1109/FOCS.2009.72
7. Goel, G., Mehta, A.: Online budgeted matching in random input models with applications to adwords. In: Proceedings of the Nineteenth Annual ACM-SIAM Symposium on Discrete Algorithms, SODA 2008, pp. 982–991. Society for Industrial and Applied Mathematics (2008)
8. Huang, Z., Shu, X.: Online stochastic matching, poisson arrivals, and the natural linear program (2021). arXiv:2103.13024
9. Huang, Z., Tang, Z.G., Wu, X., Zhang, Y.: Online vertex-weighted bipartite matching: beating $1 − 1/e$ with random arrivals. ACM Trans. Algorithms **15**(3) (2019). https://doi.org/10.1145/3326169

10. Jaillet, P., Xin, L.: Online stochastic matching: new algorithms with better bounds. Math. Oper. Res. **39**(3), 624–646 (2014). https://doi.org/10.1287/moor.2013.0621
11. Karande, C., Mehta, A., Tripathi, P.: Online bipartite matching with unknown distributions. In: Proceedings of the Forty-Third Annual ACM Symposium on Theory of Computing, STOC 2011, pp. 587–596 (2011). https://doi.org/10.1145/1993636.1993715
12. Karp, R.M., Vazirani, U.V., Vazirani, V.V.: An optimal algorithm for on-line bipartite matching. In: Proceedings of the 22nd Annual ACM Symposium on Theory of Computing - STOC 90 (1990). https://doi.org/10.1145/100216.100262
13. Mahdian, M., Yan, Q.: Online bipartite matching with random arrivals. In: Proceedings of the 43rd Annual ACM Symposium on Theory of Computing - STOC 11 (2011). https://doi.org/10.1145/1993636.1993716
14. Manshadi, V.H., Gharan, S.O., Saberi, A.: Online stochastic matching: online actions based on offline statistics. Math. Oper. Res. **37**(4), 559–573 (2012). https://doi.org/10.1287/moor.1120.0551
15. Manshadi, V.H., Gharan, S.O., Saberi, A.: Online stochastic matching: online actions based on offline statistics. Math. Oper. Res. **37**, 559–573 (2012)
16. Mehta, A.: Online matching and ad allocation. Found. Trends Theor. Comput. Sci. **8**(4), 265–368 (2013). https://doi.org/10.1561/0400000057
17. Schrijver, A.: Combinatorial Optimization: Polyhedra and Efficiency. Springer, Heidelberg (2003)

Beyond Pigouvian Taxes: A Worst Case Analysis

Moshe Babaioff[1], Ruty Mundel[2(✉)], and Noam Nisan[2]

[1] Microsoft Research, Herzliya, Israel
moshe@microsoft.com
[2] Hebrew University of Jerusalem, Jerusalem, Israel
ruth.mundel@mail.huji.ac.il, noam@cs.huji.ac.il

Abstract. In the early 20th century, Pigou observed that imposing a marginal cost tax on the usage of a public good induces a socially efficient level of use as an equilibrium. Unfortunately, such a "Pigouvian" tax may also induce other, socially inefficient, equilibria. We observe that this social inefficiency may be unbounded, and study whether alternative tax structures may lead to milder losses in the worst case, i.e. to a lower price of anarchy. We show that no tax structure leads to bounded losses in the worst case. However, we do find a tax scheme that has a lower price of anarchy than the Pigouvian tax, obtaining tight lower and upper bounds in terms of a crucial parameter that we identify. We generalize our results to various scenarios that each offers an alternative to the use of a public road by private cars, such as ride sharing, or using a bus or a train.

Keywords: Price of anarchy · Pigouvian tax · Public good · Ride sharing

1 Introduction

This paper studies the design of taxes intended to overcome the "tragedy of the commons" in the use of a shared public resource. We consider a situation with a public good and a population of users where each of them may choose to use either the public good or to use a more costly alternative instead. Users of the public good "congest" it, causing a negative externality to all others so the social planner wishes to reduce the use of the public good to the socially optimal level, by levying taxes on such use. Examples of this scenario abound, e.g.:

- Road tolls. Commuters may either use the road by driving their car to work or take public transportation. There is an inconvenience for taking public transportation, but driving a car increases the congestion on the road leading to increased commute times for everyone. Tolls on the road may incentivize taking public transportation.

Supported by the European Research Council (ERC) under the European Union's Horizon 2020 research and innovation programme (grant agreement No 740282).
Supported by the Israeli Smart Transportation Research Center (ISTRC).

© Springer Nature Switzerland AG 2022
M. Feldman et al. (Eds.): WINE 2021, LNCS 13112, pp. 226–243, 2022.
https://doi.org/10.1007/978-3-030-94676-0_13

– Carbon taxes. People that use carbon-based energy sources may instead opt to use more expensive renewable energy sources (e.g. an electric car vs. a petrol based car). The carbon-based energy sources have an externality in terms of pollution and global warming. Carbon taxes incentivize a switch to renewable energy sources.

1.1 Public-Good Congestion Games

Here is a very basic model for these situations. We capture the demand for the public good by the *individual cost function* $\alpha : [0, 1] \rightarrow \mathbb{R}_{\geq 0}$, where $\alpha(q)$ is the price at which fraction q of the population chooses to use the public good.[1] We capture the negative externality imposed by the use of the public good by a non-decreasing *externality function* $l : [0, 1] \rightarrow \mathbb{R}_{\geq 0}$, where $l(q)$ is the negative externality when fraction q of the population uses the public good. We assume that this externality is borne by every member of the population[2]. The *Social Cost* when fraction q of the population uses the public good is thus $SC(q) = l(q) + \int_q^1 \alpha(x)dx$. In this model, if the public good is offered for free then everyone will use the public good so the total social cost will be $l(1)$, which may be tragically high relative to the socially optimal usage level, the one that minimizes this social cost. Already in 1920, Pigou [7] suggested a taxation scheme that will result in an efficient equilibrium: tax each user of the public good the marginal externality that she imposes on society, i.e. the taxation function is given by $t(q) = l'(q)$, where q is the fraction of the population using the public good (and $l'(\cdot)$ is the derivative of l with respect to q). Using first order conditions to minimize the social cost $l(q) + \int_q^1 \alpha(x)dx$ we get that for an optimal q we have that $\alpha(q) = l'(q)$ which is indeed obtained as the equilibrium with this Pigouvian tax.

The starting point of this paper notes that this analysis only guarantees the existence of an efficient equilibrium[3], but it ignores the possibility of other, non-efficient equilibria. It turns out that existence of bad equilibria is indeed possible. Figure 1 shows an example where there are three equilibrium points with $\alpha(q) = l'(q)$, at least one of which has a tragically high social cost, while another has very low social cost. Furthermore, it is indeed possible that natural, best-reply, dynamics will lead to this socially-costly equilibrium.

---

[1] The individual cost function is also the inverse function of the demand function. For every $q \in [0, 1]$, q fraction of the population have disutility at least $\alpha(q)$ for not using the public good (and using the alternative instead).

[2] We also study a variant where only those that use the public good incur the cost of $l(q)$, see Sect. 5.

[3] Under appropriate continuity assumptions.

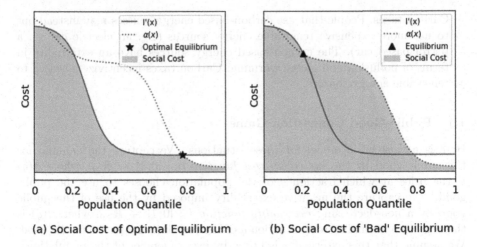

(a) Social Cost of Optimal Equilibrium   (b) Social Cost of 'Bad' Equilibrium

Fig. 1. An example in which the Pigouvian tax induces both the optimal point as an equilibrium (a), as well as a'bad' equilibrium (b). For each population quantile $q \in [0,1]$, the total social cost is given by the shaded area, combining the area under the externality derivative l' (total externality) to its left, and the area under the individual cost function α to its right.

The example in Fig. 1 demonstrates that there may exist non-optimal equilibrium points, but how bad can they be? In other words, what is the "price of anarchy" in this setting? Taking the example of Fig. 1 to its limits, we observe that the ratio between the social cost of an equilibrium and the optimal social cost may be unbounded.

Observation 1. *The price of anarchy of the Pigouvian tax is unbounded.*

Given this failure of the Pigouvian tax, the question that we ask is whether some other taxation scheme $t(\cdot)$ can guarantee a reasonably good social cost in all equilibria points. In other words, yield a better price of anarchy. Like the Pigouvian tax, the desired taxing scheme will charge a price according to the current load in the system. Our main result is negative, showing no tax function yields a small price of anarchy. To formulate this we need some definitions:

Definitions

- For a fixed externality function $l : [0,1] \longrightarrow \mathbb{R}_{\geq 0}$, a fixed tax function $t : [0,1] \longrightarrow \mathbb{R}_{\geq 0}$ and a fixed individual cost function $\alpha : [0,1] \longrightarrow \mathbb{R}_{\geq 0}$, the *Price of Anarchy PoA*(l, t, α) is the social cost of the worst equilibrium[4] divided by the cost of the social optimum, in the game induced by l, t, and α.

---

[4] If no equilibrium exists then we define the price of anarchy to be 1. An equilibrium always exists if t is continuous. None of our results rely on non-existence of equilibrium.

- For a given fixed externality function $l : [0,1] \longrightarrow \mathbb{R}_{\geq 0}$, the *Taxed Price of Anarchy for l*, $TPoA(l)$, is the price of anarchy obtained for the worst individual cost curve α under the best taxation scheme $t(\cdot)$: $TPoA(l) = \inf_t \sup_\alpha PoA(l, t, \alpha)$.
- The *Taxed Price of Anarchy of a family (set) of externality functions* $\mathcal{L} = \{l(\cdot)\}$, is the taxed price of anarchy of the worst externality function in the family: $TPoA(\mathcal{L}) = \sup_{l \in \mathcal{L}} TPoA(l)$.

Note that under the definition of $TPoA(l)$ the tax function $t(\cdot) = t_l(\cdot)$ may depend on the externality function $l(\cdot)$ as we assume it is public knowledge but may not depend on the individual cost function $\alpha(\cdot)$ which we assume is unknown to the planner and possibly varies over time.[5] This basic modeling choice, which follows Pigou, makes it impossible to analyze our problem by encoding it as a routing game, even when the number of agents is finite. This is so since this would require explicitly encoding α within the graph, while we assume α is unknown to the tax designer.

Before presenting a full version of our main negative result, we state a qualitative version, showing that no tax function has a bounded price of anarchy:

Theorem ("Lower Bound", qualitative version): The taxed-price-of-anarchy of the family of all monotone non-decreasing externality functions is unbounded.

Thus, the Pigouvian tax is not the only one susceptible to the problem of agents being trapped in a bad equilibrium given that tax, but rather, *every* tax suffers from that problem when demand is adversarial. Given this negative result we look for conditions under which the taxed price of anarchy can be bounded using some carefully chosen tax scheme.

In the classic special case of strictly convex externality functions (i.e. increasing $l'(\cdot)$) the Pigouvian tax ensures that there is only a single equilibrium point which is thus optimal and so the price of anarchy is 1. Convexity of l is, however, a strong assumption that may not always hold, e.g. when the externality function l has a sigmoid-like shape in cases where the marginal externality saturates at a certain load. We extend the classic convex case to externality functions $l(\cdot)$ that have derivative $l'(\cdot)$ that is "approximately monotone" and show that the taxed price of anarchy degrades linearly in the monotonicity approximation level. See Theorem 2.

Our main results concern general (far from convex) externality functions for which we present tight upper and lower bounds for the taxed price of anarchy as a function of a crucial parameter which we identify. This parameter is the maximal ratio between derivative of l at any two points, which we denote by H.

Definition 1. *For a fixed constant $H > 1$, let $\mathcal{L}_H$ be the* family of functions with bounded derivative ratio H: *the family of externality functions that each satisfies $l'(q_1)/l'(q_2) \leq H$ for every $0 \leq q_1, q_2 \leq 1$.*

---

[5] It is not difficult to see that the problem becomes trivial if the demand is fixed and the tax function may depend on it.

We first prove an upper bound on the taxed price of anarchy of $\mathcal{L}_H$:

Theorem ("Upper Bound"): For $\mathcal{L}_H$, the family of functions with bounded derivative ratio $H > 1$, it holds that the taxed price of anarchy of $\mathcal{L}_H$ is at most $\sqrt{H}$, i.e. $TPoA(\mathcal{L}_H) \leq \sqrt{H}$.

This theorem is obtained with a simple constant tax that does not depend on the load and is set to be the geometric mean of the maximum and minimum values of l'. In contrast we show in the full version of the paper [2] that the price of anarchy of the Pigouvian tax is much worse, $\Theta(H)$. This suggests that unless the social planner can influence the equilibrium selection process, the Pigouvian tax might not be the right choice from a worst-case perspective. The quantitative version of our main theorem shows a tight lower bound on the taxed price of anarchy of $\mathcal{L}_H$:

Theorem ("Lower Bound", quantitative version): For $\mathcal{L}_H$, the family of functions with bounded derivative ratio $H > 1$, it holds that the taxed price of anarchy of $\mathcal{L}_H$ is at least $\sqrt{H}$, i.e. $TPoA(\mathcal{L}_H) \geq \sqrt{H}$.

Combining these two results we immediately derive that the taxed price of anarchy of $\mathcal{L}_H$ is exactly $\sqrt{H}$, i.e. $TPoA(\mathcal{L}_H) = \sqrt{H}$.

We view the main message of our results to be that the Pigouvian tax does not ensure an even approximately efficient use of the public resource, and even though no other taxation scheme is good either, in certain senses, a different taxation scheme may have an advantage over the Pigouvian tax. These results were shown in a model that is obviously idealized. The next subsection considers a variety of more realistic generalizations and shows that the main conclusions hold for these generalizations as well.

1.2   Ride Sharing Games

In the second part of this paper, we look at a generalization of the basic model that captures transportation problems with private cars and ride-sharing options of various types. Here are some natural ride-sharing settings that the model captures:

- **Bus:** The basic model studied above may be viewed as capturing the situation where a commuter may take the bus instead of a private car. The public good in this case is the road, and the taxes are tolls. A bus-ride causes only negligible increased congestion on the road compared to a car ride. However, bus-riders also incur the cost of congestion. Bus passengers do not pay tolls.
- **Carpools with tolls:** In this model, K passengers may share a car, where K is some fixed constant capturing the capacity of a shared car. The load on the road when q fraction of the population take a private car and the other $1 - q$ fraction carpool is $q + (1 - q)/K$. In this model we assume that carpoolers equally split the toll.
- **Carpools without tolls:** This model is similar to the previous one, only that we assume that carpools are toll-free.

One may also think of various intermediate models, e.g. where carpoolers receive a discount on tolls. In Sect. 4 we define a parameterized common generalization of all of these models which we call *Ride Sharing Games*. We prove analogs of our public-good congestion model results for this general class of settings. The main takeaway is that no toll function provides a good price of anarchy in any of these models and that from a worst-case perspective in all of these models there are toll functions that do better than the Pigouvian tax.

In our models so far all users suffers from the externalities of the usage of the public good, i.e. of the road. One can alternatively consider a model in which the population that is not using the public good does not suffer from the externality imposed by public-good users. A simple example for that setting is a public road for which there is a parallel train track, and each person needs to decide whether to ride a private car or take the train. The train creates no congestion at all on the road, and train passengers do not suffer from the road delays at all (travel time is independent of number of cars on the road). We show that also for this model, no toll function provides a good price of anarchy.

The rest of the paper is structured as follows: We first present related work in Sect. 1.3. Section 2 formally defines the public-good congestion model. We phrase our results for this model in Sect. 3. Section 4 describes the ride sharing model and extend our results to this general model. Finally, in Sect. 5 we consider the case that the negative externality imposed by the use of the public good is only suffered by the public-good users, and present similar impossibility results for this model as well. All proofs are omitted and can be found in the full version of the paper [2].

1.3   Related Work

Marginal cost pricing for public goods was first proposed by Pigou [7]. Samuelson [24, 25] has mathematically formulated the theory of public goods. Baumol and Oats [3] and Laffont [17] provides a good introduction to the topic as well as further references.

The concept of price of anarchy analysis was introduced by Koutsoupias and Papadimitriou [16], and used by Roughgarden and Tardos [23] in the context of selfish routing games. Since then this type of analysis has drawn much research attention in routing games as well as a variety of other games (see e.g. [19, 22] as well as references within). The complementary notion of price of stability was introduced in [1].

Traffic congestion problems were first formulated and analyzed in the 1950s by Wardrop [28] and Beckman et al. [4]. A general model in this area is the *Congestion Game* model defined by Rosenthal [21] in 1972. As mentioned above, much of the early work on price of anarchy considered congestion games. Some work has also been done regarding tolls in routing games, starting with [8], where it was shown that when using taxes one can induce the optimal state as an equilibrium. Since then a lot of work has been done on tolls in congestion games [5, 6, 12, 14, 15, 26].

Our model is different from the literature on tolls in congestion games in several respects: first, our public-good model does not fall into the category of routing games, since in our model *everyone* suffers the externalities resulting from the usage of the public good, even those who do not use it. Additionally, the crux of our model is that we treat the demand function as unknown, which is conceptually different from the implicit fixed demand assumption in congestion models.

In transportation research, technologies advancements as well as the option to combine several aspects to combat congestion has been researched extensively, both empirically, e.g. [11, 27] and theoretically, e.g. [9, 10, 13, 18, 20].

2   The Model

A *Public-Good Congestion Game (PGCG)* includes a population of agents where each of them may choose to use either a public good or a more costly alternative instead. Users of the public good "congest" it, causing a negative externality to all others. We assume a large population of agents, with each agent by itself having an infinitesimal effect on the system. This is formalized by modeling the population as a continuum, who have a total mass 1. We assume that the amount that each agent pays for the alternative might vary. We denote by $x \in [0,1]$ the fraction of the population that use the public good while $1 - x$ is the fraction of the population that use the alternative. The externality of using the public good is a function of the total mass of population using it. The social planner can impose a tax on the agents that use the public good, in order to affect agents chosen strategy, aiming to minimize the social cost.

Formally, a Public-Good Congestion Game $PGCG(l, \alpha, t)$ is defined by three functions – an externality function l, a tax function t and an individual cost function α as follows:

Definition 2 (Externality Function l). *We use $l : [0,1] \longrightarrow \mathbb{R}_{\geq 0}$ to denote an externality function, with $l(x)$ being the externality, in monetary terms, experienced by every individual, when the mass of public-good users is x.*

Assumption 1. *We assume that a externality function $l(\cdot)$ satisfies the following:*

- *l is non-negative*
- *l is non-decreasing*
- *l is continuous and left differentiable with a left derivative function denoted by $l'(\cdot)$*

Definition 3 (Tax Function t). *We use $t : [0,1] \longrightarrow \mathbb{R}_{\geq 0}$ to denote a tax function, with $t(x)$ being the tax that every public-good user pays when the mass of public-good users is x.*

Note that we do not make any assumptions on $t(\cdot)$ – it can be any function (not necessarily increasing or continuous). Also note that we define the tax as non-negative (which means the social planner does not pay agents to use the public good), as a social planner that is aiming to minimize the social cost would never choose to use subsidies.

Definition 4 (Individual Cost Function α). *We use $\alpha : [0,1] \longrightarrow \mathbb{R}_{\geq 0}$ to denote an individual cost function with $\alpha(x)$ being the disutility, in monetary terms, an individual that lies on the x percentile of the population suffers by refraining from using the public good. I.e. for every $x \in [0,1]$, x fraction of the population have disutility at least $\alpha(x)$ for not using the public good (and using the alternative instead).*

Assumption 2. *We assume that an individual cost function $\alpha(\cdot)$ satisfies the following:*

- *α is non-negative*
- *α is non-increasing*

As $\alpha(\cdot)$ is a monotone function, it is integrable and additionally, for every $x \in [0,1]$ it has a limit from the left as well as a limit from the right. For a given x, we use $\alpha(x^-) = \inf\{\alpha(t) \mid t < x\}$ to denote the limit from the left at x, and $\alpha(x^+) = \sup\{\alpha(t) \mid t > x\}$ to denote the limit from the right at x.

Note our model assumes the externality, individual cost and tax functions are all measured in the same units of money. We assume agents have the same value for money, and the differences in preferences between different agents are reflected in their individual cost function. For convenience we will denote $PGCG(l, \alpha, t)$ by $G(l, \alpha, t)$.

Definition 5 (Personal Cost c_A). *We use $c_A : [0,1] \times [0,1] \longrightarrow \mathbb{R}_{\geq 0}$ to denote a personal cost function, with $c_A(i, x)$ being the personal cost (or disutility) of an agent in the $i \in [0,1]$ percentile choosing action $A \in \{PG, ALT\}$ (using the public good or using the alternative), when x fraction of the population use the public good.*

This means that the personal cost can be written as follows:

$$c_{PG}(i, x) = l(x) + t(x) \qquad \text{when } i \text{ uses public good}$$
$$c_{ALT}(i, x) = \alpha(i) + l(x) \qquad \text{when } i \text{ uses the alternative}$$

We assume agents are rational: each infinitesimal agent will take the action (use the public good or use the alternative) that has lower personal cost for the agents (breaking ties arbitrarily).

It is now possible to define means of analyzing Public-Good Congestion Games, by defining equilibrium points, as well as socially optimal points.

Definition 6 (Equilibrium Point $\hat{x}$). *Given $G(l, \alpha, t)$, $\hat{x}$ is an equilibrium point if the following hold:*

$$\forall i \leq \hat{x} : \quad c_{ALT}(i, \hat{x}) \geq c_{PG}(i, \hat{x})$$
$$\forall i > \hat{x} : \quad c_{ALT}(i, \hat{x}) \leq c_{PG}(i, \hat{x})$$

I.e. the agents that choose to use the public good are the $\hat{x}$-fraction of the agents whose cost of the alternative is highest. Note that in the case where α and t are continuous, any internal equilibrium point $\hat{x} \in (0,1)$ must satisfy $c_{ALT}(i, \hat{x}) = c_{PG}(i, \hat{x})$. That is, $\hat{x}$ is such that the agents in the $\hat{x}$ percentile are indifferent between using the public good and the alternative.

We next present a simple characterization of equilibrium points. For that characterization it is convenient to define $\alpha(0^-) = +\infty$ and $\alpha(1^+) = -\infty$.

Observation 2. *Given $G(l, \alpha, t)$, $\hat{x}$ is an equilibrium point if and only if*

$$\alpha(\hat{x}^-) \geq t(\hat{x}) \geq \alpha(\hat{x}^+) \tag{1}$$

Note that if t is continuous then an equilibrium is guaranteed to exist. We denote by $EQ(l, \alpha, t)$ the set of equilibrium points of the game $G(l, \alpha, t)$. When l, α and t are clear from context, we use EQ to denote $EQ(l, \alpha, t)$.

Definition 7 (Social Cost SC). *We use $SC_{(l,\alpha)} : [0,1] \longrightarrow \mathbb{R}_{\geq 0}$ to denote the Social Cost (or total disutility) function for the game $G(l, \alpha, t)$, with $SC_{(l,\alpha)}(x)$ being the total amount of disutility when the x fraction of the population with the highest individual cost values uses the public good in the game $G(l, \alpha, t)$. As agents are infinitesimal, the social cost can be given by integrating the personal cost over all $i \in [0,1]$, while omitting the taxes part in the cost, as taxes are paid to the social planner and do not affect the total social cost when considering the planner as part of society:*

$$SC_{(l,\alpha)}(x) = l(x) + \int_x^1 \alpha(z)dz = l(0) + \int_0^x l'(z)dz + \int_x^1 \alpha(z)dz \tag{2}$$

When l and α are clear from context we omit them in the notation and denote $SC_{(l,\alpha)}(x)$ by $SC(x)$.

As α is integrable, the function $A(x) = \int_x^1 \alpha(z)dz$ is continuous. l is continuous as well, so the function SC is continuous on the compact set $[0,1]$. This means that the minimum of the function SC is obtained for some $x \in [0,1]$. We will call such a point a social optimal point:

Definition 8 (Social Optimal Point x^*). *Given a game $G(l, \alpha, t)$, x^* is called a social optimal point if its social cost is minimal:*

$$x^* \in \arg\min_{x \in [0,1]} SC_{(l,\alpha)}(x) \tag{3}$$

Note that there might be multiple social optimal points, but every such point has the same minimal social cost. With a slight abuse of notation we denote that optimal social cost by $SC_{(l,\alpha)}(x^*)$.

Definition 9 (Price of Anarchy $PoA(l, \alpha, t)$). *Given a game $G(l, \alpha, t)$, the Price of Anarchy $PoA(l, \alpha, t)$ is given by the largest ratio between the social cost of an equilibrium, and the optimal social cost[6]:*

$$PoA\,(l, \alpha, t) = \frac{\sup\limits_{\hat{x} \in EQ(l, \alpha, t)} SC_{(l, \alpha)}(\hat{x})}{SC_{(l, \alpha)}(x^*)} \tag{4}$$

In the case that the set $EQ(l, \alpha, t) = \emptyset$ we define $PoA\,(l, \alpha, t) = 1$.

Given an externality function l, the social planner wishes to find a tax function $t = t_l$ (that may depend on l) that minimizes the social cost in the worst case over the population's individual cost function, ensuring a good outcome even if the population's individual cost function arbitrarily changes. Hence, we interpret the taxed price of anarchy as a bound ensuring the tax is good in the worst case (over preferences):

Definition 10 (Taxed Price of Anarchy $TPoA(l)$). *The Taxed Price of Anarchy for externality function l, $TPoA(l)$, is the price of anarchy of l with the best tax function $t = t_l$ for the worst individual cost function α:*

$$TPoA(l) = \inf_t \sup_\alpha PoA\,(l, \alpha, t) \tag{5}$$

Definition 11 (Taxed Price of Anarchy $TPoA(\mathcal{L})$). *The Taxed Price of Anarchy of a family (set) of externality functions $\mathcal{L} = \{l(\cdot)\}$, $TPoA(\mathcal{L})$, is the taxed price of anarchy of the worst externality function in the family:*

$$TPoA(\mathcal{L}) = \sup_{l \in \mathcal{L}} TPoA(l) \tag{6}$$

3   Bounds on TPoA

Pigou [7] has proven that the Pigouvian tax induces an efficient equilibrium. An immediate conclusion is that when the externality derivative is increasing (the externality function is convex), every equilibrium is efficient (since there is a unique equilibrium). For completeness we first present this result. After presenting this result we first extend it to externality functions that are approximately-monotone and then present our main result - a tight bound on the taxed price of anarchy as a function of the worst latency derivative ratio.

Theorem 1 (Pigou 1920 [7] - Increasing Latency Derivative). *Fix $G(l, \alpha, t)$. If the externality function l has a (left) derivative l' that is non-decreasing, then $TPoA(l) = 1$. Furthermore, for such l, setting the Pigouvian tax $t(x) = l'(x)$ yields $PoA(l, \alpha, t) = 1$ for any individual cost function α.*

---

[6] If the optimal social cost is zero then the price of anarchy is defined to be infinity, unless every equilibrium has zero social cost, in that case the PoA is defined to be 1.

Theorem 1 requires the (left) derivative of the externality function to be (weakly) increasing, which is a strong assumption. We next relax this assumption, allowing it to only be "close to" an increasing function, captured by the concept of γ-approximately increasing function:

Definition 12 (γ-approximately increasing function). Let $\gamma > 1$. Given a function $h : [0,1] \longrightarrow \mathbb{R}_{>0}$ we say $h(\cdot)$ is a γ-approximately weakly-increasing function with $H(\cdot)$, if $H : [0,1] \longrightarrow \mathbb{R}_{>0}$ is a non-decreasing function such that for every $x \in [0,1]$ it holds:

$$1 \leq \frac{h(x)}{H(x)} \leq \gamma \tag{7}$$

Note that if $h(\cdot)$ is itself a non-decreasing function, it is $1-$approximately weakly increasing (with $h(\cdot)$ itself).

Theorem 2 (γ-approximately Increasing l'). Fix $G(l, \alpha, t)$. If the (left) derivative of the externality function l' is γ-approximately weakly increasing with an integrable function L', then $TPoA(l) \leq \gamma$. Furthermore, for such l, setting the tax function to be $t(x) = L'(x)$, yields $PoA(l, \alpha, t) \leq \gamma$ for every inconvenience function α.

We now present our main result – two complementary theorems that tightly bound the TPoA when the ratio between the maximal and minimal value of the externality derivative is bounded.

Theorem 3 (Bounded Latency Derivative Upper Bound). Given $H > 0$, let l be an externality function with (left) derivative l' satisfying $\frac{\sup_{0 \leq x \leq 1} l'(x)}{\inf_{0 \leq x \leq 1} l'(x)} \leq H$. Then, $TPoA(l) \leq \sqrt{H}$. Moreover, for any such l, the guarantee is obtained by the constant tax $t(x) = \sqrt{H} \cdot \inf_{0 \leq x \leq 1} l'(x)$ for every $x \in [0,1]$.

Theorem 4 (Bounded Latency Derivative Lower Bound). For any $H > 0$ there exists an externality function l with (left) derivative l' satisfying $\frac{\sup_{0 \leq x \leq 1} l'(x)}{\inf_{0 \leq x \leq 1} l'(x)} \leq H$, for which $TPoA(l) \geq \sqrt{H}$.

4   Ride Sharing Games

The Public-Good Congestion Game model studied so far captures the situation of congestion on a road, where the public good is the road usage. A passenger that chooses to take a private car uses the public good, but may alternatively take the bus. Taking the bus is less convenient, yet passengers that ride the bus do not increase the latency on the road. In this section we expand the model to support additional ride sharing (carpooling) models in which passengers can ride in shared cars of limited capacity (unlike the bus).

Shared car models can differ not only by their carpool capacity, but also by the tolls that carpooling passengers need to pay. One possibility is that carpools are charged the same as a private car and the toll is shared among its passengers, another is that the shared ride is exempt from toll. One can also consider inter-mediate cases in between (e.g., shared cars get 50% discount on the toll which is still equally split). The general model we present captures all these models by adding two additional parameters to the game. The first is the marginal load a passenger in a shared ride adds to the road, and the other is the fraction of toll charged to a private car that a passenger sharing a ride is required to pay. We next formally define this family of ride sharing games.

4.1   Ride Sharing Game Definition

A Ride Sharing Game $RSG(l, \alpha, t, \nu)$ is defined by four parameters. The first three parameters are the same as in the definition of Public-Good Congestion Games: $l(x)$ is the latency experienced by every road-taking agent when the mass of vehicles on the road is x, $\alpha(x)$ is the marginal inconvenience of the x'th percentile of population from sharing their ride with others and $t(x)$ is the toll that every private car taking the road needs to pay when the mass of vehicles on the road is x. The fourth parameter of a Ride Sharing Game is the *Ride Sharing Technology* ν captured by a pair of parameters κ and τ:

Definition 13 (Ride Sharing Technology $\nu = \{\kappa, \tau\}$). *We use $\nu = \{\kappa, \tau\}$ to denote the set of parameters defining the* Ride Sharing Technology *used in the game:*

- *$\kappa \in [0, 1)$ is the marginal load a passenger in a shared ride adds to the road, when normalizing the load of a private car to 1.[7] Thus, with x fraction taking private cars (and $1 - x$ riding shared cars), the total load on the road is $x + \kappa(1 - x) = (1 - \kappa)x + \kappa$ which we denote by $\kappa(x)$. For example, when ride sharing passengers do not add any additional load to the road (as the case for a bus) then $\kappa = 0$, while $\kappa = 1/2$ means that a shared ride passenger add to the road half the load of a private car passenger.[8]*
- *$\tau \in [0, 1)$ is the fraction of toll charged to a private car that a passenger sharing a ride is required to pay.[9][10] For example, $\tau = 0$ means ride sharing passengers are exempt from paying any toll, while $\tau = 1/2$ means a ride sharing passenger is required to pay half the toll of a private car passenger.*

For example, in the bus model, we consider the bus as a single vehicle that has negligible impact on the congestion of the road, and bus passengers do not add

---

[7] The assumption that $\kappa < 1$ corresponds to the externality of a passenger in a private car being larger than that of a passenger riding a shared car.

[8] This formulation is general enough to capture ride sharing vehicles that impose larger load on the road than private cars, as long as the per-passenger load is at most 1: for example, a minibus with 20 passengers that have total load of 2 (like two private cars) has a per-passenger load of 1/10.

[9] The assumption that $\tau < 1$ corresponds to the natural assumption that the toll charged to a passenger of a private car is larger than the toll imposed on a passenger in a shared car.

[10] Note that the total toll paid by all carpool passengers might not necessarily be equal to the toll a private car is charged.

additional load to the road ($\kappa = 0$). Buses are assumed to be exempt from tolls ($\tau = 0$). Thus, using the above notations we denote the Bus *Technology as $BUS = \{0, 0\}$. We present additional examples in Sect. 4.2.*

Using these parameters, we redefine the personal cost of an agent.

Definition 14 (Personal Cost c_A). *We use $c_A : [0, 1] \times [0, 1] \longrightarrow \mathbb{R}_{\geq 0}$ to denote a personal cost function, with $c_A(i, x)$ being the personal cost (or disutility) of a passenger in the $i \in [0, 1]$ percentile choosing action $A \in \{CAR, RS\}$ (riding a private car or a ride sharing), when x fraction of the population rides private cars.*

This means that when the ride sharing technology is $\nu = \{\kappa, \tau\}$ the personal cost can be written as follows:

$$c_{CAR}(i, x) = l(\kappa(x)) + t(\kappa(x)) \qquad \text{when } i \text{ rides a private car}$$
$$c_{RS}(i, x) = \alpha(i) + l(\kappa(x)) + \tau \cdot t(\kappa(x)) \qquad \text{when } i \text{ rides a shared car}$$

We assume passengers are rational: each infinitesimal passenger will take the action (ride a private car or a shared car) that has lower personal cost for the passenger (breaking ties arbitrarily).

We will be interested in studying games induced by various RS technologies. For any RS technology ν, we will consider the game for that fixed ν and denote $G_\nu(l, \alpha, t) = RSG(l, \alpha, t, \nu)$. The generalization of the game affects the expressions given for equilibrium points, as well as the social cost. The generalized equilibrium means a passengers in the equilibrium percentile $\hat{x}$ is not interested in changing his choice between riding a private car and sharing his ride, so the equilibrium condition becomes:

$$\alpha(\hat{x}^-) \geq (1 - \tau) \cdot t(\kappa(\hat{x})) \geq \alpha(\hat{x}^+) \tag{8}$$

and the social cost, which is given by integrating the personal cost over all $i \in [0, 1]$, while omitting the tolls part in the cost, is given by

$$SC_{(l, \alpha, \nu)}(x) = l(\kappa(x)) + \int_x^1 \alpha(z) dz = l(\kappa) + \int_0^x l'(\kappa(z))(1 - \kappa) dz + \int_x^1 \alpha(z) dz \tag{9}$$

The reader can find a full formal definition of the generalized game in the full version of the paper [2].

4.2   Examples of Ride Sharing Games

We next highlight several *Ride Sharing Technologies* of interest: The bus model (which is equivalent to the *Public-Good Congestion Game*), a carpool model where ride sharing passengers are required to pay the same toll as a private car (and split it equally), and a carpool model in which ride sharing passengers are exempt from paying the toll. We show how the parameters of the ride sharing technology capture these three models, and give explicit formulas for the personal and social costs of these scenarios.

Bus Model. In the bus model, an agent can choose whether to ride a private car or to ride a bus with infinite capacity sharing it with other agents being exempt from toll. In our model of an "ideal bus", the bus does not increase the congestion on the road, so every bus passenger has no impact on road congestion (meaning $\kappa = 0$).[11] An agent in the i percentile who chooses to take the bus pays their marginal inconvenience $\alpha(i)$ and is exempt from paying the toll on the road, as the bus itself is exempt from paying it (meaning $\tau = 0$).

We denote by $G_{BUS}(l, \alpha, t) = RSG(l, \alpha, t, \{0, 0\})$ a bus model game in which the *personal cost* of a passenger is

$$c_{CAR}(i, x) = l(x) + t(x) \qquad \text{when } i \text{ rides a private car}$$
$$c_{BUS}(i, x) = \alpha(i) + l(x) \qquad \text{when } i \text{ rides the bus}$$

and the *social cost* $SC_{BUS}(x)$ is

$$SC_{BUS}(x) = l(x) + \int_x^1 \alpha(z)dz = l(0) + \int_0^x l'(z)dz + \int_x^1 \alpha(z)dz$$

Meaning that the bus technology defined by the parameters $\nu = \{0, 0\}$ induces the same game as the *Public-Good Congestion Game*.

Finite Capacity Ride Sharing Models. In the finite-capacity ride sharing model, an agent can choose whether to ride a private car alone or to ride a shared car with additional $K - 1$ agents (meaning $\kappa = \frac{1}{K}$) and thus the load on the road when x fraction of the population ride private cars is $\kappa(x) = \frac{K-1}{K}x + \frac{1}{K}$. Though all agents are affected by the road latency, riding a shared car decreases the load.

Remark 1. As agents 'fill up' cars up to their capacity before using an additional shared car, the only 'shared' car that has less then the maximal capacity number of passengers is the last one, and as agents are infinitesimal, this is negligible.

We highlight two finite-capacity ride sharing models: the *Non-tolled Ride Sharing* model in which agents are exempt from paying the toll (meaning $\tau = 0$) and the *Tolled Ride Sharing* model in which the carpool is charged the same as a private car and its K passengers split the toll equally (meaning $\tau = \frac{1}{K}$).

Non-tolled Ride Sharing. We denote by $G_{RS}(l, \alpha, t) = RSG(l, \alpha, t, \{\frac{1}{K}, 0\})$ a non-tolled ride sharing game in which the *personal cost* of a passenger is:

$$c_{CAR}(i, x) = l\left(\frac{K-1}{K} \cdot x + \frac{1}{K}\right) + t\left(\frac{K-1}{K} \cdot x + \frac{1}{K}\right) \qquad \text{when } i \text{ rides a private car}$$
$$c_{RS}(i, x) = \alpha(i) + l\left(\frac{K-1}{K} \cdot x + \frac{1}{K}\right) \qquad \text{when } i \text{ share a ride}$$

---

[11] The assumption that no congestion is created by an infinite-size bus is ideal, and clearly not very realistic. A more nuanced model has finite-capacity ride sharing vehicles that do increase road congestion.

Tolled Ride Sharing. We denote by $G_{RS-TOLLED}(l, \alpha, t) = RSG(l, \alpha, t, \{\frac{1}{K}, \frac{1}{K}\})$ a tolled ride sharing game in which the *personal cost* of a passenger is:

$$c_{CAR}(i, x) = l\Big(\frac{K-1}{K} \cdot x + \frac{1}{K}\Big) + t\Big(\frac{K-1}{K} \cdot x + \frac{1}{K}\Big)$$

when i rides a private car, and

$$c_{RS-TOLLED}(i, x) = \alpha(i) + l\Big(\frac{K-1}{K} \cdot x + \frac{1}{K}\Big) + \frac{1}{K} \cdot t\Big(\frac{K-1}{K} \cdot x + \frac{1}{K}\Big)$$

when i share a tolled ride.

Note that as the *social cost* $SC_{l,\alpha,\nu}(x)$ is not affected by tolls, the social cost of both these models is the same:

$$SC_{RS}(x) = SC_{RS-TOLLED}(x)$$
$$= l\Big(\frac{K-1}{K} \cdot x + \frac{1}{K}\Big) + \int_x^1 \alpha(z)dz$$
$$= l(\frac{1}{K}) + \int_0^x \frac{K-1}{K} \cdot l'\Big(\frac{K-1}{K} \cdot z + \frac{1}{K}\Big) dz + \int_x^1 \alpha(z)dz$$

Table 1 summarize the ride sharing technologies of these highlighted models.

Table 1. The ride sharing technologies parameters of the highlighted models.

Model	RS marginal road load (κ)	Passenger's share of car toll(τ)
Bus (of infinite capacity)[a]	0	0
(Non-tolled) Shared Rides with carpool capacity K	$\frac{1}{K}$	0
Tolled Shared Rides with carpool capacity K	$\frac{1}{K}$	$\frac{1}{K}$

[a]Note that one can view the bus model as the limit of the shared ride models when K goes to infinity.

4.3 Results for Ride Sharing Games

We next present our results for Ride Sharing Games, which generalize the results we have presented for Public-Good Congestion Games. We start by generalizing Theorem 1, showing that if the latency derivative is non-decreasing then every equilibrium under the Pigouvian tax is efficient.

Theorem 5 (Increasing Latency Derivative). *Fix $G_\nu(l, \alpha, t)$. If the latency function l has a (left) derivative l' that is non-decreasing, then $TPoA_\nu(l) = 1$. Furthermore, for such l, setting the tax $t(x) = \frac{1-\kappa}{1-\tau} l'(x)$ yields $PoA_\nu(l, \alpha, t) = 1$ for any inconvenience function α.*

Table 2 specify the values of t for the models highlighted in Sect. 4.2.

Table 2. The toll function that ensures $TPoA_\nu(l) = 1$ whenever l' is non decreasing, presented for the highlighted models.

Model	$t(x)$
Bus (of infinite capacity)	$l'(x)$
(Non-tolled) Shared Rides with carpool capacity K	$\frac{K-1}{K}l'(x)$
Tolled Shared Rides with carpool capacity K	$l'(x)$

We next generalize Theorem 2, showing that if the latency derivative is γ-approximately weakly increasing then the toll price of anarchy is at most γ.

Theorem 6 (γ-approximately increasing l'). *Fix $G_\nu(l, \alpha, t)$. If the (left) derivative of the externality function l' is γ-approximately weakly increasing with an integrable function L', then $TPoA_\nu(l) \leq \gamma$. Furthermore, for such l, setting the tax function to be $t(x) = \frac{1-\kappa}{1-\tau}L'(x)$, yields $PoA(l, \alpha, t) \leq \gamma$ for every inconvenience function α.*

Finally, we generalize Theorem 3 and Theorem 4, that tightly bound the taxed price of anarchy when the ratio between the maximal and minimal value of the externality derivative is bounded.

Theorem 7 (Bounded Latency Derivative Upper Bound). *Fix ν. Given $H > 0$, for latency l assume that the (left) latency function derivative l' satisfies $\frac{\sup_{0<x\leq 1} l'(\kappa(x))}{\inf_{0<x\leq 1} l'(\kappa(x))} \leq H$. Then, $TPoA_\nu(l) \leq \sqrt{H}$.*

Theorem 8 (Bounded Latency Derivative Lower Bound). *Fix ν. For any $H > 0$ there exists a latency function l with (left) derivative l' satisfying $\frac{\sup_{0<x\leq 1} l'(\kappa(x))}{\inf_{0<x\leq 1} l'(\kappa(x))} \leq H$, for which $TPoA_\nu(l) \geq \sqrt{H}$.*

5   Externality on Public Good Users only (Train)

In this section we consider settings in which the population that does not use the public good does not suffer the externality imposed by public-good users. It will be convenient to think of this model as the train model. In this model passengers may take a train instead of a private car. Train riders do not cause any congestion on the road, nor do they incur the congestion costs or pay any tolls. Under the same definitions for the latency, toll and individual cost functions as in our original 'bus' model, the personal cost in this model is

$$c_{CAR}(i, x) = l(x) + t(x) \qquad \text{when } i \text{ rides a private car}$$
$$c_{TRAIN}(i, x) = \alpha(i) \qquad \text{when } i \text{ rides the train}$$

242   M. Babaioff et al.

Which means the equilibrium condition is

$$\alpha(\hat{x}^-) \geq l(\hat{x}) + t(\hat{x}) \geq \alpha(\hat{x}^+) \tag{10}$$

and the social cost, which is given by integrating the personal cost over all $i \in [0,1]$, while omitting the tolls part in the cost, is given by

$$SC_{(l,\alpha)}(x) = x \cdot l(x) + \int_x^1 \alpha(z)dz \tag{11}$$

In this model the TPoA is unbounded:

Theorem 9. *In the train model, for any $Z > 1$ there exists a non-decreasing latency function l, for which $TPoA(l) \geq Z$.*

References

1. Anshelevich, E., Dasgupta, A., Kleinberg, J., Tardos, É., Wexler, T., Roughgarden, T.: The price of stability for network design with fair cost allocation. SIAM J. Comput. **38**(4), 1602–1623 (2008)
2. Babaioff, M., Mundel, R., Nisan, N.: Beyond Pigouvian taxes: a worst case analysis (2021). https://arxiv.org/abs/2107.12023
3. Baumol, W.J., Oates, W.E.: The Theory of Environmental Policy. Cambridge University Press (1988)
4. Beckmann, M.J., McGuire, C.B., Winsten, C.B.: Studies in the Economics of Transportation. Rand Corporation (1955)
5. Bilò, V., Vinci, C.: Dynamic taxes for polynomial congestion games. ACM Trans. Econ. Comput. **7**(3) (2019). https://doi.org/10.1145/3355946
6. Bonifaci, V., Salek, M., Schäfer, G.: Efficiency of restricted tolls in non-atomic network routing games. In: Persiano, G. (ed.) SAGT 2011. LNCS, vol. 6982, pp. 302–313. Springer, Heidelberg (2011). https://doi.org/10.1007/978-3-642-24829-0_27
7. Pigou, A.C.: The Economics of Welfare. MacMillan, New York (1920)
8. Cole, R., Dodis, Y., Roughgarden, T.: How much can taxes help selfish routing? J. Comput. Syst. Sci. **72**(3), 444–467 (2006)
9. Cramton, P., Geddes, R.R., Ockenfels, A.: Markets in road use: eliminating congestion through scheduling, routing, and real-time road pricing. Working Paper (2017)
10. Cramton, P., Geddes, R.R., Ockenfels, A., et al.: Using technology to eliminate traffic congestion. J. Inst. Theor. Econ. **175**(1), 126–39 (2019)
11. Eliasson, J., et al.: The stockholm congestion charges: an overview. Stockholm: Centre for Transport Studies CTS Working Paper 7, 42 (2014)
12. Fleischer, L., Jain, K., Mahdian, M.: Tolls for heterogeneous selfish users in multicommodity networks and generalized congestion games. In: 45th Annual IEEE Symposium on Foundations of Computer Science, pp. 277–285. IEEE (2004)
13. Hall, J.D.: Pareto improvements from Lexus lanes: the effects of pricing a portion of the lanes on congested highways. J. Public Econ. **158**, 113–125 (2018). https://doi.org/10.1016/j.jpubeco.2018.01.003. http://www.sciencedirect.com/science/article/pii/S0047272718300033

14. Harks, T., Schäfer, G., Sieg, M.: Computing flow-inducing network tolls. Technical report, Technical Report 36–2008, Institut für Mathematik, Technische Universität (2008)
15. Karakostas, G., Kolliopoulos, S.G.: Edge pricing of multicommodity networks for heterogeneous selfish users. In: FOCS, pp. 268–276 (2004)
16. Koutsoupias, E., Papadimitriou, C.: Worst-case equilibria. In: Meinel, C., Tison, S. (eds.) STACS 1999. LNCS, vol. 1563, pp. 404–413. Springer, Heidelberg (1999). https://doi.org/10.1007/3-540-49116-3_38
17. Laffont, J.J., et al.: Fundamentals of Public Economics. MIT Press Books 1 (1988)
18. Mehr, N., Horowitz, R.: Pricing traffic networks with mixed vehicle autonomy. In: 2019 American Control Conference (ACC), pp. 2676–2682 (2019). https://doi.org/10.23919/ACC.2019.8814392
19. Nisan, N., Roughgarden, T., Tardos, E., Vazirani, V.V.: Algorithmic Game Theory. Cambridge University Press, New York (2007)
20. Ostrovsky, M., Schwarz, M.: Carpooling and the economics of self-driving cars. In: Proceedings of the 2019 ACM Conference on Economics and Computation, EC 2019, pp. 581–582 (2019)
21. Rosenthal, R.W.: A class of games possessing pure-strategy Nash equilibria. Internat. J. Game Theory 2(1), 65–67 (1973)
22. Roughgarden, T.: Selfish Routing and the Price of Anarchy, vol. 174. MIT Press, Cambridge (2005)
23. Roughgarden, T., Tardos, É.: How bad is selfish routing? J. ACM (JACM) 49(2), 236–259 (2002)
24. Samuelson, P.A.: Diagrammatic Exposition of a Theory of Public Expenditure, pp. 159–171. Macmillan Education UK (1995)
25. Samuelson, P.A.: The pure theory of public expenditure. Rev. Econ. Stat. 36(4), 387–389 (1954)
26. Swamy, C.: The effectiveness of stackelberg strategies and tolls for network congestion games. In: SODA, pp. 1133–1142 (2007)
27. Varaiya, P.: What we've learned about highway congestion. Access Mag. 1(27), 2–9 (2005)
28. Wardrop, J.G.: Road paper. Some theoretical aspects of road traffic research. Proc. Inst. Civil Eng. 1(3), 325–362 (1952)

The Core of Housing Markets from an Agent's Perspective: Is It Worth Sprucing Up Your Home?

Ildikó Schlotter[1,2(✉)], Péter Biró[1,3], and Tamás Fleiner[1,2]

[1] Centre for Economic and Regional Studies, Budapest, Hungary
{schlotter.ildiko,biro.peter,fleiner.tamas}@krtk.hu
[2] Budapest University of Technology and Economics, Budapest, Hungary
[3] Corvinus University of Budapest, Budapest, Hungary

Abstract. We study housing markets as introduced by Shapley and Scarf [39]. We investigate the computational complexity of various questions regarding the situation of an agent a in a housing market H: we show that it is NP-hard to find an allocation in the core of H where (i) a receives a certain house, (ii) a does not receive a certain house, or (iii) a receives a house other than her own. We prove that the core of housing markets *respects improvement* in the following sense: given an allocation in the core of H where agent a receives a house h, if the value of the house owned by a increases, then the resulting housing market admits an allocation where a receives either h, or a house that she prefers to h; moreover, such an allocation can be found efficiently. We further show an analogous result in the STABLE ROOMMATES setting by proving that stable matchings in a one-sided market also respect improvement.

1 Introduction

Housing markets is a classic model in economics where agents are initially endowed with one unit of an indivisible good, called a *house*, and agents may trade their houses according to their preferences without using monetary transfers. In such markets, trading results in a reallocation of houses in a way that each agent ends up with exactly one house. Motivation for studying housing markets comes from applications such as kidney exchange [8,12,35] and housing programs [1,43].

In their seminal work Shapley and Scarf [39] examined housing markets where agents' preferences are weak orders. They proved that such markets always admit a *core* allocation, that is, an allocation where no coalition of agents can strictly improve their situation by trading only among themselves. They also described the Top Trading Cycles (TTC) algorithm, proposed by David Gale, and proved that the set of allocations that can be obtained through the TTC algorithm coincides with the set of competitive allocations; hence the TTC always produces an allocation in the core. When preferences are strict, the TTC produces the

Supported by the Hungarian Academy of Sciences (Momentum Programme LP2021-2) and the Hungarian Scientific Research Fund (NFKIH grants K128611, K124171).

M. Feldman et al. (Eds.): WINE 2021, LNCS 13112, pp. 244–261, 2022.
https://doi.org/10.1007/978-3-030-94676-0_14

unique allocation in the *strict core*, that is, an allocation where no coalition of agents can weakly improve their situation by trading among themselves [34].

Although the core of housing markets has been the subject of considerable research, there are still many challenges which have not been addressed. Consider the following question: given an agent a and a house h, does there exist an allocation in the core where a obtains h? Or one where a does not obtain h? Can we determine whether a may receive a house better than her own in some core allocation? Similar questions have been extensively studied in the context of the STABLE MARRIAGE and the STABLE ROOMMATES problems [20–23,31], but have not yet been considered in relation to housing markets.

Even less is known about the core of housing markets in cases where the market is not static. Although some researchers have addressed certain dynamic models, most of these either focus on the possibility of repeated allocation [28, 29,34], or consider a situation where agents may enter and leave the market at different times [13,32,42]. Recently, Biró et al. [9] have investigated how a change in the preferences of agents affects the housing market. Namely, they considered how an improvement of the house belonging to agent a affects the situation of a. Following their lead, we aim to answer the following question: if the value of the house belonging to agent a increases, how does this affect the core of the market from the viewpoint of a? Is such a change bound to be beneficial for a, as one would expect? This question is of crucial importance in the context of kidney exchange: if procuring a new donor with better properties (e.g., a younger or healthier donor) does not necessarily benefit the patient, then this could undermine the incentive for the patient to find a donor with good characteristics, damaging the overall welfare.

1.1 Our Contribution

We consider the computational complexity of deciding whether the core of a housing market contains an allocation where a given agent a obtains a certain house. In Theorem 1 we prove that this problem is NP-complete, as is the problem of finding a core allocation where a does *not* receive a certain house. Even worse, it is already NP-complete to decide whether a core allocation can assign *any* house to a other than her own. Various generalizations of these questions can be answered efficiently in both the STABLE MATCHING and STABLE ROOMMATES settings [20–23,31], so we find these intractability results surprising.

Instead of asking for a core allocation where a given agent can trade her house, one can also look at the optimization problem which asks for an allocation in the core with the maximum number of agents involved in trading. This problem is known to be NP-complete [18]. We show in Theorem 2 that for any $\varepsilon > 0$, approximating this problem with ratio $|N|^{1-\varepsilon}$ for a set N of agents is NP-hard. We complement this strong inapproximability result in Proposition 3 by pointing out that a trivial approach yields an approximation algorithm with ratio $|N|$.

Turning our attention to the question of how an increase in the value of a house affects its owner, we show the following result in Theorem 4. If the core of a housing market contains an allocation where a receives a house h, and the market changes in a way such that some agents perceive an increased value for the house owned by a (and nothing else changes in the market), then the resulting housing market admits an allocation in its core where a receives either h or a house that she prefers to h. We prove this in a constructive way, by presenting an algorithm that finds such an allocation. This settles an open question by Biró et al. [9] who ask whether the core *respects improvement* in the sense that the best allocation achievable for an agent a in a core allocation can only (weakly) improve for a as a result of an increase in the value of a's house.

It is clear that an increase in the value of a's house may not always yield a *strict* improvement for a (as a trivial example, some core allocation may assign a her top choice even before the change), but one may wonder if we can efficiently determine when a strict improvement for a becomes possible. This problem turns out to be closely related to the question whether a can obtain a given house in a core allocation; in fact, we were motivated to study the latter problem by our interest in determining the possibilities for a strict improvement. Although one can formulate several variants of the problem depending on what exactly one considers to be a strict improvement, by Theorem 11 each of them leads to computational intractability (NP-hardness or coNP-hardness).

Finally, we also answer a question raised by Biró et al. [9] regarding the property of respecting improvements in the context of the STABLE ROOMMATES problem. An instance of STABLE ROOMMATES contains a set of agents, each having preferences over the other agents; the usual task is to find a matching between the agents that is *stable*, i.e., no two agents prefer each other to their partners in the matching. It is known that a stable matching need not always exist, but if it does, then Irving's algorithm [26] finds one efficiently. In Theorem 14 we show that if some stable matching assigns agent a to agent b in a STABLE ROOMMATES instance, and the valuation of a increases (that is, if she moves upward in other agents' preferences, with anything else remaining constant), then the resulting instance admits a stable matching where a is matched either to b or to an agent she prefers to b. This result is a direct analog of the one stated in Theorem 4 for the core of housing markets; however, the algorithm we propose in order to prove it uses different techniques.

We remark that we use a model with partially ordered preferences (a generalization of weak orders), and describe a linear-time implementation of the TTC algorithm in such a model.

1.2 Related Work

Most works relating to the core of housing markets aim for finding core allocations with some additional property that benefits global welfare, most prominently Pareto optimality [4,5,27,33,37]. Another line of research comes from kidney exchange where the length of trading cycles is of great importance and often plays a role in agents' preferences [7,15–17,19] or is bounded by some constant [2,10,11,18,25]. None of these deal with problems where a core allocation is required to fulfill some constraint regarding a given agent or set of agents—that they be trading, or that they obtain (or not obtain) a certain house. Nevertheless, some of them focus on finding a core allocation where the number of agents involved in trading is as large as possible. Cechlárová and Repiský [18] proved that this problem is NP-hard in the classical housing market model, while Biró and Cechlárová [7] considered a special model where agents care first about the house they receive and after that about the length of their trading cycle (shorter being better); they prove that for any $\varepsilon > 0$, it is NP-hard to approximate the number of agents trading in a core allocation with a ratio $|N|^{1-\varepsilon}$ (where N is the set of agents).

The property of respecting improvement has first been studied in a paper by Balinski and Sönmez [6] on college admission. They proved that the student-optimal stable matching algorithm respects the improvement of students, so a better test score for a student always results in an outcome weakly preferred by the student (assuming other students' scores remain the same). Hatfield et al. [24] contrasted their findings by showing that no stable mechanism respects the improvement of school quality. Sönmez and Switzer [40] applied the model of matching with contracts to the problem of cadet assignment in the United States Military Academy, and have proved that the cadet-optimal stable mechanism respects improvement of cadets. Recently, Klaus and Klijn [30] have obtained results of a similar flavor in a school-choice model with minimal-access rights.

Roth et al. [36] deal with the property of respecting improvement in connection with kidney exchange: they show that in a setting with dichotomous preferences and pairwise exchanges priority mechanisms are donor monotone, meaning that a patient can only benefit from bringing an additional donor on board. Biró et al. [9] focus on the classical Shapley-Scarf model and investigate how different solution concepts behave when the value of agent's house increases. They prove that both the strict core and the set of competitive allocations satisfy the property of respecting improvements, however, this is no longer true when the lengths of trading cycles are bounded by some constant.

2   Preliminaries

Preferences as Partial Orders. In the majority of the existing literature, preferences of agents are usually considered to be either strict or, if the model allows for indifference, weak linear orders. Weak orders can be described as lists containing *ties*, a set of alternatives considered equally good for the agent. Partial orders are a generalization of weak orders that allow for two alternatives to be *incomparable* for an agent. Incomparability may not be transitive, as opposed to indifference in weak orders. Formally, an (irreflexive)[1] *partial ordering* $\prec$ on a set of alternatives is an irreflexive, antisymmetric and transitive relation.

Partially ordered preferences arise by many natural reasons; we give two examples motivated by kidney exchanges. For example, agents may be indifferent between goods that differ only slightly in quality. Indeed, recipients might be indifferent between two organs if their expected graft survival times differ by less than one year. However, small differences may add up to a significant contrast:

---

[1] Throughout the paper we will use the term *partial ordering* in the sense of an irreflexive (or strict) partial ordering.

an agent may be indifferent between a and b, and also between b and c, but strictly prefer a to c. Partial preferences also emerge in multiple-criteria decision making. The two most important factors for estimating the quality of a kidney transplant are the HLA-matching between donor and recipient, and the age of the donor.[2] An organ is considered better than another if it is better with respect to both of these factors, leading to partial orders.

Housing Markets. Let $H = (N, \{\prec_a\}_{a \in N})$ be a *housing market* with agent set N and with the preferences of each agent $a \in N$ represented by a partial ordering $\prec_a$ of the agents. For agents a, b, and c, we interpret $a \prec_c b$ as agent c *preferring* the house owned by agent b to the house of agent a. We will write $a \preceq_c b$ as equivalent to $b \not\prec_c a$, and we write $a \sim_c b$ if $a \not\prec_c b$ and $b \not\prec_c a$. We say that agent a finds the house of b *acceptable*, if $a \preceq_a b$, and we denote by $A(a) = \{b \in N : a \preceq_a b\}$ the set of agents whose house is acceptable for a. We define the *acceptability graph* of the housing market H as the directed graph $G^H = (N, E)$ with $E = \{(a, b) \mid b \in A(a)\}$; we let $|G^H| = |N| + |E|$. Note that $(a, a) \in E$ for each $a \in N$. The *submarket* of H on a set $W \subseteq N$ of agents is the housing market $H_W = (W, \{\prec_a^{|W}\}_{a \in W})$ where $\prec_a^{|W}$ is the partial order $\prec_a$ restricted to W; the acceptability graph of H_W is the subgraph of G^H induced by W, denoted by $G^H[W]$. For a set W of agents, let $H - W$ be the submarket $H_{N \setminus W}$ obtained by *deleting* W from H; for $W = \{a\}$ we may write simply $H - a$.

For a set $X \subseteq E$ of arcs in G^H and an agent $a \in N$ we let $X(a)$ denote the set of agents b such that $(a, b) \in X$; whenever $X(a)$ is a singleton $\{b\}$ we will abuse notation by writing $X(a) = b$. We also define $\delta_X^-(a)$ and $\delta_X^+(a)$ as the number of in-going and out-going arcs of a in X, respectively. For a set $W \subseteq N$ of agents, we let $X[W]$ denote the set of arcs in X that run between agents of W.

We define an *allocation* X in H as a subset $X \subseteq E$ of arcs in G^H such that $\delta_X^-(a) = \delta_X^+(a) = 1$ for each $a \in N$, that is, X forms a collection of cycles in G^H containing each agent exactly once. Then $X(a)$ denotes the agent whose house a obtains according to allocation X. If $X(a) \neq a$, then a is *trading* in X. For allocations X and X', we say that a *prefers* X to X' if $X'(a) \prec_a X(a)$.

For an allocation X in H, an arc $(a, b) \in E$ is X-*augmenting*, if $X(a) \prec_a b$. We define the *envy graph* $G_{X\prec}^H$ of X as the subgraph of G^H containing all X-augmenting arcs. A *blocking cycle* for X in H is a cycle in $G_{X\prec}^H$, that is, a cycle C where each agent a on C prefers $C(a)$ to $X(a)$. An allocation X is contained in the *core* of H, if there does not exist a blocking cycle for it, i.e., if $G_{X\prec}^H$ is acyclic. A *weakly blocking cycle* for X is a cycle C in G^H where $X(a) \preceq_a C(a)$ for each agent a on C and $X(a) \prec_a C(a)$ for at least one agent a on C. The *strict core* of H contains allocations that do not admit weakly blocking cycles.

Organization. Section 3 contains an adaptation of the TTC algorithm for partially ordered preferences, followed by our results on finding core allocations with various arc restrictions and on maximizing the number of agents involved in trading. In Sect. 4 we present our results on the property of respecting improvements

---

[2] In fact, these are the two factors for which acceptability thresholds can be set by the patients in the UK program [8].

in relation to the core of housing markets, including our main technical result, Theorem 4. In Sect. 5 we study the respecting improvement property in the context of the STABLE ROOMMATES problem. We conclude with some questions for future research in Sect. 6.

3 The Core of Housing Markets: Some Computational Problems

We investigate a few computational problems related to the core of housing markets. In Sect. 3.1 we describe our adaptation of TTC to partially ordered preferences. In Sect. 3.2 we turn our attention to the problem of finding an allocation in the core of a housing market that satisfies certain arc restrictions, requiring that a given arc be contained or, just the opposite, not be contained in the desired allocation. In Sect. 3.3 we look at the most prominent optimization problem in connection with the core: given a housing market, find an allocation in its core where the number of trading agents is as large as possible.

3.1 Top Trading Cycles for Preferences with Incomparability

Strict Preferences. If agents' preferences are represented by strict orders, then the TTC algorithm [39] produces the unique allocation in the strict core. TTC creates a directed graph D where each agent a points to her top choice, that is, to the agent owning the house most preferred by a. In the graph D each agent has out-degree exactly 1, since preferences are assumed to be strict. Hence, D contains at least one cycle, and moreover, the cycles in D do not intersect. TTC selects all cycles in the graph D as part of the desired allocation, deletes from the market all agents trading along these cycles, and repeats the whole process until there are no agents left.

Preferences as Partial Orders. When preferences are represented by partial orders, one can modify the TTC algorithm by letting each agent a in D point to her *undominated* choices: b is undominated for a, if there is no agent c such that $b \prec_a c$. Notice that an agent's out-degree is then *at least* 1. Thus, D contains at least one cycle, but in case it contains more than one cycle, these may overlap.

A simple approach is to select a set of mutually vertex-disjoint cycles in each round, removing the agents trading along them from the market and proceeding with the remainder in the same manner. It is not hard to see that this approach yields an algorithm that produces an allocation in the core: by the definition of undominated choices, any arc of a blocking cycle leaving an agent a necessarily points to an agent that was already removed from the market at the time when a cycle containing a got selected. Clearly, no cycle may consist of such "backward" arcs only, proving that the computed allocation is indeed in the core.

Implementation in Linear Time. Abraham et al. [3] describe an implementation of the TTC algorithm for strict preferences that runs in $O(|G^H|)$ time. We extend their ideas to the case when preferences are partial orders as follows.

For each agent $a \in N$ we assume that a's preferences are given using a *Hasse diagram* which is a directed acyclic graph H_a that can be thought of as a compact representation of $\prec_a$. The vertex set of H_a is $A(a)$, and it contains an arc (b,c) if and only if $b \prec_a c$ and there is no agent c' with $b \prec_a c' \prec_a c$. Then the description of our housing market H has length $\sum_{a \in A} |H_a|$ which we denote by $|H|$. If preferences are weak or strict orders, then $|H| = O(|G^H|)$.

Throughout our variant of TTC, we will maintain a list $U(a)$ containing the undominated choices of a among those that still remain in the market, as well as a subgraph D of G^H spanned by all arcs (a,b) with $b \in U(a)$. Furthermore, for each agent a in the market, we will keep a list of all occurrences of a as someone's undominated choice. Using H_a we can find the undominated choices of a in $O(|H_a|)$ time, so initialization takes $O(|H|)$ time in total.

Whenever an agent a is deleted from the market, we find all agents b such that $a \in U(b)$, and we update $U(b)$ by replacing a with its in-neighbors in H_b. Notice that the total time required for such deletions (and the necessary replacements) to maintain $U(b)$ is $O(|H_b|)$. Hence, we can efficiently find the undominated choices of each agent at any point during the algorithm, and thus traverse the graph D consisting of arcs (a,b) with $b \in U(a)$.

To find a cycle in D, we simply keep building a path using arcs of D, until we find a cycle (perhaps a loop). After recording this cycle and deleting its agents from the market (updating the lists $U(a)$ as described above), we simply proceed with the last agent on our path. Using the data structures described above the total running time of our variant of TTC is $O(|N| + \sum_{a \in N} |H_a|) = O(|H|)$.

3.2   Allocations in the Core with Arc Restrictions

We now focus on the problem of finding an allocation in the core that fulfills certain arc constraints. The simplest such constraints arise when we require a given arc to be included in, or conversely, be avoided by the desired allocation.

We define the ARC IN CORE problem as follows: given a housing market $H = (N, \{\prec_a\}_{a \in N})$ and an arc (a,b) in G^H, decide whether there exists an allocation in the core of H that contains (a,b), or in other words, where agent a obtains the house of agent b. Analogously, the FORBIDDEN ARC IN CORE problem asks to decide if there exists an allocation in the core of H *not* containing (a,b).

By giving a reduction from ACYCLIC PARTITION [14], we show in Theorem 1 that both of these problems are computationally intractable, even if each agent has a strict ordering over the houses. In fact, we cannot even hope to decide for a given agent a in a housing market H whether there exists an allocation in the core of H where a is trading; we call this problem AGENT TRADING IN CORE.

Theorem 1 ($\star^3$). *Each of the following problems is NP-complete, even if agents' preferences are strict orders:*

- ARC IN CORE,
- FORBIDDEN ARC IN CORE, *and*
- AGENT TRADING IN CORE.

---

[3] Proofs marked by an asterisk can be found in the full version of our paper [38].

3.3   Maximizing the Number of Agents Trading in a Core Allocation

Perhaps the most natural optimization problem related to the core of housing markets is the following: given a housing market H, find an allocation in the core of H whose *size*, defined as the number of trading agents, is maximal among all allocations in the core of H; we call this the MAX CORE problem. MAX CORE is NP-hard by a result of Cechlárová and Repiský [18]. In Theorem 2 below we show that even approximating MAX CORE is NP-hard. Our result is tight in the following sense: we prove that for any $\varepsilon > 0$, approximating MAX CORE with a ratio of $|N|^{1-\varepsilon}$ is NP-hard, where $|N|$ is the number of agents in the market. By contrast, a very simple approach yields an approximation with ratio $|N|$.

We remark that Biró and Cechlárová [7] proved a similar inapproximability result, but since they considered a special model where agents not only care about the house they receive but also about the length of their exchange cycle, their result cannot be translated to our model, and so does not imply Theorem 2. Instead, our reduction relies on the ideas we use to prove Theorem 1.

Theorem 2 ($\star$). *For any constant $\varepsilon > 0$, the* MAX CORE *problem is* NP-*hard to approximate within a ratio of $\alpha_\varepsilon(N) = |N|^{1-\varepsilon}$ where N is the set of agents, even if agents' preferences are strict orders.*

We contrast Theorem 2 with the observation that an algorithm that outputs *any* allocation in the core yields an approximation for MAX CORE with ratio $|N|$.

Proposition 3 ($\star$). MAX CORE *can be approximated with a ratio of $|N|$ in polynomial time, where $|N|$ is the number of agents in the input.*

4   The Effect of Improvements in Housing Markets

Let $H = (N, \{\prec_a\}_{a \in N})$ be a housing market containing agents p and q. We consider a situation where the preferences of q are modified by "increasing the value" of p for q without altering the preferences of q over the remaining agents. If the preferences of q are given by a strict or weak order, then this translates to *shifting* the position of p in the preference list of q towards the top. Formally, a housing market $H' = (N, \{\prec'_a\}_{a \in N})$ is called a (p,q)-*improvement* of H, if $\prec_a = \prec'_a$ for any $a \in N \setminus \{q\}$, and $\prec'_q$ is such that (i) $a \prec'_q b$ iff $a \prec_q b$ for any $a, b \in N \setminus \{p\}$, and (ii) if $a \prec_q p$, then $a \prec'_q p$ for any $a \in N$. We will also say that a housing market is a p-*improvement* of H, if it can be obtained by a sequence of (p, a_i)-improvements for a series $a_1, \dots, a_k$ of agents for some $k \in \mathbb{N}$.

To examine how p-improvements affect the situation of p in the market, one may consider several solution concepts such as the core, the strict core, and so on. We regard a solution concept as a function Φ that assigns a set of allocations to each housing market. Based on the preferences of p, we can compare allocations in Φ. Let $\Phi_p^+(H)$ denote the set containing the best houses p can obtain in $\Phi(H)$:

$$\Phi_p^+(H) = \{X(p) \mid X \in \Phi(H), \forall X' \in \Phi(H) : X'(p) \preceq_p X(p)\}.$$

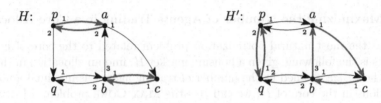

Fig. 1. The housing markets H and H' in the proof of Proposition 6. Here and everywhere else we depict markets through their acceptability graphs with all loops omitted; preferences are indicated by numbers along the arcs. For both H and H', the allocation represented by red (and bold) arcs yields the worst possible outcome for p in any core allocation of the given market. (Color figure online)

Similarly, let $\Phi_p^-(H)$ be the set containing the worst houses p can obtain in $\Phi(H)$.

Following the notation used by Biró et al. [9], we say that Φ *respects improvement for the best available house* or simply *satisfies the RI-best property*, if for any housing markets H and H' such that H' is a p-improvement of H for some agent p, $a \preceq_p a'$ for every $a \in \Phi_p^+(H)$ and $a' \in \Phi_p^+(H')$. Similarly, Φ *respects improvement for the worst available house* or simply *satisfies the RI-worst property*, if for any housing markets H and H' such that H' is a p-improvement of H for some agent p, $a \preceq_p a'$ for every $a \in \Phi_p^-(H)$ and $a' \in \Phi_p^-(H')$.

Notice that the above definition does not take into account the possibility that a solution concept Φ may become empty as a result of a p-improvement. To exclude such a possibility, we may require the condition that an improvement does not destroy all solutions. We say that Φ *strongly satisfies the RI-best (or RI-worst) property*, if besides satisfying the RI-best (or, respectively, RI-worst) property, it also guarantees that whenever $\Phi(H) \neq \emptyset$, then $\Phi(H') \neq \emptyset$ also holds where H' is a p-improvement of H for some agent p.

We prove that the core of housing markets satisfies the RI-best property. In fact, Theorem 4 (proved in Sect. 4.2) states a slightly stronger statement. By contrast, Proposition 6 shows that the core of housing markets violates the RI-worst property.

Theorem 4. *Given an allocation X in the core of the housing market H and a p-improvement H' of H, there exists an allocation X' in the core of H' such that either $X(p) = X'(p)$ or p prefers X' to X. Moreover, given H, H' and X, it is possible to find such an allocation X' in polynomial time.*

Corollary 5. *The core of housing markets strongly satisfies the RI-best property.*

Proposition 6. *The core of housing markets violates the RI-worst property.*

Proof. Let $N = \{a, b, c, p, q\}$ be the set of agents. The preferences indicated in Fig. 1 define a housing market H and a (p, q)-improvement H' of H.

We claim that in every allocation in the core of H, agent p obtains the house of a. To see this, let X be an allocation where $(p,a) \notin X$. If agent a is not trading in X, then a and p form a blocking cycle; therefore, $(b,a) \in X$. Now, if $(c,b) \notin X$, then c and b form a blocking cycle for X; otherwise, q and b form a blocking cycle for X. Hence, p obtains her top choice in all core allocations of H.

However, it is easy to verify that the core of H' contains an allocation where p obtains only her second choice (q's house), as shown in Fig. 1.   □

We describe our algorithm for Theorem 4 in Sect. 4.1, and prove its correctness in Sect. 4.2. In Sect. 4.3 we look at the problem of deciding whether a p-improvement leads to a situation strictly better for p.

4.1 Description of Algorithm HM-Improve

Before describing our algorithm for Theorem 4, we need some notation.

Pre-allocations and Their Envy Graphs. Given a housing market $H = (N, \{\prec_a\}_{a\in N})$ and two distinct agents u and v in N, we say that a set Y of arcs in $G^H = (N, E)$ is a *pre-allocation* from u to v in H, if

- $\delta_Y^-(u) = \delta_Y^+(v) = 0$,
- $\delta_Y^+(a) = 1$ for each $a \in N \setminus \{v\}$, and
- $\delta_Y^-(a) = 1$ for each $a \in N \setminus \{u\}$.

Note that Y is a collection of vertex-disjoint cycles and a unique path P in G^H, with P leading from u to v. We call u the *source* of Y and v its *sink*.

Given a pre-allocation Y from u to v in H, an arc $(a,b) \in E$ is Y-*augmenting*, if $a \neq v$ and $Y(a) \prec_a b$. We define the *envy graph* of Y as $G_{Y\prec}^H = (N, E_Y)$ where E_Y is the set of Y-augmenting arcs in E. A blocking cycle for Y is a cycle in $G_{Y\prec}^H$; notice that such a cycle cannot contain the sink v, since no Y-augmenting arc leaves v. We say that the pre-allocation Y is *stable*, if no blocking cycle exists for Y, that is, if its envy graph is acyclic.

We are now ready to propose an algorithm called HM-Improve that given an allocation X in the core of H outputs an allocation X' as required by Theorem 4. Observe that we can assume w.l.o.g. that H' is a (p,q)-improvement of H for some agent q, as we can apply such a single-agent version of Theorem 4 repeatedly to obtain the theorem for p-improvements involving multiple agents.

Algorithm HM-Improve. First, HM-Improve checks whether X belongs to the core of H', and if so, outputs $X' = X$. Hence, we may assume that X admits a blocking cycle in H'. Observe such a cycle must contain the arc (q,p), as otherwise it would block X in H as well. This implies that $X(q) \prec_q' p$.

HM-Improve proceeds by modifying the housing market: it adds a new agent $\widetilde{q}$ to H', with $\widetilde{q}$ taking the place of p in the preferences of q; the only house that agent $\widetilde{q}$ prefers to her own will be the house of p. Let $\widetilde{H}$ be the housing market obtained. Then the acceptability graph $\widetilde{G}$ of $\widetilde{H}$ can be obtained from the acceptability graph of H' by subdividing the arc (q,p) with a new vertex corresponding to agent $\widetilde{q}$. Let $\widetilde{N} = N \cup \{\widetilde{q}\}$, and let $\widetilde{E}$ be the set of arcs in $\widetilde{G}$.

Initialization. Let $Y = X \setminus \{(q, X(q))\} \cup \{(q, \widetilde{q})\}$ in $\widetilde{G}$. Observe that Y is a pre-allocation from the source $X(q)$ to the sink $\widetilde{q}$ in $\widetilde{H}$. Additionally, we define a set R of *irrelevant* agents, initially empty. We may think of irrelevant agents as temporarily deleted from the market.

Iteration. Next, algorithm HM-Improve iteratively modifies the pre-allocation Y and the set R of irrelevant agents. It will maintain the property that Y is a pre-allocation in $\widetilde{H} - R$; we denote its envy graph by $\widetilde{G}_{Y \prec}$, having vertex set $\widetilde{N} \setminus R$. While the source of Y changes during the iteration, the sink $\widetilde{q}$ remains fixed.

At each iteration, HM-Improve performs the following steps:

1. Let u be the source of Y. If $u \in \{p, \widetilde{q}\}$, then the iteration stops.
2. Otherwise, if there exists a Y-augmenting arc (s, u) in $\widetilde{G}_{Y \prec}$ entering u (note that $s \in \widetilde{N} \setminus R$), then let $u' = Y(s)$. The algorithm modifies Y by deleting the arc (s, u') and adding the arc (s, u) to Y. Note that Y thus becomes a pre-allocation from u' to $\widetilde{q}$ in $\widetilde{H} - R$.
3. Otherwise, no arc in $\widetilde{G}_{Y \prec}$ enters u; let $u' = Y(u)$. The algorithm adds u to the set R of irrelevant agents, and modifies Y by deleting the arc (u, u'). Again, Y becomes a pre-allocation from u' to $\widetilde{q}$ in $\widetilde{H} - R$.

Output. Let Y be the pre-allocation at the end of the above iteration, u its source, and R the set of irrelevant agents. HM-Improve applies the variant of the TTC algorithm described in Sect. 3.1 to the submarket H'_R of H' when restricted to the set of irrelevant agents. Let X_R denote the obtained allocation in the core of H'_R. Then HM-Improve outputs an allocation X' defined as

$$X' = \begin{cases} X_R \cup Y \setminus \{(q, \widetilde{q})\} \cup \{(q, p)\} & \text{if } u = p, \\ X_R \cup Y & \text{if } u = \widetilde{q}. \end{cases}$$

4.2 Correctness of Algorithm HM-Improve

We begin proving the correctness of algorithm HM-Improve with the following.

Lemma 7. *At each iteration, pre-allocation Y is stable in $\widetilde{H} - R$.*

Proof. The proof is by induction on the number n of iterations performed. For $n = 0$, observe that initially $Y(a) = X(a)$ for each agent $a \in N \setminus \{q\}$, and by $X(q) \prec'_q p$ we know that q prefers $Y(q) = \widetilde{q}$ to $X(q)$. Note also that neither (q, p) nor the arcs $(q, \widetilde{q})$ and $(\widetilde{q}, p)$ are contained in the envy graph $\widetilde{G}_{Y \prec}$. Thus, a cycle in $\widetilde{G}_{Y \prec}$ would be present in the envy graph of X in H as well. Since X is in the core of H, it follows that Y is stable in $\widetilde{H}$. Note that initially $R = \emptyset$.

For $n \geq 1$, assume that the algorithm has performed $n - 1$ iterations so far. Let Y and R be as defined at the beginning of the n-th iteration, with u being the source of Y, and let Y' and R' be the pre-allocation and the set of irrelevant agents obtained after the modifications in this iteration. Assume that Y is stable in $\widetilde{H} - R$, so $\widetilde{G}_{Y \prec}$ is acyclic. In case HM-Improve does not stop in Step 1 but modifies Y and possibly R, we distinguish between two cases:

(a) the algorithm modifies Y in Step 2, by using a Y-augmenting arc (s, u); then $R' = R$. Note that s prefers Y' to Y, and for any other agent $a \in N \setminus R'$ we know $Y(a) = Y'(a)$. Hence, this modification amounts to deleting all arcs (s, a) from the envy graph $\widetilde{G}_{Y \prec}$ where $Y(s) \prec_s a \preceq_s Y'(s)$.
(b) the algorithm modifies Y in Step 3, by adding the source u to the set of irrelevant agents, i.e., $R' = R \cup \{u\}$. Then $Y'(a) = Y(a)$ for each agent $a \in N \setminus R'$, so the envy graph $\widetilde{G}_{Y' \prec}$ is obtained from $\widetilde{G}_{Y \prec}$ by deleting u.

Since deleting some arcs or a vertex from an acyclic graph results in an acyclic graph, the stability of Y' is clear. □

We proceed with the observation that an agent's situation in Y may only improve, unless it becomes irrelevant: this is a consequence of the fact that the algorithm only deletes arcs and agents from the envy graph $\widetilde{G}_{Y \prec}$.

Proposition 8. *Let Y_1 and Y_2 be two pre-allocations computed by algorithm* HM-Improve, *with Y_1 computed at an earlier step than Y_2, and let a be an agent that is not irrelevant at the end of the iteration when Y_2 is computed. Then either $Y_1(a) = Y_2(a)$ or a prefers Y_2 to Y_1.*

We need an additional lemma that will be useful for arguing why irrelevant agents may not become the cause of instability in the housing market.

Lemma 9. *At the end of algorithm* HM-Improve, *there does not exist an arc $(a, b) \in \widetilde{E}$ such that $a \in N \setminus R$, $b \in R$ and $Y(a) \prec'_a b$.*

Proof. Suppose for contradiction that (a, b) is such an arc, and let Y and R be as defined at the end of the last iteration. Let us suppose that HM-Improve adds b to R during the n-th iteration, and let Y_n be the pre-allocation at the beginning of the n-th iteration. By Proposition 8, either $Y_n(a) = Y(a)$ or $Y_n(a) \prec'_a Y(a)$. The assumption $Y(a) \prec'_a b$ yields $Y_n(a) \prec'_a b$ by the transitivity of $\prec'_a$. Thus, (a, b) is a Y_n-augmenting arc entering b, contradicting our assumption that the algorithm put b into R in Step 3 of the n-th iteration. □

The following lemma, the last one necessary to prove Theorem 4, shows that HM-Improve runs in linear time; the proof relies on the fact that in each iteration but the last either an agent or an arc is deleted from the envy graph, thus limiting the number of iterations by $|E| + |N|$.

Lemma 10 (⋆). *Algorithm* HM-Improve *runs in $O(|H|)$ time.*

Proof (of Theorem 4). By Lemma 10 it suffices to show that algorithm HM-Improve is correct. Let Y and R be the pre-allocation and the set of irrelevant agents, respectively, at the end of algorithm HM-Improve, and let u be the source of Y. To begin, we prove it formally that X' is an allocation for H'.

First assume $u = \widetilde{q}$. This means that Y is the union of disjoint cycles covering each agent in $N \setminus R$ exactly once; note that no arc of Y enters or leaves $\widetilde{q}$. Hence, Y is an allocation not only in $\widetilde{H} - R$, but also in the submarket of H' on agent set $N \setminus R$, i.e., $H'_{N \setminus R}$. Second, assume that $u = p$; in this case $(q, \widetilde{q}) \in Y$, because $\widetilde{q}$ can be entered only through $(q, \widetilde{q})$. So the arc set $Y \setminus \{(q, \widetilde{q})\} \cup \{(q, p)\}$ is an allocation in $H'_{N \setminus R}$. Consequently, X' is indeed an allocation in H' in both cases.

Now, let us prove that the allocation X' is in the core of H' by showing that the envy graph $G^{H'}_{X' \prec}$ of X' is acyclic. First, the subgraph $G^{H'}_{X' \prec}[R]$ is exactly the envy graph of X_R in H'_R and hence is acyclic.

Claim. *Let $a \in N \setminus R$ and let (a, b) be an X'-augmenting arc in H'. Then (a, b) is Y-augmenting as well, i.e., $Y(a) \prec'_a b$.*

Proof (of Claim). If $(a, b) \neq (q, p)$, then (a, b) is an arc in $G^{\widetilde{H}}$, and thus the claim follows immediately from $X'(a) = Y(a)$ except for the case $a = q$ and $Y(q) = \widetilde{q}$; in this latter case $X'(q) = p \prec'_q b$ implies that q prefers b to $Y(q) = \widetilde{q}$ in $\widetilde{H}$ as well, that is, (q, b) is Y-augmenting.

We finish the proof of the claim by showing that (q, p) is not X'-augmenting if $q \notin R$. Let u be the source of Y. If $u = p$, then this is clear by $(q, p) \in X'$. If $u = \widetilde{q}$, then let us now consider the penultimate iteration in which the source of Y is moved to $\widetilde{q}$ either in Step 2 or in Step 3. Recall that the only arc entering $\widetilde{q}$ is $(q, \widetilde{q})$. If $\widetilde{q}$ became the source of Y in Step 2, then we know $\widetilde{q} \prec_q Y(q)$. By the construction of $\widetilde{H}$, this means that q prefers $Y(q) = X'(q)$ to p in H', so (q, p) is not X'-augmenting, a contradiction. Finally, if $\widetilde{q}$ became the source of Y in Step 3, then we get $q \in R$, which contradicts our assumption $q \notin R$. $\blacksquare$

As a consequence of our claim, we obtain that $G^{H'}_{X' \prec}[N \setminus R]$ is a subgraph of $\widetilde{G}_{Y \prec}$ and therefore it is acyclic by Lemma 7. Hence, any cycle in $G^{H'}_{X' \prec}$ must contain agents both in R and in $N \setminus R$ (recall that $G^{H'}_{X' \prec}[R]$ is acyclic as well). However, $G^{H'}_{X' \prec}$ contains no arcs from $N \setminus R$ to R, since such arcs cannot be Y-augmenting by Lemma 9. Thus $G^{H'}_{X' \prec}$ is acyclic and X' is in the core of H'. $\square$

4.3   Strict Improvement

Looking at Theorem 4 and Corollary 5, one may wonder whether it is possible to detect efficiently when a p-improvement leads to a situation that is strictly better for p. For a solution concept Φ and housing markets H and H' such that H' is a p-improvement of H for some agent p, one may ask the following questions:

1. POSSIBLE STRICT IMPROVEMENT FOR BEST HOUSE or PSIB:
   is it true that $a \prec_p a'$ for some $a \in \Phi(H)^+_p$ and $a' \in \Phi(H')^+_p$?
2. NECESSARY STRICT IMPROVEMENT FOR BEST HOUSE or NSIB:
   is it true that $a \prec_p a'$ for every $a \in \Phi(H)^+_p$ and $a' \in \Phi(H')^+_p$?
3. POSSIBLE STRICT IMPROVEMENT FOR WORST HOUSE or PSIW:
   is it true that $a \prec_p a'$ for some $a \in \Phi(H)^-_p$ and $a' \in \Phi(H')^-_p$?
4. NECESSARY STRICT IMPROVEMENT FOR WORST HOUSE or NSIW:
   is it true that $a \prec_p a'$ for every $a \in \Phi(H)^-_p$ and $a' \in \Phi(H')^-_p$?

Focusing on the core of housing markets, it turns out that all of the above four problems are computationally intractable, even in the case of strict preferences.

Theorem 11 ($\star$). *With respect to the core of housing markets, PSIB and NSIB are NP-hard, while PSIW and NSIW are coNP-hard, even if agents' preferences are strict orders.*

5 The Effect of Improvements in STABLE ROOMMATES

In the STABLE ROOMMATES problem we are given a set N of agents, and a preference relation $\prec_a$ over N for each agent $a \in N$; the task is to find a stable matching M between the agents. A matching is *stable* if it admits no *blocking pair*, that is, a pair of agents such that each prefers the other over her partner in the matching. Notice that an input instance for STABLE ROOMMATES is in fact a housing market. Viewed from this perspective, a stable matching in a housing market can be thought of as an allocation that (i) contains only cycles of length at most 2, and (ii) does not admit a blocking cycle of length at most 2.

For an instance of STABLE ROOMMATES, we assume mutual acceptability, that is, for any two agents a and b, we assume that $a \prec_a b$ holds if and only if $b \prec_b a$ holds. Consequently, it will be more convenient to define the acceptability graph G^H of an instance H of STABLE ROOMMATES as an undirected simple graph where agents a and b are connected by an edge $\{a, b\}$ if and only if they are acceptable to each other and $a \neq b$. A *matching* in H is then a set of edges in G^H such that no two of them share an endpoint.

Biró et al. [9] have shown the following statements.

Proposition 12 [9]. *Stable matchings in the STABLE ROOMMATES model*

- *violate the RI-worst property (even if agents' preferences are strict), and*
- *violate the RI-best property, if agents' preferences may include ties.*

Complementing Proposition 12, we show that a (p, q)-improvement can lead to an instance where no stable matching exists at all. This may happen even in the case when preferences are strict orders; hence, stable matchings do not strongly satisfy the RI-best property. For an illustration of Propositions 12 and 13 by simple examples see the full version of our paper [38].

Proposition 13 ($\star$). *Stable matchings in the STABLE ROOMMATES model do not strongly satisfy the RI-best property, even if agents' preferences are strict.*

Contrasting Propositions 12 and 13, it is somewhat surprising that if agents' preferences are strict, then the RI-best property holds for the STABLE ROOMMATES setting. Thus, the situation of p cannot deteriorate as a consequence of a p-improvement unless instability arises. The proof of Theorem 14 is provided at the end of this section.

Theorem 14. *Let $H = (N, \{\prec_a\}_{a \in N})$ be a housing market where agents' preferences are strict orders. Given a stable matching M in H and a (p, q)-improvement H' of H for two agents $p, q \in N$, either H' admits no stable matchings at all, or there exists a stable matching M' in H' such that $M(p) \preceq_i M'(p)$. Moreover, given H, H' and M it is possible to find such a matching M' in polynomial time.*

Corollary 15. *Stable matchings in the* STABLE ROOMMATES *model satisfy the RI-best property.*

Structural Ingredients. To prove Theorem 14 we are going to rely on the concept of proposal-rejection alternating sequences introduced by Tan and Hsueh [41], originally used as a tool for finding a stable partition in an incremental fashion by adding agents one-by-one to a STABLE ROOMMATES instance. We somewhat tailor their definition to fit our current purposes.

Let $\alpha_0 \in N$ be an agent in a housing market H, and let M_0 be a stable matching in $H - \alpha_0$. A sequence S of agents $\alpha_0, \beta_1, \alpha_1, \ldots, \beta_k, \alpha_k$ is a *proposal-rejection alternating sequence* starting from M_0, if there exists a sequence of matchings $M_1, \ldots, M_k$ such that for each $i \in \{1, \ldots, k\}$

(i) β_i is the agent most preferred by α_{i-1} among those who prefer α_{i-1} to their partner in M_{i-1} or are unmatched in M_{i-1},

(ii) $\alpha_i = M_{i-1}(\beta_i)$, and

(iii) $M_i = M_{i-1} \setminus \{\{\alpha_i, \beta_i\}\} \cup \{\{\alpha_{i-1}, \beta_i\}\}$ is a matching in $H - \alpha_i$.

We say that the sequence S *starts* from M_0, and that the matchings $M_1, \ldots, M_k$ are *induced* by S. We say that S *stops* at α_k, if there does not exist an agent fulfilling condition (i) in the above definition for $i = k + 1$, that is, if no agent prefers α_k to her current partner in M_k and no unmatched agent in M_k finds α_k acceptable. We will also allow a proposal-rejection alternating sequence to take the form $\alpha_0, \beta_1, \alpha_1, \ldots, \beta_k$, in case conditions (i), (ii), and (iii) hold for each $i \in \{1, \ldots, k-1\}$, and β_k is an unmatched agent in M_{k-1} satisfying condition (i) for $i = k$. In this case we define the last matching induced by the sequence as $M_k = M_{k-1} \cup \{\{\alpha_{k-1}, \beta_k\}\}$, and we say that the sequence *stops* at agent β_k.

We summarize the most important properties of proposal-rejection alternating sequences in Lemma 16 as observed and used by Tan and Hsueh.[4]

Lemma 16 ([41] ⋆)**.** *Let* $\alpha_0, \beta_1, \alpha_1, \ldots, \beta_k(, \alpha_k)$ *be a proposal-rejection alternating sequence starting from a stable matching* M_0 *and inducing the matchings* $M_1, \ldots, M_k$ *in a housing market* H. *Then the following hold.*

1. M_i *is a stable matching in* $H - \alpha_i$ *for each* $i \in \{1, \ldots, k-1(, k)\}$.
2. *If* $\beta_j = \alpha_i$ *for some* i *and* j, *then* H *does not admit a stable matching; in such a case we say that sequence* S *has a return.*
3. *If the sequence stops at* α_k *or* β_k, *then* M_k *is a stable matching in* H.
4. *For any* $i \in \{1, \ldots, k-1\}$ *agent* α_i *prefers* $M_{i-1}(\alpha_i)$ *to* $M_{i+1}(\alpha_i)$.
5. *For any* $i \in \{1, \ldots, k-1\}$ *agent* β_i *prefers* $M_i(\beta_i)$ *to* $M_{i-1}(\beta_i)$.

Description of Algorithm SR-Improve. Let $M = (N, \{\prec_a\}_{a \in N})$ be the stable matching given for the housing market H, and let $H' = (N, \{\prec'_a\}_{a \in N})$ be a (p, q)-improvement of H for two agents p and q in N (recall that $\prec'_a = \prec_a$ unless $a = q$). We now propose algorithm SR-Improve that computes a stable matching M' in H' with $M(p) \preceq_p M'(p)$, whenever H' admits some stable matching.

---

[4] The first claim of the lemma is only implicit in the paper by Tan and Hsueh [41], we prove it for the sake of completeness in the full version of our paper [38].

First, SR-Improve checks whether M is stable in H', and if so, returns the matching $M' = M$. Otherwise, $\{p, q\}$ must be a blocking pair for M in H'.

Second, the algorithm checks whether H' admits a stable matching and if so, computes *any* stable matching $M^\star$ in H' using Irving's algorithm [26]; if no stable matching exists for H', algorithm SR-Improve stops. Now, if $M(p) \preceq'_p M^\star(p)$, then SR-Improve returns $M' = M^\star$, otherwise proceeds as follows.

Let $\widetilde{H}$ be the housing market obtained from H' by deleting all agents $\{a \in N : a \preceq'_q p\}$ from the preference list of q (and vice versa, deleting q from the preference list of these agents). Notice that in particular this includes the deletion of p as well as of $M(q)$ from the preference list of q (recall that $M(q) \prec'_q p$).

Let us define $\alpha_0 = M(q)$ and $M_0 = M \setminus \{q, \alpha_0\}$. Notice that M_0 is a stable matching in $\widetilde{H} - \alpha_0$: clearly, any possible blocking pair must contain q, but any blocking pair $\{q, a\}$ that is blocking in $\widetilde{H}$ would also block H by $M(q) \prec_q a$. Observe also that q is unmatched in M_0.

Finally, SR-Improve builds a proposal-rejection alternating sequence S of agents $\alpha_0, \beta_1, \alpha_1, \ldots, \beta_k(, \alpha_k)$ in $\widetilde{H}$ starting from M_0, and inducing matchings $M_1, \ldots, M_k$ until one of the following cases occurs:

(a) $\alpha_k = p$: in this case SR-Improve outputs $M' = M_k \cup \{\{p, q\}\}$;
(b) S stops: in this case SR-Improve outputs $M' = M_k$.

Correctness of Algorithm SR-Improve. The proof that algorithm SR-Improve is correct relies on the following two facts.

Lemma 17 ($\star$). *The sequence S cannot have a return. Furthermore, if S stops, then it stops at β_k with $\beta_k = q$.*

Lemma 18 ($\star$). *If SR-Improve outputs a matching M', then M' is stable in H' and $M(p) \preceq'_p M'(p)$.*

Proof (of Theorem 14). From the description of SR-Improve and Lemma 18 it is immediate that any output the algorithm produces is correct. It remains to show that it does not fail to produce an output. By Lemma 17 we know that the sequence S built by the algorithm cannot have a return and can only stop at q, implying that SR-Improve will eventually produce an output. Considering the fifth statement of Lemma 16, we also know that the length of S is at most $2|E|$. Thus, the algorithm finishes in $O(|E|)$ time. □

6   Further Research

Even though the property of respecting improvement is important in exchange markets, many solution concepts have not been studied from this aspect. For instance, in the STABLE ROOMMATES setting with weakly or partially ordered preferences, do strongly stable matchings satisfy the RI-best property? What about stable half-matchings (or equivalently, stable partitions) in instances of STABLE ROOMMATES without a stable matching? Although the full version of our paper [38] contains an example about stable half-matchings where improvement of an agents' house damages her situation, perhaps a more careful investigation may shed light on some interesting monotonicity properties.

References

1. Abdulkadiroğlu, A., Sönmez, T.: House allocation with existing tenants. J. Econ. Theory **88**(2), 233–260 (1999)
2. Abraham, D.J., Blum, A., Sandholm, T.: Clearing algorithms for barter exchange markets: enabling nationwide kidney exchanges. In: EC'07: Proceedings of the 8th ACM Conference on Electronic Commerce, pp. 295–304 (2007)
3. Abraham, D.J., Cechlárová, K., Manlove, D.F., Mehlhorn, K.: Pareto optimality in house allocation problems. In: Fleischer, R., Trippen, G. (eds.) ISAAC 2004. LNCS, vol. 3341, pp. 3–15. Springer, Heidelberg (2004). https://doi.org/10.1007/978-3-540-30551-4_3
4. Alcalde-Unzu, J., Molis, E.: Exchange of indivisible goods and indifferences: The Top Trading Absorbing Sets mechanisms. Game. Econ. Behav. **73**(1), 1–16 (2011)
5. Aziz, H., de Keijzer, B.: Housing markets with indifferences: a tale of two mechanisms. In: AAAI'12, pp. 1249–1255 (2012)
6. Balinski, M., Sönmez, T.: A tale of two mechanisms: student placement. J. Econ. Theory **84**(1), 73–94 (1999)
7. Biró, P., Cechlárová, K.: Inapproximability of the kidney exchange problem. Inform. Process. Lett. **101**(5), 199–202 (2007)
8. Biró, P., Haase-Kromwijk, B., Andersson, T., Ásgeirsson, E.I., Baltesová, T., Boletis, I., et al.: Building kidney exchange programmes in Europe: an overview of exchange practice and activities. Transplantation **103**(7), 1514–1522 (2019)
9. Biró, P., Klijn, F., Klimentova, X., Viana, A.: Shapley-Scarf housing markets: respecting improvement, integer programming, and kidney exchange. CoRR arXiv:2102.00167 [econ.TH] (2021)
10. Biró, P., Manlove, D., Rizzi, R.: Maximum weight cycle packing in directed graphs, with application to kidney exchange programs. Discrete Math. Algorithms Appl. **1**(4), 499–517 (2009)
11. Biró, P., McDermid, E.: Three-sided stable matchings with cyclic preferences. Algorithmica **58**(1), 5–18 (2010)
12. Biró, P., van de Klundert, J., Manlove, D., et al.: Modelling and optimisation in European Kidney Exchange Programmes. Eur. J. Oper. Res. **291**(2), 447–456 (2021)
13. Bloch, F., Cantala, D.: Markovian assignment rules. Soc. Choice Welf. **40**, 1–25 (2003). https://doi.org/10.1007/s00355-011-0566-x
14. Bokal, D., Fijavž, G., Juvan, M., Kayll, P.M., Mohar, B.: The circular chromatic number of a digraph. J. Graph Theor. **46**(3), 227–240 (2004)
15. Cechlárová, K., Fleiner, T., Manlove, D.F.: The kidney exchange game. In: SOR'05, pp. 77–83 (2005)
16. Cechlárová, K., Hajduková, J.: Computational complexity of stable partitions with B-preferences. Int. J. Game Theory **31**(3), 353–364 (2003)
17. Cechlárová, K., Lacko, V.: The kidney exchange problem: how hard is it to find a donor? Ann. Oper. Res. **193**, 255–271 (2012)
18. Cechlárová, K., Repiský, M.: On the structure of the core of housing markets. Technical report, P. J. Šafárik University (2011). IM Preprint, series A, No. 1/2011
19. Cechlárová, K., Romero-Medina, A.: Stability in coalition formation games. Int. J. Game Theory **29**(4), 487–494 (2001)
20. Cseh, Á., Manlove, D.F.: Stable marriage and roommates problems with restricted edges: complexity and approximability. Discrete Optim. **20**, 62–89 (2016)

21. Dias, V., da Fonseca, G., Figueiredo, C., Szwarcfiter, J.: The stable marriage problem with restricted pairs. Theor. Comput. Sci. **306**, 391–405 (2003)
22. Fleiner, T., Irving, R.W., Manlove, D.F.: Efficient algorithms for generalized stable marriage and roommates problems. Theor. Comput. Sci. **381**(1), 162–176 (2007)
23. Gusfield, D., Irving, R.W.: The Stable Marriage Problem: Structure and Algorithms. MIT Press, Cambridge (1989)
24. Hatfield, J.W., Kojima, F., Narita, Y.: Improving schools through school choice: a market design approach. J. Econ. Theory **166**(C), 186–211 (2016)
25. Huang, C.-C.: Circular stable matching and 3-way kidney transplant. Algorithmica **58**(1), 137–150 (2010). https://doi.org/10.1007/s00453-009-9356-6
26. Irving, R.W.: An efficient algorithm for the "stable roommates" problem. J. Algorithms **6**(4), 577–595 (1985)
27. Jaramillo, P., Manjunath, V.: The difference indifference makes in strategy-proof allocation of objects. J. Econ. Theory **147**(5), 1913–1946 (2012)
28. Kamijo, Y., Kawasaki, R.: Dynamics, stability, and foresight in the Shapley-Scarf housing market. J. Math. Econ. **46**(2), 214–222 (2010)
29. Kawasaki, R.: Roth-Postlewaite stability and von Neumann-Morgenstern stability. J. Math. Econ. **58**, 1–6 (2015)
30. Klaus, B., Klijn, F.: Minimal-access rights in school choice and the deferred acceptance mechanism. Cahiers de Recherches Economiques du Département d'économie 21.11, Université de Lausanne (2021)
31. Knuth, D.E.: Mariages stables et leurs relations avec d'autres problèmes combinatoires. Les Presses de l'Université de Montréal, Montreal, Quebec (1976)
32. Kurino, M.: House allocation with overlapping generations. Am. Econ. J.-Microrecon. **6**(1), 258–289 (2014)
33. Plaxton, C.G.: A simple family of Top Trading Cycles mechanisms for housing markets with indifferences. In: ICGT 2013 (2013)
34. Roth, A.E., Postlewaite, A.: Weak versus strong domination in a market with indivisible goods. J. Math. Econ. **4**, 131–137 (1977)
35. Roth, A.E., Sönmez, T., Ünver, M.U.: Kidney exchange. Q. J. Econ. **119**, 457–488 (2004)
36. Roth, A.E., Sönmez, T., Ünver, M.U.: Pairwise kidney exchange. J. Econ. Theory **125**(2), 151–188 (2005)
37. Saban, D., Sethuraman, J.: House allocation with indifferences: a generalization and a unified view. In: EC'13: Proceedings of the 14th ACM Conference on Electronic Commerce, pp. 803–820 (2013)
38. Schlotter, I., Biró, P., Fleiner, T.: The core of housing markets from an agent's perspective: is it worth sprucing up your home? CoRR arXiv:2110.06875 [cs.GT] (2021)
39. Shapley, L., Scarf, H.: On cores and indivisibility. J. Math. Econ. **1**, 23–37 (1974)
40. Sönmez, T., Switzer, T.: Matching with (Branch-of-Choice) contracts at the United States Military Academy. Econometrica **81**, 451–488 (2013)
41. Tan, J.J.M., Hsueh, Y.-C.: A generalization of the stable matching problem. Discrete Appl. Math. **59**(1), 87–102 (1995)
42. Unver, M.U.: Dynamic kidney exchange. Rev. Econ. Stud. **77**(1), 372–414 (2010)
43. Yuan, Y.: Residence exchange wanted: a stable residence exchange problem. Eur. J. Oper. Res. **90**(3), 536–546 (1996)

Mechanisms for Trading Durable Goods

Sigal Oren[✉] and Oren Roth

Ben-Gurion University of the Negav, 8410501 Beer-Sheva, Israel

Abstract. We consider trading indivisible and easily transferable *durable goods*, which are goods that an agent can receive, use, and trade again for a different good. This is often the case with books that can be read and later exchanged for unread ones. Other examples of such easily transferable durable goods include puzzles, video games and baby clothes.

We introduce a model for the exchange of easily transferable durable goods. In our model, each agent owns a set of items and demands a different set of items. An agent is interested in receiving as many items as possible from his demand set. We consider mechanisms that exchange items in cycles in which each participating agent receives an item that he demands and gives an item that he owns. We aim to develop mechanisms that have the following properties: they are *efficient*, in the sense that they maximize the total number of items that agents receive from their demand set, they are *strategyproof* (i.e., it is in the agents' best interest to report their preferences truthfully) and they run in *polynomial time*.

One challenge in developing mechanisms for our setting is that the supply and demand sets of the agents are updated after a trade cycle is executed. This makes constructing strategyproof mechanisms in our model significantly different from previous works, both technically and conceptually and requires developing new tools and techniques. We prove that simultaneously satisfying all desired properties is impossible and thus focus on studying the tradeoffs between these properties. To this end, we provide both approximation algorithms and impossibility results.

1 Introduction

The sharing economy [7] puts on steroids the ancient idea of sharing physical assets. Instead of only sharing among friends, technology advancements facilitate the sharing of physical goods among strangers [8]: rather than booking a hotel room, we often use Airbnb to live in someone's apartment, and instead of throwing away items we no longer use, we can exchange them for others that we do need. At the heart of the sharing economy is a desire to increase social efficiency by using underutilized resources.

A perfect demonstration of the credo of the sharing economy is the efficient reallocation of *durable* goods, like books, toys and sports gear. For example,

Work partially supported by ISF grant 2167/19. Missing proofs can be found in the full version which is available on arXiv.

suppose you have just finished reading the first Harry Potter book "Harry Potter and the Philosopher's Stone". You will probably be happy to exchange it in return for the second book in the series "Harry Potter and the Chamber of Secrets". In the offline world, for a swap to occur, we must have two individuals, each of whom is interested in the other's item. However, it is implausible that a person that read the second book in the Harry Potter series will be interested in the first book. Thus, such an exchange is more likely to occur as part of a larger cycle of exchanges which might be difficult to coordinate without a central mechanism.

Online platforms (such as Swappy Books, Rehash Clothes, TradeMade and others) provide a central mechanism where each user can report the items he can give and the items he would like to receive. The platforms essentially provide infrastructure that reduces the search friction and allows to orchestrate complex exchanges. In this paper we focus on designing mechanisms for allocating easily transferable durable goods with the objective of maximizing the number of exchanges. A main challenge in designing such mechanisms is that the same item can be traded several times. This crucial difference from classic works on barter and trading (e.g., [2,15]) requires developing new algorithmic tools to ensure that these platforms will live up to their potential.

A Model of Durable Goods. We consider a stylized model for exchanging easily transferable durable goods. Our model is based on the classic work of Shapley and Scarf [15] for the house allocation problem. We consider a set N ($|N| = n$) of agents and a set M ($|M| = m$) of items. Each agent has a subset $D_i \subseteq M$ of items that he demands and a subset $S_i \subseteq M$ of items he owns. We make the simplifying assumption that each agent is willing to give any item from S_i in return for any item from D_i. This means that the agent is indifferent between all the items that he demands and also between all the items that he owns. This is a reasonable assumption for books, for example, where books that the agent already read serve as a commodity that can be exchanged to get new desired books. Moreover, we assume that the agent is unwilling to receive an item that is not in his current demand set.[1] This can be, for example, due to the physical or emotional burden of handling an unwanted item. We model the demands and endowments of the agents as a directed bipartite graph $G = (N, M, E)$ in which there is a directed edge (i, j) if agent i demands item j and a directed edge (j, i) if agent i owns item j. We refer to this graph as the *trading graph*.

We focus on mechanisms that execute exchanges according to cycles in the trading graph. In each exchange, every participating agent i receives an item in D_i and gives an item in S_i. The novelty of our model is that after agent i received item $j \in D_i$ and used it, he can trade it later for another item from D_i. Our model is dynamic in the sense that after each step, the sets D_i and S_i are updated for each agent i. We refer to a sequence of cycles as an *execution*.

---

[1] Formally, this can be modeled as part of the agent's utility as a large penalty that the agent exhibits for receiving an item not in the demand set or setting the utility to 0 if the agent receives such an item.

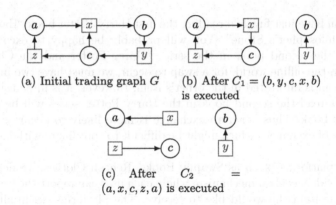

(a) Initial trading graph G (b) After $C_1 = (b, y, c, x, b)$
is executed

(c)   After C_2 =
(a, x, c, z, a) is executed

Fig. 1. An example of an execution on a trading graph. Squares denote items and circles denote agents.

It is useful to go over an example to better understand the model. Consider the instance illustrated in Fig. 1. The set of agents is $N = \{a, b, c\}$ and the set of items is $M = \{x, y, z\}$. The set of items that agent a demands is $D_a = \{x\}$ and the set of items that agent a owns is $S_a = \{z\}$. For agent b: $D_b = \{y\}$ and $S_b = \{x\}$ and for agent c: $D_c = \{x, z\}$ and $S_c = \{y\}$. The trading graph of this instance is illustrated in Fig. 1(a). In the optimal execution we first execute the cycle $C_1 = (b, y, c, x, b)$. That is, agent b receives item y from agent c and agent c receives item x from agent b. Then, after agent c receives item x the edge (c, x) is flipped. The graph after the execution of C_1 is illustrated in Fig. 1(b). Now, we can execute the cycle $C_2 = (a, x, c, z, a)$. As we see in Fig. 1(c), after executing cycle C_2 there are no more cycles that we can execute. The social welfare of this execution is 4 since overall it performs 4 exchanges, the utilities of agents a and b is one since they received one item and the utility of agent c is 2 since he received two items from D_c. This is the optimal execution (i.e., the execution that performs the maximal possible number of exchanges) for this instance. This is in contrast to the best execution that can allocate each item at most once (i.e., a "static" execution) which consists of only 3 exchanges (the cycle (a, x, b, y, c, z, a)).

Our Results. In this paper, we initiate the study of mechanisms that efficiently reallocate easily transferable durable goods. Due to the dynamic nature of this setting, designing such mechanisms is very challenging and requires developing new tools and techniques. Our goal is to develop algorithms that maximize the *social welfare* which is defined as the total number of items that all agents receive from their demand sets. Furthermore, we take a mechanism design approach and assume that for each agent i, the demand set D_i and the supply set S_i are his private information. Thus, we would like our algorithms to be *strategyproof* in the sense that each agent maximizes his utility (i.e., the number of items that he receives from his demand sets) by truthfully reporting his private information. Lastly, we would like our algorithms to run in *polynomial time*. Ideally, we would like to develop algorithms that have all these properties. Unfortunately, we will prove that simultaneously obtaining all three properties is impossible. Thus we focus on studying the tradeoffs between them.

We begin by considering simple executions that reallocate each item at most once. We refer to such executions as *static executions*. We show that an optimal static execution does not only provide a reasonable approximation to the social welfare of an optimal (dynamic) execution but can also be computed in polynomial time using a strategyproof algorithm. Formally, we show:

Theorem 1. *Let* $l = \max_{i \in N} |D_i|$ *be the maximum number of items that an agent demands. There exists a polynomial-time and strategyproof algorithm that computes an optimal static execution and provides an l-approximation to the social welfare of an optimal (possibly dynamic) execution.*

Our algorithm for computing the optimal static execution computes a maximum cycle cover of a graph by finding a maximum-weight perfect-matching in a corresponding bipartite graph. This general approach is similar to Abraham et al. [2]; however, defining the appropriate bipartite graph for our model is more intricate as each agent may own and demand multiple items. Notably, in contrast to [2], the algorithm that we devise is also strategyproof. This requires careful selection of the optimal execution that the algorithm outputs (there might be several allocations that provide the maximal welfare). Loosely speaking, we divide the proof that our algorithm is strategyproof into two parts. First, we show that an agent cannot increase his utility by not reporting some of the items that he demands. For this part of the proof, we take the edges of two maximum-weight perfect matchings for two instances: the instance in which agent i reports his true demand set and the instance in which agent i reports a subset of his demand. Using these edges, we construct two different matchings for the two instances. Since both pairs of matchings use the same set of edges, their sum of weights is identical. Hence it is impossible that both original matchings are optimal for their corresponding instances. Then, we show that an agent cannot benefit from not reporting some items that he owns. We prove this by reversing the directions of all the edges in the graph and applying our previous result showing that an agent cannot benefit from not reporting some of the items in his demand set.

We then go on to study the limits of dynamic executions. First, we show that the approximation ratio achieved by our algorithm is close to the optimal approximation ratio achievable by any strategyproof algorithm. In particular, we show that any strategyproof algorithm cannot attain an approximation ratio better than $\approx \frac{l}{2}$ (where $l = \max_{i \in N} |D_i|$). Interestingly, we show that if we consider algorithms that are both strategyproof and always return a Pareto efficient allocation, the best achievable approximation ratio is $\Theta(n)$. Finally, we drop the requirement of strategyproofness and consider the computational question of computing an optimal execution. By constructing a careful reduction from the 3D-matching problem, we show that not only computing the optimal execution is NP-hard but it also cannot be approximated within some small constant unless P = NP. We note that the fact that the trading graph changes with the execution makes the reduction quite challenging. We leave open the question of closing the gap between the l-approximation ratio of our algorithm and the impossibility result showing that the problem cannot be approximated within a small constant.[2]

This work focuses on executions that are sequences of cycles in which each participating agent receives an item in his demand set. This restriction is a result of two main assumptions. First, we assume that an agent is unwilling to give an item without immediately receiving an item in return. The reasoning behind this is that the agent views the items he has as commodities used to get items that he is interested in. Thus, he does not want to lose such a commodity without immediately getting something in return. This can lead to a problem known as the "double coincidence of wants": Bob might demand some item that Alice has while Alice may not currently demand any item of Bob. The simple solution to this classic problem is to introduce some form of money. Thus, our results can be interpreted as demonstrating the necessity of money (not necessarily fiat money[3]) when considering dynamic barter markets. The second assumption we make is that an agent is not willing to give an item and receive in return an item that is not in his demand set. The rationale is that we assume that exchanging and perhaps storing an item that the agent is not interested in may have a nonnegligible physical or emotional cost associated with it.

Related Literature. Abbassi et al. [1] consider a similar setting of barter networks with the main exception that each item may be allocated only once. This is comparable to our static executions but is very different from the more general dynamic executions. Abbassi et al. mainly focus on a different setting than ours in which the length of each trading cycle is bounded. For this setting, they present algorithms and hardness of approximation results for strategyproof mechanisms. For the setting in which the length of the trading cycles is unconstrained, they present a polynomial-time algorithm that computes an optimal static execution. It is important to note that, unlike our algorithm, this algorithm is not strategyproof.

The problem of computing an optimal static execution is very much related to the literature on matching. Our setting, in this respect, has two notable properties: 1) the agents may own and demand multiple items and 2) the agents have a high level of indifference in the sense that an agent only cares about the number of items from D_i that he receives. In contrast, till recently, previous work did not consider situations that exhibit both of these properties. In [16], for example, Sönmez considered a setting in which each agent may own and demand multiple items. He showed that when each agent has a strict preference order over subsets of items, there is no individually rational, strategyproof, and Pareto-efficient mechanism. Konishi et al. [11] demonstrated that this impossibility result holds even in cases where there are only two different types of items (e.g., houses and cars) and the agents have strict preference order over

---

[2] In the full version we show that a greedy algorithm that sequentially finds an optimal static allocation cannot get an approximation ratio better than l.

[3] [3,10] suggest that "memory" could be used instead of money.

these items. The high degree of indifference in our model allows us to escape these impossibility results. Till recently, most works in the matching literature featuring indifference considered only settings in which each agent owns and can receive exactly a single item [4,6,9,14].

In a recent working paper, Manjunath and Westkamp [13] considered a *static* setting they refer to as *Trichotomous Preferences*. Similar to our setting, each agent labels the item that he does not own as either desirable or undesirable. However, in contrast to our setting, the items that an agent *owns* are also labeled as desirable or undesirable. The only undesirable items that an agent is willing to accept as part of a bundle are those in his initial endowment. The agents rank all the acceptable bundles according to the number of desirable items in them. The authors show that there is a computationally efficient mechanism that is individually rational, Pareto efficient and strategyproof in this setting. Similarly to [1] the mechanism that Manjunath and Westkamp present is also based on a fixed ordering of the agents. In [13] this often leads to executions that are Pareto efficient but may be far from optimal. A slightly different setting was considered in a working paper by Andersson et al. [5] in which each agent is endowed with multiple copies of an item that is unique only to him. Agents label other agents' items as desirable or undesirable and for desirable items, they have a cap on the number of units they are willing to receive. Andersson et al. present a mechanism that is individually rational, strategyproof and optimal. Notice that their result can not be applied to our static model since their model is too restrictive.

Paper Outline. In Sect. 2 we formally define our model. In Sect. 3 we present a strategyproof and computationally efficient algorithm that provides an l-approximation to the optimal execution by computing an optimal static execution. In Sect. 4 we discuss the limits of dynamic executions.

2   Model

We consider a set of agents N ($|N| = n$) and a set of items M ($|M| = m$). For each agent i we denote the subset of the items that he owns by S_i and the subset of items that he demands by D_i. We assume that $(S_1, \ldots, S_n)$ is a partition of M and that no player demands an item that he also owns (i.e., $S_i \cap D_i = \emptyset$). We denote an instance by $\mathcal{G} = (N, M, S_1, D_1, \ldots, S_n, D_n)$. We denote by $l = \max_{i \in N} |D_i|$ the maximal number of items demanded by a single agent. We model the agents' preferences using a directed bipartite *trading graph* $G = (N, M, E)$. For agent i and item $j \in D_i$ we have a *demand edge* $(i, j) \in E$. Similarly, for each item $j \in S_i$ we have a *supply edge* $(j, i) \in E$.

We only allow agents to exchange items within cycles such that each participant i gives an item that is currently in S_i and receives an item that is currently in D_i.[4] After a cycle is executed we update the demand and supply sets of all

---

[4] As discussed in the introduction, we assume that the agents do not want to hold on to items that are not in their demand set nor to give items without getting anything in return.

the agents accordingly and update the graph by reversing all the demand edges of the cycle and removing all the supply edges of the cycle. Formally:

Definition 1 (Cycle Step). *Let $G = (N, M, E)$ denote the current trading graph and let $C = (i_1, j_1, i_2, j_2, ..., i_k, j_k, i_1)$ denote a cycle in G. After executing the cycle C, the edge set of the graph is updated to $E' = (E \setminus \{e \in C\}) \cup \{(j_t, i_t)|(i_t, j_t) \in C\}$. The number of exchanges in a cycle step is $|C| = |\{(i,j)|i \in N, j \in M, (i,j) \in C\}|$.*

In this paper we study executions, these are sequences of cycles that obey the conditions we previously defined. Formally:

Definition 2. *Consider an instance $G = (N, M, S_1, D_1, ..., S_n, D_n)$. An execution $r = C_1, ..., C_k$ for G is a sequence of cycles such that for each $1 \le i \le k$, C_i is a cycle in G_{i-1}, where G_{i-1} is the trading graph that results from executing cycles $C_1, ..., C_{i-1}$ sequentially on the original trading graph $G = (N, M, E)$ and $G_0 = G$.*

We denote the set of all executions by R and the number of exchanges of an execution $r = C_1, ..., C_k$ by $|r| = \sum_{i=1}^{k} |C_i|$. We denote the set of agents that participate in an execution r by $N(r)$ and the set of demand edges that are used in an execution r by $E(r)$. We refer to a demand edge that was used in an execution and then was flipped and used as a supply edge as a *dynamic edge*. For example, the edge (c, x) in the execution described in Fig. 1 is a dynamic edge. The utility of an agent i in an execution r is defined as the number of items from his demand set that i received throughout the execution. Furthermore, we assume that at each step agents are only willing to accept items that are currently in their demand set. Accepting an item currently not in their demand set results in setting their utility to $-n \cdot m$. Similarly, the utility of an agent that is asked to give an item that is not currently in his demand set is also $-n \cdot m$. Formally,

Definition 3 (Agent's utility). *Consider the instance $G = (N, M, S_1, D_1, ..., S_n, D_n)$ and an execution r for G. Let $A_i(r)$ denote the set of items that agent i received in r. The utility of agent i is: $u_i(r) = |A_i(r)|$, if for each trading cycle that i participated in, the item that he received was in his current demand set and the item that he gave was in his current supply set. Else, $u_i(r) = -n \cdot m$.*

We consider the standard objective function of maximizing the social welfare:

Definition 4 (Social Welfare). *Consider the instance $G = (N, M, S_1, D_1, ..., S_n, D_n)$ and an execution r for G. The social welfare of r is $U(r) = \sum_{i \in N} u_i(r)$.*

In other words, as we only consider executions in which agents receive items in their demand set the social welfare equals to the total number of items that the agents received (i.e., $\sum_{i=1}^{n} |A_i(r)|$). We denote the execution maximizing the social welfare (i.e., the optimal execution) by r_o.

3 A Strategyproof l-Approximation Algorithm

Most of the literature on exchange economics usually focuses on models in which each item can be reallocated at most once. In our model we allow items to be reallocated several times. In our analysis, we will refer to executions that, as in the traditional literature, allocate each item at most once as *static executions* and executions that can allocate each item more than once as *dynamic executions*.

In this section we present a polynomial time algorithm for computing the optimal static execution. Recall that l is the maximum number of items that a single agent demands. We show that our algorithm provides an l-approximation to the social welfare of an optimal *dynamic* execution. Furthermore, we show that our algorithm is strategyproof – each agent maximizes his utility by truthfully reporting the items that he demands and the items that he owns.

Proposition 1. *In any instance $\mathcal{G}$, the social welfare of an optimal static execution is at least $\frac{1}{l}$ of the social welfare of an optimal (possibly dynamic) execution.*

Proof. Consider an instance $\mathcal{G}$ and let r_o be an optimal execution. We will show that there exists a static execution r_s such that each agent that received at least one item in r_o will receive one item in r_s (formally, $N(r_o) \subseteq N(r_s)$). As the maximal number of items that an agent may receive is l, this implies that the execution r_s provides an l-approximation to the optimal (dynamic) execution. Thus, the optimal static execution also provides an l-approximation.

We now construct a static execution r_s such that $N(r_o) \subseteq N(r_s)$. Denote by $S(r_o)$ the set of supply edges that were used in r_o. Let $G_{r_o} = (N, M, E(r_o) \cup S(r_o))$ denote the trading graph that includes all the edges that were used in r_o. Since an execution is a sequence of cycles, the in-degree of each node in G_{r_o} is the same as its out-degree. Recall that a static execution can only allocate each item at most once, thus we essentially remove from G_{r_o} the edges associated with items that were allocated more than once. Formally, for each supply edge (j, i) such that $i \in N$ and $j \in M$ that was not included in the initial trading graph G (i.e., $(j, i) \in S(r_o) \setminus E$) we remove both the supply edge (j, i) and the dynamic edge $(i, j) \in E$ that was flipped to create (j, i). Denote the new graph by G'_{r_o}. Observe that the edges of G'_{r_o} are a subset of E and it is still the case that each node in G'_{r_o} has the same out-degree and in-degree. Furthermore, since for each agent that received some items in r_o the first supply edge that was used is $(j, i) \in S(r_o) \cap E$ we have that if a node was not isolated in G_{r_o} it is still not isolated in G'_{r_o}. Thus, by Euler's theorem on the connected components of G'_{r_o} we have that there exists a cycle cover of all the edges in G'_{r_o}. This cycle cover includes all the agents that received at least one item in r_o. Therefore, we have that there exists a static execution r_s for which $N(r_o) \subseteq N(r_s)$ as required. $\square$

Recall that for any instance $\mathcal{G}$ we denote by r_s an optimal static execution for $\mathcal{G}$ and by r_o an optimal execution for $\mathcal{G}$. We now show that this bound is tight:

Claim. For any $l \geq 1$ and $n > l$, there exists an instance $\mathcal{G}$ with n agents such that for each agent i $|D_i| \leq l$ and $|r_s| = \frac{1}{l}|r_o|$.

Proof. Consider an instance $\mathcal{G}$ with $l+1$ items ($M = \{1, \ldots, l+1\}$). Let $N_{l+1} = \{1, \ldots, l+1\}$ denote the set of the first $l+1$ agents. These are the only agents that own and demand items in $\mathcal{G}$. For each agent $i \in N_{l+1}$ we have that $S_i = \{i\}$ and $D_i = M \setminus \{i\}$. The optimal execution executes l cycle steps where in each step all agents in N_{l+1} give the item they own and receive a new item which they will swap in the next step. In this execution all the $(l+1)l$ demand edges are used. Note that only $l+1$ of them are not dynamic edges. On the other hand, in the optimal static execution each of the agents in N_{l+1} will receive a single item. This is optimal as in any static execution the maximal number of items that an agent that owns a single item may receive is one. Hence, there is a gap of l between the social welfare of the optimal execution and the social welfare of the optimal static execution.   □

3.1   Computing an Optimal Static Execution

We present an algorithm for computing an optimal static execution. Our algorithm computes a maximal cycle cover by finding a maximum weight perfect matching. This is similar to the algorithm of Abraham et al. [2] for the kidney exchange setting. However, since in our setting the same agent may participate in more than one cycle, if he owns several items and demands several items, we consider edge-disjoint cycles whereas [2] considers node-disjoint cycles. To handle this difference we construct a new bipartite graph where we have two copies of each edge play as the vertices of the graph. Roughly speaking, each edge is connected to its copy with weight zero and to all the edges that are adjacent to it with weight 1. Now, a perfect matching can define an execution in the following way: any edge that is matched to its copy does not take part in the execution and any edge that is matched to a different edge is included together with the edge it was matched to in some cycle in the execution. Notice that the maximum weight matching will maximize the number of edges that are not matched to their copies(and hence participate in the execution) as they are the only edges that have positive weight. We now formalize this construction:

Theorem 2. *An optimal static execution r can be computed in polynomial time* $(O(|E|^3))$.

Proof. Given an instance $\mathcal{G}$ and its corresponding trading graph $G = (N, M, E)$ we construct a new undirected weighted bipartite graph, $H(G) = (E \times \{0\}, E \times \{1\}, E_H)$ such that:

$$E_H = \underbrace{\{((e,0),(e,1)) \mid e \in E\}}_{E_1} \cup \underbrace{\{((e,0),(e',1)) \mid e = (u,v), e' = (v,z) \in E\}}_{E_2}.$$

We assign each edge in E_1 a weight of 0 and each edge in E_2 a weight of 1. We illustrate this construction in Fig. 2. Note that we can construct $H(G)$ in polynomial time. After constructing the bipartite graph $H(G)$ we compute a maximum-weight perfect matching $\mathcal{M}$ of $H(G)$. This problem is known as the "Assignment problem" and can be solved in polynomial time, for example using

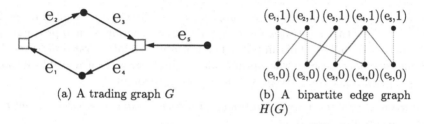

(a) A trading graph G

(b) A bipartite edge graph $H(G)$

Fig. 2. A trading graph G and the corresponding bipartite edge graph $H(G)$. In the illustration of $H(G)$ dotted edges have weight 0 and solid edges weight 1.

the Hungarian method [12]. Note that a perfect matching of the graph $H(G)$ is guaranteed to exist since the edges of E_1 by themselves form a perfect matching. We establish the correctness of our algorithm by the following two lemmas (The proofs of both lemmas can be found in the full version.). First we construct from $\mathcal{M}$ a static execution r such that $|r|$ equals half the weight of $\mathcal{M}$:

Lemma 1. *Consider a trading graph G. Given a perfect matching $\mathcal{M}$ for $H(G)$ of weight x we can compute in polynomial time a corresponding static execution $r = C_1, \ldots, C_k$ for G such that $|r| = x/2$.*

To show that r is an optimal static execution, in Lemma 2 we prove that given an execution r that makes x exchanges we can construct a perfect matching of the graph $H(G)$ of weight $2x$.

Lemma 2. *If there exists a static execution $r = C_1, \ldots, C_k$ for the trading graph G such that $|r| = x$, then a perfect matching $\mathcal{M}$ for $H(G)$ of weight $2x$ exists.*

3.2 A Strategyproof Algorithm for Computing an Optimal Static Execution

So far, we assumed that the agents truthfully report the items that they demand and own to the algorithm. In this section, we consider a mechanism design type of question and ask if indeed it is in the agents' best interest to report their true preferences. We denote the vector of the agents' reports by $\vec{x}' = (x_1', x_2', \ldots, x_n')$, where for each agent i: $x_i' = (D_i', S_i')$. We denote the trading graph constructed by the reports as $G(\vec{x}')$. The utility of each agent depends on the execution chosen by the mechanism. Recall that, roughly speaking, the utility of agent i is the number of items that he received from his demand set D_i in the execution. Formally, we denote by $A(\vec{x}')$ the execution computed by algorithm A when it gets as input the reports vector $\vec{x}'$. With this notation, an algorithm A is strategyproof (i.e., truthful) if and only if $\forall i, (D_i', S_i'), \vec{x}'_{-i} \; u_i(A((D_i, S_i), \vec{x}'_{-i})) \geq u_i(A((D_i', S_i'), \vec{x}'_{-i}))$ where D_i and S_i are the agent's private information.

In this section, we modify the l-approximation algorithm we presented in Sect. 3.1 for computing an optimal static execution to make it strategyproof. We note that the approximation ratio achieved by our algorithm is close to the optimal approximation ratio achievable by any strategyproof algorithm, as in Sect. 4 we show that no strategyproof algorithm can guarantee an approximation ratio better than $\frac{l+1}{2}$. This implies that the approximation ratio achieved by our algorithm is close to the optimal approximation ratio achievable by any strategyproof algorithm. We prove the following theorem:

Theorem 3. *There exists a strategyproof algorithm that computes an optimal static execution in poly-time.*

In many cases there is no unique optimal static execution. In particular, often there exists some agent i that in one optimal execution receives more items than in another optimal execution. In a strategyproof mechanism we need to make sure that such an agent cannot misreport the items that he demands or owns in order to get the mechanism to output an execution that is better for him. To handle this issue we apply a consistent tie-breaking role to select an optimal static execution.

In particular, recall that in the algorithm described in Theorem 2 we computed an optimal static execution by constructing a bipartite graph $H(G)$ in which the nodes of the graph are the edges of the trading graph G. Recall that in $H(G)$ the weight of each edge in E_1 is 0 and the weight of each edge in E_2 is 1. We now slightly perturb the weights of the edges in $H(G)$ to make sure that the algorithm breaks ties consistently between optimal static executions that give different utilities to the same agent. To this end, we first define a complete order π over all edges (i, j) such that $i \in N$, $j \in M$. The order assigns each possible edge a distinct natural number between 1 and $|N| \cdot |M|$. Next, we define a graph $H'(G)$ which is identical to $H(G)$ except that the weight of an edge $(((i,j), 0), ((j,k), 1)) \in E_2$ such that $i, k \in N$ and $j \in M$ is perturbed as follows:

$$w'(((i,j), 0), ((j,k), 1)) = 1 + \underbrace{2^{-\pi((i,j))}}_{\varepsilon_{(((i,j),0),((j,k),1))}}.$$

Similarly, the weight of an edge $(((j,i), 0), ((i,k), 1)) \in E_2$ such that $i \in N$ and $j, k \in M$ is: $w'(((j,i), 0), ((i,k), 1)) = 1 + 2^{-\pi((i,j))} = w'(((i,j), 0), ((j,k), 1))$.

Observe that for any matching $\mathcal{M}$ we have that $w'(\mathcal{M}) = |E_2 \cap \mathcal{M}| + \sum_{e \in E_2 \cap \mathcal{M}} \varepsilon_e$. It is not hard to see that for any matching $\mathcal{M}$ the sum of perturbations is less than 1 (i.e., $\sum_{e \in \mathcal{M}} \varepsilon_e < 1$). Thus, we have that $\lfloor w'(\mathcal{M}) \rfloor = |E_2 \cap \mathcal{M}|$. Together with the fact that $w(\mathcal{M}) = |E_2 \cap \mathcal{M}|$ this implies the following claim:

Claim. If $\mathcal{M}'$ is a maximum weight perfect matching of the graph $H'(G)$, then $\mathcal{M}'$ is also a maximum weight perfect matching of the graph $H(G)$.

Next, we show that any execution that corresponds to some maximum weight perfect matching in $H'(G)$ gives each agent the same utility:

Proposition 2. *For any two maximum weight perfect matchings of the graph $H'(G)$: $\mathcal{M}, \mathcal{M}'$ and any two executions r, r' that correspond to $\mathcal{M}$ and $\mathcal{M}'$ respectively, for any agent i, $A_i(r) = A_i(r')$. In other words, the utilities of all agents are identical in r and r'.*

Proof. Assume towards contradiction that there exist two perfect-matchings of the graph $H'(G)$: $\mathcal{M}, \mathcal{M}'$ such that $w'(\mathcal{M}) = w'(\mathcal{M}')$ but for executions r and r' that correspond to $\mathcal{M}$ and $\mathcal{M}'$ there exists an agent i and an item j such that $j \in A_i(r)$ but $j \notin A_i(r')$. This implies that there exists an edge $(((i,j),0),((j,k),1)) \in E_2 \cap \mathcal{M} \setminus \mathcal{M}'$ where $k \in N$. In this case, $w'(\mathcal{M})$ has a $2^{-\pi((i,j))}$ term that will be missing from $w'(\mathcal{M}')$. We note that this term has a unique exponent, as the only other option to achieve this term is by having an edge $(((j,i),0),((i,a),1)) \in E_2 \cap \mathcal{M}'$ for $a \in M$. However this is impossible since an agent cannot both own and demand the same item. Furthermore, since the exponent is unique this term cannot be derived by adding different perturbations. Thus, we conclude that $w'(\mathcal{M}) \neq w'(\mathcal{M}')$ in contradiction to our assumption. □

Strategyproofness. Denote by A_s the algorithm that computes an optimal execution by choosing a maximum weight perfect matching of the graph $H'(G)$ defined above. We now prove Theorem 3 and show that A_s is strategyproof. To this end, we show that $\forall i, \vec{x}'_{-i}, x'_i, u_i(A_s(x_i, \vec{x}'_{-i})) \geq u_i(A_s(x'_i, \vec{x}'_{-i}))$, where $x_i = (D_i, S_i)$ is a truthful report. We first consider reporting items that are not in D_i or S_i respectively. Observe that if as a result of this misreport agent i receives an item which is not in D_i or need to give an item which is not in S_i, then by definition his utility will be 0. To show that the agent cannot benefit from such misreport when this is not the case, we prove a type of irrelevancy property. We show that an agent cannot improve his utility by not reporting items that the he did not receive or did not give. In the full version we observe:

Observation 4. *Fix some agent i, (not necessarily truthful) report $x'_i = (D'_i, S'_i)$ and reports vector $\vec{x}'_{-i}$ for the rest of the agents. Let $r = A_s(x'_i, \vec{x}'_{-i})$. Denote by $B_i(r)$ the set of items that agent i gave in r and recall that $A_i(r)$ is the set of items that agent i received in r. For any $\tilde{D}_i \subseteq D'_i - A_i(r)$ and $\tilde{S}_i \subseteq S'_i - B_i(r)$ we have that $u_i(A_s((D'_i - \tilde{D}_i, S'_i - \tilde{S}_i), \vec{x}'_{-i})) = u_i(A_s(x'_i, \vec{x}'_{-i}))$.*

We conclude that:

Corollary 1. *Any agent i cannot increase his utility by reporting that he demands an item $j \notin D_i$ or that he owns an item $j \notin S_i$.*

The main part of the proof is showing that an agent cannot benefit from not reporting some of the items in his demand set. In Proposition 3 we show that for any report of the items that the agent owns he cannot benefit from hiding items in his demand set. Then in Proposition 5 we apply Proposition 3 on an instance in which each agent switches between the items he demands and the items that he owns. We show that for any demand report the agent cannot benefit from hiding some of the items that he owns. The two propositions together with Corollary 1 complete the proof of Theorem 3 showing that for any agent i, demand and supply reports D'_i and S'_i and reports of the other agents $\vec{x}'_{-i}$ it hold that $u_i(A_s((D'_i, S'_i), \vec{x}'_{-i})) \leq u_i(A_s((D_i, S_i), \vec{x}'_{-i}))$. First by Corollary 1 we have that an agent can never benefit from including in his demand report items that are not in his true demand set and including in his supply report items that are not in his true supply. Then, for $D_i' \subseteq D_i, S_i' \subseteq S_i$ Proposition 3 guarantees us that $u_i(A_s((D_i', S_i'), \vec{x}_{-i}')) \leq u_i(A_s((D_i, S_i'), \vec{x}_{-i}'))$. Finally we use Proposition 5 to get that for any $S_i' \subseteq S_i$: $u_i(A_s((D_i, S_i'), \vec{x}_{-i}')) \leq u_i(A_s((D_i, S_i), \vec{x}_{-i}'))$ as required. We now state and discuss Proposition 3 and Proposition 5.

Proposition 3. *For every agent i, supply report $S_i' \subseteq S_i$, reports of the other agents $\vec{x}_{-i}'$ and $X \subseteq D_i$ we have that $u_i(A_s((X, S_i'), \vec{x}_{-i}')) \leq u_i(A_s((D_i, S_i'), \vec{x}_{-i}'))$.*

Proof. We define the function $f_i(X) = u_i(A_s((X, S_i'), \vec{x}_{-i}'))$ for every $X \subseteq D_i$. We claim that f_i is a monotone set function for subsets of D_i. That is, $\forall X \subseteq D_i$ and $\forall Y \subseteq X$ we have that $f_i(X) \geq f_i(Y)$. Note that this concludes the proof of the proposition. Also note that in order to prove that f_i is monotone it is sufficient to show that for any agent i, $X \subseteq D_i$ and $j \in X$, $f_i(X) \geq f_i(X - \{j\})$.

Let $r_X = A_s((X, S_i'), \vec{x}_{-i}'))$ be the execution that the algorithm A_s outputs when agent i reports demand X. By Observation 4, for any $j \notin A_i(r_X)$ we have that $f_i(X) = f_i(X - \{j\})$ as required. Thus, for the rest of the proof we consider the case that $j \in A_i(r_X)$.

Let $r_{X-\{j\}} = A_s((X - \{j\}, S_i'), \vec{x}_{-i}'))$. Denote by $Y = A_i(r_{X-\{j\}}) - A_i(r_X)$ the set of items that agent i received when reporting $X - \{j\}$ but did not receive when reporting X. Assume towards contradiction that $f_i(X - \{j\}) > f_i(X)$. As $f_i(X - \{j\}) \leq f_i(X) - 1 + |Y|$, this implies that $|Y| \geq 2$. Let $\mathcal{M}_X$ and $\mathcal{M}_{X-\{j\}}$ be the maximum weight perfect matchings that were computed as part of A_s for demand reports X and $X - \{j\}$ respectively (these are the matchings that are used to derive the executions r_X and $r_{X-\{j\}}$). Roughly speaking, we will construct from the union of their edges two different matchings: $\mathcal{M}_X'$ and $\mathcal{M}_{X-\{j\}}'$ such that one is a valid perfect matching when agent i reports X and the other is a valid perfect matching when agent i reports $X - \{j\}$. Since those matchings cover the same edges as $\mathcal{M}_X$ and $\mathcal{M}_{X-\{j\}}$ the sum of their weights is the same. This implies that one of the matchings $\mathcal{M}_X$ or $\mathcal{M}_{X-\{j\}}$ does not have the maximum weight. The crux of the proof is the following proposition, which we prove in the full version:

Proposition 4. *If $f_i(X - \{j\}) > f_i(X)$, then, for the maximum weight perfect matchings $\mathcal{M}_X$ and $\mathcal{M}_{X-\{j\}}$ there exist matchings $\mathcal{M}_X'$ and $\mathcal{M}_{X-\{j\}}'$ such that:*

1. $\mathcal{M}_X'$ *is a valid perfect matching of the graph $H'(G((X, S_i'), \vec{x}_{-i}'))$ (i.e., when agent i reports X) and $\mathcal{M}_{X-\{j\}}'$ is a valid perfect matching of the graph $H'(G((X - \{j\}, S_i'), \vec{x}_{-i}'))$ (i.e., when agent i reports $X - \{j\}$).*
2. $w'(\mathcal{M}_X) + w'(\mathcal{M}_{X-\{j\}}) = w'(\mathcal{M}_X') + w'(\mathcal{M}_{X-\{j\}}')$
3. $w'(\mathcal{M}_X') \neq w'(\mathcal{M}_X)$.

Observe that the three statements of the proposition imply that either $w'(\mathcal{M}_X') > w'(\mathcal{M}_X)$ or $w'(\mathcal{M}_{X-\{j\}}') > w'(\mathcal{M}_{X-\{j\}})$. Since both $\mathcal{M}_X'$ and $\mathcal{M}_X$ are valid matchings of $H'(G((X, S_i'), \vec{x}_{-i}'))$ and both $\mathcal{M}_{X-\{j\}}'$ and $\mathcal{M}_{X-\{j\}}$ are valid matchings of $H'(G((X - \{j\}, S_i'), \vec{x}_{-i}'))$ this in contradiction to the assumption that $\mathcal{M}_X$ and $\mathcal{M}_{X-\{j\}}$ are maximum weight perfect matchings. Thus, we conclude that $f_i(X - \{j\}) \leq f_i(X)$. $\square$

Next, we show that an agent maximizes his utility by truthfully reporting all the items that he owns. In the following proofs we will compare the utility of agents in two different instances. For this purpose we use the notation $u_i^{\mathcal{G}}(r)$ for the utility of agent i in execution r of instance $\mathcal{G}$.

Proposition 5. *For any agent i, demand report $D_i' \subseteq D_i$, reports of the other agents $\vec{x}_{-i}'$ and $S_i' \subset S_i$ we have that $u_i(A_s((D_i', S_i'), \vec{x}_{-i}')) \leq u_i(A_s((D_i', S_i), \vec{x}_{-i}'))$.*

Proof. Denote the original instance of the problem by $\mathcal{G}$. We define the reversed instance $\bar{\mathcal{G}}$, in this instance the demand of each agent $a \in N$ is $\bar{D}_a = S_a$ and his supply is $\bar{S}_a = D_a$. For a vector of reports $\vec{x}'$ such that $x_a' = (D_a', S_a')$ for every agent $a \in N$ we define the reversed reports vector $\vec{\bar{x}}'$ such that for each agent a, $\bar{x}_a' = (S_a', D_a')$.

By Proposition 3 we have that for the instance $\bar{\mathcal{G}}$ and any reports vector of the other agents $\vec{\bar{x}}_{-i}'$, for any $S_i' \subset \bar{D}_i$ and any $D_i' \subseteq \bar{S}_i$: $u_i^{\bar{\mathcal{G}}}(A_s((S_i', D_i'), \vec{\bar{x}}_{-i}')) \leq u_i^{\bar{\mathcal{G}}}(A_s((S_i, D_i'), \vec{\bar{x}}_{-i}'))$. To prove the proposition we will show in Claim 3.2 below that $u_i^{\bar{\mathcal{G}}}(A_s((S_i', D_i'), \vec{\bar{x}}_{-i}')) = u_i^{\mathcal{G}}(A_s((D_i', S_i'), \vec{x}_{-i}'))$ and that $u_i^{\bar{\mathcal{G}}}(A_s((S_i, D_i'), \vec{\bar{x}}_{-i}')) = u_i^{\mathcal{G}}(A_s((D_i', S_i), \vec{x}_{-i}'))$. □

We now observe the strong symmetry between an instance of our game and the reversed instance in which each agent swaps between the items he receives and the items he demands. In the full version we prove:

Claim. For every agent i, reports vector $\vec{x}'$ for $\mathcal{G}$ and a reversed report vector $\vec{\bar{x}}'$ for $\bar{\mathcal{G}}$, we have that $u_i^{\mathcal{G}}(A_s(\vec{x}')) = u_i^{\bar{\mathcal{G}}}(A_s(\vec{\bar{x}}'))$.

4   Limitations of Dynamic Executions

In Sect. 3.2 we showed that the efficient algorithm that computes the optimal static execution is both strategyproof and provides an l-approximation. In this section we prove that the best approximation ratio achievable by a strategyproof algorithm is $\frac{l+1}{2}$. Then, we consider the problem of finding the optimal execution from a strictly computational perspective and prove that unless P = NP the problem cannot be approximated within some small constant.

Theorem 5. *There is no strategyproof algorithm which gives better approximation than $\frac{l+1}{2}$.*

Proof sketch: Consider the following instance $\mathcal{G}$: the set of agents contains two subsets of cardinality l: $N = \{i_1, i_2, \ldots, i_l\}$ and $N' = \{i_1', i_2', \ldots, i_l'\}$. The set of items contains two subsets of cardinality l: $M = \{j_1, j_2, \ldots, j_l\}$ and $M' = \{j_1', j_2', \ldots, j_l'\}$. For $1 \leq k \leq l$ the demand of agent i_k is $D_{i_k} = M$ and his supply is $S_{i_k} = \{j_k'\}$. For $1 \leq k \leq l$ the demand of agent i_k' is $D_{i_k'} = \{j_k'\}$ and his supply is $S_{i_k'} = \{j_k\}$. The instance $\mathcal{G}$ also includes $\frac{l+1}{2}(l^2 + l) \cdot l$ extra agents that are partitioned into l groups. There are also extra items such that the demand and

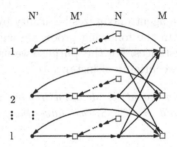

Fig. 3. The trading graph for an instance showing that no strategyproof algorithm can attain an approximation ratio better than $\frac{l+1}{2}$.

supply of each group of $\frac{l+1}{2}(l^2 + l)$ agents creates a path that ends in an item in M'. We illustrate the corresponding trading graph in Fig. 3.

The optimal execution for the instance $\mathcal{G}$ first executes the l cycles in which each pair of agents i_k and i'_k swap items j'_k and j_k between them. There are l such cycles and in each cycle there are 2 exchanges. Then, it executes $l-1$ cycles with the items of M. The number of exchanges in each cycle is l so the total number of exchanges is $(l - 1) \cdot l + 2l = l^2 + l$. Assume towards contradiction that there exists an algorithm that achieves an approximation ratio $\alpha < \frac{l+1}{2}$. Note that in $\mathcal{G}$ such an algorithm must allocate at least one agent two or more items. We conclude that there exists an agent $i_k \in N$ that is allocated by the algorithm in the instance $\mathcal{G}$ at least two items. Now consider an instance $\mathcal{G}'$ which is identical to $\mathcal{G}$ except that agent i_k also demands item j^1_k. This means that the trading graph now has a giant cycle of size $\frac{l+1}{2}(l^2 + l)$ and since the algorithm guarantees an approximation ratio better than $\frac{l+1}{2}$ it has to execute this cycle. In this case i_k can only participate in the giant cycle and hence only gets a single item. Thus, agent i_k can increase his utility by not reporting j^1_k. □

4.1   Pareto Efficiency and Strategyproofness

An execution r is Pareto efficient if for any other execution r' there exists an agent i such that $u_i(r') < u_i(r)$. We leave open the question of whether there exists a strategyproof algorithm that always returns a Pareto efficient execution. In any case, we show that even if such an algorithm exists, its performance is quite poor.

Proposition 6. *Any algorithm that is strategyproof and returns a Pareto efficient execution cannot guarantee an approximation ratio better than $\Theta(n)$.*

Proof. Consider the following n-agent instance $\mathcal{G}$. In this instance we have 3 agents i_1, i_2, i_3 such that for agent i_1, $D_{i_1} = \{a, b\}$ and $S_{i_1} = \{c\}$. For agent i_2, $D_{i_2} = \{b\}$ and $S_{i_2} = \{a\}$. For agent i_3, $D_{i_3} = \{c\}$ and $S_{i_3} = \{b\}$. The instance also include a sequence of $n - 3$ agents that in the trading graph take part in a long path that starts from item p'_1 and end in item c: $P = (p'_1, p_1, p'_2, \ldots, p_{n-3}, c)$. In Fig. 4 we illustrate the trading graph for the instance $\mathcal{G}$.

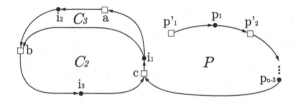

Fig. 4. The trading graph for instance $\mathcal{G}$ in the proof of Proposition 6.

In this instance the only (dynamic) Paerto efficient execution first executes the cycle $C_2 = (i_1, b, i_3, c, i_1)$ and then executes the cycle $C_3 = (i_1, a, i_2, b, i_1)$. The utility of agent i_1 in this execution is 2. Consider the case that agent i_1 also demands item p'_1. Now, the graph has a giant cycle $C_1 = (p'_1, p_1, p'_2, \ldots, p_{n-3}, c, i_1, p'_1)$ that includes $n-2$ agents and $n-2$ items. Note that it is impossible to execute both cycles C_1 and C_2 since in C_2 agent p_{n-3} receives item c and in C_2 agent i_3 receives item c and c is the only item that agent p_{n-3} and i_3 demand. This implies that the algorithm cannot execute cycle C_1 as in this case the utility of agent i_1 would be 1 and he can increase his utility by not reporting that he demands item p'_1. Thus, the algorithm has to execute first C_2 and then C_3 which accumulates to a total of 4 exchanges where the optimal execution performs $n-2$ exchanges.[5] □

4.2 Computational Hardness

In this section we discuss the problem of computing an optimal execution from a purely computational perspective (the complete proof can be found in the full version):

Theorem 6. *Unless $P = NP$ there is no polynomial time c-approximation for computing the optimal execution unless $P = NP$ where $c > 1$ is a small constant.*

Proof sketch: We reduce from the NP-Complete 3D-matching problem[6]: Let X, Y, and Z be finite, disjoint sets, of size n and let T be a subset of $X \times Y \times Z$. Does there exist a subset $S \subseteq T$ of size n such that for any two distinct triplets $(x_1, y_1, z_1), (x_2, y_2, z_2) \in S$, we have $x_1 \neq x_2, y_1 \neq y_2$, and $z_1 \neq z_2$?

Recall that computing the optimal static execution can be done in polynomial time. This means that proving the hardness of computing an optimal dynamic execution requires us to devise a very careful reduction in which the optimal execution has to execute certain cycles at the first round in order to execute

---

[5] Observe that the problem here is that because of Pareto Efficiency the algorithm has to execute in the first instance both C_2 and C_3. In comparison, an optimal static execution will only execute one cycle in this instance and hence agent i_i would not be able to benefit by misreporting in the modified instance.

[6] [2] also reduce from 3D-matching, however, our reductions are inherently different. Specifically, their hardness stems from limiting the size of the cycles whereas we have no such limitation.

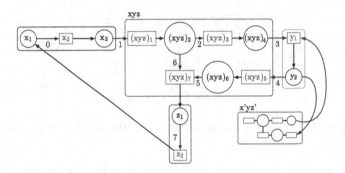

Fig. 5. An illustration for the proof Theorem 6 featuring the components of $(x, y, z) \in T$ and $(x', y, z') \in T$.

other cycles in the next rounds. In the reduction, given an instance X, Y, Z and $T \subseteq X \times Y \times Z$ of the 3D-matching problem, we construct an instance that includes a component for every triplet $(x, y, z) \in T$, $x \in X$, $y \in Y$ and $z \in Z$ (see illustration in Fig. 5). In this instance an algorithm for computing an optimal execution in the trading graph $G = (N, M, E)$ will return an execution of size $8n$ if and only if there exists a perfect 3D-matching in (X, Y, Z, T). This is done by making sure that the optimal execution executes two cycles for every triplet (x, y, z) in the perfect 3D-matching in (X, Y, Z, T): first the cycle the contains edges numbered $0, 1, 6, 7$ in Fig. 5 and then after edge 6 was flipped the cycle including edges $2, 3, 4, 5$. The complete proof requires a delicate analysis of the possible cycles that may be executed in each sequence.   □

5   Conclusion and Discussion

Our paper contributes to forming the mathematical foundations of barter markets. As such, the paper does not aim to provide a full modeling of a concrete market, but rather to mathematically capture some of the major challenges in designing them. We identify a central aspect of many barter markets that yet to be studied: the market may be dynamic in the sense that the same item can move from hand to hand several times. A main contribution of our paper is identifying this aspect and formally modeling it. The second set of contributions is in a comprehensive analysis of dynamic executions.

Our results on the approximation ratio of strategyproof mechanisms in this setting can be interpreted in two ways. First, in many cases, the approximation ratio of l (the maximal number of items an agent demands) achieved by an optimal static execution is reasonable since the number of items that an agent demands does not grow with the size of the network. This gives a justification for studying static executions even in a dynamic environment such as ours. Second, the impossibility result showing that a strategyproof mechanism cannot provide an approximation ratio better than $\approx l/2$ suggests that to increase efficiency, barter networks should include some form of money. This may explain why many barter applications indeed often involve vouchers, for example. We hope that the understanding of barter markets we gained in this paper will provide a stepping stone towards understanding markets with vouchers.

References

1. Abbassi, Z., Haghpanah, N., Mirrokni, V.: Exchange market mechanisms without money. In: Web and Internet Economics: 11th International Conference, WINE 2015, Amsterdam, The Netherlands, 9–12 December 2015, Proceedings, vol. 9470, p. 429. Citeseer (2015)
2. Abraham, D.J., Blum, A., Sandholm, T.: Clearing algorithms for barter exchange markets: enabling nationwide kidney exchanges. In: Proceedings of the 8th ACM Conference on Electronic Commerce, pp. 295–304. ACM (2007)
3. Akbarpour, M., Combe, J., He, Y., Hiller, V., Shimer, R., Tercieux, O.: Unpaired kidney exchange: overcoming double coincidence of wants without money. In: Proceedings of the 21st ACM Conference on Economics and Computation, pp. 465–466 (2020)
4. Alcalde-Unzu, J., Molis, E.: Exchange of indivisible goods and indifferences: The Top Trading Absorbing Sets mechanisms. Games Econ. Behav. **73**(1), 1–16 (2011)
5. Andersson, T., Cseh, A., Ehlers, L., Erlanson, A.: Organizing time banks: lessons from matching markets. Am. Econ. J.: Microecon. **13**(1), 338–373 (2018)
6. Aziz, H., De Keijzer, B.: Housing markets with indifferences: a tale of two mechanisms. In: AAAI, vol. 12, pp. 1249–1255 (2012)
7. Botsman, R., Rogers, R.: What's Mine is Yours: The Rise of Collaborative Consumption. Collins, London (2010)
8. Frenken, K., Schor, J.: Putting the sharing economy into perspective. Environ. Innov. Soc. Trans. **23**, 3–10 (2017)
9. Jaramillo, P., Manjunath, V.: The difference indifference makes in strategy-proof allocation of objects. J. Econ. Theory **147**(5), 1913–1946 (2012)
10. Kocherlakota, N.R.: Money is memory. J. Econ. Theory **81**(2), 232–251 (1998)
11. Konishi, H., Quint, T., Wako, J.: On the Shapley-Scarf economy: the case of multiple types of indivisible goods. J. Math. Econ. **35**(1), 1–15 (2001)
12. Kuhn, H.W.: The Hungarian method for the assignment problem. Nav. Res. Logist. (NRL) **52**(1), 7–21 (2005)
13. Manjunath, V., Westkamp, A.: Strategy-proof exchange under trichotomous preferences. Technical report, Working paper, University of Cologne (2019)
14. Saban, D., Sethuraman, J.: House allocation with indifferences: a generalization and a unified view. In: Proceedings of the Fourteenth ACM Conference on Electronic Commerce, pp. 803–820. ACM (2013)
15. Shapley, L., Scarf, H.: On cores and indivisibility. J. Math. Econ. **1**(1), 23–37 (1974)
16. Sönmez, T.: Strategy-proofness and essentially single-valued cores. Econometrica **67**(3), 677–689 (1999)

Formal Barriers to Simple Algorithms for the Matroid Secretary Problem

Maryam Bahrani[1], Hedyeh Beyhaghi[2(✉)], Sahil Singla[3], and S. Matthew Weinberg[4]

[1] Columbia University, New York, USA
m.bahrani@columbia.edu
[2] Toyota Technological Institute at Chicago, Chicago, USA
hedyeh@ttic.edu
[3] Georgia Institute of Technology, Atlanta, USA
ssingla@gatech.edu
[4] Princeton University, Princeton, USA
smweinberg@princeton.edu

Abstract. Babaioff et al. [4] introduced the matroid secretary problem in 2007, a natural extension of the classic single-choice secretary problem to matroids, and conjectured that a constant-competitive online algorithm exists. The conjecture still remains open despite substantial partial progress, including constant-competitive algorithms for numerous special cases of matroids, and an $O(\log \log \text{rank})$-competitive algorithm in the general case.

Many of these algorithms follow principled frameworks. The limits of these frameworks are previously unstudied, and prior work establishes only that a handful of particular algorithms cannot resolve the matroid secretary conjecture. We initiate the study of impossibility results for frameworks to resolve this conjecture. We establish impossibility results for a natural class of greedy algorithms and for randomized partition algorithms, both of which contain known algorithms that resolve special cases.

Keywords: Secretary problem · Matroids · Optimal stopping theory · Graph theory · Greedy algorithms

1 Introduction

The problem of finding a max-weight basis of a matroid $\mathcal{M} = (V, \mathcal{I})$[1] is central in the field of combinatorial optimization (see books [18, 21, 23]). More specifically, each element $e \in V$ has a weight $w(e) \geq 0$, and the goal is to find the set $S \in \mathcal{I}$ maximizing $w(S) := \sum_{e \in S} w(e)$. Seminal works of Rado, Gale, and Edmonds establish that the following simple greedy algorithm finds a max-weight basis of a

---

[1] Given a finite set V and a family of subsets of V called I, we say $\mathcal{M} = (V, \mathcal{I})$ is a matroid if it satisfies (i) $\emptyset \in \mathcal{I}$, (ii) *Hereditary Property (downwards closed):* $\forall T \subseteq S \subseteq V$, set $S \in \mathcal{I}$ implies $T \in \mathcal{I}$, and (iii) *Exchange Property:* For any $S, T \in \mathcal{I}$ where $|S| > |T|$, there exists some $x \in S$ such that $T \cup \{x\} \in \mathcal{I}$.

© Springer Nature Switzerland AG 2022
M. Feldman et al. (Eds.): WINE 2021, LNCS 13112, pp. 280–298, 2022.
https://doi.org/10.1007/978-3-030-94676-0_16

matroid $(V, \mathcal{I})$: Initialize $A = \emptyset$, then process the elements of V in decreasing order of $w(e)$, adding to A any element such that $A \cup \{e\} \in \mathcal{I}$ [8, 11, 22]. In fact, if for some $(V, \mathcal{I})$ this algorithm is optimal for all $w(\cdot)$, then $(V, \mathcal{I})$ must be a matroid.

While simple, this algorithm still requires knowledge of all weights up front. Motivated by applications to mechanism design and other online problems [3, 13], recent work considered the problem in an online setting: elements are still processed one at a time and are immediately and irrevocably accepted or rejected upon processing, but an element's weight remains unknown until the element is processed. In particular, the algorithm does not have control over the order of elements and therefore cannot run the simple greedy algorithm.

For a fully adversarial order, it's folklore that the best algorithm can do no better than simply selecting a random element. Babaioff et al. [4][2] therefore introduced the *Matroid Secretary Problem* (MSP), where elements arrive in a *uniformly random order* (while the weight function is still adversarial). This formulation extends the classic single-item secretary problem [7].

Consider an algorithm $\mathcal{A}$ for the matroid secretary problem on matroid $\mathcal{M}$. Let OPT be the max-weight basis of $\mathcal{M}$ under $w(\cdot)$, and let ALG be the set of elements chosen by $\mathcal{A}$ (under $w(\cdot)$). The following notion of *utility-competitiveness* for a matroid secretary algorithm was studied in Babaioff et al. [4].

Definition 1 (Utility-Competitive). *An algorithm $\mathcal{A}$ is α-utility-competitive if $\mathbb{E}[w(\mathsf{ALG})]/w(\mathsf{OPT}) \geq \alpha$, where the expectation is over the randomness of the arrivals and any internal randomness of algorithm $\mathcal{A}$.*

In the same paper that introduced the matroid secretary problem, Babaioff et al. [4] conjecture that there is a constant-utility-competitive algorithm. The stronger form of the conjecture is that this constant is $1/e$.

Conjecture 1 (Matroid Secretary). There is an $\Omega(1)$-utility-competitive algorithm for the matroid secretary problem.

Despite extensive follow-up work, this conjecture still remains open. Many constant-utility-competitive algorithms have been proposed for specific classes of matroid (see related work in Sect. 1.3). For general matroids, however, the best known algorithms are $1/O(\log\log r)$-competitive [9, 20] (here, r denotes the *rank* of the matroid, which is the size of the largest set in $\mathcal{I}$).

As the only known lower bound, even for general matroids, is the same $1/e$ from the classic single-item setting, and because Dynkin's algorithm guarantees a stronger property that the heaviest element is selected with probability $1/e$, the following stronger notion of *probability-competitive* algorithms has been also studied [14, 25].

Definition 2 (Probability-Competitive). *An algorithm $\mathcal{A}$ is α-probability-competitive if for all $i \in \mathsf{OPT}$ it satisfies that $\mathbb{P}[i \in \mathsf{ALG}] \geq \alpha$.*

Note that probability-competitiveness is a stronger notion than utility-competitiveness, since the former implies the latter with the same competitive

---
[2] Conference version [5] appeared in 2007.

ratio. Soto et al. [25] showed that many (but not all) existing utility-competitive algorithms can be extended to obtain probability-competitive algorithms. This results in the following more ambitious conjecture. Again, the stronger version conjectures that this constant is $1/e$.

Conjecture 2. There is an $\Omega(1)$-probability-competitive algorithm for the matroid secretary problem.

Progress on both conjectures has been slow. Indeed, even the strong version of Conjecture 2 remains plausible, while the best utility-competitive algorithms have stalled at $1/O(\log \log r)$ [9,20]. One thesis motivating our work is that the community currently lacks structure for narrowing a search among numerous promising approaches. Existing algorithms for special cases indeed follow principled frameworks, but these frameworks are quite flexible and it remains unknown which (if any) of them might produce a resolution to either conjecture.

One particularly enticing possibility is that a simple "greedy-like" algorithm might even work. Note that such algorithms indeed work in the Free-Order model [16], or for the related Matroid Prophet Inequality [17], or for special cases of the Matroid Secretary Problem [2,7]. There are numerous variants of "greedy" algorithms, though. While many particular variants are known to fail on the same "hat graph" [4], there is previously no approach to quickly tell whether a novel greedy variant is already known to fail.

In this work, we rigorously consider two general classes of algorithms, and prove super-constant lower bounds on what they can achieve for the matroid secretary problem. This both helps explain why these types of algorithms have faced difficulty extending beyond the special cases for which they were originally designed, and helps guide future work towards precisely the variants that merit further exploration.

1.1   Greedy Algorithms

Since finding the max-weight basis of matroids without requiring irrevocable commitments can be done exactly by the simple greedy algorithm, the class of greedy algorithms is a very natural candidate for solving the Matroid Secretary Problem. We consider a large family of "greedy-like" algorithms. We define three natural properties that a greedy algorithm might have, and establish that any algorithm satisfying these properties cannot be constant-utility-competitive (Theorem 2). We postpone formal statements of the properties until Sect. 3, but overview them here: (i) the algorithm should reject the first T fraction of elements, (ii) the algorithm at all times stores an independent set I containing all accepted elements and no elements rejected after T, (iii) an element is accepted if and only if it improves the max-weight basis of I after contracting the accepted elements.[3] Note that this a general framework rather than a fully-specified algo-

---

[3] To rephrase (iii), an element e is accepted iff after contracting the accepted elements (not including e), the max-weight basis of the restricted matroid to $I \cup \{e\}$ is heavier than the max-weight basis of the restricted matroid to I (the latter being exactly the weight of I since I is independent).

rithm, since it allows for the algorithm to choose I (it need not be the max-weight basis after contracting the accepted elements, just some independent set).

In Sect. 3 we overview several existing algorithms that fit this framework, and Theorem 2 unifies a proof that none of these algorithms (or many hypothetical ones) can be constant-utility-competitive. Our lower bound construction is a variant of the well-known "hat graph", which has been known since [4] to be problematic for greedy-like algorithms. So our main contribution is not this construction itself, but rather a formalization of precisely the class of greedy algorithms for which this graph is problematic.

Main Result 1 (Informal, see Theorem 2). *No Greedy algorithm (as per Algorithm 1) is constant-utility-competitive.*

We emphasize that while the hat graph itself is not a novel construction, our proof is quite distinct (and more involved) from prior work as it must rule out a broad class of algorithms rather than just a single one.

1.2 Randomized Partition Algorithms

Another class of particularly simple algorithms are *randomized partition algorithms*:

1. Before looking at any weights, (perhaps randomly) partition all the elements[4] into parts S_i.
2. Within each part, run Dynkin's algorithm.
3. Output the union of the selected elements.

Note that these algorithms are allowed to use *any randomized* partition. The elegant $1/(2e)$-approximation of Korula and Pal for graphic matroids[5] is a randomized partition algorithm [19]. Their algorithm is utility-competitive, but not probability-competitive. Soto et al. [25] recently designed a different constant probability-competitive algorithm for graphic matroids. While their algorithm is still quite elegant, it is perhaps not quite as simple as randomized partition algorithms. It is also worth noting that algorithms such as [9,20] follow a more general framework, where the algorithm in step one looks at the weights before partitioning and step two is not necessarily Dynkin's single-choice algorithm (but perhaps some simple greedy algorithm). This raises the question whether the novel development beyond [19] is necessary to achieve probability-competitive algorithms? Our second main result answers this question: no randomized partition algorithm can be constant-probability-competitive (or even $\omega(n^{-1/8})$-probability-competitive).

---

[4] We consider the known matroid setting where the matroid is known but the weights are revealed one-by-one.

[5] Given a graph with edges E, a graphic matroid $(E, \mathcal{I})$ is defined with $\mathcal{I}$ consisting of all subsets of edges that do not contain a cycle.

Main Result 2 (Informal, see Theorem 4). *No Randomized Partition algorithm is constant-probability-competitive.*

Our construction witnessing Theorem 4 is also a graphic matroid, although it is unrelated to the hat graph (and to the best of our knowledge, novel). Note that our proof cannot be extended to utility-competitive algorithms since we know [19] is a constant-utility-competitive randomized partition algorithm for graphic matroids.

1.3 Related Work and Brief Summary

There is a *substantial* body of work on random-order problems for matroids (the Matroid Secretary Problem [4]) and for several other discrete optimization problems; we will not attempt to overview it (e.g., see [6,12]). Here, we will briefly repeat the most related works.

Our work takes first steps towards characterizing classes of algorithms which might resolve the Matroid Secretary Problem. We focus on the simplest classes of algorithms which previously succeeded in special cases or for related problems, Greedy [16,17] or Randomized Partition [19], and study the limits of these classes. First, we consider extremely simple greedy algorithms. A specific instantiation of this class of algorithms was shown to fail on a now-canonical "hat graph" in [4], but related algorithms known to succeed in the Free-Order Model [1,16], and in the related Matroid Prophet Inequality [17]. In addition, Dynkin's algorithm and the Optimistic algorithm for k-uniform matroids of [2] fit this model. Our Theorem 2 shows that no Greedy algorithm is constant-utility-competitive for all matroids. Second, we consider probability-competitive algorithms, formally considered in [25], and related to the ordinal model considered in [14]. Soto et al. [25], in particular, develop several probability-competitive algorithms for core settings such as graphic, transversal, and laminar matroids. Our work asks whether the extremely simple algorithms previously developed in [19] can match these stronger probability-competitive guarantees, and we show in Theorem 4 that the answer is no.

2   Preliminaries

The Matroid Secretary Problem (MSP) is defined as:

1. There is a matroid $\mathcal{M} = (V, \mathcal{I})$, and weight function $w(\cdot) : V \to \mathbb{R}_{\geq 0}$. Matroid $\mathcal{M}$ is fully-known to the algorithm in advance.[6] Function $w(\cdot)$ is initially completely unknown to the algorithm.

---

[6] We are not concerned with computational efficiency of our algorithms in this work (our lower bounds are unconditional), so we will not stress about the precise format in which access to the matroid is given. To be concrete, one access model is that the algorithm has oracle access to $\mathcal{I}$ (query a set S and learn whether or not $S \in \mathcal{I}$). To the best of our knowledge, most algorithms previously considered for MSP are polytime given oracle access to $\mathcal{I}$.

2. Initially, the set of accepted elements, A, is empty. Elements of V arrive in a uniformly random order. When an element $i \in V$ arrives, the algorithm learns its weight $w(i)$, and must make an immediate and irrevocable decision whether or not to accept it (adding it to A). The algorithm must maintain $A \in \mathcal{I}$ at all times.
3. If set A is selected, the algorithm achieves payoff $\sum_{i \in A} w(i)$.

We will abuse notation and use $w(S) := \sum_{i \in S} w(i)$. Because $w(\cdot)$ is fixed, the offline optimum is the max-weight basis: $\mathrm{MWB}(\mathcal{M}) := \arg\max_{S \in \mathcal{I}} \{w(S)\}$.[7] We will also use standard matroid notation such as *restriction*: the matroid $\mathcal{M}|_S$ is the matroid $\mathcal{M}$ restricted to S, and has ground set S and independent sets $\mathcal{I}|_S := \{T \cap S \mid T \in \mathcal{I}\}$. We also discuss matroid *contractions*: the matroid $\mathcal{M} \setminus S$ is the matroid $\mathcal{M}$ contracted by S, and has ground set $V \setminus S$ and independent sets $\mathcal{I} \setminus S := \{T \mid T \cup S \in \mathcal{I}\}$. When $\mathcal{M}$ is clear from context, we will also (slightly) abuse notation and write $\mathrm{MWB}(T) := \mathrm{MWB}(\mathcal{M}|_T)$.

We will later reference Dynkin's $1/e$-probability-competitive algorithm for selecting a single item, i.e., a 1-uniform matroid: (1) Reject the first $T = \mathrm{Binom}(n, 1/e)$ elements and call this the *sampling stage*. (2) Afterwards, accept an element i iff it is the heaviest element seen so far.

Theorem 1 [7]. *Dynkin's algorithm is $1/e$-probability-competitive for 1-uniform matroids, this is optimal.*

3   Greedy Algorithms

Because matroids are exactly the constraints for which the simple greedy algorithm is optimal, greedy-like algorithms are a natural family to consider as candidates for resolving the Matroid Secretary Problem. Indeed greedy-like algorithms solve the related Matroid Prophet Inequality [17], Matroid Secretary in the free-order model [1,16], and special cases of Matroid Secretary [2,7]. In this section, we give an impossibility result for certain greedy algorithms. This helps unify counterexamples for related algorithms, and also helps narrow future research towards algorithms which have hope of resolving the Matroid Secretary Problem.

3.1   A Class of Greedy Algorithms

We now define a natural framework of greedy algorithms for the Matroid Secretary Problem (Algorithm 1). Without loss of generality, we consider the continuous arrival setting, where each element $e \in V$ arrives at a time $t(e)$ independently and uniformly drawn from $[0, 1]$. We refer by V_t to the set of elements that arrive (strictly) before t, and by A_t to the set of elements accepted by the algorithm (strictly) before time t.

---

[7] In this work, we assume for simplicity that the max-weight basis is unique. In case of ties, we tie-break by choosing the lexicographically-earlier basis.

Algorithm 1. Greedy Algorithm for the matroid secretary problem

---

We define a *greedy algorithm* as one that satisfies the following properties:

(i) Reject (but store) elements that arrive before T (sampling stage). Denote $S := V_T$ to emphasize this.

(ii) At all times t, maintain an independent set I_t such that:
- I_t contains all accepted elements and no elements which were rejected after T, i.e. $A_t \subseteq I_t \subseteq A_t \cup S$.
- At all times t, I_t spans V_t.

(iii) Accept e if and only if $e \in \mathrm{MWB}((\mathcal{M} \setminus A_{t(e)})|_{I_{t(e)} \cup \{e\}})$ (and $t(e) > T$). That is, accept e if and only if it is in the max-weight basis of $I_{t(e)} \cup \{e\}$ *after contracting by $A_{t(e)}$*.

Before getting into our results, it is helpful to understand why Algorithm 1 is a class of algorithms (rather than a fully-specified algorithm). The reason is that the algorithm has flexibility in which subset of S to include in I_t (but it must include A_t, and must span V_t). The restriction is that the algorithm does not know which element might arrive at time t, nor its weight, when setting I_t. Furthermore, the algorithm can choose the length of the sampling stage T.

It is also helpful to see how this framework captures (or doesn't capture) existing greedy-like algorithms:

- Dynkin's algorithm (with $T = 1/e$) fits this framework. But so do suboptimal algorithms (e.g., accept the first element after T which exceeds the 5th-highest sample. Or even accept an element which arrives at time $t > T$ iff it exceeds the $(\lfloor 5t/T \rfloor)^{\text{th}}$-highest sample).
- The Optimistic Algorithm for k-uniform matroids of [2] fits this framework. The algorithm maintains a list U, initially the k heaviest elements of S. If e exceeds the lightest element in U, it is accepted, and the lightest element of U is removed. In our language, this has $I_t := A_t \cup U$ at all times.
- There is a natural extension of the Optimistic Algorithm to all matroids, which was previously considered in [4].
- A related Pessimistic Algorithm (similar to the rehearsal algorithm for the related k-uniform prophet inequality of [1]) for k-uniform matroids fits this framework. The algorithm also maintains a list U, initially the k heaviest elements of S. If e exceeds the lightest element in U, it is accepted, but the *heaviest element of U lighter than e* is removed. In our language, this again has $I_t := A_t \cup U$ at all times (but U is updated differently to the previous bullet).
- The Virtual Algorithm for k-uniform matroids of [2] does *not* fit this framework. The algorithm accepts an element e if and only if e is one of the heaviest k elements so far *and* the kth-heaviest element of $V_{t(e)}$ is in S (i.e., e is accepted if and only if it "kicks out a sample" from the top k so far). This is because the algorithm needs to remember rejected elements in order to properly keep track of the kth-heaviest element so far, and whether it was a sample.

Observe finally that all of the algorithms above (which fit the framework) further have the following. First, if an element is rejected (after T), it is forgotten forever, and the algorithm proceeds as if the element had never existed in the first place.[8] Similarly, once an element e is accepted, the algorithm updates $\mathcal{M}$ by contracting by e, and then proceeds identically as if the true matroid had been $\mathcal{M} \setminus \{e\}$ the whole time.[9] These attributes are shared by the matroid prophet inequality of [17], and initially drove our formulation.

With an understanding of Greedy algorithms in hand, we now state our main result.

Theorem 2. *Any algorithm satisfying the 3 properties of Algorithm 1 cannot be constant-utility-competitive.*

3.2   Hard Instance: The Hat

In this section, we will study a *hat graph* which drives our impossibility result. The hat has a special element which is significantly heavier than the sum of all others, and thus any algorithm with a good utility-competitive ratio must accept it. Furthermore, this special element appears in many small circuits, so the algorithm must not accept the remaining elements of any of these circuits prior to the arrival of the heavy element (otherwise, the heavy element cannot be accepted when it arrives). The hat was used in [4] as a counterexample against a particular greedy algorithm; and variants of the graph have been informally known to be problematic for "greedy-like" algorithms. However, prior to our work there was no formal classification of "greedy-like".

The hat on $n + 2$ vertices is a collection of n triangles, all sharing the same edge. Formally, an undirected graph (V, E) is a hat if $V = \{a, b, v_1, \ldots, v_n\}$ for some $n > 0$, and $E = \{\{a, b\}\} \cup \{e_i = \{a, v_i\} : i \in [n]\} \cup \{e_i' = \{b, v_i\} : i \in [n]\}$. Several weight assignments to the edges of the hat can serve as counterexamples to the algorithms considered in this section, but we consider a particular weight assignment for ease of exposition (as we only need one counterexample). We define this weight function $w : E \to \mathbb{R}_{\geq 0}$ to maintain the following ordering of the edge weights: $w(e_1) > \ldots > w(e_n) > w(e_1') > \ldots > w(e_n')$. Furthermore, $w(\{a, b\})$ is much larger than the sum of the weights of all other edges. We will refer to $\{a, b\}$ as the *infinity edge*, and we refer to its arrival time as $t_\infty := t(\{a, b\})$ to emphasize this. Additionally, we consider the drawing of the hat in the plane as shown in Fig. 1, where e_i is to the left of e_j for $i < j$, and e_i is above e_i' for all i. Accordingly, we will sometimes refer to the relative position of edges to imply a relation between their relative weights.

---

[8] But, the framework is rich enough to also allow for algorithms which update I_t as they reject an element. This makes impossibility results stronger.

[9] The framework is rich enough to allow for algorithms which update I_t based on A_t, rather than just $\mathcal{M} \setminus \{A_t\}$, which again just makes impossibility results stronger.

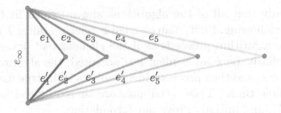

Fig. 1. A hat on seven vertices. All purple edges $(e_1, \ldots, e_5)$ are heavier than all blue edges $(e'_1, \ldots, e'_5)$, and e_∞ is significantly heavier than all other edges. Within each color, darker edges are heavier. (Color figure online)

We call the pair of edges (e_i, e'_i) the i-th *claw*. Recall that any algorithm satisfying the 3 properties listed in Sect. 3.1 has memory limited to an independent set I_t. At any time t, given the history of arrivals and the algorithm's past decisions, we can classify the claws into one of 9 kinds in $\{-, A, S\}^2$. The first character in the pair describes the state of the top edge e_i, and the second character describes the state of the bottom edge e'_i. S refers to an edge that is in I_t and arrived in the sampling stage. A refers to an edge that has been accepted by the algorithm (and is therefore in I_t). $-$ refers to any edge that is not in I_t. For example, if the i-th claw is of type $(S-)$ at some time t, it means that $t(e_i) < T$, $e_i \in I_t$, and $e'_i \notin I_t$. Figure 2 illustrates these claws.

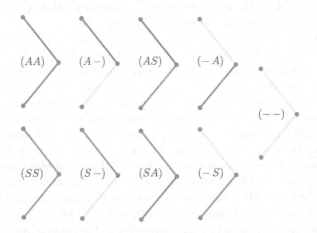

Fig. 2. All possible kinds of claws at any time t. S refers to sample edges in I_t (drawn in orange), A refers to an accepted edge in I_t (drawn in green), and $-$ refers to any other edge (drawn in gray). (Color figure online)

We next state a few lemmas about different classes of claws and their implications about the performance of the algorithm. Since the infinity edge weighs significantly more than other edges combined, we say the algorithm "loses" (i.e., fails to have a constant utility-competitive ratio) if it fails to accept the infinity edge. Conversely, the algorithm "wins" if it accepts the infinity edge. Our first observation characterizes the exact scenarios in which the algorithm loses. All missing proofs can be found in the full version.

Observation 3 (Loss condition). *The algorithm loses iff there is an (AA) claw before t_∞.*[10]

The next lemma specifies the unique blocking structure that would prevent the loss-inducing (AA) claws from forming. Our analysis focuses on the case of a $(-A)$ claw becoming a (AA) claw, as these events are significantly more likely than a $(A-)$ claw turning into an (AA) claw, and suffice for our analysis.

Lemma 1 (Blockers and Protection). *Suppose there is no (AA) claw yet. Consider a $(-A)$ claw whose upper edge is about to arrive. The upper edge is accepted iff there is no (SA) claw to its left. For this reason, we will refer to (SA) as the* blocker. *We say that the algorithm is* protected *at time t if there is a blocker in I_t.*

Importantly, note that there can be *at most one blocker in I_t*, as two blockers form a cycle. So we can unambiguously refer to *the* blocker at any time t. A blocker's effectiveness is a function of its location: Blockers far to the left "protect" more claws and are therefore more effective.

With this language in mind, we can reframe the algorithm's objective, while working within the Greedy framework. The algorithm loses whenever the upper edge of a $(-A)$ claw arrives without a blocker to its left. So the algorithm would like to maintain a blocker in I_t as far to the left as possible.[11] So the remainder of this section studies decisions the algorithm can make (again, within the Greedy framework) to include blockers far to the left. Lemma 2, however, establishes that we cannot create a new blocker without destroying our old one first (thereby going "unprotected" for some period).

Lemma 2. *If the lower edge of an $(S-)$ arrives at time t and I_t has a blocker, this edge will not be accepted.*

Lemma 2 means that the algorithm faces a tradeoff. If I_t has a blocker, it is safe from accepting the upper edge of a $(-A)$ claw *to its right* at time t. But, the algorithm *cannot move its blocker to the left, even if the lower edge of an $(S-)$ arrives during this interval*. Alternatively, the algorithm may not have a blocker during I_t. In that case, the algorithm can possibly accept a good blocker, if one happens to arrive at time t. But, the algorithm is at risk of accepting the upper edge of a $(-A)$ claw that arrives at time t *no matter its location*, because I_t has no blockers at all.

---

[10] Babaioff et al. [4] used the same graph as a counterexample to a special case of our greedy algorithm, also relying on this observation. Our lemmas are otherwise new, and necessary since we rule out a much larger class of greedy-like algorithms.

[11] Note that an arbitrary algorithm can simply decide to violate the properties defining Greedy. Our goal is to analyze Greedy algorithms, which must fit this framework.

3.3   Main Result: Ruling out all Greedy Algorithms

Armed with a better understanding of some properties of the hat structure, we are ready to prove Theorem 2, which states that greedy algorithms fail to be α-utility-competitive for any constant α.

We give a detailed proof sketch below, and defer calculations to the full version. We first repeat the main intuition: The algorithm's goal is to not accept any (AA) claw before t_∞ (Observation 3). To do so, the algorithm *must* make sure I_t includes a blocker to the left of every $(-A)$ whose upper edge arrives at time $t < t_\infty$ (Lemma 1). We can order potential blockers $(S-)$ by the arrival times of their lower edges, each of which is uniformly distributed in $[T, 1]$. Therefore, it is unlikely that a blocker far to the left arrives very early.

The algorithm can try to start with a mediocre blocker and improve it over time by accepting blockers further to the left as they arrive. The caveat is that due to Lemma 2, *blocker improvements are only possible in unprotected periods*, during which *any* arriving upper edge of $(-A)$ claws is accepted. Therefore, the algorithm faces a trade-off: Forming a more effective blocker costs more unprotected time. Importantly, the algorithm does not know whether the next arriving edge will be part of a potential blocker, or part of an $(-A)$.

In order to show that the algorithm fails, we show that with high probability there will be a (AA) claw before the arrival of the infinity edge. Specifically, we show that with high probability, an $(-A)$ claw becomes (AA) in an interval of length $\ell = n^{-0.1}$ after T, which is with high probability before the arrival of the infinity edge.

We now get into details of our proof approach. We first choose a parameter $x \in [n]$ (thinking of the claws as labeled 1 through n from left to right). We will undercount the algorithm's failure, noting that it fails whenever any of the following happens:

- The upper edge of some $(-A)$ to the left of x arrives during $[T, T + \ell]$, and I_t does not include any blocker to the left of x for any $t \in [T, T + \ell]$.
- The upper edge of some $(-A)$ arrives at an unprotected $t \in [T, T + \ell]$.

In other words, we are zeroing in on two potential sources of failure: the upper edge of *any* $(-A)$ claw could arrive during an unprotected time, or the upper edge of an $(-A)$ claw to the left of x could arrive before the algorithm accepts a blocker to the left of x. Note that these are very narrow possibilities for failure, but they suffice for our analysis.

So there are three probabilities to analyze. The first part of the first bullet is independent of the algorithm,[12] and simply considers the probability that the upper edge of a $(-A)$ to the left of x arrives during $[T, T + \ell]$.

Lemma 3. *With probability at least* $1 - 2^{-\ell^2 x/2}$, *the upper edge of a* $(-A)$ *claw to the left of* x *arrives between* T *and* $T + \ell$.

---

[12] Recall that the first edge of a $(--)$ claw to arrive must always be accepted since I_t must span V_t.

The next two probabilities are significantly more involved, as they consider decisions made by the algorithm. Note that the algorithm can decide *adaptively* when to go unprotected, based on the current ratio of $(-A)$s (potential (AA)s) versus $(S-)$s (potential blockers) to the left of x. To this end, we will let the algorithm adaptively choose any (measurable) subset of $[T, T+\ell]$ to go unprotected, and let y denote the total measure of this interval.[13] y captures the aforementioned tradeoff: small y means that the algorithm is likely to fail bullet one, while large y means the algorithm is likely to fail bullet two. Lemma 4 quantifies the cost of keeping y small, lowerbounding the probability of the second part of the first bullet.

Lemma 4. *Conditioned on the upper edge of a $(-A)$ claw to the left of x arriving between T and $T+\ell$ (i.e. Lemma 3 happening), any greedy algorithm which goes unprotected for a total measure of y during $[T, T+\ell]$ fails to accept a blocker to the left of x with probability at least:*

$$(1 - -2x\ell e^{-\frac{2x}{3}})(1-y)^{4x}.$$

Finally, we analyze the second bullet, lower bounding the probability that the upper edge of a $(-A)$ claw (anywhere) arrives during a period when the algorithm is unprotected (while the precise form is complicated, recall the intuition that as y gets larger, the probability of this particular bad event goes up, and y is at most ℓ):

Lemma 5. *Any greedy algorithm which goes unprotected for a total measure of $y \geq n^{-0.4}/2$ during $[T, T+\ell]$ has the upper edge of a $(-A)$ claw arrive during an unprotected t with probability at least:*

$$1 - \left(1 - \frac{2y - n^{-0.4}}{2\ell}\right)^{\frac{n^{0.6}(4\ell - n^{-0.4})}{32\ell^2}}.$$

Finally, we just need to combine the three bounds in Lemmas 3, 4, 5. We will choose a value of ℓ and x for the analysis, and then the algorithm (knowing x) can adaptively allocate the unprotected intervals within $[T, T+\ell]$ for a total measure of y. More formally, we let $f(y) = (1 - 2x\ell e^{-\frac{2x}{3}})(1-y)^{4x}\left(1 - (\frac{1}{2})^{\ell^2 x/2}\right)$ denote the lowerbound on failure probability derived in Lemma 4. Furthermore, we let

$$g(y) = \begin{cases} 1 - \left(1 - \frac{2y - n^{-0.4}}{2\ell}\right)^{\frac{n^{0.6}(4\ell - n^{-0.4})}{32\ell^2}}, & y \geq \frac{n^{0.4}}{2}; \\ 0, & y < \frac{n^{0.4}}{2}. \end{cases}$$

The first case follows from Lemma 5, and setting g to 0 elsewhere only strengthens our lower bound. Overall, the algorithm fails with probability at least $\min_y \{\max\{f(y), g(y)\}\}$. The next lemma sets parameters to lower bound this expression.

---

[13] The algorithm does not need to commit to the value of y in advance or choose it deterministically.

Lemma 6. *When $x = n^{0.3}$ and $\ell = n^{-0.1}$, we have*

$$\lim_{n \to \infty} \min_{y \in [0,\ell]} \{f(y), g(y)\} = 1.$$

The proof of Theorem 2 now follows from the four lemmas of this section.

4   Randomized Partition Algorithms

This section is devoted to a class of algorithms based on partition matroids. These are generalizations of an algorithm by Korula and Pal [19] for the secretary problem on graphic matroids. We show that this algorithm and natural generalizations of it fail to provide good probability-competitive performance.

4.1   Defining Randomized Partition Algorithms

The algorithm by Korula and Pal [19] was phrased in the language of graphs. Let us try to generalize it in a language applicable to all matroids. Before seeing any weights, their algorithm restricts itself (potentially randomly) to accepting only a subset of independent sets. More specifically, the algorithm will restrict its attention to the disjoint union[14] of solutions to simpler subproblems. The algorithm must ensure that for all feasible solutions to the subproblems, their union is a feasible solution to the main problem. In the case of the Korula-Pal algorithm, the smaller subproblems are instances of 1-uniform matroid secretary problems. (Several other algorithms for the Matroid Secretary Problem use similar high-level techniques, where the "simpler" matroids are not 1-uniform [9,15,20,24], and this idea is also used for the related prophet inequality [10].)

More concretely, we say that a partition is *valid* if the union of what is accepted by the instances of Dynkin's algorithm is an independent set (regardless of the weights and order of arrivals). Now we consider the following class of algorithms based on partition matroids:

Algorithm 2. Randomized Partition

1. Before looking at any weights, (perhaps randomly) validly partition the elements into parts S_i.
2. Within each part, run Dynkin's algorithm, and output the union of the selected elements.

One can ask whether any algorithm in this framework can be constant-*probability*-competitive. Theorem 4 shows that the answer is 'no'.

---
[14] This disjointness is why we refer to these generalizations as algorithms based on "partition matroids."

4.2   Randomized Partitions

In this section, we will rule out all algorithms based on partition matroids as candidates for achieving a constant probability-competitive ratio for the matroid secretary problem.

For the algorithm to always output a feasible solution, any partition it uses must be valid. Recall that a *valid* partition is one for which the union of what is accepted by the instances of Dynkin's algorithm is always independent. We say a distribution over partitions is *valid* if every partition in its support is valid.

Without loss of generality, we can assume the input graph is always complete. Otherwise, one can consider a modified weight-function that assigns a weight of zero to every edge that is not present. Since the algorithm cannot see the weights of the edges in advance, it will have to choose a partition of the complete graph at the start.

Theorem 4. *Any algorithm that draws a partition from a valid distribution $\mathcal{D}$ in Algorithm 2 is not α-probability-competitive for any $\alpha = \omega(n^{-1/8})$.*

The high-level plan in the proof of Theorem 4 is to plant a random *broom*, illustrated in Fig. 3, and show that with high probability, its handle is not accepted. We will refer to the lone neutral edge $\{u, w\}$ connecting the two stars as the *handle* of the broom. Note that the edges of non-zero weight in the broom form an acyclic subgraph and are therefore the unique max-weight basis of this graphic matroid.

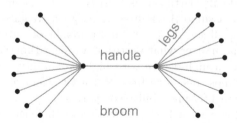

Fig. 3. Two stars connected by an edge form a broom. We call the bridge between the two stars the handle of the broom, and we the other edges of the broom as its legs.

Before proving this theorem, we characterize valid partitions.

Characterizing Valid Partitions. In this section we give a few characterizations of what valid partitions look like, which serve to provide intuition into why validity is a strong enough condition that prevents partition-based algorithms from probability-competitiveness.

We define a *valid* partition to be one where the union of what is accepted by the instances of Dynkin's algorithm is always an independent set, even for adversarial weights and arrival orders. We first give several equivalent descriptions of what valid partitions should look like in the case of graphic matroids, which provides certain structural properties enforced by validity. It will be later used to prove our Theorem 4.

Lemma 7. *Let* $\{S_1, \ldots, S_k\}$ *partition the edges of a complete graph* K_n, *and let* part(e) *denote the* S_i *containing edge* e. *The following are equivalent:*

(a) **Matroid condition:** $\{S_1, \ldots, S_k\}$ *is valid.*
(b) **Graph condition (i):** *Every cycle has at least two edges in the same part.*
(c) **Graph condition (ii):** *Every triangle has at least two edges in the same part.*

Proof of Theorem 4. We provide a counterexample in the case of graphic matroids using the broom. Consider a partition $\mathcal{S} = \{S_1, \ldots, S_k\}$ of the edges of the complete graph. We say an edge $e \in S_i$ is "high-degree" if the sum of the degrees of its endpoints within the same part S_i is large. More concretely, we define the part-i degree of a vertex v as $\deg_i(v) = |\{e = \{a, b\} \in S_i : v \in \{a, b\}\}|$. Given an edge $e = \{a, b\}$ in part S_i, its degree is given by $\deg(e) = \deg_i(a) + \deg_i(b) - 1$, which intuitively means that we are counting all the incident edges in that part and the edge itself. An edge e is said to be *high-degree* if $\deg(e) \geq C$ for some C that we will choose later.

We will show that a $1 - o(1)$ fraction of the edges are high-degree for super-constant C. Therefore, an adversary can plant a random broom by assigning weights according to the following distribution: Pick a random edge $\{u, v\}$ in the graph, and randomly partition the vertices $V \setminus \{u, v\}$ into two parts X and Y of equal size (we assume $|V|$ is even). Assign a weight of 1 to every edge $\{u, x\}$ and $\{v, y\}$ for all $x \in X, y \in Y$, and a weight of zero to everything else. We will show that no matter what partition an algorithm chooses, the random edge $\{u, v\}$ will have a high-degree with high probability. The algorithm must therefore choose at most one edge from at least C elements of OPT. Hence, it cannot be better than $1/C$-probability-competitive.

It remains to show that a $1 - o(1)$ fraction of the edges are high-degree for some super-constant C in any valid partition $\mathcal{S}$ of the edges of the complete graph. A partition of the edges of K_n can be thought of as a coloring of its adjacency matrix $A \in M^{n \times n}$ (ignoring diagonal entries) in the obvious way (i.e., assign a different color to each S_i, and the color part(e) to the entry of A corresponding to e). In this notation, an entry of A is low-degree if there are fewer than C entries of the same color in its row or column. Note that by Lemma 7, a partition is valid iff every triangle has at least two edges in the same part. In the matrix language, a partition is valid iff for every three row indices u, v, w, at least two of $A(u, v)$, $A(u, w)$ and $A(v, w)$ are the same color. We will show using this interpretation of feasibility that each row and column must mostly consist of high-degree entries. More specifically, we will fix a vertex v, and consider any other two vertices u and w.

Proposition 1. *Let* $C \leq (n-1)/2$ *and let* $T(n)$ *be the maximum possible number of low-degree edges in any valid coloring of the complete graph on n vertices. For*

any vertex v, *let* x_i *denote the number of edges adjacent to* v *in partition* i.
$T(n) \leq$

$$\max_{x \in \mathbb{N}_{\geq 0}^{n-1}, \sum_i x_i = n,} \min \left\{ \sum_i T(x_i) + 2C(n-1), \; T(n-1) + 2(n-1 - \max_i\{x_i\}) \right\}.$$

Proof. There are two steps: for any x, we show that both the left term and the right term are always upper bounds (and therefore their minimum is a valid upper bound too).

Intuitively, the left term is better when $\max_i\{x_i\}$ is not too large. To see that the left term is always an upper bound, consider the following cases. Below, let X_i denote the set of nodes z such that (z,v) is in partition i (and therefore $x_i := |X_i|$).

- First, consider each X_i, and consider the induced subgraph on just these x_i nodes. The number of low-degree edges *just counting those between two nodes in* X_i is at most $T(x_i)$, by definition of $T(\cdot)$. Clearly, a node must be low-degree in the induced subgraph to possibly be low-degree in the full graph. This means there are at most $\sum_i T(x_i)$ low-degree edges between two nodes in the same X_i.
- Next, consider an edge between two nodes x, y both $\neq v$ which are *not* in the same X_i. This means that the edges (v,x) and (v,y) are *not* colored the same, and therefore the edge (x,y) *must* share a color with one of them for A to be valid. Whichever edge shares its color, we will charge its non-v endpoint (e.g. if (x,y) shares a color with (v,x), we charge x). Observe that once a vertex is charged C times, this means there are $C+1$ edges adjacent to it which share the color of (v,x). This means that none of these edges are low-degree. Therefore, an edge can be low-degree only if its non-v endpoint is charged at most C times, and therefore there can be at most $C(n-1)$ such low-degree edges.
- Finally, consider an edge adjacent to v. We will lazily upper bound the number of low-degree edges by just the total number of edges, $n-1$, and further upper bound it by $C(n-1)$ for cleanliness of the expression.

This establishes the left term, which holds for any x. Now we establish the right term. Intuitively, the right term is a better bound whenever $\max_i\{x_i\}$ is large. Let $j := \arg\max_i\{x_i\}$. If $x_i > C$, then there can be no low-degree edges adjacent to v in X_1. Therefore, there are at most $(n-1-x_j)$ low-degree edges adjacent to v. On the subgraph induced by the $n-1$ nodes other than v, there are clearly at most $T(n-1)$ low-degree edges by definition of $T(\cdot)$, and again any edge which is low-degree in the full graph must be low-degree in every induced subgraph. On the other hand, if $x_j \leq C$, then perhaps all edges adjacent to v are low-degree, and we can only use this technique to give an upper bound of $T(n-1) + n - 1$. In both cases, our bound is at most $T(n-1) + 2(n-1 - \max_i\{x_i\})$ as long as $C \leq (n-1)/2$.

We will show inductively in Lemma 8 that $T(n) \leq b \cdot C \cdot n^{1+a}$, where a is a constant, and b and C are super-constant in n, as long as a few conditions hold. Corollary 1 lists values that satisfy these conditions, concluding that for all $0 < \varepsilon < 1/2$, there are valid assignments to the variables that achieve $T(n) \leq n^{3/2+\varepsilon}$.[15] Furthermore, Corollary 1 ensures that C is super-constant (and in particular polynomial in n), implying that with probability at least $\frac{\binom{n}{2} - n^{3/2+\varepsilon}}{\binom{n}{2}}$, the handle of the randomly planted broom will be high-degree for super-constant C. It can therefore only be selected with a sub-constant probability.

Lemma 8. *Consider the following recurrence when $C \leq (n-1)/2$. $T(n) \leq$*

$$\max_{x \in \mathbb{N}_{\geq 0}^{n-1}, \sum_i x_i = n,} \min\left\{ \sum_i T(x_i) + 2C(n-1), \ T(n-1) + 2(n-1-\max_i\{x_i\}) \right\}.$$

with a base case of $T(n) = n(n-1)/2$ when $(n-1)/2 < C$. For all N, $T(N) \leq b \cdot C \cdot N^{1+a}$, as long as

1. *$a \in (0,1)$ is a constant;*
2. *C is a super-constant function of N;*
3. *b is a super-constant function of N such that $b(N) \geq 1$ for all N;*
4. *for all $n < N$, the following is satisfied: $\frac{2(n-1)}{abn^{1+a}} < \frac{(1+a)bC}{2n^{1-a}}$.*

As an immediate corollary, we get the following.

Corollary 1. *Let $T(n)$ be defined as in Lemma 8. Then for all $0 < \varepsilon < 1/2$, $T(n) \leq n^{3/2+\varepsilon}$.*

Now we can complete the proof of Theorem 4. Corollary 1 together with Proposition 1 establishes that for any $\varepsilon > 0$, there are at most $n^{3/2+\varepsilon}$ edges with degree at most $C := n^{\varepsilon/3}$. This means that with probability $1 - n^{-1/2+\varepsilon}$, a randomly selected edge (u,v) of the complete graph has degree at least $n^{\varepsilon/3}$. Conditioned on (u,v) having high-degree, we know that $n^{\varepsilon/3}$ edges of the max-weight spanning tree are in the same partition as (u,v). Therefore, at least one of them is selected with probability at most $n^{-\varepsilon/3}$. Setting $\varepsilon = 3/8$, we conclude that except with probability $n^{-1/8}$, there is some edge selected with probability at most $n^{-1/8}$, and therefore no randomized partition algorithm can be $\omega(n^{-1/8})$-probability-competitive.

References

1. Azar, P.D., Kleinberg, R., Weinberg, S.M.: Prophet inequalities with limited information. In: Chekuri, C. (ed.) Proceedings of SODA, pp. 1358–1377. SIAM (2014)
2. Babaioff, M., Immorlica, N., Kempe, D., Kleinberg, R.: A knapsack secretary problem with applications. In: Charikar, M., Jansen, K., Reingold, O., Rolim, J.D.P. (eds.) APPROX/RANDOM -2007. LNCS, vol. 4627, pp. 16–28. Springer, Heidelberg (2007). https://doi.org/10.1007/978-3-540-74208-1_2

---

[15] It can be shown that this is in fact tight.

3. Babaioff, M., Immorlica, N., Kempe, D., Kleinberg, R.: Online auctions and generalized secretary problems. SIGecom Exch. **7**(2), 1–11 (2008)
4. Babaioff, M., Immorlica, N., Kempe, D., Kleinberg, R.: Matroid secretary problems. J. ACM **65**(6), 35:1–35:26 (2018)
5. Babaioff, M., Immorlica, N., Kleinberg, R.: Matroids, secretary problems, and online mechanisms. In: Proceedings of the Eighteenth Annual ACM-SIAM Symposium on Discrete Algorithms, pp. 434–443 (2007)
6. Dinitz, M.: Recent advances on the matroid secretary problem. ACM SIGACT News **44**(2), 126–142 (2013)
7. Dynkin, E.B.: The optimum choice of the instant for stopping a Markov process. Sov. Math. **4**, 627–629 (1963)
8. Edmonds, J.: Matroids and the greedy algorithm. Math. Program. **1**(1), 127–136 (1971)
9. Feldman, M., Svensson, O., Zenklusen, R.: A simple O (log log (rank))-competitive algorithm for the matroid secretary problem. In: Proceedings of SODA, pp. 1189–1201 (2015)
10. Feldman, M., Svensson, O., Zenklusen, R.: Online contention resolution schemes. In: Proceedings of SODA, pp. 1014–1033 (2016)
11. Gale, D.: Optimal assignments in an ordered set: an application of matroid theory. J. Comb. Theory **4**(2), 176–180 (1968)
12. Gupta, A., Singla, S.: Random-order models. In: Roughgarden, T. (ed.) Beyond the Worst-Case Analysis of Algorithms. Cambridge University Press (2020)
13. Hajiaghayi, M.T., Kleinberg, R.D., Parkes, D.C.: Adaptive limited-supply online auctions. In: Proceedings 5th ACM Conference on Electronic Commerce (EC-2004), 17–20 May 2004, pp. 71–80. ACM, New York (2004)
14. Hoefer, M., Kodric, B.: Combinatorial secretary problems with ordinal information. In: Proceedings of ICALP, pp. 133:1–133:14 (2017)
15. Huynh, T., Nelson, P.: The matroid secretary problem for minor-closed classes and random matroids. arXiv preprint arXiv:1603.06822 (2016)
16. Jaillet, P., Soto, J.A., Zenklusen, R.: Advances on matroid secretary problems: free order model and laminar case. In: Goemans, M., Correa, J. (eds.) IPCO 2013. LNCS, vol. 7801, pp. 254–265. Springer, Heidelberg (2013). https://doi.org/10.1007/978-3-642-36694-9_22
17. Kleinberg, R., Weinberg, S.M.: Matroid prophet inequalities. In: Proceedings of the Forty-Fourth Annual ACM Symposium on Theory of Computing, pp. 123–136. ACM (2012)
18. Korte, B., Vygen, J.: Combinatorial Optimization, Volume 21 of Algorithms and Combinatorics. Springer, Heidelberg (2008). https://doi.org/10.1007/978-3-662-56039-6
19. Korula, N., Pál, M.: Algorithms for secretary problems on graphs and hypergraphs. In: Albers, S., Marchetti-Spaccamela, A., Matias, Y., Nikoletseas, S., Thomas, W. (eds.) ICALP 2009. LNCS, vol. 5556, pp. 508–520. Springer, Heidelberg (2009). https://doi.org/10.1007/978-3-642-02930-1_42
20. Lachish, O.: O (log log rank) competitive ratio for the matroid secretary problem. In: 2014 IEEE 55th Annual Symposium on Foundations of Computer Science, pp. 326–335. IEEE (2014)
21. Oxley, J.G.: Matroid Theory, vol. 3. Oxford University Press, New York (2006)
22. Rado, R.: Note on independence functions. Proc. London Math. Soc. **3**(1), 300–320 (1957)
23. Schrijver, A.: Combinatorial Optimization: Polyhedra and Efficiency, vol. 24. Springer, Heidelberg (2003)

298   M. Bahrani et al.

24. Soto, J.A.: A simple PTAS for weighted matroid matching on strongly base orderable matroids. Electron. Notes Discrete Math. **37**, 75–80 (2011)
25. Soto, J.A., Turkieltaub, A., Verdugo, V.: Strong algorithms for the ordinal matroid secretary problem. In: Proceedings of SODA, pp. 715–734 (2018)

Threshold Tests as Quality Signals: Optimal Strategies, Equilibria, and Price of Anarchy

Siddhartha Banerjee[1], David Kempe[2($\boxtimes$)], and Robert Kleinberg[1]

[1] Cornell University, Ithaca, NY 14853, USA
sbanerjee@cornell.edu, rdk@cs.cornell.edu
[2] University of Southern California, Los Angeles, CA 90089, USA
David.M.Kempe@gmail.com

Abstract. We study a signaling game between two firms competing to have their product chosen by a principal. The products have (real-valued) qualities, which are drawn i.i.d. from a common prior. The principal aims to choose the better of the two products, but the quality of a product can only be estimated via a coarse-grained *threshold test*: given a threshold θ, the principal learns whether a product's quality exceeds θ or fails to do so.

We study this selection problem under two types of interactions. In the first, the principal does the testing herself, and can choose tests optimally from a class of allowable tests. We show that the optimum strategy for the principal is to administer *different* tests to the two products: one which is passed with probability $\frac{1}{3}$ and the other with probability $\frac{2}{3}$. If, however, the principal is required to choose the tests in a symmetric manner (i.e., via an i.i.d. distribution), then the optimal strategy is to choose tests whose probability of passing is drawn uniformly from $[\frac{1}{4}, \frac{3}{4}]$.

In our second interaction model, test difficulties are selected endogenously by the two firms. This corresponds to a setting in which the firms must commit to their testing (quality control) procedures before knowing the quality of their products. This interaction model naturally gives rise to a *signaling game* with two senders and one receiver. We characterize the unique Bayes-Nash Equilibrium of this game, which happens to be symmetric. We then calculate its Price of Anarchy in terms of the principal's probability of choosing the worse product. Finally, we show that by restricting both firms' set of available thresholds to choose from, the principal can lower the Price of Anarchy of the resulting equilibrium; however, there is a limit, in that for every (common) restricted set of tests, the equilibrium failure probability is strictly larger than under the optimal i.i.d. distribution.

Keywords: Signaling · Information elicitation · Price of anarchy

S. Banerjee—Supported by the NSF under grants CNS-1955997, DMS-1839346 and ECCS-1847393.
R. Kleinberg—Supported by the NSF under grant CCF-1512964.

M. Feldman et al. (Eds.): WINE 2021, LNCS 13112, pp. 299–316, 2022.
https://doi.org/10.1007/978-3-030-94676-0_17

1   Introduction

A principal wants to choose between two firms producing interchangeable products, whose qualities are drawn i.i.d. from a known prior. The principal wants to pick the product of higher quality — however, she cannot directly observe the products' qualities. In order to learn more about the products' qualities, the principal can simultaneously subject the products to *tests*. Specifically, we consider the simplest and most coarse-grained tests: binary (i.e., pass/fail) threshold tests that reveal whether the product's quality lies above or below a chosen θ. How should the principal choose the tests to administer to the two products, so as to help her maximize the probability of picking the better of the two? We refer to this as the *optimal selection* problem.

Now consider an alternative setting in which firms conduct their own quality control in-house, according to a fully disclosed and verifiable procedure. This may be necessary if the principal does not possess the expertise to conduct quality control herself. In this setting, while the principal may not be able to conduct a test, we assume that she can verify that a firm correctly followed its disclosed testing protocol; in other words, we assume that firms inherently have the power to *commit* to a test. At the time a firm commits to a testing protocol, it will not know the exact quality of each individual product — for example, due to variations across batches and over time, or because the firm acts as an intermediary (e.g. head hunters who vet candidates for a hiring firm). Indeed, such variation is the reason testing is needed in the first place. As before, we assume that firms have independent common priors for their product qualities. How will firms choose tests in such an *endogenous selection* setting, if each firm wants to maximize the probability of its own product being selected? Will competition push the firms to subject themselves to very difficult tests, or will they coordinate on easy tests at equilibrium? How much worse off is the principal due to having to outsource quality control tests, rather than conducting them herself? Can she improve her probability of choosing the better product by restricting the set of tests from which the firms can choose, e.g., by prescribing standards that such tests must adhere to?

Endogenous test selection by two firms can be naturally viewed as a form of signaling; committing to a testing procedure takes the role of committing to a signaling scheme.[1] Thus, our work can be construed as a natural *game* played between two agents whose strategies are signaling schemes from a restricted class of available schemes. This parallels several recent works on Bayesian persuasion games between multiple firms vying for customers [1,2,4,5,12]; we discuss these in detail in the full version [3]. Our high-level question is what the equilibria of such signaling games look like, and how much efficiency is lost (if any) by letting

---

[1] This is the more common view of signaling in the economics community: a signaling scheme is interpreted as a device (physical or otherwise) that maps relevant states of the world to observable signals. Fixing a device constitutes committing to a signaling scheme. In contrast, recent works in computer science apply signaling/persuasion to scenarios such as communications where it is less clear whether the sender has the ability to commit to a mapping.

the agents/firms choose their own signaling scheme rather than the principal being able to control how she receives information about the state of the world.

We investigate such questions using the following simple model (see Sect. 2). The two firms have products with real-valued *qualities* X, Y drawn randomly from a common prior with continuous cdf Ψ. The principal has at her disposal a collection of tests parametrized by a threshold $\theta \in \mathbb{R}$ which encodes the difficulty level of each test. When a firm's product with quality X is subjected to a test with threshold θ, the outcome reliably reveals whether $X \geq \theta$ (the product passes the test) or $X < \theta$ (the product fails the test). In the language of signaling, this means that we restrict to signaling schemes with binary outcomes, in which the sets mapped to each outcome are intervals.

Based on the chosen test difficulties (which are observable in both optimal and endogenous selection regimes) and their outcomes, the principal selects one of the products. Her objective is to minimize the probability of choosing the worse product, while each firm's objective is to maximize the probability of having its product chosen. We consider the following models, which endow the principal with varying degrees of control:

1. The principal must give both firms the same test.
2. The principal has full control over the difficulties θ_X, θ_Y of the tests given to the two firms.
3. The principal specifies a distribution from which both firms draw tests in an i.i.d. manner. The restriction to *identical* distributions may be required to achieve ex-post fairness, compared to, for instance, randomizing which of the two firms gets which of two non-identical tests.
4. The firms endogenously choose their own tests via equilibrium strategies.
5. The principal can restrict available tests to a set S (common to both firms), and firms endogenously choose their tests from S. Such a restriction could arise if the principal is a government agency or sufficiently powerful firm providing binding quality control guidelines.

It is clear — simply from suitable subset relationships on sets of available actions — that in terms of the principal's error probability, $\{1, 4\} \geq 5 \geq 3 \geq 2$. Our goal is to explicitly characterize the optimal or equilibrium outcomes under these five models, thereby inferring which of the preceding comparisons are strict, as well as to quantify the increase in error probability for the principal resulting from a move to a weaker model. When comparing a model in which the principal has control with one in which the agents are allowed to choose tests according to an equilibrium strategy, this ratio exactly corresponds to the Price of Anarchy.

1.1 Other Applications and Model Discussion

While we phrase our work in terms of two firms offering products, our model applies more broadly. In particular, it can be viewed as a generalization of the classic "forum shopping" model of Lerner and Tirole [13] to multiple firms (*property owners*, in their language).[2] In this model, firms can choose an external

---

[2] Lerner and Tirole [13] do briefly discuss a multi-firm setting, but only consider one extremely limited example.

certification agency to issue a recommendation on whether or not their product is "acceptable." There is a continuum of agencies, ranging from fully aligned with the firm's interests to fully independent. Under suitable parameters, this model precisely corresponds to being able to choose any quantile threshold for a test. While the model does not place the tests "in house," in terms of the firms' choices, it is equivalent to our model. The focus of [13] is on the interplay of the independence/difficulty of the agency and the owner's "concessions" — direct transfers to any user of the property, such as price reductions or additional features. As they argue, such a setup not only captures agencies certifying products, but also journals/conferences reviewing papers and similar endeavors. In addition to these applications, some of the literature on multi-sender cheap talk/Bayesian persuasion is motivated in terms of competing proposals, either to a funding agency or internal within an organization; see, e.g., [4,5].

Another application, aligned with the classic work of Spence [18] and Ostrovsky and Schwarz [16], is in the assessment of students. Here, the test is a pass/fail exam (or class) via which a student is assessed. The optimization problem may guide a teacher aiming to correctly rank the students in a class, while the endogenous test selection model roughly corresponds to students choosing the difficulty of projects to undertake or of classes to enroll in.

In the context of applications, three key assumptions in our model are worth discussing. The first is that firms are unaware of their quality when choosing tests. This power of commitment *before* the state of the world is revealed is the defining distinction between Bayesian Persuasion and Cheap Talk models, and is covered in depth in the full version [3]. As we discuss, most works on inter-firm signaling make this assumption. For example, Lerner and Tirole [13] assume that property owners do not know users' utilities for their product.[3] Similarly, Ostrovsky and Schwarz [16] consider early contracting between students and employers, in which students at the time of negotiation only have priors on their future performance. Naturally, as with all models, this assumption is a simplification, with reality lying between full and no commitment power.

The second assumption is that tests have binary and monotone (i.e., pass/fail) outcomes; in particular, we assume that no test can be passed with quality x, but failed with quality $x' > x$. Restricting to monotone information structures is quite common in the literature: for recent examples, see [9,15], and [6] and discussions therein. Other kinds of restricted signal spaces also have significant precedent in the literature. Dughmi et al. [8] analyze Bayesian Persuasion in which the sender is restricted in terms of the number of signals. Boleslavsky et al. [4] assume that the state of the world is binary (the product is good or bad) and allow each sender to only send one of two signals; nevertheless, competition between senders results in complex signal distributions at equilibrium. Similarly, the certification models of [11,13] mostly consider binary outcomes (recommend/don't recommend). As argued in [11] (see, e.g., Footnote 3 in [11] and the literature cited there), the main

---

[3] However, we note that in addressing the same real-world scenario, Gill and Sgroi [11] instead consider a model where the owner knows the state before choosing the certifier; see the full version [3] for details.

purpose of a test or evaluation is to provide a *concise* summary of the product. When the outcome of the evaluation must be concise, the number of possible signals that can be sent is necessarily bounded, and a binary signal is a clean and idealized way to capture such a desideratum. Monotonicity is natural to assume when signals should be interpretable by a decision maker. This justification is also borne out by the coarse-grained grading systems (pass/fail, grades A–F) typically used in education contexts. It also closely aligns with the argument made in [17] that there is a tradeoff between accuracy and complexity of advice (i.e., signals).

The third assumption is that there are exactly two firms (for most of our results), and that their qualities are drawn i.i.d. This assumption is very standard in the study of related questions in competitive signaling; see, e.g., the in-depth discussion of [4,5,12,14] and additional related work in the full version [3]. We discuss the difficulties with extending the result to $n > 2$ firms or non-identical priors in Sect. 7.

1.2 Our Results

As we elaborate in Sect. 2, it is equivalent — and much more convenient — to characterize tests not in terms of their thresholds, but in terms of the probability that a product will fail the test. Thus, we can view each possible test as a real number in $[0, 1]$; in this case, the products' qualities can be assumed w.l.o.g. to be drawn *uniformly* from $[0, 1]$.

When both firms' products have to be subjected to the same test, it is easy to see that the optimum test is the *median* test, passed with probability exactly $\frac{1}{2}$, which chooses the wrong product with probability $\frac{1}{4}$ (see Sect. 2). When the principal can give the firms different tests, our main result is summarized by the following theorem. (See Sects. 4 and 3 for formal statements.)

Theorem 1 (Optimal Selection of Tests by Principal: Informal).

1. *If the principal can assign arbitrary tests to the two firms, then it is optimal to give one firm a test of $\frac{1}{3}$ and the other a test of $\frac{2}{3}$. This results in a probability of $\frac{1}{6}$ of incorrect selection.*
2. *If the principal must draw i.i.d. tests for the firms, then the optimal rule draws test thresholds uniformly from the interval $[\frac{1}{4}, \frac{3}{4}]$. This results in a probability of $\frac{5}{24}$ of incorrect selection.*

The preceding theorem is rather surprising! Even though the firms' products have i.i.d. qualities, the principal can decrease her failure probability significantly (by 33%) by giving the firms very different tests. Analogously, a teacher trying to optimally rank students by ability should give the students different tests, even if their abilities share a common prior distribution.

For the case of endogenous test selection, the equilibrium and its probability of a mistake are characterized by the following result, stated formally and discussed in Sect. 5:

Theorem 2 (Equilibrium Distribution). *When firms' qualities are drawn i.i.d. uniformly from* $[0, 1]$*, and firms choose their test difficulties endogenously, there is a unique Bayes-Nash Equilibrium, which is symmetric, and consists of each firm choosing difficulty* $\theta \in [0, 1]$ *from the probability density function (pdf)* $f(\theta) = \frac{1}{2(\theta^2 + (1-\theta)^2)^{3/2}}$.

The principal's resulting probability of incorrect selection is approximately 0.23056, causing a Price of Anarchy of approximately 1.38336 compared to the optimum correlated tests and approximately 1.10653 compared to the optimum i.i.d. test distribution.

Finally, in Sect. 6, we allow the principal to set "guidelines" for the firms' quality control tests, by prescribing a set $S \subseteq [0, 1]$ from which the thresholds must be drawn.

Theorem 3 (Restricted Equilibrium Distribution). *When the firms' qualities are drawn i.i.d. uniformly from* $[0, 1]$*, and the firms choose their test difficulties endogenously from an interval* $S = [a, b] \subseteq [0, 1]$*, there is a unique Bayes-Nash Equilibrium. This unique Bayes-Nash equilibrium is symmetric and can be explicitly characterized in closed form.*

Moreover, there exist values a, b *for which the resulting probability of a mistake by the principal is strictly smaller than for the interval* $[0, 1]$*; for example, for the interval* $[0, 0.79]$*, the probability of a mistake is approximately 0.22975.*

However, even compared to a principal restricted to i.i.d. test choices, under symmetric Bayes-Nash Equilibria, the Price of Anarchy is lower-bounded by a constant strictly larger than one: for every set $S \subseteq [0, 1]$ *(not just intervals), the probability of a mistake is at least* $\frac{5}{24} + \frac{1}{82944}$.

One interesting interpretation of the preceding theorem is that a somewhat bigger part of the problem with endogenous test selection is that firms skew too much towards harder tests. Making extremely difficult tests (the top 20%) unavailable results in a (slightly) better equilibrium probability for the principal. However, as we will see in the analysis, when restricting the interval of available tests, the equilibrium distribution has non-trivial point mass at the upper end of the interval; in other words, at equilbrium, firms will still compete by choosing difficult tests.

A visual representation of our results is given in Fig. 1. Taken together, our theorems imply a strict separation of all five models of test selection, and notably show that the principal has a higher probability of incorrect selection when choosing the same test for both firms compared to when they choose tests endogenously.

Our work raises a wealth of directions for future inquiry, discussed in detail in Sect. 7. Most immediate would be extensions to more than two firms and to richer signaling schemes. For an extension to multiple firms, an important point is to decide what the principal's and the firms' objectives are. One natural generalization is to have the principal still choose one (or k) of the firms' products; this appears difficult. A "friendlier" generalization involves a principal who wants to fully rank the firms by quality (e.g., a teacher in a classroom setting), and aims

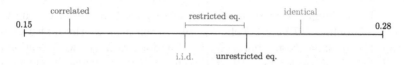

Fig. 1. The principal's failure probabilities under different models of threshold choices.

to minimize the number of inversions compared to the true order. In this setting, a firm/student may try to minimize the expected number of other firms ranked ahead of it. Because the objective functions naturally decompose into pairwise objectives by linearity of expectation, our results carry over to this setting completely. The only necessary generalization is for the case of correlated tests. In fact, in Sect. 4, we characterize the optimal choice of tests for the principal in the presence of any number of firms.

Due to space constraints, essentially all proofs, as well as a much more in-depth discussion of related work, are deferred to the full version [3].

2   Model and Preliminaries

2.1   Qualities, Tests and Selection

We consider a setting in which a principal wants to pick the better of the products provided by two firms X and Y. We will equivalently refer to this process as *selecting* or *choosing* a firm or *ranking* the firms. The two firms' products have i.i.d. *qualities* X, Y drawn from a common prior distribution with continuous cdf[4] Ψ on $\mathbb{R}$. Abusing notation, we use X, Y to refer both to the firms themselves and their products' (random) qualities.

Information about the products' qualities is revealed by means of *binary threshold tests* (henceforth simply *tests*) administered to the products. More specifically, a test is completely characterized by a threshold $\theta \in \mathbb{R}$. A product of quality X subjected to a test with threshold θ *passes* if and only if $X \geq \theta$; otherwise, we say that the product *fails* the test θ. To avoid unnecessary clutter in writing, we also refer to the *firm* X or Y as passing or failing the test (instead of its product). The larger θ, the less likely a product is to pass the test, so we can naturally think of θ as the *difficulty* of the test. When a product is subjected to a test, the outcome (pass or fail) is revealed to everyone, but no additional information can be inferred about the product. This model is mathematically equivalent to the certification model of [13].

The principal's goal is to minimize the probability of selecting the product of lower quality. We refer to this as an *incorrect selection*, or as an *error* by the principal, or — by analogy with ranking — as an *inversion*. Formally, consider

---

[4] We adopt the convention that the cumulative distribution function (cdf) of a probability measure on $\mathbb{R}$ is defined by setting $F(x)$ to be the measure of the set $(-\infty, x]$ under the distribution.

a rule $\mathcal{T}$ for assigning tests to firms and selecting a firm based on the tests' outcome. We define $I(\mathcal{T}) := \mathbb{1}[\mathcal{T}\text{chooses the wrong firm}]$ as the indicator of $\mathcal{T}$ inverting the ranking. Note that $I(\mathcal{T})$ is a random variable, with randomness arising from: (1) $\mathcal{T}$'s selection of test thresholds, (2) the firms' products' random qualities, and (3) possibly randomized aggregation of test outcomes. The principal's goal is to choose $\mathcal{T}$ so as to minimize $\mathbb{E}[I(\mathcal{T})]$.

Given a firm's test result, the principal can form a posterior belief of its product's quality. The posterior expected quality of a product passing threshold test θ is $\mathbb{E}_{X\sim\Psi}[X \mid X \geq \theta]$, while the posterior expected quality of a product failing it is $\mathbb{E}_{X\sim\Psi}[X \mid X < \theta]$. Observe that for any product quality cdf Ψ, we have that $\mathbb{E}_{X\sim\Psi}[X \mid X < \theta]$ and $\mathbb{E}_{X\sim\Psi}[X \mid X \geq \theta]$ are monotone non-decreasing in θ, and strictly increasing for θ in the support of Ψ. Furthermore

$$\mathbb{E}_{X\sim\Psi}[X \mid X < \theta] \leq \mathbb{E}_{X\sim\Psi}[X] \leq \mathbb{E}_{X\sim\Psi}[X \mid X \geq \theta],$$

and both inequalities are strict if θ is in the support of Ψ. Because both products' qualities are drawn from the same distribution, these observations imply the following proposition.

Proposition 1. *Let $\theta_X > \theta_Y$ be the thresholds of the tests to be applied to the products of firms X, Y. Assume that both θ_X, θ_Y lie in the support of Ψ.*

1. *If both firms' products pass their tests, or both fail their tests, then the principal minimizes the probability of an inversion by selecting X.*
2. *If exactly one of the products of X, Y passes its test, then the principal minimizes the probability of an inversion by selecting the firm that passed.*

Proposition 1 characterizes a rational principal's choice (once test outcomes have been revealed) almost completely. To complete the description, we assume that when there is a tie, the principal picks one of the firms uniformly at random. We will refer to this case as a coin flip, and say that X (or Y) *wins/loses the coin flip*. As an illustration, consider the following example:

Example 1 (The Median Test). Suppose that both firms' products have i.i.d. quality levels $X, Y \sim \text{Uniform}[0, 1]$ (i.e., drawn uniformly over $[0, 1]$). A natural test is the median test $\mathcal{T}_{median}$, under which both products are subjected to a test with $\theta = \frac{1}{2}$. A product's posterior expected quality upon passing is $\mathbb{E}[X \mid X \geq 1/2] = 3/4$, and upon failing $\mathbb{E}[X \mid X \leq 1/2] = 1/4$. Now w.l.o.g. suppose that the two firms' products have qualities $X < Y$. If $X < \frac{1}{2} \leq Y$, then Y passes and X fails, and the principal ranks them correctly. However, if $Y < \frac{1}{2}$, then both fail, and if $X \geq \frac{1}{2}$, then both pass. In either case, a coin flip is required, and the principal chooses correctly only with probability $\frac{1}{2}$. Thus, the median test achieves $\mathbb{E}[I(\mathcal{T}_{median})] = \frac{1}{4}$.

More generally, if the principal gives the same test θ to both agents, then an inversion happens if: (1) either both $X, Y \geq \theta$ or both $X, Y < \theta$, and (2) the coin flip determines the wrong winner. Thus, the probability of an inversion is $\frac{1}{2}(\theta^2 + (1 - \theta)^2)$. This is minimized at $\theta = \frac{1}{2}$, showing that the median test is optimal for the principal if she must give the same test to both agents.

Given complete control over the choice of testing rule $\mathcal{T}$, the principal's goal is to choose the rule that minimizes $\mathbb{E}[I(\mathcal{T})]$. This could be a single threshold for both firms (as with the median test); a distribution G over $\mathbb{R}$ such that, for each firm, the principal draws an i.i.d. threshold $\theta \sim G$; or, most generally, a joint distribution G over thresholds for both firms. The optimal i.i.d. threshold distribution and the optimal joint distribution are studied in Sects. 3 and 4.

2.2 Endogenous Test Selection and Quantile Thresholds

In many settings, firms may be better equipped than the principal to perform quality control tests in house.[5] In these cases, the firms will typically commit to a verifiable quality control procedure for their products. The principal gets to observe (only) the threshold θ and the outcome of the test. In other words, both firms commit to a signaling scheme about their products' qualities, where the space of signaling schemes is restricted to a binary signal space and threshold functions.

Each firm's goal is to maximize its probability of being selected, or — equivalently — of being ranked ahead of the other firm. Due to the competitive nature of the game, the appropriate solution concept (which we will study) is a *Bayes-Nash Equilibrium*. We refer to this setting as *endogenous test selection*. Because the firms are a priori symmetric, in any equilibrium, each firm's product must be selected with probability $\frac{1}{2}$.

In a further generalization, note that the principal may be able to rule out some types of tests. In other words, in a more general model, the principal may specify a closed set S and restrict the firms to selecting test thresholds $\theta \in S$ only. We will be primarily interested in the case when S is an interval, but also consider more general closed sets S.

Before continuing, we note that since the utilities of both the principal and the firms depend only on rankings and not actual qualities, it is convenient to work in the quantile space $[0, 1]$ rather than the quality space $\mathbb{R}$. To do this, note that for any quality $X \sim \Psi$, its corresponding (random) *quantile* $\Psi(X)$ is distributed uniformly in $[0, 1]$. Now, suppose that firm X chooses (or is assigned) a threshold $\sigma \in \mathbb{R}$ for its test; we can equivalently view this as the firm picking a *threshold quantile* $\theta = \Psi(\sigma) \in [0, 1]$. Note that a product with quality $X \sim \Psi$ passes a test with threshold quantile θ with probability $1 - \theta$; moreover, a threshold quantile $\theta \in [0, 1]$ corresponds to a threshold $\sigma = \Psi^{-1}(\theta)$ in the quality space, where $\Psi^{-1}(x) \triangleq \inf\{y \in \mathbb{R} \mid \Psi(y) \geq x\}$ is the generalized inverse function associated with the cdf Ψ. Thus, w.l.o.g., we henceforth focus on product qualities drawn from $\Psi \sim \text{Uniform}[0, 1]$, and understand "threshold" to refer to the threshold quantile $\theta \in [0, 1]$.

---

[5] Alternatively, the setting may be such that the agents naturally have the choice of test difficulty, such as in external certification of product quality [10,11,13] or students' selection of which classes to attempt [18]. In these settings, it is still frequently assumed that agents are *not* aware of their private quality value when they make their choice of difficulty; see for example [13] for a model of certification and [16] for a model of contracting between students and employers.

2.3   Extension to More Firms

While we have focused thus far on the paradigmatic case of two firms, the model can be naturally extended to $n \geq 2$ firms. Several natural generalizations suggest themselves, both in terms of the principal's objective and the firms' objective. With n firms, the principal may try to maximize the probability of choosing the best product, or try to produce a complete ranking of all firms' products, minimizing the total number of inversions.[6] For a firm, the goal might be to maximize the probability of being selected, or to be ranked as highly as possible in expectation. Our results extend naturally to the latter objectives, namely,

- The utility of a firm is proportional to the number of firms which have a lower rank.
- The disutility of the principal is proportional to the (normalized) *Kendall tau distance*[7] between the true and inferred rankings, i.e., the fraction of pairwise inversions between the two lists.

Extending our notation from the case of two firms, for a given rule T for choosing tests for firms, we denote the (random) Kendall tau distance between the resulting ranking and the correct ranking by $I(T)$. Again, the principal's goal is to minimize $\mathbb{E}[I(T)]$. Using linearity of expectations for both the firms and the principal, all of our results for two firms carry over immediately to the case of n firms, with exactly the same guarantees regarding the fraction of misranked pairs. The only exception is that for *correlated tests* (in Sect. 4), the optimal choice for the principal will depend on the number n of firms. These results do not extend to other objectives, and both optimal and equilibrium strategies will typically look different for $n \geq 3$ firms. See Sect. 7 for a discussion.

3   Optimal I.I.D. Tests

In this section, we explicitly characterize the optimal distribution from which the principal should draw thresholds if it is required that both firms' thresholds be drawn independently from the same distribution; in contrast, in the next section we consider the case of correlated thresholds.

3.1   Characterizing the Expected-Inversions Functional

Let T_G denote the test selection rule under which each firm is given a test with threshold drawn i.i.d. from G. We begin by characterizing the expected number of inversions as a functional of the cdf G from which the thresholds are drawn. In the next section, we will show how to choose G to minimize this functional. For notational convenience, we henceforth denote $I(G) = \mathbb{E}[I(T_G)]$.

---

[6] There are naturally other objectives in between these two extremes.

[7] Recall that the Kendall τ distance between two rankings is the number of inversions between the rankings, i.e., the number of pairs of elements that are in different order.

Lemma 1. *Assume that the quality distribution Ψ is uniform on $[0,1]^8$. Suppose that thresholds for both firms are drawn i.i.d. from the distribution G on $[0,1]$ (not necessarily continuous). The probability of selecting the worse product is given by the functional*

$$I(G) = \int_0^1 \int_0^x (1 - G(x) + G(y))^2 \, \mathrm{d}y \, \mathrm{d}x. \tag{1}$$

3.2 Optimizing the Objective Function

We now characterize the i.i.d. distribution H^* that minimizes $I(G)$.

Theorem 4. *Assume that the quality distribution Ψ is uniform (see Footnote 8) on $[0,1]$. Let H^* be the cdf corresponding to the uniform distribution over the interval $[\frac{1}{4}, \frac{3}{4}]$.*

The inversion probability of H^* is $\frac{5}{24}$, and this is optimal for i.i.d. distributions: for every distribution G over $[0,1]$, we have $I(G) \geq I(H^*)$.

In other words, the optimal way to pick i.i.d. tests is to sample them *uniformly from* $[\frac{1}{4}, \frac{3}{4}]$. This may seem somewhat surprising. Some intuition for this can be derived from looking at *correlated* test selection rules in the limit of infinitely many firms. (See the discussion after Theorem 5 in Sect. 4.)

4 Optimal Correlated Tests

In Sect. 3, we derived the optimal distribution to sample tests from if each firm must be assigned a test *independently* from the same distribution. Here, we consider the problem when the firms' tests can be chosen *in a correlated way*.

As we mention in Sect. 2.3, although most of our analysis looks at two firms, it extends naturally to multiple firms when the goal is to minimize the expected number of inversions. When the test assignments can be correlated, the actual number of firms affects the optimal solution. Hence, in this section, we explicitly characterize the optimal choices when there are n firms. Surprisingly, this takes the following simple form:

Theorem 5. *Assume that the quality distribution Ψ is uniform (see Footnote 8) on $[0,1]$. Recall that $I(\mathcal{T})$ denotes the (random) Kendall tau distance between the true and inferred rankings. For n firms, the expected fraction of inversions $\mathbb{E}[I(\mathcal{T})]$ is minimized over all correlated test selection rules $\mathcal{T}$ by one which assigns the test with threshold $\theta_i = \frac{n+2(i-1)}{4n-2}$ to firm i. The resulting expected fraction of inverted pairs of firms is $\frac{5n-4}{12(2n-1)}$.*

To get intuition for this result, it is instructive to consider it for $n = 2$. In this case, the optimal $\mathcal{T}$ allocates two tests at thresholds $\frac{1}{3}$ and $\frac{2}{3}$, respectively,

---

[8] Recall from Sect. 2.2 that this assumption is without loss of generality.

and this improves the fraction of misclassified pairs from $\frac{5}{24}$ to $\frac{1}{6}$. The main reason behind this improvement is that the ability to give different tests to the two firms allows the principal to choose tests to maximally split up the space $[0, 1]$, such that the only way the principal makes a mistake is if the products' qualities X, Y are in the same interval.

Theorem 5 is also instructive in the limit as $n \to \infty$. Here, one sees that the optimal test distribution converges to uniformly spaced tests over the interval $[\frac{1}{4}, \frac{3}{4}]$ (and leads to a $\frac{5}{24}$ fraction of pairs being inverted). This suggests that a uniform distribution of tests over $[\frac{1}{4}, \frac{3}{4}]$ should be the optimal distribution for i.i.d. tests for any number of firms, since drawing n tests from a continuous distribution results in all n tests being unique almost surely, and close to the optimal correlated tests. This intuition is indeed confirmed by the earlier Theorem 4.

5   Endogenous Test Selection and Price of Anarchy

In this and the next section, we turn to the question of endogenous test selection. Here, we consider the setting where the principal makes all threshold tests in $[0, 1]$ available to the firms for selection; in the next section, we consider the benefits of being able to restrict the set of offered tests.

The equilibrium concept we study for endogenous test selection is Bayes-Nash Equilibria. A pair of distributions (F_X, F_Y) supported on (a subset of) $[0, 1]$ constitutes a Bayes-Nash Equilibrium of the *endogenous test selection game* if, given that X chooses a random test from $[0, 1]$ according to F_X, choosing a test from F_Y is a *best response* for firm Y (i.e., in the set of strategies that maximize Y's selection probability), and similarly with the roles of X and Y reversed. The case when F_X and F_Y are identical is referred to as a symmetric Bayes-Nash Equilibrium. As discussed in Sect. 2.2, even though we focus on quality distributions being Uniform$[0, 1]$, the results extend naturally to any distribution Ψ which is absolutely continuous.

Theorem 6. *There is a unique Bayes-Nash Equilibrium of the endogenous test selection game when firms have access to all tests in $[0, 1]$. The unique Bayes-Nash Equilibrium is symmetric, and its equilibrium distribution $F_X = F_Y = F_{eq}$ has the following cdf F_{eq} and pdf f_{eq}.*

$$F_{eq}(\theta) = \frac{1}{2} \cdot \left(1 - \frac{1 - 2\theta}{\sqrt{\theta^2 + (1 - \theta)^2}} \right) \qquad f_{eq}(\theta) = \frac{1}{2} \cdot \frac{1}{(\theta^2 + (1 - \theta)^2)^{3/2}}.$$

The proof of this theorem is fairly technical. It utilizes heavily that at equilibrium, each agent can ensure to be ranked first with probability $\frac{1}{2}$, simply by copying the other agent's strategy. As is the case with many equilibrium proofs, the bulk of the technical work goes into proving that each player's distribution has full support and is continuous. Once this fact is established, the equilibrium indifference condition (i.e., the fact that each point in $[0, 1]$ yields probability $\frac{1}{2}$ of being ranked first) can be used to derive a differential equation for the cdf. This differential equation is then solved explicitly to yield the theorem.

Figure 2a shows the cdf and pdf of the equilibrium distribution of Theorem 6. Observe that the cdf is continuous and has support $[0, 1]$. Moreover, note that the pdf f_{eq} is also symmetric about $\frac{1}{2}$. (This is not a priori obvious, and indeed, will not be the case when we consider restricted test sets in the next section). Finally, as discussed before, observe that if quality levels X, Y are drawn from any absolutely continuous distribution Ψ, then the unique equilibrium distribution for thresholds $\sigma \in \mathbb{R}$ is given by $F_{eq, \Psi}(\sigma) = F_{eq}(\Psi(\sigma)) = \frac{1}{2} \left(1 - \frac{1 - 2\Psi(\sigma)}{\sqrt{\Psi(\sigma)^2 + (1 - \Psi(\sigma))^2}} \right)$.

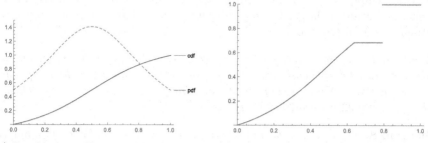

(a) Equilibrium cdf and pdf for firms with unrestricted choice of tests.

(b) Equilibrium cdf for firms restricted to choosing tests in $[0, 0.79]$.

Fig. 2. Examples of equilibrium cdfs for unrestricted and restricted sets of tests.

We are now in a position to combine Theorem 4 and Theorem 6 to determine the Price of Anarchy (in terms of the principal's probability of selecting the wrong firm) of allowing firms to choose their own tests. Substituting the characterizations into the functional (Eq. (1)), the resulting expression unfortunately does not lend itself to closed-form evaluation. However, a numerical calculation establishes the following.

Corollary 1. *The equilibrium cdf F_{eq} satisfies that $I(F_{eq}) \approx 0.23056$. Consequently, compared to the optimal i.i.d. test selection rule, endogenous test selection over unrestricted tests has a Price of Anarchy of roughly 1.10653 for any number of firms. Compared to the optimal correlated test selection rule, it has a Price of Anarchy of approximately 1.38336 for two firms, decreasing to 1.10653 as the number of firms $n \to \infty$.*

6 Endogenous Test Selection with Restricted Tests

We now consider a more general treatment: the principal restricts the firms to choose tests from a non-empty closed set $S \subseteq [0, 1]$, and the firms will play according to equilibrium distributions F_X, F_Y supported on subsets of S. Note that although the firms' tests are restricted to the set S, their products' qualities are still drawn uniformly from the entire interval $[0, 1]$; this is reflected in the probabilities of passing/failing tests.

The existence of a (mixed, symmetric) Bayes-Nash Equilibrium follows from Lemma 7 of [7]. However, we note that in general, the Nash equilibrium may not be unique; for example, when $S = \{1 - \frac{\sqrt{2}}{2}, \frac{\sqrt{2}}{2}\}$, every pair of probability distributions on S constitutes an equilibrium. To see that this is the case, observe that conditional on the firms choosing *any* ordered pair of tests in the product set $S \times S$, each firm's probability of being selected is $\frac{1}{2}$.

The following pair of theorems shows that by restricting the set S available to the firms, even to an interval, the principal can achieve a strictly smaller inversion probability than under the equilibrium for $S = [0,1]$; however, for every non-empty set S, the inversion probability under every symmetric Bayes-Nash Equilibrium is larger by some absolute constant than the one under the optimum i.i.d. distribution.

Theorem 7. *Let $F_{[0,0.79]}$ be the unique[9] symmetric Bayes-Nash equilibrium distribution when firms choose from the interval $[0,0.79]$, and $F_{[0,1]}$ the unique and symmetric Bayes Nash equilibrium distribution for unrestricted firms. Then,[10] $I(F_{[0,0.79]}) < 0.22975 < 0.23052 < I(F_{[0,1]})$.*

Theorem 8. *Let $S \subseteq [0,1]$ be an arbitrary non-empty set, and F any symmetric Bayes-Nash equilibrium distribution of firms restricted to choosing tests from S. The expected probability of choosing the wrong firm under F is $I(F) \geq \frac{5}{24} + \frac{1}{82944}$.*

We emphasize that Theorem 8 establishes a lower bound only for *symmetric* equilibria. For general S, there may be asymmetric equilibria, and they may achieve error probabilities strictly smaller than $\frac{5}{24}$. For example, as observed above, when tests are restricted to the set $S = \{1 - \frac{\sqrt{2}}{2}, \frac{\sqrt{2}}{2}\}$, there is an asymmetric equilibrium in which firm X always chooses $\theta_X = 1 - \frac{\sqrt{2}}{2}$, Y always chooses $\theta_Y = \frac{\sqrt{2}}{2}$, and the inversion probability is $\frac{1}{2}(\theta_X^2 + (\theta_Y - \theta_X)^2 + (1 - \theta_Y)^2) = 3 - 2\sqrt{2} \approx 0.17157$, whereas $\frac{5}{24} \approx 0.2083$.

The key to proving Theorem 7 is the following complete characterization of the unique Bayes-Nash equilibrium when S is restricted to intervals.

Theorem 9. *Let $S = [a, b]$ be a non-empty interval, and consider the game when both firms are restricted to choosing tests from S. There is a unique Bayes-Nash equilibrium, which is symmetric. Its cdf F_{eq} is given by the following:*

1. *If $(1 - a) \cdot b \leq \frac{1}{2}$, then F_{eq} is a step function at b, i.e., both firms deterministically choose b.*
2. *Otherwise, let*

$$\delta_b = \frac{1 - a(1 - b) - b(1 - a)}{(1 - a)((1 - b)^2 + b^2)} \qquad \gamma = \frac{1 - a - 2b + 4ab - 2ab^2}{1 - 4(1 - a)b + 2(1 - 2a)b^2}.$$

---

[9] As will be established in Theorem 9.
[10] Recall that we write $I(F) = \mathbb{E}[I(\mathcal{T}_F)]$.

The equilibrium cdf F_{eq} is given by:

$$F_{eq}(\theta) = \begin{cases} \frac{1}{2(1-a)} \cdot \left((1-2a) + \sqrt{a^2 + (1-a)^2} \cdot \frac{2\theta-1}{\sqrt{\theta^2 + (1-\theta)^2}} \right) & \textit{for } a \leq \theta < \gamma \\ 1 - \delta_b & \textit{for } \gamma \leq \theta < b \\ 1 & \textit{for } \theta = b. \end{cases}$$

(2)

Figure 2b illustrates the equilibrium cdf for firms restricted to the interval $[0, 0.79]$.

6.1 Suboptimality of All Symmetric Equilibria

We next outline the proof of Theorem 8. Recall that we use H^* to denote the cdf of the optimal distribution, i.e., the uniform distribution on $[\frac{1}{4}, \frac{3}{4}]$. We begin with an easy proposition, capturing that a sufficient condition for H^* and an arbitrary cdf G to differ by at least ε at z is for G to be "sufficiently discontinuous" at some point θ.

Proposition 2. *If $G(\theta) \geq \varepsilon + \lim_{t \uparrow \theta} G(t)$, then $|H^*(z) - G(z)| \geq \frac{\varepsilon}{2}$ for some z.*

Proposition 2 is the key ingredient to proving Lemma 2, which shows that symmetric equilibrium distributions deviate far from the optimal distribution,

Lemma 2. *Let F be the cdf of an equilibrium distribution for some non-empty closed set S. There exists a $z \in (0, 1)$ with $|F(z) - H^*(z)| \geq \frac{1}{24}$.*

The second key lemma shows that a large deviation at even one point implies a significantly larger error probability.

Lemma 3. *Let G be any distribution such that $|G(z) - H^*(z)| \geq \varepsilon$ for some $z \in (0, 1)$ and $\varepsilon > 0$. Then, $I(G) \geq I(H^*) + \frac{1}{6}\varepsilon^3$.*

Combining Lemmas 2 and 3, with $\varepsilon = \frac{1}{24}$, immediately implies Theorem 8.

7 Conclusions

We introduced and studied a problem of optimal and endogenous test selection in a setting where a principal wants to select the product of higher quality from one of two firms, but the products' qualities can only be measured through *threshold tests* which reveal whether a product's quality lies above or below a threshold θ. We explicitly characterized the optimal correlated and i.i.d. distributions for the principal, as well as the equilibrium distribution when the firms can choose their own thresholds from an interval $[a, b]$ (in particular including the case of the interval $[0, 1]$). Using these characterizations, we showed that the principal can do strictly better by giving the firms different tests than drawing their tests i.i.d. The best i.i.d. distribution is better than any symmetric equilibrium for any set S offered to the firms (including sets S that are not intervals), and the equilibrium under the *best interval* gives the principal strictly higher probability of selecting the best product than the equilibrium for the interval $[0, 1]$.

Our work raises a wealth of questions for future work. An immediate question implicitly raised in Sect. 6 is which set of tests a principal should offer to achieve the smallest probability of selecting the wrong product at equilibrium.[11] There are two variants to this question: when the principal is interested only in symmetric equilibria, or also in asymmetric (non-unique) ones. For the former version, a natural conjecture would be that the optimal set for the principal is an interval, in which case our numerical calculations from Sect. 6 would imply that the optimum set would be the interval $[0, 0.79\ldots]$. While we cannot prove or disprove this conjecture at this point, a similar-looking stronger conjecture is false: there are discrete sets S the principal can offer under which the unique equilibrium is strictly better than if the principal instead offered the smallest interval containing all of S. For the latter case, we conjecture optimality of the set $\{1 - \sqrt{2}/2, \sqrt{2}/2\}$, discussed in Sect. 6.

The endogenous test selection game between the firms can be viewed as a natural instance of a *signaling game*, in which each firm's strategy is a signaling scheme. Our problem setup severely restricted the signaling schemes the firms could choose from, to binary threshold tests. Naturally, it would be desirable to extend the results to broader classes of signaling schemes. At the full extreme, when firms may choose any signaling scheme, the unique equilibrium of our game is full disclosure. This follows from Corollary 1 of [12]. However, an analysis of the intermediate regime, in which the number of signals is still constrained (as in [8]), would still be of interest.

Perhaps the most immediate next step along these lines would be signaling schemes in which firms can choose an arbitrary mapping from qualities to $\{\text{pass}, \text{fail}\}$. It is not hard to show that w.l.o.g., it suffices to consider signaling schemes specified by an interval $[\theta_1, \theta_2]$ such that the firm passes the test iff its quality lies in the interval. A natural conjecture would be that at equilibrium, firms given access to such tests would always choose threshold tests only, i.e., set $\theta_1 = 0$. This conjecture is false! If such a symmetric equilibrium existed, it must be the equilibrium we derived in Sect. 5 — however, against this strategy, there are responses yielding firm X a selection probability strictly larger than $\frac{1}{2}$. Explicitly characterizing the equilibrium distribution appears difficult.

Another natural version is to require threshold tests, but allow multiple thresholds $\theta_1 \leq \theta_2 \leq \cdots \leq \theta_k$. This naturally corresponds to the type of tests encountered in classes, where cutoffs are defined between multiple grades. Even for two thresholds, characterizing the equilibrium outcomes appears difficult – a firm with a difficult-to-attain 'B' grade may have to be ranked ahead of a firm with an easy-to-attain 'A' grade (similarly between easy 'B' and difficult 'C'...).

---

[11] Of course, if the principal can choose different sets for different firms, then she can choose $S_X = \{\frac{1}{3}\}$ and $S_Y = \{\frac{2}{3}\}$, which would implement the optimal strategy for her. The more interesting question is to find one set S to restrict *all* firms to, which naturally corresponds to prescribing standards for quality control.

This is different from the pass-fail model, where every firm that passes a test is ranked ahead of every firm that fails a test, regardless of the tests' difficulties.

A very interesting direction for future work is considering firms whose product qualities are drawn independently from *different* distributions. If one distribution stochastically dominates the other, it would be interesting to see if the weaker firm may at equilibrium follow "moon shot" strategies of taking very hard tests and hoping that this will allow it to win some of the time. Characterizing the equilibrium again appears to be quite challenging, because when both firms pass tests of the same difficulty, their posterior quality distributions will be different — as a result, the principal will not simply rank passing firms by their thresholds, and this results in a possibly infinite-dimensional system of differential equations characterizing the equilibrium distribution.

There are several open directions in terms of alternate objectives when extending the model to $n > 2$ firms. When the principal's goal is to obtain a complete ranking minimizing the Kendall tau distance, and the firms' goal is to be ranked as highly as possible in expectation, we argued that our results carry over immediately; and for correlated tests, we explicitly characterized the optimum distribution. However, when the objectives are changed, this ceases to be true. A natural objective is for the principal to maximize the probability of selecting the best product, and for each firm to maximize the probability of being selected. Even for $n = 3$ firms, it appears difficult to characterize the equilibria of the endogenous test selection game, or the principal's optimal test distribution.

Instead of having the principal try to maximize the probability of selecting the better firm, an alternative objective would be for the principal to maximize the expected quality of the selected firm. While this is a natural objective, it requires the model to ascribe meaning to the concrete quality values, rather than using them only for comparison, in contrast to a viewpoint where utilities predominantly encode preferences. Nonetheless, the optimization and equilibrium questions would likely yield a rich set of questions.

Finally, we note a possibly interesting connection to a very different setting.[12] One can interpret our setting as a principal trying to allocate an item to one of two agents X, Y via a price-discriminating posted-price mechanism. Different from standard such setups, the natural correspondence has a *welfare-maximizing* (rather than *revenue-maximizing*) principal. The mechanism corresponding to our testing setting then has the principal offer the two agents possibly different posted prices. If exactly one agent is interested in buying the item at his posted price, that agent is given the item at the posted price. If both agents are interested in buying at their respective prices, the agent with higher price obtains the item at his posted price. If neither agent is interested in buying, then again, the agent with higher price obtains the item, and pays 0. This model raises the issue of strategic manipulation: an agent might decline the item at his posted price, hoping that his price is higher and he will get the item for free. A natural question is whether the principal can price-discriminate in a way that

---

[12] We thank Nicole Immorlica for suggesting this interpretation.

316   S. Banerjee et al.

will provide higher social welfare than offering both agents the same price (and choosing randomly which agent obtains the item if both accept/decline).

Acknowledgement. We would like to thank Odilon Camara, Peter Frazier, Moshe Hoffman, Nicole Immorlica, Jonathan Libgober, Erez Yoeli, and Christina Lee Yu for useful discussions and pointers.

References

1. Au, P.H., Kawai, K.: Competitive information disclosure by multiple senders. Games Econ. Behav. **119**, 56–78 (2020)
2. Au, P.H., Kawai, K.: Competitive disclosure of correlated information. Econ. Theory **72**(3), 767–799 (2019). https://doi.org/10.1007/s00199-018-01171-7
3. Banerjee, S., Kempe, D., Kleinberg, R.: Threshold tests as quality signals: optimal strategies, equilibria, and price of anarchy. arXiv preprint arXiv:2110.10881 (2021)
4. Boleslavsky, R., Carlin, B.I., Cotton, C.S.: Bayesian exaggeration and the (mis)allocation of scarce resources (2016, under submission)
5. Boleslavsky, R., Cotton, C.S.: Limited capacity in project selection: competition through evidence production. Econ. Theory **65**(2), 385–421 (2018)
6. Candogan, O.: On information design with spillovers (2020). SSRN 3537289
7. Dasgupta, P., Maskin, E.: The existence of equilibrium in discontinuous economic games, I: Theory. Rev. Econ. Stud. **53**(1), 1–26 (1986)
8. Dughmi, S., Kempe, D., Qiang, R.: Persuasion with limited communication. In: Proceedings of 17th ACM Conference on Economics and Computation, pp. 663–680 (2016)
9. Dworczak, P., Martini, G.: The simple economics of optimal persuasion. J. Polit. Econ. **127**(5), 1993–2048 (2019)
10. Gill, D., Sgroi, D.: Sequential decisions with tests. Games Econ. Behav. **63**(2), 663–678 (2008)
11. Gill, D., Sgroi, D.: The optimal choice of pre-launch reviewer. J. Econ. Theory **147**(3), 1247–1260 (2012)
12. Hwang, I., Kim, K., Boleslavsky, R.: Competitive advertising and pricing (2019, under submission)
13. Lerner, J., Tirole, J.: A model of forum shopping. Am. Econ. Rev. **96**(4), 1091–1113 (2006)
14. Li, Z., Rantakari, H., Yang, H.: Competitive cheap talk. Games Econ. Behav. **96**, 65–89 (2016)
15. Onuchic, P.F., Ray, D.: Conveying value via categories. arXiv preprint arXiv:2103.12804 (2019)
16. Ostrovsky, M., Schwarz, M.: Information disclosure and unraveling in matching markets. Am. Econ. J.: Microecon. **2**(2), 34–63 (2010)
17. Sobel, J.: Giving and receiving advice. In: Acemoglu, D., Arellano, M., Dekel, E. (eds.) Advances in Economics and Econometrics: Tenth World Congress (Econometric Society Monographs), pp. 305–341 (2013)
18. Spence, M.: Job market signaling. Q. J. Econ. **87**(3), 355–374 (1973)

The Platform Design Problem

Christos Papadimitriou, Kiran Vodrahalli$^{(\boxtimes)}$, and Mihalis Yannakakis

Department of Computer Science, Columbia University, New York, NY, USA
{christos,kiran.vodrahalli}@columbia.edu,
mihalis@cs.columbia.edu

Abstract. On-line firms deploy suites of software platforms, where each platform is designed to interact with users during a certain activity, such as browsing, chatting, socializing, emailing, driving, etc. The economic and incentive structure of this exchange, as well as its algorithmic nature, have not been explored to our knowledge. We model this interaction as a Stackelberg game between a Designer and one or more Agents. We model an Agent as a Markov chain whose states are activities; we assume that the Agent's utility is a linear function of the steady-state distribution of this chain. The Designer may design a platform for each of these activities/states; if a platform is adopted by the Agent, the transition probabilities of the Markov chain are affected, and so is the objective of the Agent. The Designer's utility is a linear function of the steady state probabilities of the accessible states, minus the platform development costs. The underlying optimization problem of the Agent—how to choose the states for which to adopt the platform—is an MDP. If this MDP has a simple yet plausible structure (the transition probabilities from one state to another only depend on the target state and the recurrent probability of the current state) the Agent's problem can be solved by a greedy algorithm. The Designer's optimization problem (designing a custom suite for the Agent so as to optimize, through the Agent's optimum reaction, the Designer's revenue), is in general NP-hard to approximate within any finite ratio; however, in the special case, while still NP-hard, has an FPTAS. These results generalize, under mild additional assumptions, from a single Agent to a distribution of Agents with finite support, as well as to the setting where other Designers have already created platforms. We discuss directions of future research.

Keywords: Theory of the online firm · Markov decision process · Bi-level optimization · Approximation algorithms · Stackelberg games

1 Introduction

In economics, the creation of wealth happens through markets: environments in which firms employ land, labor, capital, raw materials, and technology to produce new goods for sale, at equilibrium prices, to consumers and other firms. Since all agents in this scenario participate voluntarily, wealth must be created.

© Springer Nature Switzerland AG 2022
M. Feldman et al. (Eds.): WINE 2021, LNCS 13112, pp. 317–333, 2022.
https://doi.org/10.1007/978-3-030-94676-0_18

Accordingly, markets have been the focus of a tremendous intellectual effort by economists, mathematicians, and, more recently, computer scientists.

Over the past three decades the global information environment has spawned novel business models seemingly beyond the reach of the extant theory of markets, and which, arguably, account for a large part of present-time wealth creation, chief among them a new kind of software company that can be called *platform designer*. On-line platforms are created with which consumers interact during certain activities: search engines facilitate browsing, social networks host social interactions, movie, music, and game sites provide entertainment, chatting and email apps mediate communication. Shopping platforms, navigation maps, tax preparation sites, and many more platforms bring convenience and therefore value to consumers' lives. Increasingly during these past two decades, on-line firms have created comprehensive *suites* of platforms, covering many such life activities. Platform designers draw much of their revenue through the data that they collect about the users interacting with their platforms, which data they either sell to other firms or use to further fine tune and enhance their own business. In this paper we point out that, in the case of platform designers, the most elementary aspects of markets, for example the theory of production and consumption, are quite nontrivial. We focus on a restricted case of the problem corresponding to the "substitutes" case, having proved that the case with complements (when platforms are allowed to feed into one another) is hopeless. Note that this reflects the history of the search, in the market context, for conceptually, and implicitly computationally, tractable cases. (Recall the fruitful early work by Arrow and other economists on the identification of classes of markets with good structural properties, such as the gross substitutes case [2], and the extensive more recent work in computer science developing algorithms for special cases, like the case of linear utilities [25].) We also note that while the problem of platform design is most naturally considered in a learning setting, where user behavior and revenues are unknown and must be learned from interaction and data, our focus in this paper is on computational tractability, and our results can be interpreted as demonstrating that even for the easier setting where the parameters of the user and the amount of revenue received for various design decisions is known, the problem is not easy to solve. The tractable cases we discover can then serve as an initial point to base further investigations of the tractability of more general learning settings with interaction between the platform designer and the agents where both the designer and the agents have unknowns they need to learn.

Our Model and Results. We model the platform design problem as a Stackelberg game (that is, a game where one player goes first and the others react optimally) with two players, a Designer and an Agent (the extension to many Agents is also studied, and the case of many competing Designers is also discussed). Here, the Designer plays first, and the Agent responds. The Agent is modeled as an ergodic Markov chain on a set of states $\mathcal{A}$, representing the Agent's life activities. We assume that the Agent receives a fixed payoff per unit of time spent at each state. The Designer has the opportunity to design a platform for each state in $\mathcal{A}$, which the Agent may or may not choose to adopt. There is a one-time cost for the Designer to build a platform for a given state. If the Agent adopts the platform, the transitions of the Agent's life change at that state, and the Agent's utility at that state may increase or decrease as a result of adoption[1]. In return, the Designer gets to observe the Agent at that state and derives a fixed utility payoff for the fraction of time the Agent spends in that state (modeling the Designer's collection of the relevant Agent data). We assume that platform revenue is proportional to the time users spend on the platform, which strikes us as a reasonable first approximation.

We note immediately that the Agent's optimization problem, once the Designer has deployed a set S of platforms, is a Markov decision process (MDP), and it follows from MDP theory that the Agent will adopt some of the platforms offered and reject the rest and the optimum set of adopted platforms can be computed by linear programming (and other methods).

Now the *platform design problem* (PDP) is the following: Given the Markov chain, all utility coefficients for both the Agent and the Designer, and the development costs of the platforms, choose a set of states S for which to create platforms, so as to maximize the Designer's utility; namely, the utility to the Designer of the Markov chain that results from the optimum response by the Agent to the platforms in S, minus the development costs of the platforms in S. It is immediate that, since the Designer can anticipate Agent's optimal response, at optimality all platforms in the optimum set S will be adopted.

We show that PDP is NP-hard to approximate within *any* finite ratio (Theorem 1). The proof of this result is quite instructive, because it relies almost exclusively on the fact that introduced platforms can modify the Markov chain so as to funnel traffic from one platform to the other, and therefore create the stark choices necessary for this level of complexity. The construction has the property that offering a platform in one state can make it more attractive for the Designer to offer a platform also at another state (if the adoption of the platform in the first state increases the transition probability to the second state) In economic terms, the platforms offered by the Designer can be *complementary goods*, and making decisions for such goods tend to be difficult.

In view of this obstacle, we next turn to a special kind of Markov chain, for which platforms are essentially *substitute goods;* generally, substitution is known to lead to better behaving markets. A Markov chain of this sort, called *the flower* (see Fig. 1), has a number of transition parameters that is linear in $|\mathcal{A}|$. At each state i, the transition leads back to the state with some probability q_i, while the rest of the probability $(1 - q_i)$ is split among the other states *in proportions that are fixed*. Evidently, this is equivalent to a chain that has an extra "rest state" 0 with $q_0 = 0$, that is, a purely transitional state (see Fig. 1). Adopting a platform now increases or decreases the transition probability of the state to itself, decreasing or increasing, respectively, the transition probability to the other states. We show that, in this case, the MDP optimizing the Agent's

---

[1] One possible reason for diminished utility is the aversion of the Agent to the Designer's access to personal information pertaining to that state.

objective, given the available platforms, becomes a quasiconcave combinatorial optimization problem with special structure (Lemma 1), which can be solved by a greedy algorithm (Theorem 2). The algorithm can be extended to a setting where there are multiple available platforms for each state in $\mathcal{A}$, and the agent can choose to adopt one or none of these options for each state (see Sect. 6 and Appendix D of [20]).

The PDP in the flower specal case is still NP-hard (Theorem 4), but has a dynamic programming FPTAS if one parameter—the expected time spent at each state—is quantized (Theorem 3). The dynamic programming algorithm can be extended, through some further quantization, to the case of many agents— except that the number of agents is now in the exponent (see Sect. 5 of [20]). Given that the number of agents is likely to be very large, the best way to think of this algorithm is as an algorithm for the case in which one is given a *distribution* of agents of finite support—that is, with a small number of *agent types*—an essential step towards a model of platform design where we must learn an optimal design given (perhaps indirect) sample access to such a distribution. Similarly, essentially the same algorithm can be adapted to the competitive setting, where a Designer enters a field where many Designers have already built existing platforms, and must now decide which platforms to build (Sect. 6 and Appendix D of [20]).

Related Work. We are not aware of past research on the production and consumption of online platforms. Computational aspects of Stackelberg games between consumers and firms designing or packaging on-line products have been explored to a small degree, see e.g. [13,14]. There has been work on online decision making, where at each round the Designer gets to select from some set of options (e.g., which is the best ad to display to the user of a website) and receives a reward after deployment for that round, as well as additional information about the performance of the other options [6]; see also [9,16–18,22]. This line of research is of obvious relevance to the present one, even though our Agent model is far more complex. More recently, trade-offs in on-line activity by consumers, for example between effectiveness of browsing and privacy, have been discussed [23,24]. The ways in which on-line firms profit from data has been somewhat explored, see e.g. [1] but not in any manner that can be used in our model; here we consider it a given parameter.

Our Contributions. Our main contributions are: the articulation of the Platform Design Problem, the observation that it is profoundly intractable in its generality, the identification of the tractable class of flower Markov chains, roughly corresponding to substitution in markets, the solution of the Agent's and the Designer's problems through the Agent's greedy algorithm and dynamic programming, the generalizations of these algorithms to multiple Agents and Designers, and the many directions for further research opened (see the discussion in Sect. 5).

2   PDP: Intractability of the General Case

The platform design problem (PDP) is a Stackelberg game between a Designer and an Agent[2]. The Agent inhabits a discrete state space with transitions and rewards. The Designer moves first by building, at some fixed cost and for certain states, one platform per state. Each platform, if adopted by the Agent, changes the Agent's transitions and rewards at that state, and also yields to the Designer a reward rate (modeling the Designer's utility from learning about the Agent) per unit of time the Agent spends in the platform for each platform the Agent accepts. The Agent adopts platforms to optimize its expected reward in the resulting Markov Decision Process (MDP). The Designer's goal is to build platforms so that the Agent behaves in a way that optimizes the Designer's total reward. Formally:

Definition 1 (PDP). *The Agent's environment is an irreducible Markov chain with state space $\mathcal{A} = [n]$ with n states. At each state i, the transition probabilities out of i are a vector of probabilities T_i^{life} and the reward coefficient is a real number c_i^{life}.*

The Designer chooses a set $S \subseteq [n]$ of these states for which to build platforms. The Designer pays a fixed $\text{cost}_i > 0$ to build a platform at state i, and receives reward rate d_i per unit of time the Agent spends at state i, provided the Agent opts in to the platform at state i.

After the Designer's move, the Agent faces a Markov Decision Problem (see [21] for an introduction to Markov decision theory). At each state $i \in S$, adoption of the platform will result in the transition probabilities changing to T_i^{platform} and the reward coefficient changing to c_i^{platform}. We assume that these changes in the transition probabilities are such that the reachable part of the Markov chain is irreducible[3].

The Agent's optimal decision in response to the Designer's move S is a set $S' \subseteq S$ of states on which to adopt the platform (recall that in MDPs, it is well known that we can restrict the possible policies, without loss of optimality, to deterministic, Markovian, stationary policies computed by linear programming). Let $M(S')$ be the Markov chain resulting from adopting the subset S' of the platforms offered by the Designer.

Coming now back to the Designer's first move, and since the Designer can fully anticipate the Agent's response S' to S and every extra platform has a positive cost, the Designer omits any platforms that would not be adopted—that is, makes sure that $S = S'$. Among all such sets, the Designer chooses the one that optimizes the Designer's profit

$$\text{profit}(S) := \sum_{i \in S} d_i \cdot \pi_i(S) - \sum_{i \in S} \text{cost}_i$$

---

[2] We later consider the case with multiple Agents and multiple Designers, as well as multiple platforms per state.

[3] Irreducibility can be guaranteed by maintaining a cycle of tiny probability around the states; it will never be a problem in our arguments and constructions.

$\pi_i(S)$ *denotes the steady state distribution at state* i *of the Markov* $M(S)$.

We prove that the PDP in its generality is as severely intractable as any optimization problem can be: It is NP-hard to approximate *within any finite approximation ratio*.

Theorem 1. *It is strongly NP-hard to decide whether the optimum solution to a PDP instance has zero or positive profit for the designer.*

Proof. We reduce from the Set Cover problem. Given a family F of m subsets of a set U of n elements and an integer k, we want to determine if there is a subfamily of F with k sets whose union is U. We define an instance of the PDP problem as follows. There are $m + n + 1$ states, one for each set of F and each element of U, and an additional 'bad' state. For each set-state S_i, there is one potential platform $p(S_i)$ that the Designer may decide to offer at the state S_i. For each element state u_j and every set S_i of F that contains element u_j there is a platform $p(u_j, S_i)$ that the Designer may offer at state u_j; the Designer will offer at most one of these platforms at state u_j.[4] The Designer has no platform for the last 'bad' state.

The Agent likes all the platforms: that is, the Agent's rewards are such that he will adopt every platform that is offered by the Designer. Initially the MDP is at any element-state u_j with uniform probability $1/n$. The transition probabilities of the Agent's MDP are as follows. An element-state u_j with platform $p(u_j, S_i)$ (if adopted) transitions with probability 1 to the set-state S_i. An element state u_j with no adopted platform transitions with probability 1 to the bad state. A set-state S_i with adopted platform $p(S_i)$ self-loops with probability $1 - 1/k^2$ and transitions with the remaining probability to a uniformly random element-state. A set-state S_i with no (adopted) platform transitions with probability 1 to the bad state. The bad state self-loops with probability $1 - 1/nk^4$ and transitions with the remaining probability to a uniformly random element-state.

The Designer's rewards and costs are as follows. The reward rate for each set-state platform $p(S_i)$ is set to $r = k^2 + k$, i.e. the Designer receives revenue equal to r times the fraction of the time that the Agent spends in platform $p(S_i)$; the cost of building the platform is k. The reward rates and costs of the platforms $p(u_j, S_i)$ are set to 0. The objective of the Designer is to select a set of platforms to offer that maximizes the total profit, which is the total reward minus the total cost.

We claim that the optimal profit for the Designer is positive if and only if the Set Cover instance has a solution with at most k sets. Intuitively, the goal of the Designer is to keep the Agent at all times within her "ecosystem", i.e. in states with her platforms, while making a profit.

First, suppose that there is a set cover C with at most k sets. The Designer offers the platform $p(S_i)$ for every $S_i \in C$ at the set-state S_i, and for each

---

[4] We allow here the Designer to have a choice among several platforms in a state; it is easy to modify the construction, by using additional states, so that in each state the Designer has only one potential platform, which she may choose to build.

element-state u_j, the Designer offers a platform $p(u_j, S_i)$ for some $S_i \in C$ that contains u_j. The cost of building the platforms is $k|C| \leq k^2$. The Agent adopts all the offered platforms, and because of the transition probabilities, spends almost all the time at the set-states corresponding to sets in C, specifically a fraction $\frac{k^2}{1+k^2}$ of the time. Therefore, the profit of the Designer is at least $r\frac{k^2}{1+k^2} - k^2 > 0$.

Conversely, suppose that the Designer has a solution with positive profit. Suppose that some element-state u_j does not have a platform, or u_j has a platform $p(u_j, S_i)$ but the corresponding set-state S_i does not have the corresponding platform $p(S_i)$. Then, every time the MDP visits u_j will then move subsequently to the bad state. Therefore, the MDP will spend most of the time (specifically at least $1 - 1/k^2$ fraction of the time) in the bad state, which does not provide any revenue to the Designer. Thus, the total revenue to the Designer is at most r/k^2 which is less than the cost of a set platform. We conclude that, if the profit is positive, then every element state u_j must have a platform $p(u_j, S_i)$ and the corresponding state S_i must have the corresponding platform $p(S_i)$. This implies that the collection C of set-states S_i with a platform forms a set cover. The Designer's profit is at most $r - k|C| = k^2 + k - k|C|$. Since the profit is positive, $|C| \leq k$.

3 Flower Case: The Agent's Problem

The intractability proof of the general PDP in the previous section relies on the complementary nature of the construction: offering a platform in one state can make it more attractive for the designer to also offer a platform in certain other states. We will next define a special case of the PDP which is much better behaved, and in economic terms roughly corresponds to substitution.

An agent divides her time among the different states. If the designer offers a platform at a state s and the agent adopts it, she spends more time at s, and hence has less time to spend in the rest of the states. In the absence of complementarity, this means that it is now less beneficial for the designer to offer a platform in another state. In other words, platforms at different states compete for the attention (and the time) of the agent, and it is the agent's time spent on the platforms that determines their contribution to the profit of the designer.

We define now formally the model in the special case, which will be our focus in the rest of the paper.

Definition 2 (Flower MDP). *We have the same setup as defined in Sect. 2, with some added constraints on what the possible transitions can be. We also add in a dummy state 0 with no reward or platforms[5]. In Fig. 1, we define the transitions of the Markov chain T_{life} to represent the Agent's life, and T_{platform} to represent the Agent's life when the platform is adopted at all states. Here,*

---

[5] Note that the rest state 0 is for convenience and is not necessary in our model. We could equivalently have a graph where each node i transitions to node j with probability $(1 - q_i - y_i) \cdot p_j$, and self-transitions with probability $q_i + y_i + p_i(1 - q_i - y_i)$.

324   C. Papadimitriou et al.

p_i, q_i, y_i satisfy $\sum_i p_i = 1$, $0 < p_i$, $0 < q_i < 1$, and $0 < y_i < 1 - q_i$ for all $i \in [n]$. In words, p_i denote transition probabilities to different states from the rest state, q_i denote the self-transition probabilities, and y_i denote the modification to the self-transitions due to the Agent accepting the platform at state i. At state 0, the action chosen by the Agent does not affect the transitions, since the Designer never builds a platform there.

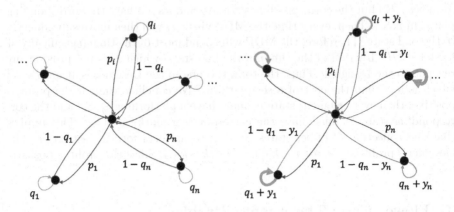

Fig. 1. T_{life} (left) and T_{platform} (right).

3.1 The Greedy Algorithm

Irreducible average-reward MDPs are efficiently solvable via linear programming, value and policy iteration, etc. [5]. Here, we reformulate the Agent's problem as a combinatorial optimization problem with special structure, and solve it through a greedy algorithm. The following is straightforward:

Lemma 1. *The agent's objective for an optimal policy defined in Sect. 2 can be re-written as the following optimization in the special case of the flower MDP (Definition 2):*

$$\underset{S \subseteq [n]}{\operatorname{argmax}} \frac{A + \sum_{j \in S} z_j \phi(j)}{B + \sum_{j \in S} z_j} \tag{1}$$

where

$$A := \sum_{i=1}^n \lambda_i c_i^{\text{life}}; \quad B := 1 + \sum_{i=1}^n \lambda_i; \quad \lambda_i = \frac{p_i}{1 - q_i}; \quad z_i = \frac{p_i}{1 - q_i - y_i} - \frac{p_i}{1 - q_i} \geq 0;$$

$$\phi(i) := \begin{cases} c_i^{\text{platform}} + \frac{\lambda_i}{z_i}\left(c_i^{\text{platform}} - c_i^{\text{life}}\right) & \text{if } z_i > 0 \\ 0 & \text{if } z_i = 0 \end{cases};$$

We therefore define

$$\text{utility}^{\text{Agent}}(S) := \frac{A + \sum_{j \in S} z_j \phi(j)}{B + \sum_{j \in S} z_j}$$

Proof. See Appendix A of [20].

We note that here we assume that $y_i > 0$ at each state—that is to say, adopting the platform increases a state's recurrence probability. This assumption is not necessary, and the general case can be handled in a similar way by modifying the greedy algorithm to pay attention to signs (see Appendix A of [20]). We also reiterate that the solution to the original average case MDP problem need not be unique. Therefore, the argmax solution to Eq. 1 has many potential solutions.

The optimization problem formulated in Lemma 1 can be solved in polynomial time. The intuitive reason is this: Looking at the fractional objective function, we note that it is the ratio of two linear functions of the combinatorial (integer) variables implicit in S, and such functions are known to be quasiconvex. It is therefore no huge surprise that a greedy algorithm solves it—however, the details are rather involved. Incidentally, one could arrive at the same algorithm by tracing the simplex algorithm on the MDP linear program.

ALGORITHM 1: GREEDY ALGORITHM

Input: Parameters of the Agent's problem: transition probabilities and utility coefficients in and out of the platform.
Output: An optimal subset $S \subseteq [n]$ of states where the Agent accepts the platform.
Initialize $S := \{\}$
for $k \in [n]$ *sorted[6] from largest to smallest* $\phi(k)$ **do**
    if utility$^{\text{Agent}}(S) < \phi(k)$ **then**
        Update $S := S \cup \{k\}$
    else
        return S
    end
end
return S

Theorem 2. *Algorithm 1 returns*

$$S^* \in \underset{S \subseteq [n]}{\text{argmax}} \ \text{utility}^{\text{Agent}}(S)$$

That is, the policy

$$\pi(s) = \begin{cases} a^1 & \text{if } s \in S^* \\ a^0 & \text{o.w.} \end{cases}$$

is an optimal policy. Here, a^1 and a^0 refer to the actions available to the Agent's MDP: "accept platform" is a^1 and "do not accept platform" is a^0.

---

[6] Note the sort order may not be unique in case of ties.

Before we prove the theorem, we give a useful definition and a lemma.

Definition 3 (Prefix policy). *We say a policy S is prefix if the states in the policy are the first m states in order sorted by ϕ, for some value of m.*

Lemma 2 (Mediant Inequality).

$$\frac{x}{y} < \frac{r}{s} \iff \frac{x}{y} < \frac{x+r}{y+s} < \frac{r}{s} \quad \text{where } y, s > 0.$$

Proof. Since $y, s > 0$ and thus $y + s > 0$, cross-multiply and simplify to get the desired inequalities. □

With this lemma in hand, we prove Theorem 2.

Proof (Proof of Theorem 2). We can prove optimality in two steps.

1. First we will show that any non-prefix policy is dominated by a prefix policy. Thus an optimal policy must be prefix.
2. Then, we show that the greedy algorithm necessarily finds a best prefix policy (e.g., an optimal stopping point).

We begin with the first step. Suppose we have a non-prefix policy S. Let state $\ell \in [n]$ be a "missing piece" (e.g., if we index by sorted order and S contained $1, 2, 4, 5, 7$, missing pieces would be 3 and 6). This $\ell \in [n]$ necessarily exists since S is non-prefix. Now there are two cases.

Case 1:

$$\text{utility}^{\text{Agent}}(S) < \phi(\ell)$$

We apply Lemma 2 to show that adding state ℓ results in improvement in the objective.

Case 2:

$$\text{utility}^{\text{Agent}}(S) \geq \phi(\ell)$$

Here we show that removing all states $k \in S$ where $\phi(k) < \phi(\ell)$ improves the objective. If equality holds, then it does not matter whether we add the state to the objective, so for simplicity, we terminate at equality. From the assumption and the definition of ϕ, we have

$$\text{utility}^{\text{Agent}}(S) \geq \phi(\ell) > \phi(k)$$

for all such k. By Lemma 2, removing state k increases the ratio, e.g.

$$\text{utility}^{\text{Agent}}(S \setminus \{k\}) > \text{utility}^{\text{Agent}}(S) \geq \phi(\ell) > \phi(k)$$

The same argument applies to all $k' \in S \setminus \{k\}$ such that $\phi(k') \leq \phi(k)$ as well. Therefore, we can remove all the k' with score less than the score of ℓ and improve the objective.

After a single round of considering a missing state ℓ (where either case 1 or case 2 applies), we produce a new S', which can again be non-prefix. However, the maximum index present in the new S' has either decreased (if the second case happened and we deleted everything worse than ℓ) or we filled in the missing state ℓ. Either way, the number of missing pieces has strictly decreased and we have added no new missing states. Using induction on the number of missing pieces proves that iterating over all original missing pieces will "fill in all the gaps" and produce a prefix policy S^* which is strictly better than the original non-prefix policy S.

Finally, we show the greedy algorithm selects an optimal prefix policy. Let the output of the greedy algorithm be $\hat{S}$. The desired result directly follows since if the next state ℓ satisfies

$$\text{utility}^{\text{Agent}}(\hat{S}) \geq \phi(\ell)$$

and is not selected, since all smaller states (sorted by ϕ) are less than or equal to $\phi(\ell)$, any prefix subset of the smaller states is an effective average which is $\leq \phi(\ell)$, and any prefix subset of future states is worse off. Thus, the greedy algorithm produces a maximal solution.

4 Flower Case: The Designer's Problem

We now consider the Designer's problem (Definition 1) in the special case where the Agent lives in the flower MDP (Definition 2). Under this assumption on the Agent's MDP, it will be possible to give an FPTAS for the Designer's problem, due to the additional structure imposed in this setting. Let $\text{Agent}(S)$ denote the subset of states that the Agent adopts when the Designer offers platforms for the subset S of states. Given the results of Sect. 3, the fraction of the time that the Agent spends in state $i \in \text{Agent}(S)$ is

$$\frac{\frac{p_i}{1-q_i-y_i}}{B + \sum_{i \in \text{Agent}(S)} z_i}$$

using the notation of Sect. 2 given in Definition 2 and Lemma 1 (see Appendix A of [20] for the stationary distribution of the Markov chain). Thus, we can simplify the expression for the Designer's profit function:

$$\text{profit}(S) := \frac{\sum_{i \in \text{Agent}(S)} d_i \cdot \frac{p_i}{1-q_i-y_i}}{B + \sum_{i \in \text{Agent}(S)} z_i} - \sum_{i \in S} \text{cost}_i$$

Call a set S of states *feasible* if $\text{Agent}(S) = S$. Since the Agent's response is completely anticipated by the Designer (Agent's parameters are known to the Designer, and the Designer can therefore simulate the greedy algorithm from Sect. 3), only feasible sets S need be considered.

A few additional properties result after we specialize to the flower MDP setting. It is easy to see from the definition and the greedy algorithm of Sect. 3

that if a set S is feasible then so are all its subsets. If for some state i, profit$(\{i\}) \leq 0$, then it follows that for all sets S that contain i we have profit$(S) \leq$ profit$(S - \{j\})$. Hence, there is no reason to build a platform at i, and we can ignore i. Thus, we may restrict our attention to the states i such that profit$(\{i\}) > 0$. We may assume also without loss of generality that every state i by itself is feasible: If $\{i\}$ is not feasible, then neither is any set that contains i, therefore we can ignore i.

We now add a few more assumptions to ensure tractability. Let $K = \max_i$ profit$(\{i\})$. It is easy to see that for any set S, profit$(S) \leq \sum_{i \in S}$ profit$(\{i\})$. Therefore, the optimal profit OPT is at most nK and at least K. We will also assume that the cost cost$_i$ of building a platform at any site i is not astronomically larger than the anticipated optimal profit, specifically we assume cost$_i \leq rK$ for some polynomially bounded factor r. Furthermore, and importantly, for our dynamic programming FPTAS to work in polynomial time, a discretization assumption is necessary. For each state i, the platform available will change (increase or decrease) the term $\frac{p_i}{1-q_i}$ appearing in the numerator and the denominator of the Agent's objective by an additive z_i. *We assume that all these z_i's are multiples of the same small constant δ* (think of δ as 1%), and moreover that the z_i's are polynomially bounded in n. This assumption means that there are $O\left(\frac{1}{\delta}\right)$ possible values of the denominator, and ensures the dynamic programming is polynomial-time. One should think of this maneuver as one of the compromises (in addition to accepting a slightly suboptimal solution) for the approximation of the whole problem. We suspect that the problem has no FPTAS without this assumption, although there is a pseudo-polytime algorithm.

4.1   FPTAS for the PDP

The Platform Design Problem is approximately solvable in polynomial time; in this section we present a FPTAS which returns a $(1 - \epsilon)$-approximate solution. Our approach is inspired by the FPTAS for knapsack presented in [12]. We also note that our algorithm relies on the structure of the greedy algorithm presented in Algorithm 1.

The algorithm uses dynamic programming. It employs a 3-dimensional hash table, called SET, into which the sets under consideration are being hashed. The hash function has three components that correspond to the following components of the profit of the set, scaled and rounded appropriately to integers: (1) the whole profit profit(S), (2) the first term in the profit, denoted $P_1(S)$, and (3) the denominator of the first term, denoted $\mathbf{D}(S)$ (which note, is also the denominator of the Agent's objective function). We use $\mathbf{N}(S)$ to denote the numerator of the Agent's objective function.

Lemma 3. *Let $S, S' \subseteq [k]$ be two sets that hash in the same bin and suppose that $\mathbf{N}(S) \leq \mathbf{N}(S')$. Then for every set $T \subseteq \{k+1, \ldots, n\}$, if $S' \cup T$ is feasible then $S \cup T$ is also feasible, and profit$(S \cup T) \geq$ profit$(S' \cup T) - \epsilon K/n$.*

Proof. Proved in Appendix B of [20]. ∎

ALGORITHM 2: DESIGNER'S FPTAS FOR THE PDP

Input: The parameters of the PDP: transition probabilities, utility and cost
        coefficients for the Agent and the Designer, and small positive reals ϵ, δ

Output: A $(1 - \epsilon)$-approximately optimal subset of states S^* for which to deploy
        platforms.

$\mathbf{N}(S)$ and $\mathbf{D}(S)$ denote the numerator and the denominator of the Agent's objective
function, with the constant terms omitted

$P_1(S)$ denotes the first term in the Designer's profit function

SET is a hash table of subsets of $[n]$ indexed by triples of integers

The hash function is $\text{hash}(S) := \left(\lceil \frac{\text{profit}(S)}{\epsilon K / 2n} \rceil, \lceil \frac{P_1(S)}{\epsilon K / 2n} \rceil, \mathbf{D}(S)/\delta \right)$

Initialize the hash table SET to contain only the empty set in the bin $(0, 0, 0)$

for $k \in [n]$ **do**

    for $S \in$ SET in lexicographic order **do**

        $S' := S \cup \{k\}$

        if *Agent will adopt all platforms in S' and* $\text{profit}(S') > 0$ **then**

            if $\text{hash}(S') \in$ SET **then**

                $\hat{S} :=$ SET$[\text{hash}(S')]$

                if $\mathbf{N}(\hat{S}) > \mathbf{N}(S')$ **then**

                    SET$[\text{hash}(S')] := S'$

                end

            else

                SET$[\text{hash}(S')] := S'$

        end

    end

end

return *the set S in the hash table with largest first hash value*

Lemma 4. *For every $k = 0, 1, \ldots, n$, after the kth iteration of the loop, there is a set S in the hash table that can be extended with elements from $\{k + 1, \ldots, n\}$ to a feasible set that has profit $\geq$ OPT $- \epsilon k \cdot K/n$.*

Proof. Proved in Appendix B of [20]. ∎

Theorem 3. *Algorithm 2 is a FPTAS for the Platform Design Problem.*

Proof. Lemma 4 for $k = n$ tells us that at the end, the table contains a set S whose profit is within ϵK of OPT. Since $OPT \geq K$, the profit of S is at least $(1 - \epsilon)OPT$.

Regarding the complexity of the algorithm, note that the three dimensions of the hash table have respectively size $O(n^2/\epsilon)$ (since the maximum profit is at most nK), $O(rn^2/\epsilon)$, and n/δ. In every iteration the algorithm spends time proportional to the number of sets stored in the table. In particular, the algorithm only needs linear time to check the feasibility of each S' as well as calculate $\mathbf{N}(S')$ and $\text{profit}(S')$. Thus, the total time is polynomial in n and $\frac{1}{\epsilon}$. ∎

It turns out that an FPTAS is the best we could hope for, even if all $z_i = 1$:

Theorem 4. *The PDP in the flower case is NP-complete.*

Proof. Proved in Appendix B of [20].

5   Discussion and Future Work

We believe that we have barely scratched the surface of a very important subject: the economic, mathematical, and algorithmic modeling of the inter-actions between Designers of on-line platforms and the consumers of on-line services/producers of data. Our model captures a few of the important aspects of this complex environment: the way adoption of services affects both the user's activities and the user's enjoyment of these activities, while it enhances the Designer's revenue in ways that depend on the time spent and activities per-formed on the platform; the nature of the Designer's profit (revenue from the acquisition of data pertaining to the user minus the significant development costs); the fact that multiple platforms, even by the same Designer, compete for the user's attention and use; the nature of some of the user's dilemmas (chief among them: surrender privacy for increased efficiency and/or enjoyment?). A simplified model of these aspects (the flower chain, linearity of utilities) is a tractable bi-level optimization problem. However, there are many effects that our current model does not capture, which are quite interesting for future research: for instance, as a sample, we may want to model time dependencies in profits, rewards, and costs, scaling effects for the platform designers due to increasing numbers of users, synergistic effects for the agents who may adopt suites of plat-forms (for instance, adopting all of the Google suite of products may provide more benefit than using different providers for each service), and potential net-work effects involved in influencing agent behavior when there are many agents.

Several generalizations of our results are possible: in Sect. 5 of [20], we gen-eralize the FPTAS for the flower to the case with k Agents, each with their own flower Markov chain. Our algorithm is polynomial runtime for constant k, and is exponential if k is allowed to vary—hence, it is perhaps more natural to think of k as the number of types in a finite-support distribution of agents, where the number of types may naturally be a small constant in settings of interest. In Sect. 6 of [20] (and continuing in the Appendix D of [20] in full detail), we gener-alize the setting to the interaction of two or more platform designers with agents. There we confront the algorithmic problems involved with designer competition, such as the *best response problem:* if a Designer is confronted with a situation in which other designers have already deployed several platforms at various states, which platforms should this Designer deploy? It turns out that Agent behavior in this setting is still governed by a greedy algorithm, but the potential is differ-ent. Furthermore, the Designer has a polynomial time algorithm for the platform design problem with a constant number of agents – however, the runtime is still exponential in the number of agents. The algorithm can also be extended to the case where the Designer can choose between multiple possible platforms for each state.

We believe that intractability (both analytical and computational) lurks in many of the further possible immediate generalizations of this model—for example, to undiscretized coefficients, to Markov chains more general than the flower, or to more complex objectives than linear (such as the addition of an entropy regularizer to the objectives of both the Agent and the Designer—an especially tempting variant to consider in this particular problem). We believe that more ambitious problem formulations in these directions may need to further simplify the other aspects of the model in this paper to become tractable.

On the other hand, we also believe that any form of intractability of the Designer's problem is arguably *affordable*. Our dynamic programming FPTAS would likely not generalize to more general contexts—such as those involving complex chains, nonlinear objectives, many Designers, learning of the statistics of the Agents' parameters etc., see below—but the alternative exhaustive algorithm, with its rather benign exponential dependency on n, the number of platforms, is extremely realistic in this context. We believe that the true challenges in generalizing our results are challenges of formulation and modeling.

Superficially, platform design resembles Mechanism Design (MD) [19], but the essence of much of MD is that the Designer knows only *statistics* of the Agent's characteristics and designs the mechanism to optimize revenue over all possible eventualities by incentivizing the Agent to implicitly reveal their type; and this essence is missing in the PDP. In the on-line platform environment, the subject of incentives for type revelation and truthfulness is rather clearly related to the *personalization* of the platform, and we believe that a generalization of our model will have to address this important issue and aspect of platform design.

In the present first brush at platform design, we have abstracted the PDP in terms of a single Agent—a maneuver and methodology familiar from Economics —, and next ventured to the case of a few Agent types. But of course, the motivating environment involves myriads of atypical Agents whose parameters are unknown. Moreover, the rewards for the Designer may also not be completely known, and the Designer problem can be modeled as an interactive game between Designer and (multiple) Agent(s). Thus, the learning nature of platform design resembles interactive learning (the learner and the Agent whose parameters are being learned interact closely, and the learner can easily experiment with variants of the platform), and also has certain characteristics of learning from revealed preferences, see e.g. [26]. We believe that a wealth of novel and intriguing technical problems within Learning Theory and Machine Learning lie in this direction, and can build on recent work in the intersection of these areas with Algorithmic Mechanism Design, Learning in Games, and Reinforcement Learning [8,11]. In situations where the Agent rewards are unknown to the Designer, it may also be possible for the Agent to behave strategically – understanding these impacts on platform design is but one interesting direction to pursue. Of course, recent cautionary results on the limitations of optimization by samples [3,4] come to mind as well.

Regarding the important subject of strategic interactions between designers, we have not addressed the equilibrium problem—beyond the best-response algorithm. We can show (see Appendix F of [20]) that a pure Nash equilibrium may not exist even in the flower setting, and we conjecture that finding a pure equilibrium is Σ_2-complete.

But perhaps the most interesting strategic questions go beyond the model of this paper: How are Designers incentivized by the competition to design and deploy platforms that are more beneficial to the Agents than in the monopolistic situation?

Finally—and almost needless to say—the subject of platform design, as circumscribed in this paper, is crying out for treatment from the point of view of the exploding literature on *ethics, fairness, and privacy* in algorithm design—see for example [7,10,15] among many other important works—and exposes new aspects of today's algorithmic environment to these important considerations and emerging methodologies. The PDP defines an environment where privacy and fairness concerns are ubiquitous and paramount. Understanding what kinds of social, economic, regulatory, and technological interventions may result in fairer outcomes of platform design is an important direction of future work.

Acknowledgement. We thank John Tsitsklis, Eva Tardos, and Yang Cai for helpful conversations during the development of this work. K. Vodrahalli acknowledges support from an NSF Graduate Fellowship. C. Papadimitriou and M. Yannakakis acknowledge support from NSF Grant CCF-1763970.

References

1. Agarwal, A., Dahleh, M., Sarkar, T.: A marketplace for data: an algorithmic solution. In: Proceedings of the 2019 ACM Conference on Economics and Computation, EC 2019, pp. 701–726. Association for Computing Machinery, New York (2019). https://doi.org/10.1145/3328526.3329589
2. Arrow, K.J., Block, H.D., Hurwicz, L.: On the stability of the competitive equilibrium, ii. Econometrica **27**(1), 82–109 (1959), http://www.jstor.org/stable/1907779
3. Balkanski, E., Rubinstein, A., Singer, Y.: The limitations of optimization from samples. In: Proceedings of the 49th Annual ACM SIGACT Symposium on Theory of Computing, STOC 2017, pp. 1016–1027. Association for Computing Machinery, New York (2017). https://doi.org/10.1145/3055399.3055406
4. Balkanski, E., Singer, Y.: The sample complexity of optimizing a convex function. In: Kale, S., Shamir, O. (eds.) Proceedings of the 2017 Conference on Learning Theory. Proceedings of Machine Learning Research, vol. 65, pp. 275–301. PMLR, Amsterdam, 07–10 July 2017. http://proceedings.mlr.press/v65/balkanski17a.html
5. Bertsekas, D.P.: Dynamic Programming and Optimal Control, vol. I, 4th edn.. Athena Scientific, Belmont (2017)
6. Cesa-Bianchi, N., Lugosi, G.: Prediction, Learning, and Games. Cambridge University Press, New York (2006)
7. Dwork, C., Hardt, M., Pitassi, T., Reingold, O., Zemel, R.: Fairness through awareness. In: Proceedings of the 3rd Innovations in Theoretical Computer Science Conference, pp. 214–226 (2012)
8. Foster, D.J., Li, Z., Lykouris, T., Sridharan, K., Tardos, E.: Learning in games: robustness of fast convergence. In: Advances in Neural Information Processing Systems, pp. 4734–4742 (2016)

9. Frazier, P., Kempe, D., Kleinberg, J., Kleinberg, R.: Incentivizing exploration. In: Proceedings of the Fifteenth ACM Conference on Economics and Computation, pp. 5–22 (2014)

10. Gemici, K., Koutsoupias, E., Monnot, B., Papadimitriou, C., Piliouras, G.: Wealth inequality and the price of anarchy (2018)

11. Haghtalab, N.: Foundation of machine learning, by the people, for the People. Ph.D. thesis, Carnegie Mellon University (2018). http://reports-archive.adm.cs.cmu.edu/anon/anon/usr0/ftp/usr/ftp/2018/CMU-CS-18-114.pdf

12. Ibarra, O.H., Kim, C.E.: Fast approximation algorithms for the knapsack and sum of subset problems. J. ACM **22**(4), 463–468 (1975)

13. Kleinberg, J., Papadimitriou, C., Raghavan, P.: A microeconomic view of data mining. Data Min. Knowl. Discov. **2**(4), 311–324 (1998)

14. Kleinberg, J., Papadimitriou, C., Raghavan, P.: Segmentation problems. In: Proceedings of the Thirtieth Annual ACM Symposium on Theory of Computing, STOC 1998, pp. 473–482. Association for Computing Machinery, New York (1998). https://doi.org/10.1145/276698.276860

15. Kleinberg, J., Papadimitriou, C.H., Raghavan, P.: On the value of private information. In: Proceedings of the 8th Conference on Theoretical Aspects of Rationality and Knowledge, TARK 2001, pp. 249–257. Morgan Kaufmann Publishers Inc., San Francisco (2001)

16. Liu, Y., Ho, C.: Incentivizing high quality user contributions: new arm generation in bandit learning. In: McIlraith, S.A., Weinberger, K.Q. (eds.) Proceedings of the Thirty-Second AAAI Conference on Artificial Intelligence, (AAAI-18), New Orleans, Louisiana, USA, 2–7, February 2018, pp. 1146–1153. AAAI Press (2018). https://www.aaai.org/ocs/index.php/AAAI/AAAI18/paper/view/16879

17. Lykouris, T., Tardos, E., Wali, D.: Feedback graph regret bounds for thompson sampling and ucb (2019)

18. Mansour, Y., Slivkins, A., Syrgkanis, V.: Bayesian incentive-compatible bandit exploration. In: Proceedings of the Sixteenth ACM Conference on Economics and Computation, EC 2015, pp. 565–582. Association for Computing Machinery, New York (2015). https://doi.org/10.1145/2764468.2764508

19. Myerson, R.B.: Mechanism design by an informed principal. Econometrica J. Econ. Soc. **51**, 1767–1797 (1983)

20. Papadimitriou, C., Vodrahalli, K., Yannakakis, M.: The platform design problem (2021)

21. Puterman, M.L.: Markov Decision Processes: Discrete Stochastic Dynamic Programming, 1st edn. John Wiley & Sons Inc, USA (1994)

22. Roughgarden, T., Wang, J.R.: Minimizing regret with multiple reserves. In: Proceedings of the 2016 ACM Conference on Economics and Computation, EC 2016, pp. 601–616. Association for Computing Machinery, New York (2016). https://doi.org/10.1145/2940716.2940792

23. Tsitsiklis, J.N., Xu, K.: Delay-predictability trade-offs in reaching a secret goal. Oper. Res. **66**(2), 587–596 (2018). https://doi.org/10.1287/opre.2017.1682

24. Tsitsiklis, J.N., Xu, K., Xu, Z.: Private sequential learning (2018)

25. Vazirani, V.V.: Combinatorial Algorithms for Market Equilibria, pp. 103–134. Cambridge University Press, Cambridge (2007). https://doi.org/10.1017/CBO9780511800481.007

26. Zadimoghaddam, M., Roth, A.: Efficiently Learning from Revealed Preference. In: Goldberg, P.W. (ed.) WINE 2012. LNCS, vol. 7695, pp. 114–127. Springer, Heidelberg (2012). https://doi.org/10.1007/978-3-642-35311-6_9

A Consumer-Theoretic Characterization of Fisher Market Equilibria

Denizalp Goktas[✉], Enrique Areyan Viqueira, and Amy Greenwald

Brown University, 115 Waterman Street, Providence, RI 02906, USA
{denizalp_goktas,eareyan,amy_greenwald}@brown.edu

Abstract. In this paper, we bring consumer theory to bear in the analysis of Fisher markets whose buyers have arbitrary continuous, concave, homogeneous (CCH) utility functions representing locally non-satiated preferences. The main tools we use are the dual concepts of expenditure minimization and indirect utility maximization. First, we use expenditure functions to construct a new convex program whose dual, like the dual of the Eisenberg-Gale program, characterizes the equilibrium prices of CCH Fisher markets. We then prove that the subdifferential of the dual of our convex program is equal to the negative excess demand in the associated market, which makes generalized gradient descent equivalent to computing equilibrium prices via tâtonnement. Finally, we run a series of experiments which suggest that tâtonnement may converge at a rate of $O((1+E)/t^2)$ in CCH Fisher markets that comprise buyers with elasticity of demand bounded by E. Our novel characterization of equilibrium prices may provide a path to proving the convergence of tâtonnement in Fisher markets beyond those in which buyers utilities exhibit constant elasticity of substitution.

Keywords: Market equilibrium · Market dynamics · Fisher market

1 Introduction

One of the seminal achievements in mathematical economics is the proof of existence of equilibrium prices in **Arrow-Debreu competitive economies** [1]. This result, while celebrated, is non-constructive, and thus provides little insight into the computation of equilibrium prices. The computational question dates back to Léon Walras, a French economist, who in 1874 conjectured that a decentralized price-adjustment process he called **tâtonnement**, which reflects market behavior, would converge to equilibrium prices [29]. An early positive result in this vein was provided by Arrow, Block and Hurwicz, who showed that a continuous version of tâtonnement converges in markets with an aggregate demand function that satisfies the **weak gross substitutes (WGS)** property [2]. Unfortunately, following this initial positive result, Herbert Scarf provided his eponymous example of an economy for which the tâtonnement process does not converge, dashing all hopes of the tâtonnement process justifying the concept

© Springer Nature Switzerland AG 2022
M. Feldman et al. (Eds.): WINE 2021, LNCS 13112, pp. 334–351, 2022.
https://doi.org/10.1007/978-3-030-94676-0_19

of market equilibria in general [3]. Nonetheless, further study of tâtonnement in simpler models than a full-blown Arrow-Debreu competitive economy remains important, as some real-world markets are indeed simpler [17].

For market equilibria to be justified, not only should they be backed by a natural price-adjustment process such as tâtonnement, as economists have long argued, they should also be computationally efficient. As Kamal Jain put it, "If your laptop cannot find it, neither can the market" [25]. A detailed inquiry into the computational properties of market equilibria was initiated by Devanur et al. [14,15], who studied a special case of the Arrow-Debreu competitive economy known as the **Fisher market** [5]. This model, for which Irving Fisher computed equilibrium prices using a hydraulic machine in the1890s, is essentially the Arrow-Debreu model of a competitive economy in which there are no firms, and buyers are endowed with an artificial currency [25]. Devanur et al. [14] discovered a connection between the **Eisenberg-Gale convex program** and Fisher markets in which buyers have linear utility functions, thereby providing a (centralized) polynomial time algorithm for equilibrium computation in these markets [14,15].

Their work was built upon by Jain, Vazirani, and Ye [21], who extended the Eisenberg-Gale program to all Fisher markets whose buyers have **continuous, concave, and homogeneous (CCH)** utility functions. Further, they proved that the equilibrium of Fisher markets for buyers with CCH utility functions can be computed in polynomial time by interior point methods.Gao and Kroer [19] go beyond interior point methods to develop algorithms that converge in **linear, quasilinear**, and **Leontief** Fisher markets. However, unlike tâtonnement, these methods provide little insight into how markets reach equilibria.

More recently, Cole and Fleischer [9,11], and Cheung, Cole, and Devanur [8] showed the fast convergence of tâtonnement in Fisher markets where the buyers' utility functions satisfy weak gross substitutes with bounded elasticity of demand, and the **constant elasticity of substitution (CES)** properties respectively, the latter of which is a subset of the class of CCH utility functions [8,9,11]. Aside from tâtonnement being a plausible model of real-world price movements due to its decentralized nature, Cole and Fleischer argue for the plausibility of tâtonnement by proving that it is an abstraction for in-market processes in a real-world-like model called the ongoing market model [9,11]. The plausibility of tâtonnement as a natural price-adjustment process has been further supported by Gillen et al. [20], who demonstrated the predictive accuracy of tâtonnement in off-equilibrium trade settings [20]. This theoretical and empirical evidence for tâtonnement makes it even more important to understand its convergence properties, so that we can better characterize those markets for which we can predict price movements and, in turn, equilibria.

Our Approach and Findings. In consumer theory [23], consumers/buyers are assumed to solve the **utility maximization problem (UMP)**, in which each buyer maximizes its utility constrained by its budget, thereby discovering its optimal demand. Dual to this problem is the **expenditure minimization problem (EMP)**, in which each buyer minimizes its expenditure constrained by its desired utility level, an alternative means of discovering its optimal demand. These two problems are intimately connected by a deep mathematical structure, yet most existing approaches to computing market equilibria focus on UMP only.

In this paper, we exploit the relationship between EMP and equilibrium prices to provide a new convex program, which like the seminal Eisenberg-Gale program characterizes the equilibrium prices of Fisher markets assuming buyers with arbitrary CCH utility functions. Additionally, by exploiting the duality structure between UMP and EMP, we provide a straightforward interpretation of the dual of our program, which also sheds light on the dual of Eisenberg-Gale program. In particular, while it is known that an equilibrium allocation that solves the Eisenberg-Gale program is one that maximizes the buyers' utilities given their budgets at equilibrium prices (UMP; the primal), we show that equilibrium prices are those that minimize the buyers' expenditures at the utility levels associated with their equilibrium allocations (EMP; the dual).

Our characterization of CCH Fisher market equilibria via UMP and EMP also allows us to prove that the subdifferential of the dual of our convex program is equal to the negative excess demand in the corresponding market [8].[1] Consequently, solving the dual of our convex program via generalized gradient descent is equivalent to tâtonnement (just as generalized gradient descent on the dual of the Eisenberg-Gale program is equivalent to tâtonnement [15]).

Finally, we run a series of experiments which suggest that tâtonnement may converge at a rate of $O((1+E)/t^2)$ in CCH markets where buyers have **bounded elasticity of demand (BED)** with elasticity parameter E, a class of markets that includes CES Fisher markets. Assuming bounded elasticity of demand, bounded changes in prices result in bounded changes in demand. A summary of all known tâtonnement convergence rate results, as well as this conjecture, appears in Fig. 1.

Roadmap. In Sect. 2, we introduce essential notation and definitions, and summarize our results. In Sect. 3, we derive the dual of the Eisenberg-Gale program and propose a new convex program whose dual characterizes equilibrium prices in CCH Fisher markets via expenditure functions. In Sect. 4, we show that the subdifferential of the dual of our new convex program is equivalent to the negative excess demand in the market, which implies an equivalence between generalized gradient descent and tâtonnement. In Sect. 5, we include an empirical analysis of tâtonnement in CCH Fisher markets.

---

[1] Similarly, it is known that the subdifferential of the dual of the Eisenberg-Gale program is equal to the negative excess demand in the corresponding market [8]. Our result also implies this known result, since the two programs' objective functions differ only by a constant.

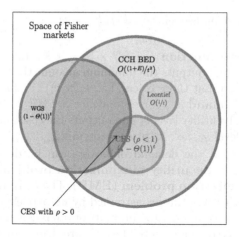

Fig. 1. The convergence rates of tâtonnement in different Fisher markets. We color previous contributions blue, and our conjecture in red. We note that the convergence rate for WGS markets does not apply to markets where the elasticity of demand is unbounded, e.g., linear Fisher markets; likewise, the convergence rate for CES Fisher markets does not apply to linear Fisher markets. (Color figure online)

2   Preliminaries and an Overview of Results

We use Roman uppercase letters to denote sets (e.g., X), bold uppercase letters to denote matrices (e.g., $\boldsymbol{X}$), bold lowercase letters to denote vectors (e.g., $\boldsymbol{p}$), and Roman lowercase letters to denote scalar quantities (e.g., c). We denote the ith row vector of any matrix (e.g., $\boldsymbol{X}$) by the equivalent bold lowercase letter with subscript i (e.g., $\boldsymbol{x}_i$). Similarly, we denote the jth entry of a vector (e.g., $\boldsymbol{p}$ or $\boldsymbol{x}_i$) by the corresponding Roman lowercase letter with subscript j (e.g., p_j or x_{ij}). We denote the set of numbers $\{1, \ldots, n\}$ by $[n]$, the set of natural numbers by $\mathbb{N}$, the set of real numbers by $\mathbb{R}$, the set of non-negative real numbers by $\mathbb{R}_+$ and the set of strictly positive real numbers by $\mathbb{R}_{++}$. We denote by Π_X the Euclidean projection operator onto the set $X \subset \mathbb{R}^n$: i.e., $\Pi_X(\boldsymbol{x}) = \arg\min_{\boldsymbol{z} \in X} \|\boldsymbol{x} - \boldsymbol{z}\|_2$. We also define some set operations. Unless otherwise stated, the sum of a scalar by a set and of two sets is defined as the Minkowski sum, and the product of a scalar by a set and two sets is defined as the Minkowski product.

2.1   Consumer Theory

In this paper, we consider the general class of utility functions $u_i : \mathbb{R}^m \to \mathbb{R}$ that are continuous, concave and homogeneous. The **indirect utility function** $v_i : \mathbb{R}_+^m \times \mathbb{R}_+ \to \mathbb{R}_+$ takes as input prices $\boldsymbol{p}$ and a budget b_i and outputs the maximum utility the buyer can achieve at those prices given that budget, i.e., $v_i(\boldsymbol{p}, b_i) = \max_{\boldsymbol{x} \in \mathbb{R}_+^m : \boldsymbol{p} \cdot \boldsymbol{x} \leq b_i} u_i(\boldsymbol{x})$

The **Marshallian demand** is a correspondence $d_i : \mathbb{R}_+^m \times \mathbb{R}_+ \rightrightarrows \mathbb{R}_+^m$ that takes as input prices $\boldsymbol{p}$ and a budget b_i and outputs the utility maximizing allocation of goods at budget b_i, i.e., $d_i(\boldsymbol{p}, b_i) = \arg\max_{\boldsymbol{x} \in \mathbb{R}^m_+ : \boldsymbol{p} \cdot \boldsymbol{x} \leq b_i} u_i(\boldsymbol{x})$: i.e., $d_i(\boldsymbol{p}, b_i) = \arg\max_{\boldsymbol{x}_i \in \mathbb{R}^m_+ : \boldsymbol{p} \cdot \boldsymbol{x}_i \leq b_i} u_i(\boldsymbol{x}_i)$.

The **expenditure function** $e_i : \mathbb{R}^m_+ \times \mathbb{R}_+ \to \mathbb{R}_+$ takes as input prices $\boldsymbol{p}$ and a utility level ν_i and outputs the minimum amount the buyer must spend to achieve that utility level at those prices, i.e., $e_i(\boldsymbol{p}, \nu_i) = \min_{\boldsymbol{x} \in \mathbb{R}^m_+ : u_i(\boldsymbol{x}) \geq \nu_i} \boldsymbol{p} \cdot \boldsymbol{x}$

The **Hicksian demand** is a correspondence $h_i : \mathbb{R}^m_+ \times \mathbb{R}_+ \rightrightarrows \mathbb{R}_+$ that takes as input prices $\boldsymbol{p}$ and a utility level ν_i and outputs the cost-minimizing allocation of goods at utility level ν_i, i.e., $h_i(\boldsymbol{p}, \nu_i) = \arg\min_{\boldsymbol{x} \in \mathbb{R}^m_+ : u_i(\boldsymbol{x}) \geq \nu_i} \boldsymbol{p} \cdot \boldsymbol{x}$

In consumer theory, the demand of buyers can be determined by studying two dual problems, the **utility maximization problem (UMP)** and **the expenditure minimization problem (EMP)**. The UMP refers to the buyer's problem of maximizing its utility constrained by its budgets (i.e., optimizing its indirect utility function) in order to obtain its optimal demand (i.e., Marshallian demand), while the EMP refers to the buyer's problem of minimizing its expenditure constrained by its desired utility level (i.e., optimizing its expenditure function) in order to obtain its optimal demand (i.e., Hicksian demand). When the utilities are continuous, concave and represent locally non-satiated preferences the UMP and EMP are related through the following identities, which we use throughout the paper:

$$\forall b_i \in \mathbb{R}_+ \qquad\qquad e_i(\boldsymbol{p}, v_i(\boldsymbol{p}, b_i)) = b_i \qquad\qquad (1)$$

$$\forall \nu_i \in \mathbb{R}_+ \qquad\qquad v_i(\boldsymbol{p}, e_i(\boldsymbol{p}, \nu_i)) = \nu_i \qquad\qquad (2)$$

$$\forall b_i \in \mathbb{R}_+ \qquad\qquad h_i(\boldsymbol{p}, v_i(\boldsymbol{p}, b_i)) = d_i(\boldsymbol{p}, b_i) \qquad\qquad (3)$$

$$\forall \nu_i \in \mathbb{R}_+ \qquad\qquad d_i(\boldsymbol{p}, e_i(\boldsymbol{p}, \nu_i)) = h_i(\boldsymbol{p}, \nu_i) \qquad\qquad (4)$$

A good $j \in [m]$ is said to be a **gross substitute (complement)** for a good $k \in [m] \setminus \{j\}$ if $\sum_{i \in [n]} d_{ij}(\boldsymbol{p}, b_i)$ is increasing (decreasing) in p_k. If the aggregate demand, $\sum_{i \in [n]} d_{ij}(\boldsymbol{p}, b_i)$, for good k is instead weakly increasing (decreasing), good j is said to be a **weak gross substitute (complement)** for good k.

The class of homogeneous utility functions includes the well-known **linear**, **Cobb-Douglas**, and **Leontief** utility functions, each of which is a special case of the **Constant Elasticity of Substitution (CES)** utility function family, parameterized by $-\infty \leq \rho \leq 1$, and given by $u_i(\boldsymbol{x}_i) = \sqrt[\rho]{\sum_{j \in [m]} w_{ij} x_{ij}^\rho}$. Linear utility functions are obtained when ρ is 1, while Cobb-Douglas and Leontief utility functions are obtained when ρ approaches 0 and $-\infty$, respectively. For $0 < \rho \leq 1$, goods are gross substitutes, e.g., Sprite and Coca-Cola, for $\rho = 1$; goods are perfect substitutes, e.g., Pepsi and Coca-Cola; and for $\rho < 0$, goods are complementary, e.g., left and right shoes.

The **(price) elasticity of demand** reflect how demand varies in response to a change in price. More specifically, buyer i's elasticity of demand for good $j \in [m]$ with respect to the price of good $k \in [m]$ is defined as $\frac{\partial d_{ij}(\boldsymbol{p}, b_i)}{\partial p_k} \frac{p_k}{d_{ij}(\boldsymbol{p}, b_i)}$. A buyer is said to have **bounded elasticity of demand** with elasticity parameter E if $\min_{\boldsymbol{p} \in \mathbb{R}^m_+, j, k \in [m]} \left\{ -\frac{\partial d_{ij}(\boldsymbol{p}, b_i)}{\partial p_k} \frac{p_k}{d_{ij}(\boldsymbol{p}, b_i)} \right\} = E < \infty$.

2.2 Fisher Markets

A **Fisher market** comprises n buyers and m divisible goods [5]. As is usual in the literature, we assume that there is one unit of each good available [25]. Each buyer $i \in [n]$ has a budget $b_i \in \mathbb{R}_+$ and a utility function $u_i : \mathbb{R}_+^m \to \mathbb{R}$. An instance of a Fisher market is thus given by a tuple $(n, m, U, \boldsymbol{b})$ where $U = \{u_1, \ldots, u_n\}$ is a set of utility functions, one per buyer, and $\boldsymbol{b} \in \mathbb{R}_+^n$ is the vector of buyer budgets. We abbreviate as $(U, \boldsymbol{b})$ when n and m are clear from context.

When the buyers' utility functions in a Fisher market are all of the same type, we qualify the market by the name of the utility function, e.g., a Leontief Fisher market. Considering properties of goods, rather than buyers, a (Fisher) market satisfies **gross substitutes** (resp. **gross complements**) if all pairs of goods in the market are gross substitutes (resp. gross complements). A Fisher market is **mixed** if all pairs of goods are either gross complements or gross substitutes. A Fisher market exhibits **bounded elasticity of demand** with parameter E, if the elasticity of demand of the buyer with highest elasticity of demand is $E < \infty$.

An **allocation** $\boldsymbol{X}$ is a map from goods to buyers, represented as a matrix s.t. $x_{ij} \geq 0$ denotes the quantity of good $j \in [m]$ allocated to buyer $i \in [n]$. Goods are assigned **prices** $\boldsymbol{p} \in \mathbb{R}_+^m$. A tuple $(\boldsymbol{X}^*, \boldsymbol{p}^*)$ is said to be a **competitive (or Walrasian) equilibrium** of Fisher market $(U, \boldsymbol{b})$ if 1. buyers are utility maximizing constrained by their budget, i.e., for all $i \in [n], \boldsymbol{x}_i^* \in d_i(\boldsymbol{p}^*, b_i)$; and 2. the market clears, i.e., for all $j \in [m], p_j^* > 0$ implies $\sum_{i \in [n]} x_{ij}^* = 1$; and $p_j^* = 0$ implies $\sum_{i \in [n]} x_{ij}^* \leq 1$.

If $(U, \boldsymbol{b})$ is a CCH Fisher market, then the optimal solution $\boldsymbol{X}^*$ to the **Eisenberg-Gale program** constitutes an equilibrium allocation, and the optimal solution to the Lagrangian that corresponds to the allocation constraints (Eq. (6)) are the corresponding equilibrium prices [10,14,18]:

Primal

$$\max_{\boldsymbol{X} \in \mathbb{R}_+^{n \times m}} \quad \sum_{i \in [n]} b_i \log\left(u_i(\boldsymbol{x}_i)\right) \tag{5}$$

$$\text{subject to} \quad \sum_{i \in [n]} x_{ij} \leq 1 \qquad \forall j \in [m] \tag{6}$$

We define the **excess demand** correspondence $z : \mathbb{R}^m \rightrightarrows \mathbb{R}^m$, of a Fisher market $(U, \boldsymbol{b})$, which takes as input prices and outputs a set of excess demands at those prices, as the difference between the demand for each good and the supply of each good: $z(\boldsymbol{p}) = \sum_{i \in [n]} d_i(\boldsymbol{p}, b_i) - \boldsymbol{1}_m$. where $\boldsymbol{1}_m$ is the vector of ones of size m.

The **discrete tâtonnement process** for Fisher markets is a decentralized, natural price adjustment, defined as:

$$p(t+1) = p(t) + G(g(t)) \qquad \text{for } t = 0, 1, 2, \ldots \qquad (7)$$
$$g(t) \in z(p(t)) \qquad\qquad (8)$$
$$p(0) \in \mathbb{R}_+^m , \qquad\qquad (9)$$

where $G : \mathbb{R}^m \to \mathbb{R}^m$ is a monotonic function s.t. for all $j \in [m], \boldsymbol{x}, \boldsymbol{y} \in \mathbb{R}^m$, if $x_j \geq y_j$, then $G_j(\boldsymbol{x}) \geq G_j(\boldsymbol{y})$. Intuitively, tâtonnement is an auction-like process in which the seller of $j \in [m]$ increases (resp. decreases) the price of a good if the demand (resp. supply) is greater than the supply (resp. demand).

2.3 Subdifferential Calculus and Generalized Gradient Descent

We say that a vector $\boldsymbol{g} \in \mathbb{R}^n$ is a **subgradient** of a continuous function $f :$ $U \to \mathbb{R}$ at $\boldsymbol{a} \in U$ if for all $\boldsymbol{x} \in U$, $f(\boldsymbol{x}) \geq f(\boldsymbol{a}) + \boldsymbol{g}^T(\boldsymbol{x} - \boldsymbol{a})$. The set of all subgradients $\boldsymbol{g}$ at a point $\boldsymbol{a} \in U$ for a function f is called the **subdifferential** and is denoted by $\partial_x f(\boldsymbol{a}) = \{\boldsymbol{g} \mid f(\boldsymbol{x}) \geq f(\boldsymbol{a}) + \boldsymbol{g}^T(\boldsymbol{x} - \boldsymbol{a})\}$. If f is convex, then its subdifferential exists everywhere. If additionally, f is differentiable at $\boldsymbol{a}$, so that its subdifferential is a singleton at $\boldsymbol{a}$, then the subdifferential at $\boldsymbol{a}$ is equal to the gradient. In this case, we write $\partial_x f(\boldsymbol{a}) = \boldsymbol{g}$; in other words, we take the subdifferential to be vector-valued rather than set-valued.

Consider the optimization problem $\min_{\boldsymbol{x} \in V} f(\boldsymbol{x})$, where $f : \mathbb{R}^n \to \mathbb{R}$ is a convex function that is not necessarily differentiable and V is the feasible set of solutions. Let $\ell_f(\boldsymbol{x}, \boldsymbol{y})$ be the **linear approximation** of f at $\boldsymbol{y}$, that is $\ell_f(\boldsymbol{x}, \boldsymbol{y}) = f(\boldsymbol{y}) + \boldsymbol{g}^T(\boldsymbol{x} - \boldsymbol{y})$, where $\boldsymbol{g} \in \partial_x f(\boldsymbol{y})$. A standard method for solving this problem is the **mirror descent** [24] update rule is as follows:

$$\boldsymbol{x}(t+1) = \arg\min_{\boldsymbol{x} \in V} \{\ell_f(\boldsymbol{x}, \boldsymbol{x}(t)) + \gamma_t \delta_h(\boldsymbol{x}, \boldsymbol{x}(t))\} \qquad \text{for } t = 0, 1, 2, \ldots \qquad (10)$$
$$\boldsymbol{x}(0) \in \mathbb{R}^n \qquad\qquad (11)$$

Here, as above, $\gamma_t > 0$ is the step size at time t and, $\delta_h(\boldsymbol{x}, \boldsymbol{x}(t))$ is the **Bregman divergence** of a convex differentiable **kernel** function $h(\boldsymbol{x})$ defined as $\delta_h(\boldsymbol{x}, \boldsymbol{y}) = h(\boldsymbol{x}) - \ell_h(\boldsymbol{x}, \boldsymbol{y})$ [6]. When the kernel is the scaled weighted entropy $h(\boldsymbol{x}) = c \sum_{i \in [n]} (x_i \log(x_i) - x_i)$, given $c > 0$, then the Bregman divergence reduces to the **scaled generalized Kullback-Leibler divergence**: $\delta_{\mathrm{KL}}(\boldsymbol{x}, \boldsymbol{y}) = c \sum_{i \in [n]} \left[x_i \log\left(\frac{x_i}{y_i}\right) - x_i + y_i \right]$, which, when $V = \mathbb{R}_+^n$, yields the following simplified update rule, where as usual $g(t) \in \partial_x f(\boldsymbol{x}(t))$:

$$\forall j \in [m] \qquad x_j(t+1) = x_j(t) \exp\left\{ \frac{-g_j(t)}{\gamma_t} \right\} \qquad \text{for } t = 0, 1, 2, \ldots \qquad (12)$$
$$x_j(0) \in \mathbb{R}_{++} \qquad\qquad (13)$$

Equations (12) and (13) do not include a projection step, because when the initial iterate is within $\mathbb{R}^n_+$, subsequent iterates remain within this set.

2.4   A High-level Overview of Our Contributions

In this paper, we bring consumer theory to bear in the analysis of CCH Fisher markets. In so doing, we first derive the dual of the Eisenberg-Gale program for arbitrary CCH Fisher markets, generalizing the special cases of linear and Leontief markets, which are already understood [15]. We then provide a new convex program whose dual also characterizes equilibrium prices in CCH Fisher markets via expenditure functions. This program is of interest because the subdifferential of the objective function of its dual is equal to the negative excess demand in the market, which implies that mirror descent on this objective is equivalent to solving for equilibrium prices in the associated market via tâtonnement. Finally, we conjecture a convergence rate of $O((1+E)/t^2)$ for CCH Fisher markets in which the elasticity of buyer demands is bounded by E.

Although the Eisenberg-Gale convex program dates back to 1959, its dual for arbitrary CCH Fisher markets is still not yet well understood. Our first result is to derive the Eisenberg-Gale program's dual, generalizing the two special cases identified by Cole et al. [12] for linear and Leontief utilities.

Theorem 1. *The dual of the Eisenberg-Gale program for any CCH Fisher market $(U, \boldsymbol{b})$ is given by:*

$$\min_{\boldsymbol{p} \in \mathbb{R}^m_+} \sum_{j \in [m]} p_j + \sum_{i \in [n]} \left[b_i \log\left(v_i(\boldsymbol{p}, b_i)\right) - b_i \right] \tag{14}$$

We then propose a new convex program whose dual characterizes the equilibrium prices of CCH Fisher markets via expenditure functions. We note that the optimal value of this convex program differs from the optimal value of the Eisenberg-Gale program by a constant factor.

Theorem 2. *The optimal solution $(\boldsymbol{X}^*, \boldsymbol{p}^*)$ to the primal and dual of the following convex programs corresponds to equilibrium allocations and prices, respectively, of the CCH Fisher market $(U, \boldsymbol{b})$:*

Primal

$$\max_{\boldsymbol{X} \in \mathbb{R}^{n \times m}_+} \sum_{i \in [n]} b_i \log\left(u_i\left(\frac{\boldsymbol{x}_i}{b_i}\right)\right)$$

$$\text{subject to} \quad \forall j \in [m], \sum_{i \in [n]} x_{ij} \leq 1$$

Dual

$$\min_{\boldsymbol{p} \in \mathbb{R}^m_+} \sum_{j \in [m]} p_j - \sum_{i \in [n]} b_i \log\left(\partial_{\nu_i} e_i(\boldsymbol{p}, \nu_i)\right)$$

This convex program formulation for CCH Fisher markets is of particular interest because its subdifferential equals the negative excess demand in the market. As a result, solving this program via (sub)gradient descent is equivalent to solving the market via tâtonnement.

Theorem 3. *The subdifferential of the objective function of the dual of the program given in Theorem 2 for a CCH Fisher market $(U, \boldsymbol{b})$ at any price $\boldsymbol{p}$ is equal to the negative excess demand in $(U, \boldsymbol{b})$ at price $\boldsymbol{p}$:*

$$\partial_{\boldsymbol{p}} \left(\sum_{j \in [m]} p_j - \sum_{i \in [n]} b_i \log \partial_{\nu_i} e_i(\boldsymbol{p}, \nu_i) \right) = -z(\boldsymbol{p}) \tag{15}$$

To prove Theorem 3, we make use of standard consumer theory, specifically the duality structure between UMP and EMP, as well as a generalized version of Shepherd's lemma [26,27]. We also provide a new, simpler proof of this generalization of Shepherd's lemma via Danskin's theorem [13].

Finally, we conduct an experimental investigation of the convergence of the tâtonnement process defined by the mirror descent rule with KL-divergence and fixed step sizes in CCH Fisher markets. This particular process was previously studied by Cheung, Cole, and Devanur [8] in Leontief Fisher markets. They showed a worst-case *lower* bound of $\Omega(1/t^2)$ to complement an $O(1/t)$ worst-case upper bound. These results suggest a possible convergence rate of $O(1/t^2)$ or $O(1/t)$ for entropic tâtonnement in a class of Fisher markets that includes Leontief Fisher markets. Our experimental results support the conjecture that a worst-case convergence rate of $O(1/t^2)$ might hold, not only in Leontief and CES Fisher markets, but in CCH Fisher markets where buyers' elasticity of demand is bounded by E.

3   A New Convex Program for CCH Fisher Markets

In this section, we provide an alternative convex program to the Eisenberg-Gale program, which also characterizes the equilibria of CCH Fisher markets. Of note, our program characterizes equilibrium prices via expenditure functions. For CCH Fisher markets, the Eisenberg-Gale program's primal allows us to calculate the equilibrium allocations, while its dual yields the corresponding equilibrium prices [7]. Cole et al. [12] provide dual formulations of the Eisenberg-Gale program for linear and Leontief utilities [12], and in unpublished work, Cole and Tao [10] present a generalization of the Eisenberg-Gale dual for arbitrary CCH utility functions. However, as we show in the full version, the optimal value of the objective of the Eisenberg-Gale program's primal differs from the optimal value of the dual provided by Cole and Tao [10] by a constant factor, despite their dual characterizing equilibrium prices accurately. Hence, their dual is technically not the dual of the Eisenberg-Gale program for which strong duality holds. The proof of the following theorem stating the Eisenberg-Gale program's dual can be found in the full version.

Theorem 1. *The dual of the Eisenberg-Gale program for any CCH Fisher market $(U, \boldsymbol{b})$ is given by:*

$$\min_{\boldsymbol{p} \in \mathbb{R}_+^m} \sum_{j \in [m]} p_j + \sum_{i \in [n]} [b_i \log (v_i(\boldsymbol{p}, b_i)) - b_i] \tag{14}$$

Before presenting our program, we present several preliminary lemmas. All omitted proofs can be found in the full version.

The next lemma establishes an important property of the indirect utility and expenditure functions in CCH Fisher markets that we heavily exploit in this work, namely that the derivative of the indirect utility function with respect to b_i—the bang-per-buck—is constant across all budget levels. Likewise, the derivative of the expenditure function with respect to ν_i—the buck-per-bang— is constant across all utility levels. In other words, both functions effectively depend only on prices. Not only are the bang-per-buck and the buck-per-bang constant, they equal $v_i(\boldsymbol{p}, 1)$ and $e_i(\boldsymbol{p}, 1)$, respectively, namely their values at exactly one unit of budget and one unit of (indirect) utility.

An important consequence of this lemma is that, by picking prices that maximize a buyer's bang-per-buck, we not only maximize their bang-per-buck at all budget levels, but we further maximize their total indirect utility, given their *known* budget. In particular, given prices $\boldsymbol{p}^*$ that maximize a buyer's bang-per-buck at budget level 1, we can easily calculate the buyer's total (indirect) utility at budget b_i by simply multiplying their bang-per-buck by b_i: i.e., $v_i(\boldsymbol{p}^*, b_i) = b_i v_i(\boldsymbol{p}^*, 1)$. Here, we see the homogeneity assumption at work.

Analogously, by picking prices that maximize a buyer's buck-per-bang, we not only maximize their buck-per-bang at all utility levels, but we further maximize the buyer's total expenditure, given their *unknown* optimal utility level. In particular, given prices $\boldsymbol{p}^*$ that minimize a buyer's buck-per-bang at utility level 1, we can easily calculate the buyer's total expenditure at utility level ν_i by simply multiplying their buck-per-bang by ν_i: i.e., $e_i(\boldsymbol{p}^*, \nu_i) = \nu_i e_i(\boldsymbol{p}^*, 1)$. Thus, solving for optimal prices at any budget level, or analogously at any utility level, requires only a single optimization, in which we solve for optimal prices at budget level, or utility level, 1.

Lemma 1. *If u_i is continuous and homogeneous of degree 1, then $v_i(\boldsymbol{p}, b_i)$ and $e_i(\boldsymbol{p}, \nu_i)$ are differentiable in b_i and ν_i, resp. Further, $\partial_{b_i} v_i(\boldsymbol{p}, b_i) = \{v_i(\boldsymbol{p}, 1)\}$ and $\partial_{\nu_i} e_i(\boldsymbol{p}, \nu_i) = \{e_i(\boldsymbol{p}, 1)\}$.*

The next lemma provides further insight into why CCH Fisher markets are easier to solve than non-CCH Fisher markets. The lemma states that the bang-per-buck, i.e., the marginal utility of an additional unit of budget, is equal to the inverse of its buck-per-bang, i.e., the marginal cost of an additional unit of utility. Consequently, by setting prices so as to minimize the buck-per-bang of buyers, we can also maximize their bang-per-buck. Since the buck-per-bang is a function of prices only, and not of prices and allocations together, this lemma effectively decouples the calculation of equilibrium prices from the calculation of equilibrium allocations, which greatly simplifies the problem of computing equilibria in CCH Fisher markets.

Corollary 1. *If buyer i's utility function u_i is CCH, then*

$$\frac{1}{e_i(\boldsymbol{p}, 1)} = \frac{1}{\partial_{\nu_i} e_i(\boldsymbol{p}, \nu_i)} = \partial_{b_i} v_i(\boldsymbol{p}, b_i) = v_i(\boldsymbol{p}, 1) \ . \tag{16}$$

We can now present our characterization of the dual of the Eisenberg-Gale program via expenditure functions. While Devanur et al. [16] provided a method to construct a similar program to that given in Theorem 2 for specific utility functions, their method does not apply to arbitrary CCH utility functions. The proof of this theorem can be found in the full version.

Theorem 2. *The optimal solution* (X^*, p^*) *to the primal and dual of the following convex programs corresponds to equilibrium allocations and prices, respectively, of the CCH Fisher market* (U, b):

Primal

$$\max_{X \in \mathbb{R}_+^{n \times m}} \sum_{i \in [n]} b_i \log\left(u_i\left(\frac{x_i}{b_i}\right)\right)$$

$$\text{subject to} \quad \forall j \in [m], \sum_{i \in [n]} x_{ij} \leq 1$$

Dual

$$\min_{p \in \mathbb{R}_+^m} \sum_{j \in [m]} p_j - \sum_{i \in [n]} b_i \log\left(\partial_{\nu_i} e_i(p, \nu_i)\right)$$

Our new convex program for CCH Fisher markets, which characterizes equilibrium expenditure functions, makes plain the duality structure between utility functions and expenditure functions that is used to compute "shadow" prices for allocations. In particular, $e_i(p, \nu_i)$ is the Fenchel conjugate of the indicator function $\chi_{\{x : u_i(x_i) \geq \nu_i\}}$, meaning the utility levels and expenditures are dual (in a colloquial sense) to one another. Therefore, equilibrium utility levels can be determined from equilibrium expenditures, and vice-versa, which implies that allocations and prices can likewise be derived from one another through this duality structure.[2]

4   Equivalence of Mirror Descent and Tâtonnement

Cheung, Cole, and Devanur [8] have shown via the Lagrangian of the Eisenberg-Gale program, i.e., without constructing the precise dual, that the subdifferential of the dual of the Eisenberg-Gale program is equal to the negative excess demand in the associated market, which implies that mirror descent equivalent to a subset of tâtonnement rules. In this section, we use a generalization of Shephard's lemma to prove that the subdifferential of the dual of our new convex program is equal to the negative excess demand in the associated market. Our proof also applies to the dual of the Eisenberg-Gale program, since the two duals differ only by a constant factor.

While Shephard's lemma is applicable to utility functions with singleton-valued Hicksian demand (i.e., strictly concave utility functions), we require a generalization of Shephard's lemma that applies to utility functions that are not strictly concave and that could have set-valued Hicksian demand. An early proof of this generalized lemma was given by Tanaka in a discussion paper [27]; a more modern perspective can be found in a recent survey by Blume [4]. For

---

[2] A more in-depth analysis of this duality structure can be found in Blume [4].

completeness, we also provide a new, simple proof of this result via Danskin's theorem (for subdifferentials) [13] the full version.

Lemma 2. *Shephard's lemma, generalized for set-valued Hicksian demand [4, 26, 27] Let $e_i(\boldsymbol{p}, \nu_i)$ be the expenditure function of buyer i and $h_i(\boldsymbol{p}, \nu_i)$ be the Hicksian demand set of buyer i. The subdifferential $\partial_{\boldsymbol{p}} e_i(\boldsymbol{p}, \nu_i)$ is the Hicksian demand at prices $\boldsymbol{p}$ and utility level ν_i, i.e., $\partial_{\boldsymbol{p}} e_i(\boldsymbol{p}, \nu_i) = h_i(\boldsymbol{p}, \nu_i)$.*

The next lemma plays an essential role in the proof that the subdifferential of the dual of our convex program is equal to the negative excess demand. Just as Shephard's Lemma related the expenditure function to Hicksian demand via (sub)gradients, this lemma relates the expenditure function to Marshallian demand via (sub)gradients. One way to understand this relationship is in terms of **Marshallian consumer surplus**, the area under the Marshallian demand curve, i.e., the integral of Marshallian demand with respect to prices.[3] Specifically, by applying the fundamental theorem of calculus to the left-hand side of Lemma 3, we see that the Marshallian consumer surplus equals $b_i \log (\partial_{\nu_i} e_i(\boldsymbol{p}, \nu_i))$. The key takeaway is thus that any objective function we might seek to optimize that includes a buyer's Marshallian consumer surplus is thus optimizing their Marshallian demand, so that optimizing this objective yields a utility-maximizing allocation for the buyer, constrained by their budget.

Lemma 3. *If buyer i's utility function u_i is CCH, then $\partial_{\boldsymbol{p}} \left(b_i \log \left(\partial_{\nu_i} e_i(\boldsymbol{p}, \nu_i) \right) \right) = d_i(\boldsymbol{p}, b_i)$.*

Remark 1. Lemma 3 makes the dual of our convex program easy to interpret, and thus sheds light on the dual of the Eisenberg-Gale program.

Specifically, we can interpret the dual as specifying prices that minimize the distance between the sellers' surplus and the buyers' Marshallian surplus.

Remark 2. The lemmas we have proven in this section and the last provide a possible explanation as to why no primal-dual type convex program is known that solves Fisher markets when buyers have *non-homogeneous* utility functions, in which the primal describes optimal allocations while the dual describes equilibrium prices. By the homogeneity assumption, a CCH buyer can increase their utility level (resp. decrease their spending) by $c\%$ by increasing their budget (resp. decreasing their desired utility level) by $c\%$. This observation implies that the marginal expense of additional utility, i.e., "bang-per-buck", and the marginal utility of additional budget, i.e., "buck-per-bang", are constant (Lemma 1). Additionally, optimizing prices to maximize buyers' "bang-per-buck" is equivalent to optimizing prices to minimize their "buck-per-bang" (Corollary 1). Further, optimizing prices to minimize their "buck-per-bang" is equivalent to maximizing their utilities constrained by their budgets (Lemma 3).

---

[3] We note that the definition of Marshallian consumer surplus for multiple goods requires great care and falls outside the scope of this paper. More information on consumer surplus can be found in Levin [22], and Vives [28].

Thus, the equilibrium prices computed by the dual of our program, which optimize the buyers' buck-per-bang, simultaneously optimize their utilities constrained by their budgets. In particular, equilibrium prices can be computed without reference to equilibrium allocations (Corollary 1 + Lemma 3). In other words, assuming homogeneity, the computation of the equilibrium allocations and prices can be isolated into separate primal and dual problems.

Next, we show that the subdifferential of the dual of our convex program is equal to the negative excess demand in the associated market.

Theorem 3. *The subdifferential of the objective function of the dual of the program given in Theorem 2 for a CCH Fisher market $(U, \boldsymbol{b})$ at any price $\boldsymbol{p}$ is equal to the negative excess demand in $(U, \boldsymbol{b})$ at price $\boldsymbol{p}$:*

$$\partial_{\boldsymbol{p}} \left(\sum_{j \in [m]} p_j - \sum_{i \in [n]} b_i \log \partial_{\nu_i} e_i(\boldsymbol{p}, \nu_i) \right) = -z(\boldsymbol{p}) \tag{15}$$

Cheung, Cole, and Devanur [8] define a class of markets called **convex potential function (CPF)** markets. A market is a CPF market, if there exists a convex potential function φ such that $\partial_{\boldsymbol{p}} \varphi(\boldsymbol{p}) = -z(\boldsymbol{p})$. They then prove that Fisher markets are CPF markets by showing, through the Lagrangian of the Eisenberg-Gale program, that its dual is a convex potential function [8]. Likewise, Theorem 3 implies the following:

Corollary 2. *All CCH Fisher markets are CPF markets.*

Proof. A convex potential function $\phi : \mathbb{R}^m \to \mathbb{R}$ for any CCH Fisher market $(U, \boldsymbol{b})$ is given by:

$$\varphi(\boldsymbol{p}) = \sum_{j \in [m]} p_j - \sum_{i \in [n]} b_i \log \left(\partial_{\nu_i} e_i(\boldsymbol{p}, \nu_i) \right) \tag{17}$$

Fix a kernel function h for the Bregman divergence δ_h. If the mirror descent procedure given in Eqs. (10) and (11) is run on Eq. (17) (i.e., choose $f = \varphi$), it is then equivalent to the tâtonnement process for some monotonic function of the excess demand [8].

Thus, by varying the kernel function h of the Bregman divergence we can obtain different tâtonnement rules. For instance, if $h = \frac{1}{2}\|x\|_2^2$, the mirror descent process reduces to the classic tâtonnement rule given by $G(\boldsymbol{x}) = \gamma_t \boldsymbol{x}$, for $\gamma_t > 0$ and for all $t \in \mathbb{N}$, in Eqs. (7) to (9).

5   Convergence of Discrete Tâtonnement

In this section, we conduct an experimental investigation[4] of the rate of convergence of **entropic tâtonnement**, which corresponds to the tâtonnement

---

[4] Our code can be found on https://github.com/denizalp/fisher-tatonnement.git.

process given by mirror descent with the scaled generalized Kullback-Leibler (KL) divergence, specifically $6\delta_{\mathrm{KL}}(\boldsymbol{p}, \boldsymbol{q})$, as the Bregman divergence, and a fixed step size γ. This particular update rule, which reduces to Eqs. (12) and (13), has been the focus of previous work [8]. Interest in this update rule stems from the fact that prices can never reach 0, which ensures that demands, and as a consequence, excess demands, are bounded throughout the tâtonnement process. This is because the demand for any good j is always upper bounded by $\frac{\sum_{i \in [n]} b_i}{p_j}$. Before presenting experimental results for entropic tâtonnement, we note that the process is not guaranteed to converge in all CCH Fisher markets. It does not converge, for example, in linear Fisher markets. An example of such a market can be found in the full version.

Cheung, Cole, and Devanur [8] proved a worst-case *lower* bound of $\Omega(1/t^2)$ to complement their $O(1/t)$ worst-case upper bound for the convergence rate of entropic tâtonnement in Leontief markets. These results suggest a possible convergence rate of $O(1/t^2)$ or $O(1/t)$ for entropic tâtonnement for a class of Fisher markets that includes Leontief markets. The goal of our experiments is to better understand the class of Fisher markets for which entropic tâtonnement converges, and to see if a worst-case convergence rate of $O(1/t^2)$ or $O(1/t)$ might hold, not only for Leontief, but for a larger class of CCH Fisher, markets.

In all our experiments, we randomly generated mixed CES Fisher markets, each with 70 buyers and 30 goods. The buyers' values for goods, and their budgets, were drawn uniformly between 2 and 3. We drew initial prices uniformly in the range $[2, 3]$. In our first two experiments, we initialized 10,000 mixed CES markets, and we chose the ρ parameter uniformly at random with $1/2$ probability in the range $[1/4, 3/4]$ and with $1/2$ probability in the range $[-1, -101]$.[5] Note that this range for ρ ensures that the elasticity of demand E of the market is bounded above by 4. Under these conditions, we ran the entropic tâtonnement process with a step size of 2 in each market.

In our first set of experiments, we assigned each buyer, uniformly at random, either CES, Cobb-Douglas, or Leontief utilities, with $E \leq 4$. We observed convergence in all experiments, at the rate depicted in Fig. 2a. These results suggest that the sublinear convergence rate of $O(1/t)$ could be improved to $O(1/t^2)$ for entropic tâtonnement in Leontief markets, and could perhaps even be extended to a larger class of Fisher markets, beyond Leontief. (The inner frame in Fig. 2a is a closeup of iterations 0 to 10, intended to highlight that the average trajectory of the objective value throughout entropic tâtonnement decreases at a rate faster than $O(1/t^2)$.)

We then ran the same experiment with buyers with linear utilities included (i.e., unbounded elasticity of demand), and found that out of 10,000 experiments, 9889 of them did *not* converge. This result is unsurprising in light of the fact that tâtonnement is not guaranteed to converge in linear markets, since, in expectation, buyers with linear utilities make up a quarter of this market.

Finally, we ran experiments in which we varied the elasticity of demand. To do so, we ran tâtonnement in markets with elasticities of demand $E \in$

---

[5] We ruled out values of ρ close to 0 and 1 to ensure numerical stability.

$\{0.1, 0.2, \ldots, 0.9\}$, and we varied the step size $\gamma \in \{1, 2, \ldots, 9\}$. The results are presented in Fig. 2b. In this heat map, purple signifies that all experiments converged, while yellow signifies that no experiments converged. Interestingly, as the elasticity of demand of the market increased, prices still converged, albeit only with a sufficiently large step size, thus at a slower rate.

In the light of the results of our experiments, we conjecture that tâtonnement converges at a rate of $O((1+E)/t^2)$ in CCH Fisher markets. We recall that for Leontief utilities $E = 0$, for weak gross complements markets $E \leq 1$, for weak gross substitutes markets $E \geq 1$, and for linear utilities $E = \infty$. Our conjecture thus implies that a convergence rate of $O(1/t^2)$ applies for Leontief Fisher markets, i.e., perfect complements, and that this rate deteriorates as the market's elasticity of demand increases, ultimately leading to non-convergence in markets of perfect substitutes, i.e., linear Fisher markets. That is, the convergence rate of tâtonnement in CCH Fisher markets can be seen as a combination of the convergence rates of two types of extreme markets: perfect complements, i.e., Leontief, and perfect substitutes, i.e., linear, Fisher markets.

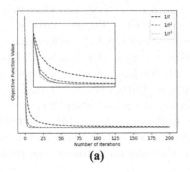
(a)
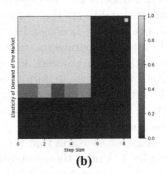
(b)

Fig. 2. (a) Average trajectory of the value of the objective function throughout tâtonnement with KL divergence for mixed CES Fisher markets with $E \leq 4$ is drawn in red. The predicted worst case sublinear convergence rate is depicted by a dashed blue line. A convergence rate of $1/t^2$ and $1/t^3$ are denoted in green and orange, respectively. (b) Percentage of experiments that con- verge as a function of step size and elasticity of demand. Purple signfies that all experiments converged; yellow signfies that no experiments converged. For sufficiently low values of E, we see convergence regardless of step size; and for sufficiently large step sizes, we see convergence regardless of E. (Color figure online)

6   Conclusion

In this paper, we introduced a new convex program whose dual characterizes the equilibrium prices of CCH Fisher markets via expenditure functions. We also related this dual to the dual of the Eisenberg-Gale program. The dual of our program is easily interpretable, and thus allows us to likewise interpret the Eisenberg-Gale dual. In particular, while it is known that an equilibrium allocation that solves the Eisenberg-Gale program (the primal) is one that maximizes the Nash social welfare, we show that equilibrium prices—the solution to the dual—minimize the distance between the sellers' surplus and the buyers' Marshallian surplus. Building on the results of Cheung, Cole, and Devanur [8], who showed that the subdifferential of the dual of the Eisenberg-Gale program is equal to the negative excess demand, we show the same for the dual of our convex program, which implies that solving our convex program via generalized gradient descent is equivalent to solving a Fisher market by means of tâtonnement.

The main technical innovation in this work is to express equilibrium prices via expenditure functions. This insight could allow us to prove the convergence of tâtonnement for more general classes of CCH utility functions, beyond CES. To this end, we ran experiments that supported the conjecture that tâtonnement converges at a rate of $O\left((1+E)/t^2\right)$ in CCH Fisher markets with elasticity of demand bounded by E. If this result holds in general, it would improve upon and generalize prior results for Leontief markets to a larger class of CCH markets, which includes nested and mixed CES utilities. In future work, we plan to continue to investigate this conjecture, using the insights gained from our consumer-theoretic characterization of the equilibrium prices of Fisher markets.

We believe that our analysis offers important insights about the Eisenberg-Gale program. We observe that in CCH markets, maximizing the bang-per-buck is equivalent to minimizing the buck-per-bang, and moreover, the buck-per-bang and bang-per-buck are constant across utility levels and budgets. Additionally, optimizing prices to minimize buyers' buck-per-bang is equivalent to maximizing their utilities constrained by their budgets. As a result, equilibrium prices can be determined by minimizing the buck-per-bang of buyers, which depends only on prices. In other words, the computation of equilibrium prices can be decoupled from the computation of equilibrium allocations. Indeed, there exists a primal-dual convex program for these markets. The challenge in solving Fisher markets where buyers' utility functions can be non-homogeneous seems to stem from the fact that the buck-per-bang and bang-per-buck vary across utility levels and budget, which in turn means that the computation of prices and allocations cannot be decoupled. As a result, we suspect that a primal-dual convex program formulation that solves Fisher markets for buyers with non-homogeneous utility functions may not exist.

Acknowledgements. We would like to thank Richard Cole, Yun Kuen Cheung, and Yixin Tao for feedback on an earlier version of this paper. This work was partially supported by NSF Grant CMMI-1761546.

References

1. Arrow, K.J., Debreu, G.: Existence of an equilibrium for a competitive economy. Econometrica J. Econ. Soc. **22**, 265–290 (1954)

2. Arrow, K.J. Hurwicz, L.: On the stability of the competitive equilibrium, i. Econometrica **26**(4), 522–552 (1958). ISSN 00129682, 14680262. http://www.jstor.org/stable/1907515

3. Arrow, K.J., Kehoe,T.J.: Distinguished fellow: Herbert scarf's contributions to economics. J. Econ. Perspect. **8**(4), 161–181 (1994). ISSN 08953309, http://www.jstor.org/stable/2138344

4. Blume, L.E.: Duality. The New Palgrave Dictionary of Economics, pp. 1–7 (2017). https://doi.org/10.1057/978-1-349-95121-5_285-2

5. Brainard, W.C., Scarf, H.E. et al.: How to Compute Equilibrium Prices in 1891. Citeseer, Princeton (2000)

6. Bregman, L.M.: The relaxation method of finding the common point of convex sets and its application to the solution of problems in convex programming. USSR Comput. Math. Math. Phys. **7**(3), 200–217 (1967)

7. Chen, L., Ye, Y., Zhang, J.: A note on equilibrium pricing as convex optimization. In: Deng, X., Graham, F.C. (eds.) WINE 2007. LNCS, vol. 4858, pp. 7–16. Springer, Heidelberg (2007). https://doi.org/10.1007/978-3-540-77105-0_5

8. Cheung, Y.K., Cole, R., Devanur, N.: Tatonnement beyond gross substitutes? gradient descent to the rescue. In Proceedings of the Forty-Fifth Annual ACM Symposium on Theory of Computing, STOC 2013, pp. 191–200. Association for Computing Machinery, New York (2013). ISBN 9781450320290, https://doi.org/10.1145/2488608.2488633

9. Cole, R., Fleischer, L.: Fast-converging tatonnement algorithms for one-time and ongoing market problems. In Proceedings of the Fortieth Annual ACM Symposium on Theory of Computing, pp. 315–324 (2008)

10. Cole, R., Tao, Y.: Balancing the robustness and convergence of tatonnement (2019)

11. Cole, R., Fleischer, L., Rastogi, A.: Discrete price updates yield fast convergence in ongoing markets with finite warehouses (2010)

12. Cole, R., et al.: Convex program duality, fisher markets, and nash social welfare (2016)

13. Danskin, J.M.: The theory of max-min, with applications. SIAM J. Appl. Math. **14**(4), 641–664 (1966). ISSN 00361399, http://www.jstor.org/stable/2946123

14. Devanur, N.R., Papadimitriou, C.H., Saberi, A., Vazirani, V.V.: Market equilibrium via a primal-dual-type algorithm. In: The 43rd Annual IEEE Symposium on Foundations of Computer Science, 2002. Proceedings, pp. 389–395 (2002). https://doi.org/10.1109/SFCS.2002.1181963

15. Devanur, N.R., Papadimitriou, C.H., Saberi, A., Vazirani, V.V.: Market equilibrium via a primal-dual algorithm for a convex program. J. ACM (JACM) **55**(5), 1–18 (2008)

16. Devanur, N.R., Jain, K., Mai, T., Vazirani, V.V., Yazdanbod, S.: New convex programs for fisher's market model and its generalizations. arXiv preprint arXiv:1603.01257 (2016)

17. Duffie, D., Sonnenschein, H.: Arrow and general equilibrium theory. J. Econ. Lit. **27**(2), 565–598 (1989). ISSN 00220515, http://www.jstor.org/stable/2726689

18. Eisenberg, E., Gale, D.: Consensus of subjective probabilities: the pari-mutuel method. Annal. Math. Stat. **30**(1), 165–168 (1959)

19. Gao, Y., Kroer, C.: First-order methods for large-scale market equilibrium computation. In: NeurIPS (2020). https://arxiv.org/abs/2006.06747

20. Gillen, B.J., Hirota, M., Hsu, M., Plott, C.R., Rogers, B.W.: Divergence and convergence in scarf cycle environments: experiments and predictability in the dynamics of general equilibrium systems. Econ. Theor. **71**, 1–52 (2020)

21. Jain, K., Vazirani, V.V., Ye, Y.: Market equilibria for homothetic, quasi-concave utilities and economies of scale in production. In: SODA, vol. 5, pp. 63–71 (2005)
22. Levin, J.: Lecture notes on consumer theory, October 2004
23. Mas-Colell, A., Whinston, M.D., Green, J.R.: Microeconomic theory. Number 9780195102680 in OUP Catalogue. Oxford University Press (1995). https://ideas. repec.org/b/oxp/obooks/9780195102680.html
24. Nemirovskij, A.S., Yudin, D.B.: Problem complexity and method efficiency in optimization (1983)
25. Nissan, N., Roughgarden, T.: Algorithmic Game Theory. Cambridge University Press, Cambridge (2007). https://doi.org/10.1017/CBO9780511800481
26. Shephard, R.W.: Theory of Cost and Production Functions. Princeton University Press, Princeton (2015)
27. Tanaka, Y.: Nonsmooth optimization for production theory. Hokkaido University, Sapporo (2008)
28. Vives, X.: Small income effects: a marshallian theory of consumer surplus and downward sloping demand. Rev. Econ. Stud. **54**(1), 87–103 (1987). ISSN 00346527, 1467937X, http://www.jstor.org/stable/2297448
29. Walras, L.: Elements of Pure Economics; or, The Theory of Social Wealth. In: Jaffé., W. (ed.)Orion Editions, vol. 2 (1969)

Learning, Fairness, Privacy
and Behavioral Models

A Tight Negative Example for MMS Fair Allocations

Uriel Feige[1], Ariel Sapir[2(✉)], and Laliv Tauber[1]

[1] Weizmann Institute, Rehovot, Israel
{uriel.feige,laliv.tauber}@weizmann.ac.il
[2] Ribbon Communications, Petah Tikva, Israel
arielsa@cs.bgu.ac.il

Abstract. We consider the problem of allocating indivisible goods to agents with additive valuation functions. Kurokawa, Procaccia and Wang [JACM, 2018] present instances for which every allocation gives some agent less than her maximin share. We present such examples with larger gaps. For three agents and nine items, we design an instance in which at least one agent does not get more than a $\frac{39}{40}$ fraction of her maximin share. Moreover, we show that there is no negative example in which the difference between the number of items and the number of agents is smaller than six, and that the gap (of $\frac{1}{40}$) of our example is worst possible among all instances with nine items.

For $n \geq 4$ agents, we show examples in which at least one agent does not get more than a $1 - \frac{1}{n^4}$ fraction of her maximin share. In the instances designed by Kurokawa, Procaccia and Wang, the gap is exponentially small in n.

Our proof techniques extend to allocation of chores (items of negative value), though the quantitative bounds for chores are different from those for goods. For three agents and nine chores, we design an instance in which the MMS gap is $\frac{1}{43}$.

Keywords: Fair division · Indivisible items · Maximin share · Goods · Chores

1 Introduction

We consider allocation problems with m items, n agents, and nonnegative additive valuation functions. The *maximin share* (MMS) of an agent i is the highest value w_i, such that if all agents have the same valuation function v_i that i has, there is an allocation in which every agent gets value at least w_i. An allocation is *maximin fair* if every agent gets a bundle that she values at least as much as her MMS. Hence if all agents have the same valuation function, a maximin fair allocation exists.

Perhaps surprisingly, if agents have different additive valuation functions, then a maximin fair allocation need not exist. Kurokawa, Procaccia and Wang [10] present negative examples showing that for every $n \geq 3$, there are

© Springer Nature Switzerland AG 2022
M. Feldman et al. (Eds.): WINE 2021, LNCS 13112, pp. 355–372, 2022.
https://doi.org/10.1007/978-3-030-94676-0_20

instances for which in every allocation, at least one agent does not receive her MMS. The *gap* (namely, the fraction of MMS lost by some agent) is not stated explicitly in [10]. However, one can derive explicit gaps from their examples by substituting values for certain parameters that are used in the examples. Doing so gives gaps that are exponentially small in n. Even for small n, the gaps shown by these examples are orders of magnitude smaller than the positive results that are known, where these positive results show that (for additive valuations) there always is an allocation that gives every agent at least a $\frac{3}{4} + \Omega(\frac{1}{n})$ of her MMS [3,6,8,10].

We present negative examples with substantially larger gaps than those shown in [10]. The motivation for designing such examples is that they are needed if one is to ever establish tight bounds on the fraction of the MMS that can be guaranteed to be given to agents. Though we are far from establishing tight bounds for general instances, our bounds are tight in special cases. In particular, when there are at most nine items and the additive valuation functions are integer valued, our results imply the following tight threshold phenomenon. If for every agent, her valuation function is such that the sum of all item values is at most 119, then a maximin fair allocation always exists. If the sum of item values is 120, then there are instances in which a maximin fair allocation does not exist, and then the gap is $\frac{1}{40}$. If the sum of item values is larger than 120, the gap cannot be larger then $\frac{1}{40}$.

1.1   Our Results

The term *negative example* will refer to an allocation instance with additive valuation functions in which there is no allocation that gives every agent her MMS. The term $Gap(n,m)$ refers to the largest possible value $\delta \geq 0$, such that there is an allocation instance with m items and n agents with additive valuations, such that in every allocation there is an agent that gets at most a $1 - \delta$ fraction of her MMS.

In our work, we find negative examples with the smallest possible number of items. The number of items turns out to be nine. Among allocation instances with nine items, we find the allocation instance with the largest gap. Theorem 1 is based on this allocation instance.

Theorem 1. *There is an allocation instance with three agents and nine items for which in every allocation, at least one of the agents does not get more than a $\frac{39}{40}$ fraction of her MMS. In other words,*

$$Gap(n = 3 \, , \, m = 9) \geq \frac{1}{40}.$$

The minimality of the number items in Theorem 1 is implied by Theorem 2, together with Proposition 1 that implies that $n \geq 3$ in every negative example.

Theorem 2. *For every n, every allocation instance with n agents and $m \leq n+5$ items has an allocation in which every agent gets her MMS. In other words,*

$$Gap(n \geq 1 \, , \, m \leq n + 5) = 0$$

A weaker version of Theorem 2 (with $m \leq n+3$) was previously proved in [4].

The maximilaty of the gap in Theorem 1 (when there are nine items) is implied by Theorem 3.

Theorem 3. *Every allocation instance with three agents and nine items has an allocation in which every agent gets at least a $\frac{39}{40}$ fraction of her MMS. In other words,*

$$Gap(n = 3 \, , \, m = 9) \leq \frac{1}{40}$$

The proof of Theorem 3 is based on analysis that reduces the infinite space of possible negative examples into a finite number of classes. For each class, the negative example with largest possible gap within the class can be determined by solving a linear program. For every class we solved the respective linear program using a standard LP solver, and verified that there is no negative example (with 9 items) for which the gap is larger than $\frac{1}{40}$.

Theorem 1 is concerned with three agents. We also provide negative examples for every number of agents $n \geq 4$. The gaps in these negative examples deteriorate at a rate that is polynomial in $\frac{1}{n}$.

Theorem 4. *For every $n \geq 4$, there is an allocation instance with n agents and at most $3n + 3$ items for which in every allocation, at least one of the agents does not get more than a $1 - \frac{1}{n^4}$ fraction of her MMS. In other words,*

$$Gap(n \geq 4 \, , \, m \leq 3n + 3) \geq \frac{1}{n^4}$$

Our negative examples that prove Theorem 4 are inspired by, and contain ingredients from, the negative examples presented in [10]. The new aspect in our constructions is the formulation of Lemma 1, the observation that this lemma suffices for the proofs to go through (a related but more demanding property was used in [10]), and a design, based on modular arithmetic, that satisfies the Lemma.

The techniques of this paper extend from allocation of goods to allocation of chores (items of negative value, or equivalently, positive dis-utility). We find that results for chores are qualitatively similar to those for goods, though quantitative values of the gaps are different from those values for goods. Likewise, the proof techniques for the case of chores are similar to those shown in this paper for goods, though some of the details change. To simplify the presentation in this paper, all sections of the paper refer only to allocation of goods, except for Sect. 6 that refers only to allocation of chores. Section 6 is kept short, and presents only the adaptation of Theorem 1 to the case of chores.

Theorem 5. *There is an allocation instance with three agents and nine chores for which in every allocation, at least one of the agents does not get less than a $\frac{44}{43}$ fraction of her MMS (of dis-utility). In other words, the instance has an MMS gap of $\frac{1}{43}$.*

We have verified that with eight chores, there always is an allocation giving every agent no more dis-utility than her MMS, and (using a computer assisted proof) that for nine items, $\frac{1}{43}$ is the largest possible gap. However, we omit details of this verification from this manuscript.

1.2   Related Work

In this section we review related work that is most relevant to the current paper. In particular, we shall only review papers that concern the maximin share (there are numerous papers considering other fairness notions), and only in the context of nonnegative additive valuation functions (some of the works we cite consider also other classes of valuation functions).

The maximin share was introduced by Budish [5]. The fact that there are allocations instances with additive valuations in which no MMS allocation exists was shown in [10]. That paper presents an instance with three agents and twelve items that has no MMS allocation. The gap in that instance as presented in that paper is around 10^{-6}, though by optimizing parameters associated with the instance it is possible to reduce the gap to the order of 10^{-3}. The paper also shows that for every $n \geq 4$ there are instances with $3n + 4$ items and no MMS allocation. The gaps in these instances are exponentially small in n, and this is inherent in the construction given in that paper.

Work on proving the existence of allocations that give a large fraction of the MMS was initiated in [10]. The largest fraction currently known is $\frac{3}{4} + \frac{1}{12n}$ [6].

For the case of three agents, it was shown in [7] that there is an allocation that gives every agent at least a $\frac{8}{9}$ fraction of her MMS. Our Theorem 1 shows that one cannot guarantee more than a $\frac{39}{40}$ fraction in this case. For the case of four agents, it was shown in [8] that there is an allocation that gives every agent at least a $\frac{4}{5}$ fraction of her MMS.

In [4] it was shown that an MMS allocation always exists if $m \leq n + 3$. We improve the bound to $m \leq n+5$, and show that this is best possible when $n = 3$.

In [1] it was shown that if all items have values in $\{0, 1, 2\}$ then an MMS allocation exists. Our negative example in Theorem 1 uses integer values as high as 26.

1.3   Preliminaries

An allocation instance has a set $M = \{e_1, \ldots, e_m\}$ of m items and a set $\{1, \ldots n\}$ of n agents. The term *bundle* will always denote a set of items. Every agent i has a valuation function $v_i : 2^M \rightarrow R$ that assigns a value to every possible bundle of items. We assume throughout that valuation functions v are normalized ($v(\emptyset) = 0$) and monotone ($v(S) \leq v(T)$ for all $S \subset T \subseteq M$). An n-partition of M is a partition of M into n disjoint bundles. $P_n(M)$ denotes the set of all n-partitions of M. An allocation $A = (A_1, \ldots, A_n)$ is an n-partition of M, with the interpretation that for every $1 \leq i \leq n$, agent i receives bundle A_i. The utility that agent i derives from this allocation is $v_i(A_i)$.

Definition 1. *Consider an allocation instance with a set $M = \{e_1, \ldots, e_m\}$ of m items and a set $\{1, \ldots n\}$ of n agents. Then the maximin share of agent i, denoted by MMS_i, is the maximum over all n-partitions of M, of the minimum value under v_i of a bundle in the n-partition.*

$$MMS_i = \max_{(B_1,\ldots,B_n)\in P_n(M)} \min_j [v_i(B_j)]$$

An n-partition that maximizes the above expression will be referred to as an MMS_i-partition.

An allocation that gives every agent at least her MMS is referred to as an MMS allocation.

A valuation function v is additive if $v(S) = \sum_{e\in S} v(e)$. Though Definition 1 applies to arbitrary valuation functions, in this paper we shall only consider additive valuation functions.

By convention, in all remaining parts of the paper, all valuation functions are additive, unless explicitly stated otherwise.

We now review some known propositions concerning the MMS (with additive valuations). For completeness, we also sketch the proofs of these propositions, though we emphasize that all propositions in this section were known and are not original contributions of the current paper.

Proposition 1. *Every allocation instance in which either all agents or all agents but one have the same valuation function has an MMS allocation.*

Proof. Let $v = v_1 = \ldots = v_{n-1}$ be the valuation function shared by all agents but one, and let v_n be the valuation function of agent n who may have a different valuation function. Let $B_1, \ldots, B_n$ be an MMS partition with respect to v. For every agent i with $1 \leq i \leq n-1$, every one of these bundles has value at least MMS_i. Allocate to agent n the bundle j that maximizes $v_n(B_j)$, and allocate the remaining bundles to the other agents. Additivity of v_n implies that $v_n(B_j) \geq MMS_n$, and hence every agent gets at least her MMS.

The following three propositions concern reduction steps that allow us to replace an allocation instance by a simpler one.

An allocation instance with additive valuations and m items $\{e_1, \ldots, e_j\}$ is *ordered* if for every agent i and every two items e_j and e_k with $j < k$ we have that $v_i(e_j) \geq v_i(e_k)$. Given an unordered allocation instance with additive valuations and m items, its *ordered version* is obtained by replacing the valuation function v_i of each agent i by a new additive valuation function v_i' in which item values are non-increasing. That is, let σ denote a permutation over m items with respect to which the values of items are non-increasing under v_i. Then for every $1 \leq j \leq m$ we have that $v_i'(e_j) = v_i(e_{\sigma^{-1}(j)})$. The following proposition is due to [4].

Proposition 2. *For every instance I with additive valuations, every allocation A' for its ordered version I' can be transformed to an allocation A for I, while ensuring that every agent derives at least as high utility from A in I as derived from A' in I'.*

Proof. A choosing sequence is a sequence of names of agents (repetitions are allowed). The choosing sequence induces an allocation by the following procedure. Starting from round 1, in each round r, the agent whose name appears in the rth location in the choosing sequence receives the item of highest value for the agent (ties can be broken arbitrarily), among the yet unallocated items. The allocation A' for I' induces a choosing sequence, where for every r, the agent in location r is the one to which A' allocated the rth most valuable item in I'. Using this choosing sequence for the instance I, in every round r, the respective agent gets an item that she values at least as her rth most valuable item, which is the value of the item that she got under A'.

Proposition 2 implies that when searching for a negative example with the maximum possible gap, it suffices to restrict attention to ordered instances.

The following two propositions are helpful for arguments that are based on induction on n. As each such proposition concerns two instances, in the MMS notation we shall specify which instance we refer to.

Proposition 3. *Let I be an arbitrary allocation instance with a set M of items and n agents. Let I' be an allocation instance derived from I by removing an arbitrary item e from M, and removing one arbitrary agent. Then for each of the remaining agent q, $MMS_q(I') \geq MMS_q(I)$.*

Proof. Let $(B_1, \ldots, B_n)$ be an $MMS_q(I)$ partition. By renaming bundles, we may assume without loss of generality that $e \in B_n$. Then $(B_1, \ldots, B_{n-2}, (B_{n-1} \cup B_n \backslash \{e\}))$ is an $(n-1)$ partition for $M \backslash \{e\}$ that certifies that $MMS_q(I') \geq MMS_q(I)$.

Proposition 4. *Let I be an arbitrary allocation instance with a set M of $m \geq 2$ items, and n agents. Let I' be an allocation instance derived from I by removing two items e_i and e_j from M, and removing one arbitrary agent. Then for every remaining agent q, if either the $MMS_q(I)$ partition has a bundle that contains both e_i and e_j, or $v_q(e_i) + v_q(e_j) \leq MMS_q(I)$, then $MMS_q(I') \geq MMS_q(I)$.*

Proof. Let $(B_1, \ldots, B_n)$ be an $MMS_q(I)$ partition for I. If both e_i and e_j belong to the same bundle, then the proof is as in that for Proposition 3. If e_i and e_j are in different bundles, by renaming bundles, we may assume without loss of generality that $e_i \in B_{n-1}$ and $e_j \in B_n$. Then $(B_1, \ldots, B_{n-2}, (B_{n-1} \backslash \{e_i\}) \cup (B_n \backslash \{e_j\}))$ is an $(n-1)$ partition for $M \backslash \{e_i, e_j\}$ that certifies that $MMS_q(I') \geq MMS_q(I)$. This is because $v_q((B_{n-1} \backslash \{e_i\}) \cup (B_n \backslash \{e_j\})) = v_q(B_{n-1}) + v_q(B_n) - (v_q(e_i) + v_q(e_j)) \geq 2MMS_q(I) - MMS_q(I) = MSS_q(I)$.

2   An MMS Gap of $\frac{1}{40}$

In this section we prove Theorem 1, showing an allocation instance for which in every allocation, at least one of the agents gets at most a $\frac{39}{40}$ fraction of her MMS.

Proof. To present the instance that proves Theorem 1, we think of the nine items as arranged in a three by three matrix, with rows r_1, r_2, r_3 (starting from the top) and columns c_1, c_2, c_3 (starting from the left).

$$\begin{pmatrix} e_1 & e_2 & e_3 \\ e_4 & e_5 & e_6 \\ e_7 & e_8 & e_9 \end{pmatrix}$$

There are three agents, referred to as R (the *row* agent), C (the *column* agent), and U (the *unbalanced* agent). The MMS of every agent is 40. When depicting valuation functions, for each agent, we present the items in one of her MMS bundles in boldface.

Every row in the valuation function of R has value 40 and gives R her MMS. Her valuation function is:

$$\begin{pmatrix} 1 & 16 & 23 \\ 26 & 4 & 10 \\ \mathbf{12} & \mathbf{19} & \mathbf{9} \end{pmatrix}$$

Every column in the valuation function of C has value 40 and gives C her MMS. Her valuation function is:

$$\begin{pmatrix} 1 & 16 & \mathbf{22} \\ 26 & 4 & \mathbf{9} \\ 13 & 20 & \mathbf{9} \end{pmatrix}$$

The bundles that give U her MMS are $p = \{e_2, e_4\}$ (the *pair*, in boldface), $d = \{e_3, e_5, e_7\}$ (the *diagonal*), and $q = \{e_1, e_6, e_8, e_9\}$ (the *quadruple*). The valuation function of U is:

$$\begin{pmatrix} 1 & \mathbf{15} & 23 \\ \mathbf{25} & 4 & 10 \\ 13 & 20 & 9 \end{pmatrix}$$

It remains to show that no allocation gives every agent her MMS. An allocation is a partition into three bundles. As a sanity check, let us first consider the three partitions that each give one of the agents her MMS. For the partition (r_1, r_2, r_3), both C and U want only r_3, and hence one of them does not get her MMS. For the partition (c_1, c_2, c_3), both R and U want only c_3, and hence one of them does not get her MMS. For the partition (p, d, q), both R and C want only p, and hence one of them does not get her MMS.

To analyse all possible partitions in a systematic way, we consider a valuation function M that values each item as the maximum value given to the item by the three agents. Hence M is:

$$\begin{pmatrix} 1 & 16 & 23 \\ 26 & 4 & 10 \\ 13 & 20 & 9 \end{pmatrix}$$

Every allocation that gives every agent her MMS partitions M into three bundles, where the sum of values in each bundle is at least 40, but not more than 42 (as the sum of all values of M is $40 * 3 + 2$). If one of the bundles has two items, then this bundle must be $\{e_2, e_4\} = p$, whose value under M is 42. Hence each of the two remaining bundles must have value 40 under M. The unique way of partitioning the remaining items into two bundles of value 40 is to have the bundles $\{e_3, e_5, e_7\} = d$ and $\{e_1, e_6, e_8, e_9\} = q$. (The only way of reaching a value 40 in a bundle that contains item e_3 of value 23 is to include the two items of values 4 and 13.) But we already saw (in the sanity check) that the partition (p, d, q) is not a valid solution.

It follows that the partition must be into three bundles, each of size three. The bundle containing e_9 must have value between 40 and 42. There are only two such bundles of size three, namely r_3 and c_3. Each of them has value 42. If one of them is chosen, the remaining two bundles in the partition must then each be of value 40. For e_4, the only two bundles of value 40 are r_2 and c_1. Hence we get only two possible partitions, (r_1, r_2, r_3) and (c_1, c_2, c_3), and both were already excluded in our sanity check.

3   MMS Gaps that Are Inverse Polynomial in the Number of Agents

We present examples that apply for every $n \geq 4$. The initial design of our examples will include $5n - 7$ items, but for $n \geq 6$, this number will be reduced later. It will be convenient to think of the items as being arranged as selected entries in an n by n matrix, along the perimeter of the matrix, and along its main diagonal. We will construct two valuation functions, where a set R of at least two agents have valuation function V_R, and a set C of at least two agents have valuation function V_C (here R stands for *row* and C stands for *column*, and $|R| + |C| = n$). We will start with a base matrix B, and then modify B so as to obtain V_R and V_C.

In the base matrix B, the items have only seven different values, regardless of the value of n. We shall partition the items into groups of items of equal value, and give an informative name to each group.

Rows are numbered from top down, and columns from left to right. We use the convention that the index j specifies an arbitrary value in the range $2 \leq j \leq n - 1$.

The value of items in each group, and the locations of the groups in B, are as follows.

- $B_{1j} = (n - 2)n$. (Top row, excluding corners).
- $B_{1n} = 1$. (Top-right corner.)
- $B_{j1} = (n - 2)(n - 1)$. (Left column, excluding corners.)
- $B_{jj} = (n - 2)(n^2 - 4n + 2)$. (Main diagonal, excluding corners.)
- $B_{jn} = (n - 2)(n - 1) + 1$. (Right column, excluding corners.)
- $B_{n1} \cup B_{nj} = (n - 2)^2 + 1$. (Bottom row, excluding bottom-right corner.)
- $B_{nn} = (n - 2)(n - 3)$. (Bottom-right corner.)

For $n = 4$, this gives the following matrix.

$$\begin{pmatrix} 0 & 8 & 8 & 1 \\ 6 & 4 & 0 & 7 \\ 6 & 0 & 4 & 7 \\ 5 & 5 & 5 & 2 \end{pmatrix}$$

Observe that all entries of B are nonnegative. Moreover, All row sums and all column sums have the same value $t_B = n(n-2)^2 + 1$

A bundle of items will be called *good* if the sum of its values is t_B. Hence all rows and all columns are good, but there are also other bundles that are good. A partition of all items into n bundles is *good* if every bundle in the partition is good. For example, a partitioning of the items into row bundles is good, and likewise, a partitioning into column bundles is good. The following lemma constrains the structure of good partitions of B.

Lemma 1. *In every partitioning of the items of B into n good bundles, the structure of the good partition is such that at least one of the following three conditions hold:*

1. *The bottom row is split among the n good bundles (one item in each bundle).*
2. *The right column is split among the n good bundles (one item in each bundle).*
3. *At least one of the bundles contains at least one item from the bottom row and at least one item from the right column, but does not contain the item B_{nn}.*

Proof. Observe that $t_B = n(n-2)^2 + 1 = 1$ modulo $n-2$. There are exactly $2n-2$ items that have value 1 modulo $n-2$ (the bottom row and the right column, excluding the bottom-right corner). We refer to these items as *special*. The remaining items have value 0 modulo $n-2$, and are not special. In every good partition, it must be the case that one good bundle has $n-1$ special items, and each other good bundle has one special item.

Consider the good bundle with $n-1$ special items.

If the $n-1$ special items are all in the bottom row (or all in the right column), then item B_{nn} must be the remaining item in the bundle (that is the only way to reach t_B), and then the right column (or bottom row) must be split.

If the $n-1$ odd items include at least one from the bottom row and at least one from the right column, then we may assume that B_{nn} is also in the bundle (as otherwise condition 3 of the Lemma holds). This accounts for n items in the bundle. The sum of values of these n items cannot possibly be equal to t_B. This can be verified by a case analysis. If B_{1n} is among these items, then the only way to reach t_B with $n-2$ additional special items is to add all items of B_{jn} (as special items in the bottom row have strictly smaller value than items in B_{jn}), but then the bundle has no special items from the bottom row. Alternatively, if B_{1n} is not among these items, then the only way to reach t_B with $n-1$ special items is to add all special items of the bottom row (as special items in B_{jn} have strictly larger value than special items in the bottom row), but then the bundle has no special items from the right column.

Consequently, the sum values of these n items needs to be strictly smaller than t_B. Their total value is minimized if they are $B_{1n} \cup B_{nj} \cup B_{nn}$, giving a value of $1 + (n-2)((n-2)^2 + 1) + (n-2)(n-3) = (n-1)(n-2)^2 + 1$. Hence a value of $(n-2)^2$ is missing in order to complete the sum of values to $t_B = n(n-2)^2 + 1$. For $n \geq 5$, none of the remaining items has such small value, and hence such a good bundle cannot be formed at all. The only case that remains to be considered is $n = 4$, because for $n = 4$ the value of diagonal items B_{jj} happens to satisfy $(n-2)(n^2 - 4n + 2) = (n-2)^2$.

Recall the matrix for $n = 4$ depicted above. The composition of values in a good bundle that has two special items from the bottom row, the special item B_{14}, the item B_{44}, and one diagonal item, is $(5, 5, 1, 2, 4)$. But then one of the two items of value 8 does not have a good bundle. (An item of value 8 needs an additional value of 9 to reach 17. However, of the items that remain, there is only one combination of items that gives value 9, namely, as $4 + 5$.)

Remark 1. Our proof for Theorem 4 follows a pattern used in [10]. In their construction, the base matrix B was required to have the property that it has only two good partitions: the row partition and the column partition. In contrast, we allow B to have many more good partitions (as specified in Lemma 1), and show that even with this extra flexibility, the proof pattern of [10] still works. Given this extra flexibility in the properties of B, we design such matrices (one for each value of n) with much smaller integer entries than the corresponding matrices designed in [10].

Using the matrix B, we shall now create two matrices, one for V_R and one for V_C. First, every entry of B is multiplied by n. Then, for V_R, subtract 1 from the value of each special item in the bottom row, and add $n - 1$ to the value of the bottom-right corner. For $n = 4$, the matrix for V_R is:

$$\begin{pmatrix} 0 & 32 & 32 & 4 \\ 24 & 16 & 0 & 28 \\ 24 & 0 & 16 & 28 \\ 19 & 19 & 19 & 11 \end{pmatrix}$$

The maximin share of every agent in R is 68 (each row is a bundle). For general $n \geq 4$, this maximin share is $t_V = nt_B = n^2(n-2)^2 + n$.

For V_C, subtract 1 from the value of each special item in the right column, and add $n - 1$ to the value of the bottom-right corner. For $n = 4$, the matrix for V_C is:

$$\begin{pmatrix} 0 & 32 & 32 & 3 \\ 24 & 16 & 0 & 27 \\ 24 & 0 & 16 & 27 \\ 20 & 20 & 20 & 11 \end{pmatrix}$$

Similar to agents in R, the maximin share of every agent in C is 68 (each column is a bundle). For general $n \geq 4$, this maximin share is $t_V = nt_B = n^2(n-2)^2 + n$.

Proposition 5. *If $|R| + |C| = n$ and $|R|, |C| \geq 2$, then in every allocation, at least one player gets a bundle that he values as at most $t_V - 1$.*

Proof. The allocation partitions the items into n bundles. If at least one of the bundles has value less than t_B in B, then the same bundle has value at most $n(t_B - 1) + (n - 1) = t_V - 1$ for the agent who receives it. Hence we may assume that every bundle has value t_B in B. By Lemma 1, there are only three possibilities for this.

1. *The bottom row is split.* Then every agent in R receives a bundle that contains a single item from the bottom row. As $|R| \geq 2$, for at least one row agent, this single item lost a value of 1 in the process of constructing V_R. Consequently, the value received by this agent is $nt_B - 1 = t_V - 1$.

2. *The right column is split.* Then every agent in C receives a bundle that contains a single item from the right column. As $|C| \geq 2$, for at least one column agent, this single item lost a value of 1 in the process of constructing V_C. Consequently, the value received by this agent is $nt_B - 1 = t_V - 1$.

3. *At least one of the bundles contains at least one item from the bottom row and at least one item from the right column, but does not contain the item B_{nn}. Such a bundle has value at most $t_V - 1$ for every agent.*

We can now prove Theorem 4. In fact, we state a somewhat stronger version of it in which the gap is improved from $\frac{1}{n^4}$ to a somewhat larger value.

Theorem 6. *For given N, let $n = \lceil \frac{N+4}{2} \rceil$. Then for every $N \geq 4$, there is an allocation instance with N agents and at most $N + 4n - 7$ items (which gives $3N + 1$ when N is even and $3N + 3$ when N is odd) for which in every allocation, at least one of the agents does not get more than a $1 - \frac{1}{f(n)}$ fraction of her MMS. Here, the function $f(n)$ has value $f(n) = n^2(n-2)^2 + n$. In other words,*

$$Gap(N \geq 5 \, , \, m \leq 3N + 3) \geq \frac{1}{f(n)}$$

Proof. For $4 \leq N \leq 5$ we have that the corresponding value of $n = \lceil \frac{N+4}{2} \rceil = N$, and hence the corresponding instances was described above. (Observe that $f(n)$ equals the corresponding value of t_V in these instances.)

For $N \geq 6$, we have that $n = \lceil \frac{N+4}{2} \rceil < N$. In this case we construct an instance as above for the corresponding value of n (with value $t_V = n^2(n-2)^2 + n$). We add to this instance $N - n$ agents so that the number of agents becomes N. Among the agents, we set $\lfloor \frac{N}{2} \rfloor$ agents to be row agents, and the remaining agents to be column agents. We also add to the instance $N - n$ auxiliary items, each of value t_V, and so the total number of items is $(5n - 7) + (N - n) = N + 4n - 7$.

For each of the N agents, the MMS is t_v (by partitioning the set of items into the $N - n$ auxiliary items, and either the n rows or the n columns). $N - n$ agents get their MMS by getting an auxiliary item. However, among the n agents that remain, at least two are row agents (because $|R| - (N-n) = \lfloor \frac{N}{2} \rfloor - N + \lceil \frac{N+4}{2} \rceil = 2$) and at least two are column agents, and this suffices for Proposition 5 to apply.

4 An MMS Allocation Whenever $m \leq n + 5$

In this section we prove Theorem 2, that if $m \leq n + 5$ there always is an MMS allocation. The proof makes use of the following two lemmas.

Lemma 2. *Let I be an allocation instance with n agents and m items, and assume that for every instance with $n - 1$ agents and $m - 1$ items there is an MMS allocation. If there is an agent i and item e for which $v_i(e) \geq MMS_i(I)$, then I has an MMS allocation.*

Proof. Remove item e and agent i, resulting in an instance I' with $n-1$ agents and $m-1$ items. By Proposition 3, for every agent $j \neq i$ it holds that $MSS_j(I') \geq MSS_j(I)$. By the assumption of the lemma, there is an MSS allocation A' for I'. Extend A' to an allocation A for I, by giving item e to agent i. Allocation A is an MSS allocation for I.

Lemma 3. *Let I be an allocation instance with n agents and m items, and assume that for every instance with $n-1$ agents and $m-2$ items there is an MMS allocation. Suppose that there is an agent q and a bundle B containing two items such that $v_q(B) \geq MMS_q(I)$, and moreover, for every agent $j \neq q$, at least one of the following conditions hold:*

1. *B is small: $v_j(B) \leq MSS_j(I)$.*
2. *B is directly dominated: B is equal to or contained in one of the bundles of the MMS_j partition.*
3. *B is indirectly dominated: the MMS_j partition contains a bundle B' such that $v_j(B') \geq v_j(B)$ and $|B' \cap B| = 1$.*

Then I has an MMS allocation.

Proof. Remove bundle B and agent q, resulting in an instance I' with $n-1$ agents and $m-2$ items. We claim that $MSS_j(I') \geq MSS_j(I)$ for every agent $j \neq q$. For agents for which either condition 1 or condition 2 hold, this follows by Proposition 4.

For an agent j for which only condition 3 holds, let B_1' denote the other bundle intersected by B. Replace the two bundles B' and B_1' in the MMS_j partition by the two bundles B and $B_1 = (B' \cup B_1') \setminus B$. We have that $v_j(B) \geq MMS_j(I)$ (as condition 1 is assumed not to hold) and $v_j(B_1) \geq MMS_j(I)$. (The last inequality can be verified as follows. Condition 3 holding implies that $v_j(B') \geq v_j(B)$. This together with $(B \cup B_1) = (B' \cup B_1')$ implies that $v_j(B_1) \geq v_j(B_1')$. The fact that B_1' is a bundle in the original MSS_j partition implies that $v_j(B_1') \geq MMS_j$.) Hence we get an MMS_j partition in which B is one of the bundles, and now we can apply condition 2 to conclude that $MSS_j(I') \geq MSS_j(I)$.

By the assumption of the lemma, there is an MSS allocation A' for I'. Extend A' to an allocation A for I, by giving bundle B to agent q. Allocation A is an MSS allocation for I.

We now prove Theorem 2.

Proof. The proof is by induction on n. The theorem trivially holds for $n=1$, and holds for $n=2$ by Proposition 1. The case $n=2$ serves as the base case of the induction, and it remains to prove the theorem for $n \geq 3$. In all cases with $n \geq 3$ we assume without loss of generality:

- The theorem has already been proved for all $n' < n$ (the inductive hypothesis).
- $m = n + 5$ (because if $m < n + 5$, we may add $n + 5 - m$ auxiliary items that have 0 value to all agents).

– All bundles in the MMS partition of every agent are of size at least 2.

The third assumption can be made without loss of generality, as otherwise there is an agent i and item e for which $v_i(e) \geq MMS_i$, and then Lemma 2 allows us to reduce the instance to one in which the induction hypothesis already holds.

Observe that the third assumption implies (among other things) that it suffices to consider only $n \leq 5$, because for $n \geq 6$ we have that $m = n + 5 < 2n$, and the third assumption cannot hold.

Using these assumptions, the cases $n = 3$, $n = 4$, and $n = 5$ are proved in Lemma 4, Lemma 5, and Lemma 6, respectively.

4.1 Three Agents, Eight Items

Lemma 4. *Every allocation instance with $n = 3$ agents and $m = n + 5$ items has an MMS allocation.*

Proof. By Proposition 2 we may assume that the instance is ordered (for every $1 \leq i < j \leq m$ and every agent q, $v_q(e_i) \geq v_q(e_j)$).

Recall (see the proof of Theorem 2) that we may assume that the MMS partition of an agent contains only bundles of size at least 2. Consequently, for every agent j, her MMS_j partition contains at least one bundle (call it B_j) that has exactly two items.

If the three bundles B_1, B_2 and B_3 are disjoint, give each agent her respective bundle, and allocate the two remaining items arbitrarily.

It remains to consider the case that at least two of these bundles intersect. W.l.o.g., let these bundles be B_1 and B_2.

Suppose that $|B_1 \cap B_2| = 1$. Then as the instance is ordered, all agents agree that one of the two bundles, B_1 or B_2, is not more valuable than the other. W.l.o.g., let this bundle be B_1. Likewise, if $B_1 = B_2$, then also in this case B_1 is not more valuable than B_2.

There are two cases to consider:

– $v_3(B_1) \geq MMS_3$. In this case Lemma 3 applies with agent 3 serving as agent q, and with B_1 serving as B. Hence an MMS allocation exists.
– $v_3(B_1) < MMS_3$. In this case Lemma 3 applies with agent 1 serving as agent q, and with B_1 serving as B. Hence an MMS allocation exists.

4.2 Four Agents, Nine Items

Lemma 5. *Every allocation instance with $n = 4$ agents and $m = n + 5$ items has an MMS allocation.*

Proof. Consider an allocation instance I with four agents and a set M of at most nine items. Recall (see the proof of Theorem 2) that we may assume that the MMS partition of an agent contains only bundles of size at least 2. Consequently, for every agent i, her MMS_i partition contains three bundles of size two, and one bundle of size three.

Let $(B_{1,1}, B_{1,2}, B_{1,3}, B_{1,4})$ denote the MMS_1 partition of agent 1, with $|B_{1,1}| = |B_{1,2}| = |B_{1,3}| = 2$ and $|B_{1,4}| = 3$. Suppose that for some $k \leq 3$, there is exactly one agent $i \geq 2$ for which $v_i(B_{1,k}) \geq MMS_i$. Then Lemma 3 applies with agent i serving as agent q, and $B_{1,k}$ serving as bundle B. Hence an MMS allocation exists.

Likewise, if for some $k \leq 3$ there is no agent $i \geq 2$ for which $v_i(B_{1,k}) \geq MMS_i$, Lemma 3 applies with agent 1 serving as agent q, and $B_{1,k}$ serving as bundle B. Hence also in this case an MMS allocation exists.

It follows that we can assume that for each of the bundles $\{B_{1,1}, B_{1,2}, B_{1,3}\}$ there is at most one agent $2 \leq i \leq 4$ that values it less than her MSS.

Consider now a bipartite graph G. Its left hand side contains four vertices, corresponding to the four agents $\{1, 2, 3, 4\}$. Its right hand side has four vertices, corresponding to the four bundles $\{B_{1,1}, B_{1,2}, B_{1,3}, B_{1,4}\}$. For every $1 \leq i, j \leq 4$ there is an edge between agent i and bundle $B_{1,j}$ if $v_i(B_{1,j}) \geq MMS_i$. Observe that a perfect matching in G induces an MMS allocation, giving every agent her matched bundle. Hence it suffices to show that G has a perfect matching.

Each of the right hand side vertices $B_{1,k}$ for $1 \leq k \leq 3$ has degree at least 3 (as at most one agent values it less than her MMS), and $B_{1,4}$ has degree at least 1 (as agent 1 values it at least as MSS_1). Hence for every $k \leq 3$, every set of k right hand side vertices has at least k left hand side neighbors. Moreover, the set of all right hand side vertices has four left hand side neighbors, as for every agent i, at least one of the four bundles has value at least $\frac{1}{4}v_i(M) \geq MMS_i$. Hence by Hall's condition, G has a perfect matching.

4.3   Five Agents, Ten Items

Lemma 6. *Every allocation instance with $n = 5$ agents and $m = n + 5$ items has an MMS allocation.*

Proof. Let I be an arbitrary allocation instance with 5 agents and 10 items. Recall (see the proof of Theorem 2) that we may assume that the MMS partition of an agent contains only bundles of size at least 2. As $m = 2n$, this implies that for every agent i, all bundles of her MMS_i partition are of size two.

By Proposition 2 we may assume that the instance is ordered (for every $1 \leq i < j \leq m$ and every agent q, $v_q(e_i) \geq v_q(e_j)$). For every agent i, consider the bundle B_i in her MMS partition that contains the item e_1. This gives five bundles (not necessarily all distinct). Among these bundles, consider the bundle B in which the second item of the bundle has highest index (lowest value). Then for every agent i we have that $v_i(B) \leq v_i(B_i)$, because the instance is ordered. Let q be an agent that has B as a bundle in her MMS partition (if there is more that one such agent, pick one arbitrarily). Lemma 3 (condition 3 in the lemma) implies I has an MMS allocation.

5   Tightness of MMS Ratio for Nine Items

Theorem 3 claims that every allocation instance with three agents and nine items has an allocation that gives each agent at least a $\frac{39}{40}$ of her MMS. Its proof has three steps.

1. The proof of Theorem 7 that shows that a negative example can have only one of two possible structures.
2. Each structure induces linear constraints on the valuation functions of the agents. For each structure, we set up a linear program that finds a solution that satisfies all linear constraints implied by the corresponding structure, while maximizing the MMS gap in that solution. These LPs are underconstrained, and the optimal feasible solutions of these LPs turn out not to correspond to true negative examples. Hence we need to add additional constraints to the LPs, preventing the LPs from producing solutions that are not true negative examples.
3. For each of the two structures, we partition all potential negative examples that have this structure into a finite number of classes, where each class offers some refinement of the structure. The classes need not be disjoint. The refined structure of a class gives rise to additional constraints to the LP. Thus we end up with a finite number of different LPs, one for each class. We then verify that none of these LPs generates a negative example with MMS gap larger than $\frac{1}{40}$ (this is done by having a computer program solve the corresponding LPs), and this proves Theorem 3.

Before stating Theorem 7, we introduce some notation and terminology. Recall that we may assume that instance I is ordered. We still do so, but we no longer assume that the order is from item e_1 to item e_9. Instead the order is left unspecified at this point. Our naming convention for items is based on arranging the items in a three by three matrix, and naming the items according to their location in the matrix, as specified below.

$$\begin{pmatrix} e_1 & e_2 & e_3 \\ e_4 & e_5 & e_6 \\ e_7 & e_8 & e_9 \end{pmatrix}$$

The rows of the matrix are referred to as R_1, R_2, R_3, starting from the top row, and the columns are referred to as C_1, C_2, C_3, starting from the left column. The two main diagonals of the matrix are $\{e_1, e_5, e_9\}$ and $\{e_3, e_5, e_7\}$.

We say that a bundle is *good* for an agent if she values it at least as her MMS, and *bad* otherwise.

Theorem 7. *A negative example for three players and nine items must have the following structure (after appropriately renaming the items). For one agent R, the MMS partition is into the three rows (R_1, R_2 and R_3), for one agent C, the MMS partition is into the three columns (C_1, C_2 and C_3), and for one agent U, the MMS partition is to a bundle $P = \{e_2, e_4\}$ (P stands for pair), a bundle D that is one of the two main diagonals (D stands for diagonal), and a bundle Q with the remaining four items (Q stands for quadruple). Bundle P is good* for all agents, whereas D and Q are bad for agents R and C. The row and the column that do not intersect P are good for all agents (these are R_3 and C_3), whereas the remaining rows and columns are good only for the agents that have them in their partition, and bad for the other agents.

As there are two main diagonals, Theorem 7 offers two possible structures. We refer to them as the parallel diagonals structure (bundle d runs in parallel to bundle p), and the crossing diagonals structure (bundle d crosses bundle p). They are depicted in Figs. 1 and 2, respectively. Within each figure, for every item e_i, the entry r_i (c_i, u_i, respectively) denotes its value to agent R (C, U, respectively).

Fig. 1. The parallel diagonals structure, with MMS partitions for players R, C and U respectively. The good bundles, marked by a $\star$ above the item, are $R_3 = \{e_7, e_8, e_9\}$, $C_3 = \{e_3, e_6, e_9\}$ and $P = \{e_2, e_4\}$.

Fig. 2. The crossing diagonals structure, with MMS partitions for players R, C and U respectively. The good bundles, marked by a $\star$ above the item, are $R_3 = \{e_7, e_8, e_9\}$, $C_3 = \{e_3, e_6, e_9\}$ and $P = \{e_2, e_4\}$.

Due to space limitations, we omit the proof of Theorem 7, as well as the rest of the proof of Theorem 3. They can be found in the full version of our paper[1].

6   Extension to Chores

Chores are items of negative value, or equivalently, positive dis-utility. In allocation problems involving only chores, the convention is that all items must be

---

[1] Uriel Feige, Ariel Sapir, Laliv Tauber: A tight negative example for MMS fair allocations. CoRR abs/2104.04977 (2021).

allocated. In analogy to Definition 1, MMS_i is the minimum over all n-partitions of M, of the maximum dis-utility under v_i of a bundle in the n-partition. (Note that as dis-utility replaces value, maximum and minimum are interchanged in this definition, compared to Definition 1.) It is known that for agents with additive dis-utility functions over chores, there are allocation instances in which in every allocation some agent gets a bundle of dis-utility higher than her MMS [2], and that there always is an allocation giving every agent a bundle of dis-utility at most $\frac{11}{9}$ times her MMS [9].

We now prove Theorem 5, that there is an instance with three agents and nine chores that has an MMS gap of $\frac{1}{43}$.

Proof. We present an example with an MMS gap of $\frac{1}{43}$, using notation as in Sect. 2.

Every row in the dis-utility function of R has value 43 and gives R her MMS. Her dis-utility function is:
$$\begin{pmatrix} 6 & 15 & 22 \\ 26 & 10 & 7 \\ \mathbf{12} & \mathbf{19} & \mathbf{12} \end{pmatrix}$$
Every column in the dis-utility function of C has value 43 and gives C her MMS. Her dis-utility function is:
$$\begin{pmatrix} 6 & 15 & \mathbf{23} \\ 26 & 10 & \mathbf{8} \\ 11 & 18 & \mathbf{12} \end{pmatrix}$$
The bundles that give U her MMS are $p = \{e_2, e_4\}$ (the *pair*, in boldface), $d = \{e_3, e_5, e_7\}$ (the *diagonal*), and $q = \{e_1, e_6, e_8, e_9\}$ (the *quadruple*). The dis-utility function of U is:
$$\begin{pmatrix} 6 & \mathbf{16} & 22 \\ \mathbf{27} & 10 & 7 \\ 11 & 18 & 12 \end{pmatrix}$$
In analogy to Sect. 2, to analyse all possible allocations in a systematic way, it is convenient to consider a dis-utility function M in which the dis-utility of each chore as the minimum (rather than maximum, as we are dealing with chores) dis-utility given to the chore by the three agents. Hence M is:
$$\begin{pmatrix} 6 & 15 & 22 \\ 26 & 10 & 7 \\ 11 & 18 & 12 \end{pmatrix}$$
Adaptation of the analysis of Sect. 2 shows that in every allocation, some agent gets chores of dis-utility at least 44, whereas the MMS is 43. Further details of the proof are omitted.

7   Discussion

The open questions below refer to allocations of goods. Questions of a similar nature can be asked for chores, though the quantitative bounds in these questions would be different from those mentioned below.

Let δ_n denote the largest value such that for n agents, there is an allocation instance with additive valuations for which no allocation gives every agent more than a $1 - \delta_n$ fraction of her MMS. We have that $\delta_1 = \delta_2 = 0$. As to δ_3, the combination of our Theorem 1 and the results of [7] imply that $\frac{1}{40} \leq \delta_3 \leq \frac{1}{9}$. It would be interesting to determine the exact value of δ_3, or at least to narrow the gap between its lower bound and upper bound. Computer assisted techniques, such as those used in the proof of Theorem 3, may turn out useful for this purpose.

For general $n \geq 3$, the combination of our Theorem 4 and the results of [6] imply that $\frac{1}{n^4} \leq \delta_n \leq \frac{1}{4} + \frac{1}{12n}$. We do not know whether δ_n tends to 0 as n grows. Determining whether this is the case remains as an interesting open question. The known results do not exclude the possibility that δ_n tends to $\frac{1}{4}$ as n grows, but we would be very surprised if this turns out to be true.

Acknowledgement. This research benefited from the use of automated solvers for mixed integer programs. Specifically, we have used solvers of https://online-optimizer. appspot.com/ and https://www.lindo.com/. We would like to thank Shai Keidar and Orin Munk for helpful discussions.

References

1. Amanatidis, G., Markakis, E., Nikzad, A., Saberi, A.: Approximation algorithms for computing maximin share allocations. ACM Trans. Algorithms **13**(4), 52:1–52:28 (2017)
2. Aziz, H., Rauchecker, G., Schryen, G., Walsh, T.: Algorithms for max-min share fair allocation of indivisible chores. In: AAAI, pp. 335–341 (2017)
3. Barman, S., Krishnamurthy, S.K.: Approximation algorithms for maximin fair division. ACM Trans. Econ. Comput. **8**(1), 5:1–5:28 (2020)
4. Bouveret, S., Lemaître, M.: Characterizing conflicts in fair division of indivisible goods using a scale of criteria. Auton. Agents Multi-Agent Syst. **30**(2), 259–290 (2015). https://doi.org/10.1007/s10458-015-9287-3
5. Budish, E.: The combinatorial assignment problem: approximate competitive equilibrium from equal incomes. J. Polit. Econ. **119**(6), 1061–1103 (2011)
6. Garg, J., Taki, S.: An improved approximation algorithm for maximin shares. In: EC, pp. 379–380 (2020)
7. Gourves, L., Monnot, J.: On maximin share allocations in matroids. Theor. Comput. Sci. **754**, 50–64 (2019)
8. Ghodsi, M., Hajiaghayi, M.T., Seddighin, M., Seddighin, S., Yami, H.: Fair allocation of indivisible goods: improvements and generalizations. In: EC, pp. 539–556 (2018)
9. Huang, X., Lu, P.: An algorithmic framework for approximating maximin share allocation of chores. In: EC, pp. 630–631 (2021)
10. Kurokawa, D., Procaccia, A.D., Wang, J.: Fair enough: guaranteeing approximate maximin shares. J. ACM **65**(2), 8:1–8:27 (2018)

Approximating Nash Social Welfare Under Binary XOS and Binary Subadditive Valuations

Siddharth Barman[1] and Paritosh Verma[2(✉)]

[1] Indian Institute of Science, Bengaluru 560012, Karnataka, India
barman@iisc.ac.in
[2] Purdue University, West Lafayette, Indiana 47907, USA
verma136@purdue.edu

Abstract. We study the problem of allocating indivisible goods among agents in a fair and economically efficient manner. In this context, the Nash social welfare—defined as the geometric mean of agents' valuations for their assigned bundles—stands as a fundamental measure that quantifies the extent of fairness of an allocation. Focusing on instances in which the agents' valuations have binary marginals, we develop essentially tight results for (approximately) maximizing Nash social welfare under two of the most general classes of complement-free valuations, i.e., under binary XOS and binary subadditive valuations.

For binary XOS valuations, we develop a polynomial-time algorithm that finds a constant-factor (specifically 288) approximation for the optimal Nash social welfare, in the standard value-oracle model. The allocations computed by our algorithm also achieve constant-factor approximation for social welfare and the groupwise maximin share guarantee. These results imply that—in the case of binary XOS valuations—there necessarily exists an allocation that simultaneously satisfies multiple (approximate) fairness and efficiency criteria. We complement the algorithmic result by proving that Nash social welfare maximization is APX-hard under binary XOS valuations.

Furthermore, this work establishes an interesting separation between the binary XOS and binary subadditive settings. In particular, we prove that an exponential number of value queries are necessarily required to obtain even a sub-linear approximation for Nash social welfare under binary subadditive valuations.

Keywords: Discrete fair division · Nash social welfare · Binary marginals

1 Introduction

At the core of discrete fair division lies the problem of fairly allocating *indivisible* goods among agents with equal entitlements, but distinct preferences. In this context, the Nash social welfare [38]—defined as the geometric mean of agents'

© Springer Nature Switzerland AG 2022
M. Feldman et al. (Eds.): WINE 2021, LNCS 13112, pp. 373–390, 2022.
https://doi.org/10.1007/978-3-030-94676-0_21

valuations for their assigned bundles—stands as a fundamental measure that quantifies the extent of fairness of an allocation. This welfare function achieves a balance between the extremes of social welfare and egalitarian welfare. The relevance of Nash social welfare is further substantiated by the fact that it satisfies key fairness axioms, including the Pigou-Dalton transfer principle [37]. Furthermore, Nash social welfare is indifferent to individual scales of the valuations: multiplicatively scaling any agent's valuation by a positive number does not alter the relative ordering of the allocations (induced by this welfare objective) and, in particular, keeps the Nash optimal allocation unchanged. In terms of practical applications, Nash social welfare is used as an optimization criterion by the widely-used platform `Spliddit.org` for finding fair allocations [29].

With these considerations in hand, a substantial body of research in recent years has been directed towards maximizing the Nash social welfare in settings with indivisible goods; see, e.g., [1,2,6,10,18,19,24]. This maximization problem is known to be APX-hard, even when the agents have additive valuations [33].[1] Hence, in general, algorithmic results for this problem aim for approximation guarantees. A key focus of this line of research has been on the hierarchy of complement-free valuations, which includes the following valuation classes, in order of containment: additive, submodular, XOS (fractionally subadditive), and subadditive. Recall that submodular functions satisfy a diminishing returns property, XOS functions are pointwise maximizers of additive functions, and subadditive functions constitute the most general class in this hierarchy. These valuation classes have also been extensively studied in the literature on combinatorial auctions, wherein the focus is primarily on maximizing social welfare [39].

In the context of maximizing Nash social welfare, the best-known approximation algorithm for additive valuations achieves an approximation ratio of $e^{1/e}$ (in polynomial time) [6]. Furthermore, for submodular valuations, a recent result of Li and Vondrák [35] obtains a constant-factor (specifically 380) approximation ratio; see also [25]. In contrast to these constant-factor bounds, the problem of maximizing Nash social welfare under (general) XOS and subadditive valuations has a linear (in the number of agents) approximation guarantee [4,16,26]. This approximation ratio is in fact tight in the standard value-oracle model: under general XOS (and, hence, subadditive) valuations, a sub-linear approximation of the optimal Nash social welfare necessarily requires exponentially many value queries [4].

The current work contributes to this thread of research with a focus on valuations that have binary (dichotomous) marginals. Formally, a valuation v is said to bear the binary-marginals property iff, for every subset of goods S and each good g, the marginal value of g relative to S is either zero or one, $v(S \cup \{g\}) - v(S) \in \{0,1\}$. Such valuations capture preferences in many real-world domains and have received significant attention in the fair division literature; see, e.g., [12,13,23,31,40,42]. Our results address the two most general valuation classes

---

[1] Recall that a valuation v is said to be additive iff the value of any subset of goods, S, is the sum of values of the goods in it, $v(S) = \sum_{g \in S} v(\{g\})$.

in the above-mentioned hierarchy, i.e., we study XOS and subadditive valuations in conjunction with the binary-marginals property. This meaningfully extends prior work on binary additive and binary submodular valuations. Throughout, we will say that a valuation is binary additive (submodular/XOS/subadditive) iff it is additive (submodular/XOS/subadditive) and also has binary marginals.

In particular, under binary additive valuations, Nash optimal allocations can be computed in polynomial time [7,20]. The work of Halpern et al. [30] shows that—in the case of binary additive valuations—maximizing Nash social welfare (with a lexicographic tie-breaking rule) provides a truthful and fair[2] mechanism. For the broader class of binary submodular valuations,[3] a truthful, fair, and polynomial-time mechanism was obtained by Babaioff et al. [3]; this result considers *Lorenz domination* as a notion of fairness and, hence, ensures that the computed allocation maximizes the Nash social welfare and satisfies other fairness criteria. These works identify multiple domains wherein binary additive and binary submodular functions are applicable; see also [11].

The current work moves up in the hierarchy of complement-free valuations and develops essentially tight results for both binary XOS and binary subadditve valuations. Before detailing our results, we provide a stylized example that illustrates the relevance of such a generalization: consider a spectrum-allocation setting wherein transmission rights of distinct (frequency) bands have to be fairly allocated among different agents. Here, each band is an indivisible good of unit value, to every agent. However, an agent can utilize a subset of bands only if the frequencies across the allocated bands are close enough. In particular, say the bands, $B_1, B_2, \ldots, B_m$, are indexed such that an agent can use a subset of bands only if their indices are within a parameter $\Delta \in \mathbb{Z}_+$ of each other. For instance, if $\Delta = 3$, then an agent will have value two for the bundle $\{B_1, B_4, B_{10}\}$ and value the bundle $\{B_1, B_4, B_5, B_7\}$ at three, by transmitting on B_4, B_5, and B_7. We note that such valuations can be expressed as binary XOS functions (and not as submodular functions). Hence, in such a resource-allocation setting, finding a fair, or economically efficient, allocation falls under the purview of the current work. For a realistic treatment of spectrum allocations, and accompanying high-stakes auctions, see [34] and references therein.

Our Results. The algorithmic results in this work only require access to standard value queries: given any subset of goods S and an agent i, the value oracle returns the value that i has for S.[4] The following list summarizes our results.

1. We develop a polynomial-time 288-approximation algorithm for maximizing Nash social welfare under binary XOS valuations (Theorem 2). We

---

[2] Specifically, the mechanism of Halpern et al. [30] ensures envy-freeness up to one good.

[3] Binary submodular functions admit the following characterization: every binary submodular function is necessarily a rank function of a matroid [43, Chapter 39].

[4] That is, the developed algorithms do not require an explicit description of the valuations. Note that in the current context the agents' valuations are combinatorial set functions, hence explicitly representing the valuations might be prohibitive, i.e., require one to specify exponential (in the number of goods) values.

also complement this algorithmic result by proving that—under binary XOS valuations—Nash social welfare maximization is APX-hard (Theorem 4).

To obtain the approximation guarantee, we consider allocations wherein, for each agent i, the envy is multiplicatively bounded towards the entire set of goods, G_i, allocated to agents with bundle size at least four times that of i. Specifically, for an allocation, write H_i to denote the set of agents who have received a bundle of size at least four times that of i, and let G_i denote the set of goods allocated among all the agents in H_i (along with unallocated goods, if any). We show that, under binary XOS valuations, if, in an allocation $\mathcal{A}$, each agent i's value for her own bundle is at least $1/2$ times her value for G_i, then $\mathcal{A}$ achieves a constant-factor approximation guarantee for Nash social welfare. Our algorithm (Algorithm 1) finds such an allocation by iteratively updating the agents' bundles towards the desired property. The algorithm also maintains an analytically useful property that each agent's value for her bundle is equal to the cardinality of the bundle, i.e., the bundles are *non-wasteful*. For binary XOS valuations, one can show that multiplicatively bounding envy between pairs of agents does not, by itself, provide a constant-factor approximation guarantee (see the full version of this paper [8]). Hence, bounding envy of every agent i against all of G_i is a crucial extension. It is relevant to observe that while the algorithm is simple, its analysis is based on novel counting arguments (Lemma 2 and Proposition 2). Notably, the combinatorial nature of the algorithm makes it amenable to large-scale implementations.

2. Furthermore, our algorithm (Algorithm 1) achieves an approximation ratio of $(3 + 2\sqrt{2})$ for the problem of maximizing social welfare under binary XOS valuations (the proof is deferred to the full version of the paper [8]). That is, the computed allocation simultaneously provides approximation guarantees for Nash social welfare (a fairness metric) and social welfare (a measure of economic efficiency).

   In addition, the allocation (approximately) satisfies the fairness notion of *groupwise maximin shares* (GMMS); see full version of this paper for definitions and proof [8]. GMMS is a stronger criterion than the well-studied fairness concept of maximin shares (MMS). Specifically, an allocation $\mathcal{A}$ is said to be α-GMMS iff $\mathcal{A}$ is α-approximately MMS for every subgroup of agents. The allocations computed by our algorithm are $1/6$-GMMS [8].

3. Complementing the above-mentioned positive results, we prove that, under binary subadditive valuations, an exponential number of value queries are necessarily required to obtain a sub-linear approximation for the Nash social welfare (Theorem 5). Indeed, this query complexity bound identifies an interesting dichotomy between the binary subadditive and the binary XOS settings: while for binary XOS valuations a polynomial-number of value queries suffice for approximating the optimal Nash social welfare within a constant factor, binary subadditive valuations are essentially as hard as general subadditive (or general XOS) valuations.

Additional Related Work. The current work provides a single algorithm that achieves constant-factor approximation guarantees for both Nash social welfare and social welfare, under binary XOS valuations. Focusing solely on social welfare maximization, one can compute an $(e/(e-1))$-approximation (of the optimal social welfare) under general XOS valuations, using the algorithm of Feige [22]. This result, however, requires oracle access to *demand queries*, which is a more stringent requirement than one used in the current work (of value queries).[5]

We obtain the query complexity, under binary subadditive valuations, by utilizing a lower-bound framework of Dobzinski et al. [21]. In [21], a lower bound was obtained—for social welfare maximization—under general XOS and subadditive valuations. The notable technical contribution of the current work is to establish the query complexity with valuations that in fact have binary marginals. Furthermore, one can show that our lower bound (Theorem 5) holds more broadly for maximizing *p-mean welfare*, for any $p \leq 1$; this includes social welfare maximization as a special case. Hence, we also strengthen the negative result of [4], by showing that it continues to hold even if the marginals of the subadditive valuations are binary.

Maximin share (MMS) is a prominent fairness notion in discrete fair division [15]. For an agent i, this fairness threshold is defined as the maximum value that i can guarantee for herself by partitioning the set of goods into n bundles and receiving the minimum valued one; here, n denotes the total number of agents. While MMS allocations (i.e., allocations that provide each agent a bundle of value at least as much as her maximin share) are not guaranteed to exist [32,41], this fairness notion is quite amenable to approximation guarantees across the hierarchy of complement-free valuations; see, [27], [28], and references therein. In the binary-marginals case, we note that MMS allocations are guaranteed to exist and can be computed efficiently for binary additive [14] and binary submodular [9] valuations. By contrast, such an existential result does not hold for binary XOS valuations [9]. For such valuations, however, the work of Li and Vetta [36] provides a polynomial-time algorithm that finds allocations wherein each agent receives a bundle of value at least 0.367 times her maximin share.[6] The current work addresses the stronger notion of groupwise maximin shares [5] under binary XOS valuations.

2 Notation and Preliminaries

We study the problem of allocating m indivisible goods among n agents in a fair and economically efficient manner. Throughout, we will use $[m] := \{1, 2, \ldots, m\}$ to denote the set of goods and $[n] := \{1, 2, \ldots, n\}$ to denote the set of agents. The cardinal preference of the agents $i \in [n]$, over subsets of goods, are expressed via

---

[5] The question of whether sub-linear approximation bounds can be achieved for Nash social welfare with demand-oracle access to general XOS, and subadditive, valuations remains an interesting direction of future work.

[6] The result of Li and Vetta [36] holds for a somewhat more general valuation class, which are defined via *hereditary set systems*.

valuations $v_i : 2^{[m]} \mapsto \mathbb{R}_+$. Specifically, $v_i(S) \in \mathbb{R}_+$ denotes the value that agent $i \in [n]$ has for subset of goods $S \subseteq [m]$. We represent fair division instances by the triple $\langle [m], [n], \{v_i\}_{i=1}^n \rangle$.

An allocation $\mathcal{A} = (A_1, A_2, \ldots, A_n)$ is a collection of n pairwise disjoint subsets of goods, $A_i \cap A_j = \emptyset$ for all $i \neq j$. Here, the subset of goods $A_i \subseteq [m]$ is assigned to agent $i \in [n]$ and will be referred to as a bundle. For ease of presentation and analysis, we do not force the requirement that, in an allocation, all the goods are assigned, i.e., the allocations can be partial with $\cup_{i=1}^n A_i \neq [m]$. Write $A_0 := [m] \setminus (\cup_{i=1}^n A_i)$ to denote the subset of unassigned goods in an allocation $\mathcal{A} = (A_1, \ldots, A_n)$.[7]

Valuation Classes. This work focuses on valuations that have the binary-marginals property, i.e., are dichotomous. Formally, a valuation v is said to have binary marginals iff $v(S \cup \{g\}) - v(S) \in \{0, 1\}$ for any subset of goods $S \subseteq [m]$ and any good $g \in [m]$. As a direct consequence, the valuations we consider are monotonic: $v(S) \leq v(T)$ for any subsets $S \subseteq T \subseteq [m]$. In addition, we assume that the valuations are normalized, $v_i(\emptyset) = 0$ for each $i \in [n]$.

A set of goods $S \subseteq [m]$ is said to be *non-wasteful*, with respect to a valuation v, iff $v(S) = |S|$. Note that, under valuations with binary marginals, subsets of non-wasteful sets are also non-wasteful; a proof of the following proposition is provided in the full version of this paper [8].

Proposition 1. *Let $v : 2^{[m]} \mapsto \mathbb{Z}_+$ be a function with binary marginals and $S \subseteq [m]$ be a non-wasteful set (with respect to v), then each subset of S is non-wasteful as well.*

We consider valuations that—in conjunction with satisfying the binary-marginals property—belong to the following classes of complement-free functions, presented in order of containment.

(i) *Additive*: A valuation $v : 2^{[m]} \mapsto \mathbb{R}_+$ is said to be additive iff the value of any subset of goods $S \subseteq [m]$ is equal to the sum of values of the goods in it, $v(S) = \sum_{g \in S} v(\{g\})$.

(ii) *Submodular*: A valuation $v : 2^{[m]} \mapsto \mathbb{R}_+$ is said to be submodular iff $v(S \cup \{g\}) - v(S) \geq v(T \cup \{g\}) - v(T)$ for every $S \subseteq T \subset [m]$ and good $g \in [m] \setminus T$.

(iii) XOS: A valuation $v : 2^{[m]} \mapsto \mathbb{R}_+$ is said to be XOS iff it can be expressed as a pointwise maximum over a collection of additive functions, i.e., there exists a collection of additive functions $\{\ell_t\}_{t=1}^L$ such that, for every subset $S \subseteq [m]$, we have $v(S) = \max_{t \in [L]} \ell_t(S)$. Here, the number of additive functions, L, can be exponentially large in m.

---

[7] Note that one can always allocate the subset of unassigned goods A_0 arbitrarily among the agents without reducing the Nash social welfare of $\mathcal{A} = (A_1, A_2, \ldots, A_n)$.

(iv) *Subadditive*: A valuation $v : 2^{[m]} \mapsto \mathbb{R}_+$ is said to be subadditive iff it does not admit any complementary subset of goods: $v(S \cup T) \le v(S) + v(T)$, for every pair of subsets $S, T \subseteq [m]$.

As mentioned previously, our algorithmic results hold in the standard value-oracle model, wherein, given any subset of goods $S \subseteq [m]$ and an agent $i \in [n]$, the value oracle returns $v_i(S) \in \mathbb{R}_+$ in unit time.

We will use the prefix *binary* before the names of function classes to denote that the valuation additionally has binary marginals, e.g., a function v is binary XOS iff it is XOS and has binary marginals. The following theorem (proved in the full version of this paper [8]) provides useful characterizations of binary XOS valuations.

Theorem 1. *A valuation $v : 2^{[m]} \mapsto \mathbb{R}_+$ is binary XOS iff it satisfies anyone of the following equivalent properties*

(P$_1$): Function v is XOS and it has binary marginals.

(P$_2$): Function v has binary marginals and for every set $S \subseteq [m]$ there exists a subset $X \subseteq S$ with the property that $v(X) = |X| = v(S)$.

(P$_3$): Function v can be expressed as a pointwise maximum of binary additive functions $\{\ell_t : 2^{[m]} \mapsto \mathbb{R}_+\}_{t=1}^{L}$, i.e., $v(S) = \max_{1 \le t \le L} \ell_t(S)$ for every $S \subseteq [m]$. Here, each function ℓ_t is additive and $\ell_t(g) \in \{0, 1\}$ for every $g \in [m]$.

(P$_4$): There exists a family of subsets $\mathcal{F} \subseteq 2^{[m]}$ such that $v(S) = \max_{F \in \mathcal{F}} |S \cap F|$ for every set $S \subseteq [m]$.

Social welfare and Nash social welfare. The social welfare $\mathrm{SW}(\cdot)$ of an allocation $\mathcal{A} = (A_1, \ldots, A_n)$ is defined as the sum of the values that the agents derive from their bundles in $\mathcal{A}$, i.e., $\mathrm{SW}(\mathcal{A}) := \sum_{i=1}^{n} v_i(A_i)$. The Nash social welfare $\mathrm{NSW}(\cdot)$ of an allocation $\mathcal{A} = (A_1, \ldots, A_n)$ is defined as the geometric mean of the agents' values in $\mathcal{A}$, i.e., $\mathrm{NSW}(\mathcal{A}) := (\prod_{i=1}^{n} v_i(A_i))^{\frac{1}{n}}$. An allocation $\mathcal{N} = (N_1, \ldots, N_n)$ with the maximum possible Nash social welfare (among the set of all allocations) is referred to as a Nash optimal allocation.

For binary XOS valuations, one can assume, without loss of generality, that welfare-maximizing allocations solely consist of non-wasteful bundles; the proof of this lemma is deferred to the full version of this paper [8].

Lemma 1. *For any allocation $\mathcal{P} = (P_1, \ldots, P_n)$ among agents with binary XOS valuations, there exists an allocation $\mathcal{P}' = (P'_1, \ldots, P'_n)$ of non-wasteful bundles that has the same valuation profile as $\mathcal{P}$, i.e., $v_i(P'_i) = |P'_i| = v_i(P_i)$ for all agents $i \in [n]$.*

3   Approximation Algorithm for Nash Social Welfare

Our algorithm (Algorithm 1) computes an allocation $\mathcal{A} = (A_1, \ldots, A_n)$ in which, for each agent i, the envy is multiplicatively bounded towards the entire set of goods, G_i, allocated to agents with bundle size at least four times that of i.

Specifically, with respect to allocation $\mathcal{A}$, write H_i to denote the set of agents who have received a bundle of size at least four times that of i, and let $G_i = \left(\cup_{j \in H_i} A_j \right) \cup A_0 \cup A_i$; recall that A_0 denotes the set of unassigned goods in $\mathcal{A}$. We show that, under binary XOS valuations, if $v_i(A_i) > \frac{1}{2} v_i(G_i)$ for all agents i, then $\mathcal{A}$ achieves a constant-factor approximation guarantee for Nash social welfare.

Algorithm 1 finds such an allocation by iteratively updating the agents' bundles. In particular, if for an agent i the envy requirement is not met (i.e., we have $v_i(A_i) \leq \frac{1}{2} v_i(G_i)$), then the algorithm finds a non-wasteful subset $X \subset G_i$ with twice the value of A_i, i.e., finds a subset $X \subset G_i$ with the property that $v_i(X) = |X| = 2v_i(A_i)$. The algorithm then assigns X to agent i, and updates the remaining bundles accordingly. Note that, under binary XOS valuations, such a subset X can be computed efficiently (in Line 5 of the algorithm): one can initialize $X = G_i$ and iteratively remove goods from X until the desired property is achieved; recall (P2) in Theorem 1. Also, with these updates, the algorithm maintains the invariant that the bundles assigned to the agents are non-wasteful. Indeed, the value of agent i doubles after receiving subset X, and we show that the algorithm necessarily finds the desired allocation after at most a polynomial number of such value increments, i.e., the algorithm runs in polynomial time (Lemma 3).

Algorithm 1. ALG

Input: Fair division instance $\langle [m], [n], \{v_i\}_{i=1}^n \rangle$ with value-oracle access to the binary XOS valuations v_is

Output: Allocation $\mathcal{A} = (A_1, \ldots, A_n)$

1: Compute an allocation $\mathcal{A} := (A_1, \ldots, A_n)$ with $v_i(A_i) = |A_i| = 1$, for every agent $i \in [n]$. {Such an allocation $\mathcal{A}$ can be computed by finding a perfect matching between the agents i and the goods valued by i.}

2: Initialize $A_0 = [m] \setminus (\cup_{j=1}^n A_j)$

3: For each agent $i \in [n]$, initialize subset of agents $H_i := \{j \in [n] \ : \ |A_j| > 4|A_i|\}$ and subset of goods $G_i := \left(\cup_{j \in H_i} A_j \right) \cup A_0 \cup A_i$

4: **while** there exists agent $i \in [n]$ such that $v_i(A_i) \leq \frac{1}{2} v_i(G_i)$ **do**

5:    Find subset $X \subseteq G_i$ with the property that $v_i(X) = |X| = 2v_i(A_i)$ {Such a non-wasteful subset X can be computed efficiently for binary XOS valuations}

6:    Set $A_i = X$, and update $A_j \leftarrow A_j \setminus X$ for each $j \in H_i$

7:    Set $A_0 = [m] \setminus (\cup_{j=1}^n A_j)$

8:    Set $H_k = \{j \in [n] \ : \ |A_j| > 4|A_k|\}$ and $G_k = \left(\cup_{j \in H_k} A_j \right) \cup A_0 \cup A_k$, for each agent $k \in [n]$

9: **end while**

10: **return** $\mathcal{A} = (A_1, \ldots, A_n)$

Write $\mathcal{N} = (N_1, \ldots, N_n)$ to denote a Nash optimal allocation for the given fair division instance. We will throughout assume that the optimal Nash welfare is positive, $\text{NSW}(\mathcal{N}) > 0$. In the complementary case, wherein $\text{NSW}(\mathcal{N}) = 0$, returning an arbitrary allocation suffices.[8] Note that the assumption $\text{NSW}(\mathcal{N}) > 0$ and the fact that the valuations have binary marginals ensure that, for each agent i, the bundle N_i contains a unit valued (by i) good. Hence, in Line 1 of the algorithm we are guaranteed to find a matching wherein each agent is assigned a good of value one.

The following lemma establishes an interesting property of the allocation $\mathcal{A} = (A_1, \ldots, A_n)$ returned by our algorithm. In particular, the lemma shows that—for any integer $\alpha \in \mathbb{Z}_+$—at most n/α agents i receive a bundle A_i of value less than $\frac{1}{18\alpha}$ times $v_i(N_i)$. That is, in allocation $\mathcal{A}$, for any $\alpha \in \mathbb{Z}_+$, the number of (18α)-suboptimal agents is at most n/α. We will establish the approximation ratio of Algorithm 1 (in Theorem 2 below) by invoking the lemma with dyadic values of α.

Lemma 2. *Let $\mathcal{A} = (A_1, \ldots, A_n)$ be the allocation returned by Algorithm 1 and $\mathcal{N} = (N_1, \ldots, N_n)$ be a Nash optimal allocation, with $\text{NSW}(\mathcal{N}) > 0$. Also, for any integer $\alpha \in \mathbb{Z}_+$, let $X_\alpha := \left\{ i \in [n] : v_i(A_i) < \frac{1}{18\alpha} v_i(N_i) \right\}$. Then,*

$$|X_\alpha| \leq \frac{n}{\alpha}$$

The proof of the above lemma uses a counting argument that is described as the following proposition, its proof is presented in the full version of this paper [8].

Proposition 2. *Let $D_0, D_1, \ldots, D_\ell$ be a collection of pairwise disjoint subsets of goods such that $|D_k|$ is an integer multiple of 4^{k+2}, for each $0 \leq k \leq \ell$. Also, let $\mathcal{B} = (B_1, B_2, \ldots, B_t)$ be any t-partition of the set $\cup_{k=0}^{\ell} D_k$ with the property that*

(P): For any index k and each good $g \in D_k$, if $g \in B_b$, then $|B_b| \leq 4^{k+2}$ (i.e., goods in D_k must be assigned among bundles of size at most 4^{k+2}).
Then, in the partition, the number of bundles $t \geq \sum_{k=0}^{\ell} \frac{|D_k|}{4^{k+2}}$.

We now give a proof of Lemma 2.

Proof of Lemma 2. Throughout its execution ALG assigns a non-wasteful bundle to every agent (see Lines 5 and 6) and, hence, for the returned allocation $\mathcal{A} = (A_1, \ldots, A_n)$ we have $v_i(A_i) = |A_i|$, for all agents i. Also, we assume, without loss of generality, that the bundles in the Nash optimal allocation $\mathcal{N} = (N_1, \ldots, N_n)$ are non-wasteful; see Lemma 1.

Fix an integer $\alpha \in \mathbb{Z}_+$ and consider any agent $i \in X_\alpha$. Recall that G_i contains the set of goods that (under allocation $\mathcal{A}$) are assigned among agents

---

[8] Here, in fact, one can also maximize the Nash social welfare subject to the constraint that the maximum possible number of agents receive a good: write n' to denote the size of the maximum-cardinality matching between the agents i and the goods valued by i, and introduce $(n - n')$ "dummy" goods, any nonempty subset of which gives unit value to any agent. Approximating Nash social welfare in this modified instance (with binary XOS valuations) addresses the constrained version of the problem.

in $H_i := \{j \in [n] \ : \ |A_j| > 4|A_i|\}$. We begin by upper bounding the size of the intersection between N_i and G_i; in particular, this bound shows that, in $\mathcal{A}$, not too many goods from the optimal bundle N_i can get assigned among agents in H_i. Towards this, note that the termination condition of the while-loop (Line 4) ensures that the returned non-wasteful bundle A_i satisfies $v_i(G_i) < 2v_i(A_i) = 2|A_i|$. Therefore, using Proposition 1 and the fact that N_i is non-wasteful we get

$$|N_i \cap G_i| = v_i(N_i \cap G_i) \leq v_i(G_i) < 2|A_i| \qquad (1)$$

Also, since agent $i \in X_\alpha$, the cardinality of N_i is more than 18α times that of A_i: $|N_i| = v_i(N_i) > 18\alpha \, v_i(A_i) = 18\alpha \, |A_i|$. This observation and inequality (1) imply that N_i has a sufficiently large intersection with $G_i^c := [m] \setminus G_i$

$$|N_i \cap G_i^c| = |N_i| - |N_i \cap G_i| > 18\alpha|A_i| - 2|A_i| \geq 16\alpha|A_i| \qquad (2)$$

Indeed, G_i^c is the set of goods that, in allocation $\mathcal{A}$, are assigned among the agents $j \in [n] \setminus (H_i \cup \{i\})$, i.e., among the agents $j \neq i$ with bundles of value $v_j(A_j) = |A_j| \leq 4|A_i|$.

To establish the desired upper bound on the size of X_α, we partition it into subsets. Specifically, for each $0 \leq k \leq \lfloor \log_4 m \rfloor$, define set

$$X_\alpha^k := \{i \in X_\alpha \ : \ 4^k \leq v_i(A_i) < 4^{k+1}\}.$$

That is, X_α^k is the set of agents for whom the ratio between assigned value and the optimal value is less than $\frac{1}{18\alpha}$ (i.e., $i \in X_\alpha$) and the assigned value is in the range $[4^k, 4^{k+1})$. We note that with k between 0 and $\lfloor \log_4 m \rfloor$, the subsets X_α^ks, partition X_α. In particular, initially in ALG (see Line 1) each agent achieves a value of one; recall the assumption that $\mathrm{NSW}(\mathcal{N}) > 0$ and, hence, there exists a matching wherein each agent is assigned a nonzero valued good. Furthermore, during the execution of ALG the agents' valuations inductively continue to be at least one: consider any iteration of the while-loop and let $\hat{i}$ be the agent that receives a new bundle X in the iteration (see Lines 5 and 6). The selection criterion of X ensures that the valuation of $\hat{i}$ in fact doubles. For any agent $j \in H_{\hat{i}}$, before the update in Line 6 we have $v_j(A_j) = |A_j| > 4|A_{\hat{i}}| = 2|X|$ and, hence, even after the update $(A_j \leftarrow A_j \setminus X)$ agent j's value continues to be at least one. Finally, for each remaining agent (in the set $[n] \setminus \left(H_{\hat{i}} \cup \{\hat{i}\}\right)$) its bundle remains unchanged. Hence, for the returned allocation $\mathcal{A} = (A_1, \ldots, A_n)$ we have $v_i(A_i) = |A_i| \geq 1$. Also, the fact that the marginals of the valuation v_i are binary implies $v_i([m]) \leq m$, i.e., $v_i(A_i) \leq m$. Therefore, the bounds $1 \leq v_i(A_i) \leq m$ (for all agents i) imply that the subsets X_α^ks, with $0 \leq k \leq \lfloor \log_4 m \rfloor$, partition X_α; in particular, $\sum_k |X_\alpha^k| = |X_\alpha|$.

Furthermore, for each agent $i \in X_\alpha^k$ we have

$$|N_i \cap G_i^c| > 16\alpha \, |A_i| \qquad \text{(via inequality (2))}$$
$$\geq 16\alpha \, 4^k \qquad \text{(since } i \in X_\alpha^k)$$
$$= \alpha \, 4^{k+2}$$

Therefore, for each k, the set of goods $D_k := \bigcup_{i \in X_\alpha^k}(N_i \cap G_i^c)$ satisfies $|D_k| \geq \alpha 4^{k+2} |X_\alpha^k|$. That is, for each k, and whenever $X_\alpha^k \neq \emptyset$, the size of set D_k is at least a positive integer multiple of 4^{k+2}. Also, note that for each good $\bar{g} \in D_k$, we have (by definition of D_k) that $\bar{g} \in N_{\bar{i}} \cap G_{\bar{i}}^c$ for some $\bar{i} \in X_\alpha^k$. These containments (and the definition of $G_{\bar{i}}^c$) ensure that $\bar{g} \in A_j$,[9] for some agent $j \in [n]$ with the property that $|A_j| \leq 4|A_{\bar{i}}| < 4^{k+2}$. That is, for each k (with $X_\alpha^k \neq \emptyset$), the cardinality of D_k is a positive integer multiple of 4^{k+2} and (under $\mathcal{A}$) the goods in D_k must be assigned to agents with bundles of size at most 4^{k+2}. These two properties ensure that (in allocation $\mathcal{A}$) a sufficiently large number of bundles are necessarily required to cover the set of goods $\bigcup_k D_k = \bigcup_k \bigcup_{i \in X_\alpha^k}(N_i \cap G_i^c) = \bigcup_{i \in X_\alpha}(N_i \cap G_i^c)$. Specifically, write $t \in \mathbb{Z}_+$ to denote the number of agents that have been assigned (under allocation $\mathcal{A}$) at least one good from $\bigcup_k D_k$ (i.e., $t := |\{j \in [n] : A_j \cap (\cup_k D_k) \neq \emptyset\}|$), then Proposition 2 (proved in the full version of this paper [8]) gives us

$$t \geq \sum_{k=0}^{\lfloor \log_4 m \rfloor} \frac{|D_k|}{4^{k+2}} \geq \sum_{k=0}^{\lfloor \log_4 m \rfloor} \frac{\alpha 4^{k+2} \cdot |X_\alpha^k|}{4^{k+2}} = \alpha \sum_{k=0}^{\lfloor \log_4 m \rfloor} |X_\alpha^k| = \alpha |X_\alpha|.$$

However, the number of agents t cannot be more than n. Hence, the stated claim follows $n \geq t \geq \alpha |X_\alpha|$.

The allocation $\mathcal{A} = (A_1, A_2, \ldots, A_n)$ returned by Algorithm 1 can be made complete by allocating the (unassigned) goods in $[m] \setminus \cup_{i=1}^n A_i$ arbitrarily. Doing this would not affect the approximation guarantee.

The following lemma establishes the time complexity of ALG and its proof is delegated to the full version of the paper [8].

Lemma 3. *For any given fair division instance with n agents, m goods, and value-oracle access to the binary XOS valuations, Algorithm 1 (ALG) returns an allocation in time that is polynomial in n and m.*

The following theorem is our main result for Nash social welfare.

Theorem 2. *For binary XOS valuations and in the value-oracle model, there exists a polynomial-time 288-approximation algorithm for the Nash social welfare maximization problem.*

Proof. Let $\mathcal{A} = (A_1, A_2, \ldots, A_n)$ be the (non-wasteful) allocation returned by ALG, and $\mathcal{N} = (N_1, N_2, \ldots, N_n)$ be a (non-wasteful) Nash optimal allocation. As mentioned previously, under the condition that optimal Nash welfare NSW($\mathcal{N}$) > 0, every agent necessarily receives a value of at least one in the allocation $\mathcal{A}$, i.e., $v_i(A_i) \geq 1$ for all i. In addition, $v_i(N_i) \leq v_i([m]) \leq m$; the last inequality follows from the fact the valuations have binary marginals. Hence, for all agents i, we have $v_i(A_i) \geq \frac{1}{m} v_i(N_i)$.

---

[9] Recall that $A_0 \subseteq G_{\bar{i}}$ and, hence, $A_0 \cap G_{\bar{i}}^c = \emptyset$.

Next, we partition the set of agents based on the ratio of their assigned value, $v_i(A_i)$, and their optimal value, $v_i(N_i)$. Specifically, define set $Y_{2^d} := \left\{ i \in [n] : \frac{1}{2^{d+1}} \frac{v_i(N_i)}{18} \leq v_i(A_i) < \frac{1}{2^d} \frac{v_i(N_i)}{18} \right\}$, for each integer $d \in \{0, 1, \ldots, \lceil \log m \rceil\}$. Since $v_i(A_i) \geq \frac{1}{m} v_i(N_i)$ for all i, the remaining agents $i' \in Y' := [n] \setminus \left(\bigcup_{d=0}^{\lceil \log m \rceil} Y_{2^d} \right)$ satisfy $v_{i'}(A_{i'}) \geq \frac{1}{18} v_{i'}(N_{i'})$. Indeed, the subsets Y_{2^d}s and Y' partition the set of agents, and $|Y'| + \sum_{d=0}^{\lceil \log m \rceil} |Y_{2^d}| = n$.

Note that, with $\alpha = 2^d$, we have $Y_\alpha \subseteq X_\alpha$; here, set X_α is defined as in Lemma 2. Hence, invoking this lemma we get

$$|Y_{2^d}| \leq \frac{n}{2^d} \qquad \text{for all } 0 \leq d \leq \lceil \log m \rceil \tag{3}$$

Write $\pi(S) := \prod_{i \in S} \frac{v_i(A_i)}{v_i(N_i)}$, if the subset of agents $S \neq \emptyset$, and 1 otherwise. Now, towards establishing the approximation ratio, consider

$$\frac{\text{NSW}(\mathcal{A})}{\text{NSW}(\mathcal{N})} = \left(\prod_{i=1}^{n} \frac{v_i(A_i)}{v_i(N_i)} \right)^{1/n} = \left(\pi(Y') \prod_{d=0}^{\lceil \log m \rceil} \pi(Y_{2^d}) \right)^{1/n}$$

$$\geq \left(\left(\frac{1}{18} \right)^{|Y'|} \prod_d \pi(Y_{2^d}) \right)^{1/n}$$

$$\text{(since } v_i(A_i) \geq \tfrac{1}{18} v_i(N_i) \text{ for all } i \in Y')$$

$$\geq \left(\left(\frac{1}{18} \right)^{|Y'|} \prod_d \left(\frac{1}{18 \, 2^{d+1}} \right)^{|Y_{2^d}|} \right)^{1/n}$$

$$\text{(since } v_i(A_i) \geq \tfrac{1}{18 \, 2^{d+1}} v_i(N_i) \text{ for all } i \in Y_{2^d})$$

$$= \frac{1}{18} \left(\prod_d \left(\frac{1}{2^{d+1}} \right)^{|Y_{2^d}|} \right)^{1/n} \qquad \text{(since } |Y'| + \sum_k |Y_{2^d}| = n)$$

$$= \frac{1}{18} \left(\prod_d \left(\frac{1}{2^{d+1}} \right)^{|Y_{2^d}|/n} \right)$$

$$\geq \frac{1}{18} \left(\prod_d \left(\frac{1}{2^{d+1}} \right)^{\frac{1}{2^d}} \right) \qquad \text{(via inequality (3))}$$

We can show that the product $\prod_d \left(\frac{1}{2^{d+1}} \right)^{\frac{1}{2^d}} \geq \frac{1}{16}$; see, full version of the paper [8]. Therefore, the stated approximation bound follows

$$\frac{\text{NSW}(\mathcal{A})}{\text{NSW}(\mathcal{N})} \geq \frac{1}{18} \prod_{d=0}^{\lceil \log m \rceil} \left(\frac{1}{2^{d+1}} \right)^{\frac{1}{2^d}} \geq \frac{1}{18} \cdot \frac{1}{16} = \frac{1}{288}.$$

4   Hardness of Approximation for Binary XOS Valuations

This section establishes the APX-hardness of maximizing Nash social welfare in fair division instances with binary XOS valuations. This inapproximability holds even if the agents' (binary XOS) valuations are identical and admit a succinct representation. We obtain the hardness result by developing an approximation preserving reduction from the following gap version of the independent set problem in 3-regular graphs.

Theorem 3 ([17]). *Given a 3-regular graph $\mathcal{G}$ and a threshold τ, it is* NP-*hard to distinguish between*

- YES *Instances: The size of the maximum independent set in $\mathcal{G}$ is at least τ.*
- NO *Instances: The size of the maximum independent set in $\mathcal{G}$ is at most $\frac{94}{95}\tau$.*

Our hardness result is presented in the following theorem; refer to the full version of this paper for its proof [8]. Notably, the hardness result is obtained by reducing the gap problem described above to a gap version of computing maximum Nash social welfare for instances with binary XOS valuations.

Theorem 4. *For fair division instances with (identical) binary* XOS *valuations, it is* NP-*hard to approximate the maximum Nash social welfare within a factor of* 1.0042.

5   Lower Bound for Binary Subadditive Valuations

In this section we prove that, under binary subadditive valuations, an exponential number of value queries are required to obtain a sub-linear approximation for the Nash social welfare.

Theorem 5. *For fair division instances $\langle [m], [n], \{f_i\}_{i=1}^n \rangle$ with binary subadditive valuations and a fixed constant $\varepsilon \in (0, 1]$, exponentially many value queries are necessarily required for finding any allocation with Nash social welfare at least $\frac{1}{n^{1-\varepsilon}}$ times the optimal.*

Towards establishing this theorem, we define two (families of) fair division instances, each with n agents, $m = n^2$ goods, and binary subadditive valuations. In the first instance, all the agents will have the same binary subadditive valuation, $f : 2^{[m]} \mapsto \mathbb{R}_+$, while in the second instance, the valuations of the agents will be non-identical, $f_i' : 2^{[m]} \mapsto \mathbb{R}_+$ for each agent $i \in [n]$. In particular, we will construct the valuations, f and $\{f_i'\}_i$, such that (i) distinguishing whether the agents' valuations are $\{f_i'\}_i$ or f requires an exponential number of value queries (Lemma 6) and (ii) the optimal Nash social welfare of the two instances differ multiplicatively by a linear factor. Since the second property implies that one can use any sub-linear approximation of the optimal Nash social welfare to distinguish between the two instances (i.e., between the two valuation settings), these properties will establish the stated query lower bound.

To specify the valuations, fix a small constant $\delta \in \left(0, \frac{1}{16}\right)$ and write integers $p := \lfloor (1 + \delta) n^{4\delta} \rfloor$ along with $q := \lfloor n^{1+2\delta} \rfloor$. We will assume, throughout, that n is large enough to ensure that the integers $p, q \in \mathbb{Z}_+$ satisfy $p < q$. With these parameters in hand, define valuation $f : 2^{[m]} \mapsto \mathbb{Z}_+$ as follows

$$f(S) := \begin{cases} |S| & \text{if } |S| \le p, \\ p & \text{if } p < |S| \le q, \\ \left\lceil \frac{p |S|}{q} \right\rceil & \text{otherwise, if } |S| > q \end{cases}$$

For constructing valuations $\{f_i'\}_i$, consider a random n-partition, $T_1, T_2, \ldots, T_n$, of the set of goods $[m]$, with $|T_i| = n$ for each $i \in [n]$. Now, for every $i \in [n]$ and subset $S \subseteq [m]$, define $f_i'(S) := \max\{f(S), |S \cap T_i|\}$. The following two lemmas show that the constructed valuations are binary subadditive; their proofs appear in the full version of this paper [8].

Lemma 4. *The valuation f (as defined above) is subadditive and has binary marginals.*

Lemma 5. *The valuations $\{f_i'\}_{i \in [n]}$ (as defined above) are subadditive and have binary marginals.*

The following lemma shows that the functions f and f_i' (for any $i \in [n]$) cannot be distinguished from each other using polynomial number of values queries; its proof is deferred to the full version of the paper [8].

Lemma 6. *An exponential number of value queries are required to distinguish between the functions f and f_i', for any $i \in [n]$.*

5.1   Proof of Theorem 5

Here, we establish Theorem 5, our main negative result for binary subadditive valuations.

With n agents and $m = n^2$ goods, we consider two families of instances with binary subadditive valuations (see Lemmas 4 and 5): the first one in which all the agents have the same valuation f, and the other wherein the agents' valuations are $\{f_i'\}_{i=1}^n$. Lemma 6 shows that exponentially many value value queries are required to distinguish between these two cases, i.e., to determine whether the agents' valuations are f or $\{f_i'\}_i$.

We will next establish that such a distinction can be made via an $n^{1-\varepsilon}$ approximation to the optimal Nash social welfare and, hence, obtain the stated query complexity of approximating the Nash social welfare. Note that, under valuations $\{f_i'\}_i$, the optimal Nash welfare is equal to n. In particular, allocating bundle T_i to agent i leads to $f_i'(T_i) = n$, for each $i \in [n]$, i.e., here the Nash social welfare of the allocation $(T_1, \ldots, T_n)$ is n.

By contrast, under valuation f, the optimal Nash social welfare is at most $(2n^{4\delta} + 1)$. In fact, the following argument shows that, for any allocation $(A_1, \ldots, A_n)$, the average social welfare $\frac{1}{n} \sum_{i=1}^n f(A_i) \le 2n^{4\delta} + 1$. Hence, via the AM-GM inequality, this upper bound holds for the optimal Nash social welfare as well.

Let $\mathcal{A} = (A_1, A_2, \ldots, A_n)$ be the allocation that maximizes the average social welfare under f. We can assume, without loss of generality, that for each agent $i \in [n]$, either $|A_i| > q$ or $|A_i| \leq p$. Otherwise, if for some agent $j \in [n]$, we have $p < |A_j| \leq q$, then we can iteratively remove goods from A_j until $|A_j| = p$. This update will not decrease $f(A_j)$ (this value will continue to be p) and, hence, the social welfare remains unchanged as well. For ease of analysis, we further modify allocation $\mathcal{A}$: while there are two agents $j, k \in [n]$ with $|A_k| \geq |A_j| > q$, we iteratively transfer goods from A_j to A_k until $|A_j| = p$. Note that after each transfer the social welfare decreases by at most one (i.e., the drop in average social welfare is at most $1/n$):[10] for any $s \leq |A_j|$, we have $\left\lceil \frac{p}{q} (|A_k| + s) \right\rceil +$ $\left\lceil \frac{p}{q} (|A_j| - s) \right\rceil \geq \left\lceil \frac{p}{q} (|A_k| + s) + \frac{p}{q} (|A_j| - s) \right\rceil = \left\lceil \frac{p}{q} |A_k| + \frac{p}{q} |A_j| \right\rceil \geq \left\lceil \frac{p}{q} |A_k| \right\rceil +$ $\left\lceil \frac{p}{q} |A_j| \right\rceil - 1$.

Since at most n such transfers can occur (between pairs of agents), the *average* social welfare of allocation $\mathcal{A}$ decreases by at most one after all the transfers. Now, allocation $\mathcal{A}$ has exactly one agent with bundle size greater than p. This observation gives us the following upper bound

$$\frac{1}{n} \sum_{i=1}^{n} f(A_i) \leq \frac{1}{n} \left((n-1)p + \left\lceil \frac{p}{q} m \right\rceil \right)$$

$$\leq \frac{1}{n} \left(np + \frac{p}{q} m \right)$$

$$\leq p + \frac{np}{q} \qquad \text{(since } m = n^2)$$

$$\leq 2n^{4\delta} \qquad \text{(since } p = \lfloor (1+\delta)n^{4\delta} \rfloor \text{ and } q = \lfloor n^{1+2\delta} \rfloor)$$

Therefore, under valuation f, the average social welfare is at most $2n^{4\delta} + 1$. As mentioned previously, this implies that the ratio of the optimal Nash welfares under f_i's and f is $O(n^{1-4\delta})$. This, overall, establishes that any sub-linear approximation would differentiate between the valuations and, hence, require exponentially many value queries. The theorem stands proved.

6   Conclusion and Future Work

We develop algorithmic and hardness result for Nash social welfare maximization under binary XOS and binary subadditive valuations. Our algorithm provides (under binary XOS valuations) constant-factor approximations simultaneously for Nash social welfare, social welfare, and GMMS. It would be interesting to extend the positive result for Nash social welfare to the asymmetric version, wherein each agent has an associated weight (entitlement) $e_i \in$

---

[10] Recall that for any $a, b \in \mathbb{R}_+$, the following inequalities hold: $\lceil a \rceil + \lceil b \rceil - 1 \leq \lceil a + b \rceil \leq \lceil a \rceil + \lceil b \rceil$.

$\mathbb{R}_+$, and the objective is to find an allocation $(X_1, \ldots, X_n)$ that maximizes $(\prod_i (v_i(X_i))^{e_i})^{\frac{1}{\sum_i e_i}}$. Another interesting direction for future work is to develop, under binary XOS valuations, constant-factor approximation algorithms for p-mean welfare maximization, with $p \leq 1$.

References

1. Anari, N., Gharan, S.O., Mai, T., Vazirani, V.V.: Nash Social Welfare for Indivisible Items under Separable, Piecewise-Linear Concave Utilities. In: Proceedings of the 29th Annual ACM-SIAM Symposium on Discrete Algorithms (SODA), pp. 2274–2290 (2018)
2. Anari, N., Gharan, S.O., Saberi, A., Singh, M.: Nash social welfare, matrix permanent, and stable polynomials. In: Proceedings of the 8th Conference on Innovations in Theoretical Computer Science (ITCS) (2017)
3. Babaioff, M., Ezra, T., Feige, U.: Fair and truthful mechanisms for dichotomous valuations. In: Proceedings of the AAAI Conference on Artificial Intelligence. vol. 35. AAAI (2021)
4. Barman, S., Bhaskar, U., Krishna, A., Sundaram, R.G.: Tight approximation algorithms for p-mean welfare under subadditive valuations. In: 28th Annual European Symposium on Algorithms (ESA 2020). Schloss Dagstuhl-Leibniz-Zentrum für Informatik (2020)
5. Barman, S., Biswas, A., Krishnamurthy, S., Narahari, Y.: Groupwise maximin fair allocation of indivisible goods. In: Proceedings of the AAAI Conference on Artificial Intelligence. vol. 32 (2018)
6. Barman, S., Krishnamurthy, S.K., Vaish, R.: Finding fair and efficient allocations. In: Proceedings of the 2018 ACM Conference on Economics and Computation, pp. 557–574 (2018)
7. Barman, S., Krishnamurthy, S.K., Vaish, R.: Greedy algorithms for maximizing Nash social welfare. In: Proceedings of the 17th International Conference on Autonomous Agents and MultiAgent Systems, pp. 7–13 (2018)
8. Barman, S., Verma, P.: Approximating nash social welfare under binary XOS and binary subadditive valuations. CoRR abs/2106.02656 (2021)
9. Barman, S., Verma, P.: Existence and computation of maximin fair allocations under matroid-rank valuations. In: Proceedings of the 20th International Conference on Autonomous Agents and MultiAgent Systems, pp. 169–177. AAMAS '21 (2021)
10. Bei, X., Garg, J., Hoefer, M., Mehlhorn, K.: Earning limits in fisher markets with spending-constraint utilities. In: Proceedings of the International Symposium on Algorithmic Game Theory (SAGT), pp. 67–79 (2017)
11. Benabbou, N., Chakraborty, M., Igarashi, A., Zick, Y.: Finding fair and efficient allocations when valuations don't add up. In: International Symposium on Algorithmic Game Theory, pp. 32–46. Springer (2020). https://doi.org/10.1007/978-3-030-57980-7_3
12. Bogomolnaia, A., Moulin, H., Stong, R.: Collective choice under dichotomous preferences. J. Econ. Theory **122**(2), 165–184 (2005)
13. Bouveret, S., Lang, J.: Efficiency and envy-freeness in fair division of indivisible goods: logical representation and complexity. J. Artif. Intell. Res. **32**, 525–564 (2008)

14. Bouveret, S., Lemaître, M.: Characterizing conflicts in fair division of indivisible goods using a scale of criteria. Auton. Agents Multi-Agent Syst. **30**(2), 259–290 (2016)
15. Budish, E.: The combinatorial assignment problem: approximate competitive equilibrium from equal incomes. J. Polit. Econ. **119**(6), 1061–1103 (2011)
16. Chaudhury, B.R., Garg, J., Mehta, R.: Fair and efficient allocations under subadditive valuations. In: Proceedings of the AAAI Conference on Artificial Intelligence. vol. 35. AAAI (2021)
17. Chlebík, M., Chlebíková, J.: Inapproximability results for bounded variants of optimization problems. In: International Symposium on Fundamentals of Computation Theory, pp. 27–38. Springer (2003). https://doi.org/10.1007/978-3-540-45077-1_4
18. Cole, R., et al.: Convex program duality, Fisher markets, and Nash social welfare. In: Proceedings of the 2017 ACM Conference on Economics and Computation, pp. 459–460 (2017)
19. Cole, R., Gkatzelis, V.: Approximating the Nash social welfare with indivisible items. In: Proceedings of the forty-seventh annual ACM symposium on Theory of computing, pp. 371–380 (2015)
20. Darmann, A., Schauer, J.: Maximizing Nash product social welfare in allocating indivisible goods. European J. Oper. Res. **247**(2), 548–559 (2015)
21. Dobzinski, S., Nisan, N., Schapira, M.: Approximation algorithms for combinatorial auctions with complement-free bidders. Math. Oper. Res. **35**(1), 1–13 (2010)
22. Feige, U.: On maximizing welfare when utility functions are subadditive. In: Proceedings of the Thirty-Eighth Annual ACM Symposium on Theory of Computing. STOC '06, Association for Computing Machinery, pp. 41–50. New York, NY, USA (2006)
23. Freitas, G.: Combinatorial assignment under dichotomous preferences (2010)
24. Garg, J., Hoefer, M., Mehlhorn, K.: Approximating the Nash social welfare with budget-additive valuations. In: Proceedings of the Twenty-Ninth Annual ACM-SIAM Symposium on Discrete Algorithms, pp. 2326–2340. SIAM (2018)
25. Garg, J., Husić, E., Végh, L.A.: Approximating Nash social welfare under Rado valuations. In: Proceedings of the 53rd Annual ACM SIGACT Symposium on Theory of Computing, pp. 1412–1425 (2021)
26. Garg, J., Kulkarni, P., Kulkarni, R.: Approximating Nash social welfare under submodular valuations through (un)matchings. In: Proceedings of the fourteenth annual ACM-SIAM symposium on discrete algorithms, pp. 2673–2687. SIAM (2020)
27. Garg, J., Taki, S.: An improved approximation algorithm for maximin shares. In: Proceedings of the 21st ACM Conference on Economics and Computation, pp. 379–380 (2020)
28. Ghodsi, M., HajiAghayi, M., Seddighin, M., Seddighin, S., Yami, H.: Fair allocation of indivisible goods: improvements and generalizations. In: Proceedings of the 2018 ACM Conference on Economics and Computation, pp. 539–556 (2018)
29. Goldman, J., Procaccia, A.D.: Spliddit: unleashing fair division algorithms. ACM SIGecom Exchanges **13**(2), 41–46 (2015)
30. Halpern, D., Procaccia, A.D., Psomas, A., Shah, N.: Fair division with binary valuations: one rule to rule them all. In: International Conference on Web and Internet Economics, pp. 370–383. Springer (2020)
31. Kurokawa, D., Procaccia, A.D., Shah, N.: Leximin allocations in the real world. ACM Trans. Econ. Comput. (TEAC) **6**(3–4), 1–24 (2018)

32. Kurokawa, D., Procaccia, A.D., Wang, J.: When can the maximin share guarantee be guaranteed? In: Proceedings of the Thirtieth AAAI Conference on Artificial Intelligence, pp. 523–529 (2016)
33. Lee, E.: Apx-hardness of maximizing Nash social welfare with indivisible items. Inf. Proc. Lett. **122**, 17–20 (2017)
34. Leyton-Brown, K., Milgrom, P., Segal, I.: Economics and computer science of a radio spectrum reallocation. Proc. National Acad. Sci. **114**(28), 7202–7209 (2017)
35. Li, W., Vondrák, J.: A constant-factor approximation algorithm for Nash social welfare with submodular valuations. arXiv preprint arXiv:2103.10536 (2021)
36. Li, Z., Vetta, A.: The fair division of hereditary set systems. ACM Trans. Econ. Comput. (TEAC) **9**(2), 1–19 (2021)
37. Moulin, H.: Fair division and collective welfare. MIT press (2004)
38. Nash, J.: The bargaining problem. Econometrica **18**(2), 155–162 (1950)
39. Nisan, N., Roughgarden, T., Tardos, E., Vazirani, V.V.: Algorithmic game theory. Cambridge University Press (2007)
40. Ortega, J.: Multi-unit assignment under dichotomous preferences. Math. Soc. Sci. **103**, 15–24 (2020)
41. Procaccia, A.D., Wang, J.: Fair enough: guaranteeing approximate maximin shares. In: Proceedings of the fifteenth ACM conference on Economics and computation, pp. 675–692 (2014)
42. Roth, A.E., Sönmez, T., Ünver, M.U.: Pairwise kidney exchange. J. Econ. Theory **125**(2), 151–188 (2005)
43. Schrijver, A.: Combinatorial optimization: polyhedra and efficiency, vol. 24. Springer Science and Business Media (2003)

Default Ambiguity: Finding the Best Solution to the Clearing Problem

Pál András Papp[✉] and Roger Wattenhofer

ETH Zürich, Zürich, Switzerland
{apapp,wattenhofer}@ethz.ch

Abstract. We study financial networks with debt contracts and credit default swaps between specific pairs of banks. Given such a financial system, we want to decide which of the banks are in default, and how much of their liabilities can these defaulting banks pay. There can easily be multiple different solutions to this problem, leading to a situation of *default ambiguity*, and a range of possible solutions to implement for a financial authority.

In this paper, we study the properties of the solution space of such financial systems, and analyze a wide range of reasonable objective functions for selecting from the set of solutions. Examples of such objective functions include minimizing the number of defaulting banks, minimizing the amount of unpaid debt, maximizing the number of satisfied banks, and many others. We show that for all of these objectives, it is NP-hard to approximate the optimal solution to an $n^{1-\epsilon}$ factor for any $\epsilon > 0$, with n denoting the number of banks. Furthermore, we show that this situation is rather difficult to avoid from a financial regulator's perspective: the same hardness results also hold if we apply strong restrictions on the weights of the debts, the structure of the network, or the amount of funds that banks must possess. However, if we restrict both the network structure and the amount of funds simultaneously, then the solution becomes unique, and it can be found efficiently.

Keywords: Financial network · Default ambiguity · Clearing problem · Credit default swap

1 Introduction

Financial systems are often called "highly complex", suggesting that relations and contracts between different financial institutions such as banks form a networked system that is basically impossible to understand. In order to model this phenomenon, there is a recent line of work that aims to describe this complexity in terms of computational complexity.

At the core of understanding financial systems is the *clearing problem*: given a system of banks with (conditional or unconditional) debt contracts between specific banks, we need to decide which of the banks are in default due to these

© Springer Nature Switzerland AG 2022
M. Feldman et al. (Eds.): WINE 2021, LNCS 13112, pp. 391–409, 2022.
https://doi.org/10.1007/978-3-030-94676-0_22

debts, and how much of their liabilities can these defaulting banks pay. This is a fundamental problem in a financial system, and an essential task for a financial regulator after a shock, with the 2008 financial crisis as a recent example.

Earlier results show that the clearing problem is computationally easy if all contracts between the banks are unconditional debts, or more generally, if the contracts in the network represent *"long" positions*; that is, a better outcome for one bank ensures a better (or the same) outcome for other banks. However, this is not always the case in practice: banks often have *"short" positions* on each other, when it is more favorable for a bank if another bank is in a worse situation. Typical short positions are credit default swaps (CDSs), short-selling options and other types of derivatives.

This suggests that a realistic analysis of financial systems requires a model that can capture both long and short positions. However, with both long and short positions in the network, financial systems exhibit significantly richer behavior: we can easily have situations of *default ambiguity* when there are multiple solutions in the system, and none of these solutions is obviously superior to the others in terms of clearing.

In practice, a clearing authority has to make a choice among these different solutions of the system, yielding an outcome that is more favorable to some banks and less favorable to others. In this paper, we focus on such cases of default ambiguity; we study the different solutions of the system, and various criteria to evaluate these solutions and select one of them to implement.

We begin with some fundamental observations about the solution space of financial systems. We then introduce a wide range of problems that aim to find the best solution according to a specific objective. These include finding e.g. the solution with the smallest number of defaults, the solution preferred by the largest number of banks, the best solution for a specific bank, and many others.

Our first main contribution is negative, showing that all these problems are not only NP-hard to solve, but also NP-hard to approximate to any $n^{1-\epsilon}$ factor (for any $\epsilon > 0$). This shows that even if the clearing authority has a well-defined objective to select among the solutions, finding a reasonably good solution is still not viable in practice.

We then study the same problem from a financial regulator's perspective, showing that it is rather difficult to come up with restrictions on the network to prevent this situation. In particular, we show that the same hardness results still hold in many restricted variants of the model: with unit-weight contracts, with severe restrictions on the network structure, and also if we require banks to own a positive amount of funds.

However, on the positive side, we also show that if we restrict both the network structure and the funds of banks simultaneously, then the resulting financial networks have a unique solution, and this solution can be found efficiently.

2   Related Work

The fundamental model of financial systems was introduced by Eisenberg and Noe [9], which only assumes simple debt contracts between the banks. Following works have also extended this model by e.g. default costs [23], cross-ownership relations [10, 26] or so-called covered CDSs [16]. However, these model variants can only describe long positions in a network. This means that there is always a *maximal* solution in the system that is simultaneously the best for all banks, and thus the clearing problem is not particularly interesting in this setting.

In contrast to this, the recent work of Schuldenzucker *et al.* [24, 25] introduces a model which also allows CDSs in the network, i.e. conditional debt contracts where the payment obligation depends on the default of a specific third bank. While a CDS is still a very simple contract, it can already capture short positions in the network. Moreover, CDSs are a prominent kind of derivative in real-world financial systems that also played a major role in the 2008 financial crisis [12].

We use this model of Schuldenzucker *et al.* as the base model for our findings. With both debts and CDSs, the clearing problem suddenly becomes significantly more challenging. The work of [24, 25] mostly focuses on the existence of a solution in this model, and the complexity of finding an arbitrary solution; we summarize these results in Sect. 4.

However, in the general case, these financial networks do not have a maximal solution, and thus an authority has to select from a set of solutions that represent a trade-off between the interests of different banks. The work of [24, 25] does not study this situation, describing it as unwanted since it is prone to the lobbying activity of banks in the system. Our work analyzes the clearing problem in this general case; to our knowledge, the problem has not been studied from this perspective before.

In general, there are many previous works that study the propagation of shocks in financial networks, and its dependence on the connectivity of the network [1, 3, 6, 11]. There are also several results that study the topic from a computational complexity perspective; however, they mostly assume a simple debt-only model, and focus on more complex questions, such as sensitivity to shocks or bailout policies [8, 15, 18, 20]. Other works introduce more substantial changes into these models, e.g. time-dependent clearing mechanisms [2, 22] or game-theoretic aspects [4, 19].

There is also a wide literature on different financial derivatives, and CDSs in particular [7, 12, 17]. On the more practical side, the clearing problem also plays a central role in stress tests to evaluate the sensitivity of financial systems, e.g. in the European Central Bank's stress test framework [5].

3   Model Definition

3.1   Banks and Contracts

A financial network consists of a set of *banks* B. Individual banks are mostly denoted by u, v or w, the number of banks by $n = |B|$. Each bank v has a certain amount of *funds* (in financial terms: external assets) available to the bank, denoted by e_v.

We assume that there are contracts for payments between given pairs of banks in the system. Each such contract is between two specific banks u and

v, and obliges u (the debtor) to pay a specific amount of money (known as the *notional*) to the other bank v (the creditor), either unconditionally or based on a specific condition.

These contracts result in a specific amount of payment obligation for each bank v. If v cannot fulfill these obligations, then we say that v is *in default*. In this case, the *recovery rate* of v, denoted by r_v, is the proportion of liabilities that v is able to pay. Note that $r_v \in [0,1]$, and v is in default exactly if $r_v < 1$.

The model allows two kinds of contracts between banks. Debt contracts (or simply *debts*) oblige bank u to pay a specific amount to v unconditionally, i.e. in any case. On the other hand, we also allow *credit default swaps (CDSs)* between u and v in reference to a third bank w. A CDS represents a conditional debt that obliges u to pay a specific amount to v only in case if bank w is in default. More specifically, if the weight of the CDS is δ and the recovery rate of bank w is r_w, then the CDS incurs a payment obligation of $\delta \cdot (1 - r_w)$ from node u to v. In practice, CDSs are often used as an insurance policy against the default of the debtors of the bank, or as a speculative bet based on insights into the market.

Before a formal definition, let us consider the example in Fig. 1. In this system, bank u has a total liability of 4 due to the 2 outgoing debts, but it only has funds of 2; hence it is in default, and its recovery rate is $r_u = \frac{2}{4} = \frac{1}{2}$. In accordance with earlier works (such as [9,23]), the model assumes that in this case, u has to make payments *proportionally to the respective liabilities* in the contracts; hence it transfers 1 unit of money to w and 1 unit to v.

Since u has a recovery rate of $r_u = \frac{1}{2}$, the CDS from w to v translates to a liability of $2 \cdot (1 - r_u) = 1$. Although w has no funds, it receives 1 unit of money from u, so it can fulfill this payment obligation and narrowly avoids default, $r_w = 1$. Finally, v has no liabilities at all, so $r_v = 1$. Since it receives 1 unit of money from both u and w, and has $e_v = 1$, it has 3 units of money after the clearing of the system.

3.2   Assets and Liabilities

Formally, our systems are defined by a vector $e = (e_v)_{v \in B}$, the matrix $D = (\delta_{u,v})_{u,v \in B}$, where $\delta_{u,v}$ denotes the weight of debt from u to v (interpreted as $\delta_{u,v} = 0$ if there is no such debt), and the matrix $C = (\delta_{u,v}^w)_{u,v,w \in B}$, where $\delta_{u,v}^w$ denotes the weight of the CDS from u to v in reference to w. We assume that no bank enters into a contract with itself or in reference to itself. Given a financial system on B by (e, D, C), we are interested in the recovery rates r_v of banks, which can also be represented as a vector $r = (r_v)_{v \in B}$.

Given a recovery rate vector r, the *liability* of u to v is formally defined as

$$l_{u,v}(r) = \delta_{u,v} + \sum_{w \in B} \delta_{u,v}^w \cdot (1 - r_w).$$

The total liability of bank u is $l_u(r) = \sum_{v \in B} l_{u,v}(r)$, i.e. the sum of payment obligations for u. However, the actual payment from u to v can be lower than $l_{u,v}(r)$ if u is in default. The model assumes that defaulting banks always use all their assets to pay for liabilities, and they make payments proportionally to the respective liabilities. With a recovery rate of r_u, u can pay an r_u portion of each liability, so the payment from u to v is $p_{u,v}(r) = r_u \cdot l_{u,v}(r)$.

On the other hand, the *assets* of v are defined as

$$a_v(r) = e_v + \sum_{u \in B} p_{u,v}(r).$$

Given the assets and liabilities of v, the recovery rate r_v has to satisfy $r_v = 1$ if $a_v(r) \geq l_v(r)$ (i.e. if v is not in default), and $r_v = \frac{a_v(r)}{l_v(r)}$ if $a_v(r) < l_v(r)$ (if v is in default). If a vector r is an equilibrium point of these equations, i.e. it satisfies this condition on $a_v(r)$ and $l_v(r)$ for every bank v, then r is a *clearing vector* of the system. Our main goal is to analyze the different clearing vectors.

The *equity* of v in a solution is defined as

$$q_v(r) = \max\left(a_v(r) - l_v(r), 0\right),$$

i.e. the amount of money available to v after clearing. In the example of Fig. 1, we have $q_u = 0$, $q_w = 0$ and $q_v = 3$. We assume that the main goal of banks is to maximize their equity. Note that we have written q_u instead of $q_u(r)$ in order to simplify notation; we often do not show the dependence on r when r is clear from the context.

Previous works also consider an extension of this base model with *default costs* [23–25]; we also refer to this setting as systems with loss. In this case, the financial network has two more parameters $\alpha, \beta \in [0, 1]$, and when a bank goes into default, it loses a specific fraction of its assets. More specifically, if v is in default, then its assets are defined as

$$a_v(r) = \alpha \cdot e_v + \beta \cdot \sum_{u \in B} p_{u,v}(r).$$

Throughout the paper, we mostly focus on the base model without loss, i.e. we assume $\alpha = \beta = 1$ unless specified otherwise. However, we also discuss the extension of our proofs to systems with loss, and we briefly study some questions that only arise in case we have default costs.

In the rest of the paper, we switch to a computer science terminology: we refer to the banks as *nodes*, clearing vectors as *solutions* (with the set of solutions denoted by S), and the notions of contracts as the *weight* of the contracts.

4   Properties of the Solution Space

Previous Work. The work of Schuldenzucker *et al.* focuses on the existence and computability of solutions [24,25]. Their results can be summarized as follows:

- *Loss-free systems* ($\alpha = \beta = 1$): in this case, there always exists a solution; however, this proof of existence is non-constructive. Finding an (approximate) solution is PPAD-hard.

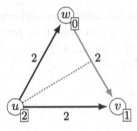

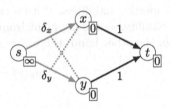

Fig. 1. Example system on 3 banks. External assets are shown in rectangles besides the bank, simple debts are shown as blue arrows, and CDSs are shown as brown arrows with a dotted line to the reference bank. (Color figure online)

Fig. 2. Branching gadget consisting of two nodes x and y, both having an outgoing debt to a sink t and an incoming CDS from a source s.

– *Systems with loss* ($\alpha < 1$ or $\beta < 1$): in this case, a solution might not exist at all. Deciding if a system has an (approximate) solution is NP-hard.

Once we know that a solution exists, another natural question is if there is a *maximal* solution, i.e. a solution r such that $q_v(r) \geq q_v(r')$ for every node v and every solution r'. If such a maximal solution exists, then we can assume that an authority always prefers to implement this solution. However, in both settings, a system can easily have multiple solutions with none of them being maximal.

Branching Gadget. A basic building block in our constructions is the *branching gadget* shown in Fig. 2, which has already been used with some parametrizations in the works of [24,25], e.g. as an example system with no maximal solution. For the weight parameters δ_x and δ_y, we always assume $\delta_x \geq \delta_y \geq 1$.

Since the source and sink nodes can never go into default, we only analyze the recovery rate subvector (r_x, r_y). First, observe that we cannot have both nodes surviving, i.e. $(1, 1)$ as a solution: both nodes only receive any funds if the other node is in default. However, if either $r_x = 0$ or $r_y = 0$, then the other node can already pay its debt, thus $(0, 1)$ and $(1, 0)$ are always solutions in this system.

Besides this, there may be other solutions when both nodes are in default with a positive recovery rate; these depend on the concrete values of δ_x and δ_y. If x and y are in default, then their assets are equal to the amount of debt they can pay, so the remaining solutions are obtained from the equations $r_x = \delta_x \cdot (1 - r_y)$ and $r_y = \delta_y \cdot (1 - r_x)$.

However, there are also choices of δ_x, δ_y for which these equations confirm that $(0, 1)$ and $(1, 0)$ are indeed the only solutions. One such example is $\delta_x = 2$, $\delta_y = 1$; we refer to this case as the *clean branching gadget*, and we assume this parametrization unless specified otherwise. This gadget is a natural candidate for representing a binary choice: x is a *binary node* in the sense that r_x is either 0 or 1 in any solution, and r_y offers a convenient representation of its negation.

Number of Solutions. We now discuss the size of the solution space in our systems.

Lemma 1. *There exists a financial system with infinitely many solutions.*

Proof. Consider the branching gadget of Fig. 2 with $\delta_x = \delta_y = 1$. For any $\rho \in [0, 1]$, the vector $(\rho, 1-\rho)$ satisfies the equations above, thus it is a solution of the system.

While this shows that the number of solutions is potentially unlimited, the difference between most of these vectors is only the extent of the defaults. Thus it is also natural to study another concept of difference between solutions: we say that two solutions r and r' are *essentially different* if there is a node v such that either $r_v = 1$ but $r'_v < 1$, or $r'_v = 1$ but $r_v < 1$. Since we only consider a boolean value for each node in this definition, the number of pairwise essentially different solutions is at most 2^n.

Lemma 2. *There exists a system with $2^{\Omega(n)}$ solutions that are pairwise essentially different.*

Proof. Let us take $\frac{n}{4}$ independent copies of the clean branching gadget. In each gadget, there are two possible subsolutions: (0,1) or (1,0). Over the distinct gadgets, these can be combined in any way, adding up to $2^{n/4}$ solutions that are pairwise essentially different.

Better and Worse Solutions. While our systems may not always have a maximal solution, it is still reasonable to say that some solutions are better than others.

Definition 1. *Given two solutions r and r', we say that r' is* strictly better *than r if $q_v(r') \geq q_v(r)$ for every node v, and there exists a node u such that $q_u(r') > q_u(r)$. A solution r is* Pareto-optimal *if there is no solution r' that is strictly better than r (otherwise, r is* Pareto-suboptimal).

A financial authority might want to avoid implementing Pareto-suboptimal solutions, and prefer a strictly better solution instead. However, selecting among Pareto-optimal solutions is more difficult, since they represent a trade-off between the preferences of different nodes.

In our base financial system model without loss, every solution is Pareto-optimal. However, if we also have default costs, then some funds are lost when a node goes into default, and thus this is not the case anymore. While the proofs of these claims are simple, we defer them to the full version of the paper [21].

Lemma 3. *In loss-free financial systems, every solution is Pareto-optimal.*

Lemma 4. *In systems with loss, there can be solutions that are Pareto-suboptimal.*

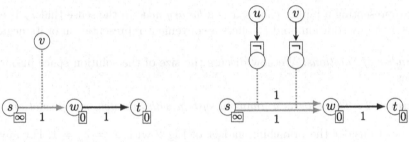

Fig. 3. NOT gate   **Fig. 4.** OR gate

5   Finding the "Best" Solution

In this section, we discuss a wide range of realistic objective functions for selecting a solution in out networks. We show that for these objectives, the optimal solution is even hard to reasonably approximate. The details of these proofs are discussed in the full version of the paper.

5.1   Tools and Gadgets

Besides the branching gadget, our constructions also apply gadgets that simulate boolean operations on nodes. Note that most of these gadgets have already been used before in the work of [24, 25], sometimes in a slightly different form.

    More specifically, given two binary nodes u and v, we can construct the following gadgets:

- NOT gate: a node w such that $r_w = 1$ if $r_v = 0$, and $r_w = 0$ if $r_v = 1$,
- OR gate: a node w such that $r_w = 0$ if $r_u = r_v = 0$, and $r_w = 1$ otherwise,
- AND gate: a node w such that $r_w = 1$ if $r_u = r_v = 1$, and $r_w = 0$ otherwise.

    We demonstrate the NOT and OR gates in Figs. 3 and 4, and discuss the behavior of these gadgets in the full version. Note that Fig. 4 already uses the NOT gate as black box, denoted by a ¬ symbol. In a similar fashion, we can also create AND and OR gates on multiple inputs.

    Finally, when adding incoming or outgoing contracts to a bank v, our main goal is often to establish a specific behavior for v, and thus it is unimportant where these contracts come from/go to. Hence for simplicity, we add a specific *source node* s with $e_s = \infty$ and a *sink node* t to our constructions, which act as the source/recipient of all such contracts.

5.2   Example: Maximizing the Equity of a Node

To demonstrate the main idea behind our constructions, we first discuss the problem of maximizing the equity of a specific node. That is, given a node v, we define the value of a solution r as the equity q_v, and we denote the search problem of finding the highest-value solution by MaxEQUITY(v). This is a very natural problem, and a crucial question for v if it wants to understand its situation in the network. However, this problem is already hard to solve in our model.

Theorem 1. *The problem MaxEQUITY(v) is NP-hard to approximate to any $n^{1-\epsilon}$ factor.*

Proof. We use a reduction from the boolean satisfiability (SAT) problem, which is known to be NP-complete [13]. Given an input boolean formula ϕ on N variables and M clauses, we transform this into a financial system representation by creating N distinct branching gadgets, each corresponding to a specific variable. Recall that if we understand r_x to be the value of the variable in an assignment, then r_y represents its negation.

Given these variables, we can use our logical gates to compute the value of ϕ for a specific assignment: we first combine each clause into a node with an OR gate, and then combine all these nodes with an AND gate. This provides a binary indicator node v_I which describes the value of ϕ under a specific assignment. We also add a further NOT gate on top of v_I to obtain a convenient representation of its negation in a new bank $\overline{v_I}$.

Most of our hardness results will use this *base construction*, extended by further gadgets representing the specific objective function. For the MaxEQUITY(v) objective, we only add a node v that has $e_v = 0$, and an incoming CDS of weight n in reference to $\overline{v_I}$.

If there exists a satisfying assignment to ϕ, then there is a solution in this system that has $r_{\overline{v_I}} = 0$, and thus $q_v = n$. As such, any $n^{1-\epsilon}$ approximation algorithm must return a solution in this case with $q_v \geq n^\epsilon > 0$. On the other hand, if ϕ is unsatisfiable, then every solution of the system has $q_v = 0$. Hence a polynomial-time approximation would also allow us to decide whether ϕ is satisfiable, which completes the reduction.

Note that the branching gadgets already determine the recovery rate of all other nodes, so the system has exactly the 2^N solutions that correspond to the different variable assignments. This allows us to easily characterize the entire solution space, so the source of computational hardness is not the fact that we cannot even find a single solution, as described in [25] before.

With a slightly different gadget appended to the base construction, we can present a similar reduction for the problem of *minimizing* the equity of a bank v.

Theorem 2. *The problem MinEQUITY(v) is NP-hard to approximate to any $n^{1-\epsilon}$ factor.*

5.3   Global Objective Functions

Given a financial system with many solutions, there are various objectives that an authority could follow when choosing the solution to implement. Some of the most natural objective functions are as follows:

- MinDEFAULT: minimize the number of defaulting nodes, i.e. $|\{v \in B \mid r_v < 1\}|$
- MaxPREFER: find the solution that is the primary preference of most nodes, i.e. define the maximal equity of bank v as $q_v^{(\max)} = \max_{r \in S} q_v(r)$, and then maximize $|\{v \in B \mid q_v(r) = q_v^{(\max)}\}|$,
- MinUNPAID: minimize the amount of unpaid liabilities, i.e. $\sum_{u,v \in B} l_{u,v} - p_{u,v}$.

One can show that these are indeed different problems: they can obtain their optimum in distinct solutions, and the optimum for one objective might give a very low-quality solution in terms of another one.

Theorem 3. *For any objectives f_1, f_2 from above, there is a system such that in the optimal solution for f_1, the value of f_2 is a $\Theta(n)$ factor worse than the optimum value of f_2.*

We provide example constructions for these claims in the full version of the paper. In fact, one can even combine these examples into a single system with a very different optimum for each function.

Theorem 4. *There exists a financial system such that the optima for the objective functions above are all obtained in different solutions, and in terms of the respective metrics, each of these optima are a factor of $\Omega(\sqrt{n})$ better than any other solution in the system.*

Now let us analyze these problems from a complexity perspective. We can apply a similar technique to Theorem 1 to show that it is hard to approximate any of these objectives.

Theorem 5. *The problem MinDEFAULT is NP-hard to approximate to any $n^{1-\epsilon}$ factor.*

Proof sketch. Given a fixed constant ϵ, let us select an ϵ' such that $0 < \epsilon' < \epsilon$. Also, given a formula ϕ on N variables and M clauses, let us introduce $m := N + M$. We extend the base construction of Sect. 5.2 by introducing $m^{1/\epsilon'}$ distinct new banks u_i to the system that all have $e_{u_i} = 0$, and an outgoing CDS of weight 1 in reference to the indicator node v_I.

For every variable assignment that evaluates to false, we have $r_{v_I} = 0$, so all the new nodes are in default; as such, the number of defaulting nodes is $m^{1/\epsilon'} + O(m)$. On the other hand, if there is a satisfying assignment, then the banks u_i have no liability in the corresponding solution, so the number of defaulting banks is only $O(m)$. Since $n = \Theta(m^{1/\epsilon'})$ in this system, the best solution has either $\Theta(n)$ or $O(n^{\epsilon'}) < n^\epsilon$ defaults, depending on whether ϕ is satisfiable; this shows an inapproximability to any $n^{1-\epsilon}$ factor.

We can also rephrase the MinDEFAULT problem as maximizing the number of surviving (non-defaulting) nodes; the two problems clearly have the same optimal solution. However, this MaxSURVIVING problem is defined by a different metric in its objective function, so it could behave very differently in terms of approximability (see e.g. the minimum vertex cover and maximum independent set problems, which are also complements [13,14]). However, it turns out that in our case, the problem is hard to approximate in both metrics.

Theorem 6. *The problem MaxSURVIVING is NP-hard to approximate to any $n^{1-\epsilon}$ factor.*

We can use different variants of the same proof technique to show the same hardness result for the other two objectives. Furthermore, similar to MaxSURVIVING, we can also define dual problems for these objectives, which are also hard to approximate.

Theorem 7. *The problems MaxPREFER and MinUNPAID (as well as their dual problems MinLEASTPREFER and MaxPAID) are NP-hard to approximate to any $n^{1-\epsilon}$ factor.*

5.4   More Complex Objectives

Most Balanced Solution. In a different setting, an authority might want to find a solution where the distribution of equity is balanced in some sense. E.g. if we have two larger alliances of banks (sets of nodes), then our goal might be to find a solution that distributes the total equity evenly between these alliances.

We show our hardness result for the simplest case of only two nodes v_1 and v_2, and the problem MinDIFF(v_1, v_2) of minimizing $|q_{v_1} - q_{v_2}|$ is minimal. It follows that the more general problem of minimizing $|\sum_{v_1 \in V_1} q_{v_1} - \sum_{v_2 \in V_2} q_{v_2}|$ for two sets of nodes V_1 and V_2 is also hard.

Theorem 8. *The problem MinDIFF(v_1, v_2) is NP-hard to approximate to any $n^{1-\epsilon}$ factor.*

Proof. We can simply consider the MinEQUITY(v) construction with $v_1 := v$, and add an extra bank v_2 such that $q_{v_2} = 0$. This system has $|q_{v_1} - q_{v_2}| = q_{v_1}$, so we can apply the same reduction as in the MinEQUITY case.

Most Representative Solution. It could also be a reasonable goal to select a solution that is somehow representative of the whole solution space S. Assuming a fixed distance metric between two solutions (e.g. let $d(r, r') := \sum_{v \in B} |r_v - r'_v|$), there are many natural ways to define a metric of centrality for a solution $r \in S$.

We only discuss one natural approach here: let us define

$$cent(r) = \frac{1}{|S|} \sum_{r' \in S} d(r, r'),$$

as the centrality of a solution r, and let MinDIST denote the problem of finding the solution r with the lowest $cent(r)$ value.

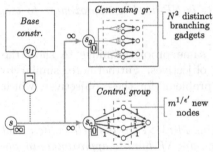

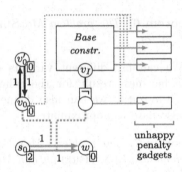

Fig. 5. Construction of Theorem 9   **Fig. 6.** Construction of Theorem 10

Note that our result essentially shows that the solution space can exhibit a threshold behavior between two very different shapes, and it is already hard to decide which of the two shapes is obtained. This suggests that the problem is also hard in any other reasonable formulation, i.e. for other distance functions or centrality metrics.

Theorem 9. *The problem MinDist is NP-hard to approximate to any $n^{1-\epsilon}$ factor.*

Proof sketch. The main idea is to add two large sets of nodes to our construction, as sketched in Fig. 5. The *generating group* consists of N^2 independent branching gadgets, while the *control group* has $m^{1/\epsilon'}$ single nodes with an outgoing debt (where m denotes the size of ϕ and $\epsilon' < \epsilon$ as before). We ensure that both groups only receive funds if $r_{v_I} = 1$; otherwise, all the new nodes are in default.

Since the control group contains almost all of the nodes asymptotically, the centrality of a solution is essentially defined by the recovery rates of the nodes in the control group. If ϕ is unsatisfiable, then every assignment produces $r_{v_I} = 0$, and thus the control nodes have recovery rates of 0 in every solution. On the other hand, if ϕ is satisfiable, then the branching gadgets in the generating group will introduce 2^{N^2} new solutions (for each satisfying assignment), which reduces the at most 2^N unsatisfying solutions to an asymptotically irrelevant part of S. In this case, the control nodes have a recovery rate of 1 in almost every solution.

Hence the two cases are very different in terms of solution space. An approximation algorithm would always need to find a satisfying assignment if one exists; otherwise, it returns a solution with an average distance of at least $m^{1/\epsilon'} \approx n$, while the optimum has a distance of only $O(m) \approx n^{1-\epsilon'}$.

Strictly Better Solution. Recall that in systems with loss, we can have Pareto-suboptimal solutions, so it is natural to ask if a specific solution can be improved: if there is a solution r' strictly better than r, then we would probably want to implement r' instead of r. If such an r' was easy to find, then we could iteratively improve an initial solution until we eventually find a Pareto-optimal one.

Theorem 10. *It is NP-hard to decide if a given solution r is Pareto-suboptimal.*

Proof sketch. The construction is built around a binary node v_0 (see Fig. 6). To each node u of our base construction, we add a so-called *unhappy penalty gadget*. This essentially means that if $r_{v_0} = 0$, then u pays a large penalty to a special sink t_0; however, t_0 has further gadgets attached to ensure that t_0 is still worse off if $r_{v_0} = 0$, even though it receives money from this penalty. As such, the default of v_0 is not favorable to any node in the system.

The base idea then is to add another node w, which, on the other hand, receives 1 unit of money if either $r_{v_0} = 0$, or $r_{v_I} = 1$. Let r be the solution where $r_{v_0} = 0$, and thus all nodes in the base construction are in default, but $q_w = 1$. Any solution strictly better than r must also have $q_w \geq 1$. If v_0 is not in default, this is only possible if we find a satisfying assignment of ϕ, thus ensuring $r_{v_I} = 1$.

6 Restricted Financial Networks

Our final goal in the paper is to understand the key reasons behind this computational complexity, and whether we can introduce some restrictions to our network model to eliminate this phenomenon. In particular, we show that the same hardness results also hold in many severely restricted variants of our financial system model, and it takes a combination of multiple restrictions to ensure that the solution space is sufficiently simple.

Before considering restrictions to the network, let us first briefly discuss a familiar extension of the model: default costs. We point out that while our hardness results were mostly shown for systems without loss, they can also be extended to systems with loss with some minor modifications.

Theorem 11. *Theorems 1–2 and 5–10 also hold for any $\alpha, \beta \in (0, 1]$.*

6.1 Unweighted Networks

For convenience, we have sometimes used rather large edge weights in our constructions; one could argue that this is unrealistic in practice. As such, we first show that our hardness results also carry over to the setting when each contract in the network has the same weight.

Theorem 12. *Theorems 1–2 and 5–10 also hold in unit-weight networks.*

Proof sketch. The modifications required for this setting are rather straightforward: most edges in our constructions have unit weight to begin with. Whenever the weight is a larger integer k, we can split this into k distinct contracts that come from/go to k distinct source/sink nodes. The only cases that require some extra consideration are the gadgets used in Theorems 9 and 10.

6.2   Restricted Network Structure

In their work, Schuldenzucker *et al.* also discuss several restrictions to the network structure [24,25]. While they study these restrictions from a different perspective (their goal is to ensure that the system always has a solution, even with default costs), it is natural to ask whether our hardness results still hold in these restricted network models.

In particular, the authors define the so-called *dependency graph* to express the relations of banks in a directed graph with edges of two colors:

- *Green edges*: intuitively, these indicate long positions. For example, there is a green edge from u to v if u has a contract towards v (debt or CDS), or if v has an outgoing CDS in reference to u.
- *Red edges*: intuitively, these indicate short positions. There is a red edge from w to v if v has an incoming CDS in reference to w (unless there is a debt of even larger weight from w to v).

For details on the dependency graph, we refer the reader to the full version of the paper or the work of [24].

The work of [24] studies different restrictions to the network based on this dependency graph. In the most restricted case, they study systems where the dependency graph contains exclusively (or almost exclusively) green edges, so short positions are essentially banned.

Definition 2. *We say that a financial network is a* green system *if its dependency graph only contains green edges.*

Using a fixed-point theorem, one can show that green systems are similar to debt-only networks in the sense that they always contain a maximal solution. As such, this simpler case is not so interesting in terms of default ambiguity.

However, [24] also studies a more general setting where short positions are still allowed in the network, but only in a structurally restricted fashion.

Definition 3. *A financial network is an* RFC (red-free cycle) system *if no directed cycle of the dependency graph contains a red edge.*

The authors show that in RFC systems, one can always find a solution efficiently. Intuitively, one can iterate through the strongly connected components (SCCs) of the dependency graph in topological order, since every SCC is only dependent on the preceding ones. Since each SCC is a green system, there is always a maximal subsolution in the current SCC (if the subsolutions in previous SCCs are already fixed), and we can find this efficiently.

In contrast to this, our goal of finding the best solution is still not straightforward in these RFC systems. In particular, selecting a different (non-maximal) solution in the first SCC could allow us to find a different solution in the second SCC; while this is unfavorable to banks in the first SCC, it might be much better in terms of our global objective. In fact, our hardness results even hold in this heavily restricted class of networks.

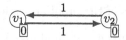

Fig. 7. A simple debt-only network with multiple solutions

Theorem 13. *Theorems 1–2 and 5–10 also hold in RFC systems.*

Proof sketch. The key observation is that the dependency graphs of our constructions are already very close to DAGs: the only directed cycles occur within the branching gadgets, where banks x and y have short position on each other. As such, it is sufficient to come up with an alternative branching gadget design that satisfies the RFC property.

The main idea of this gadget is to consider two banks v_1 and v_2 as in Fig. 7; for any $\rho \in [0, 1]$, $r_{v_1} = r_{v_2} = \rho$ is a solution of this system. We can then essentially use the small and large ρ values in this system (e.g. $\rho < \frac{1}{3}$ and $\rho > \frac{2}{3}$) as the new representations of the two binary states. We exclude the intermediate ρ values by artificially making the solution significantly worse (in terms of our objective function) whenever we have $\rho \in [\frac{1}{3}, \frac{2}{3}]$.

We note that the situation in Fig. 7 seems rather artificial. However, recall that default ambiguity often arises after an external shock hits the market; as such, one should imagine this as a situation where banks in a cycle have lost all their funds due to such an event.

6.3 Green Systems and Regularity

Our alternative construction in Theorem 13 uses the fact that a debt-only network can still have multiple solutions in some special edge cases, thus allowing us to create a large solution space. To prevent this phenomenon, we first need a deeper understanding of these cases when green system have multiple solutions.

The work of [9] already studies this question in debt-only networks, showing that the solution is unique if from any bank there is a directed path to another bank with positive funds. We prove a more general version of this result, extending the theorem to any green system, and using a weaker assumption on the topology. In particular, we show that green systems can only have multiple solutions in a special edge case: when we have a network segment with no funds and no incoming assets at all.

Theorem 14. *Let G be a green system, and assume that v is a bank that has two distinct recovery rates $r_v \neq r'_v$ in two solutions. Let C be the set of nodes reachable from v on a path of simple debts. Then the following must hold:*

- *for all $u \in C$ we have $e_u = 0$,*
- *if there is a path of contracts from a bank $w \in G$ to a bank $u \in C$, then $e_w = 0$.*

Proof sketch. The main steps of the proof are as follows:

- Recall from before that a green system always has a maximal solution r (and also a minimal solution r'); these assigns the highest/lowest recovery rate to all banks.
- In such a setting, all banks must have the same equity in any solution. Intuitively, in systems without loss, if a bank had less equity in a solution r_0 than in the maximal solution r, then some other bank would need to have more equity in r_0 than in r.
- If $r_v > r'_v$ (i.e. v can have different recovery rates), then v makes strictly more payment on its outgoing debts in r than in r'. In a loss-free system, these extra payments traverse the network in until they either (i) reach a node u with no more unfulfilled liabilities, or (ii) they arrive back at v. However, the first option is not possible, since this would mean $q_u > q'_u$; hence all such payments must ultimately arrive back at v.
- This means that from v, any path of contracts (with positive liability) must eventually lead back to v, implying that these contracts form an SCC C.
- Finally, no node $u \in C$ can have $e_u > 0$, and also no node $w \in G$ can have a positive payment towards a bank $u \in C$. This is because C is closed under outgoing payments, so if any funds arrive in C, then the loss-free property implies that some $u \in C$ must have $q'_u > 0$; hence we already have $r'_u = 1$ in the solution r'. However, if $r_v > r'_v$, then in r there is a strictly positive extra payment arriving at u; this implies $q_u > q'_u$, which is again a contradiction.

Note that the proof also makes a structural observation that the banks reachable from v must form a SCC in the graph of "meaningful" contracts (which induce a positive liability in some solution). However, since it is not immediately clear whether a CDS is meaningful, we expressed Theorem 14 in a weaker form, stating the restrictions only for the set of nodes C that are reachable from v on simple debts.

The situation described in Theorem 14 is a very special case, so there are various ways to ensure that we exclude such networks. One natural approach is to restrict the amount of funds that banks must possess, since this is usually strictly supervised in practice.

Definition 4. *We say that a financial system G is* regular *if we have $e_v > 0$ for all $v \in B$.*

This assumption is realistic in many legal frameworks: financial regulations usually require banks to possess enough funds to cover at least a specific portion of their liabilities. Considering that default ambiguity often happens after a shock hits the market, an alternative (more practical) interpretation of this property is that all banks must keep at least some of their funds in a format that is resilient to external shocks.

Note that there are various other options to exclude the edge case of Theorem 14 with weaker conditions; however, most of these approaches are difficult to enforce from a regulator's perspective.

On the other hand, note that Theorem 14 only applies to green systems. If our network is not a green system, then even this rather strong condition is not sufficient to ensure that the solution is unique.

Theorem 15. *Theorems 1–2 and 5–10 also hold in regular financial systems.*

Proof sketch. The main idea is to consider a new representation of the binary states in our gadgets: instead of $r_v = 0$ and $r_v = 1$, the two binary states will be represented by $r_v = 0.5$ and $r_v = 1$. This allows us to give some funds to every node in our construction, thus fulfilling the regularity condition.

Most of our gadgets are actually rather easy to adapt to this setting; it is only the constructions of Theorems 9 and 10 where this is more technical.

6.4   Combined Restrictions: a Unique Solution

This shows that we need both the RFC property and regularity together to ensure that the solution of the system is unique, and thus our hardness results can be avoided. This provides an interesting final message from our analysis: it suggests that financial regulators might need to use both topological and fund-based restrictions simultaneously in order to eliminate the computational problems arising from default ambiguity.

Theorem 16. *If a system is both regular and RFC, then it has a unique solution. This solution can be efficiently approximated in polynomial time.*

Proof. We can now apply the approach of [24] for RFC systems, computing a solution by visiting the SCCs in topological order. The payments coming from the previous SCCs can simply be considered as extra funds at the bank when processing the current SCC of the network.

Due to the RFC property, the current SCC is always a green system. Regularity implies that every node u in the SCC has $e_u > 0$; this is only further increased by the payments from previous SCCs. As such, Theorem 14 shows that there is always a unique subsolution in the current SCC. Altogether, this implies that the solution r is unique in the whole network; as such, we can indeed simply apply the algorithm of [24] for RFC systems, which always finds an arbitrary solution.

Note, however, that the solution of our networks can also be irrational in some cases, so we can only claim that it is efficiently approximated with this method. It is already discussed in [24,25] that given an error margin $\epsilon > 0$, this algorithm finds a recovery rate vector r^ϵ such that $|r_v - r_v^\epsilon| \leq \epsilon$ for all $v \in B$, and its running time is polynomial in n and $1/\epsilon$.

Finally, we point out that if we have default costs, then our hardness results still hold even in the setting of Theorem 16. This is because with default costs, a green system can still have multiple solutions even if it is regular. If we modify Fig. 7 to have $e_u = e_v = \frac{1}{3}$ and we assume $\alpha = \beta = \frac{1}{2}$, then both $r_u = r_v = 1$ and $r_u = r_v = \frac{1}{3}$ are solutions; while the former is clearly better for u and v, the latter might be superior in terms of our objective.

Acknowledgements. We would like to thank Steffen Schuldenzucker for his very valuable feedback and improvement ideas.

References

1. Acemoglu, D., Ozdaglar, A., Tahbaz-Salehi, A.: Systemic risk and stability in financial networks. Am. Econ. Rev. **105**(2), 564–608 (2015)
2. Banerjee, T., Feinstein, Z.: Impact of contingent payments on systemic risk in financial networks. Math. Financ. Econ. **13**(4), 617–636 (2019). https://doi.org/10.1007/s11579-019-00239-9
3. Bardoscia, M., Battiston, S., Caccioli, F., Caldarelli, G.: Pathways towards instability in financial networks. Nat. Commun. **8**, 14416 (2017)
4. Bertschinger, N., Hoefer, M., Schmand, D.: Strategic payments in financial networks. In: 11th Innovations in Theoretical Computer Science Conference (ITCS 2020). LIPIcs, vol. 151, pp. 1–16. Dagstuhl, Germany (2020)
5. Dees, S., Henry, J., Martin, R.: Stamp€: stress-test analytics for macroprudential purposes in the euro area. ECB, Frankfurt a. M. (2017)
6. Demange, G.: Contagion in financial networks: a threat index. Manage. Sci. **64**(2), 955–970 (2016)
7. D'Errico, M., Battiston, S., Peltonen, T., Scheicher, M.: How does risk flow in the credit default swap market?, J. Financ. Stab. **35**, 53–74 (2018)
8. Egressy, B., Wattenhofer, R.: Bailouts in financial networks. Arxiv preprint arxiv:2106.12315 (2021)
9. Eisenberg, L., Noe, T.H.: Systemic risk in financial systems. Manage. Sci. **47**(2), 236–249 (2001)
10. Elliott, M., Golub, B., Jackson, M.O.: Financial networks and contagion. Am. Econ. Rev. **104**(10), 3115–53 (2014)
11. Elsinger, H., Lehar, A., Summer, M.: Risk assessment for banking systems. Manage. Sci. **52**(9), 1301–1314 (2006)
12. Fender, I., Gyntelberg, J.: Overview: global financial crisis spurs unprecedented policy actions. BIS Quarter. Rev. **13**(4), 1–24 (2008)
13. Garey, M.R., Johnson, D.S.: Computers and intractability: a guide to the theory of NP-Completeness. Freeman, W.H. and Co. (1979)
14. Hastad, J.: Clique is hard to approximate within $n^{1-\epsilon}$. In: Proceedings of 37th Conference on Foundations of Computer Science, pp. 627–636. IEEE (1996)
15. Hemenway, B., Khanna, S.: Sensitivity and computational complexity in financial networks. Algorith. Finance **5**(3–4), 95–110 (2016)
16. Leduc, M.V., Poledna, S., Thurner, S.: Systemic risk management in financial networks with credit default swaps. In: SSRN, p. 2713200 (2017)
17. Loon, Y.C., Zhong, Z.K.: The impact of central clearing on counterparty risk, liquidity, and trading: evidence from the credit default swap market. J. Financial Econ. **112**(1), 91–115 (2014)
18. Papachristou, M., Kleinberg, J.: Allocating stimulus checks in times of crisis. Arxiv preprint arxiv:2106.07560 (2021)
19. Papp, P.A., Wattenhofer, R.: Network-aware strategies in financial systems. In: 47th International Colloquium on Automata, Languages, and Programming (ICALP 2020). LIPIcs, vol. 168, pp. 1–17. Dagstuhl, Germany (2020)

20. Papp, P.A., Wattenhofer, R.: Debt swapping for risk mitigation in financial networks. In: Proceedings of the 22nd ACM Conference on Economics and Computation. EC '21, Association for Computing Machinery, pp. 765–784. New York, NY, USA (2021)

21. Papp, P.A., Wattenhofer, R.: Default ambiguity: finding the best solution to the clearing problem (full version). Arxiv preprint arxiv:2002.07741v2 (2021)

22. Papp, P.A., Wattenhofer, R.: Sequential defaulting in financial networks. In: 12th Innovations in Theoretical Computer Science Conference (ITCS 2021). LIPIcs, vol. 185, pp. 52:1–52:20. Dagstuhl, Germany (2021)

23. Rogers, L.C., Veraart, L.A.: Failure and rescue in an interbank network. Manage. Sci. **59**(4), 882–898 (2013)

24. Schuldenzucker, S., Seuken, S., Battiston, S.: Clearing payments in financial networks with credit default swaps. In: Proceedings of the 2016 ACM Conference on Economics and Computation, p. 759. EC '16, ACM, New York, NY, USA (2016)

25. Schuldenzucker, S., Seuken, S., Battiston, S.: Finding clearing payments in financial networks with credit default swaps is PPAD-complete. In: 8th Innovations in Theoretical Computer Science Conference (ITCS 2017). LIPIcs, vol. 67, pp. 1–20. Dagstuhl, Germany (2017)

26. Vitali, S., Glattfelder, J.B., Battiston, S.: The network of global corporate control. PloS one **6**(10), e25995 (2011)

Planning on an Empty Stomach:
On Agents with Projection Bias

Sigal Oren[✉] and Nadav Sklar

Ben-Gurion University of the Negav, 8410501 Beer-Sheva, Israel
sklar@post.bgu.ac.il

Abstract. People often believe that their future preferences will be similar to their current ones. For example, people who go hungry to the supermarket, often buy less healthy food items than when they go on a full stomach. Loewenstein et al. [10] coined the term *projection bias* to capture this and similar behaviors.

Our first contribution is a generalization of the restricted model of Loewenstein et al. by considering agents with projection bias that traverse a state graph for time horizon t. Our generalization allows us to capture more complex planning scenarios, such as a student that plans his occupational path. We analyze the planning behavior of biased agents and show that their loss due to their projection bias may be unbounded. Obviously, agents who do not suffer from projection bias at all will be able to traverse the graph optimally. We show–perhaps surprisingly–that agents that exhibit a strong projection bias sometimes fare better than agents that exhibit projection bias to a smaller extent. Similarly, we show that agents that plan for a longer time horizon do not necessarily fare better than agents that plan for a shorter time horizon. We then provide bounds on the number of these "non-monotonicity" points in a given state graph. Among other results, we prove a hardness result for computing a subgraph that maximizes the utility of the biased agent.

Keywords: Projection bias · Planning · Behavioral bias

1 Introduction

It is well documented that people who go grocery shopping on an empty stomach make much less healthy-conscious choices and might buy more food [3,12,15]. The underlying reason is that when we are in a "hungry state" it is difficult for us to imagine which types of food we would like to eat later when we are in a "less satiated state". A crisp demonstration of this is an experiment by Read and van Leeuwen [13] in which office workers had to choose whether they will get a healthy or an unhealthy snack in a week from now. It turned out that workers who were hungry when deciding were more likely to choose an unhealthy snack

Work supported by BSF grant 2018206 and ISF grant 2167/19. The full version can be found on arXiv.

for next week. Loewenstein et al. [10] coined the term *projection bias* to describe the behavior in such situations. Individuals exhibiting projection bias believe that their preferences in future states will be similar to their current preferences. In particular, they know whether the change will be positive or negative with respect to their current state, but they misestimate the magnitude of the change.

There is ample evidence in the Psychology and Behavioral Economics literature (see [4,10,11] for surveys) of people exhibiting behaviors that could be explained by projection bias. Many of those have to do with *underappreciation of adaptation* [4] – individuals fail to realize how well they will adapt to a new situation and hence exaggerate the implications of a change. For example, patients with renal failure expect that a kidney transplant will dramatically change their lives, but when asked after the transplant, they report a milder change [14]; Academic faculty exaggerate the effect that getting or being denied tenure will have on their quality of life [4].

Many of the studies considering projection bias rely on self-reporting, which may hinder the robustness of the results. Conlin et al. [2] test projection bias in field data. They analyze data of catalog orders and returns of weather-related items. They show that people who order cold-weather items on a cold day are more likely to return them. Similarly, if the temperature on the day after they receive the item is high, they are more likely to return the item.

In their influential paper, Loewenstein et al. [10] suggest a formal model for capturing projection bias. The rough idea is that the utility of an agent from consuming some good depends not only on the amount it consumes but also on the agent's state. The state can change as a function of the consumption and hence the agent has to predict his utility in different states to make intertemporal decisions. Formally, an agent currently at state s has a utility $u(c, s)$ for consuming c units. The agent's projected utility for a state $s' \neq s$ is $\tilde{u}(c, s'|s) = \alpha \cdot u(c, s) + (1 - \alpha) \cdot u(c, s')$ where $0 \leq \alpha < 1$ is a parameter that specifies the extent to which the agent's prediction is biased. They demonstrate their model by considering settings of endowment effect, impulse buying of durable goods and addiction formation. For the latter, the conceptual claim is that novice smokers fail to understand that quitting will be much harder after a year of smoking. Moreover, they do not realize that in the future, they might need to smoke more to attain a "pleasure level" similar to their initial "pleasure level".

The focus of Loewenstein et al. [10] was on settings in which the state essentially changes as a simple function of the consumption. This modeling fails to capture situations in which the state changes in a more elaborate manner. As an example, consider a high school graduate planning her career path: the different states correspond to her occupational status (e.g., *completed B.A in Biology, first year of business school*) and the utility at each step can equal to some measure of her well being for staying at this state or her yearly earning ability if she stayed at this state. It is often the case that there are multiple career paths leading to the same state. We suggest overcoming this limitation by introducing a state graph to capture how the state changes as a function of the agent's action.

Model. We consider an agent with projection bias traversing a directed acyclic state graph G for $t > 0$ discrete time steps[1], starting from an initial state s. (see Fig. 1 for an illustration). Each state v is defined a non-negative payoff, $u(v)$, that the agent collects for each time step it stays in v. Each edge (v_1, v_2) is assigned a cost that corresponds to the cost of transitioning from state v_1 to state v_2. We assume that transitioning from a state to a neighboring state takes exactly one time step.[2] The payoffs and costs can be of either monetary value or more abstract measures such as a measure of well-being.

We model the behavior of an agent with projection bias similarly to Loewenstein et al. [10]. An agent is characterized by a parameter $0 \le \alpha < 1$ that captures the extent of the bias. As greater α is, the agent believes that its payoff in any future state will resemble its current payoff more. Formally, the projected payoff of an α-biased agent currently at v for any future state w is

$$u^\alpha(w|v) = \alpha \cdot u(v) + (1 - \alpha)u(w)$$

At each state v the α-biased agent can either stay at v and collect a payoff of $u(v)$ or continue to a neighboring state. The agent's objective is to maximize its total utility (i.e., the sum of payoffs the agent collected minus the cost it paid for the edges that it traversed). However, due to its projection bias, the agent at each state will aim to maximize its total *projected* utility instead. We observe that after the agent decides to stay at some state v it will choose to stay at v for all remaining time steps. This observation implies that at each state an α-biased agent plans to take some path on the state graph and then stay at the last state on the path for the remaining time steps. It is not hard to see that at each state the α-biased agent can compute a path maximizing its total projected utility in polynomial time by a dynamic program.

It is instructive to consider the behavior of a biased agent in an example. Consider a 1/2-biased agent traversing the state graph G in Fig. 1. For $t = 6$ the agent at s will plan to follow the path (s, v_1, v_4) as the projected utility of this path is $(1/2 \cdot 1 + 1/2 \cdot 8) \cdot (6 - 2) - 0 = 18$. This is greater than the utility of staying at s which is $1 \cdot 6 = 6$, the utility of continuing to v_1 which is $(1/2 \cdot 1 + 1/2 \cdot 6) \cdot (6 - 1) - 0 = 17.5$ and the utility of continuing to v_3 which is $(1/2 \cdot 1 + 1/2 \cdot 10) \cdot (6 - 3) - 0 = 16.5$. At v_1 the agent, learns that the payoff at v_1 is higher than it expected and decides to abandon its original plan and instead stay at v_1. To see why this is the case, observe that the total utility for staying at v_1 is 30 while the projected utility for continuing to v_4 is 28. The changing payoffs may not only cause the agent to abandon a plan but may also make it to completely change its plan. To see this, consider again the state graph in Fig. 1 where now $t = 9$. In this case an 1/2-biased agent at s will plan to take a path to v_3. However, once it arrives to v_1 it will change its plan and decide to continue to v_4 instead.

---

[1] To focus on the effects of the projection bias we assume that future costs and payoffs are not discounted.

[2] Essentially, this is without loss of generality since longer transition periods can be handled by adding more edges of cost 0.

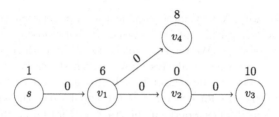

Fig. 1. An example of a state graph G.

Overview of Results. We provide a thorough analysis of the behavior of an α-biased agent traversing a direct acyclic state graph. We first present some insights into the agent's planning behavior. When discussing the instance in Fig. 1 we observed that an agent might change its plan. Such a change can only occur when the agent moves between states that have different payoffs. When the agent moves from a low-payoff state to a high-payoff state, the projected payoffs of future states increase. As a result, the agent will favor shorter paths, as each time step can be used either to traverse an edge or collect a payoff and a higher payoffs make the second option more appealing. In the complementary case of moving from a high payoff state to a low payoff state, we see the opposite phenomenon as the agent now favors longer paths. We show that the number of changes in plan that a biased agent makes may be linear in the size of the graph. In fact, there are instances in which the α-biased agent changes its plan after each step it takes.

Non-Monotonicity. Next, we consider the effects of the different parameters of the model on the (actual) total utility of the α-biased agent. First, we discuss how the agent's utility changes with the time horizon t. For an unbiased agent, the total utility may only increase with t, since the unbiased agent can always increase its utility by taking the same path it took for $t-1$. This is not necessarily the case for an α-biased agent. Roughly speaking, as t increases, an α-biased agent would plan to follow a path with a higher total utility (both actual and projected). However, due to its projection bias, it may reach a state it would not have reached otherwise. At this state, the agent may change its plan to one in which the total utility is smaller. To bound the number of non-monotonicity points (i.e., t is a non-monotonicity point if the total utility of an α-biased agent for $t+1$ is lower than its total utility for t), we bound the number of possible paths that an α-biased agent, for a fixed value of α, may take for any value of t. We provide a lower bound of n and an upper bound of n^3 on this quantity. We obtain orthogonal results when considering the behavior of the agent as a function of α. Perhaps counterintuitively, we show that as the agent's bias decreases, its total utility may also decrease. We also bound the number of non-monotonicity points in α by the number of paths that α-biased agents with different values may take. We show a lower bound of n and an upper bound of n^4 on this number.

Performance Ratio and Hardness of Computing an Optimal Subgraph. A natural question in this setting is to understand how much an agent can lose from its bias. Specifically, we ask how large can be the gap between the utility of an unbiased agent and the utility of an α-biased agent traversing a state graph for the same number of steps. We refer to this ratio as the performance ratio. Note that, since the α-biased agent always knows which states have a higher payoff than others and is only wrong about how high the payoff is, for any instance, there exists a t such that for every $t' > t$ the agent takes the same path as an unbiased agent. This implies a performance ratio of 1 for large values of t. On the other hand, for a fixed time horizon t, we show that for any value of $0 < \alpha < 1$, the performance ratio can be made arbitrarily high.

The possible high loss of an agent due to its bias motivates finding ways to increase its utility. One possible way of doing this is by removing states and edges from the graph. Consider the state graph in Fig. 1. For $t = 10$ and $\alpha = 1/2$, assuming ties are broken in favor of shorter paths, the agent will take the path (s, v_1, v_4) for a total (actual) utility of $(10 - 2) \cdot 8 = 64$. However, if we remove v_4 then the agent will now take the path (s, v_1, v_2, v_3) that has a total utility of $(10 - 3) \cdot 10 = 70$. This leads us to the computational question of whether we can efficiently compute a subgraph maximizing the total utility of an α-biased agent. We show that this problem is NP-hard.

Discussion. The current paper is situated in the growing literature on the planning of agents with behavioral biases such as present bias and sunk cost bias (e.g., [1,5–9,16]). In contrast to other papers in this line of work, the modeling of projection bias requires a fully-fledged model of planning that includes costs on the edges, payoffs on the nodes and a time horizon t that the agent plans for. Moreover, projection bias operates on a more global level than present bias which makes planning for agents with projection bias and agents with present bias considerably different, technically and conceptually. While all of these biases imply some form of time-inconsistent planning, the type of inconsistency and its implications differ. For example, the loss of an agent with projection bias due to its bias may be unbounded, whereas the loss of an agent with present bias is bounded by a function of the graph's size. The non-monotonicity results we present are also unique to agents with projection bias.[3]

Our work leaves several open questions, including:

- Are the upper bounds we presented on the number of non-monotonicity points in the agent's utility as a function of α or t tight?
- We show that computing a subgraph that maximizes the total utility of an α-biased agent is an NP-hard problem. One can consider other methods to increase its utility. For example, instead of removing edges from the underlying graph, by how much can decreasing the cost of different paths help to increase the agent's utility? Can this be done optimally in polynomial time?
- On a more high level, in the spirit of [8], it could be interesting to explore the behavior of agents that exhibit projection bias simultaneously with other planning related biases such as present bias and sunk cost bias.

---

[3] [7] considers a different type of monotonicity results for sophisticated agents in terms of other parameters of the graph.

Structure of the Paper. In Sect. 2 we formally describe the model and analyze the behavior of the agent. In Sect. 3 we discuss the performance ratio and computing a subgraph maximizing the agent's utility. In Sect. 4 we consider the effects of the parameters t and α on the choices of the biased agent. Finally, in Sect. 5 we discuss an extension of our model for state graphs that may have cycles.

2 Model and Preliminaries

We consider an agent traversing a directed acyclic *state graph*, that has n states. The initial state of the agent is s and the agent traverses the graph for $t \geq 0$ time steps. At every time step, an agent currently at state v can either stay and collect a payoff of $u(v)$ (positive) or continue to a different state. An agent who chooses to continue to an adjacent state, by taking an edge e, will not collect any payoff at this time step and will pay a cost of $c(e)$ (negative). An agent with projection bias currently at a state v, believes that its payoff in a different state w will be similar to its payoff at v. Formally, we assume that an agent with projection bias currently at v believes its payoff in a future state w will be $u^{\alpha}(w|v) = \alpha \cdot u(v) + (1 - \alpha) \cdot u(w)$. We refer to this as the agent's projected payoff. Where $0 \leq \alpha < 1$ denotes the extent of projection bias that the agent exhibits. For $\alpha = 0$ we have an unbiased agent and as α increases, the agent believes that its payoff in future states will be more similar to the payoff in its current state.

The goal of the agent is to maximize its utility in the t steps it traverses the graph. The utility equals the sum of payoffs the agent collected minus the cost it paid for the edges that it traversed. Observe that if an agent decides to stay at a state v, then at the next time step, it will also decide to remain at v. The reason for this is that in both time steps the projected utilities are identical. In this case, it is easy to observe that if an agent decides to stay at some state v when it has t remaining time steps, it will also choose to stay at v when planning for $0 \leq t' < t$ time steps. We proved the following observation:

Observation 1. *If an α-biased agent, currently at v, decides to stay at v then it will stay there for all remaining time steps.*

The observation allows us to specify the agent's behavior concisely: an α-biased agent currently at state v will plan to follow a path P to a state w which maximizes its *projected utility*: $(t - |P|) \cdot u^{\alpha}(w|v) - C(P)$, where $C(P) = \sum_{e \in P} c(e)$. If there are several paths with the same utility, we assume the agent breaks ties in favor of shorter paths. It is not hard to see that the agent can compute the path maximizing its projected utility in polynomial time using a simple dynamic program (see the full version for a formal proof).

We note that for any state graph G and any $0 \leq \alpha \leq 1$, if t is large enough, then the α-biased agent will behave as an optimal (unbiased) agent and will take the minimal cost path that reaches the state of maximal payoff. When planning for a shorter time horizon, it is no longer the case that the agent behaves

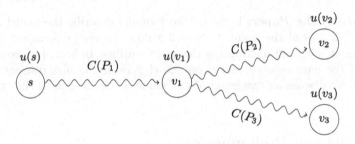

Fig. 2. Illustration for proposition 2.

optimally or follows a consistent plan. The following observation formalizes under which circumstances the α-biased agent will change its plan.

Observation 2. *Consider an α-biased agent that at state s plans to follow path P_1 to v_1 and then to follow P_2 to v_2. When reaching v_1 the agent changes its plan and plans to follow P_3 to v_3 instead (see Fig. 2 for an illustration). Then, $u(v_1) > u(s)$ and $|P_2| > |P_3|$ or $u(v_1) < u(s)$ and $|P_2| < |P_3|$.*

Proof. Since the plan of the α-biased agent at s was to reach v_2 and not v_3 we have that:

$$(t - |P_1| - |P_2|)u^\alpha(v_2|s) - C(P_1) - C(P_2) > (t - |P_1| - |P_3|)u^\alpha(v_3|s) - C(P_1) - C(P_3)$$

Similarly, since the plan of the α-biased agent at v_1 was to reach v_3 and not v_2 we have that:

$$(t - |P_1| - |P_3|)u^\alpha(v_3|v_1) - C(P_3) > (t - |P_1| - |P_2|)u^\alpha(v_2|v_1) - C(P_2)$$

Putting this together we get that $\alpha \cdot u(v_1)(|P_2| - |P_3|) > \alpha \cdot u(s)(|P_2| - |P_3|)$. We may assume that $\alpha > 0$ since the planning of an unbiased agent is always consistent. Hence we can divide by α and get that $u(v_1)(|P_2|-|P_3|) > u(s)(|P_2|-|P_3|)$. This implies that $u(v_1) > u(s)$ if and only if $|P_2| > |P_3|$. We conclude that if an α-biased agent changes its plan at a higher payoff state, it will choose a shorter path. Conversely, when the agent changes its plan at a lower payoff state, it will choose a longer path. $\square$

Notice that since we consider a directed acyclic graph, the number of times an agent may formulate a new plan is bounded by $n-1$, where n is the number of states in the graph. This is because, by Observation 1, once an agent decides to stay at a state, it will remain there for all remaining time steps. This implies that the number of plan changes is bounded by the number of states with outgoing edges, which is at most $n-1$. This bound is tight, as we show in the full version that for any $0 \le \alpha < 1$ there exists a fan graph (illustrated in Fig. 3) in which the α-biased agent ends up taking a path of $k + 1 = n$ states and at each state except for the last two states the agent formulates a new plan.

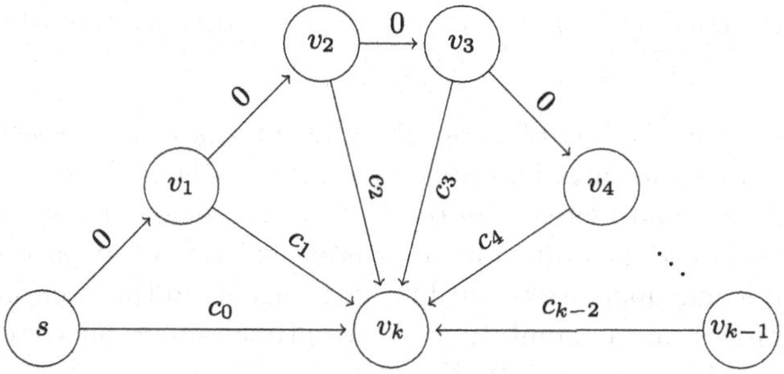

Fig. 3. A state graph in which the biased agent changes its plan $k - 1 = n - 2$ times.

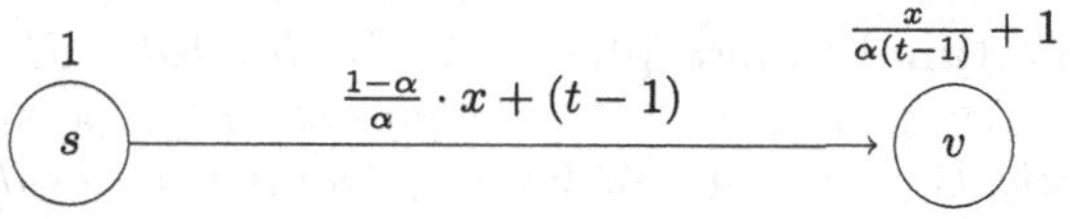

Fig. 4. An example with unbounded performance ratio

3 On the Utility of the α-Biased Agent

It is clear that an agent with projection bias may obtain a smaller utility than an optimal agent due to its failure to accurately predict its payoff in different states. A natural question in our model is to quantify how much an α-biased agent can lose due to its bias. To this end, we attempt to bound the ratio between the total utility of an optimal (unbiased) agent and the (actual) utility of an α-biased traversing a state graph G for t time steps. We refer to this ratio as the performance ratio and show that this ratio can be arbitrarily large. Perhaps surprisingly, the instance demonstrating this has only two states.

Theorem 3. *For any* $0 < \alpha < 1$, $t > 1$ *and* $x > t$, *there exists a state graph* G *in which the performance ratio of an* α*-biased agent traversing* G *for time horizon* t *is* $\frac{x}{t}$.

Proof. Consider the instance illustrated in Fig. 4. An optimal agent in this instance will go to v for a total utility of

$$\frac{x}{\alpha} + t - 1 - \left(\frac{1-\alpha}{\alpha}x + (t-1)\right) = x.$$

We claim that an α-biased agent stays at s and hence its total utility is t. This implies that the performance ratio is $\frac{x}{t}$, as required. To see why the α-biased agent will stay at s, observe that:

$$u^{\alpha}(v|s) = \alpha + (1-\alpha)\left(\frac{x}{\alpha(t-1)} + 1\right) = \frac{x}{t-1} \cdot \frac{1-\alpha}{\alpha} + 1.$$

Thus, we have that $(t-1) \cdot u^{\alpha}(v|s) - c(s,v) = 0$ and hence the α-biased chooses to stay at s.   □

As we just seen, the loss of an α-biased agent due to its projection bias can be very large. In some cases an outside planner can help the α-biased agent to take a better path and increase its total utility by removing some of the states in the graph.[4] In the introduction we give an example of an instance in which removing states may increase the utility of the agent. In this section we consider the characteristics and computation of an optimal subgraph which is a graph that maximizes the agent's utility. Formally, let $U^{\alpha,t}(G')$ denote the total utility of an α-biased agent traversing G for t time steps. With this notation we consider the following problem:

Definition 1 (Optimal Subgraph). *Given a directed acyclic state graph G, an initial state s, a time horizon $t > 0$ and a projection bias parameter $0 < \alpha < 1$, compute a subgraph G' of G (that includes s) that maximizes the total utility of the α-biased agent over all subgraphs (i.e., $U^{\alpha,t}(G') = \max_{H \subseteq G} U^{\alpha,t}(H)$).*

To better understand the possible structure of optimal subgraphs, we consider minimal subgraphs that only include states and edges that are necessary to maximize the agent's utility. Formally,

Definition 2 (Minimal Optimal Subgraph). *Given a directed acyclic state graph G, an initial state s, a time horizon $t > 0$ and a projection bias parameter $0 < \alpha < 1$, a minimal optimal subgraph G' is an optimal subgraph such that for any $H \subset G'$ the utility of the agent is smaller than its utility in G'.*

In the next section, we make several observations on the structure of minimal optimal subgraphs and observe that their structure is inherently different than the structure of optimal subgraphs for agents exhibiting other biases (e.g., present bias [6]).

3.1   The Structure of a Minimal Optimal Subgraph

We begin by observing that the subgraph maximizing the total utility of the agent may require more states than just the states that the agent actually visits (i.e., it is not necessarily a path). This is illustrated in the instance in Fig. 5. In this instance, the path that has the maximal utility when planning for $t = 10$ time steps is $P = (s, v_1, v_2, v_3, v_4)$ that has a utility of $220 \cdot 7 - 115 = 1425$. However, its projected utility for a $1/2$-biased agent at s is only $(0.5 \cdot 100 + 0.5 \cdot 250) \cdot 6 - 70 = 980$. This is smaller than the utility of staying at s which is 1000, which means that an $1/2$-biased agent traversing P in isolation will choose to stay at s. If we add the edge (v_2, v_5) to the path, then the $1/2$-biased agent will follow

---

[4] Unfortunately, even removing states and edges cannot circumvent the unbounded loss, in the worst case. This is demonstrated by Claim 3 as the utility of the agent in the only strict subgraph that contains only state s is the same as its utility on the two states graph.

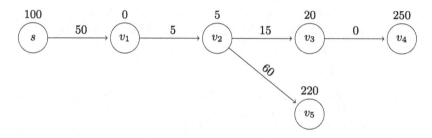

Fig. 5. For $t = 10, \alpha = 0.5$ the agent will not traverse $P = (s, v_1, v_2, v_3, v_4)$ in isolation but if we include the edge (v_1, v_5) the agent will traverse P.

P since at s it will plan to reach v_5, since the path (s, v_1, v_2, v_5) has projected utility of $(0.5 \cdot 100 + 0.5 \cdot 220) \cdot 7 - 115 = 1005$. Once the agent reaches v_1 it will plan to follow the path (v_1, v_2, v_3, v_4) since $125 \cdot 6 - 20 = 730 > 110 \cdot 7 - 65 = 705$ and it will continue to follow this path till it reaches v_5. Thus, the path P by itself does not maximize the α-biased agent's utility.

After observing that the optimal subgraph may not include only the path that the agent traverses, we look for characterization of minimal optimal subgraphs. Observe that if the α-biased agent traverses a path P in a minimal optimal subgraph, then, except for $P = (s, v_1, \ldots, v_k)$ the graph includes at most k paths that start from a state on P and end in some other state. These paths are paths that the biased agent plans to take at some node v_i and they are required to get the agent to continue from v_i to v_{i+1}. For an agent with present bias, Kleinberg and Oren [6] proved that each state v_i might be the source of at most one such path. In the full version, we observe that this is not the case for agents with projection bias.

3.2   Hardness of Computing an Optimal Subgraph

As discussed, limiting the agent's options by removing states and edges may increase the agent's utility. Unfortunately, we show that computing a subgraph that maximizes the utility of the agent is NP-hard:

Theorem 4. *The Optimal Subgraph Problem is NP-hard.*

To prove the theorem we define the following more concrete decision problem:

Definition 3 (Bias Mitigating Subgraph Problem). *Consider a directed acyclic state graph G, a time horizon $t > 0$ and an α-biased agent where $0 \leq \alpha < 1$. Let $\mathcal{P}_o$ denote the set of paths that maximize the utility of an unbiased agent in G. Does there exist a subgraph $G' \subseteq G$ in which the α-biased agent follows a path $P_o \in \mathcal{P}_o$?*

Observe that the hardness of the Bias Mitigating Subgraph Problem implies the hardness of the optimal subgraph problem. If there exists a subgraph G'

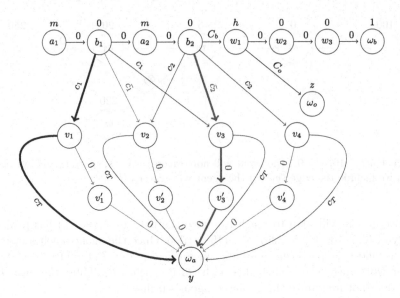

Fig. 6. Corresponding graph for $(x_1 \lor \neg x_2 \lor x_3) \land (x_2 \lor \neg x_3 \lor x_4)$. The bold edges represent a bias mitigating subgraph which corresponds to the assignment $x_1 = TRUE, x_3 = FALSE$. (Color figure online)

in which the α-biased agent follows a path $P \in \mathcal{P}_o$, then an algorithm solving the optimal subgraph problem has to return such subgraph. Furthermore both the path that an optimal agent takes in G and the path that an α-biased agent takes G' can be computed in polynomial time. Thus, to prove the hardness of the Optimal Subgraph problem, we show that:

Theorem 5. *The Bias Mitigating Subgraph Problem (a.k.a BMS) is NP-complete.*

It is easy to see that the BMS problem is in NP since in the full version we show that computing the total utility of an unbiased agent in G and the path taken by an α-biased agent in G' can be done in polynomial time. We show that BMS is NP-hard by reducing from 3-SAT. In the classic 3-SAT problem, we are given a conjunctive normal form formula φ with r variables and k clauses such that each clause has exactly three variables. We need to return yes if and only if there exists an assignment to the variables that satisfies the φ. Since the reduction is quite involved, we leave the formal construction and proof to the full version and only provide a sketch here.

Given a 3-CNF formula φ with r variables $x_1, x_2, \ldots, x_r$ and k clauses $C_1, C_2, \ldots, C_k$, we construct a directed acyclic state graph $G(\varphi)$ for the BMS problem. The graph is illustrated in Fig. 6. Roughly speaking, each clause C_i in the formula corresponds to two states a_i, b_i. Each state a_i is a medium-payoff state and each state b_i is a low-payoff state with fixed utilities. Each variable x_j in the formula corresponds to two nodes v_j, v'_j, where each clause is connected to the variables it contains by an edge (b_i, v_j). The weight and color of such an edge differs between the case where x_j is in C_i (blue) and the case where $\neg x_j$ is in C_i (red). For each state v_j we have a blue edge (v_j, w_a) and red edges (v_j, v_j') and (v_j', w_a). The edges are colored such that there exists a satisfying assignment to φ if and only if there exists a subgraph where each state b_i has a unique path to w_a and in this path all the edges have the same color. Roughly speaking, in the subgraph, an edge (a_i, v_j) correspond to clause C_i being satisfied by variable x_j and the path chosen from v_j to w_a correspond to assignment to variable x_j. In addition there are three states w_o, w_a, w_b that serve as "planning targets" at different states. The utilities of these states are chosen such that $u(w_o) > u(w_a) > u(w_b)$. The following proposition establishes the correctness of the reduction:

Proposition 1. *For any $\frac{2k-1.5}{2k-1} < \alpha < 1$ and $t > 2k + 4$, the state graph $G(\varphi)$, constructed from the formula φ, has a bias mitigating subgraph if and only if there exists a subgraph in which for every $1 \le i \le k$ all direct paths from b_i to w_a are unicolored and for every b_i there is at least one direct unicolored path.*

We now sketch the proof. We first show that an optimal agent takes the path $P_o = (a_1, b_1, \ldots, a_k, b_k, w_1, w_o)$ and this is the unique path maximizing its utility. Thus, in any bias mitigating subgraph the α-biased agent should follow P_o. Then, we argue that in any subgraph $G' \subseteq G(\varphi)$ which includes (w_1, w_o), an α-biased agent at w_1 will continue to w_o. This implies that to prove the proposition we need to show that there exists a subgraph in which the α-biased agent will reach w_1 if and only if there exists a subgraph in which for every $1 \le i \le k$ all direct paths from b_i to w_a are unicolored and there is at least one such path.

For the first direction, we show that an α-biased agent will reach w_1 when traversing a subgraph G' in which for every $1 \le i \le k$ all direct paths from b_i to w_a are unicolored and there is at least one such path. The idea here is the agent would have different types of plans at low-payoff states and at medium utility states. In a medium-payoff state the α-biased agent will continue from a_i to b_i as part of a plan to get to w_a. In a low-payoff state b_i will continue to a_{i+1}, whenever the mixed-colored path (b_i, v_j, v_j', w_a) (the first edge is blue) does not exist.

As for the second direction, first we show that for every $1 \le i \le k$ if the agent at a_i continues to b_i then it is either the case that one of the paths $(a_i, b_i, v_j, v_j', w_a)$ or (a_i, b_i, v_j, w_a) is in G' or the mixed-color path $(a_l, b_l, v_j, v_j', w_a)$ for some $l > i$ is in G'. Then we show that if the agent plans to follow a direct path from b_i to w_a then this path cannot be a mixed colored path (b_i, v_j, w_a) (first edge is red) since it is too expensive. Moreover, we have that, in any case, the path cannot be mixed colored path (b_l, v_j, v_j', w_a) (i.e., the edge (b_i, v_j) is blue) for any $l \ge i$, since in this case an α-biased agent at b_l will continue to v_j instead of a_{l+1}. This is because the cost of this path is too low.

Putting this together, we get that, for any $1 \le i \le k$, there exists at least one unicolored path from b_i to w_a. Moreover, if the bias mitigating subgraph includes two paths connecting v_j and w_a, then any state b_i that is connected to v_j is connected by a red edge. This implies that we can remove all blue (v_j, w_a) edges without revoking the property that there exists at least one unicolored path from every state b_i to w_a.

We note that the structure of the reduction is similar to the reduction in [16] proving that finding a minimal subgraph in which a present biased agent will reach the target is NP-hard. We note that while the structure is similar and both proofs reason about the behavior of time-inconsistent agents, the planning of agents with these two biases is significantly different. In particular, the current reduction attains an extra level of complexity as the planning of an α-biased agent is not constrained to choosing a path to a specific state. Moreover, the choice of the path for an agent with projection bias is more nuanced as it depends not only on the cost of the path but also on the utility of the current node and the number of remaining time steps.

4   Non-monotonicity and the Number of Different Paths

In this section, we consider how the total utility of an α-biased agent changes as a function of the time horizon t and the value of α. We show that the total utility of the α-biased agent is non-monotonic in both of these parameters. For each one of the parameters, we present a polynomial upper bound on the number of non-monotonicity points. Instead of bounding the number of non-monotonicity points directly, we show a polynomial bound on the number of *transition points*. These are points in which the path that the agent takes changes. The polynomial upper bound on the number of transition points also implies an upper bound on the number of different paths that the agent can take as a function of t or α. This is a question of independent interest and for both parameters, we also observe a lower bound of n on the number of different paths. This is done by constructing and analyzing an out-directed star instance with carefully selected costs and payoffs.

The following piece of notation will be useful for proving our claims. Let $P_{\alpha,t}(v)$ denote the path *followed* by an α-biased agent traversing the graph for t time steps, starting from state v and let $\tilde{P}_{\alpha,t}(v)$ denote the path that the agent *plans* to follow. We are now ready to specify and prove our claims.

4.1   Planning for Different Time Horizons

Recall that for any state graph G and a bias parameter $0 \leq \alpha < 1$, an α-biased agent planning for a large enough time horizon t^* will take the same path as an optimal agent. This implies that for any $t > t^*$, the total utility of the agent is monotonically increasing in t. This is not necessarily the case for intermediate values of t. The reason for this is that while the initial plan that an agent planning for a larger time horizon makes is better, the agent may change its plan later and as a result, its actual utility may decrease. As an illustration, consider a 3/4-biased agent traversing the state graph illustrated in Fig. 7. For $t = 2$ the agent arrives at v_1 and will end up with a total utility of 210. However, for $t = 3$, the 3/4-biased agent plans to take the path (s, v_2, v_3) since the projected utility for taking it is $(1/4 \cdot 440 + 3/4 \cdot 0) \cdot 1 = 110$ and the projected utility for going to v_1 is $(1/4 \cdot 210 + 3/4 \cdot 0) \cdot 2 = 105$. When the agent reaches v_2 it will choose to

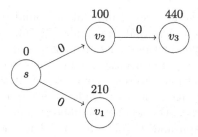

Fig. 7. An example for non-monotonicity point

stay as the projected utility from reaching v_3 is now $(3/4 \cdot 100 + 1/4 \cdot 440) = 185$ where the utility from staying at the same state is 200. The value $t = 3$ is a non-monotonicity point in the utility function of the $3/4$-biased agent. These are points in which the total utility of an α-biased planning for time horizon t is lower than its utility for planning for time horizon $t - 1$. We show that the maximal number of non-monotonicity points is polynomial in the number of nodes in G:

Theorem 6. *For any acyclic state graph G with n states, an initial state s and projection bias $0 \leq \alpha < 1$, the number of non-monotonicity points in the utility function of an α-biased agent traversing the graph from s is at most n^3.*

As discussed in the introduction, instead of tackling this problem directly, we compute a bound on the number of time horizons for which the path that the α-biased agent takes changes. Notice that this quantity is necessarily bounded since the α-biased agent will always take the optimal path for a large enough time horizon.

Proposition 2. *For every directed acyclic state graph G with n states, an initial state s and $0 \leq \alpha < 1$, the number of transition points t such that the path the agent takes from s when planning for $t - 1$ time steps is different than the path it takes when planning for t (i.e., $P_{\alpha,t-1}(s) \neq P_{\alpha,t}(s)$) is at most n^3.*

Proof. Since the α-biased agent takes a different path when planning for $t - 1$ and for t, there has to be a state v on the path $P_{\alpha,t-1}(s)$ at which the α-biased agent that plans for t steps makes a different plan (this state can also be s). Let v denote the first such state. We label the transition point t by $(v, k, \tilde{P})$, where $0 \leq k \leq n - 1$ is the length of the path that the agent followed from s to v and $\tilde{P} = \tilde{P}_{\alpha,t-1-k}(v)$ is the path that the agent that initially planned for $t - 1$ steps plans to take. We observe that each label can appear at most once. This is because the *planning* of the agent is consistent in the time horizon in the sense that if a path is optimal for t_1 and suboptimal for $t_2 > t_1$ it will be suboptimal for any $t_3 > t_2$. This implies that if $\tilde{P}_{\alpha,t-1-k}(v) \neq \tilde{P}_{\alpha,t-k}(v)$ (i.e., an α-biased agent with time horizon $t - k - 1$ plans to follow $\tilde{P}$ and an α-biased agent initially planning for time horizon $t - k$ plans to follow a different path), then for any $t' > t$, $\tilde{P}_{\alpha,t-1-k}(v) \neq \tilde{P}_{\alpha,t'-k}(v)$.

We conclude that the number of intervals is bounded by the number of different labels $(v, k, \tilde{P})$. Observe that the number of different *plans* that α-biased agents planning for different time horizons makes is at most n. The reason for this is that since the projected utilities remain the same for any time horizon, then whenever the agent plans to reach some state w it will take the same path. This implies that if any agent changes its plan it has to be the case that it plans to a different state. Hence, the number of different labels is at most n^3 as there are n nodes in the graph and $0 \leq k \leq n - 1$. Thus, the number of transition points is bounded by n^3. ☐

In the full version we show that there exists instances that have $n - 1$ transition points. We derive this simple bound by considering an out-directed star with s as the center and choose the costs and payoffs such that for every value of $1 \leq t \leq n$ the agent plans to reach a different state.

4.2 Agents with Different Values of α

In this section we take a deeper look on how different values of α affect the α-biased agent behavior. Given a state graph G and time horizon t, an agent with $\alpha = 0$ (e.g., an optimal agent) would take an optimal path and an agent with $\alpha = 1$ would stay at the initial state. What would agents with an intermediate value of α do? For intermediate values, the following example demonstrates that the agent's utility function does not necessary increase as the extent of the biases decreases: Consider the graph illustrated in Fig. 8. A $1/2$-biased agent traversing the graph for $t = 3$ time steps would traverse the edge (s, v_1) as the projected utility for reaching v_1 is $(1/2 \cdot 0 + 1/2 \cdot 50) \cdot 2 = 50$ while the projected utility for reaching v_3 is $(1/2 \cdot 0 + 1/2 \cdot 105) \cdot 1 - 3 = 49.5$. For the same graph, a $1/3$-biased agent would traverse the edge (s, v_2) as the projected utility for reaching v_3 is $(1/3 \cdot 0 + 2/3 \cdot 105) \cdot 1 - 3 = 67$ while the projected utility for reaching v_1 is $(1/3 \cdot 0 + 2/3 \cdot 50) \cdot 2 \approx 66.7$. When the agent reaches v_2 it chooses to stay as the projected utility for reaching v_3 is $(1/3 \cdot 45 + 2/3 \cdot 105) \cdot 1 = 85$ and the total utility from staying at v_2 is 87.

We establish an upper bound of n^4 on the number of non-monotonicity points in α. We do this by bounding the number of different transition points in the interval $[0, 1)$. α^* is a transition point if for any $\alpha < \alpha^*$ an α-biased agent takes a different path than an α^*-biased agent.

Proposition 3. *For any state graph G, an initial state s and time horizon $t \geq 0$, the number of transition points as a function of α is at most n^4.*

Proof. Similar to the proof of the bound on the number transition points for a time horizon t, we label each transition point by the label $(v, k, \tilde{P_1})$. Where v is the first node that the agent changed its plan, k is the length of the path it took to get to v and $\tilde{P_1}$ is the path that it planned to take at v right before the transition point. We observe that, when labeling all transition points in the interval $[0, 1)$, each label may appear at most once. The reason for this is that Lemma 1 below shows that if α_1-biased with time horizon $t - k$ plans to follow

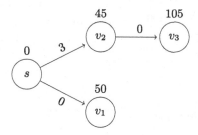

Fig. 8. An example for non-monotonic planning in α. When $t = 3$, a 1/2-biased agent would end up with total utility of 100 while a 1/3-biased agent would end up with total utility of 87.

$\tilde{P}_1$ and for $\alpha_2 > \alpha_1$ an α_2-biased agent plans to follow $\tilde{P}_2 \neq \tilde{P}_1$, then for any $\alpha \geq \alpha_2$, an α-biased agent will not plan to follow $\tilde{P}_1$. Hence it is impossible that for $\alpha > \alpha_2$, an α-biased agent that reaches v with a path of length k will plan to take $\tilde{P}_1$ again. To complete the proof, in Lemma 2 below we show that the number of different *plans* that biased agents with different values of α, that are currently at a state v with time horizon t make is at most n^2. This implies that the number of different labels is at most n^4, as required. □

Next, we state and prove the two lemmas we used to prove Proposition 3:

Lemma 1. *Consider $0 \leq \alpha_1 < \alpha_2$. If $\tilde{P}_{\alpha_1,t}(v) \neq \tilde{P}_{\alpha_2,t}(v)$ then for any $\alpha_3 \geq \alpha_2$, $\tilde{P}_{\alpha_3,t}(v) \neq \tilde{P}_{\alpha_1,t}(v)$.*

Proof. For ease of notation let $\tilde{P}_1 = \tilde{P}_{\alpha_1,t}(v)$, $\tilde{P}_2 = \tilde{P}_{\alpha_2,t}(v)$ and assume that v_1 and v_2 are the last states in these two paths respectively. By applying Lemma 3, we get that $(t-|\tilde{P}_1|)(u(v_1)-u(v)) > (t-|\tilde{P}_2|)(u(v_2)-u(v))$ Assume towards contradiction that α_3-biased agent plans to take the path $\tilde{P}_1$. By applying Lemma 3 for α_2, α_3, we get that $(t - |\tilde{P}_2|)(u(v_2) - u(v)) > (t - |\tilde{P}_1|)(u(v_1) - u(v))$ and reach a contradiction to our assumption. □

Our second lemma bounds the number of paths that agents with different bias parameters can plan for:

Lemma 2. *Consider a directed acyclic state graph G and let $\tilde{\mathcal{P}}_t(v) = \{\tilde{P}_{\alpha,t}(v)|0 \leq \alpha < 1\}$, then $|\tilde{\mathcal{P}}_t(v)| \leq n^2$.*

Proof. We first claim that the number of different paths that agents with different values of α may take to get from a state v to a state w is at most n. To see why this is the case, consider $\alpha_2 > \alpha_1$ such that an α_1-biased agent takes the path $\tilde{P}_1$ to w and an α_2-biased agent takes the path $\tilde{P}_2$ to w. By applying Lemma 3 we have that $(t - |\tilde{P}_1|)(u(w) - u(v)) > (t - |\tilde{P}_2|)(u(w) - u(v))$.

Observe that $u(w) > u(v)$, as otherwise the agent would have stayed at v. This implies that $t - |\tilde{P}_1| > t - |\tilde{P}_2|$ and hence $|\tilde{P}_2| > |\tilde{P}_1|$. As the number of different paths lengths is bounded by n, we have that the number of different

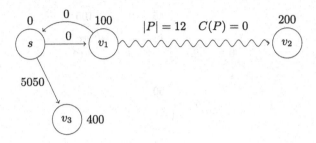

Fig. 9. Instance in which a 0.5-biased agent traverses a cycle.

paths reaching w that the agent takes is at most n. Since there are n nodes in the graph we conclude that the number of different paths that agents with different values of α currently at v may plan to follow is bounded by n^2.   □

In the full version we prove the following auxiliary lemma:

Lemma 3. *Consider a directed acyclic state graph G. If an α_1-biased agent plans to follow a path $\tilde{P}_1$ and an α_2-biased agent for $\alpha_2 > \alpha_1$ plans to follow a path $\tilde{P}_2$, such that $\tilde{P}_1 \neq \tilde{P}_2$, then $(t-|\tilde{P}_1|)(u(v_1)-u(v)) > (t-|\tilde{P}_2|)(u(v_2)-u(v))$.*

Finally, to complete the picture we show in the full version that there are instances in which the number of paths taken by agents with different values of α is n.

5   Extensions - State Graphs with Cycles

So far, our focus was on state graphs that do not have any cycles, as such graphs represent realistic scenarios, and it is easier to reason about them. However, projection bias is not limited to misestimating the payoffs of states we have not experienced yet. For example, consider the different food choices of people with varying levels of hunger. This motivates the analysis of graphs that do have cycles. Intuitively one might suspect that since the agent can tell which of two states has the higher payoff, it will be impossible for it to traverse a cycle. In this section we show that this is not the case.

Consider the graph illustrated in Fig. 9 where $t = 39$. A 1/2-biased agent will traverse the edge (s, v_1) since the projected utility for reaching v_2 is $(1/2 \cdot 0 + 1/2 \cdot 200) \cdot (39 - 13) = 2600$ while the projected utility for reaching v_3 is $(1/2 \cdot 0 + 1/2 \cdot 400) \cdot (39 - 1) - 5050 = 2550$. When the agent reaches v_1 he chooses to traverse the edge (v_1, s) as the projected utility for reaching v_3 is $(1/2 \cdot 100 + 1/2 \cdot 400) \cdot (38 - 2) - 5050 = 3950$ while the projected utility for reaching v_2 is $(1/2 \cdot 100 + 1/2 \cdot 200) \cdot (38 - 12) = 3900$ and the total utility for staying in v_1 is 3800. This concludes that the agent will traverse the cycle (s, v_1, v) in its first two steps.

In the full version we generalize the instance in Fig. 9 to show that any extent of projection bias suffices to get the agent to follow a cycle.

References

1. Albers, S., Kraft, D.: Motivating time-inconsistent agents: a computational approach. In: Proceedings of 12th Workshop on Internet and Network Economics (2016)
2. Conlin, M., O'Donoghue, T., Vogelsang, T.J.: Projection bias in catalog orders. Am. Econ. Rev **97**(4), 1217–1249 (2007)
3. Gilbert, D.T., Gill, M.J., Wilson, T.D.: The future is now: temporal correction in affective forecasting. Organ. Behav. hHuman Decision Proc. **88**(1), 430–444 (2002)
4. Gilbert, D.T., Pinel, E.C., Wilson, T.D., Blumberg, S.J., Wheatley, T.P.: Immune neglect: a source of durability bias in affective forecasting. J. Personal. Social Psychol. **75**(3), 617 (1998)
5. Gravin, N., Immorlica, N., Lucier, B., Pountourakis, E.: Procrastination with variable present bias. In: Proceedings of the 2016 ACM Conference on Economics and Computation. EC '16, ACM, p. 361. New York, NY, USA (2016)
6. Kleinberg, J., Oren, S.: Time-inconsistent planning: a computational problem in behavioral economics. In: Proceedings of the Fifteenth ACM Conference on Economics and Computation. EC '14, ACM, pp. 547–564. New York, NY, USA (2014)
7. Kleinberg, J., Oren, S., Raghavan, M.: Planning problems for sophisticated agents with present bias. In: Proceedings of the 2016 ACM Conference on Economics and Computation. EC '16, ACM, pp. 343–360. New York, NY, USA (2016)
8. Kleinberg, J., Oren, S., Raghavan, M.: Planning with multiple biases. In: Proceedings of the 2017 ACM Conference on Economics and Computation. EC '17, ACM, pp. 567–584. New York, NY, USA (2017)
9. Kleinberg, J., Oren, S., Raghavan, M., Sklar, N.: Stochastic model for sunk cost bias. In: To appear in proceedings of the 37th Conference on Uncertainty in Artificial Intelligence (2021)
10. Loewenstein, G., O'Donoghue, T., Rabin, M.: Projection bias in predicting future utility. Quarter. J. Econ. **118**(4), 1209–1248 (2003)
11. Loewenstein, G., Schkade, D.: Wouldn't it be nice? predicting future feelings. wellbeing: the foundations of hedonic psychology, pp. 85–105 (1999)
12. Muellbauer, J.: Habits, rationality and myopia in the life cycle consumption function. Annales d'Economie et de Statistique, pp. 47–70 (1988)
13. Read, D., Van Leeuwen, B.: Predicting hunger: the effects of appetite and delay on choice. Organ. Behav. Human Decision Proc. **76**(2), 189–205 (1998)
14. Smith, D., Loewenstein, G., Jepson, C., Jankovich, A., Feldman, H., Ubel, P.: Mispredicting and misremembering: patients with renal failure overestimate improvements in quality of life after a kidney transplant. Health Psychol. **27**(5), 653 (2008)
15. Tal, A., Wansink, B.: Fattening fasting: hungry grocery shoppers buy more calories, not more food. JAMA Internal Med. **173**(12), 1146–1148 (2013)
16. Tang, P., Teng, Y., Wang, Z., Xiao, S., Xu, Y.: Computational issues in time-inconsistent planning. In: Proceedings of the Thirty-First AAAI Conference on Artificial Intelligence, pp. 3665–3671 (2017)

Eliciting Social Knowledge for Creditworthiness Assessment

Mark York[1(✉)], Munther Dahleh[2], and David C. Parkes[1]

[1] Harvard John A. Paulson School of Engineering and Applied Sciences,
Allston, MA 02134, USA
markyork@g.harvard.edu, parkes@eecs.harvard.edu
[2] MIT Laboratory for Information and Decision Systems,
Cambridge, MA 02139, USA
dahleh@mit.edu

Abstract. Access to capital is a major constraint for economic growth in the developing world. Yet lenders often face high default rates due to their inability to distinguish creditworthy borrowers from the rest. In this paper, we propose two novel scoring mechanisms that incentivize community members to truthfully report their signal on the creditworthiness of others in their community. We first design a truncated asymmetric scoring rule for a setting where the lender has no liquidity constraints. We then derive a novel, strictly-proper Vickrey-Clarke-Groves (VCG) scoring mechanism for the liquidity-constrained setting. Whereas Chen et al. [7] give an impossibility result for an analogous setting in which sequential reports are made in the context of decision markets, we achieve a positive result through appeal to interim uncertainty about the reports of others. Additionally, linear belief aggregation methods integrate nicely with the VCG scoring mechanism that we develop.

Keywords: Information elicitation · Scoring rules · Mechanism design

1 Introduction

Access to capital has become the primary anti-poverty tool in development. Global microfinance grew from 13 million borrowers and $7 billion in loans in 1995 to 140 million borrowers and $129 billion in loans in 2019 [14,20]. A particular challenge with microfinance is that the unbanked have minimal credit history, creating an information asymmetry problem between lenders and borrowers.

Muhammad Yunus launched microfinance in 1976 with the Grameen bank. They lend to groups of people who are jointly-liable to repay the loan. This creates self-selection based on community information [10], but it also imposes significant cost on lenders and borrowers through bi-weekly meetings, the risk of default by fellow group members, and administration.

Another solution is the advent of data-analytics based lenders. These lenders typically give loans to individuals, and they leverage demographic or other information to select borrowers. Branch, operating in Kenya, requires users to own a

© Springer Nature Switzerland AG 2022
M. Feldman et al. (Eds.): WINE 2021, LNCS 13112, pp. 428–445, 2022.
https://doi.org/10.1007/978-3-030-94676-0_24

smartphone with their app installed and runs analytics on the calls, text messages, emails, and other usage data from the phone. Based on the performance of past borrowers, these companies determine how likely a new potential borrower is to repay. Loans are as small as five USD, and interest rates start at 18% monthly (199% APR) [5]. While this expands credit access, it excludes people who do not have smartphones and the interest rates are high. Another issue is that un-creditworthy borrowers learn which factors the algorithm considers, and they can modify their behavior to receive loans [4].

Fortunately, research shows that community members are knowledgeable about the creditworthiness of people in their community. Maitra et al. [16] deployed an agent-intermediated lending scheme in West Bengal, India through which they appointed agents to select borrowers and administer loans. These agents were compensated based on repayment rates, and the repayment rates were higher than those for group lending schemes in the same region. Hussam et al. [12] went one step further and deployed a community-recommendation scheme employing the *Robust Bayesian Truth Serum* (RBTS) to reward recommenders for giving reports that conform closely to those of their peers [28]. RBTS was found to partially nullify the incentives of recommenders to lie on behalf of family members. Of note, RBTS does not reward recommenders based on repayment outcomes.

We propose a new information elicitation system that incentivizes community members to report their true beliefs about the likelihood that others will repay a loan.[1] The goal is to support a lender who wants to lend to the borrowers who are most likely to repay. We design incentives in a way that recommenders strictly prefer to report their true beliefs about borrower creditworthiness.

The main results are the following:

1. The *truncated Winkler mechanism*, which is strictly proper for a lender without liquidity constraints and with a monotone non-decreasing belief aggregation rule, for a technical *grain-of-no-veto* condition on beliefs. The mechanism is not incentive compatible for a liquidity-constrained borrower.
2. The *Vickrey-Clarke-Groves (VCG) scoring mechanism*, which is strictly proper for a lender with or without liquidity constraints and with a weighted-linear belief aggregation rule, for beliefs with full support and a requirement that the weight on any single recommender is not too extreme.
3. The VCG scoring mechanism also aligns incentives with recommenders wanting to receive larger weights in the aggregator, and thus long-run incentives to provide higher-quality predictions.

In regard to the truncated Winkler mechanism, we use asymmetric scoring rules so that the minimum expected score is associated with the lender's threshold on minimum probability of repayment at which making a loan is profitable. We also develop the *grain-of-no-veto* condition, which provides strict incentives, by reasoning about the interim utility and uncertainty faced by a recommender.

---

[1] An initial deployment of the scheme, conducted under Harvard University's IRB, is underway in Uganda with 100 agricultural borrowers, thanks to a partnership with Makere University.

The application of the VCG mechanism to this context is novel and non-standard. In particular, we use outcome-contingent payments to construct the *valuation functions* of recommenders for different loan outcomes; in effect, this gives a recommender a valuation function for making a loan to a borrower that is proportional to their belief as to the repayment probability of the borrower. By folding the typical VCG-style payments on top, we take a constant (trivially proper but not strictly proper) scoring rule and generate an elicitation mechanism that is strictly proper. Moreover, the allocation rule of the mechanism corresponds to a belief aggregation model, and can embed weights assigned to recommenders in an incentive-compatible way. This use of linear-weighted aggregators corresponds naturally to well-studied belief aggregation systems [6, 23].

While our work is inspired by lending, the two mechanisms that we introduce are broadly applicable to decision settings under information asymmetries and where the decision maker can elicit information from different parties; e.g., employee screening, tenant screening, insurance underwriting, contractor selection, and service provider ratings.

1.1   Related Work

One way to formulate the problem of gathering community lending recommendations is as a *peer prediction* problem, i.e., as a problem of information elicitation without verification. The approach in peer prediction is to leverage correlation and mutual-information structure between reports to promote incentive alignment around true reports. A number of peer prediction mechanisms have been proposed, each requiring varying levels of knowledge on the part of the designer, on the kinds of reports, and on the task [1, 13, 15, 17, 21, 22, 24, 25, 27, 28].

While peer prediction has been used for belief elicitation for microfinance in [12], this is more naturally a problem of information elicitation with verification, where a lender will observe whether or not a borrower makes a repayment or defaults in the future, making this setting well-suited for the methods of scoring rules, prediction markets, and decision markets. *Scoring rules* are methods to elicit beliefs about uncertain future events where the outcome will be later observed [9]. Agents make reports are compensated for their level of accuracy after the event. A scoring rule is *strictly proper* if agents uniquely maximize their expected compensation by reporting their true belief. The *logarithmic scoring rule* and *Brier* or *quadratic scoring rules* are symmetric, in the sense that the expected score when truthfully reporting is minimized when the true belief $p = 0.5$ and symmetric about that point. The class of *Winkler scoring rules* [26], allow designers to set the minimum score point at any arbitrary location $c \in (0, 1)$.

Prediction markets can be used in a way that combines scoring rules with sequential elicitation from multiple participants, with the current market price reflecting the aggregate belief of the population about the outcome of an uncertain event [8]. In a prediction market with an automated market maker, for example the logarithmic market-scoring rule, one agent's report is in effect scored relative to the preceding report [11].

In our setting, reports also affect whether an event is observed, where reports determine who gets a loan. This relates to the *decision market* framework, where a principal makes a decision based on market prices. This creates new incentive challenges, and Chen et al. [7] provide a characterization of strict properness that requires randomization over decisions. This may not be credible behavior in real-world settings where the stakes are high. As we discuss in Sect. 4, we provide a counterpoint to this requirement, by considering *interim* analysis, when a recommender knows their own belief but is uncertain about the reports of others. This interim uncertainty enables strict incentive alignment for a deterministic decision rule that lends to the borrowers who are most likely not to default.

Zermeño [30] proposed a piecewise mechanism that can be viewed as a special case of Theorem 1 for the setting of a single recommender, though he focused on incentivizing effort in contrast to our focus on eliciting existing knowledge. VCG concepts have been used together with scoring rules, but not in the way described here [19]. Whereas we assume the set of recommenders is disjoint from the set of potential borrowers, Alon et al. [3] focus on the incentive issues that arise when this is not the case.

2 Preliminaries

A lender has a set of candidate *borrowers* $M = \{1, \ldots, m\}$ and recruits a set $N = \{1, \ldots, n\}$ of *recommenders* who know the candidate borrowers personally. Each recommender i has a subjective belief $p_{iq} \in [0, 1]$ of the likelihood with which candidate borrower q will repay a loan. We write $p_i = (p_{i1}, \ldots, p_{im})$, $p_q = (p_{1q}, \ldots, p_{nq})$, and $p = (p_1, \ldots, p_n)$. We also refer to belief p_i as the *type* of the recommender. We let $\mathcal{D}$ denote a *prior* on beliefs, such that $p \sim \mathcal{D}$. We write $p_{-i} = (p_1, \ldots, p_{i-1}, p_{i+1}, \ldots, p_n)$, and write $p_{-i} \sim \mathcal{D}_{-i}$, marginalizing out over recommender i (beliefs p_i may be correlated between recommenders). We assume $\mathcal{D}$ and $\mathcal{D}_{-i}$ are common knowledge.

Recommender i makes a *report* $\hat{p}_{iq} \in [0, 1]$ to the lender (principal), and we allow $\hat{p}_{iq} \neq p_{iq}$. We use $\hat{p}_i$ to denote the profile of all belief reports of recommender i. In our mechanisms, recommenders make reports independently with no knowledge of other recommenders' reports. The *repayment outcome* for a borrower q who receives a loan is a binary variable, $o_q \in \{0, 1\}$, with 1 representing repayment and 0 representing default. We sometimes write $o = \{o_1, \ldots, o_m\}$.

The lender makes a decision about which borrowers will receive a loan. We assume that the lender has a *profit threshold* $c \in [0, 1]$, such that the lender makes profit when making a loan where the repayment probability is c or higher. The lender forms a *belief* about the repayment probability of a borrower $q \in M$ with an aggregation function $B_q(p_q)$, which represents the lender's belief where $p_q = (p_{1q}, \ldots, p_{nq})$. We assume this belief is weakly monotone increasing in p_{iq}, for each i and q. We also sometimes work with a linear aggregator, $B_q(p_q) = \sum_{i \in N} w_i p_{iq}$, with *weight* $w_i > 0$ on recommender i and $\sum_{i \in N} w_i = 1$. The lender also has a *liquidity constraint* $K \leq m$, and if $K < m$ then can only make loans to a limited number of borrowers. We assume a uniform loan size and uniform interest rate; using ratings to both make lending decisions and optimize loan size and interest rates is important future work.

After observing repayment outcomes, the mechanism compensates recommenders using a scoring rule (potentially negative) of $s_{iq}(\hat{p}_q, o_q)$, where s_{iq} maps the report $\hat{p}_q$ and outcome o_q to reward. A *proper scoring rule* is one in which the expected score from a truthful report is at least as great as the expected score from any non-truthful report, i.e.,

$$\mathbb{E}_{o_q \sim p_{iq}}[s_{iq}(p_{iq}, \hat{p}_{-iq}, o_q)] \geq \mathbb{E}_{o \sim p_{iq}}[s_{iq}(\hat{p}_{iq}, \hat{p}_{-iq}, o_q)] \; ; \; \forall \hat{p} \neq p \tag{1}$$

A *strictly proper scoring rule* replaces this inequality with a strict inequality.

Definition 1 (Elicitation mechanism). *We design an* elicitation mechanism $\mathcal{M} = (x, t, s)$:

1. *Elicit belief reports* $\hat{p} = (\hat{p}_1, \ldots, \hat{p}_n)$ *from recommenders*
2. *Determine the set of borrowers,* $x(\hat{p}) \in \{0, 1\}^m$, *that will receive a loan, such that* $\sum_{q \in M} x_q(\hat{p}) \leq K$, *and define two-part payments:*
   (a) *A* fixed payment $t_i(\hat{p}) \in \mathbb{R}$ **made by each** *recommender* $i \in N$
   (b) *An outcome-contingent payment* $s_{iq}(\hat{p}_q, o_q) \in \mathbb{R}$ **made to each** *recommender* i *for each borrower* $q \in x(\hat{p})$, *i.e., for each borrower for which* $x_q(\hat{p}) = 1$.

Given reports $\hat{p}$ and outcome profile $o = (o_1, \ldots, o_m)$, the *realized utility* to recommender i is

$$u_i(\hat{p}_i, \hat{p}_{-i}, o) = \sum_{q \in x(\hat{p})} s_{iq}(\hat{p}_q, o_q) - t_i(\hat{p}). \tag{2}$$

Here, $\hat{p}_{-i} = (\hat{p}_1, \ldots, \hat{p}_{i-1}, \hat{p}_{i+1}, \ldots, \hat{p}_n)$. The *utility* for recommender i with belief p_i is

$$U_i(p_i, \hat{p}_i, \hat{p}_{-i}) = \sum_{q \in x(\hat{p})} (p_{iq} s_{iq}(\hat{p}_q, 1) + (1 - p_{iq}) s_{iq}(\hat{p}_q, 0)) - t_i(\hat{p})$$

$$= \sum_{q \in x(\hat{p})} \mathbb{E}_{o_q \sim p_{iq}}[s_{iq}(\hat{p}_q, o_q)] - t_i(\hat{p}) \tag{3}$$

This quantity is ex post with respect to the reports of others, and takes an expectation over borrower outcomes with respect to the beliefs of recommender i. It is useful to interpret the outcome-contingent payment to the recommender as inducing a term that plays a similar role as an agent's valuation in mechanism design, where $\mathbb{E}_{o_q \sim p_{iq}}[s_{iq}(\hat{p}_q, o_q)]$ is the recommender's "value" for the lender's decision to lend to borrower q. There are a number of possible desiderata for an elicitation mechanism.

– *Allocative efficiency,* so that the mechanism allocates to the borrowers with the maximum probability of repayment amongst those better than the profit threshold c. For lender belief $B_q(\hat{p}_q)$, this requires that

$$x(\hat{p}) \in \underset{a \in \{0,1\}^m}{\arg\max} \sum_{q \in M : B_q(\hat{p}_q) > c} B_q(\hat{p}_q) \times a_q$$

$$\text{s.t.} \sum_{q \in M} a_q \leq K \tag{4}$$

- *Weak ex post incentive compatibility* (weak EPIC), so that each recommender's utility is weakly maximized by reporting truthfully, regardless of the reports of others (*ex post proper* in the language of scoring rules); i.e.,

$$U_i(p_i, p_i, \hat{p}_{-i}) \geq U_i(p_i, \hat{p}_i, \hat{p}_{-i}) \; ; \forall i, \; \forall p_i, \; \forall \hat{p}_i, \; \forall \hat{p}_{-i} \tag{5}$$

- *Strict ex post incentive compatibility* (strict EPIC), so that each recommender's utility is strictly maximized by reporting truthfully, regardless of the reports of others (*ex post strict proper* in the language of scoring rules); i.e.,

$$U_i(p_i, p_i, \hat{p}_{-i}) > U_i(p_i, \hat{p}_i, \hat{p}_{-i}) \; ; \forall i, \; \forall p_i, \; \forall \hat{p}_i \neq p_i, \; \forall \hat{p}_{-i} \tag{6}$$

- *Strict interim incentive compatibility* (strict IIC), so that each recommender's interim utility, considering the distribution on reports of others, is strictly maximized by reporting truthfully (*strict properness* in the language of scoring rules); i.e.,

$$\mathbb{E}_{p_{-i} \sim \mathcal{D}_{-i}}[U_i(p_i, p_i, p_{-i})] > \mathbb{E}_{p_{-i} \sim \mathcal{D}_{-i}}[U_i(p_i, \hat{p}_i, p_{-i})] \; ; \forall i, \; \forall p_i, \; \forall \hat{p}_i \neq p_i \tag{7}$$

- *Ex post Individually Rational* (IR), so that all recommenders that make a truthful report have non-negative expected utility once loans are allocated, but before repayment outcomes are observed; i.e.,

$$U_i(p_i, p_i, \hat{p}_{-i}) \geq 0 \; ; \forall i, \; \forall p_i, \; \forall \hat{p}_{-i}. \tag{8}$$

- *Strong ex post IR*,[2] so that all recommenders that make a truthful report have a non-negative realized utility after repayment outcomes are observed; i.e.,

$$u_i(p_i, \hat{p}_{-i}, o) \geq 0 \; ; \forall i, \; \forall p_i, \; \forall \hat{p}_{-i}, \; \forall o. \tag{9}$$

3 Unconstrained-Liquidity Setting: Truncated Winkler Mechanism

In this section, we design incentive-compatible elicitation mechanisms for the unconstrained-liquidity setting. We truncate strictly-proper scoring rules to provide a constant score when reports are too low to make a loan; this constant

---

[2] The first concept of IR is ex post with respect to the reports of others. For this reason, we adopt the phrasing *strong ex post IR* here, since this holds once outcomes are observed.

score is the expected score for truthful reports just above the lending threshold. We use the *Winkler scoring rule transformation* to ensure a convex expected score (or generalized entropy) for any valid threshold. We then show that this mechanism is weak EPIC (proper) in all cases and strict IIC (strictly proper) when no one recommender has the ability to unilaterally deny a loan (the *Grain-of-no-veto* condition). Finally, we show how the truncated Winkler mechanism fails to be weak EPIC in the liquidity-constrained setting. Relative to the otherwise more general VCG scoring mechanism, the truncated Winkler mechanism accommodate non-linear belief aggregation rules.

3.1   The Truncated Winkler Mechanism

We lend to borrower q if and only if $B_q(\hat{p}_q) > c$. Since $B_q(\hat{p}_q)$ is monotonic non-decreasing, we can define a *marginal threshold* $c_{iq} \in [0,1]$ for each borrower-recommender combination above which recommender i must report for borrower q to receive a loan. This threshold depends on the reports of others and is defined while holding others' reports constant. Formally, let c_{iq} denote the *minimum report by recommender i on borrower q such that q receives a loan*, i.e.,

$$c_{iq} \triangleq \inf_{p' \in [0,1]} p' \text{ s.t. } B_q(\hat{p}_{iq} = p', \hat{p}_{-iq}) > c, \tag{10}$$

where $\hat{p}_{-iq} = (\hat{p}_{1q}, \ldots, \hat{p}_{i-1,q}, \hat{p}_{i+1,q}, \ldots, \hat{p}_{nq})$. We leave the dependence of c_{iq} on the reports of others implicit in the notation. By weak monotonicity of the belief aggregator, for any $\hat{p}_{iq} > c_{iq}$ we have that borrower q receives a loan. With a linear aggregator $B_q(\hat{p}_q) = \sum_{i \in N} w_i \hat{p}_{iq}$, for example, we have

$$c_{iq} = \min(1, \max(0, \frac{1}{w_i}(c - \sum_{j \neq i} w_j \hat{p}_{jq}))). \tag{11}$$

We truncate each recommender's payout when $\hat{p}_{iq} < c_{iq}$, to make them indifferent about the lending decision when their belief p_{iq} equals their threshold c_{iq}. It is necessary that the expected utility function when reporting truthfully $U_i(p_i, p_i, \hat{p}_{-i})$ is convex for weak EPIC [9]. Common strictly proper scoring rules for binary outcomes have their minimum U_i for truthful reports when $p = .5$, so we use the Winkler scoring rule to ensure convexity when even when $c_{iq} < .5$ (see Fig. 1).

Let s be any symmetric proper scoring rule, and consider the marginal lending threshold $c_{iq} \in (0,1)$. The *Winkler scoring rule* s^W [26] is defined as

$$s_{iq}^W(\hat{p}_q, o_q) = \frac{s(\hat{p}_{iq}, o_q) - s(c_{iq}, o_q)}{T(c_{iq}, \hat{p}_{iq})}, \quad T(c_{iq}, \hat{p}_{iq}) = \begin{cases} s(0,0) - s(c_{iq}, 0) & \text{if } \hat{p}_{iq} \leq c_{iq} \\ s(1,1) - s(c_{iq}, 1) & \text{otherwise.} \end{cases} \tag{12}$$

The Winkler scoring rule is (strictly) proper when s is (strictly) proper.

Definition 2 (Truncated Winkler elicitation mechanism). *The* Truncated Winkler elicitation mechanism *with unconstrained lender liquidity, lender profit threshold c, and monotone belief aggregation B_q, is defined as following:*

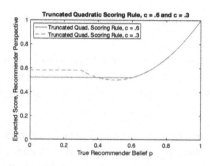 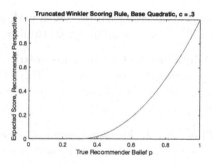

Fig. 1. Left: Expected utility with truthful reporting when truncating the familiar quadratic scoring rule [9] with threshold values $c_{iq} = 0.6$ and $c_{iq} = 0.3$; note convexity when $c_{iq} \geq .5$ and the lack of convexity when $c_{iq} < .5$. **Right**: Expected utility with truthful reporting for the truncated Winkler elicitation mechanism and lender threshold $c_{iq} = 0.3$. Note that we omit subscripts iq.

– *Allocation: for each borrower q, $x_q(\hat{p}) = 1$ if $B_q(\hat{p}_q) > c$ and $x_q(\hat{p}) = 0$ otherwise*
– *Payment:*
  • *Immediate payment: zero*
  • *Outcome-contingent payment: for each borrower that receives a loan, $s_{iq}(\hat{p}_q, o_q) = s_{iq}^W(\hat{p}_q, o_q)$ as per Eqs. (10) and (12). There is no outcome-contingent payment when $\hat{p}_{iq} \leq c_{iq}$.*

Theorem 1. *For unconstrained liquidity and a weak monotone-increasing aggregation function B_q, the truncated Winkler mechanism is ex post individually rational, and weak EPIC (proper) for a report of recommender i on borrower q for marginal lending threshold c_{iq} when recommender belief $p_{iq} \leq c_{iq}$ and strict EPIC (strictly proper) when $p_{iq} > c_{iq}$.*

Proof. The expected score from the Winkler rule under truthful reporting $U_i(p_i, p_i, \hat{p}_{-i})$ is strictly positive when $p_{iq} > c_{iq}$ [26]. The fact that $B_q(\hat{p}_q)$ is weak monotone-increasing ensures that for $\hat{p}_{iq} > c_{iq}$ we have $B_q(\hat{p}_q) > c$. Since reports of $\hat{p}_{iq} \leq c_{iq}$ yield payment of 0, risk-neutral agents will always prefer $\hat{p}_{iq} > c_{iq}$ when $p_{iq} > c_{iq}$ (since they prefer a positive expected value to zero expected value). Given that s_{iq}^W is strict EPIC, the full mechanism is also strict EPIC (strictly proper) when $p_{iq} > c_{iq}$. When $p_{iq} \leq c_{iq}$, the convexity of U_i and the fact that $U_i(c_{iq}, c_{iq}, \hat{p}_{-i}) = 0$ guarantees that a recommender weakly maximizes its expected payment from truthful reporting when $p_{iq} \leq c_{iq}$. This gives weak EPIC for all beliefs. Individual rationality is immediate, since $U_i(p_i, p_i, \hat{p}_{-i}) \geq 0$. □

3.2   Strict IIC with the Grain-of-no-veto Condition

We next extend Theorem 1 to achieve strict IIC under certain conditions. For this, define *no veto* for i and q at threshold c to mean $B_q(0, p_{-i,q}) > c$, i.e., the beliefs of others such that even with $p_{iq} = 0$ from recommender i the borrower q will receive a loan under truthful reports.

Definition 3 (Grain-of-no-veto). *The distribution on beliefs satisfies a* grain-of-no-veto *at c when $\mathbb{P}[p_{-i,q} \sim \mathcal{D}_{-i} : B_q(0, p_{-i,q}) > c] > 0$, for all recommenders i, all borrowers q, i.e., no veto is satisfied with positive measure of the type distribution.*

Theorem 2. *For unconstrained liquidity, multiple recommenders $(n > 1)$, and a weak monotone-increasing aggregation function B_q, the truncated Winkler mechanism is ex post IR, and also strictly IIC (strict proper) when the distribution on beliefs satisfies grain-of-no-veto.*

Proof. Per Theorem 1, when $p_{iq} > c_{iq}$, the Winkler scoring mechanism is strictly ex post proper with regard to i's report on q. Otherwise, it is weakly ex post proper. By grain-of-no-veto, for any belief of i on q, p_{iq}, there is non-zero probability that $p_{iq} > c_{iq}$. This implies $\mathbb{E}_{p_{-i} \sim \mathcal{D}_{-i}}[U_i(p_i, p_i, p_{-i})] > \mathbb{E}_{p_{-i} \sim \mathcal{D}_{-i}}[U_i(p_i, \hat{p}_i, p_{-i})]$, for all $\hat{p}_i \neq p_i$. IR follows immediately from Theorem 1. $\qquad\square$

For a weighted linear aggregator, grain-of-no-veto requires (for all i, all q) that $\mathbb{P}[p_{-i,q} \sim \mathcal{D}_{-i} : \sum_{j \neq i} w_j p_{jq} > c] > 0$. That is, there is non-zero probability that the weighted sum over reports of all but one recommender is large enough. We can also state a corollary for the case that beliefs have full support on $[0, 1]$.

Corollary 1. *For unconstrained liquidity and multiple recommenders, and a belief distribution with full support, the truncated Winkler mechanism with the weighted linear aggregator is ex post IR, and also strictly IIC (strict proper) for lender profit threshold c when $\max_i[w_i] < 1 - c$, which requires $c < (n-1)/n$.*

Proof. Since $\max_i[w_i] < 1 - c$, then $\sum_{j \neq i} w_j > c$, for all $j \in N$. From this, we have $\mathbb{P}[p_{-i,q} \sim \mathcal{D}_{-i} : \sum_{j \neq i} w_j p_{jq} > c] > 0$, by full support, and thus grain-of-no-veto. Moreover, since $\max_i[w_i] \geq 1/n$, we need $1/n < 1 - c$ and thus $c < (n-1)/n$. $\qquad\square$

As the threshold increases, the system needs more recommenders to provide strict properness and the lender becomes less able to put a very large weight on any single recommender.

3.3 Failure of Truncated Winkler with Constrained Liquidity

When there are a limited number of loans that can be made, lending decisions are no longer independent between borrowers. This opens up new manipulations, where recommenders can prioritize a higher-payoff borrower over a lower one to improve their expected utility.

Theorem 3. *In the constrained-liquidity setting with more than one recommender, the Truncated Winkler mechanism is not weakly EPIC (ex post proper).*

Proof. Consider 3 recommenders, 2 borrowers, budget $K = 1$, profit threshold $c = 0.5$, s_{iq}^W based on the logarithmic scoring rule, an unweighted aggregator, and recommenders with beliefs in Table 1. Under truthful reports, the lender's belief will be approximately 0.57 and 0.55, for borrowers 1 and 2 respectively, and borrower 1 will be allocated. The expected utility of recommender 2 will be $p_{2,1}$ · $s_{2,1}^W(p_1, 1) + (1 - p_{2,1})s_{2,1}^W(p_1, 0) = 0.4\frac{\ln(0.4) - \ln(0.2)}{-\ln(0.2)} + (1 - 0.4)\frac{\ln(1-0.4) - \ln(1-0.2)}{-\ln(0.2)} = 0.4 * 0.43 + 0.6 * (-0.18) = 0.07$. If, recommender 2 misreports $\hat{p}_{2,1} = 0$, then the lender's beliefs will be 0.43 and 0.55 for borrowers 1 and 2, and borrower 2 will be allocated. In this case, recommender 2's expected utility will be $p_{2,2} \cdot s_{2,1}^W(p_2, 1) + (1 - p_{2,2})s_{2,1}^W(p_2, 0) = 0.85\frac{\ln(0.85) - \ln(0.7)}{-\ln(0.7)} + (1 - 0.85)\frac{\ln(1-0.85) - \ln(1-0.7)}{-\ln(0.7)} = 0.85 * 0.54 + 0.15 * (-1.94) = 0.17$. □

Table 1. Perverse incentives with Truncated Winkler and constrained liquidity.

	Recommender 1	Recommender 2	Recommender 3
Belief on Borrower 1	0.7	0.4	0.6
Belief on Borrower 2	0.4	0.85	0.4
Borrower 1 threshold c_{i1}	0.5	0.2	0.4
Borrower 2 threshold c_{i2}	0.25	0.7	0.25
Expected utility, Honest	0.12	0.07	0.09
Expected utility, Recommender 2 Misreport	0.04	0.17	0.04

4   The VCG Scoring Mechanism

In this section, we introduce a mechanism that provides strict properness for both the unconstrained and constrained-liquidity cases, i.e., handling $K < m$ in addition to $K = m$. The mechanism combines scoring rules and the VCG mechanism to achieve strict properness together with a linear belief aggregator. We define the outcome-contingent payment to be the non-standard *constant scoring rule*, i.e.,

$$s_{iq}^{VCG}(\hat{p}_q, o_q) = \begin{cases} w_i, \text{if } o_q = 1 \\ 0, \text{otherwise.} \end{cases} \tag{13}$$

This scoring rule is trivially proper, but not strictly proper (the payment does not depend on the report). This will provide strictly proper incentives when embedded within the framework of the VCG mechanism. In the context of the VCG scoring mechanism, it is useful to define the *value function* of a recommender for loan decisions $a \in \{0, 1\}^m$ as

$$v_i(a) \triangleq \sum_{q \in M} a_q \times \mathbb{E}_{o_q \sim p_{iq}}[s_{iq}^{VCG}(p_q, o_q)] = \sum_{q \in M} a_q w_i p_{iq}. \tag{14}$$

Similarly, we define the *reported value function* as

$$\hat{v}_i(a) \triangleq \sum_{q \in M} a_q \times \mathbb{E}_{o_q \sim \hat{p}_{iq}}[s_{iq}^{VCG}(\hat{p}_q, o_q)] = \sum_{q \in M} a_q w_i \hat{p}_{iq}. \qquad (15)$$

These play the typical role of valuations and reported valuations in the analysis of the incentive properties of a VCG mechanism (we could also use other monotonically-increasing scoring rules in the value function, but then we would sometimes end up allocating to borrowers who do not have the highest weighted-average report). This mechanism is not the same as weighted VCG because the weight w_i *directly impacts agents' value functions.* In weighted VCG, the weights do not affect the intrinsic value of an allocation to the agents.[3]

By defining value in this way, the allocation of loans that maximizes the total, weighted reported value is also the allocation that lends to the borrowers for which the aggregate belief of repayment is largest. The value-maximizing allocation given reports $\hat{p}$ is

$$\max_{a \in \{0,1\}^m} \sum_{q \in M} a_q \left(\sum_{i \in N} w_i \hat{p}_{iq} \right)$$

$$\text{s.t.} \quad \sum_{q \in M} a_q \leq K. \qquad (16)$$

To introduce a profit threshold $c > 0$ for the lender, we can add K imaginary *reserve borrowers* to the system and a *reserve recommender* with weight 1 who reports c for a loan decision to each of these borrowers and 0 for other borrowers (a weight of 1 makes this equivalent to all actual recommenders making a report of c for each of the reserve borrowers). We leave the weights to other recommenders unchanged. Adopting R to represent the *reserve borrowers*, the modified allocation rule is:

$$x^{VCG}(\hat{p}) \in \arg\max_{a \in \{0,1\}^m} \left[\sum_{q \in M} a_q \left(\sum_{i \in N} w_i \hat{p}_{iq} \right) + \sum_{q \in R} a_q \cdot c \right]$$

$$\text{s.t.} \quad \sum_{q \in M \cup R} a_q \leq K. \qquad (17)$$

Going forward we will incorporate the reserve recommender into the set of N recommenders and the K reserve borrowers into the set M of borrowers.

Definition 4 (VCG scoring mechanism). *The* VCG scoring mechanism *with possibly constrained lender liquidity, lender profit threshold c, and linear weighted belief aggregation B_q with weights $w = (w_1, \ldots, w_n)$, is defined as following:*

– *Allocation: adopt $x^{VCG}(\hat{p})$*

---

[3] In particular, our payment function in Eq. 18 does not involve a $1/w_i$ term as in weighted VCG mechanisms [18]. We discuss this further in the extended version [29].

– *Payment (no payments are made by or collected from the reserve recommender):*
  • *Immediate payment:*

$$t_i^{VCG}(\hat{p}_i, \hat{p}_{-i}) = \sum_{j \neq i} \hat{v}_j(x^{-i}(\hat{p}_{-i})) - \sum_{j \neq i} \hat{v}_j(x(\hat{p}_i, \hat{p}_{-i})), \qquad (18)$$

  where x^{-i} is the allocation decision that would be made without i present, i.e., ignoring the reports from recommender i.
  • *Outcome-contingent payment: for each borrower that receives a loan, $s_{iq}^{VCG}(\hat{p}_q, o_q) = w_i$ if $o_q = 1$ and 0 otherwise.*

The realized utility of recommender i after the repayment outcomes are known by the lender is

$$u_i(\hat{p}_i, \hat{p}_{-i}, o) \triangleq \sum_{q \in x^{VCG}(\hat{p})} w_i \cdot o_q - t_i^{VCG}(\hat{p}_i, \hat{p}_{-i}). \qquad (19)$$

4.1 Strict Properness of the VCG Scoring Mechanism

Theorem 4. *The VCG Scoring Mechanism is efficient, satisfies weak EPIC (ex post proper), and is ex post individually rational.*

Once valuation functions are set-up to correspond to aggregate belief reports, this proof follows the standard recipe for the incentive compatible and IR properties of a VCG mechanism (see the extended version [29]). However, this only gives weak EPIC and we want strict IIC (strict properness), so that it is a unique best response of a recommender to report their true beliefs.

For our first of two main theorems, we define an *equal-shift misreport* as a misreport $\hat{p}_i \neq p_i$ for which $p_{iq} - p_{iq'} = \hat{p}_{iq} - \hat{p}_{iq'}$ for every q, every q'. We say a mechanism is *strictly proper up to equal-shift misreports* if truthful reporting is a unique best response, maximizing interim utility except for possible tie-breaking amongst equal-shift misreports.

Theorem 5. *For constrained liquidity ($K < m$), two or more borrowers, three or more recommenders, and a belief distribution with full support, the VCG Scoring mechanism without a lender profit threshold (i.e., $c = 0$) is strict IIC (strictly proper) up to equal-shift misreports when $\max_i[w_i] < 1/2$.*

Proof. Three or more recommenders are required for $\max_{i'}[w_{i'}] < 1/2$. Two or more borrowers allows for constrained liquidity. We consider recommender i, belief p_i, any $\hat{p}_i \neq p_i$ that is not an equal-shift misreport, and establish a non-zero measure on the beliefs p_{-i} of others such that the allocation changes in a way that reduces the total value (i.e., not selecting the borrowers with the top K aggregate belief of repayment). Since VCG is weakly EPIC (Theorem 4), this establishes strict IIC up to equal-shift misreports.

For any q, let $B(q)$ denote the aggregate belief on q at p_i and $\hat{B}(q)$ at report $\hat{p}_i$. If p_{-i} satisfies $p_{jq} = \frac{1/2 - w_i p_{iq}}{\sum_{j' \neq i} w_{j'}} = p_q^*$, $\forall j \neq i$, then $B(q) = 1/2$. This belief p_{-i} is feasible by full support, and since for $p_{iq} = 0$ we have $p_q^* = (1/2)/\sum_{j' \neq i} w_{j'} < 1$ since $\sum_{j' \neq i} > 1/2$ from $w_i < 1/2$. For $p_{iq} = 1$ we have $p_q^* = (1/2 - w_i)/\sum_{j' \neq i} w_{j'} > 0$ since $w_i < 1/2$.

For a non equal-shift misreport, there are borrowers q and q', such that $p_{iq} - p_{iq'} = \hat{p}_{iq} - \hat{p}_{iq'} + \epsilon$, for $\epsilon > 0$; i.e., with the relatively disadvantaged borrower labeled q. Consider a profile p_{-i} that satisfies the following properties:

1. K borrowers, including borrower q, are allocated:
   - For $q'' \neq q$, set $p_{jq''} \in (p_{q''}^*, 1]$, for $j \neq i$, such that $B(q'') > 1/2$, where this belief of others is feasible since $p_{q''}^* < 1$.
   - For borrower q, set $p_{jq} \in (p_q^*, \min(1, p_q^* + \frac{1}{\sum_{j' \neq i} w_{j'}} \frac{w_i \epsilon}{2}))$, so that $B(q) \in (\frac{1}{2}, \frac{1}{2} + \frac{w_i \epsilon}{2})$, where this belief of others is feasible since $p_q^* < 1$.
2. $K - m$ borrowers, including borrower q', are not allocated.
   - For $q'' \neq q'$, set $p_{jq''} \in [0, p_{q''}^*)$, for $j \neq i$, such that $B(q'') < 1/2$, where this belief of others is feasible since $p_{q''}^* > 0$.
   - For borrower q', set $p_{jq'} \in (\max(0, p_{q'}^* - \frac{1}{\sum_{j' \neq i} w_{j'}} \frac{w_i \epsilon}{2}), p_{q'}^*)$, so that $B(q') \in (\frac{1}{2} - \frac{w_i \epsilon}{2}, \frac{1}{2})$, where this belief of others is feasible since $p_{q'}^* > 0$.

There is a non-zero measure on beliefs p_{-i} satisfying these properties by the full support assumption. For any such p_{-i}, at misreport $\hat{p}_i$ we have $\hat{B}(q') > \hat{B}(q)$, since $\hat{B}(q') - B(q') = \hat{B}(q) - B(q) + w_i \epsilon$ and $B(q) - B(q') < w_i \epsilon$. By the monotonicity of the VCG allocation rule, this implies one of the following at this misreport:

1. Borrower q' but not q is allocated, which is an outcome with lower total value since $B(q') < B(q)$.
2. Neither q nor q' are allocated, which is an outcome with lower total value since $B(q) > 1/2$ and only $K - 1$ other borrowers q'' have true aggregate belief $B(q'') > 1/2$.
3. Both q and q' are allocated, which is an outcome with lower total value since $B(q') < 1/2$ while K borrowers q'' (including q) have aggregate belief $B(q'') > 1/2$. □

Moreover, an equal-shift misreport does not change the loan allocation, as it does not advantage any one borrower over another. Thus, the allocation remains efficient, payments are unaffected, and the outcome of the mechanism is substantially equivalent to that under truthful reporting.

Theorem 6. *For possibly constrained liquidity ($K \leq m$), one or more borrowers, three or more recommenders, and a belief distribution with full support, the VCG Scoring mechanism with a lender profit threshold c, with $0 < c < 1$, is strict IIC (strictly proper) when $\max_i[w_i] < \min(1 - c, c)$ (which requires $n > 1/\min(1 - c, c)$ recommenders).*

Proof. We need three or more recommenders because $\min(1 - c, c) \leq 1/2$, and thus $n > 1/(1/2) = 2$. We consider recommender i, belief p_i, any $\hat{p}_i \neq p_i$, and establish a non-zero measure on the beliefs p_{-i} of others such that the allocation changes in a way that reduces the total value (i.e., not selecting the top borrowers amongst those with aggregate belief at least c). Since VCG is weakly EPIC (Theorem 4), this establishes strict IIC up to equal-shift misreports.

For any q, let $B(q)$ denote the aggregate belief on q at p_i and $\hat{B}(q)$ at report $\hat{p}_i$. If p_{-i} satisfies $p_{jq} = \frac{c - w_i p_{iq}}{\sum_{j' \neq i} w_{j'}} = p_q^*$, $\forall j \neq i$, then $B(q) = c$. This belief p_{-i} is feasible by full support, and since for $p_{iq} = 0$ we have $p_q^* = c / \sum_{j' \neq i} w_{j'} < 1$ since $\sum_{j' \neq i} w_{j'} > c$ from $w_i < 1 - c$. For $p_{iq} = 1$, $p_q^* = (c - w_i) / \sum_{j' \neq i} w_{j'} > 0$ since $w_i < c$.

For misreport $\hat{p}_i$, consider borrower q with $\hat{p}_{iq} \neq p_{iq}$.

(**Case 1:** $\hat{p}_{iq} < p_{iq}$) Let $\hat{p}_{iq} = p_{iq} - \epsilon$, some $\epsilon > 0$. Consider a profile p_{-i} that satisfies the following properties:

1. $B(q) \in (c, c + w_i \epsilon)$, by setting $p_{jq} \in (p_q^*, \min(1, p_q^* + \frac{1}{\sum_{j' \neq i} w_{j'}} w_i \epsilon))$, all $j \neq i$, where this belief of others is feasible since $p_q^* < 1$ and $c < 1$.
2. At least $m - K$ (≥ 0) other borrowers $q' \neq q$ have $B(q') < c$, by setting $p_{jq'} \in [0, p_{q'}^*)$, all $j \neq i$, where this belief of others is feasible since $p_{q'}^* > 0$.

There is a non-zero measure on beliefs p_{-i} satisfying these properties by the full support assumption. Given (1) and (2), at true beliefs we have borrower q allocated since $B(q) > c$ and at least $m - K$ others cannot be allocated, so q is in the top K of those with aggregate belief above the threshold c. For any such p_{-i}, at misreport $\hat{p}_i$ we have $\hat{B}(q) = B(q) - w_i \epsilon < c$, since $B(q) \in (c, c + w_i \epsilon)$ and $\hat{p}_{iq} = p_{iq} - \epsilon$. This implies that q is not allocated, resulting in an outcome with lower total value since q was in the top K and with true aggregate belief above the threshold.

(**Case 2:** $\hat{p}_{iq} > p_{iq}$) Let $\hat{p}_{iq} = p_{iq} + \epsilon$, some $\epsilon > 0$.
Consider a profile p_{-i} that satisfies the following properties:

1. $B(q) \in (c - w_i \epsilon, c)$, by setting $p_{jq} \in (\max(0, p_q^* - \frac{1}{\sum_{j' \neq i} w_{j'}} w_i \epsilon), p_q^*)$, all $j \neq i$, where this belief of others is feasible since $p_q^* > 0$ and $c > 0$.
2. At least $m - K$ (≥ 0) other borrowers $q' \neq q$ have $B(q') < c$, by setting $p_{jq'} \in [0, p_{q'}^*)$, all $j \neq i$, where this belief of others is feasible since $p_{q'}^* > 0$.

Given (1), at true beliefs borrower q is not allocated. At misreport $\hat{p}_i$, we have $\hat{B}(q) = B(q) + w_i \epsilon > c$, since $B(q) \in (c - w_i \epsilon, c)$ and $\hat{p}_{iq} = p_{iq} + \epsilon$. This implies one of the following at this misreport:

1. Borrower q is allocated, resulting in an outcome with lower total value since the true aggregate belief on q is below the threshold (that is, by causing q to be allocated, i displaces a reserve borrower q'' with $B(q'') = c$, and i must pay this difference to the system).
2. If q is not allocated, then since $\hat{B}(q) > c$ there must be K others allocated, by the definition of the VCG outcome rule. At least $m - K$ others have $B(q') < c$, and thus at most $(m - 1) - (m - K) = K - 1$ others have $B(q') \geq c$. This means that at least one other borrower with $B(q') < c$ is allocated, and the outcome has lower total value. $\square$

4.2   Strong Ex Post IR

We can also achieve strong ex post IR by ensuring that the immediate payment by each agent is weakly negative, and noting that the outcome-contingent payments to each recommender are weakly-positive. Define $tcomp_i(\hat{p}_{-i})$ as the worst-case immediate payment in VCG given reports of others. This quantity is independent of the recommender's own report. At the same time, we introduce a multiplier $\alpha > 0$ to the outcome-contingent payments, so that $s_{iq}^{aVCG}(\hat{p}_{iq}, o_q) = \alpha w_i$ if $o_q = 1$, and 0 otherwise. Neither change affects the incentive analysis. Modifying the definition of reported valuations accordingly, for example with $\hat{v}_i(a) = \sum_{q \in M} a_q \cdot \alpha w_i \cdot \hat{p}_{iq}$, we have

$$tcomp_i(\hat{p}_{-i}) = \max_{\hat{p}_i} \left(\sum_{j \neq i} \hat{v}_j(x_{-i}(\hat{p}_{-i})) - \sum_{j \neq i} \hat{v}_j(x^*(\hat{p}_i, \hat{p}_{-i})) \right). \tag{20}$$

We refer to this as the *rescaled VCG scoring mechanism*. By *worst-case deficit* we mean the worst-case, total payment made by the mechanism to the agents, considering both the immediate and outcome-contingent payments.

Theorem 7. *In the possibly constrained liquidity setting, and with multiple recommenders and multiple borrowers, there is some value of $\alpha_0 > 0$ such that for any $\alpha < \alpha_0$ the rescaled VCG scoring mechanism is strong ex-post IR and worst-case deficit at most $\epsilon > 0$.*

Proof. For strong ex post IR, this follows from the definition of $tcomp_i(\hat{p}_{-i})$ and outcome-contingent payments being non-negative. For the strict properness, this follows from the invariance of incentive analysis to scaling payments by any $\alpha > 0$ and that $tcomp_i(\hat{p}_{-i})$ is independent of recommender i's reports. The claim of deficit smaller than ϵ for any $\alpha < \alpha_0$, for some $\alpha_0 > 0$ follows from linearity, recognizing that α scales all payments. □

4.3   Incentive Alignment with Better Reporting Quality

The VCG scoring mechanism also aligns incentives with recommenders preferring to have larger weights in the aggregator, this providing long-run incentives for a recommender to improve its reporting quality and thus attain a higher weight over time in the aggregation rule.

Theorem 8. *Whatever the reports of others, for any recommender i, increasing the weight w_i to $w_i' > w_i$, fixing the weights of others, increases the utility $U_i(p_i, p_i, \hat{p}_{-i})$ to the recommender from truthful participation in the VCG scoring mechanism.*

Proof. The utility to recommender i is

$$U_i(p_i, p_i, \hat{p}_{-i}) = v_i(x^{VCG}(p)) + \sum_{j \neq i} \hat{v}_j(x^{VCG}(p_i, \hat{p}_{-i})) - \sum_{j \neq i} \hat{v}_j(x^{-i}(\hat{p}_{-i})),$$

where x^{-i} is the allocation decision that would be made in the VCG scoring mechanism without i. Let v_i and v_i' denote the recommender's valuation for weight w_i and w_i', respectively. The third term does not depend on its weight. Consider the first two terms, and let a and a' denote the allocation for w_i and $w_i' > w_i$, respectively. We have $v_i'(a') + \sum_{j \neq i} \hat{v}_j(a') \geq v_i'(a) + \sum_{j \neq i} \hat{v}_j(a) > v_i(a) + \sum_{j \neq i} \hat{v}_j(a)$. The first inequality holds trivially when $a' = a$, and if $a' \neq a$ then by the optimizing property of the VCG allocation rule. The second inequality holds since $v_i'(a) = \sum_q a_q w_i' p_{iq} > \sum_q a_q w_i p_{iq} = v_i(a)$, and since reported values of others are unchanged.   □

4.4   Relation to Chen et al.'s Impossibility Result

Theorem 2 [7] states that a decision market, which uses belief reports as reflected in market prices to make a decision, is strictly proper if and only if the decision is randomized and the distribution has full support. That is, for strict properness every decision must be taken with non-zero probability. The key difference in the present model is that agents report their beliefs simultaneously and without awareness of the reports of others. This creates interim uncertainty about the allocation, given the common prior $\mathcal{D}$ and the technical conditions stated in Theorems 2, 5 and 6. In contrast, the agents in Chen et al. [7] have certainty about the way belief aggregation will proceed (since they know current prices in the decision market). In effect, we achieve strict properness together with a deterministic decision rule by leveraging full *interim* support.

4.5   Linear Belief Aggregation

From Theorems 5 and 6, the VCG scoring mechanism is IIC when $max_{i \in N}[w_i]$ is sufficiently small. There are many linear aggregators in the literature that can be adjusted to satisfy this condition. See the extended version [29], where we describe a linear aggregator [6] that is based on reports from previous rounds and therefore does not affect current-round incentives.

5   Conclusion

Our formulation of the creditworthiness problem as a social-knowledge elicitation problem brings a novel source of information and a rich body of mechanism design literature to the task. We have developed a class of truncated, asymmetric scoring rules that are ex-post proper in the sufficient-liquidity case. We have also connected scoring rules and VCG-based mechanism design in a novel way, creating the *VCG scoring mechanism* through which we set agents' values via scoring rules. This mechanism is strictly IIC (and thus strictly proper) with sufficiently-distributed weights in both liquidity-constrained and liquidity-unconstrained settings, and both with or without a lender profit threshold. Given impossibility results in the adjacent setting of decision markets [7], these results expand the range of settings in which information can be elicited and paid for based on outcomes, in this case by leveraging agents' interim uncertainty. We have also connected these mechanisms with the belief aggregation literature, allowing us to retain incentive compatibility properties along with linear aggregation techniques.

Turning back to application, one concern is that of collusion, where recommenders may misreport to help friends. Clever work may be brought to bear from adjacent settings, such as *Sum of Us* by Alon et al. [3] and collusion detection [2]. It is also of interest to study online-learning, together with navigating the exploration-exploitation tradeoff for lenders who are building client bases in new communities. We may also seek to motivate recommenders to invest appropriately in providing good information; this is an area that we have seen as important from an ongoing field study in Uganda. Finally, we may consider non-binary settings, where repayments may be partial, and interest rates and loan sizes may vary according to a borrower's creditworthiness assessment.

Acknowledgements. This project was funded in part through the generous support of the OCP Group in Casablanca, Morocco and the Global Challenges in Economics and Computation Grant. The authors would like to thank Yiling Chen, Ariel Procaccia, Pablo Ducru, and Mutembesa Daniel for their detailed feedback.

References

1. Agarwal, A., Mandal, D., Parkes, D.C., Shah, N.: Peer prediction with heterogeneous users. In: Proceedings of the 2017 ACM Conference on Economics and Computation, EC 2017, pp. 81–98. ACM (2017)
2. Allahbakhsh, M., Ignjatovic, A., Benatallah, B., Beheshti, S.-M.-R., Bertino, E., Foo, N.: Collusion detection in online rating systems. In: Ishikawa, Y., Li, J., Wang, W., Zhang, R., Zhang, W. (eds.) APWeb 2013. LNCS, vol. 7808, pp. 196–207. Springer, Heidelberg (2013). https://doi.org/10.1007/978-3-642-37401-2_21
3. Alon, N., Fischer, F., Procaccia, A., Tennenholtz, M.: Sum of us: strategyproof selection from the selectors. In: Proceedings of the 13th Conference on Theoretical Aspects of Rationality and Knowledge, pp. 101–110. TARK XIII, ACM (2011)
4. Björkegren, D., Grissen, D.: Behavior revealed in mobile phone usage predicts credit repayment. World Bank Econ. Rev. **34**(3), 618–634 (2020)
5. Branch.co: Branch.co homepage (2021). https://branch.co/. Accessed June 2021
6. Budescu, D.V., Chen, E.: Identifying expertise to extract the wisdom of crowds. Manag. Sci. **61**(2), 267–280 (2015)
7. Chen, Y., Kash, I., Ruberry, M., Shnayder, V.: Decision markets with good incentives. In: Chen, N., Elkind, E., Koutsoupias, E. (eds.) WINE 2011. LNCS, vol. 7090, pp. 72–83. Springer, Heidelberg (2011). https://doi.org/10.1007/978-3-642-25510-6_7
8. Chen, Y., Pennock, D.M.: Designing markets for prediction. AI Mag. **31**(4), 42–52 (2010)
9. Gneiting, T., Raftery, A.E.: Strictly proper scoring rules, prediction, and estimation. J. Am. Stat. Assoc. **102**(477), 359–378 (2007)
10. Grameen: Grameen bank homepage (2020). http://www.grameen.com/. Accessed Feb 2020

11. Hanson, R.D.: Logarithmic market scoring rules for modular combinatorial information aggregation. J. Prediction Markets **1**(1), 3–15 (2007)
12. Hussam, R., Rigol, N., Roth, B.: Targeting high ability entrepreneurs using community information: mechanism design in the field. Am. Econ. Rev. (2021, forthcoming)
13. Jurca, R., Faltings, B.: Mechanisms for making crowds truthful. J. Artif. Intell. Res. **34**, 209–253 (2009)
14. Kassim, S.H., Rahman, M.: Handling default risks in microfinance: the case of Bangladesh. Qual. Res. Financ. Mark. **10**(4), 363–380 (2018)
15. Kong, Y., Schoenebeck, G.: An information theoretic framework for designing information elicitation mechanisms that reward truth-telling. ACM Trans. Econ. Comput. **7**(1), 2:1–2:33 (2019)
16. Maitra, P., Mitra, S., Mookherjee, D., Motta, A., Visaria, S.: Agent intermediated lending: a new approach to microfinance. Monash University, Department of Economics (2013)
17. Miller, N., Resnick, P., Zeckhauser, R.: Eliciting informative feedback: the peer-prediction method. Manag. Sci. **51**(9), 1359–1373 (2005)
18. Nisan, N., Ronen, A.: Computationally feasible VCG mechanisms. J. Artif. Intell. Res. **29**, 19–47 (2007)
19. Papakonstantinou, A., Rogers, A., Gerding, E.H., Jennings, N.R.: Mechanism design for the truthful elicitation of costly probabilistic estimates in distributed information systems. Artif. Intell. **175**(2), 648–672 (2011)
20. Porter, K.: Microcredit summit (2020). https://www.microcreditsummit.org/microfinance-statistics/. Accessed Sept 2020
21. Radanovic, G., Faltings, B., Jurca, R.: Incentives for effort in crowdsourcing using the peer truth serum. ACM Trans. Intell. Syst. Technol. **7**(4), 48:1–48:28 (2016)
22. Shnayder, V., Agarwal, A., Frongillo, R.M., Parkes, D.C.: Informed truthfulness in multi-task peer prediction. In: Proceedings of the 2016 ACM Conference on Economics and Computation, EC 2016, pp. 179–196. ACM (2016)
23. Soule, D., Grushka-Cockayne, Y., Merrick, J.R.: A heuristic for combining correlated experts. SSRN Electron. J. (2020)
24. Waggoner, B., Chen, Y.: Output agreement mechanisms and common knowledge. In: Proceedings of the Second AAAI Conference on Human Computation and Crowdsourcing (2014)
25. Wang, J., Liu, Y., Chen, Y.: Forecast aggregation via peer prediction. CoRR abs/1910.03779 (2019)
26. Winkler, R.L.: Evaluating probabilities: asymmetric scoring rules. Manag. Sci. **40**(11), 1395–1405 (1994)
27. Witkowski, J., Parkes, D.C.: Peer prediction without a common prior. In: Proceedings of the 13th ACM Conference on Electronic Commerce, EC, pp. 964–981. ACM (2012)
28. Witkowski, J., Parkes, D.C.: A robust Bayesian truth serum for small populations. In: Proceedings of the Twenty-Sixth AAAI Conference on AI (2012)
29. York, M., Dahleh, M., Parkes, D.C.: Eliciting social knowledge for creditworthiness assessment. arXiv (2108.09289) (2021)
30. Zermeño, L.: A principal-expert model and the value of menus. Technical report, MIT Economics (2011)

Social Choice and Cryptocurrencies

Decentralized Asset Custody Scheme with Security Against Rational Adversary

Zhaohua Chen[1,4] and Guang Yang[2,3,4(✉)]

[1] Peking University, Beijing, China
chenzhaohua@pku.edu.cn
[2] Shanghai Tree-Graph Blockchain Research Institute, Shanghai, China
[3] Tree-Graph Blockchain Innovation Center of Xiang River Hunan, Changsha, China
[4] Conflux Foundation, Beijing, China
guang.yang@confluxnetwork.org

Abstract. Asset custody is a core financial service in which the custodian holds in-safekeeping assets on behalf of the client. Although traditional custody service is typically endorsed by centralized authorities, decentralized custody scheme has become technically feasible since the emergence of digital assets, and furthermore, it is greatly needed by new applications such as blockchain and DeFi (Decentralized Finance).

In this work, we propose a framework of decentralized asset custody scheme that is able to support a large number of custodians and safely hold customer assets of multiple times the value of the total security deposit. The proposed custody scheme distributes custodians and assets into many custodian groups via combinatorial designs, where each group fully controls the assigned assets. Since every custodian group is small, the overhead cost is significantly reduced. The liveness is also improved because even a single alive group would be able to process transactions.

The security of this custody scheme is guaranteed under the rational adversary model, i.e. any adversary corrupting a bounded fraction of custodians cannot move assets more than the security deposit paid. We further analyze the security and performance of our constructions from both theoretical and experimental sides, and provide explicit examples with concrete numbers and figures for a better understanding.

Keywords: Blockchain application · Decentralized asset custody · Rational adversary

1 Introduction

Asset custody is a core financial service in which an institution, known as the custodian, holds in-safekeeping assets such as stocks, bonds, precious metals,

This work is supported by Shanghai Committee of Science and Technology, China (Grant No. 20511102300, 20DZ2221800) and Innovation-led High-tech Industry Programs of Department of Science and Technology of Hunan Province (Grant No. 2020GK2006, 2020GK2007). Z. Chen is also supported by NSFC under project No. 62172012.

M. Feldman et al. (Eds.): WINE 2021, LNCS 13112, pp. 449–466, 2022.
https://doi.org/10.1007/978-3-030-94676-0_25

and currency on behalf of the client. Custody service reduces the risk of clients losing their assets or having them stolen, and in many scenarios, a third-party custodian is required by regulation to avoid systematic risk. In general, security is the most important reason why people use custody services and place their assets for safekeeping in custodian institutions.

The security of traditional asset custody service is usually endorsed by the reputation of the custodian, together with the legal and regulatory system. Such centralized endorsement used to be the only viable option until the emergence of blockchain and cryptocurrencies. Cryptocurrencies enjoy two major advantages over their physical counterparts: (1) they are intrinsically integrated with information technology such as the Internet and modern cryptography, which technically enables multiple custodians to safeguard assets collectively; (2) with the underlying blockchain as a public ledger, the management of cryptocurrencies becomes transparent to everyone and hence any fraud behavior will be discovered immediately, which makes prosecution much easier.

From a systematic point of view, asset custody service provided by a federation of multiple independent custodians has better robustness and resistance against single-point failure, and hence achieves a higher level of security. Such credit enhancement is especially important for the safekeeping of cryptoassets on decentralized blockchains such as Bitcoin [16] and Ethereum [27], where the legal and regulatory system is absent or at least way behind the development of applications. For example, in the year 2019 alone, at least 12 cryptocurrency exchanges claimed being hacked and loss of cryptoassets totaled to around 2.9 billion dollars [21]. However, it is difficult for customers to distinguish that whether the claimed loss was caused by a hacker attack or internal fraud and embezzlement, and therefore raises the need for decentralized asset custody.

Decentralized asset custody finds applications in many scenarios related to blockchain and digital finance. A motivating example is the *cross-chain assets mapping* service (a.k.a. *cross-chain portable assets* [3, 28]) which maps cryptoassets on one blockchain to tokens on another blockchain for inter-chain operability. For instance, the mapping from Bitcoin to Ethereum enables usage of tokens representing bitcoins within Ethereum ecosystem, and in the meanwhile, the original bitcoins must be safeguarded so that the bitcoin tokens are guaranteed redeemable for real bitcoins in full on the Bitcoin network. Nowadays the volume of cryptoassets invested into Ethereum DeFi applications is massive, and the highest point in history almost reaches 90 billion dollars [9], among which a significant fraction (e.g. H-Tokens [13], imBTC [24], tBTC [23], WBTC [25], renBTC [19], etc.) is mapped from Bitcoin. Due to the reality that most of those DeFi applications and tokens remain in a gray area of regulation, decentralized cryptoassets custody turns out an attractive approach for better security and credit enhancement.

In this work, we propose a framework of decentralized asset custody scheme designed for cross-chain assets mapping (especially from blockchains with poor programmability, e.g. Bitcoin). More specifically, custodians and assets are distributed into multiple custodian groups, where each group consists of few custodians as its members and fully controls a small portion of all assets under custody. The authentication of each custodian group requires the consent of sufficiently many group members, which can be implemented with voting or threshold signature. Under this framework, transactions can be processed more efficiently within the very few group members, since the computational and communicational cost is significantly reduced. The liveness and robustness are also improved since even a single alive custodian group can process transactions.

The security of our proposed asset custody scheme is guaranteed against a rational adversary: every custodian in this scheme must offer a fund as the security deposit, which is kept together with the asset under custody and will be used to compensate for any loss caused by misbehaving custodians. The system remains secure as long as an adversary cannot steal more assets than the deposit paid, i.e. comparing to launching an attack the adversary would be better off by just withdrawing the security deposit of custodians controlled. Furthermore, we prove that for an adversary who corrupts a limited fraction of custodians, our scheme can safeguard customer assets of multiple times the value of the total security deposit under suitable construction. This approach significantly reduces the financing cost of a collateralized custody service.

1.1 Related Works

The prototype of decentralized asset custody scheme first appears in Bitcoin as multisignature (multisig) [2], where the authentication requires signatures from multiple private keys rather than a single signature from one key. For example, an M-of-N address requires signatures by M out of totally N predetermined private keys to move the money. This naïve scheme works well for small M and N but can hardly scale out, because the computational and communicational cost of authenticating and validating each transaction grows linearly in M. Both efficiency and liveness of the scheme are compromised for large M and N, especially in the sleepy model proposed by Pass and Shi [17] where key holders do not always respond in time. In practice, a multisignature scheme is typically used at the wallet level rather than as a public service, since the scheme becomes costly for large N and most Bitcoin wallets only support $N \leq 7$. We remark that multisignature schemes may be coupled with advanced digital signature techniques such as threshold signature [8,10,11] or aggregate signature [1,15,20] to reduce the cost of verifying multi-signed signatures.

As for the cross-chain asset mapping service, existing solutions mainly include the following types:

- Centralized: custody in a trusted central authority, with the endorsement fully from that authority, e.g. H-Tokens [13], WBTC [25] and imBTC [24];
- Consortium: custody in multisignature accounts controlled by an alliance of members, and endorsed by the reputation of alliance members, e.g. cBTC [7] (in its current version) and Polkadot [26];
- Decentralized (with deposit/collateral): custody provided by permissionless custodians, with security guaranteed by over-collateralized cryptoassets, e.g. tBTC [23] and renBTC [19] (in its future plan).

The last type seems satisfiable in decentralization and security against single-point failure and collusion. Meanwhile, existing solutions (e.g. tBTC and renBTC) have security guaranteed in the sense that an adversary will not launch a non-profitable attack. However, for these solutions, significant drawbacks exist as well. The first drawback is the inefficiency caused by over-collateralization, e.g. tBTC requires the custodian to provide collateral worth of 150% value of customer's assets, and renBTC requires 300%. The second drawback is that these solutions cannot support homogeneous collateral as the assets under custody, and hence breaking the safety of the custody service in market volatility. We remark that [12] considers the dynamic adjustment of the deposit of custodians in the long run. However, this work implicitly assumes that the security of the system is irrelevant with the behavior of custodians (e.g. by introducing cryptographic methods like in Bitcoin). Such assumption is inapplicable in the game-theoretic setting we discuss here.

1.2   Our Contributions

Our contributions lie in the following parts:

- In literature, we are the first to consider the possibility of homogeneously keeping exterior assets and custodians' deposit in the scenario of decentralized asset custody. To model such feasibility, we formalize the concept of custody scheme and further propose the concept of efficiency factor of a custody scheme for any adversary power (Sect. 2). The latter captures the maximal ratio of capable exterior assets to deposit that the underlying custody scheme can safely handle against a rational adversary.
- We propose a series of evaluation criteria to specify the performance of a custody scheme (Sect. 2). Combining with the previous point, we give a complete framework for analyzing a custody scheme and comparing different custody schemes. We point out that the underlying group assignment scheme is the core of a custody scheme.
- We present four kinds of concrete construction of group assignment schemes (three of them shown in Sect. 3). For each of them, we theoretically give an exact value/a lower bound on the efficiency factor of the custody scheme they induce. Some results turn out to be magnificent. For example, we show that we can assign 24 custodians to 759 groups such that as long as the adversary corrupts $\gamma \leq 1/4$ fraction of all custodians, the custody scheme is capable of safekeeping assets worthy of $\eta > 30.62$ times of total collateral.
- We prove that the random sampling trick significantly reduces the size of group assignment scheme without losing too much in the efficiency factor (Sect. 4). Therefore, random sampling resolves the problem of too many groups inside a custody scheme. More specifically, suppose we have a custody scheme consisting of n participants and its efficiency factor is η against some adversary. By randomly sampling $O(\eta n)$ many groups, the newly induced custody scheme would have efficiency factor $\eta' \geq \sqrt{\eta + 1} - 2$ against the same adversary with high probability. An important corollary shows that we

can construct $\Theta(n)$ groups with identical size $\Theta(1)$ to obtain an efficiency factor of $\Theta(1)$ against an adversary with constant power.

In the full version of this paper [4], we further study the computational complexity of finding the optimal corrupting strategy in general. Meanwhile, we conduct extensive experiments to validate the real-world performance of proposed group assignment scheme designs. The full version also contains more details of construction and analysis of multi-layer sharding design and full proofs of all propositions and theorems.

2 Model

Our goal is to implement the decentralized custody scheme without relying on any trusted party. More specifically, we investigate the feasibility that n custodians (a.k.a. n nodes) jointly provide the custody service, such that the security is guaranteed as long as a bounded fraction of custodians are corrupted, e.g. no more than $n/3$ nodes are corrupted simultaneously. This assumption of an honest majority is much milder than assuming a single party trusted by everyone, and hence likely leads to a better security guarantee in practice.

The decentralized custody scheme is based on overlapping group assignments. That is, custodians are assigned to overlapping groups, and each group is fully controlled by its members and holds a fraction of the total assets under custody, including both deposit from custodians and assets from customers. In what follows we assume that the in-safekeeping assets are *evenly* distributed to custodian groups, since an uneven distribution naturally leads to degradation of security and capital efficiency.

Furthermore, we consider the security of a custody scheme against a *rational adversary*: the adversary may corrupt multiple nodes, but will not launch an attack if the potential profit does not exceed the cost. To achieve security under such a model, every custodian in our scheme must provide an equal amount of deposit, which will be confiscated and used for compensation in case of misbehavior. Thus, if misbehavior can be detected in time, no rational adversary would ever launch an attack as long as the deposit paid outweighs the revenue of a successful attack. Here, we emphasize that instead of resorting to another level of collateral custody service, the deposit from custodians is maintained as a part of the total assets under custody, together with assets from external customers.

As a remark, we assume that attacks in the decentralized custody scheme can be detected immediately. If the decentralized custody service is for cryptoassets and deployed on a blockchain, then all instructions from customers and transfer of assets are transparent to everyone, and hence any malicious transaction will be caught immediately. Alternatively, the detection may be implemented with the periodic examination which ensures that misbehavior is discovered before the adversary can exit or change the set of corrupted nodes. In other scenarios, detecting corrupted behavior may be a non-trivial problem, but for the sake of this study we will leave it out to avoid another layer of complication.

The incentive of agents participating in this collateralized custody scheme is also indispensable for a full-fledged decentralized custody service. A reasonable rate of the commission fee and/or inflation tax would be sufficient to compensate the cost of agents providing such custody service. In the blockchain scenario, an extra per-transaction fee is also an option. Overall we believe that the mechanism design to incentivize custodians is essentially another topic, which is beyond the scope of this work and should be left for future study.

A Trivial but Useless Solution. In the most trivial solution, the asset under custody can only be moved when approved by all custodians or at least a majority of them. However, as n grows getting such an approval becomes expensive and even infeasible in practice, especially when honest participants may go off-line (as in the sleepy model [17]), which renders the trivial scheme useless.

Although the above solution is not satisfactory, it does provide enlightening ideas for designing a better custody scheme. The threshold authorization scheme guarantees that the adversary cannot move any assets under custody if not a sufficient number of nodes are corrupted. More generally, this is a specific case of security against the rational adversary, where with bounded power, the adversary's deposit outweighs the revenue of launching an attack. Again, as long as this property is satisfied the custody scheme is secure in our model.

In particular, the following toy example shows the feasibility of implementing our idea with multiple overlapping subsets of S as custodian groups. In this example, *each* 3-subset of S controls a certain fraction of the total assets under custody. Here S is the set of all custodians.

Example 1 (Toy example). Consider the case when 10 units of exterior assets are under custody. Assume there are $n = 5$ custodians, each paying a deposit of 6 units of assets, amounting to 30 units. Let each of the 10 3-subsets of S form a custodian group, and assign all 40 units of assets equally to all groups, i.e. each custodian group controls 4 units. If the asset controlled by each group can be moved with approval of 2 out of 3 members in that group, then an adversary controlling 2 nodes can corrupt exactly 3 custodian groups. However, by controlling 3 groups the adversary can only move $4 \times 3 = 12$ units, which is no more than the deposit of corrupted nodes (also 12 units). Thus such a custody scheme for $n = 5$ is secure against adversaries controlling up to two nodes.

In what follows, we will formalize the model of a decentralized custody scheme with assets evenly distributed among custodian groups. To start with, we introduce a formal definition of the custody scheme we consider in this work.

Definition 1 (Custody scheme). *A custody scheme $(S, \mathcal{A}, \mu)$ consists of the following three parts:*

- $S = \{1, 2, \cdots, n\}$ *denotes the set of all custodians (or simply nodes);*
- $\mathcal{A}$ *denotes a family of m k-subsets of S, such that each element in $\mathcal{A}$ (i.e. a k-subset of S) represents a custodian group under the given custody scheme;*
- $\mu \in [1/2, 1)$ *denotes a universal authentication threshold for all custodian groups, i.e. the asset controlled by that group can be settled arbitrarily with approval of strictly above μk group members.*

We emphasize that the elements in $\mathcal{A}$ do not have to be disjoint. In fact, it is imperative to use overlapping subsets in any meaningful solution. In certain cases, there might even exist repeated elements in $\mathcal{A}$.

In this work, we focus on the symmetric setting where every node provides the same amount of deposit and every custodian group has the same fraction of total assets in custody. At the same time, our discussion of the authentication threshold μ mainly focuses on $\mu = 1/2$ and $\mu = 2/3$.[1] We let $r = \lceil \mu k + \epsilon \rceil$ denote the smallest integer greater than μk, and hence the authentication of every custodian group is essentially an r-of-k threshold signature scheme.

We represent the adversary power with $\gamma \in (0, 1)$, which refers to the fraction of corrupted nodes in S. Specifically, we let $s = \lfloor \gamma n \rfloor$ denote the number of corrupted nodes in S.[2] The adversary is allowed to adaptively select corrupted nodes and then get all information and full control of those nodes thereafter, as long as the number of corrupted nodes does not exceed s. In case a group in $\mathcal{A}$ contains *at least* r corrupted nodes, we say that group is *corrupted*. Furthermore, we remark that the adversary has reasonably bounded computing power, so that cryptographic primitives such as digital signatures are not broken.

Given a custody scheme $(S, \mathcal{A}, \mu)$, together with γ for the adversary power, we use the function $f(\gamma; S, \mathcal{A}, \mu)$ to denote the maximal number of groups that may be corrupted. Formally,

$$f(\gamma; S, \mathcal{A}, \mu) := \max_{B \subseteq S : |B| = \lceil \gamma n \rceil} |\{ A \in \mathcal{A} \mid |A \cap B| > \mu k \}|. \tag{1}$$

Recall that as all assets under custody are equally distributed to all custodian groups, each corrupted group values equal to the adversary. Therefore, $f(\gamma; S, \mathcal{A}, \mu)$ directly resembles the maximal gain of the adversary.

We further define the efficiency factor of a custody scheme, which captures the ability to securely holding exterior assets.

Definition 2 (Efficiency factor of a custody scheme). *Given a custody scheme $(S, \mathcal{A}, \mu)$ and adversary power γ defined as above, the* efficiency factor *of this scheme against γ-adversary, denoted by η, is defined as:*

$$\eta := \frac{\gamma \cdot m}{f(\gamma; S, \mathcal{A}, \mu)} - 1.$$

where m is the total number of custodian groups induced by $\mathcal{A}$.

---

[1] In a synchronous network, $\mu \geq 1/2$ is a sufficient condition for the existence of expected-constant-round Byzantine agreement protocols in the authenticated setting (i.e., with digital signature and public-key infrastructure) [14], whereas $\mu \geq 2/3$ is necessary and sufficient for the existence of Byzantine agreement protocols in the unauthenticated setting [18]. We further remark that larger μ implies higher security but worse liveness, for example, when $\mu \to 1$, even a single corrupted member can block a custodian group from confirming any transaction. However, the discussion of liveness is beyond the scope of this work.

[2] In most parts of the paper, we slightly abuse the notation and assume that γn is always a natural number, i.e. $s = \gamma n \in \mathbb{N}$.

The efficiency factor η indeed equals the maximal ratio of capable exterior assets to deposit that the underlying custody scheme can handle. Specifically, suppose that u units of assets are deposited in total, and v units of exterior assets are in custody. According to (1), by launching an attack the adversary is able to seize the funds of $f(\gamma; S, \mathcal{A}, \mu)$ custodian groups, which amounts to $(u + v) \cdot f(\gamma; S, \mathcal{A}, \mu)/m$ units of assets, at the cost of losing deposit worthy of value $\gamma \cdot u$ units. Recall that in our model, collateral and exterior assets are homogeneous and kept together by the custodian groups, therefore, the custody scheme is secure as long as $f(\gamma; S, \mathcal{A}, \mu)/m \cdot (u + v) \leq \gamma \cdot u$, or equivalently, $v/u \leq \eta$ according to Definition 2.

As an example for the definition, $\eta = 1$ implies that the system is secure when the total value of exterior assets is no more than the total value of deposit.

Notice that when the efficiency factor $\eta < 0$ for some γ, the custody scheme against that γ-adversary is always insecure, regardless of the amount of deposit. To capture such property, we further define the reliability and safety of a custody scheme based on the Definition 2.

Definition 3 (Reliability and safety of custody scheme). *For a custody scheme $(S, \mathcal{A}, \mu)$ and adversary power γ, we say that the custody scheme is γ-reliable if the efficiency factor η of the scheme is non-negative against γ-adversary, i.e. $f(\gamma; S, \mathcal{A}, \mu) \leq \gamma \cdot m$. Furthermore, the scheme is secure against γ-adversary (or simply secure) if it is γ'-reliable for every $\gamma' \in (0, \gamma]$.*[3]

Putting into our formal definition, the trivial solution with only one custodian group (i.e. $k = n$, $m = 1$) has efficiency factor $\eta = \infty$ for $\gamma \leq \mu$ and $\eta < 0$ for $\gamma > \mu$; the custody scheme in Example 1 has its efficiency factor η changing according to the adversary power γ as summarized in Table 1. In particular, for $\gamma = 1/5$ and $\gamma = 2/5$, the scheme is reliable with $\eta = \infty$ and $\eta = 1/3$ respectively. For $\gamma \geq 3/5$ the scheme is unreliable with $\eta < 0$.

From the formalization of our decentralized custody scheme, it is clear that the custodian group assignment $\mathcal{A}$ is the core of the whole custody scheme. In particular, for a fixed n, every specific group assignment $\mathcal{A}$ and fixed constant μ (say, $\mu \in \{1/2, 2/3\}$), as the parameters m and k are already specified in $\mathcal{A}$, the maximal number of corrupted groups and the efficiency factor η are functions solely depending on the adversary power γ.[4]

Therefore, in the rest of this paper, we will focus on the construction and analysis of custodian group assignment schemes. In the meantime, we point out that it is meaningless merely to study a single group assignment scheme. Even in real life, the group assignment scheme should be adjusted with the joining and leaving of custodians. Instead, we focus on the systematic construction methods which lead to group assignment scheme *families*.

---

[3] Notice that although $f(\gamma; S, \mathcal{A}, \mu)$ increases with γ, $f(\gamma; S, \mathcal{A}, \mu)/\gamma$ may not be a increasingly-monotone function in γ, and as a result, γ-reliability does not necessarily lead to γ-security.

[4] We remark that the number of custodians n is not always extractable from the group assignment scheme $\mathcal{A}$, as in some cases, especially when we consider random sampling in Sect. 4), some custodians may belong to no group.

Table 1. The efficiency factor of the custody scheme under different adversary power in Example 1.

Parameters\adversary power (γ)	1/5	2/5	3/5	4/5
# corrupted nodes (s)	1	2	3	4
# corrupted custodian groups ($f(\gamma; S, \mathcal{A}, \mu)$)	0	3	7	10
Efficiency factor (η)	∞	1/3	$-1/7$	$-1/5$

The authentication threshold is realized as $r = 2$ and $\mu = 1/2$ (in this example equivalent to have $\mu \in [1/3, 2/3)$).

Definition 4 (Group assignment scheme family). *We say $\mathcal{C} = \{\mathcal{A}^n\}_{n \in \mathcal{I}}$ is a group assignment scheme family, if*

- *$\mathcal{I}$ is an index set;*
- *$\mathcal{A}^n$ is a group assignment scheme with n nodes;*
- *all group assignment schemes in $\mathcal{C}$ imply an identical group size.*

Evaluation Criteria. In this work, we use the following evaluation criteria when comparing two group assignment scheme families with the same group size:

1. *Efficiency factor.* Firstly, we consider the efficiency factor η of schemes in two families with the same number of nodes under adversary power $\gamma = 1/2 \cdot \mu, 2/3 \cdot \mu$. We prefer the family with a higher efficiency factor of group assignment schemes.
2. *Number of groups.* Secondly, we consider the size m of schemes in two families with the same number of nodes. We prefer the family with less size of group assignment schemes. In real life, a large amount of groups leads to a high maintenance cost of the custody scheme.

3   Constructions of Group Assignment Schemes

In this section, we propose three types of group assignment schemes and analyze the performance of resultant custody schemes. We also provide empirical analysis of these schemes with concrete numbers for a better understanding.

3.1   Symmetric Design

Definition 5 (Symmetric design). *Given n and k, let $\mathcal{A}_{sym}$ be a family consisting of all k-subsets of S as custodian groups, i.e. $\mathcal{A}_{sym}$ is an assignment with $m = \binom{n}{k}$ different groups where each group has k nodes. For every authentication threshold μ, a custody scheme is induced by $\mathcal{A}_{sym}$ and μ.*

Due to the perfect symmetry of $\mathcal{A}_{sym}$, it immediately follows that the number of corrupted groups in the above custody scheme only depends on the number of corrupted nodes. For the adversary corrupts any set of γn nodes, the number of corrupted groups can be calculated as follows:

$$f(\gamma; S, \mathcal{A}_{sym}, \mu) = \sum_{r \leq t \leq k} \binom{\gamma n}{t} \binom{n - \gamma n}{k - t}. \tag{2}$$

The efficiency factor turns out to be $\eta = \gamma \cdot \binom{n}{k} / \sum_{t=r}^{k} \binom{\gamma n}{t} \binom{n-\gamma n}{k-t} - 1$. When $\mu \geq \gamma$,[5] according to the tail bound of hypergeometric distribution [5], we have

$$\eta = \gamma \cdot \binom{n}{k} / \sum_{t=r}^{k} \binom{\gamma n}{t} \binom{n - \gamma n}{k - t} - 1 \geq \gamma \cdot e^{2(\gamma - \mu)^2 k} - 1, \tag{3}$$

which establishes a good lower bound on the efficiency factor of the symmetric design under appropriate γ.

In the following proposition, we demonstrate that for appropriately large k, $\mathcal{A}_{sym}$ is secure for γ close to μ.

Proposition 1. *For any k and n, given μ and corresponding $r = \lceil \mu k + \epsilon \rceil$, if $\sqrt{2(r-1) \ln \frac{k-1}{r-1}} < \min\{r-1, k-r\}$, then the custody scheme induced by $\mathcal{A}_{sym}$ and μ is secure against γ_{sym}-adversary, for γ_{sym} defined as follows:*

$$\gamma_{sym} := \frac{r - 1 - \sqrt{2(r-1) \ln \frac{k-1}{r-1}}}{k - 1}.$$

For the special case when n is even, k is odd, $n \geq 2k$ and $\mu = 1/2$, the security threshold of custody scheme induced by symmetric design can be enhanced to $1/2$, as shown in the following proposition.

Proposition 2. *For any odd k and even n with $n \geq 2k$, the custody scheme derived from $\mathcal{A}_{sym}$ and $\mu = 1/2$ is secure against $1/2$-adversary.*

Figure 1 depicts the relation between efficiency factor η and adversary power γ, for $n \in \{20, 60\}$, $k \in \{5, 7\}$, and $\mu \in \{1/2, 2/3\}$. Basically, we see that with fixed n, k and μ, the efficiency factor η of the custody scheme induced by symmetric design decreases as γ grows. Further, for combinations of reasonably large n and k, the efficiency factor η can be above 10 when γ is roughly $1/2 \cdot \mu$. For instance, when $n = 20$, $k = 5$ and $\mu = 2/3$, we have $m = \binom{20}{5} = 15,504$ and the efficiency factor $\eta = 10.4$ against adversary with power $\gamma = 0.35$.

Figure 2 illustrates the behavior of the efficiency factor η versus the custodian group size k, for $n \in \{20, 60\}$, $\mu \in \{1/2, 2/3\}$ and $\gamma \in \{1/3 \cdot \mu, 1/2 \cdot \mu, 2/3 \cdot \mu\}$.

---

[5] We mention that in this work, when considering the reliability of a custody scheme, we tacitly approve that $\mu \geq \gamma$. For a better understanding, consider the first example with only one group consisting of all custodians. Under such group assignment, when $\gamma > \mu$, the scheme is surely γ-unreliable.

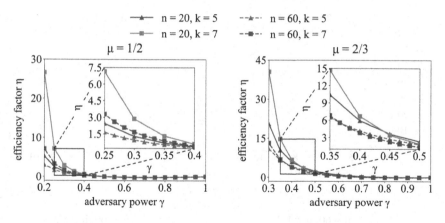

Fig. 1. The efficiency factor η against adversary power γ for $\mathcal{A}_{sym}$. In particular, $\eta < 0$ iff the custody scheme is not secure for the corresponding γ.

The figure shows that in general, η increases with k for custody schemes induced by $\mathcal{A}_{sym}$. The sawteeth appearing on the curves are due to the rounding of r and s, i.e. the authentication threshold and the number of corrupted nodes.

Finally we remark that the construction of $\mathcal{A}_{sym}$ by itself is mainly a theoretical result. Because the size of such group assignment $m = \binom{n}{k}$ grows too fast and hence n and k must be severely bounded in practice, e.g. $n \sim 20$ and $k \sim 5$, in order to keep m reasonable. One solution to mitigate the above issues is by random sampling, as exhibited in Sect. 4.

3.2 Polynomial Design

The following construction of group assignments relies on polynomial-based combinatorial designs.

Definition 6 (Polynomial design). *For given k, let $q \geq k$ be a prime and the number of custodians be $n = kq$. Let $T = \{(a, b) \mid 0 \leq a \leq k-1, 0 \leq b \leq q-1\}$ be a set of size kq, therefore, there is a bijection from S to T. (For simplicity, we use an element in T to represent the unique corresponding element S.) At last, let $0 < d < k$ be a integer. The polynomial design $\mathcal{A}_{poly}$ is a family of $m = q^d$ k-subsets of S defined as $\mathcal{A}_{poly} := \{A(p) \mid p \text{ is a degree-d monic polynomial over } \mathbb{Z}/q\mathbb{Z}\}$, where $\forall p$, $A(p) := \{(i, p(i)) \mid 0 \leq i \leq k-1\}$. Then, for every authentication threshold μ, a custody scheme can be induced by $\mathcal{A}_{poly}$ and μ.*

It is easy to verify that $\mathcal{A}_{poly}$ consists of m distinct groups, and the intersection of any two distinct groups in $\mathcal{A}_{poly}$ is strictly bounded by d by the Fundamental Theorem of Algebra, i.e.:

$$\forall A_p, A_q \in \mathcal{A}_{poly}, A_p \neq A_q \implies |A_p \cap A_q| < d. \tag{4}$$

Hence, the efficiency factor η of the custody scheme induced by polynomial design is lower bounded as below.

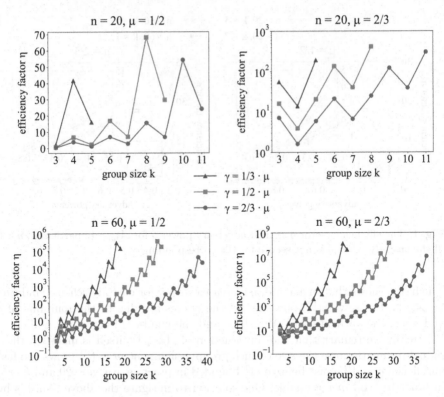

Fig. 2. The efficiency factor η against group size k for $\mathcal{A}_{sym}$. Blank points on the right side refer to $\eta = \infty$ when adversary cannot corrupt even a single custodian group.

Theorem 1. *Given parameters k, q, d, $n = kq$, μ and corresponding r, the efficiency factor η of the custody scheme induced by $\mathcal{A}_{poly}$ and μ against a γ-adversary is lower bounded as follows:*

$$\eta \geq \gamma^{1-d} \cdot \binom{r}{d} \bigg/ \binom{k}{d} - 1.$$

Surprisingly, the lower bound of η given by Theorem 1 does not rely on the selection of q.

From Theorem 1, we immediately obtain the following proposition:

Proposition 3. *Given parameters k, q, d, $n = kq$, μ and corresponding r, the custody scheme induced by $\mathcal{A}_{poly}$ and μ is secure against γ_{poly}-adversary for γ_{poly} defined as:*

$$\gamma_{poly} := \left(\binom{r}{d} \bigg/ \binom{k}{d} \right)^{\frac{1}{d-1}}.$$

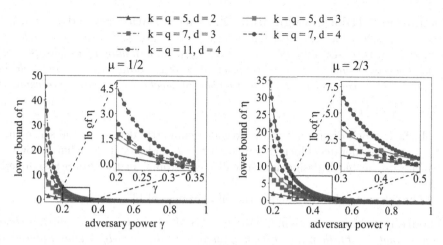

Fig. 3. The lower bound of efficiency factor η (by Theorem 1) against adversary power γ for $\mathcal{A}_{poly}$. Recall that $n = kq$ in $\mathcal{A}_{poly}$.

Figure 3 depicts the relation between the lower bound of η following Theorem 1 against the adversary power γ, for $\mu \in \{1/2, 2/3\}$ and k, q, d as shown in the figure. Note that $n = kq$. It is easy to see that the lower bound of η increases as k and d become larger with fixed corrupted fraction γ. For specific choices we get $\eta \geq 9.45$ against adversary with $\gamma = 3/11$, when $\mu = 2/3$ and $\mathcal{A}_{poly}$ is parameterized by $n = 121$, $k = q = 11$ and $d = 4$, with totally $m = 14,641$ groups. Furthermore, we remark that under the estimation of Theorem 1, the efficiency factor η increases rapidly as γ decreases since $\eta \sim \gamma^{1-d}$. For instance, the lower bound for η is improved to no less than 34.29 when γ is reduced from $3/11$ to $2/11$ in the above example.

The polynomial design only implies a group number of $k^d = O(n^{d/2})$, which is far smaller than the group number of $\binom{n}{k}$ given by symmetric design. Our subsequent experiments show that considering efficiency factor, polynomial design behaves a bit worse than symmetric design. Nevertheless, the result is pleasing enough for a realization in practice.

3.3 Block Design

One may notice that the previous two constructions give custody schemes with a rather large number of groups. For symmetric design, we have $\binom{n}{k}$ groups; and for polynomial design, we have $k^d = \Theta(n^{d/2})$ groups. In this section, we consider block designs, which lead to a smaller number of groups.

A block design is a particular combinatorial design consisting of a set of elements and a family of subsets (called blocks) whose arrangements satisfy generalized concepts of balance and symmetry.

Definition 7 (Block design, from [22], with notation revised). *Let n, k, λ and t be positive integers such that $n > k \geq t$. $(S, \mathcal{A}_{blck})$ is called a t-(n, k, λ)- design if S is a set with $|S| = n$ and $\mathcal{A}_{blck}$ is a family of k-subsets of S (called blocks), such that every t-subset of S is contained in exactly λ blocks in $\mathcal{A}_{blck}$. One can verify that the number of blocks of a t-(n, k, λ)-design is $m = \lambda \cdot \binom{n}{t} \big/ \binom{k}{t}$.*

In fact, block design naturally extends symmetric design, in the sense that $\mathcal{A}_{sym}$ is a degenerated block design with $t = k$ and $\lambda = 1$. In what follows, a "block" in the block design is also called a "group" in the group assignment scheme.

The following theorem, shows the effectiveness of block designs:

Theorem 2. *For every t-(n, k, λ)-design $(S, \mathcal{A}_{blck})$, let $\mu \geq (t-1)/k$ (which implies that $r \geq t$), then the efficiency factor η of the custody scheme induced by $\mathcal{A}_{blck}$ and μ against a γ-adversary (i.e., the adversary corrupting $s = \gamma n$ nodes) is lower bounded as follows:*

$$\eta \geq \gamma \cdot \frac{\binom{n}{t}}{\binom{k}{t}} \cdot \frac{\binom{r}{t}}{\binom{s}{t}} - 1.$$

The following proposition further shows that the custody scheme induced by block design is secure with proper γ.

Proposition 4. *When $n \geq 3k - 3$, and $\mu \geq 1/2$, $r \geq \max\{t, 3\}$, the custody scheme induced by an r-(n, k, λ)-design with μ is secure against γ_{blck}-adversary, for γ_{blck} defined as follows:*

$$\gamma_{blck} := \frac{1}{k} \cdot \mu^{\frac{1}{t-1}} + \frac{t-1}{n}.$$

When $t = 2$, according to Theorem 2, we have $\eta \geq \frac{n-1}{s-1} \cdot \frac{r(r-1)}{k(k-1)} - 1 \approx \frac{\mu^2}{\gamma} - 1$, which implies that the efficiency factor is at least $\Omega(1)$ when $\gamma \geq 1/2 \cdot \mu$. When k and λ are constant, the corresponding number of groups is $\lambda \cdot \binom{n}{2} \big/ \binom{k}{2} = \Theta(n^2)$. With larger t, the result given by Theorem 2 is even more inspiring.

Figure 4 shows the lower bound of η obtained by Theorem 2 versus the adversary's power γ for different block designs with $\mu \in \{1/2, 2/3\}$. We clearly observe that the lower bound of η significantly increases with the value of t under fixed corrupted fraction γ. Further, although Theorem 2 only provides a lower bound estimation for large γ, we still achieve satisfying numerical results. For instance, using the custody scheme induced from the 5-$(24, 8, 1)$-design (see [6,22] for the construction) with $m = 759$ custodian groups and $\mu = 1/2$, the efficiency factor η is no less than 30.62 when $\gamma \leq 1/4$.

Meanwhile, our further experimental results demonstrate that block design has a comparable performance with polynomial design, which indicates that block design finds its application in constructing custodian groups under the scenario of decentralized asset custody in our model.

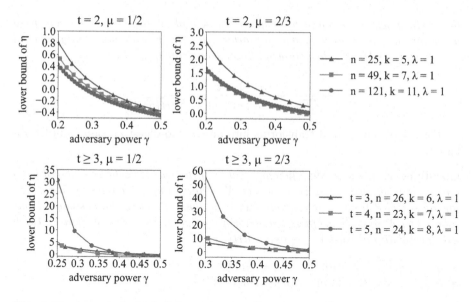

Fig. 4. The lower bound of efficiency factor η (by Theorem 2) against adversary power γ for $\mathcal{A}_{blck}$. All six concrete block designs shown in this figure have explicit constructions [6, 22].

4   Compressing Group Assignment Schemes via Random Sampling

We notice that under symmetric design and polynomial design, a group assignment scheme $\mathcal{A}$ may contain too many custodian groups, which renders the induced custody scheme almost impossible to manage in practice. To mitigate this problem, we propose a randomized sampling technique to construct compact custody schemes with a smaller number of custodian groups sampled from $\mathcal{A}$ as representatives.

Definition 8 (Random Sampling). *Given a group assignment scheme $\mathcal{A}$ consisting of m groups, as well as a sampling rate $\beta \in (0,1)$, we uniformly sample a subset of βm elements from $\mathcal{A}$ at random as the new assignment scheme $\mathcal{A}'$, and then construct a custody scheme based on $\mathcal{A}'$. The sampling process does not affect on the authentication threshold μ.*

In what follows we analyze the efficiency of $\mathcal{A}'$ comparing to $\mathcal{A}$. For a given corrupted fraction γ, let $H(\gamma)$ be a function of γ defined as $H(\gamma) := -(\gamma \ln \gamma + (1 - \gamma) \ln(1 - \gamma))$. Then the efficiency factor of custody scheme induced by $\mathcal{A}'$ is lower bounded as in the following theorem:

Theorem 3. *Let $\mathcal{A}$ and $\mathcal{A}'$ be defined as above, and suppose the corrupted fraction γ satisfies $n\gamma(1 - \gamma) \geq 1$.[6] Let η and η' be the efficiency factor of*

---

[6] This is trivial if $n > 4$ and $\gamma n \geq 2$.

the custody scheme induced by respectively $\mathcal{A}$ and $\mathcal{A}'$ together with some fixed μ against a γ-adversary. Then, for arbitrary $c \geq 0$, with probability at least $1 - \frac{e}{2\pi} \exp(-cnH(\gamma))$, the following lower bound for η' holds:

$$\eta' \geq \frac{\gamma(\eta + 1) \cdot \sqrt{\beta m}}{\gamma \cdot \sqrt{\beta m} + (\eta + 1) \cdot \sqrt{(1 + c)nH(\gamma)/2}} - 1.$$

For $c = 1$, Theorem 3 transforms into an easy-to-digest version as in Corollary 1.

Corollary 1. *Let η be the efficiency factor of the custody scheme induced by $\mathcal{A}$ and some μ against a γ-adversary. Let $\mathcal{A}'$ be the group assignment scheme uniformly sampled from $\mathcal{A}$ at random with m' groups. Suppose the custody scheme induced by $\mathcal{A}'$ and μ has efficiency factor η' against the same γ-adversary. Then, with probability at least $1 - \frac{e}{2\pi} \exp(-nH(\gamma))$,*

- $\eta' \geq \sqrt{\eta + 1} - 2$, *with* $m' = (\eta + 1)nH(\gamma)/\gamma^2$;
- $\eta' \geq (\eta - 1)/2$, *with* $m' = (\eta + 1)^2 nH(\gamma)/\gamma^2$.

To better illustrate the effect of Corollary 1, we consider the symmetric design in Sect. 3.1. (3) shows that the efficiency factor of the custody scheme induced by symmetric design reaches $\Theta(n)$ with $k = \Theta(\log n)$, and $\Theta(1)$ with $k = \Theta(1)$. Combining with Corollary 1, we further obtain the following important corollary:

Corollary 2. *For fixed $\gamma < \mu$, we can uniformly choose m different k-subsets of S at random, where $|S| = n$, such that with probability $1 - O(\exp(-nH(\gamma)))$, the efficiency factor η of the custody scheme induced by these subsets and μ against a γ-adversary satisfies:*

- $\eta = \Omega(1)$, *with* $k = \Theta(1)$ *and* $m = \Theta(n)$;
- $\eta = \Omega(\sqrt{n})$, *with* $k = \Theta(\log n)$ *and* $m = \Theta(n^2)$;
- $\eta = \Omega(n)$, *with* $k = \Theta(\log n)$ *and* $m = \Theta(n^3)$.

5   Summary and Discussion

In this work we propose a framework of decentralized asset custody schemes based on overlapping group assignments. The custody scheme reaches high efficiency, with security guaranteed against any rational adversary that corrupts a bounded fraction of custodians.

Explicit constructions of compact assignments with much less custodian groups, efficient approximation algorithms for estimating the actual efficiency factor of a given custody scheme in our framework, and more rigorous analysis of liveness guarantee as well as the trade-off between liveness and security are of independent interest, which we left for future work. Meanwhile, how the scheme handles a change in the total amount of assets in custody is also an interesting and realistic question.

Acknowledgments. We thank Jihan Wu, Yuesheng Ge, and Bitmain Technologies Ltd. for supporting the initial study of this work. We also thank Ge Chen and Tian Lin for their helpful discussion, and thank the anonymous reviewers for their constructive comments. Zhaohua Chen further thanks Xiaotie Deng for supporting his internship in Conflux Foundation.

References

1. Bellare, M., Neven, G.: Multi-signatures in the plain public-key model and a general forking lemma. In: Juels, A., Wright, R.N., di Vimercati, S.D.C. (eds.) Proceedings of the 13th ACM Conference on Computer and Communications Security, CCS 2006, Alexandria, VA, USA, 30 October–3 November 2006, pp. 390–399. ACM (2006). https://doi.org/10.1145/1180405.1180453
2. BitcoinWiki: Multisignature. https://en.bitcoin.it/wiki/Multisignature
3. Buterin, V.: Chain interoperability. Technical report, R3 (2016)
4. Chen, Z., Yang, G.: Decentralized asset custody scheme with security against rational adversary. CoRR abs/2008.10895 (2020)
5. Chvátal, V.: The tail of the hypergeometric distribution. Discret. Math. **25**(3), 285–287 (1979)
6. Colbourn, C.J., Dinitz, J.H.: Handbook of Combinatorial Designs. CRC Press, Boca Raton (2007)
7. Conflux: Conflux ShuttleFlow: A cross-chain asset protocol. https://medium.com/conflux-network/conflux-shuttleflow-a-cross-chain-asset-protocol-15ad6b2a9539
8. Damgård, I., Jakobsen, T.P., Nielsen, J.B., Pagter, J.I., Østergaard, M.B.: Fast threshold ECDSA with honest majority. In: Galdi, C., Kolesnikov, V. (eds.) SCN 2020. LNCS, vol. 12238, pp. 382–400. Springer, Cham (2020). https://doi.org/10.1007/978-3-030-57990-6_19
9. DeFi Pulse: DeFi - the decentralized finance leaderboard at DeFi Pulse. https://defipulse.com/
10. Doerner, J., Kondi, Y., Lee, E., Shelat, A.: Threshold ECDSA from ECDSA assumptions: the multiparty case. In: 2019 IEEE Symposium on Security and Privacy, SP 2019, San Francisco, CA, USA, 19–23 May 2019, pp. 1051–1066. IEEE (2019). https://doi.org/10.1109/SP.2019.00024
11. Gennaro, R., Goldfeder, S.: Fast multiparty threshold ECDSA with fast trustless setup. In: Lie, D., Mannan, M., Backes, M., Wang, X. (eds.) Proceedings of the 2018 ACM SIGSAC Conference on Computer and Communications Security, CCS 2018, Toronto, ON, Canada, 15–19 October 2018, pp. 1179–1194. ACM (2018). https://doi.org/10.1145/3243734.3243859
12. Harz, D., Gudgeon, L., Gervais, A., Knottenbelt, W.J.: Balance: dynamic adjustment of cryptocurrency deposits. In: Cavallaro, L., Kinder, J., Wang, X., Katz, J. (eds.) Proceedings of the 2019 ACM SIGSAC Conference on Computer and Communications Security, CCS 2019, London, UK, 11–15 November 2019, pp. 1485–1502. ACM (2019). https://doi.org/10.1145/3319535.3354221
13. Huobi Blockchain Team: H-Tokens white paper: bridge between centralized and DeFi markets. https://www.htokens.finance/static/pdf/whitepaper-en.pdf
14. Katz, J., Koo, C.: On expected constant-round protocols for Byzantine agreement. J. Comput. Syst. Sci. **75**(2), 91–112 (2009). https://doi.org/10.1016/j.jcss.2008.08.001

15. Maxwell, G., Poelstra, A., Seurin, Y., Wuille, P.: Simple Schnorr multi-signatures with applications to Bitcoin. Des. Codes Crypt. **87**(9), 2139–2164 (2019). https://doi.org/10.1007/s10623-019-00608-x
16. Nakamoto, S.: Bitcoin: a peer-to-peer electronic cash system (2008)
17. Pass, R., Shi, E.: The sleepy model of consensus. In: Takagi, T., Peyrin, T. (eds.) ASIACRYPT 2017, Part II. LNCS, vol. 10625, pp. 380–409. Springer, Cham (2017). https://doi.org/10.1007/978-3-319-70697-9_14
18. Pease, M.C., Shostak, R.E., Lamport, L.: Reaching agreement in the presence of faults. J. ACM **27**(2), 228–234 (1980). https://doi.org/10.1145/322186.322188
19. Ren: Ren. https://renproject.io/
20. Schnorr, C.P.: Efficient signature generation by smart cards. J. Cryptol. **4**(3), 161–174 (1991). https://doi.org/10.1007/BF00196725
21. SelfKey: A comprehensive list of cryptocurrency exchange hacks. https://selfkey.org/list-of-cryptocurrency-exchange-hacks/
22. Stinson, D.: Combinatorial Designs: Constructions and Analysis. Springer, New York (2007). https://doi.org/10.1007/b97564
23. tBTC: tBTC: A decentralized redeemable BTC-backed ERC-20 token. https://docs.keep.network/tbtc/index.pdf
24. Tokenlon: imBTC - more accessible Bitcoin. https://tokenlon.im/imBtc
25. WBTC: Wrapped tokens: A multi-institutional framework for tokenizing any asset. https://wbtc.network/assets/wrapped-tokens-whitepaper.pdf
26. Wood, G.: Polkadot: Vision for a heterogeneous multi-chain framework. White Paper (2016)
27. Wood, G., et al.: Ethereum: a secure decentralised generalised transaction ledger. Ethereum Project Yellow Paper **151**(2014), 1–32 (2014)
28. Zamyatin, A., et al.: SoK: communication across distributed ledgers. IACR Cryptology ePrint Archive, p. 1128 (2019)

The Distortion of Distributed Metric Social Choice

Elliot Anshelevich[1], Aris Filos-Ratsikas[2], and Alexandros A. Voudouris[3(✉)]

[1] Computer Science Department, Rensselaer Polytechnic Institute, Troy, USA
eanshel@cs.rpi.edu
[2] Department of Computer Science, University of Liverpool, Liverpool, UK
aris.filos-ratsikas@liverpool.ac.uk
[3] School of Computer Science and Electronic Engineering, University of Essex,
Colchester, UK
alexandros.voudouris@essex.ac.uk

Abstract. We consider a social choice setting with agents that are partitioned into disjoint groups, and have metric preferences over a set of alternatives. Our goal is to choose a single alternative aiming to optimize various objectives that are functions of the distances between agents and alternatives in the metric space, under the constraint that this choice must be made in a distributed way: The preferences of the agents within each group are first aggregated into a representative alternative for the group, and then these group representatives are aggregated into the final winner. Deciding the winner in such a way naturally leads to loss of efficiency, even when complete information about the metric space is available. We provide a series of (mostly tight) bounds on the distortion of distributed mechanisms for variations of well-known objectives, such as the (average) total cost and the maximum cost, and also for new objectives that are particularly appropriate for this distributed setting and have not been studied before.

1 Introduction

The main goal of social choice theory [31] is to come up with outcomes that accurately reflect the collective opinions of individuals within a society. A prominent example is that of elections, where the preferences of voters over different candidates are aggregated into a single winner, or a set of winners in the case of committee elections. Besides elections, the abstract social choice theory setting, where a set of *agents* express preferences over a set of possible *alternatives* captures very broad decision-making applications, such as choosing public policies, allocations of resources, or the most appropriate position to locate a facility.

In the field of computational social choice, Procaccia and Rosenschein [30] defined the notion of *distortion* to measure the loss in an aggregate objective (typically the utilitarian welfare), due to making decisions whilst having access

This work was partially supported by NSF awards CCF-1527497 and CCF-2006286.

M. Feldman et al. (Eds.): WINE 2021, LNCS 13112, pp. 467–485, 2022.
https://doi.org/10.1007/978-3-030-94676-0_26

to only *ordinal* information about the preferences of the agents, rather than their exact values (or costs). Following their work, a lot of effort has been put forward to bound the distortion of social choice rules, with Anshelevich *et al.* [5,7] being the first to consider settings with *metric* preferences. In such settings, agents and alternatives are points in a metric space, and the distances between them (which define the agent costs) satisfy the triangle inequality. The metric space can be thought of as evaluating the proximity between agents and alternatives for different political issues or ideological axes (e.g., liberal to conservative, or libertarian to authoritarian). The distortion in metric social choice has received significant attention, with many variants being considered over the recent years.

In contrast to the *centralized* decision-making settings considered in the papers mentioned above, there are cases where it is logistically too difficult to aggregate the preferences of the agents directly, or different groups of agents play inherently different roles in the process. In such scenarios, the collective decisions have to be carried out in a *distributed* manner, as follows. The agents are partitioned into groups (such as electoral districts, focus groups, or subcommittees), and the members of each group locally decide a single alternative that is representative of their preferences, without taking into account the agents of different groups. Then, the final outcome is decided based on properties of the group representatives, and not on the underlying agents within the groups; for example, the representatives act as agents themselves and choose an outcome according to their own preferences. However, since the representatives cannot perfectly capture all the information about the preferences of the agents (even when it is available in the group level), it is not surprising that choosing the final outcome this way may lead to loss of efficiency.

Motivated by this, Filos-Ratsikas *et al.* [20] studied the deterioration of the social welfare in general normalized distributed settings, by extending the notion of distortion to account for the information about the agents' preferences that is lost after the local decision step. They showed bounds on the distortion of max-weight mechanisms when the number of groups is given. Very recently, Filos-Ratsikas and Voudouris [21] considered the metric distributed distortion problem, and showed tight bounds on the distortion of mechanisms under several restrictions: (a) the metric is a line, (b) their objective is the (average) social cost objective (total distance between agents and the chosen alternative), and (c) the groups are mainly limited to be of the same size.

In this paper, we extend the results of Filos-Ratsikas and Voudouris in all three axes: We provide bounds for (a) general metrics (including refined bounds for the line metric), (b) four different objectives (including the average total cost, the maximum cost, and two new objectives that clearly motivated by the distributed nature of the setting), and (c) groups of agents that could vary in size. We paint an almost complete picture of the distortion landscape of distributed mechanisms when the agents have metric preferences.

1.1   Our Contributions

We consider a distributed, metric social choice setting with a set of agents and a set of alternatives, all of whom are located in a metric space. The preferences of the agents for the alternatives are given by their distances in the metric space, and as such they satisfy the triangle inequality. Furthermore, the agents are partitioned into a given set of *districts* of possibly different sizes. A distributed mechanism selects an alternative based on the preferences of the agents in two steps: first, each district selects a *representative* alternative using some local aggregation rule, and then the mechanism uses *only* information about the representatives to select the final winning alternative.

The goal is to choose the alternative that optimizes some aggregate objective that is a function of the distances between agents and alternatives. In the main part of the paper we consider the following four cost minimization objectives, which can be defined as compositions of objectives applied over and within the districts:

– The *average of the average agent distance in each district*[1] (AVG ∘ AVG);
– The *average of the maximum agent distance in each district* (AVG ∘ MAX);
– The *maximum agent distance in any district* (MAX ∘ MAX);
– The *maximum of the average agent distance in each district* (MAX ∘ AVG).

While AVG ∘ AVG and MAX ∘ MAX are adaptations of objectives that have been considered in the centralized setting, AVG∘MAX and MAX∘AVG are only meaningful in the context of distributed social choice. In particular, MAX∘AVG can be thought of as a fairness-inspired objective guaranteeing that no district has a very large cost, where the cost of a district is the average cost of its members. Similarly, AVG ∘ MAX guarantees that the average district cost is small, where the cost of a district is now defined as the egalitarian (maximum) cost of any of its members. We consider the introduction and study of these objectives as one of the major contributions of our work.

We measure the performance of a distributed mechanism by its *distortion*, defined as the worst-case ratio (over all instances of the problem) between the objective value of the alternative chosen by the mechanism and the minimum possible objective value achieved over all alternatives. The distortion essentially measures the deterioration of the objective due to the fact that the mechanism must make a decision via a distributed two-step process, on top of other possible informational limitations related to the preferences of the agents. We consider *deterministic* mechanisms that are either *cardinal* (in which case they have access to the exact distances between agents and alternatives), or *ordinal* (in which case they have access only to the rankings that are induced by the distances). Table 1 gives an overview of our bounds on the distortion of distributed mechanisms,

---

[1] Note that this objective is not exactly equivalent to the well-known (average) *social cost* objective, defined as the (average) total agent distance over all districts. All of our results extend for this objective as well, by adapting our mechanisms to weigh the representatives proportionally to the district sizes. When all districts have the same size, AVG∘AVG coincides with the average social cost.

for the four objectives defined above. We provide bounds that hold for general metric spaces, and then more refined bounds for the fundamental special case where the metric is a line.

Table 1. An overview of the distortion bounds for the various settings studied in this paper. Each entry consists of an interval showing a lower bound on the distortion of all distributed mechanisms for the corresponding setting, and an upper bound that is achieved by some mechanism; when a single number is presented, the bound is tight. The results marked with a (*) for the line metric and the AVG ∘ AVG objective (as well as the corresponding lower bounds for general metrics) follow from the work of Filos-Ratsikas and Voudouris [21]; all other results in the table were not known previously.

	General metric		Line metric	
	Cardinal	Ordinal	Cardinal	Ordinal
AVG ∘ AVG	3^*	$[7^*, 11]$	3^*	7^*
AVG ∘ MAX	3	$[2 + \sqrt{5}, 11]$	3	$[2 + \sqrt{5}, 5]$
MAX ∘ MAX	$[1 + \sqrt{2}, 3]$	$[3, 5]$	$1 + \sqrt{2}$	3
MAX ∘ AVG	$[1 + \sqrt{2}, 3]$	$[2 + \sqrt{5}, 5]$	$1 + \sqrt{2}$	$[2 + \sqrt{5}, 5]$

Several of our bounds for general metric spaces are based on a novel composition technique for designing distributed mechanisms. In particular, we prove a rather general *composition theorem*, which appears in many versions throughout our paper, depending on the objective at hand. Roughly speaking, the theorem relates the distortion of a distributed mechanism to the distortion of the centralized voting rules it uses for the local (in-district) and global (over-districts) aggregation steps. In particular, for two such voting rules with distortion bounds α and β, the distortion of the composed mechanism is at most $\alpha + \beta + \alpha\beta$. This effectively enables us to plug in voting rules with known distortion bounds, and obtain distributed mechanisms with low distortion. The theorem is also robust in the sense that the AVG and MAX objectives can be substituted with more general objectives satisfying specific properties, such as monotonicity and subadditivity; we provide more details on that in Sect. 5.2.

To demonstrate the strength of the theorem, consider the objective AVG∘AVG. For cardinal mechanisms and general metrics, the bound of 3 follows by using optimal centralized voting rules (with distortion 1) for both steps. Similarly, for ordinal mechanisms, the bound of 11 follows by using the PLURALITYMATCHING rule of Gkatzelis *et al.* [22] in both steps; this rule is known to have distortion at most 3 for general instances, and at most 2 when all agents are at distance 0 from their most-preferred alternative (which is the case when the representatives are thought of as agents in the second step of the mechanism).

Even though the composition theorem is evidently very powerful, it comes short of providing tight bounds in some cases. To this end, we design explicit mechanisms with improved distortion guarantees, both for general metrics as well as the fundamental special case where the metric is a line. A compelling highlight of our work is a novel mechanism for objectives of the form $\mathtt{MAX} \circ G$, to which we refer as λ-ACCEPTABLE-RIGHTMOST-LEFTMOST (λ-ARL). While this mechanism has the counter-intuitive property of *not* being *unanimous* (i.e., there are cases where all agents agree on the best alternative, but the mechanism does not choose this alternative as the winner), it achieves the best possible distortion of $1 + \sqrt{2}$ among all distributed mechanisms on the line. In contrast, we prove that unanimous mechanisms cannot achieve distortion better than 3. To the best of our knowledge, this is the first time that not satisfying unanimity turns out to be a necessary ingredient for achieving the best possible distortion in the metric social choice literature.

1.2 Related Work

The distortion of social choice voting rules has been studied extensively for many different settings. For a comprehensive introduction to the distortion literature, we refer the reader to the survey of Anshelevich et al. [6].

After the work of Procaccia and Rosenschein [30], a series of papers adopted their normalized setting, where the agents have unit-sum values, and proved asymptotically tight bounds on the distortion of ordinal single-winner rules [13,15], multi-winner rules [14], rules that choose rankings of alternatives [10], and strategyproof rules [11]. Recent papers considered more general questions related to how the distortion is affected by the amount of available information about the values of the agents [3,27,28]. The normalized distortion has also been investigated in other related problems, such as participatory budgeting [9], and one-sided matching [2,19].

The metric distortion setting was first considered by Anshelevich et al. [5] who, among many results, showed a lower bound of 3 for deterministic single-winner ordinal rules for the social cost, and an upper bound of 5, achieved by the Copeland rule. Following their work, many papers were devoted to bridging this gap (e.g., see [25,29]) until, finally, Gkatzelis et al. [22] designed a rule with distortion at most 3; in fact, this bound holds for the more general *fairness ratio* [23] (which captures various different objectives, including the social cost and the maximum cost). Besides the main setting, many other works have shown bounds on the metric distortion for randomized rules [7,18], rules that use less than ordinal information [4,17,26], committee elections [16,24], primary elections [12], and for many other problems [1,8].

Most related to our work are the recent papers of Filos-Ratsikas et al. [20] and Filos-Ratsikas and Voudouris [21], who initiated the study of the distortion in distributed normalized and metric social choice settings, respectively. As already previously discussed, we improve the results of Filos-Ratsikas and Voudouris by extending them to hold for general metrics and asymmetric districts, and also show bounds for many other objectives. In our terminology, Filos-Ratsikas and Voudouris showed a tight bound of 3 for cardinal distributed mechanisms and a tight distortion of 7 for ordinal mechanisms, when the metric is a line, the districts have the same size, and the objective is the social cost (which is equivalent to our $\mathtt{AVG} \circ \mathtt{AVG}$ objective when the districts are symmetric). Interestingly, not only do we generalize these results to hold for asymmetric districts and other objectives, but our composition theorem also provides easier proofs, compared to the characterizations of worst-case instances used in their paper.

2 Preliminaries

An instance of our problem is defined as a tuple $I = (N, A, D, \delta)$, where

- N is a set of n *agents*.
- A is a set of m *alternatives*.
- D is a collection of k *districts*, which define a partition of N (i.e., each agent belongs to a single district). Let N_d be the set of agents that belong to district $d \in D$, and denote by $n_d = |N_d|$ the size of d.
- δ is a *metric space* that contains points representing the agents and the alternatives. In particular, δ defines a *distance* $\delta(i, j)$ between any $i, j \in N \cup A$, such that the *triangle inequality* is satisfied, i.e., $\delta(i, j) \leq \delta(i, x) + \delta(x, j)$ for every $i, j, x \in N \cup A$.

A *distributed* mechanism takes as input information about the metric space, which can be of cardinal or ordinal nature (e.g., agents and alternatives could specify their exact distances between them, or the linear orderings that are induced by the distances), and outputs a single winner alternative $w \in A$ by implementing the following two steps:

- Step 1: For every district $d \in D$, the agents therein decide a *representative* alternative $y_d \in A$.
- Step 2: Given the district representatives, the output is an alternative $w \in A$.

In both steps, the decisions are made by using *direct voting rules*, which map the preferences of a given subset of agents to an alternative. To be more specific, in the first step, an *in-district* direct voting rule is applied for each district $d \in D$ with input the preferences of the agents in the district (set N_d) to decide its representative $y_d \in A$. Then, in the second step, the district representatives can be thought of as pseudo-agents, and an *over-districts* direct voting rule is applied with input their preferences to decide the final winner $w \in A$. In the special case of instances consisting of a single district, the process is not distributed, and thus the two steps collapse into one: The final winner is the alternative chosen to be the district's representative.

2.1 Objectives

We consider standard minimization objectives that have been studied in the related literature, and also propose new ones that are appropriate in the context distributed setting. Each objective assigns a value to every alternative as a cost function composition $F \circ G$ of an objective function F that is applied over the districts and an objective function G that is applied within the districts. Our main four objectives are defined by considering all possible combinations of $F, G \in \{\text{AVG}, \text{MAX}\}$, where the functions AVG and MAX define an average and a max over districts or agents within a district, respectively. In particular, we have:

- $(\text{AVG} \circ \text{AVG})(j|I) = \frac{1}{k} \sum_{d \in D} \left(\frac{1}{n_d} \sum_{i \in N_d} \delta(i,j) \right).$
- $(\text{MAX} \circ \text{MAX})(j|I) = \max_{d \in D} \max_{i \in N_d} \delta(i,j).$
- $(\text{AVG} \circ \text{MAX})(j|I) = \frac{1}{k} \sum_{d \in D} \max_{i \in N_d} \delta(i,j).$
- $(\text{MAX} \circ \text{AVG})(j|I) = \max_{d \in D} \left\{ \frac{1}{n_d} \sum_{i \in N_d} \delta(i,j) \right\}.$

The AVG∘AVG objective is similar to the well-known utilitarian *average social cost* objective measuring the average total distance between all agents and alternative j; actually, AVG∘AVG coincides with the average social cost when the districts are symmetric (i.e., have the same size), but not in general. The MAX∘MAX objective coincides with the egalitarian *max cost* measuring the maximum distance from j among all agents. The new objectives AVG∘MAX and MAX∘AVG make sense in the context of distributed voting, and can be thought of as measures of fairness between districts. For example, minimizing the MAX∘AVG objective corresponds to making sure that the final choice treats each district fairly so that the average social cost of each district is almost equal to that of any other district. Of course, besides combinations of AVG and MAX, one can define many more objectives; we consider such generalizations in Sect. 5.2.

2.2   Distortion of Voting Rules and Distributed Mechanisms

Direct voting rules can be suboptimal, especially when they have limited access to the metric space (for example, when ordinal information is known about the preferences of the agents over the alternatives). This inefficiency is typically captured in the related literature by the notion of distortion, which is the worst-case ratio between the objective value of the optimal alternative over the objective value of the alternative chosen by the rules. Formally, given a minimization cost objective $F \in \{\text{AVG}, \text{MAX}\}$, the *F-distortion* of a voting rule V is

$$\text{dist}_F(V) = \sup_{I=(N,A,\delta)} \frac{F(V(I)|I)}{\min_{j \in A} F(j|I)},$$

where $V(I)$ denotes the alternative chosen by the voting rule when given as input the (single-district) instance I consisting of a set of agents N, a set of alternatives A, and a metric space δ.

The notion of distortion can be naturally extended for the case of distributed mechanisms. Given a composition objective $F \circ G$, the *($F \circ G$)-distortion* of a distributed mechanism M is

$$\text{dist}_{F \circ G}(M) = \sup_{I=(N,A,D,\delta)} \frac{(F \circ G)(M(I)|I)}{\min_{j \in A}(F \circ G)(j|I)},$$

where $M(I)$ is the alternative chosen by the mechanism when given as input an instance I consisting of a set of agents N, a set of alternatives A, a set D of districts, and a metric space δ. Our goal is to bound the distortion of distributed mechanisms for the different objectives we consider. To this end, we will either show how known results from the literature about the distortion of direct voting rules can be composed to yield distortion bounds for distributed mechanisms, or design explicit mechanisms with low distortion. Due to lack of space, some proofs are omitted.

3   Composition Results for General Metric Spaces

In this section we consider general metric spaces, and show how known distortion bounds for direct voting rules can be composed to yield distortion bounds for distributed mechanisms that rely on those voting rules. Given an objective $F \circ G$, we say that a distributed mechanism is α-in-β-over if it uses an in-district rule with G-distortion at most α and an over-districts rule with F-distortion at most β. Our first technical result is an upper bound on the $(F \circ G)$-distortion of α-in-β-over mechanisms, for any $F, G \in \{\texttt{AVG}, \texttt{MAX}\}$. The proof of the following theorem also follows from the more general Theorem 12 for objectives that are compositions of functions satisfying particular properties, such as monotonicity and subadditivity.

Theorem 1. *For any $F, G \in \{\texttt{AVG}, \texttt{MAX}\}$, the $(F \circ G)$-distortion of any α-in-β-over mechanism is at most $\alpha + \beta + \alpha\beta$.*

Proof. Here, we present a proof only for the $\texttt{AVG} \circ \texttt{AVG}$ objective; the proof for the other objectives is similar. Consider an arbitrary α-in-β-over mechanism M and an arbitrary instance $I = (N, A, D, \delta)$. Let w be the alternative that M outputs as the final winner when given I as input, and denote by o an optimal alternative. By the definition of M, we have the following two properties:

$$\forall j \in A, d \in D : \sum_{i \in N_d} \delta(i, y_d) \le \alpha \sum_{i \in N_d} \delta(i, j) \tag{1}$$

$$\forall j \in A : \sum_{d \in D} \delta(y_d, w) \le \beta \sum_{d \in D} \delta(y_d, j) \tag{2}$$

By the triangle inequality, we have $\delta(i, w) \le \delta(y_d, w) + \delta(i, y_d)$ for any agent $i \in N$. Using this, we obtain

$$(\texttt{AVG} \circ \texttt{AVG})(w|I) = \frac{1}{k} \sum_{d \in D} \left(\frac{1}{n_d} \sum_{i \in N_d} \delta(i, w) \right)$$

$$\le \frac{1}{k} \sum_{d \in D} \left(\frac{1}{n_d} \sum_{i \in N_d} \delta(y_d, w) \right) + \frac{1}{k} \sum_{d \in D} \left(\frac{1}{n_d} \sum_{i \in N_d} \delta(i, y_d) \right)$$

$$= \frac{1}{k} \sum_{d \in D} \delta(y_d, w) + \frac{1}{k} \sum_{d \in D} \left(\frac{1}{n_d} \sum_{i \in N_d} \delta(i, y_d) \right).$$

By (1) and (2) for $j = o$, we obtain

$$(\text{AVG} \circ \text{AVG})(w|I) \leq \beta \cdot \frac{1}{k} \sum_{d \in D} \delta(y_d, o) + \alpha \cdot \frac{1}{k} \sum_{d \in D} \left(\frac{1}{n_d} \sum_{i \in N_d} \delta(i, o) \right)$$

$$= \beta \cdot \frac{1}{k} \sum_{d \in D} \left(\frac{1}{n_d} \sum_{i \in N_d} \delta(y_d, o) \right) + \alpha \cdot (\text{AVG} \circ \text{AVG})(o|I).$$

By the triangle inequality, we have $\delta(y_d, o) \leq \delta(i, y_d) + \delta(i, o)$ for any agent $i \in N$. Using this and (1) for $j = o$, we can upper bound the first term of the last expression above as follows:

$$\beta \cdot \frac{1}{k} \sum_{d \in D} \left(\frac{1}{n_d} \sum_{i \in N_d} \delta(y_d, o) \right) \leq \beta \cdot \frac{1}{k} \sum_{d \in D} \left(\frac{1}{n_d} \sum_{i \in N_d} \delta(i, y_d) \right)$$

$$+ \beta \cdot \frac{1}{k} \sum_{d \in D} \left(\frac{1}{n_d} \sum_{i \in N_d} \delta(i, o) \right)$$

$$\leq (\beta + \alpha\beta) \cdot \frac{1}{k} \sum_{d \in D} \left(\frac{1}{n_d} \sum_{i \in N_d} \delta(i, o) \right)$$

$$= (\beta + \alpha\beta) \cdot (\text{AVG} \circ \text{AVG})(o|I).$$

Putting everything together, we obtain the desired bound.   □

By applying Theorem 1 using known distortion results, we can show bounds on the $(F \circ G)$-distortion of distributed mechanisms, for any $F, G \in \{\text{AVG}, \text{MAX}\}$. Specifically, if the whole metric is known (we have access to the exact distances between agents and alternatives), we can compute the alternative that optimizes F and G, thus obtaining a 1-in-1-over distributed mechanism.

Corollary 1. *For any* $F, G \in \{\text{AVG}, \text{MAX}\}$, *there exists a cardinal distributed mechanism with* $(F \circ G)$-*distortion at most 3.*

If only ordinal information is available about the distances between agents and alternatives, then we can employ the PLURALITYMATCHING rule of Gkatzelis *et al.* [22] both within and over the districts. This rule is known to achieve the best possible distortion of 3 among all ordinal rules, for any $F, G \in \{\text{AVG}, \text{MAX}\}$. In fact, this rule achieves a distortion bound of 2 when all agents are at distance 0 from their top alternative; this is the case when the agents are a subset of the alternatives as in the second step of a distributed mechanism.[2] Hence, we have a 3-in-2-over mechanism, and Theorem 1 yields the following statement.

Corollary 2. *For any* $F, G \in \{\text{AVG}, \text{MAX}\}$, *there exists an ordinal distributed mechanism with* $(F \circ G)$-*distortion at most 11.*

---

[2] See Theorem 1 and Proposition 6 in the arxiv version of [22].

Corollaries 1 and 2 demonstrate the power of Theorem 1. However, this method does not always lead to the best possible distributed mechanisms. In particular, let us consider the objectives MAX ∘ G, for $G \in \{\text{AVG}, \text{MAX}\}$, and the class of ordinal mechanisms. We can improve upon the bound of 11 implied by Corollary 2 using a much simpler class of ordinal mechanisms. A distributed mechanism is *α-in-arbitrary-over* if it chooses the district representatives using an ordinal in-district rule with G-distortion at most α, and then outputs an arbitrary representative as the final winner.

Theorem 2. *For any $G \in \{\text{AVG}, \text{MAX}\}$, the (MAX ∘ G)-distortion of any α-in-arbitrary-over mechanism is at most $2 + \alpha$.*

Using again the rule of Gkatzelis *et al.* [22] as an in-district rule, we obtain a 3-in-arbitrary-over mechanism, and Theorem 2 implies the following result.

Corollary 3. *For any $G \in \{\text{AVG}, \text{MAX}\}$, there exists an ordinal distributed mechanism with (MAX ∘ G)-distortion at most 5.*

4   Improved Results on the Line Metric

We now focus on the line metric, where agents and alternatives are points on the line of real numbers. Exploiting this structure, there are classes of mechanisms for which we can obtain significantly improved bounds compared to those implied by the general composition Theorem 1, as well as Theorem 2.

4.1   Ordinal Mechanisms

We start with ordinal distributed mechanisms and the two objectives AVG ∘ G for $G \in \{\text{AVG}, \text{MAX}\}$. Recall that Corollary 2 implies a distortion bound of 11 for these objectives. However, when the metric is a line, we can do much better by observing that there is an ordinal over-districts voting rule with AVG-distortion of 1. In particular, we can identify the median district representative and choose it as the final winner. Using the rule of Gkatzelis *et al.* [22] as the in-district voting rule, we obtain a distributed 3-in-1-over mechanism with distortion at most 7 due to Theorem 1.

Corollary 4. *When the metric is a line, there exists an ordinal distributed mechanism with (AVG ∘ G)-distortion at most 7, for any $G \in \{\text{AVG}, \text{MAX}\}$.*

Corollary 4 recovers the tight distortion bound of 7 by Filos-Ratsikas and Voudouris [21] for AVG ∘ AVG when the districts are symmetric, and also extends it to the case of asymmetric districts. For AVG ∘ MAX, this bound of 7 is a first improvement, but we can do even better with the following ARBITRARY-MEDIAN mechanism:

1. For every district $d \in D$, choose its representative to be the favorite alternative of an arbitrary agent $j_d \in N_d$.

2. Output the median representative as the final winner.

Theorem 3. *When the metric is a line,* ARBITRARY-MEDIAN *has* (AVG ∘ MAX)-*distortion at most* 5.

Next, we show an almost matching lower bound of approximately 4.23. Before going through with the proof, we argue that ordinal distributed mechanisms with finite distortion must be unanimous. Formally, a distributed mechanism is *unanimous* if it chooses the representative of a district d to be an alternative a whenever all agents in N_d prefer a over all other alternatives.

Lemma 1. *For any* $F, G \in \{\text{AVG}, \text{MAX}\}$, *every ordinal distributed mechanism with finite* $(F \circ G)$-*distortion must be unanimous.*

We are now ready to present the ordinal lower bound for AVG ∘ MAX.

Theorem 4. *The* (AVG ∘ MAX)-*distortion of any ordinal distributed mechanism is at least* $2 + \sqrt{5} - \varepsilon$, *for any* $\varepsilon > 0$, *even when the metric is a line.*

Proof. Suppose towards a contradiction that there is an ordinal distributed mechanism M with distortion strictly smaller than $2 + \sqrt{5} - \varepsilon$, for any $\varepsilon > 0$. We will define instances with two alternatives a and b (located at 0 and 1, respectively), and districts consisting of the same size. Without loss of generality, we assume that M chooses alternative a as the final winner when given as input any instance with only two districts, such that both alternatives are representative of some district. Let x and y be two integers such that $\phi > y/x \geq \phi - \varepsilon/2$, where $\phi = (1 + \sqrt{5})/2$ is the golden ratio.

First, consider an instance I_1 consisting of the following two districts:

- The first district consists of two agents, such that one of them prefers alternative a and the other prefers alternative b.
- The second district consists of two agents, such that both of them prefer alternative b. Due to unanimity (Lemma 1), the representative of this district must be b.

Suppose that M chooses a as the representative of the first district, in which case both alternatives are representative of some district, and thus M chooses a as the final winner. Consider the following metric:

- In the first district, the agent that prefers alternative a is positioned at $1/2$, whereas the agent that prefers alternative b is positioned at $3/2$.
- In the second district, both agents are positioned at 1.

Then, we have that $(\text{AVG} \circ \text{MAX})(a|I_1) = \frac{1}{2}\left(\frac{3}{2} + 1\right) = 5/4$ and $(\text{AVG} \circ \text{MAX})(b|I_1) = \frac{1}{2}\left(\frac{1}{2} + 0\right) = 1/4$, leading to a distortion of 5. Consequently, M must choose b as the representative of the first district (where one agent prefers a and one prefers b).

Next, we argue that for instances with $x + y$ districts such that a is the representative of x districts and b is the representative of y districts, M must choose b as the final winner. Assume otherwise that M chooses a in such a situation, and consider an instance I_2 consisting of the following $x + y$ districts:

- Each of the first x districts consists of a single agent that is positioned at $1/2$ and prefers alternative a; thus, a is the representative of all these districts.
- Each of the next y districts consists of a single agent that is positioned at 1 and prefers alternative b; thus, b is the representative of all these districts.

Since $(\text{AVG} \circ \text{MAX})(a|I_2) = \frac{1}{x+y}\left(\frac{x}{2}+y\right) = \frac{1}{x+y}\cdot\frac{x+2y}{2}$ and $(\text{AVG}\circ\text{MAX})(b|I_2) = \frac{1}{x+y}\left(\frac{x}{2}+0\right) = \frac{1}{x+y}\cdot\frac{x}{2}$, the distortion is $\frac{x+2y}{x} = 1+2\frac{y}{x} \geq 1+2\phi-\varepsilon = 2+\sqrt{5}-\varepsilon$. Consequently, M must choose b, whenever there are $x+y$ districts such that a is the representative of x districts and b is the representative of the remaining y districts.

Finally, consider an instance I_3 with the following $x+y$ districts:

- Each of the first x districts consists of two agents that are positioned at 0 and prefer alternative a. Due to unanimity, a must be the representative of all these districts.
- Each of the next y districts consists of two agents, such that one of them is positioned at $-1/2$ and prefers alternative a, while the other is positioned at $1/2$ and prefers alternative b. By the discussion above (about instance I_1), b must be the representative of these districts.

As a is the representative of x districts and b is the representative of y districts, by the discussion above (about instance I_2), M chooses b as the final winner. Since $(\text{AVG}\circ\text{MAX})(a|I_3) = \frac{1}{x+y}\left(0+\frac{y}{2}\right) = \frac{1}{x+y}\cdot\frac{y}{2}$ and $(\text{AVG}\circ\text{MAX})(b|I_3) = \frac{1}{x+y}\left(x+\frac{3y}{2}\right) = \frac{1}{x+y}\cdot\frac{2x+3y}{2}$, the distortion is $\frac{2x+3y}{y} = 3+\frac{2x}{y} > 3+\frac{2}{\phi} = 1+2\phi = 2+\sqrt{5}$. This contradicts our assumption that M has distortion smaller than $2+\sqrt{5}-\varepsilon$. $\square$

Next, we consider the two objectives $\text{MAX}\circ G$ for $G \in \{\text{AVG},\text{MAX}\}$. For both of them, Corollary 3 implies a distortion bound of at most 5. When $G = \text{MAX}$ and the metric is a line, we can get an improved bound of 3 using a rather simple ARBITRARY-DICTATOR mechanism:

1. For each district $d \in D$, choose its representative to be the favorite alternative of an arbitrary agent in N_d.
2. Output an arbitrary district representative as the final winner.

Theorem 5. *When the metric is a line,* ARBITRARY-DICTATOR *has* $(\text{MAX}\circ\text{MAX})$-*distortion at most 3.*

The following theorem shows that the bound of 3 is the best possible we can hope for the $\text{MAX} \circ \text{MAX}$ objective using a unanimous distributed mechanism, even when the metric is a line. This lower bound directly extends to ordinal mechanisms, as any such mechanism with finite distortion has to be unanimous (Lemma 1).

Theorem 6. *The* $(\text{MAX} \circ \text{MAX})$-*distortion of any unanimous distributed mechanism is at least 3, even when the metric is a line.*

Finally, let us focus on the objective MAX ∘ AVG, for which we show a lower bound of approximately 4.23 on the distortion of all ordinal distributed mechanisms, thus almost matching the upper bound of 5 implied by Corollary 3.

Theorem 7. *The* (MAX ∘ AVG)*-distortion of any ordinal distributed mechanism is at least* $2 + \sqrt{5} - \varepsilon$, *for any* $\varepsilon > 0$, *even when the metric is a line.*

4.2   Cardinal Mechanisms

We now turn out attention to distributed mechanisms that have access to the distances between agents and alternatives. Recall that for such mechanisms, Corollary 1 implies a distortion bound of 3 for all the objectives we have considered so far. As in the case of ordinal mechanisms, when the metric is a line, Filos-Ratsikas and Voudouris [21] showed a matching lower bound of 3 for AVG ∘ AVG (when the districts are symmetric), which extends for AVG ∘ MAX as the construction also works for instances with single-agent districts, in which case MAX = AVG.

For objectives of the form MAX ∘ G, we design a novel distributed mechanism that is tailor-made for the line metric and achieves distortion at most $1 + \sqrt{2} \le$ 2.42. This mechanism is particularly interesting as it is *not* unanimous: Even when all agents in a district prefer an alternative a to everyone else (i.e., a is the closest alternative to all agents), the mechanism may end up choosing a different representative. In fact, by Theorem 6, we have that any unanimous mechanism cannot achieve a (MAX ∘ G)-distortion better than 3, and so to break this barrier, our mechanism *has* to be non-unanimous.

For a given $\lambda \ge 1$, we say that an alternative is λ-*acceptable* for a district $d \in D$ if her G-value for the agents in N_d is at most λ times the G-value of any other alternative for the agents in N_d. Given an objective G, we define a class of distributed mechanisms parameterized by λ that work as follows:

- For each district d, choose its representative to be the *rightmost* λ-acceptable alternative for the district.
- Output the leftmost district representative as the final winner.

We refer to this class of mechanisms as λ-ACCEPTABLE-RIGHTMOST-LEFTMOST (or λ-ARL, for short).

Theorem 8. *For any* $G \in \{\text{AVG}, \text{MAX}\}$, *the* (MAX ∘ G)*-distortion of* $(1+\sqrt{2})$*-ARL is at most* $1 + \sqrt{2}$.

We conclude this section by presenting a lower bound of $1 + \sqrt{2}$ on the (MAX∘G)-distortion of distributed mechanisms, which holds even when the metric is a line, thus showing that $(1 + \sqrt{2})$-ARL is the best possible on the line.

Theorem 9. *For any* $G \in \{\text{AVG}, \text{MAX}\}$, *the* (MAX∘$G$)*-distortion of any distributed mechanism is at least* $1 + \sqrt{2}$, *even when the metric is a line.*

5   Extensions and Generalizations

5.1   Mechanisms that Select from the Set of Representatives

In the previous sections we looked at distributed mechanisms, which can choose any alternative as the final winner by essentially considering the district representatives as proxies. We now focus on the case where the final winner can only be chosen from among the district representatives (as in the work of [21]). To make the distinction between general mechanisms and those that select from the pool of district representatives clear, we will use the term *representative-selecting* to refer to the latter.

It is not hard to see that, with the exception of the bounds implied by Theorem 1 and its corollaries for general metric spaces, the rest of our results follow by representative-selecting mechanisms. In particular, every α-in-arbitrary-over mechanism, as well as ARBITRARY-DICTATOR, choose some arbitrary representative; the 1-in-1-over cardinal mechanism and the 3-in-1-over ordinal mechanism for AVG $\circ$ G in the line metric, as well as ARBITRARY-MEDIAN, choose the median representative; every λ-ARL mechanism chooses the leftmost representative (Theorem 8). It is also not hard to see that all our lower bounds also extend for the class of representative-selecting mechanisms: some representative is always chosen as the final winner in all instances used in the constructions. Based on all of the above discussion, we have the following corollary, which collects the best distortion bounds for the different objectives we consider.

Theorem 10. *We can form ordinal representative-selecting mechanisms with distortion at most 5 for* MAX $\circ$ G *and general metric spaces. When the metric space is a line, the worst-case distortion of ordinal mechanisms is exactly 7 for* AVG $\circ$ AVG, *between* $2 + \sqrt{5}$ *and 5 for* AVG $\circ$ MAX, *exactly 3 for* MAX $\circ$ MAX, *and at least* $2 + \sqrt{5}$ *for* MAX $\circ$ AVG. *When the metric is a line, the distortion of cardinal representative-selecting mechanisms is exactly 3 for* AVG $\circ$ G, *and exactly* $1 + \sqrt{2}$ *for* MAX $\circ$ G.

Now, let us see how choosing only from the district representatives affects the bounds implied by Theorem 1 for general metric spaces. For clarity, we focus on objectives of the form AVG $\circ$ G; our discussion can easily be adapted for objectives of the form MAX $\circ$ G. Let M be a representative-selecting mechanism that uses some in-district and over-districts direct voting rules. Given an instance $I = (N, M, D, \delta)$, let $R = R_M(I)$ be the set of district representatives chosen by M, and denote by $w = M(I) \in R$ the final winner. Clearly, Theorem 1 would hold without any modifications if, for any instance I, w satisfies inequality (2) in the proof of the theorem. However, the distortion guarantees of direct voting rules used by M in and over the districts are usually only with respect to the set of alternatives from which they are allowed to choose. So, if M uses an over-districts rule that has AVG-distortion at most γ, we have that, for any instance I, w is such that

$$\forall j \in R : \sum_{i \in R} \delta(i, w) \leq \gamma \cdot \sum_{i \in R} \delta(i, j). \tag{3}$$

Inequality (3) cannot directly substitute inequality (2) in the proof of Theorem 1 as it may be the case that the optimal alternative is not included in the set of representatives. So, we need to understand the relation between β and γ.

Lemma 2. *For any mechanism M, it holds that $\beta \leq 2\gamma$.*

Due to Lemma 2, Theorem 1 implies the following distortion bounds for general metric spaces and α-in-γ-over representative-selecting mechanisms.

Theorem 11. *For general metric spaces and any $F, G \in \{\text{AVG}, \text{MAX}\}$, the $(F \circ G)$-distortion of any α-in-γ-over representative-selecting mechanism is at most $\alpha + 2\gamma + 2\alpha\gamma$.*

Using appropriate rules in and over the districts, we can now again obtain concrete upper bounds on the distortion of cardinal and ordinal representative-selecting mechanisms.

Corollary 5. *For general metric spaces and any $F, G \in \{\text{AVG}, \text{MAX}\}$, there is a representative-selecting mechanism with $(F \circ G)$-distortion at most 5, and an ordinal representative-selecting mechanism with $(F \circ G)$-distortion at most 19.*

5.2   More General Objectives

We now consider again mechanisms that can choose the final winner from the set of all alternatives, and discuss how some of our results can be extended for objectives $F \circ G$ beyond the cases where $F, G \in \{\text{AVG}, \text{MAX}\}$.

Generalizing Theorem 1. We previously showed in Theorem 1 that the $(F \circ G)$-distortion of distributed mechanisms can be bounded in terms of the F- and G-distortion of the voting rules used in and over the districts. Here, we show that this theorem still holds for a much more general class of functions. To define this properly, we should think of F and G as functions that take as input vectors of distances. More precisely, given an instance $I = (N, A, D, \delta)$, let f and g be functions so that the *cost* of any alternative $j \in A$ is

$$(F \circ G)(j|I) = f\Big(g\big(\vec{\delta}_1(j)\big), \ldots, g\big(\vec{\delta}_k(j)\big)\Big),$$

where $\vec{\delta}_d(j)$ is the vector consisting of the distances $\delta(i, j)$ between every agent $i \in N_d$ and alternative j. To give a few examples, $g\big(\vec{\delta}_d(j)\big) = \frac{1}{n_d} \sum_{i \in N_d} \delta(i, j)$ if $G = \text{AVG}$, and $g\big(\vec{\delta}_d(j)\big) = \max_{i \in N_d} \delta(i, j)$ if $G = \text{MAX}$. More generally, we consider functions f and g which satisfy the following properties:

- **Monotonicity**: A function f is *monotone* if $f(\vec{v}) \leq f(\vec{u})$, for any two vectors $\vec{v}$ and $\vec{u}$ such that $v_\ell \leq u_\ell$ for every index ℓ.

- **Subadditivity**: A function f is *subadditive* if $f(\vec{v} + \vec{u}) \leq f(\vec{v}) + f(\vec{u})$, for any two vectors $\vec{v}$ and $\vec{u}$. Moreover, for any scalar $c \geq 1$, it must be that $f(c \cdot \vec{v}) \leq c \cdot f(\vec{v})$.[3]
- **Consistency**: A function f is *consistent* if $f(\vec{v}) = c$, for any vector $\vec{v}$ such that $v_\ell = c$ for every index ℓ.

Note that both AVG and MAX, as well as many other functions, obey all of the above properties.

Theorem 12. *The distortion of any α-in-β-over mechanism is at most $\alpha + \beta + \alpha\beta$, for any objective $F \circ G$ defined by functions f and g which are monotone, subadditive, and consistent.*

Generalizing Theorem 8. When the metric space is a line, Theorem 8 holds for more general objectives of the form MAX $\circ\, G$. In particular, we are aiming to minimize the maximum cost of any district, which for an alternative $j \in A$ is given by a function $g\left(\vec{\delta}_d(j)\right)$. We again assume that g is monotone, subadditive, and consistent. In addition, we require that g is *single-peaked*: for any district d, there is a there is a unique alternative j that minimizes $g\left(\vec{\delta}_d(j)\right)$, and g increases monotonically as we move from the location of j (to the left or the right).

As in Sect. 4.2, the upper bound on the distortion is due to the λ-ARL mechanism (for a specific value of λ), which chooses the representative of each district to be the rightmost λ-acceptable alternative for the district, and then outputs the leftmost representative as the final winner. Recall that the set of λ-acceptable alternatives for a district d contains all the alternatives x such that $g\left(\vec{\delta}_d(x)\right) \leq \lambda \cdot \min_{j \in A} g\left(\vec{\delta}_d(j)\right)$.

Theorem 13. *The distortion of $(1 + \sqrt{2})$-ARL is at most $1 + \sqrt{2}$ for any objective of the form MAX $\circ\, G$, where G is defined by a monotone, subadditive, consistent, and single-peaked function g.*

6   Open Problems

In this paper, we showed bounds on the distortion of single-winner distributed mechanisms for many different objectives, some of which are novel and make sense only in this particular setting. Still, there are several challenging open questions, as well as new directions for future research. Starting with our results, it would be interesting to close the gaps between the lower and upper bounds presented in Table 1 for the various scenarios we considered. For cases where our bounds for general metrics and the line differ significantly, such as for ordinal

---

[3] The latter condition, sometimes known as sub-homogeneity, is not usually included in the standard definition of subadditive functions. It is easily implied by the first subadditivity condition when c is an integer.

mechanisms and the AVG∘MAX objective, one could focus on other well-structured metrics, like the Euclidean space or generalizations of it.

Since we focused exclusively on deterministic mechanisms, a possible direction could be to consider randomized mechanisms and investigate whether better distortion bounds are possible. Note that our composition theorem (Theorem 1 and its variants) already provide randomized bounds by plugging in appropriate randomized in-district and over-districts direct voting rules. However, these bounds seem extremely loose, and different techniques are required to obtain tight bounds. Going beyond the single-winner setting, one could study the distortion of distributed mechanisms that output committees of a given number of alternatives, or rankings of all alternatives. Finally, another interesting direction would be to study what happens when agents act strategically, and either understand how this behavior affects given distributed mechanisms, or aim to design strategyproof mechanisms that are resilient to manipulation and at the same time achieve low distortion.

References

1. Abramowitz, B., Anshelevich, E.: Utilitarians without utilities: maximizing social welfare for graph problems using only ordinal preferences. In: Proceedings of the 32nd AAAI Conference on Artificial Intelligence (AAAI), pp. 894–901 (2018)
2. Amanatidis, G., Birmpas, G., Filos-Ratsikas, A., Voudouris, A.A.: A few queries go a long way: information-distortion tradeoffs in matching. In: Proceedings of the 35th AAAI Conference on Artificial Intelligence (AAAI), pp. 5078–5085 (2021)
3. Amanatidis, G., Birmpas, G., Filos-Ratsikas, A., Voudouris, A.A.: Peeking behind the ordinal curtain: improving distortion via cardinal queries. Artif. Intell. **296**, 103488 (2021)
4. Anagnostides, I., Fotakis, D., Patsilinakos, P.: Metric-distortion bounds under limited information. CoRR arXiv:2107.02489 (2021)
5. Anshelevich, E., Bhardwaj, O., Elkind, E., Postl, J., Skowron, P.: Approximating optimal social choice under metric preferences. Artif. Intell. **264**, 27–51 (2018)
6. Anshelevich, E., Filos-Ratsikas, A., Shah, N., Voudouris, A.A.: Distortion in social choice problems: the first 15 years and beyond. In: Proceedings of the 30th International Joint Conference on Artificial Intelligence (IJCAI), pp. 4294–4301 (2021)
7. Anshelevich, E., Postl, J.: Randomized social choice functions under metric preferences. J. Artif. Intell. Res. **58**, 797–827 (2017)
8. Anshelevich, E., Zhu, W.: Ordinal approximation for social choice, matching, and facility location problems given candidate positions. In: Proceedings of the 14th International Conference on Web and Internet Economics (WINE), pp. 3–20 (2018)
9. Benadè, G., Nath, S., Procaccia, A.D., Shah, N.: Preference elicitation for participatory budgeting. In: Proceedings of the 31st AAAI Conference on Artificial Intelligence (AAAI), pp. 376–382 (2017)
10. Benadè, G., Procaccia, A.D., Qiao, M.: Low-distortion social welfare functions. In: Proceedings of the 33rd AAAI Conference on Artificial Intelligence (AAAI), pp. 1788–1795 (2019)
11. Bhaskar, U., Dani, V., Ghosh, A.: Truthful and near-optimal mechanisms for welfare maximization in multi-winner elections. In: Proceedings of the 32nd AAAI Conference on Artificial Intelligence (AAAI), pp. 925–932 (2018)

12. Borodin, A., Lev, O., Shah, N., Strangway, T.: Primarily about primaries. In: Proceedings of the 33rd AAAI Conference on Artificial Intelligence (AAAI), pp. 1804–1811 (2019)
13. Boutilier, C., Caragiannis, I., Haber, S., Lu, T., Procaccia, A.D., Sheffet, O.: Optimal social choice functions: a utilitarian view. Artif. Intell. **227**, 190–213 (2015)
14. Caragiannis, I., Nath, S., Procaccia, A.D., Shah, N.: Subset selection via implicit utilitarian voting. J. Artif. Intell. Res. **58**, 123–152 (2017)
15. Caragiannis, I., Procaccia, A.D.: Voting almost maximizes social welfare despite limited communication. Artif. Intell. **175**(9–10), 1655–1671 (2011)
16. Chen, X., Li, M., Wang, C.: Favorite-candidate voting for eliminating the least popular candidate in a metric space. In: Proceedings of the 34th AAAI Conference on Artificial Intelligence (AAAI), pp. 1894–1901 (2020)
17. Fain, B., Goel, A., Munagala, K., Prabhu, N.: Random dictators with a random referee: constant sample complexity mechanisms for social choice. In: Proceedings of the 33rd AAAI Conference on Artificial Intelligence (AAAI), pp. 1893–1900 (2019)
18. Feldman, M., Fiat, A., Golomb, I.: On voting and facility location. In: Proceedings of the 2016 ACM Conference on Economics and Computation (EC), pp. 269–286 (2016)
19. Filos-Ratsikas, A., Frederiksen, S.K.S., Zhang, J.: Social welfare in one-sided matchings: random priority and beyond. In: Lavi, R. (ed.) SAGT 2014. LNCS, vol. 8768, pp. 1–12. Springer, Heidelberg (2014). https://doi.org/10.1007/978-3-662-44803-8_1
20. Filos-Ratsikas, A., Micha, E., Voudouris, A.A.: The distortion of distributed voting. Artif. Intell. **286**, 103343 (2020)
21. Filos-Ratsikas, A., Voudouris, A.A.: Approximate mechanism design for distributed facility location. In: Caragiannis, I., Hansen, K.A. (eds.) SAGT 2021. LNCS, vol. 12885, pp. 49–63. Springer, Cham (2021). https://doi.org/10.1007/978-3-030-85947-3_4
22. Gkatzelis, V., Halpern, D., Shah, N.: Resolving the optimal metric distortion conjecture. In: Proceedings of the 61st IEEE Annual Symposium on Foundations of Computer Science (FOCS), pp. 1427–1438 (2020)
23. Goel, A., Krishnaswamy, A.K., Munagala, K.: Metric distortion of social choice rules: lower bounds and fairness properties. In: Proceedings of the 2017 ACM Conference on Economics and Computation (EC), pp. 287–304 (2017)
24. Jaworski, M., Skowron, P.: Evaluating committees for representative democracies: the distortion and beyond. In: Proceedings of the 29th International Joint Conference on Artificial Intelligence (IJCAI), pp. 196–202 (2020)
25. Kempe, D.: An analysis framework for metric voting based on LP duality. In: Proceedings of the 34th AAAI Conference on Artificial Intelligence (AAAI), pp. 2079–2086 (2020)
26. Kempe, D.: Communication, distortion, and randomness in metric voting. In: Proceedings of the 34th AAAI Conference on Artificial Intelligence (AAAI), pp. 2087–2094 (2020)
27. Mandal, D., Procaccia, A.D., Shah, N., Woodruff, D.P.: Efficient and thrifty voting by any means necessary. In: Proceedings of the 32nd Annual Conference on Neural Information Processing Systems (NeurIPS), pp. 7178–7189 (2019)
28. Mandal, D., Shah, N., Woodruff, D.P.: Optimal communication-distortion tradeoff in voting. In: Proceedings of the 21st ACM Conference on Economics and Computation (EC), pp. 795–813 (2020)

29. Munagala, K., Wang, K.: Improved metric distortion for deterministic social choice rules. In: Proceedings of the 2019 ACM Conference on Economics and Computation (EC), pp. 245–262 (2019)
30. Procaccia, A.D., Rosenschein, J.S.: The distortion of cardinal preferences in voting. In: Klusch, M., Rovatsos, M., Payne, T.R. (eds.) CIA 2006. LNCS (LNAI), vol. 4149, pp. 317–331. Springer, Heidelberg (2006). https://doi.org/10.1007/11839354_23
31. Sen, A.: Social choice theory. In: Handbook of Mathematical Economics, vol. 3, pp. 1073–1181 (1986)

Maximal Information Propagation
via Lotteries

Jing Chen[1,2] and Bo Li[3(✉)]

[1] Algorand Inc., Boston, MA 02116, USA
jing@algorand.com
[2] Department of Computer Science, Stony Brook University,
Stony Brook, NY 11794, USA
[3] Department of Computing, Hong Kong Polytechnic University,
Hong Kong, China
comp-bo.li@polyu.edu.hk

Abstract. Propagating information to more people through their friends is becoming an increasingly important technology used in domains such as blockchain, advertising, and social media. To incentivize people to broadcast the information, the designer may use a monetary rewarding scheme, which specifies who gets how much, to compensate for the propagation. Several properties are desirable for the rewarding scheme, such as budget feasible, individually rational, incentive compatible and Sybil-proof. In this work, we design a free market with lotteries, where every participant can decide by herself how much of the reward she wants to withhold before propagating to others. We show that in the free market, the participants have a strong incentive to maximally propagate the information and all the above properties are satisfied automatically.

Keywords: Information propagation · Nash equilibrium · Free market design

1 Introduction

Propagating information to more people through their friends is becoming an increasingly important technique used in many fields including advertising, social media [13] and blockchain [3,15]. To incentivize people to broadcast the information, the information holder, i.e., the mechanism designer, may use a monetary rewarding scheme $\mathbf{r} = (r_1, \cdots, r_n)$ to compensate people, where r_i is the reward assigned to player i. The rewarding scheme is expected to satisfy several properties, such as being *incentive compatible, (strongly) budget feasible, individual rational,* and *Sybil-proof.* Informally, incentive compatibility requires that each

The authors thank Costantinos Daskalakis, Yossi Gilad, Victor Luchangco, and Silvio Micali for helpful discussions in the early stage of this work, and several anonymous reviewers for helpful comments. Part of this work was done when Bo Li was an intern at Algorand. Bo Li is partially supported by the Hong Kong Polytechnic University (No. P0034420).

M. Feldman et al. (Eds.): WINE 2021, LNCS 13112, pp. 486–503, 2022.
https://doi.org/10.1007/978-3-030-94676-0_27

player does not decrease her utility by propagating the information, budget feasibility requires $\sum_i r_i \leq B$ when the mechanism designer has a budget of B, strong budget feasibility requires $\sum_i r_i = B$, individual rationality requires $r_i \geq 0$ for all i, and Sybil-proofness requires the players do not benefit by making fake copies in the information propagation process. Accordingly, designing a rewarding scheme to satisfy all or some of the above properties establishes a large research agenda. For example, maximal information propagation with budgets is studied in [27], where a rewarding scheme that is incentive compatible, strongly budget feasible and individual rational is proposed. But the scheme is not Sybil-proof. A Sybil-proof rewarding scheme is designed in [6], but it is not strongly budget feasible, where a small portion of the claimed reward is distributed among the players. In this work, instead of designing a centralized rewarding scheme, we propose a free market with lotteries where everyone can decide how much of the reward she wants to withhold before propagating to others. We show that in such a market, the players have strong incentives to fully propagate the information and all above properties are satisfied automatically.

To illustrate our design, let us first consider a toy game. Initially, a seller sends her promotion information to a small number of players she is able to reach, denoted by N and $n = |N|$ who are called *aware players*. The information is associated with a lottery such that one *winner* among aware players will be uniformly and randomly selected to get a reward normalized to \$1. When nobody propagates the information, everyone's expected reward is $1/n$. For player $i \in N$, she has a set of friends F_i such that $f_i = |F_i| \geq 1$ and $N \cap F_i = \emptyset$. If i signs an agreement with F_i such that if anyone in F_i is selected to be the winner, the reward is given i, then by propagating the information to F_i, player i's reward, which equals to the probability that the winner is selected from $F_i \cup \{i\}$, is

$$\frac{1+f_i}{n+f_i} > \frac{1}{n}. \tag{1}$$

That is i can increase her reward by propagating the information to her friends, and thus no propagation for N is not a Nash equilibrium. Actually, since Inequality 1 holds for any $n \geq 2$, everyone's dominate strategy is to propagate the formation to their friends. But to what extent will they propagate?

Assuming all players in N except i inform the information to their friends and withhold the complete reward, let us see what i will do. If i does not withhold the complete reward but shares a small amount, say $0 < c < 1$, with F_i, i can again improve her utility. For $j \in F_i$, let F_j be j's friends who are not in the game. Supposing all $j \in F_j$ adopts the same strategy with $N \setminus \{i\}$, i.e., fully propagating the information to their friends and withholding the complete reward c, we compare two cases for player i: (1) i does not leave any reward to j, and (2) i leaves c to j. To ease the notation, suppose there are n' players who are in the game except j's friends F_j for every $j \in F_i$. It is not hard to see that for case (1), player i's reward is $(1 + f_i)/n'$ and for case (2), her reward is

$$\frac{1 + (1 - c) \cdot \sum_{j \in F_i} f_j}{\sum_{j \in F_i} f_j + n'} > \frac{1 + f_i}{n'},$$

as long as for all $j \in F_i$,

$$f_j > \frac{n+1}{n(1-c)-1} \to 1 \text{ if } c \to 0.$$

Thus we conclude that withholding the complete reward from propagation is not a Nash equilibrium, and every player wants to leave partial reward to her friends for incentivizing them to further propagate the information.

We note that the initial n aware players are in the *Prisoner's dilemma*. If they do not propagate the information, each of them has expected utility of $1/n$. However, it is dominant for each of them to refer their friends by sacrificing partial reward, resulting the expected utility strictly smaller than $1/n$. This dilemma actually motivates the information propagation in the free market.

A more practical example for the above scenario is the mining game of Bitcoin [24]. In Bitcoin, when a user makes a transaction (the information sender), she wants the transaction to be broadcasted (with other necessary information such as account information, transfer amount, crypto signature and etc.) in the network so that the miners can authorize the validity of the transaction and assemble newly verified transactions into blocks. The miners compete to propose their blocks to the public chain by solving a computationally hard puzzle, and the winning probability is proportional to the share of each miner's computation power in the system. Accordingly, the transaction maker can reward the winning miner who authorizes her transaction a fixed amount of Bitcoins. At first glance, it seems that the miners may not want to broadcast the transaction since only the miners who know the transaction can be rewarded. A centralized rewarding scheme is proposed in [3] which not only rewards the winner but also other miners who helped broadcast the transaction. By carefully designing who gets much, their scheme is Sybil-proof in tree networks. However, as we will show in this work, the design of free market with lotteries automatically incentivizes the miners to propagate the transaction and satisfies all other desired properties as well. Thus the take-home message of this work is that

> the mechanism designer does not need to specify each player's reward, the market itself already provides incentives for maximal propagation.

1.1   Our Contribution

We model the problem, and the Bitcoin example, as an information propagation game in a free market, where a sender has a single piece of information to be broadcasted. For simplicity, we first assume the players are connected by a complete d-ary tree with $d \geq 3$, and all players' winning probabilities are the same. If there is an edge between two players, they are friends and one can be informed the information by the other. A strategy profile is called *full propagation* if every player withholds a minimum charge and leaves the remaining reward to all her friends, so that a maximum number of people could be aware of the information. We show that full propagation forms a Nash equilibrium which satisfies extra properties and thus is more stable than an arbitrary one.

First, full propagation is robust to collective deviations of friends, i.e., *connected coalition-proof* [4]. In the seminal work by Myerson [23], the communication game is proposed where the network on players represents the possible communication between them. Originally, it is associated with cooperative games and only coalitions formed by players who can communicate with each other (i.e., connected subgraphs in the network) are concerned. We adapt this principle to our problem and show that any deviation of a connected coalition of players from full propagation makes at least one of them worse off.

Second, full propagation survives in any order of iterative elimination of dominated strategies, and uniquely survives in a particular one. This result coincides with and generalizes the result of [3], where a centralized rewarding scheme is designed. In a centralized rewarding scheme, a player can only misbehave by withholding the information and claiming Sybil copies, while in a free market a player can arbitrarily claim how much of the reward she wants to deduct before propagating to others. We formally discuss the difference between our work and [3] at the end of Sect. 3. Recall that strategy s (weakly) dominating s', denoted by $s \preceq s'$, means choosing s always gives at least as good an outcome as choosing s', no matter what the other players do, and there is at least one profile of opponents' actions for which s gives a strictly better outcome than s'. We prove that full propagation is the unique strategy profile that survives in an interval-based monotone elimination of dominated strategies, coinciding with the players' reasoning process as illustrated in the introduction.

Our main results can be summarized as follows.

Main Result 1. (Theorems 1 and 2) *In the tree-structured free market with lotteries, full propagation achieves maximal propagation and satisfies the following:*

1. *Full propagation is a Nash equilibrium;*
2. *Any deviation of a coalition of friends hurts at least one of them;*
3. *Full propagation survives in any order of iterative elimination of dominated strategies;*
4. *There is an order of iterative elimination of dominated strategies such that full propagation is the unique surviving strategy.*

We then extend the above results to non-tree networks. For arbitrary networks, we introduce stronger relationships than friends, *good friends* and *best friends*, using shortest paths from the information sender to players. Although properties 3 and 4 in Main Result 1 do not hold, we show that if every player has at least three good friends, full propagation is a Nash equilibrium that is also connected coalition-proof on the induced *good-friendship subgraph*. It is noted that d-ary tree with $d \geq 3$ is a special case satisfying this condition.

Main Result 2. (Theorem 3) *If every player has at least 3 good friends, then full propagation is a Nash equilibrium that is connected coalition-proof on the good-friendship subgraph.*

In conclusion, if the network is well structured, the free market with lotteries is incentive compatible where players are willing to fully propagate the

490   J. Chen and B. Li

information. It is not hard to check that our scheme also satisfies other properties mentioned in the introduction. It is strongly budget feasible and individual rational, as the full reward will be given to the players and none of them needs to pay. Our scheme is also Sybil-proof, as in our game, the players are required to provide certificates on how many people they have referred to. For example, in Bitcoin each miner needs to contribute their computing power on authorizing the transaction. Therefore, as everyone in our game can arbitrarily decide how much reward she wants to withhold by informing her friends, making Sybil identities is the same as withholding a higher fraction of the reward.

Finally, it is an interesting future direction to generalize our results to random networks. We believe full propagation brings players high utility in a broader class of networks. To shed more light in this direction, we conduct experiments in Sect. 5 for our scheme under general random networks. In all experiments, full propagation brings a player the maximum utility. Moreover, the utility gap between full propagation and other strategies is actually large.

1.2 Related Works

Our work is partially motivated by the extensive study of incentivizing relays in a blockchain network to propagate transactions. While this has been studied in the literature, most of them focus on centralized algorithms where each relay's reward is fixed and decided by the algorithm. With these algorithms, to gain higher utility, a strategic relay may claim fake copies, i.e., *Sybil attack* [12]. Accordingly, in works such as [3,16], *Sybil-proof* algorithms are studied. Since in our free market each relay is able to arbitrarily decide how much reward she is willing to withhold, it is superfluous for them to make fake copies, and thus, our results directly imply Sybil-proofness. The free market ideas have also been discussed independently in [1] and [7] without a systematic analysis.

Our work also aligns with the fundamental study of what the optimal way is to reward miners for their work on authorizing transactions. Currently, the most popular rewarding scheme, as adopted by Bitcoin, is to reward miners proportionally to their share of the total contributed computational power. As shown in [10], the proportional allocation rule is the unique rule that is simultaneously non-negative, budget-balanced, symmetric, Sybil-proof, and collusion-proof. In reality, however, to earn steady rewards, miners pool themselves together, and the pools are vulnerable to security attacks, such as selfish mining attack [11,17,19,21], block withholding attack [25,26], and denial of service attack [18]. Cooperative games are used in [22] to show that under high transaction loads, it is hard for managers to distribute rewards in a stable way.

Outside the scope of blockchain, information propagation has also been studied in multi-level marketing [13,14] and query incentive networks [2,5,6,9,20]. There are major differences between the transaction propagation in a blockchain network and the query retrieval in peer-to-peer network. The players in a query incentive network do not compete with the ones who forwarded the message to them, and cannot generate an answer that they do not have. Whereas in a blockchain network, every aware player is a potential authorizer with probability proportional to their stakes or computational powers.

2 Game Theoretic Model and Preliminaries

For technical simplicity, we first assume all players are connected by a tree $\mathcal{G} = (V, E)$. Let s be the root of $\mathcal{G}$ who is the sender of the information. Note that s is not a player in our game. Suppose s has f children and thus excluding s from G, there are f subtrees denoted by $\{T_1, \cdots, T_f\}$. When there is no confusion, we also use each T_i to denote the set of nodes in it. Assume all these subtrees are complete d-ary and $f \geq d \geq 3$. Call $N = V \setminus \{s\}$ the set of *players* and denote by $n = |N|$. To make the players distinguishable from the sender, we stop calling s the root and only call her *sender*. Instead, we call the roots of these subtrees *roots on depth 0*, who are the initial players of the game, as shown in the following figure. In a similar fashion, the children of these subtree roots are viewed to be on depth 1 and so on. For each node $i \in V$, let NB_i be the set of her children in $\mathcal{G}$, and thus NB_s contains all initial players (Fig. 1).

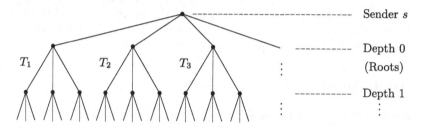

Fig. 1. An illustration of the tree structure.

We next define the propagation game $\Gamma = (N, \mathbf{S}, \mathbf{u})$ on $\mathcal{G}$. Without loss of generality, we assume the initial reward set by the sender is 1. Denote by S_i the strategy space of each player $i \in N$. Let $\mathbf{S} = S_1 \times \cdots \times S_n$ be the strategy profiles. Any strategy $s_i \in S_i$ is a mapping from $\mathbb{R}^+$ to $(\mathbb{R}^+ \cup \{\perp\})^{|NB_i|}$, where $\mathbb{R}^+ = [0, +\infty)$. That is, upon receiving the information with remaining reward x_i, player i decides how much reward $s_i(x_i)_j$ to leave to her child j by sending the information to j or not to inform j if $s_i(x_i)_j = \perp$. Denote by $s_i(x_i) = (s_i(x_i)_1, \cdots, s_i(x_i)_{|NB_i|})$. A *feasible* strategy is that for all $j \in NB_i$, if $x_i \geq x_{\min}$, $0 \leq s_i(x_i)_j \leq x_i - x_{\min}$ or $s_i(x_i)_j = \perp$; otherwise, $s_i(x_i)_j \equiv \perp$. Here $x_{\min} > 0$ is a sufficiently small number prefixed by the system. Denote by $\mathbf{s} = s_1 \times \cdots \times s_n$ a strategy profile. For any strategy profile $\mathbf{s} \in \mathbf{S}$ and an initial reward, there is a fixed set of players who can be informed the information, called *aware players* and denoted by $N^*(\mathbf{s})$. When $\mathbf{s}$ is clear in the context, we simply write N^*. To avoid cumbersome calculations, all aware players will get the reward with equal probability of $\frac{1}{|N^*|}$, and the winner needs to share the reward with her ancestors from her to the sender as committed. Thus, a player i's (expected) utility consists of two parts: *authorizing utility* and *referring utility*,

$$u_i(\mathbf{s}; x_i) = x_i \cdot \frac{1}{|N^*|} + \sum_{j \in NB_i, s_i(x_i)_j \neq \perp} (x_i - s_i(x_i)_j) \cdot \frac{n_{ij}}{|N^*|},$$

where n_{ij} is the number of aware players following j and including j.

Next we formulate the intuition in Inequality 1 by the following lemma, which shows it is always (strictly) better for a player to propagate the information.

Lemma 1. *For any player i, let $\mathbf{s}_{-i}$ be $-i$'s strategy profile such that the remaining reward x_i to i is at least $x_{\min}$. Let $s_i(x_i) = (z_1, \cdots, z_j, \cdots, z_{|NB_i|})$ be a strategy with $z_j = \bot$ for some j and $s_i'(x_i) = (z_1, \cdots, z_j', \cdots, z_{|NB_i|})$ be a new strategy by changing j's action from $\bot$ to 0. Then $u_i(s_i, \mathbf{s}_{-i}) < u_i(s_i', \mathbf{s}_{-i})$.*

We prove Lemma 1 in the full version [8]. By Lemma 1, every aware player does not get hurt by propagating the information to all her neighbours, and thus in what follows, we assume without loss of generality $s_i(x_i)_j \geq 0$ for any $x_i \geq x_{\min}$ and $j \in NB_i$. Moreover, in the following sections, we will see that the aware players actually want to maximally propagate the information. A feasible strategy s_i is called *full propagation* (FP for short) if $s_i(x_i)_j = (k-1)x_{\min}$ for all j and $kx_{\min} \leq x_i < (k+1)x_{\min}$ with $k \geq 1$. That is in an FP strategy, a player wants to inform all her neighbours by leaving the maximal reward to them. A feasible strategy profile $\mathbf{s} = (s_i)_{i \in N}$ is called FP if every s_i is FP. Denote by $\mathbf{s}^* = (s_i^*)_{i \in N}$ the FP strategy profile.

3   Main Results

To ease formulas, we define the following notations. For each subtree T, let

$$G(k) = \sum_{j=1}^{k} d^{j-1} = \frac{d^k - 1}{d - 1}$$

be the number of nodes from depth 0 to depth $k - 1$. Note that when $d \geq 3$ and $k \geq 1$, $G(k) \geq 2k - 1$. Moreover, for any $k \geq 1$,

$$(f-1)G(k+1) \geq d^{k+1} - 1 > d(d^k - 1) \geq \frac{d+1}{d-1}(d^k - 1)$$

$$= (d+1)G(k) \geq dG(k) + 2k - 1. \tag{2}$$

Denote by $H \cdot x_{\min} = 1$ the sender's initial reward, where $H > 1$ is an integer. Then H is the maximal depth that the information can reach. For any player i in depth $j \geq 0$, no matter what i's previous players do, i is not able to receive the information with remaining reward more than $(H - j) \cdot x_{\min}$ since each of i's ancestors needs to withhold at least $x_{\min}$. Moreover, all the users in and below depth H are not considered as strategic players. Finally, under FP strategies, the number of aware players is $f \cdot G(H + 1)$.

3.1   Technical Lemmas

We next introduce two technical lemmas which are crucial to prove the main results. For a subtree T, let $\pi_0(T)$ be the number of players outside of T who are aware of the information. If T is clear from the context, we write π_0 for short.

Lemma 2. *Consider the sub-game induced on a single subtree T, where r is the root player and g is one of r's descendants. Assume r receives the information with reward x_r and $k \cdot x_{\min} \leq x_r < (k+1) \cdot x_{\min}$ for some $k \geq 1$. If $d \geq 3$ and $\pi_0 \geq d \cdot G(k) + 2k - 1$, r's utility is maximized when g plays FP strategy, taking the actions of the others as given.*

Proof. Note that if $x_r < 2x_{\min}$ or the information cannot reach g, r's utility does not depend on g's action, thus the statement is trivially true. In the following we assume $x_r \geq 2x_{\min}$.

Taking the actions of the players (including r) except g as given, let Δ_j be the total number of r's referred players from r's child j and $\Delta = \sum_{j \in NB_r} \Delta_j$. Let g' be g's ancestor who is r's direct child or $g' = g$ when g is r's child. Thus g's strategy can only change $\Delta_{g'}$, which is the single variable of r's utility. Formally, r's utility can be written as (assuming π_0 includes r to simplify notions)

$$u_r(\Delta_{g'}) = x_r \frac{1}{\pi_0 + \Delta} + \sum_{j \in NB_r} (x_r - s_r(x_r)_j) \frac{\Delta_j}{\pi_0 + \Delta}$$

$$= \frac{x_r + \sum_{j \in NB_r}(x_r - s_r(x_r)_j)\Delta_j}{\pi_0 + \Delta}.$$

Calculating the derivative of $u_r(\Delta_{g'})$, we have

$$u_r'(\Delta_{g'}) = \frac{(x_r - s_r(x_r)_{g'})(\pi_0 + \Delta) - (x_r + \sum_{j \in NB_r}(x_r - s_r(x_r)_j)\Delta_j)}{(\pi_0 + \Delta)^2}$$

$$= \frac{(x_r - s_r(x_r)_{g'})\pi_0 - x_r - \sum_{j \neq g'}(s_r(x_r)_{g'} - s_r(x_r)_j)\Delta_j}{(\pi_0 + \Delta)^2}$$

$$\geq \frac{x_{\min}\pi_0 - x_r - \sum_{j \neq g'}(x_r - s_r(x_r)_j)\Delta_j}{(\pi_0 + \Delta)^2},$$

where the inequality is because $s_r(x_r)_j \leq x_r - x_{\min}$ for any j. Let

$$f_j(y_j) = (k + 1 - y_j)G(\lfloor y_j \rfloor) = (k + 1 - y_j)\frac{d^{\lfloor y_j \rfloor} - 1}{d - 1}$$

$$\leq (k + 1 - y_j)\frac{d^{y_j} - 1}{d - 1}.$$

Claim 1. *For $d \geq 3$ and $0 \leq y_j < k$, $\bar{f}_j(y_j) = (k + 1 - y_j) \cdot \frac{d^{y_j}-1}{d-1}$ monotone increases with respect to y_j.*

To prove the above claim, it suffices to calculate the derivative of $\bar{f}_j(y_j)$, and we omit the details. By Claim 1, $f_j(y_j) \leq \bar{f}_j(y_j) \leq G(k)$ for any $0 \leq y_j < k$. Thus,

$$u_r'(\Delta_{g'}) \geq \frac{x_{\min}\pi_0 - x_r - \sum_{j \neq g'}(x_r - s_r(x_r)_j)\Delta_j}{(\pi_0 + \Delta)^2}$$

$$\geq \frac{x_{\min}\pi_0 - x_r - G(k)(d-1)x_{\min}}{(\pi_0 + \Delta)^2}$$

$$\geq \frac{x_{\min}(\pi_0 - k - 1 - G(k)(d-1))}{(\pi_0 + \Delta)^2} \geq 0.$$

The last inequality is because $\pi_0 \geq dG(k) + 2k - 1$ and $k \geq 2$.

In conclusion, if g receives the information with remaining reward at least $x_{\min}$, $u'_r(\Delta_{g'}) > 0$, which means r's utility is maximized when g plays FP strategy, which finishes the proof. $\square$

Lemma 2 implies that, if the information is known to a sufficiently large number of players, for an arbitrary player, her utility is maximized when all her descendants play FP strategies. Next we show that the other direction of Lemma 2 also holds: for an arbitrary player, if the information is aware to a sufficiently large number of players and all her descendants play FP strategies, her utility is maximized when she plays FP strategy.

Fixing a tree T, let i be some player in depth j of T. Assume the remaining reward i has received is $kx_{\min} \leq x_i < (k+1)x_{\min}$ for some $0 \leq k \leq H - j - 1$, and all her descendants play FP strategies. Given any strategy $s_i(x_i)$, define

$$s'_i(x_i) = \left(\left\lfloor \frac{s_i(x_i)_1}{x_{\min}} \right\rfloor \cdot x_{\min}, \cdots, \left\lfloor \frac{s_i(x_i)_d}{x_{\min}} \right\rfloor \cdot x_{\min} \right).$$

Note that given her descendants playing FP strategies, $s'_i(x_i)$ brings i utility at least as much as $s_i(x_i)$ does. The set of all possible $s'_i(x_i)$ is called *reasonable strategies*. Thus given all i's descendants playing FP strategies, to study i's best response, it suffices to consider reasonable strategies. For convenience, we use $(d_0, d_1, \cdots, d_k)$ to represent a reasonable strategy, where $d_l \in [d]$ means i selects d_l children to propagate to next l depths by leaving $(l-1)x_{\min}$ to these children, and d_0 means i selects d_0 children to not propagate. Thus $\sum_{l=0}^{k} d_l = d$. Note that it does not matter which d_l children are selected since all i's children are symmetric. In the following, we use $u_i((d_0, d_1, \cdots, d_k); x_i)$ to denote i's utility when she receives reward x_i and her action is $(d_0, d_1, \cdots, d_k)$, given all other players adopting FP strategies.

Lemma 3. *Consider the sub-game on a single subtree T, where r is the root player. Assume r receives the information with reward x_r and $k \cdot x_{\min} \leq x_r < (k+1) \cdot x_{\min}$ for some $k \geq 1$. If $d \geq 3$, $\pi_0 \geq d \cdot G(k) + 2k - 1$, and all r's descendants adopting FP strategies, for any reasonable strategy $(d_0, d_1, \cdots, d_k)$, r's utility increases by moving a unit from $0 \leq l < k$ to $l + 1$. Formally, if for some $0 \leq l < k$ such that $d_l > 0$, then*

$$u_r((d_0, \cdots, d_l, \cdots, d_k); x_r) < u_r((d_0, \cdots, d_l - 1, d_{l+1} + 1, \cdots, d_k); x_r).$$

Proof. To simplify our notions, we ignore the index r for the root player, and let x be the propagation reward that the root receives. When $kx_{\min} \leq x < (k+1)x_{\min}$ and $k \geq 1$, the root is able to leave a proper reward to the players in depth 1 so that the information can be reached to at most depth k. When it is convenient, we write $x = (k + \epsilon)x_{\min}$ and $0 \leq \epsilon < 1$. Given any strategy $(d_0, \cdots, d_k)$ such that $\sum_{j=0}^{k} d_j = d$ and $d_i > 0$ for some $0 \leq i < k$, we compare $u((d_0, \cdots, d_l, \cdots, d_k); x)$ and $u((d_0, \cdots, d_l - 1, d_{l+1} + 1, \cdots, d_k); x)$.

If $l = 0$, Lemma 3 degenerates to Lemma 1 restricted to trees, which is trivially true. Thus in the following, we assume $l \geq 1$.

If $x_{\min} \le x < 2x_{\min}$, the remaining reward a player in depth 1 receives is always smaller than $x_{\min}$; that is there is no way the information can reach depth 2. Let $1 < d_0 \le d$ and $d_1 = d - d_0$. It is not hard to check that by notifying one more player,

$$u((d_0, d_1); x) = \frac{(d_1 + 1)x}{\pi_0 + 1 + d_1} < \frac{(d_1 + 2)x}{\pi_0 + 1 + d_1 + 1} = u((d_0 - 1, d_1 + 1); x),$$

where the inequality is because $\pi_0 \ge F(1) > 2$.

Next we assume $k \ge 2$. Denote

$$Q = x + \sum_{j=1}^{k} d_j G(j)(x - (j - 1)x_{\min}),$$

and

$$W = \pi_0 + 1 + \sum_{j=1}^{k} d_j G(j).$$

Thus the utility for $(d_0, \cdots, d_l, \cdots, d_k)$ is

$$U = u((d_1, \cdots, d_k); x) = \frac{Q}{W},$$

and the utility for $(d_0, \cdots, d_l - 1, d_{l+1} + 1, \cdots, d_k)$ is

$$U' = u((d_0, \cdots, d_l - 1, d_{l+1} + 1, \cdots, d_k); x)$$
$$= \frac{Q + G(i+1)(x - ix_{\min}) - G(i)(x - (i-1)x_{\min}))}{W + G(i+1) - G(i)}$$
$$= \frac{Q + xd^i - x_{\min}(iG(i+1) - (i-1)G(i))}{W + d^i}.$$

To show $U' > U$, it equivalent to show

$$\frac{xd^i - x_{\min}(iG(i+1) - (i-1)G(i))}{d^i} > \frac{Q}{W},$$

or

$$W\left(xd^i - x_{\min}\left(iG(i+1) - (i-1)G(i)\right)\right) > d^i Q.$$

Note that

$$W\left(xd^i - x_{\min}(iG(i+1) - (i-1)G(i))\right)$$
$$= W\left(xd^i - x_{\min}(G(i+1) - (i-1)d^i)\right)$$
$$> W\left(kx_{\min}d^i - x_{\min}(\frac{d^{i+1}}{d-1} - (i-1)d^i)\right)$$
$$= Wx_{\min}d^i\left(k + \epsilon - \frac{d}{d-1} - (i-1)\right)$$
$$\ge (\frac{1}{2} + \epsilon)Wx_{\min}d^i,$$

where the first inequality is because $x = (k + \epsilon)x_{\min}$ and

$$G(i+1) = \sum_{j=0}^{i+1} d^j = \frac{d^{i+1} - 1}{d - 1} < \frac{d^{i+1}}{d - 1};$$

the second inequality is because $k \geq i + 1$ and $d \geq 3$. Thus to show Theorem 3, it suffices to show $(1 + 2\epsilon)Wx_{\min} > 2Q$.

Claim 2. $(1 + 2\epsilon)Wx_{\min} > 2Q$.

To prove Claim 2, we note that

$$(1 + 2\epsilon)W - \frac{2Q}{x_{\min}}$$

$$= (1 + 2\epsilon)\left(\pi_0 + 1 + \sum_{j=1}^{k} d_j G(j)\right) - 2\left(k + \epsilon + \sum_{j=1}^{k} d_j G(j)(k + \epsilon - j + 1)\right)$$

$$\geq \pi_0 - 2k + 1 - \sum_{i=1}^{k} d_i G(i)(2k - 2i + 1)$$

$$> \pi_0 + 1 - 2k - dG(k) \geq 0.$$

The first equation is because $x = (k+\epsilon)x_{\min}$; the first inequality is because $\epsilon \geq 0$; the second inequality is because the following Claim 3; and the last inequality is because $\pi_0 \geq dG(k) + 2k - 1$.

Claim 3. $\sum_{i=1}^{k} d_i G(i)(2k - 2i + 1) < dG(k)$.

We prove Claim 3 in the full version [8].   □

Lemma 3 is essentially a generalization of Lemma 1 to tree structures, which means if the information is already known to a sufficiently large number of players, it is always beneficial for a player to make the information reach players in one deeper level. By induction, we have the following corollary.

Corollary 1. *Consider the sub-game on a single subtree T, where r is the root player. Assume r receives the information with reward x_r and $k \cdot x_{\min} \leq x_r < (k+1) \cdot x_{\min}$ for some $k \geq 1$. If $d \geq 3$, $\pi_0 \geq dG(k)+2k-1$, and all r's descendants adopt FP strategies, r's (unique) best strategy is $((k-1) \cdot x_{\min}, \cdots, (k-1) \cdot x_{\min})$.*

Note that Lemmas 2, 3 and Corollary 1 do not only hold for the root players, but also for any player in a subtree T.

3.2   Main Results

Given Lemma 2 and Corollary 1, it is not hard to verify that FP strategy profile s^* is a Nash equilibrium, as for each fixed tree, the information is known to the other $f - 1$ trees, and thus the number of aware players is at least

$$(f - 1)G(H + 1) \geq (d - 1)G(H + 1) \geq (d + 1)G(H).$$

However, $\mathbf{s}^*$ is not the unique Nash equilibrium. Consider the following strategy profile. Denote by 0 the root player of an arbitrary subtree T and by $1, \cdots, d$ her children. Consider the strategy profile $\mathbf{s}'$: for all $x \geq x_{\min}$, $s_0(x) = (0, \cdots, 0)$ and $s_i(x) = (\bot, \cdots, \bot)$ for all $i \in \{1, \cdots, d\}$; all the other players in T and all players not in T play arbitrary strategies. That is by notifying the players in depth 1, the root 0 withholds the complete reward no matter how much the initial reward is, and the players in depth 1 do not propagate the information at all no matter how much reward the root player leaves for them. Next, we claim strategy profile $\mathbf{s}'$ is a Nash equilibrium and for simplicity, we assume $d \geq 4$. First, for each player not in tree T, as there are at least $f - 2$ other trees play FP, $\pi_0 \geq (f - 2)G(H + 1) > dG(k) + 2k - 1$ for any $k \geq 0$. By Corollary 1, the best response of them is FP, thus all these players will not deviate. Second, the root player 0 does not deviate, as all her children do not propagate the information and her best strategy is to withhold all the reward; Finally, no player in $\{1, \cdots, d\}$ deviates, as the remaining reward for the information is 0. We can observe that in $\mathbf{s}'$, all $\{0, 1, \cdots, d\}$ played "bad" strategies. By deviating to FP strategies simultaneously, all of them can improve their utilities, which means $\mathbf{s}^*$ is more stable equilibrium than $\mathbf{s}'$.

We next define *connected coalition-proof Nash equilibria on graphs*, which are stronger than an arbitrary Nash equilibrium. Let $\Gamma = (N, \mathbf{S}, u)$ be any game with n players, and for any $C \subseteq N$, denote $\mathbf{S}_C = \times_{i \in C} S_i$. Let $G = (N, E)$ be a graph defined on players with $(i, j) \in E$ meaning players i and j can communicate with each other directly. A Nash equilibrium $\mathbf{s} \in \mathbf{S}$ is called *connected coalition-proof on G* if there is no coalition $C \subseteq N$ with the induced subgraph of C on G being connected such that by deviating to $\mathbf{s}'_C$, $u_i(\mathbf{s}'_C, \mathbf{s}_{N \setminus C}) \geq u_i(\mathbf{s})$ for any $i \in C$ and there is one $j \in C$ such that $u_j(\mathbf{s}'_C, \mathbf{s}_{N \setminus C}) > u_j(\mathbf{s})$. It is easy to see that any strong Nash equilibrium is connected coalition-proof on a complete graph and any Nash equilibrium is connected coalition-proof on an empty graph. As pointed out by Myerson in [23], the requirement of connected coalition-proof is practical as the players in any deviating coalition should be able to communicate.

Theorem 1. *For $f \geq d \geq 3$, full propagation strategy profile $\mathbf{s}^*$ is a connected coalition-proof Nash equilibrium on $\mathcal{G}$.*

Proof. For any coalition $C \subseteq N$, denote by $\mathcal{G}(C)$ the induced subgraph of C in $\mathcal{G}$. We first observe that $\mathcal{G}(C)$ being connected implies $\mathcal{G}(C)$ being a subtree in some T. Then to prove the theorem, it suffices to show for any deviation $\mathbf{s}'_C$ from $\mathbf{s}^*_C$, there is at least one player whose utility is smaller than the case when all of them play FP. Let r be the root of $\mathcal{G}(C)$. Without loss of generality, we reorder the players in $C \setminus \{r\}$ by $\{1, 2, \cdots, c\}$ where $c = |C \setminus \{r\}|$.

Given the other players not in T play FP strategies, the information will be known to at least π_0 players, and when $d \geq 3$, by Eq. 2,

$$\pi_0 \geq (f - 1)G(H + 1) > dG(H) + 2H - 1.$$

If r is in depth $H - 1$, all players in $C \setminus \{r\}$ are dummy, whose actions do not affect r's utility. Thus by Lemma 1, any deviation from $\mathbf{s}^*_C$ will strictly decreases r's utility. In the following, we assume r is above depth $H - 1$.

Since $\pi_0 \geq dG(H) + 2H - 1$, by Lemma 2,

$$u_r(\mathbf{s}'_C, \mathbf{s}^*_{N\setminus C}) \leq u_r(\mathbf{s}'_{C\setminus\{1\}}, \mathbf{s}^*_{\{1\}\cup(N\setminus C)}).$$

That is by changing player 1's strategy to FP, r's utility can only increase. We continue this procedure for player $i = 2, \cdots, c$ by changing her strategy from s_i to s_i^*, and we have

$$u_r(\mathbf{s}'_{\{r,i,\cdots,c\}}, \mathbf{s}^*_{\{1,\cdots,i-1\}\cup N\setminus C}) \leq u_r(\mathbf{s}'_{\{r,i+1,\cdots,c\}}, \mathbf{s}^*_{\{1,\cdots,i\}\cup N\setminus C}).$$

Eventually, we obtain

$$u_r(\mathbf{s}'_C, \mathbf{s}^*_{N\setminus C}) \leq u_r(\mathbf{s}'_r, \mathbf{s}^*_{N\setminus\{r\}}).$$

By Corollary 1, when all i's descendants play FP strategies,

$$u_r(\mathbf{s}'_r, \mathbf{s}^*_{N\setminus\{r\}}) < u_r(\mathbf{s}^*).$$

Thus, $u_r(\mathbf{s}'_C, \mathbf{s}^*_{N\setminus C}) < u_r(\mathbf{s}^*)$, which finishes the proof.   □

By Theorem 1, we have shown that $\mathbf{s}^*$ is more stable than an ordinary Nash equilibrium. But $\mathbf{s}^*$ is not a strong Nash equilibrium, because all the f root players can form a deviating coalition such that none of them propagates the information, which brings each root player utility $\frac{1}{f}$. However, as we have seen in the introduction, no propagation of these f players is not an equilibrium.

Next we show that $\mathbf{s}^*$ survives in any possible order of elimination of dominated strategies, and uniquely survives in an almost monotonic order of elimination of dominated strategies, which surpasses the result in [3]. We call an elimination order *monotonic* if for any player i and any two eliminated strategies s_i and s'_i, $\min_j s_i(x_i^*)_j < \min_j s'_i(x_i^*)_j$ implies that s_i is not eliminated after s'_i, where x_i^* is the minimum x_i such that $\min_j s_i(x_i)_j \neq \min_j s'_i(x_i)_j$. Assume $\perp < 0$. We call an elimination order *almost monotonic* if the condition is relaxed to $s_i(x_i^*)_{j^*} < s'_i(x_i^*)_{j^*} + x_{\min}$ implying s_i is not eliminated after s'_j.

Theorem 2. *For $f > d \geq 3$, $\mathbf{s}^*$ survives in any possible order of elimination of dominated strategies. Moreover, $\mathbf{s}^*$ is the unique Nash equilibrium survives in an almost monotonic order of elimination of dominated strategies.*

We prove Theorem 2 in the full version [8]. Note that all our results in this section hold as long as each player has at least d children, where $d \geq 3$. Actually, similar results also hold for the case of $d = 1, 2$, but the players may withhold multiples of $x_{\min}$.

Remark. Theorem 2 is similar to the result in [3]. They designed a hybrid rewarding scheme, which combines two nearly-uniform algorithms, that is Sybil-proof. Each player in a nearly-uniform algorithm A nearly-uniform algorithm specifies a maximal length H of rewarding path and rewards winning player in a chain of length h a reward of $1 + \beta \cdot (H - h + 1)$. All the players between the sender and the winner are rewarded β. Though β plays a similar role with our x_{min}, there are several differences. One of the main differences is that, by the design of the free market, the players can arbitrarily decide her own charge as long as the charge is at least of x_{min}, which is not restricted to be integer multiples of x_{min} (i.e., Sybil copies). A second difference is that we do not combine two different schemes. The advantage of the free market design is that the system does not need to (carefully) specify who gets how much, and the players themselves are already motivated to maximally propagate the information, which is also the main take-home message of the current paper.

4   A Class of Non-tree Networks

In this section, we investigate the extent to which our results for trees can be extended to general networks. In a non-tree network, a player may get the information from multiple neighbours and she will claim the one who leaves the highest reward to her, where tie is broken arbitrarily but consistently. Again let $\mathcal{G} = (V, E)$ be an arbitrary network and $s \in V$ be the initial sender. We first introduce the notions of *good friends* and *best friends*. For two players i and j, i is j's best friend if (1) i and j are connected, and (2) every shortest path from sender s to j passes i. If i is j's best friend, then j is called i's good friend. Note that each player can have at most one best friend but multiple good friends. For example, in Fig. 2(a), $a, b,$ and d are the best friends of $c, e,$ and g, respectively. However, d and f do not have any best friend as each of them has two disjoint shortest paths to s. Let $\mathcal{T} = (V, E')$ be the subgraph of $\mathcal{G}$, where $E' \subseteq E$ and $(i, j) \in E'$ if i is j's good or best friend. Note that $\mathcal{T}$ is a spanning forest of $\mathcal{G}$, and the root of each tree is either s or a player who does not have best friend. As an example, the solid lines in Fig. 2(a) form a *good-friendship* graph.

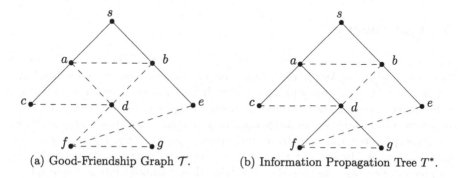

(a) Good-Friendship Graph $\mathcal{T}$.   (b) Information Propagation Tree T^*.

Fig. 2. Illustration of Good-Friendship Graph and Information Propagation Tree.

Next we prove that if every player has at least three good friends, then full propagation is a Nash equilibrium. Moreover, this equilibrium is robust to collective deviations of good friends. Note that the case of d-ary tree with $d \geq 3$ in Sect. 3 is a special case of this situation where $\mathcal{G} = \mathcal{T}$.

Theorem 3. *If every player has at least 3 good friends, then FP strategies* s^* *is a Nash equilibrium. Moreover, it is connected coalition-proof on* $\mathcal{T}$.

Proof. If every player has at least 3 good friends, then every node in $\mathcal{T}$ has at least 3 children in its corresponding tree. Let T^* be the corresponding information propagation tree under s^*, where i is connected to j if j claims i as the ancestor. Essentially, T^* connects all the trees in $\mathcal{T}$ as shown in Fig. 2(b). We first prove s^* is a Nash equilibrium. Suppose, for the sake of contradiction, some player i wants to deviate from s^*. We partition i's neighbours into two sets S_1 and S_2, where S_1 contains the players whose distance to s is at most i's distance to s, and S_2 contains the players whose distance to s is longer than i's distance to s. Note that player i's action does not affect players in S_1 and their descendants in T^*. We further partition S_2 into S_{21} and S_{22}, where S_{21} contains all i's good friends and $S_{22} = S_2 \setminus S_{21}$. If i does not fully propagate to players in S_{22}, i loses all the referring utility from S_{22} because they can still be informed via other paths, which decreases i's utility. By the condition of the lemma, $|S_{21}| \geq 3$, and only players in S_{21} are connected with i in $\mathcal{T}$. Moreover, under s^*, i will be claimed as the ancestor with higher priority by all her descendants in $\mathcal{T}$. Then our problem degenerates to the case of trees in the previous section, and by Lemma 3, i's utility is maximized by FP, which means s^* is a Nash equilibrium. Moreover, the reason why s^* is connected coalition-proof on $\mathcal{T}$ is in the same with Theorem 1: Any connected subgraph in $\mathcal{T}$ forms a subtree with a root player i, whose utility decreases by deviating from s^*.   □

Here we note that s^* may not be connected coalition-proof on the original network $\mathcal{G}$. To see this, if all senders' direct neighbours are connected with each other, they can form a connected coalition and do not propagate the information to others, which brings each of them higher utility than full propagation.

5   Experiments

Finally, we conduct experiments to confirm the validity of the free market design in random networks. We first introduce the parameters in our experiments. Given a parameter d, there are n players and each player is randomly connected to $\frac{d}{2}$ other players, so that the expected degree of each player is d. We study the utility of a fixed player and the sender is randomly selected from the other players. Fix all other players' strategies to be full propagation. Set the initial reward to 1 and $x_{\min} = \frac{1}{H}$. In each experiment, we randomly generate K networks. For simplicity, we only consider the strategies that withhold integral multiples of $x_{\min}$, and calculate the expected utility of the studied player for each strategy.

In Fig. 3, we set $n = 1000$ and $H = 6$. For each $d = 6/10/14$, we randomly construct $K = 100$ networks. We observe that full propagation (by withholding $x_{\min}$) brings a significantly higher utility on average than the other strategies

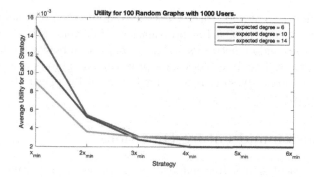

Fig. 3. When all players have the same expected degree.

(by withholding $kx_{\min}$ for $k > 1$). Then we revise the experiments by making the studied player more powerful or less powerful than the others, where a powerful player has higher degree than the others and a powerless player has lower degree than the others. In Fig. 4, the degree for the powerful player is set to $2d$ and for the powerless player it is set to be $\frac{d}{2}$. Similarly, we observe that in both cases, full propagation brings significantly higher utility than the other strategies.

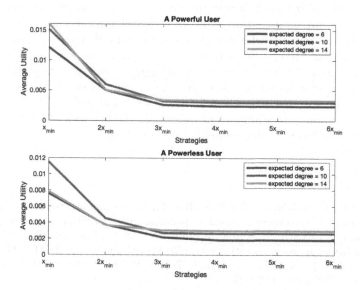

Fig. 4. When the player is powerful or powerless.

Note that full propagation is not always optimal. If a player i exclusively controls some players (i.e., the sender's information can only be reached to them via i) and these players form a complete graph, the optimal local strategy for i is to leave 0 to all of them instead of full propagation. Thus, if the network structure is known to players, the players can compute their optimal propagating strategies, which may not be full propagation. However, such "bad" networks barely happen in random networks and never happen in reality. To further avoid security threats, such as in a blockchain protocol, the players can be randomly re-connected periodically so that it is not beneficial for players to spend effort on learning network structures any more.

6   Conclusion and Future Directions

In this work, we design a free market with lotteries to incentivize players to maximally propagate information to their friends. For trees and a large class of non-tree networks, we prove that full propagation is robust against connected coalitions of players. Particularly, full propagation in tree networks uniquely survives in an interval-based monotone iterative elimination of dominated strategies. For future directions, from a theoretical perspective, it is intriguing to analyze players' behaviours in the free market within arbitrary or random networks and consider non-uniform lotteries. For example, if the property of every node having 3 good friends holds for certain random graphs, then Theorem 3 holds for such graphs as well. In addition, in the current market only the final winner and the players on the propagation path reaching her get rewarded. It would be interesting to see if there is a way to select and reward more than one winners and corresponding propagation paths. In a blockchain, this could correspond to multiple block proposers whose blocks are not eventually finalized, and may help reducing the players' risk, so as to encourage more to participate in a risk-avert model. Moreover, it is interesting and challenging to design new reward schemes to incentivize information propagation in more complex networks. Finally, from a practical perspective, experiments on synthetic and real-world data for complex networks may reveal important insights on the behavior of a free market in such networks, further enabling market design for them.

References

1. Abraham, I., Malkhi, D., Nayak, K., Ren, L., Spiegelman, A.: Solidus: an incentive-compatible cryptocurrency based on permissionless byzantine consensus (2016)
2. Arcaute, E., Kirsch, A., Kumar, R., Liben-Nowell, D., Vassilvitskii, S.: On threshold behavior in query incentive networks. In: EC, pp. 66–74. ACM (2007)
3. Babaioff, M., Dobzinski, S., Oren, S., Zohar, A.: On bitcoin and red balloons. In: EC, pp. 56–73. ACM (2012)
4. Bernheim, B.D., Peleg, B., Whinston, M.D.: Coalition-proof Nash equilibria i. concepts. J. Econ. Theory **42**(1), 1–12 (1987)
5. Cebrián, M., Coviello, L., Vattani, A., Voulgaris, P.: Finding red balloons with split contracts: robustness to individuals' selfishness. In: STOC, pp. 775–788. ACM (2012)
6. Chen, B.B., Chan, M.C.: Mobicent: a credit-based incentive system for disruption tolerant network. In: INFOCOM, pp. 875–883. IEEE (2010)
7. Chen, J., Gilad, Y.: A relay market for the algorand blockchain. Manuscript (2018)

8. Chen, J., Li, B.: Maximal information propagation via lotteries. CoRR abs/2110.10606 (2021)
9. Chen, W., Wang, Y., Yu, D., Zhang, L.: Sybil-proof mechanisms in query incentive networks. In: EC, pp. 197–214. ACM (2013)
10. Chen, X., Papadimitriou, C.H., Roughgarden, T.: An axiomatic approach to block rewards. In: AFT, pp. 124–131. ACM (2019)
11. Cossío, F.J.M., Brigham, E., Sela, B., Katz, J.: Competing (semi-)selfish miners in bitcoin. In: AFT, pp. 89–109. ACM (2019)
12. Douceur, J.R.: The Sybil attack. In: Druschel, P., Kaashoek, F., Rowstron, A. (eds.) IPTPS 2002. LNCS, vol. 2429, pp. 251–260. Springer, Heidelberg (2002). https://doi.org/10.1007/3-540-45748-8_24
13. Drucker, F., Fleischer, L.: Simpler Sybil-proof mechanisms for multi-level marketing. In: EC, pp. 441–458. ACM (2012)
14. Emek, Y., Karidi, R., Tennenholtz, M., Zohar, A.: Mechanisms for multi-level marketing. In: EC, pp. 209–218. ACM (2011)
15. Ersoy, O., Erkin, Z., Lagendijk, R.L.: Decentralized incentive-compatible and Sybil-proof transaction advertisement. In: Pardalos, P., Kotsireas, I., Guo, Y., Knottenbelt, W. (eds.) Mathematical Research for Blockchain Economy. SPBE, pp. 151–165. Springer, Cham (2020). https://doi.org/10.1007/978-3-030-37110-4_11
16. Ersoy, O., Ren, Z., Erkin, Z., Lagendijk, R.L.: Transaction propagation on permissionless blockchains: incentive and routing mechanisms. In: CVCBT, pp. 20–30. IEEE (2018)
17. Eyal, I., Sirer, E.G.: Majority is not enough: Bitcoin mining is vulnerable. In: Christin, N., Safavi-Naini, R. (eds.) Financial Cryptography and Data Security, pp. 436–454 (2014)
18. Johnson, B., Laszka, A., Grossklags, J., Vasek, M., Moore, T.: Game-theoretic analysis of DDoS attacks against bitcoin mining pools. In: Böhme, R., Brenner, M., Moore, T., Smith, M. (eds.) FC 2014. LNCS, vol. 8438, pp. 72–86. Springer, Heidelberg (2014). https://doi.org/10.1007/978-3-662-44774-1_6
19. Kiayias, A., Koutsoupias, E., Kyropoulou, M., Tselekounis, Y.: Blockchain mining games. In: EC, pp. 365–382. ACM (2016)
20. Kleinberg, J.M., Raghavan, P.: Query incentive networks. In: FOCS, pp. 132–141. IEEE Computer Society (2005)
21. Koutsoupias, E., Lazos, P., Ogunlana, F., Serafino, P.: Blockchain mining games with pay forward. In: WWW, pp. 917–927. ACM (2019)
22. Lewenberg, Y., Bachrach, Y., Sompolinsky, Y., Zohar, A., Rosenschein, J.S.: Bitcoin mining pools: a cooperative game theoretic analysis. In: AAMAS, pp. 919–927. ACM (2015)
23. Myerson, R.B.: Graphs and cooperation in games. Math. Oper. Res. 2(3), 225–229 (1977)
24. Nakamoto, S.: Bitcoin: a peer-to-peer electronic cash system (2008)
25. Rosenfeld, M.: Analysis of bitcoin pooled mining reward systems. CoRR abs/1112.4980 (2011)
26. Schrijvers, O., Bonneau, J., Boneh, D., Roughgarden, T.: Incentive compatibility of bitcoin mining pool reward functions. In: Grossklags, J., Preneel, B. (eds.) FC 2016. LNCS, vol. 9603, pp. 477–498. Springer, Heidelberg (2017). https://doi.org/10.1007/978-3-662-54970-4_28
27. Shi, H., Zhang, Y., Si, Z., Wang, L., Zhao, D.: Maximal information propagation with budgets. In: ECAI. Frontiers in Artificial Intelligence and Applications, vol. 325, pp. 211–218. IOS Press (2020)

Envy-free Division of Multi-layered Cakes

Ayumi Igarashi[1(⊠)] and Frédéric Meunier[2]

[1] National Institute of Informatics, Tokyo, Japan
`ayumi_igarashi@nii.ac.jp`
[2] École Nationale des Ponts et Chaussées, Champs-sur-Marne, France
`frederic.meunier@enpc.fr`

Abstract. We study the problem of dividing a multi-layered cake among hetero-geneous agents under non-overlapping constraints. This problem, recently proposed by Hosseini et al. (2020), captures several natural scenarios such as the allocation of multiple facilities over time where each agent can utilize at most one facility simultaneously, and the allocation of tasks over time where each agent can perform at most one task simultaneously. We establish the existence of an envy-free multi-division that is both non-overlapping and contiguous within each layered cake when the number n of agents is a prime power and the number m of layers is at most n, thus providing a positive partial answer to a recent open question. To achieve this, we employ a new approach based on a general fixed point theorem, originally proven by Volovikov (1996), and recently applied by Jojić, Panina, and Živaljević (2020) to the envy-free division problem of a cake. We further show that for a two-layered cake division among three agents with monotone preferences, an ε-approximate envy-free solution that is both non-overlapping and contiguous can be computed in logarithmic time of $1/\varepsilon$.

Keywords: Envy-freeness · Multi-layered cakes · Volovikov's theorem · FPTAS

1 Introduction

Imagine a group of n researchers deciding how to allocate m meeting rooms over a given period of time. They may have different opinions about the time slot and facility to which they would like to be assigned; for instance, some may prefer to schedule a seminar in a small room in the morning, while others may prefer to use a large room to give a lecture in the afternoon.

The problem of fairly distributing a resource has been often studied in the classical cake-cutting model [24], where the cake, represented by the unit interval, is to be divided among heterogenous agents. A variety of cake-cutting techniques have been developed over the past decades; in particular, the existence and algorithmic questions concerning an *envy-free* division, where each agent receives her first choice under the given division, have turned out to involve highly non-trivial arguments [9,25,27,29].

In the above example of sharing multiple facilities, however, one cannot trivially reduce the problem to the one-dimensional case. Indeed, if we merely divide the m time intervals independently, this may result in a division that is not feasible, i.e., it may

© Springer Nature Switzerland AG 2022
M. Feldman et al. (Eds.): WINE 2021, LNCS 13112, pp. 504–521, 2022.
https://doi.org/10.1007/978-3-030-94676-0_28

assign overlapping time intervals to the same agent who can utilize at most one room at a given time. In order to capture such constraints, Hosseini, Igarashi, and Searns [12] have recently introduced the multi-layered cake-cutting problem. There, n agents divide m different cakes under the *feasibility* constraint: the pieces of different layers assigned to the same agent should be non-overlapping, i.e., these pieces should have disjoint interiors. Besides an application to share meeting rooms, the model can capture a plethora of real-life situations; for instance, scheduling fitness lessons at a gymnasium, where each person can take at most one lesson at a time.

In this paper, we study the multi-layered cake-cutting problem, initiated by Hosseini et al. [12]. Our focus is on envy-free divisions of a multi-layered cake under feasible and contiguous constraints. For the special case of two layers and two agents, Hosseini et al. showed that the cut-and-choose protocol using one long knife produces envy-free diagonal shares that are both feasible and contiguous within each layer. For a more general combination of positive integers m and n with $m \leqslant n$, it remained an open question whether there is an envy-free multi-division that is both contiguous and feasible. In fact, the existence question is open even for $n = 3$ and $m = 2$ though, when the contiguity requirement is relaxed, they showed the existence of envy-free feasible multi-divisions. Note that when the number of layers strictly exceeds the number of agents, i.e., $m > n$, there is no way to allocate the entire cake while satisfying feasibility.

We settle this open question by Hosseini et al. affirmatively when n is a prime power and $m \leqslant n$. We show that in such cases, an envy-free multi-division, which is both feasible and contiguous, exists under mild conditions on preferences that are not necessarily monotone. Further, such divisions can be obtained using $n - 1$ long knives that hover over the entire cake. As an example of such constraints, one may consider a gymnasium where all classes begin and end at the same time.

We note that the standard proof showing the existence of an envy-free division via Sperner's lemma [27] may not work in the multi-layered setting. In the model of standard cake-cutting, the divisions into n parallel pieces of lengths x_i $(i = 1, 2, \ldots, n)$ can be represented by the points of the standard simplex Δ_{n-1}, which is then triangulated with the vertices of each triangle being labeled with distinct owner agents, and colored in such a way that each "owner" agent colors the vertex with the index of her favorite bundle of the "owned" division. If agents always prefer non-degenerate pieces to the degenerate one, a colorful triangle, which corresponds to an approximate envy-free division, is guaranteed to exist by Sperner's lemma.

In the same spirit of Su's approach [27], one may attempt to encode diagonal shares using $n - 1$ long knives, by the points of the standard simplex Δ_{n-1} and apply the usual method by using Sperner's lemma to show the existence of an envy-free division. For instance, each n-tuple $(x_1, x_2, \ldots, x_n)$ can represent a feasible and contiguous multi-division $(\mathcal{A}_1, \ldots, \mathcal{A}_n)$ where the ℓ-th layered piece of the i-th bundle $\mathcal{A}_i$ is given by the $\sigma(i, \ell)$-piece of length $x_{\sigma(i,\ell)}$ where $\sigma(i, \ell) = i + \ell - 1$ (modulo m). Unfortunately, this approach may fail to work for the multi-layered cake-cutting: even when the agents have monotone preferences over the pieces, the agents may choose the degenerate part as the most preferred one.

We thus employ a new approach of using a general Borsuk–Ulam-type theorem, originally proven by Volovikov [28], and recently applied by Jović, Panina, and

Živaljević [14,21] on the envy-free division of a cake. Volovikov's theorem ensures that for any G-equivariant coloring of the vertices of a triangulated simplicial complex, there is a fully colored simplex under assumptions on the connectivity of the domain, the co-domain on which a group of the form $(\mathbb{Z}_p)^k$ acts, and fixed-point freeness of the action. We consider a configuration space whose points encode not only diagonal shares using $n - 1$ long knives but also possible permutations of indices. Such points can be represented by the so-called *chessboard complex* $\Delta_{2n-1,n}$, which is guaranteed to be $(n - 2)$-connected, and for which Volovikov's theorem applies. With this new technique, the existence of an envy-free multi-division is shown for a general class of preferences that are not necessarily monotone. Note that this is the best one could hope to obtain, under the general preference model with choice functions: Avvakumov and Karasev [2] and Panina and Živaljević [21] provided counterexamples of a cake-cutting instance with choice functions for which no envy-free division exists, for every choice of n that is not a prime power.

Our existence result concerning envy-freeness also answers another open question raised by Hosseini et al.: It implies the existence of a *proportional* multi-division that is feasible and contiguous when $m \leqslant n$, m is a prime power, and agents have non-negative additive valuations, and thus properly generalizes the known existence result when m is a power of 2 and $m \leqslant n$ [12]; see Sect. 5 for details.

Having established that there always exists an envy-free multi-division using $n - 1$ long knives when n is a prime power, we consider a different type of division of a multi-layered cake. For instance, one may divide a two-layered cake among three agents, by cutting the top layer with one short knife and dividing the rest with one long knife. For this particular problem, one can encode such divisions by the points of the unit square. This way, a Sperner-type argument turns out to be applicable: there is an envy-free multi-division that can be constructed from only one short knife and one long knife when agents have monotone preferences. By exploiting the monotonicity that arises when fixing a long knife position, we further devise a fully polynomial-time algorithm (FPTAS) to compute an approximate envy-free multi-division. As a byproduct of our proof technique, we show that both the existence and algorithmic results extend to a *birthday cake multi-division* of a two-layered cake among three agents, i.e., a division of the cake into three parts such that whichever piece a *birthday agent* selects, there is an envy-free assignment of the remaining pieces to the remaining agents. Note that this also establishes an FPTAS to compute an approximate birthday cake multi-division for the standard one-layered cake-cutting problem among three agents with monotone preferences. Proofs omitted due to space restrictions can be found in the full version [13].

Related Work. Early literature on cake-cutting has established the existence of an envy-free contiguous division under *closed* (if the i-th piece is preferred in a convergent sequence of divisions, it is preferred in the limit) and *hungry* preferences (pieces of nonzero-length are preferred over pieces of zero-length). While classical works [25,29] applied non-constructive topological proofs, Su [27] provided a more combinatorial argument by explicitly using Sperner's lemma. Recently, the problem of dividing a partially unappetizing cake has attracted a great deal of attention. Here, some agents may find that a part of the cake is unappetizing and prefer nothing, while others may find it tasty. Even in such cases, an envy-free division only using $n - 1$ cuts has been shown to exist for a particular number n of agents under the assumption of closed preferences [2, 19, 21, 22]. The most general result obtained so far is the one by Avvakumov and Karasev [2], who showed the existence of an envy-free division for the case when n is a prime power. Panina and Živaljević [21] gave an alternative proof of the result of Avvakumov and Karasev, by using Volovikov's theorem [28].

In his classical work on cake-cutting (where he discusses the division of wine as well), Woodall [29] also proved that an envy-free division can be obtained without knowing one agent's preference. For example, the cut-and-choose protocol does not need the chooser's preference to obtain an envy-free division. More generally, for any number n of agents, there is a division of the cake into n contiguous pieces such that whichever piece a birthday agent selects, there is an envy-free assignment of the remaining pieces to the remaining agents. Asada et al. [1] gave a simple combinatorial proof that shows the existence of such division. Meunier and Su [18] showed the dual theorem, stating that there is a division of the cake into $n - 1$ contiguous pieces such that whichever agent leaves the game, there is an envy-free assignment of the remaining agents to the remaining pieces.

In general, there is no finite protocol that computes an exact envy-free division even for three agents [26], though such protocol exists when relaxing the contiguity requirement [4]. Nevertheless, for three agents with monotone valuations, Deng et al. [8] proved that an ε-approximate envy-free division can be computed in logarithmic time of $\frac{1}{\varepsilon}$, while obtaining PPAD-hardness of the same problem for choice functions whose choice is given explicitly by polynomial time algorithms. Our method for establishing FPTAS is reminiscent of that of Deng et al. A difference that is worth mentioning here is the way the triangulation together with its agent labeling is built: while Deng et al. used a special triangulation of a triangle due to Kuhn [15] and a quite complicated agent labeling given by a formula encoding the length of each piece, we subdivide the unit squares into small squares, take its barycentric subdivision, and build directly from the barycenters the agent labeling. The latter is technically light and naturally extends to a way to compute a birthday cake multi-division.

Several papers also studied the fair division problem in which agents divide multiple cakes [6, 16, 20, 23]. This model requires each agent to receive at least one non-empty piece of each cake [6, 16, 20] or receive pieces on as few cakes as possible [23]. On the other hand, our setting requires the allocated pieces to be non-overlapping. Thus, the existence/non-existence of envy-free divisions in one setting do not imply those for another.

2 Preliminaries

We consider the setting of Hosseini, Igarashi, and Searns [12], except that we allow slightly more general preferences. We are given m layers and n agents. A *cake* is the unit interval $[0, 1]$. A *piece* of cake is a union of finitely many disjoint closed subintervals of $[0, 1]$. We say that a subinterval of $[0, 1]$ is a *contiguous piece* of cake. An *m-layered cake* is a sequence of m cakes $[0, 1]$. A *layered piece* is a sequence $\mathcal{L} = (L_\ell)_{\ell \in [m]}$ of pieces of each layer ℓ; a layered piece is *contiguous* if each L_ℓ is a contiguous piece of layer $\ell \in [m]$. A layered piece $\mathcal{L}$ is *non-overlapping* if no two pieces from different layers overlap, i.e., $L_\ell \cap L'_\ell$ is formed by finitely many points or the empty set for any pair of distinct layers ℓ, ℓ'. The *length* of a layered piece is the sum of the lengths of its pieces in each layer. A *multi-division* $\mathcal{A} = (\mathcal{A}_1, \mathcal{A}_2, \ldots, \mathcal{A}_n)$ is an n tuple forming a partition of the m-layered cake into n layered pieces. (Here, "partition" is used in a slightly abusive way: while the collection covers the layered cake and the layered pieces have disjoint interiors, we allow the latter to share endpoints.) A multi-division $\mathcal{A}$ is

- *contiguous* if $\mathcal{A}_i$ is contiguous for each $i \in [n]$;
- *feasible* if $\mathcal{A}_i$ is non-overlapping for each $i \in [n]$.

We focus in this work on *complete* multi-divisions where the entire cake must be allocated.

Each agent i has a *choice function* c_i that, given a multi-division, returns the set of *preferred* layered pieces (among which the agent is indifferent). This function returns the same set of pieces over all permutations of the entries of the multi-division.

The choice function model is used in [18,27] and more general than the valuation model while the latter is more standard in fair division.

An agent i *weakly prefers* a layered piece $\mathcal{L}$ to another layered piece $\mathcal{L}'$ if there is no collection of layered pieces such that c_i selects $\mathcal{L}'$ but does not select $\mathcal{L}$. Where necessary, we impose the following additional assumption on the choice function: An agent has *hungry preferences* if she weakly prefers a layered piece of nonzero length to any layered piece of zero-length. An agent has *monotone preferences* if she always weakly prefers a layered piece $\mathcal{L} = (L_\ell)_{\ell \in [m]}$ to another $\mathcal{L}' = (L'_\ell)_{\ell \in [m]}$ whenever L'_ℓ is contained in L_ℓ for each $\ell \in [m]$. An agent has *closed preferences* if the following holds: for every sequence $(\mathcal{A}^{(t)})_{t \in \mathbb{Z}_+}$ of multi-divisions converging to a multi-division $\mathcal{A}^{(\infty)}$, we have

$$\mathcal{A}_j^{(t)} \in c_i(\mathcal{A}^{(t)}) \quad \forall t \in \mathbb{Z}_+ \quad \implies \quad \mathcal{A}_j^{(\infty)} \in c_i(\mathcal{A}^{(\infty)}) .$$

The convergence of layered pieces is considered according to the pseudo-metric $d(\mathcal{L}, \mathcal{L}') = \mu(\mathcal{L} \triangle \mathcal{L}')$; a sequence of multi-divisions is converging if each of its layered pieces converges. Here, μ is the Lebesgue measure and $\mathcal{L} \triangle \mathcal{L}' = ((L_\ell \setminus L'_\ell) \cup (L'_\ell \setminus L_\ell))_{\ell \in [m]}$.

A multi-division $\mathcal{A}$ is *envy-free* if there exists a permutation $\pi: [n] \to [n]$ such that $\mathcal{A}_{\pi(i)} \in c_i(\mathcal{A})$ for all $i \in [n]$. A *birthday cake multi-division* for agent i^* is a multi-division $\mathcal{A}$ where no matter which piece agent i^* selects, there is an envy-free assignment of the remaining pieces to the remaining agents, i.e., for every $j \in [n]$, there exists a bijection $\pi: [n] \setminus \{i^*\} \to [n] \setminus \{j\}$ such that $\mathcal{A}_{\pi(i)} \in c_i(\mathcal{A})$ for all $i \in [n] \setminus \{i^*\}$.

We also consider a setting where each agent can specify the valuation of each layered piece. Each agent i has a *valuation function* v_i that assigns a real value $v_i(\mathcal{L})$ to any layered piece $\mathcal{L}$. A valuation function naturally gives rise to a choice function that among several layered pieces, returns the most valuable layered pieces. A valuation function v_i satisfies

- *monotonicity* if $v_i(\mathcal{L}') \leqslant v_i(\mathcal{L})$ for any pair of layered pieces $\mathcal{L}, \mathcal{L}'$ such that $L'_\ell \subseteq L_\ell$ for any $\ell \in [m]$.
- *the Lipschitz condition* if there exists a fixed constant K such that for any pair of layered pieces $\mathcal{L}, \mathcal{L}'$, $|v_i(\mathcal{L}) - v_i(\mathcal{L}')| \leqslant K \times \mu(\mathcal{L} \triangle \mathcal{L}')$.

It is easy to see that monotonicity along with the Lipschitz condition implies that the hungry assumption is satisfied: the Lipschitz condition implies that all layered pieces of zero-length have the same value and thus in particular the value of the empty set; monotonicity then implies that every layered piece has a value at least the value of the empty set.

For an instance with agents' valuation functions, one can define concepts of approximate envy-freeness. A multi-division $\mathcal{A}$ is ε-envy-free if different agents approximately prefer different layered pieces, i.e., there exists a permutation $\pi: [n] \to [n]$ such that for all $i \in [n]$, $v_i(\mathcal{A}_{\pi(i)}) + \varepsilon \geq \max_{i' \in [n]} v_i(\mathcal{A}_{i'})$. An ε-birthday cake multi-division for agent i^* is a multi-division $\mathcal{A}$ where for any $j \in [n]$, there exists a bijection $\pi: [n]\setminus\{i^*\} \to [n]\setminus\{j\}$ such that for all $i \in [n]\setminus\{i^*\}$, $v_i(\mathcal{A}_{\pi(i)}) + \varepsilon \geq \max_{i' \in [n]} v_i(\mathcal{A}_{i'})$. For an instance with agents' valuation functions, we will assume that $v_i(\mathcal{L})$ can be accessed in constant time for any agent i and layered piece $\mathcal{L}$.

There are several types of divisions of a multi-layered cake that satisfy feasibility and contiguity constraints: one that is obtained by a *long knife* that hovers over the whole layered cake, and another that is obtained by a *short knife* that hovers over a single layered cake. See Fig. 1 for examples.

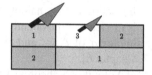 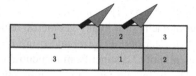

Fig. 1. Multi-divisions of a two-layered cake, obtained by one long knife and one short knife and by two long knives (pictured left-to-right).

We assume basic knowledge in algebraic topology. Definitions of abstract and geometric simplicial complexes, the fact that they are somehow equivalent, and other related notions are reminded in Appendix A of the full version [13]. The reader might consult the book by De Longueville [7] or the one by Matoušek [17], especially Chaps. 1 and 6 of the latter book, for complementary material. In the sequel, we will identify geometric and abstract simplicial complexes without further mention.

3 Envy-Free Division Using $n-1$ Long Knives

Now, we formally present the first main result of this paper, stating that an envy-free multi-division using $n-1$ long knives exists when n is a prime power.

Theorem 1. *Consider an instance of the multi-layered cake-cutting problem with m layers and n agents, $m \leq n$, with closed preferences. If n is a prime power, then there exists an envy-free multi-division that is feasible and contiguous. Moreover, it can be achieved with exactly $n-1$ long knives.*

Let us comment briefly on the special case when $m = 1$. Since the agents might prefer zero-length pieces in our setting, our theorem boils down then to a recent result of Avvakumov and Karasev [2]. They showed that when n is a prime power, there always exists an envy-free division, even if we do not assume that the agents are hungry, and that this is not true anymore if n is not a prime power. (The Avvakumov–Karasev theorem was first proved for $n = 3$ by Segal-Halevi [22]—who actually initiated the study of envy-free divisions with non-necessarily hungry agents—and for prime n by Meunier and Zerbib [19]).

3.1   Tools from Equivariant Topology

We introduce now a specific abstract simplicial complex that will play a central role in the proof of Theorem 1. The *chessboard complex* $\Delta_{m,n}$ is an abstract simplicial complex whose ground set is $[m] \times [n]$ and whose simplices are the subsets $\sigma \subseteq [m] \times [n]$ such that for every two distinct pairs (i, j) and (i', j') in σ we have $i \neq i'$ and $j \neq j'$. The name comes from the following: If we interpret $[m] \times [n]$ as an $m \times n$ chessboard, the simplices are precisely the configurations of pairwise non-attacking rooks.

Given an additive group G of order n, we get a natural action $(\varphi_g)_{g \in G}$ of G on $\Delta_{m,n}$ by identifying $[n]$ with G via a bijection $\eta \colon [n] \to G$: this natural action is defined by $\varphi_g(i, j) = g \cdot (i, j) := (i, \eta^{-1}(g + \eta(j)))$. This action is *free*, namely the orbit of each point in any geometric realization of $\Delta_{m,n}$ is of size n. Equivalently, the relative interiors of $\varphi_g(\sigma)$ and σ are disjoint for every simplex σ of $\Delta_{m,n}$ and every element g of G distinct from its neutral element; see [17, Chapter 6].

The following lemma is an immediate consequence of Volovikov's theorem [28], which has found many applications in topological combinatorics. It uses the notions of G-invariant triangulation and G-equivariant coloring we define now.

Let K be a simplicial complex on which a group G acts. This action induces an action on the underlying space $\|K\|$. Consider now a triangulation T of K. The action on $\|K\|$ induces in turn an action $(\varphi_g)_{g \in G}$ on $\|T\|$. The triangulation T is *G-invariant* if, seeing it as a geometric simplicial complex, we have $\varphi_g(\sigma) \in T$ for all $\sigma \in T$ and all $g \in G$. A coloring $\lambda \colon V(K) \to G$ is *G-equivariant* if $\lambda(g \cdot v) = g + \lambda(v)$ for all $v \in V(K)$ and $g \in G$ (considering G as an additive group).

Lemma 1. *Let $n = p^k$, where p is a prime number and k a positive integer. Denote by G the additive group $\left((\mathbb{Z}_p)^k, +\right)$. Consider a G-invariant triangulation T of $\Delta_{2n-1,n}$. For any G-equivariant coloring of the vertices of T with elements of G, there is a simplex in T whose vertices are colored with all elements of G.*

When G does not satisfy the condition of the lemma, its conclusion does not necessarily hold. Already for $G = \mathbb{Z}_6$, counterexamples are known; see [30].

Proof (of Lemma 1). Volovikov's theorem states the following; see [17, Section 6.2, Notes]. *Let G be the additive group $\left((\mathbb{Z}_p)^k, +\right)$ and let X and Y be two topological spaces on which G acts in a fixed-point free way. If X is d-connected and Y is a d-dimensional sphere, then there is no G-equivariant map $X \to Y$.* An action is *fixed-point free* if each orbit has at least two elements, and a map $f \colon X \to Y$ is *G-equivariant* if it is continuous and $f(g \cdot x) = g \cdot f(x)$ for all $x \in X$ and all $g \in G$.

Consider now any G-equivariant coloring of the vertices of T with elements of G. Suppose for a contradiction that there is no simplex in T whose vertices are colored with all elements of G. Then the coloring induces a G-equivariant simplicial map ψ from T to $\partial\Delta^G$, where this latter simplicial complex is the boundary of the $(n-1)$-dimensional standard simplex Δ^G whose vertices have been identified with G. (The action of G on Δ^G is the natural one: $g \cdot g' = g + g'$, where g is an element of G and g' is seen as a vertex of Δ^G).

The simplicial complex $\Delta_{2n-1,n}$ is $(n-2)$-connected (see [5]), and so T is $(n-2)$-connected as well. The simplicial complex $\partial\Delta^G$ is an $(n-2)$-dimensional sphere. We have already noted that the action of G on $\Delta_{2n-1,n}$, and thus on T, is free. Let us check that the action of G on $\partial\Delta^G$ is fixed-point free.

The action of G on $\partial\Delta^G$ is fixed-point free if the orbit of every point in a geometric realization of $\partial\Delta^G$ has at least two elements. Consider a point x in the underlying space $\|\partial\Delta^G\|$ of an arbitrary geometric realization. Let $\sigma \in \partial\Delta^G$ be the simplex of minimal dimension containing x. By definition of $\partial\Delta^G$, there is at least one vertex that is missed by σ. Thus, it is possible to find $g \in G$ such that $g \cdot \sigma$ is distinct from σ. The point x is not in $g \cdot \sigma$ since otherwise it would be in the intersection of σ and $g \cdot \sigma$, which is a simplex of $\partial\Delta^G$ of smaller dimension, contradicting the assumption on σ. The point $g \cdot x$ is thus distinct from x, which means that the orbit of x is of size at least two, as desired. (The action is however in general not free: for example, when $G = (\mathbb{Z}_2)^2$, the orbit of the center of the edge $\{(0,0),(1,1)\}$ in a geometric realization is of size exactly two).

Volovikov's theorem with $X = \mathsf{T}$ and $Y = \partial\Delta^G$ shows then that such a map ψ cannot exist; a contradiction.   □

The proof of Theorem 1 uses also the idea of "owner labeling" from [27]. An "owner labeling," which we call *agent labeling* in the present paper, is a labeling of the vertices of a simplicial complex with the agents as labels, such that every $(n-1)$-dimensional simplex gets pairwise distinct labels on its vertices (n being the number of agents).

Lemma 2. *There exists a G-invariant triangulation T of $\Delta_{2n-1,n}$ of arbitrary small mesh size, refining $\Delta_{2n-1,n}$, and with an agent labeling $a\colon V(\mathsf{T}) \to [n]$ that satisfies $a(g \cdot v) = a(v)$ for all $v \in V(\mathsf{T})$ and $g \in G$.*

Proof. Take any G-invariant triangulation T'. The complex $\Delta_{2n-1,n}$ itself is such a triangulation. Any barycentric subdivision of T' admits an agent labeling as desired (this is a classical idea, already present in [27]), and is still G-invariant. So, repeated barycentric subdivisions allow to get a triangulation T with the required property, of arbitrary small mesh size.   □

3.2 Proof of Theorem 1

The proof uses a "configuration space" encoding some possible contiguous and feasible multi-divisions with $n-1$ long knives. We will show that at least one of the multi-divisions corresponding to the points in such configuration space is envy-free. This configuration space is the simplicial complex $\Delta_{2n-1,n}$ introduced in Sect. 3.1, with $n = p^k$ and $G = \left((\mathbb{Z}_p)^k, +\right)$ acting on it. The elements in G will be used to identify the layered pieces. We choose an arbitrary bijection $\eta\colon [n] \to G$ to ease this identification. (In case $k = 1$, it is certainly most intuitive to set $\eta(i) = i$; note that when $k \neq 1$, this definition does not make sense.) Moreover, we fix an arbitrary injective map $h\colon [m] \to G$ and a geometric realization of $\Delta_{2n-1,n}$. We denote by $v_{i,j}$ the realization of the vertex (i, j). We explain now how each point of $\|\Delta_{2n-1,n}\|$ encodes a multi-division with $n - 1$ long knives.

Let x be a point of $\|\Delta_{2n-1,n}\|$. It is contained in (at least) one $(n - 1)$-dimensional simplex. Let $v_{i_1,1}, v_{i_2,2}, \ldots, v_{i_n,n}$ be the vertices of this simplex. We fix an arbitrary procedure $\mathfrak{P}$ to choose the simplex in case of a tie and to ease the proof of Lemma 4 below we fix it so that all points with same support provide the same simplex. A tie can occur, e.g., for $n = 3$, when x belongs to the interior of the edge $v_{3,1}$-$v_{1,2}$; there are three triangles containing this edge in $\Delta_{5,3}$; each of them contains the vertices $v_{3,1}$ and $v_{1,2}$; the third vertex can be any of $v_{2,3}$, $v_{4,3}$, and $v_{5,3}$. Moreover, we fix $\mathfrak{P}$ in an "equivariant" way: given any $g \in G$, the simplex chosen by $\mathfrak{P}$ for $g \cdot x$ is the image by φ_g of the simplex chosen by $\mathfrak{P}$ for x. Such a way to define $\mathfrak{P}$ is possible because the action of G on $\Delta_{2n-1,n}$ is free.

We write then x as $\sum_{j=1}^{n} x_{i_j} v_{i_j,j}$. We set $x_k = 0$ for every $k \notin \{i_1, \ldots, i_n\}$. The values of $x_1, \ldots, x_{2n-1}$ do not depend on the choice made by $\mathfrak{P}$: only the vertices $v_{i,j}$ spanning the minimal simplex of $\Delta_{2n-1,n}$ containing x get non-zero coefficients, and these latter are then the barycentric coordinates in this face.

Choose the permutation $\rho \in \mathcal{S}_n$ such that $i_{\rho(1)} < i_{\rho(2)} < \cdots < i_{\rho(n)}$. We interpret $x_{i_{\rho(j)}}$ as the length of the j-th piece: in a way similar to the traditional encoding of the divisions (see, e.g., [27]), the j-th piece in any layer is of length $x_{i_{\rho(j)}}$. We give then the j-th piece of the ℓ-th layer the element $\eta(\rho(j)) + h(\ell)$ of G as its "bundle-name." We get a non-overlapping layered piece by considering all pieces with a same bundle-name: two overlapping pieces have names of the form $\eta(\rho(j)) + h(\ell)$ and $\eta(\rho(j)) + h(\ell')$, and $\eta(\rho(j)) + h(\ell) = \eta(\rho(j)) + h(\ell')$ if and only if $\ell = \ell'$. The non-overlapping layered pieces obtained this way form the multi-division encoded by x, which we denote by $\mathcal{A}(x) = (\mathcal{A}_1(x), \mathcal{A}_2(x), \ldots, \mathcal{A}_n(x))$, where $\mathcal{A}_i(x)$ is the layered piece formed by the pieces with bundle-name $\eta(i) \in G$. Clearly, each $\mathcal{A}(x)$ is a feasible and contiguous multi-division that uses $n - 1$ long knives. Figure 2 depicts examples of multi-divisions corresponding to points of $\|\Delta_{2n-1,n}\|$ for $n = 2$ and $n = 3$, whereas Figs. 3 and 4 illustrate the associated chessboard complex $\Delta_{2n-1,n}$.

Here, we state two important lemmas concerning the multi-divisions $\mathcal{A}(x)$ whose proofs can be found in [13]. First, the multi-divisions $\mathcal{A}(x)$ enjoy some "equivariant" property.

Lemma 3. *We have $\mathcal{A}_i(x) = \mathcal{A}_{\eta^{-1}(g+\eta(i))}(g \cdot x)$ for all $x \in \|\Delta_{2n-1,n}\|$, $i \in [n]$, and $g \in G$.*

Another important point is that $\mathcal{A}(x)$ depends continuously on x as stated by the following lemma. The convergence of multi-divisions is defined according to the pseudo-metric $d(\cdot, \cdot)$.

Lemma 4. *Let $\left(x^{(t)}\right)_{t \in \mathbb{Z}_+}$ be a sequence of points of $\|\Delta_{2n-1,n}\|$ converging to some limit point $x^{(\infty)}$. Then $\left(\mathcal{A}(x^{(t)})\right)_{t \in \mathbb{Z}_+}$ converges to $\mathcal{A}(x^{(\infty)})$.*

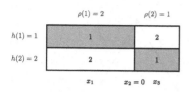

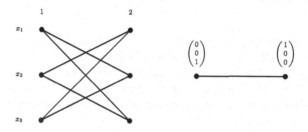

Fig. 2. Multi-divisions represented by the points of the configuration space when $h(\ell) = \ell$ for the ℓ-th layer. In the left figure, we have $G = (\mathbb{Z}_2, +)$ and $i_{\rho(1)} = i_2 = 1 < i_{\rho(2)} = i_1 = 3$. In the right figure, we have $G = (\mathbb{Z}_3, +)$ and $i_{\rho(1)} = i_2 = 3 < i_{\rho(2)} = i_3 = 4 < i_{\rho(3)} = i_1 = 5$. The number in the j-th piece of the ℓ-th layer corresponds to $\eta(\rho(j)) + h(\ell)$, where $\eta(i) = i$ (because when $k = 1$, we can identify $(\mathbb{Z}_p)^k$ and $[p]$). (Color figure online)

Fig. 3. Illustration of the chessboard complex $\Delta_{2n-1,n}$ when $n = 2$. The blue edge corresponds to a standard simplex, depicted on the right, which represents the set of multi-divisions of the form described in the left division of Fig. 2. (Color figure online)

We are now ready to prove Theorem 1.

Proof (of Theorem 1). Take a triangulation T and an agent labeling a as in Lemma 2. We partition the vertices of T into their G-orbits. From each orbit, we pick a vertex x. We ask agent $a(x)$ the index i of the non-overlapping layered piece $\mathcal{A}_i(x)$ she prefers in $\mathcal{A}(x)$. (In case of a tie, she makes an arbitrary choice.) We define $\lambda(x)$ to be $\eta(i)$. We extend λ on each orbit in an equivariant way: $\lambda(g \cdot x) := g + \lambda(x)$. This is done unambiguously because the action of G on T is free. The map λ is a G-equivariant coloring of the vertices of T with elements of G. The invariance of a along the orbits combined with Lemma 3 implies that, for every vertex x of the triangulation T, the integer $\eta^{-1}(\lambda(x))$ is the index of a layered piece preferred by agent $a(x)$ in the multi-division $\mathcal{A}(x)$.

According to Lemma 1, there exists a simplex of T whose vertices are colored with all elements of G. For every integer $N > 0$, we can choose T $:= $ T$_N$ so that it has a mesh size upper bounded by $1/N$. Denote by $x^{i,N}$ the vertex of this simplex in T$_N$ that has $a(x^{i,N}) = i$, and by $\pi_N(i)$ the integer $\eta^{-1}(\lambda(x^{i,N}))$. For each N, the map π_N is a permutation of $[n]$ (assigning the agents to their preferred layered pieces, in the multi-divisions encoded by the vertices of the simplex). When N increases, there is at least one permutation occurring infinitely many times among the π_N. Denote by π such a permutation.

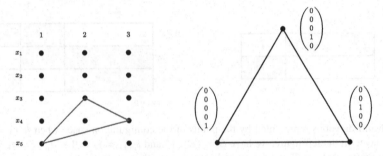

Fig. 4. Illustration of the chessboard complex $\Delta_{2n-1,n}$ when $n = 3$. The blue edge corresponds to a standard simplex, depicted on the right, which represents the set of multi-divisions of the form described in the right division of Fig. 2. (Color figure online)

Consider arbitrary large N for which $\pi_N = \pi$. We have $\mathcal{A}_{\pi(i)}(\boldsymbol{x}^{i,N}) \in c_i(\mathcal{A}(\boldsymbol{x}^{i,N}))$ for all $i \in [n]$ and all such N. Compactness implies that we can select among these arbitrarily large N an infinite sequence such that every $(\boldsymbol{x}^{i,N})_N$ converges to a same point $\boldsymbol{x}^*$ for every i. The sequence $\mathcal{A}(\boldsymbol{x}^{i,N})$ converges then to $\mathcal{A}^* := \mathcal{A}(\boldsymbol{x}^*)$ for every i (Lemma 4). Thus, by the closed preferences assumption, we have $\mathcal{A}^*_{\pi(i)} \in c_i(\mathcal{A}^*)$ for all $i \in [n]$. $\qquad\square$

4   Envy-Free Division Using One Short Knife and One Long Knife

We saw that there always exists an envy-free multi-division using $n - 1$ long knives when n is a prime power. In general, there are other types of multi-divisions that can achieve both feasibility and contiguity. For example, for a two-layered cake with three agents, one may use a short knife over one layer to obtain the first piece, and use a long knife to divide the rest in order to produce a multi-division.

In this section, we show that an envy-free multi-division of such form indeed exists for a two-layered cake division among three agents with closed and monotone preferences. Interestingly, by allowing for only such divisions, the proof for the existence of an envy-free multi-division turns out to be more elementary and constructive, in contrast with the proof based on Volovikov's theorem in the previous section. We further observe that the agents' preferences exhibit monotonicity when fixing one long knife and moving another short knife from left to right. By exploiting monotonicity, we show that an approximate envy-free division using one short knife and one long knife can be computed efficiently for three agents with valuations satisfying the Lipschitz condition and monotonicity. In fact, we prove a stronger statement where both existence and computational results extend to those for birthday cake multi-divisions:

Theorem 2. *In the case with two layers and three agents with closed, monotone, and hungry preferences, a birthday cake multi-division that is feasible and contiguous exists. Moreover, it requires only one long knife.*

Theorem 3. *In the case with two layers and three agents whose valuation functions satisfy the Lipschitz condition and monotonicity, for any $\varepsilon \in (0, 1)$, an ε-birthday cake* *multi-division that is feasible and contiguous can be found in* $O(\log^2 \frac{1}{\varepsilon})$ *time. Moreover, it requires only one long knife.*

In order to establish the above theorems, we will encode by the points of the unit square $[0,1]^2$ the divisions of a two-layered cake using one short knife and one long knife. The position of a long knife corresponds to the x-axis and the position of a short knife corresponds to the y-axis. The crucial observation is that the two vertical boundaries when $x = 0$ and $x = 1$ enjoy a certain symmetry: the divisions that appear on these boundaries are the same. By exploiting this symmetry, one can apply a Sperner-type lemma to show the existence of an envy-free division.

To formally explain the combinatorial lemma which we will use, we introduce the notion of *degree*, also mentioned in Deng et al. [8] where it was called the *index*. For a triangulated rectangle M, we consider a coloring $\lambda\colon V(M) \to \{1,2,3\}$. For such rectangle and coloring, the *degree* d(M) of M is the mod 2 number of simplices of M whose label set is exactly $\{1,2,3\}$. For subset S of $\{1,2,3\}$ of cardinality 2, let $d_S(\partial M)$ denote the mod 2 number of simplices of the boundary ∂M of M whose label set is exactly S. For a triangulated line L with coloring $\lambda\colon V(L) \to \{1,2,3\}$, the *degree* d(L) of L is the mod 2 number of simplices of L whose label set is exactly $\{1,2\}$.

The following lemma is common knowledge in combinatorial topology; see [10, Corollary 2].

Lemma 5. *Consider a triangulated rectangle* M. *Suppose that we are given a coloring* $\lambda\colon V(M) \to \{1,2,3\}$. *Then* $d(M) = d_S(\partial M)$ *for any* $S \subseteq \{1,2,3\}$ *of cardinality* 2.

4.1 Existence of an Envy-Free Multi-division

First, we will establish the existence of an envy-free multi-division using one short knife and one long knife. Suppose that there are three agents, A, B, and C, with closed, monotone, and hungry preferences. We show first the existence of a birthday cake multi-division for C: namely, there is a multi-division of the layered cake into three layered pieces such that no matter which piece C chooses, there is an envy-free assignment of the remaining pieces to the remaining agents A and B. Without loss of generality, assume that the top layer is weakly preferred to the bottom layer by at least one agent different from C, say A. The two-layered cake is divided as in Fig. 5 where x corresponds to the long knife over the two layers and y corresponds to the short knife over the top layer; such divisions can be represented by the points (x, y) of the unit square. We denote by $\mathcal{A}(v) = (\mathcal{A}_1(v), \mathcal{A}_2(v), \mathcal{A}_3(v))$ the multi-division in Fig. 5 represented by $v = (x, y) \in [0,1]^2$ where the agents first cut the cake via the short knife position y, and then cut the rest via the long knife position x. Namely, $\mathcal{A}_1(v)$ consists the $[0, y]$ segment of the top layer, $\mathcal{A}_2(v)$ consists of the $[\max\{x, y\}, 1]$ of the top layer and the $[0, x]$ segment of the bottom layer, and $\mathcal{A}_3(v)$ contains the remaining pieces. Such multi-division $\mathcal{A}(v)$ is feasible and contiguous, only using one short knife and one long knife.

1. Triangulation. To show the existence of an envy-free solution, we consider a particular triangulation T of the unit square (see Fig. 6). Let $\varepsilon \in [0,1]$. Without loss of generality, assume that $N = 1/2\varepsilon$ is an integer. Partition the unit square $[0,1]^2$ into N^2

smaller squares of side length 2ε; we call each of these squares a *basic square*. Then, divide each square into small triangles as follows: a new vertex is added in the middle of each of the four edges, and another new vertex is added at the barycenter of the square; new edges are added from this barycenter to the original vertices as well as to added vertices on the edges. We denote the vertices of the triangulation by $V(\mathsf{T})$. We call each vertical line $\mathsf{L}(x) = \{(x,y): y \in [0,1]\}$ of Fig. 6 with $x = k\varepsilon$ for $k \in \{0,2,\ldots,2N\}$ a *basic vertical line*.

2. *Labeling and coloring*. We construct the following agent labeling $a\colon V(\mathsf{T}) \to \{A,B,C\}$:

- every corner vertex of the basic squares receives label A;
- every middle vertex of the edges of the basic squares receives label B;
- every barycentric vertex of the basic squares receives label C.

To construct a coloring $\lambda\colon V(\mathsf{T}) \to \{1,2,3\}$, each owner agent $a(v)$ of $v = (x,y)$ colors the point with the index of the favorite layered piece, i.e., $\mathcal{A}_{\lambda(v)} \in c_{a(v)}(\mathcal{A}(v))$. We consider the following tie-breaking rule: a zero-length piece is never chosen and for pieces of nonzero length, in case of tie with piece 1, the owner agent $a(v)$ colors $v = (x,y)$ with 1; in case of tie between pieces 2 and 3 only, $a(v)$ colors $v = (x,y)$ with 2 if $x \leqslant y$; $a(v)$ colors $v = (x,y)$ with 3 if $x > y$.

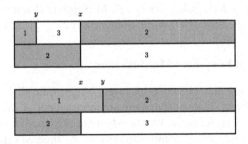

Fig. 5. Three-agent multi-divisions $\mathcal{A}(v)$. Note that the short knife y is prioritized over the long knife x.

We start by showing the following lemma: the number of 1-2 edges on the boundary of the unit square must be nonzero.

Lemma 6. *For the triangulation T of the unit square and coloring $\lambda\colon V(\mathsf{T}) \to \{1,2,3\}$, the following hold:*

(i) *the number of edges colored with 1 and 2 on $\{(x,1)\colon x_1 \leqslant x \leqslant x_2\}$ is even for any pair of $x_1, x_2 \in \{0, 2\varepsilon, 4\varepsilon, \ldots, 2N\varepsilon\}$ with $x_1 \leqslant x_2$,*

(ii) *the number of edges colored with 1 and 2 on $\mathsf{L}(0) \cup \mathsf{L}(1)$ is odd, and*

(iii) *the number of edges colored with 1 and 2 on $\{(x,0)\colon x \in [0,1]\}$ is 0.*

Proof. Agent A always answers 1 as a preferred piece at any vertex $(x,1)$ with $x \in [0,1]$ by the assumptions that A weakly prefers the top layer over the bottom layer and

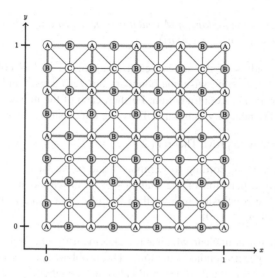

Fig. 6. Triangulation of the unit square with agent labeling.

that the preference is monotone. Hence, by the tie-breaking rule, A colors any of her owned vertex of form $(x, 1)$ with 1, which proves the claim (i).

To approach (ii), consider the boundary where $x = 1$. Let a_{12} and a_{13} denote the numbers of edges colored with $1, 2$, and $1, 3$ on $\mathsf{L}(1)$, respectively. Let b_{12} denote the number of edges colored with $1, 2$ on $\mathsf{L}(0)$. Since agent A weakly prefers the top layer to the bottom layer and owns the vertices $(1, 1)$ and $(1, 0)$, the endpoint $(1, 1)$ receives color 1 and the other endpoint $(1, 0)$ receives color 3 by the tie-breaking rule. Thus, the total number $a_{12} + a_{13}$ of edges one of whose endpoints is colored with 1 is odd. Further, for every y such that $(0, y)$ and $(1, y)$ are vertices of T, since both $(0, y)$ and $(1, y)$ receive the same agent label, we have the following symmetry:

- if one of $(0, y)$ and $(1, y)$ is colored with 1, the other is also colored with 1, and
- if one of $(0, y)$ and $(1, y)$ is colored with 2, the other is colored with 3.

Indeed, when $0 \leqslant y < 1$, the above symmetry holds because of the tie-breaking rule; in case of ties between 2 and 3, 2 is chosen at $(0, y)$ since $0 \leqslant y$ while 3 is chosen at $(1, y)$ since $1 > y$. When $y = 1$, we have already observed in the proof for 6 that, $(0, 1)$ receives color 1 and $(1, 1)$ receives color 1. Thus, we have $a_{13} = b_{12}$. We conclude that the total number $a_{12} + b_{12}$ of edges colored with 1 and 2 on $\mathsf{L}(0) \cup \mathsf{L}(1)$ is odd.

To see (iii), the piece 1 cannot be strictly preferred on the boundary of $y = 0$ since it is of zero-length. Thus, by the tie-breaking rule, every vertex on the edge $\{(x, 0) : x \in [0, 1]\}$ is colored with 2 or 3.   □

Further, each basic square with odd degree induces an approximate birthday multi-division.

Lemma 7. *Let* M *be a basic square with nonzero degree. Then, for each* $j \in \{1, 2, 3\}$, *there is a bijection* $\pi \colon \{A, B\} \to \{1, 2, 3\} \backslash \{j\}$ *such that for each* $i \in \{A, B\}$, *there* *exists a vertex v on the boundary of* M *which is owned by* i *and colored with* $\pi(i)$ *by* i, *i.e.,* $a(v) = i$ *and* $\mathcal{A}_{\pi(i)}(v) \in c_i(\mathcal{A}(v))$.

Proof. On the boundary of the square, each of agents A and B colors at least two different layered pieces; otherwise, the boundary has zero degree and by Lemma 5, so does the square, which is a contradiction. Further, each of the indices $1, 2, 3$ appears at least once on the boundary by Lemma 5. Thus, the conclusion follows. □

Now, we are ready to prove Theorem 2.

Proof (of Theorem 2). According to Lemmas 5 and 6, we have $d(\mathsf{T}) = 1$. Since the basic squares form a partition of the triangles in T, at least one of them must have nonzero degree. The conclusion follows then from Lemma 7 and a standard compactness argument (making $\varepsilon \to 0$). □

We remark that it is unclear whether the statement of Theorem 2 remains true if we consider more general choice functions. The problem appears from the fact that otherwise the side $\{(x, 1) \colon x \in [0, 1]\}$ could also receive labels 2 or 3.

4.2 FPTAS

For three agents with monotone valuations, Deng et al. [8] exploited monotonicity to design an FPTAS to compute an approximate envy-free division. We show that in the context of two-layered cake-cutting, an approximate envy-free multi-division of three agents can be computed in logarithmic time of $\frac{1}{\varepsilon}$, in a similar manner to the work of Deng et al. [8]. Note that in contrast to Deng et al. [8], the valuations do not increase monotonically when moving one knife from left to right while fixing another one. We will, nevertheless, show that a kind of monotonicy arises when moving the short knife from left to right while fixing the long one, which then enables us to design an efficient algorithm. By the previous proof of Theorem 3, there is a basic square with a nonzero degree whose corresponding a multi-division is the desired division. In order to design an FPTAS, a key observation is that fixing the position x of a long knife, the colors along the divisions are monotone. The monotonicity of valuations then enables us to calculate the degree of each interval of the vertical line, in logarithmic time of N. The above fact enables us to compute an approximate birthday multi-division of a two-layered cake among three agents in $O(\log^2 \frac{1}{\varepsilon})$ time. The full proof of Theorem 3 can be found in [13].

5 Concluding Remarks

We established the existence of an envy-free multi-division of a multi-layered cake as well as an efficient procedure for computing an approximate solution to it in certain cases. We showed that an envy-free multi-division using $n - 1$ long knives exists for n agents when n is a prime power. Further, we proved the existence of an envy-free multi-division using one short knife and one long knife for three agents with monotone preferences, and we showed that its ε-approximation can be computed in logarithmic time of $\frac{1}{\varepsilon}$. We discuss below implications of our results and some limitations of our approach.

Envy-Free Cake Division of the Same Size. The former result has an interesting implication to the standard cake-cutting problem: by aligning a single-layered cake on n parallel layers, one can use the result to show that when n is a prime power, there always exists an envy-free division of the same size, meaning that each agent receives a piece of an equal total length. This may be relevant when agents' bundles fulfill some restrictions, e.g., the employees of companies may have maximum weekly working hours. More precisely, suppose that $m = 1$ and there are n agents with closed preferences where n is a prime power. We divide the single-layered cake into n layers of equal size to create an instance of a multi-layred cake cutting. By applying the proof of Theorem 1, we obtain an envy-free division where each agent i receives a piece $(L_\ell^{(i)})_{\ell \in [n]}$ that contains n subintervals, where it satisfies the property $(\star)$ that the length of each subinterval is the same across all agents, i.e., $\mu(L_\ell^{(i)}) = \mu(L_\ell^{(j)})$ for any pair of agents $i, j \in [n]$ and $\ell \in [n]$. We note that Theorem 6.14 of Jojić et al. [14] also implies the existence of an envy-free division of an equal total length with the same number $n(n-1)$ of cuts, although such division may not satisfy the additional property $(\star)$ that the division obtained by the above procedure satisfies.

Proportional Multi-division. Another implication of our result is the existence of a proportional multi-division that is feasible and contiguous when m is a prime power, $m \leqslant n$, and each agent i has a non-negative additive valuation. Formally, a valuation function v_i is *additive* if for any pair of layered pieces $\mathcal{L}$ and $\mathcal{L}'$ whose interiors are disjoint, we have $v_i(\mathcal{L} \cup \mathcal{L}') = v_i(\mathcal{L}) + v_i(\mathcal{L}')$. Consider an instance of the multi-layered cake-cutting problem with m layers and n agents, $m \leqslant n$, with additive valuations. A *proportional fair share* of agent i is $\frac{1}{n}$ of i's valuation for the entire cake. A multi-division is *proportional* if each agent receives a layered piece of value at least her proportional fair share. Under additive valuations, it is easy to see that envy-freeness implies proportionality.

Proposition 1. *Consider an instance of the multi-layered cake-cutting problem with m layers and n agents, $m \leqslant n$, with additive valuations. Any envy-free multi-division is proportional.*

By Proposition 1, Theorem 1 implies the existence of a proportional multi-division that is feasible and contiguous when n is a prime power, $m \leqslant n$, and agents have additive valuations. We can also show that such division exists when m is a prime power, $m \leqslant n$, and agents have non-negative additive valuations in a similar manner to the proof of Theorem 5 of Hosseini et al., who recursively apply a moving-knife algorithm over a *valuable* layer that has value at least some agent's proportional fair share. The formal proof can be found in [13].

Proposition 2. *Consider an instance of the multi-layered cake-cutting problem with m layers and n agents, $m \leqslant n$, with non-negative additive valuations. If m is a prime power, then there exists a proportional multi-division that is feasible and contiguous.*

Limitation of the Approach Based on Equivariant Topology. For a number n of agents not equal to a prime power, it is known that Volovikov's theorem does not hold. Further, as already noted in the introduction, there is a counterexample of a cake cutting instance with choice functions for which no envy-free division exists [2]. The counterexamples show some limitations of the approach of equivariant topology, but do not prohibit the existence of an envy-free division in the standard valuation function model. Indeed, Avvakumov and Karasev [3] showed that an envy-free division among any number n of agents still exists when agents have identical valuations that are not necessarily monotone; this result cannot be obtained by the method of equivariant topology. See also the last paragraph of Sect. 1 in [3]. In the context of multi-layered cake-cutting, the counterexamples of [2,21] imply that an envy-free multi-division using $n - 1$ long knives may not exist under the choice function model. However, it is still an open question whether or not a counterexample exists for some natural valuation functions.

Computational Complexity. While we prove the existence of an envy-free multi-division using $n - 1$ long knives, we do not settle the precise complexity class to which the problem belongs. On a related note, the computational problem of the BSS theorem, which is a special case of the Volovikov's theorem, has been recently shown to be PPA-p-complete [11]. Due to the relation that PPA-p = PPA-p^k (see Proposition 2.2 of [11]), it would be quite surprising if our problem belongs to PPA-p^k, which would then imply that our existential result can be proven via the BSS theorem, a less powerful statement than that by Volovikov. Hence, we expect that a new complexity class, encompassing the Volovikov theorem, should probably be introduced, though the challenge lies in the fact that there is no constructive proof of Volovikov's theorem.

Acknowledgments. Ayumi Igarashi was supported by JST, PRESTO Grant Number JPMJPR 20C1.

References

1. Asada, M., et al.: Fair division and generalizations of Sperner- and KKM-type results. SIAM J. Discrete Math. **32**(1), 591–610 (2018)
2. Avvakumov, S., Karasev, R.: Envy-free division using mapping degree. Mathematika **67**(1), 36–53 (2020)
3. Avvakumov, S., Karasev, R.: Equipartition of a segment. arXiv preprint arXiv:2009.09862 (2020)
4. Aziz, H., Mackenzie, S.: A discrete and bounded envy-free cake cutting protocol for any number of agents. In: Proceedings of the 57th Symposium on Foundations of Computer Science (FOCS), pp. 416–427 (2016)
5. Björner, A., Lovász, L., Vrecica, S.T., Živaljević, R.T.: Chessboard complexes and matching complexes. J. London Math. Soc. **49**(1), 25–39 (1994)
6. Cloutier, J., Nyman, K.L., Su, F.E.: Two-player envy-free multi-cake division. Math. Soc. Sci. **59**(1), 26–37 (2010)
7. De Longueville, M.: A Course in Topological Combinatorics. Springer, New York (2012). https://doi.org/10.1007/978-1-4419-7910-0
8. Deng, X., Qi, Q., Saberi, A.: Algorithmic solutions for envy-free cake cutting. Oper. Res. **60**(6), 1461–1476 (2012)

9. Dubins, L.E., Spanier, E.H.: How to cut a cake fairly. Am. Math. Mon. **68**(1), 1–17 (1961)
10. Fan, K.: Simplicial maps from an orientable n-pseudomanifold into S^m with the octahedral triangulation. J. Comb. Theor. **2**, 588–602 (1967)
11. Filos-Ratsikas, A., Hollender, A., Sotiraki, K., Zampetakis, M.: A topological characterization of modulo-p arguments and implications for necklace splitting. In: Proceedings of the 32nd Annual ACM-SIAM Symposium on Discrete Algorithms (SODA), pp. 2615–2634 (2020)
12. Hosseini, H., Igarashi, A., Searns, A.: Fair division of time: multi-layered cake cutting. In: Proceedings of the 29th International Joint Conference on Artificial Intelligence (IJCAI), pp. 182–188 (2020)
13. Igarashi, A., Meunier, F.: Envy-free division of multi-layered cakes. CoRR, arXiv: 2106.02262 (2021)
14. Jojić, D., Panina, G., Živaljević, R.: Splitting necklaces, with constraints. SIAM J. Discret. Math. **35**(2), 1268–1286 (2021)
15. Kuhn, H.W.: Some combinatorial lemmas in topology. IBM J. Res. Dev. **4**(5), 518–524 (1960)
16. Lebert, N., Meunier, F., Carbonneaux, Q.: Envy-free two-player m-cake and three-player two-cake divisions. Oper. Res. Lett. **41**(6), 607–610 (2013)
17. Matoušek, J.: Using the Borsuk-Ulam Theorem: Lectures on Topological Methods in Combinatorics and Geometry. Springer, Heidelberg (2003). https://doi.org/10.1007/978-3-540-76649-0
18. Meunier, F., Su, F.E.: Multilabeled versions of Sperner's and Fan's lemmas and applications. SIAM J. Appl. Algebra and Geom. **3**(3), 391–411 (2019)
19. Meunier, F., Zerbib, S.: Envy-free cake division without assuming the players prefer nonempty pieces. Israel J. Math. **234**(2), 907–925 (2019). https://doi.org/10.1007/s11856-019-1939-6
20. Nyman, K., Su, F.E., Zerbib, S.: Fair division with multiple pieces. Discret. Appl. Math. **283**, 115–122 (2020)
21. Panina, G., Živaljević, R.T.: Envy-free division via configuration spaces. arXiv preprint arXiv:2102.06886 (2021)
22. Segal-Halevi, E.: Fairly dividing a cake after some parts were burnt in the oven. In: Proceedings of the 17th International Conference on Autonomous Agents and Multiagent Systems (AAMAS), pp. 1276–1284 (2018)
23. Segal-Halevi, E.: Fair multi-cake cutting. Discret. Appl. Math. **291**, 15–35 (2021)
24. Steinhaus, H.: Sur la division pragmatique. Econometrica **17**, 315–319 (1949)
25. Stromquist, W.: How to cut a cake fairly. Am. Math. Mon. **87**(8), 640–644 (1980)
26. Stromquist, W.: Envy-free cake divisions cannot be found by finite protocols. Electron. J. Comb. **15**(1), R11 (2008)
27. Su, F.E.: Rental harmony: Sperner's lemma in fair division. Am. Math. Mon. **106**(10), 930–942 (1999)
28. Volovikov, A.Y.: On a topological generalization of the Tverberg theorem. Math. Not. **59**(3), 324–326 (1996)
29. Woodall, D.R.: Dividing a cake fairly. J. Math. Anal. Appl. **78**(1), 233–247 (1980)
30. Živaljević, R.T.: User's guide to equivariant methods in combinatorics. II. Publications de l'Institut Mathématique. Nouvelle Série **64**, 107–132 (1998)

Computing Envy-Freeable Allocations with Limited Subsidies

Ioannis Caragiannis[1]($\boxtimes$) (iD) and Stavros D. Ioannidis[2]

[1] Department of Computer Science, Aarhus University, Aarhus, Denmark
iannis@cs.au.dk
[2] Department of Informatics, King's College London, London, UK
stavros.ioannidis@kcl.ac.uk

Abstract. Fair division has emerged as a very hot topic in EconCS research, and envy-freeness is among the most compelling fairness concepts. An allocation of indivisible items to agents is envy-free if no agent prefers the bundle of any other agent to his own in terms of value. As envy-freeness is rarely a feasible goal, there is a recent focus on relaxations of its definition. An approach in this direction is to complement allocations with payments (or subsidies) to the agents. A feasible goal then is to achieve envy-freeness in terms of the total value an agent gets from the allocation and the subsidies.

We consider the natural optimization problem of computing allocations that are *envy-freeable* using the minimum amount of subsidies. As the problem is NP-hard, we focus on the design of approximation algorithms. On the positive side, we present an algorithm which, for a constant number of agents, approximates the minimum amount of subsidies within any required accuracy, at the expense of a graceful increase in the running time. On the negative side, we show that, for a superconstant number of agents, the problem of minimizing subsidies for envy-freeness is not only hard to compute exactly (as a folklore argument shows) but also, more importantly, hard to approximate.

Keywords: Fair division · Indivisible goods · Subsidy minimization · Approximation algorithms

1 Introduction

Fairly dividing goods among people is an extremely important quest since antiquity. Today, fair division is a flourishing area of research in computer science, economics, and political science and *envy-freeness* is considered as the ultimate fairness concept [26]. Following a research trend that is very popular recently, we consider allocation problems with indivisible items. An allocation of items to agents is envy-free if no agent prefers the bundle of items allocated to some other agent to her own. Traditionally, agents' preferences are based on cardinal valuations they have for the items.

© Springer Nature Switzerland AG 2022
M. Feldman et al. (Eds.): WINE 2021, LNCS 13112, pp. 522–539, 2022.
https://doi.org/10.1007/978-3-030-94676-0_29

Unfortunately, with indivisible items, envy-freeness is rarely a feasible goal. For example, no such allocation exists in the embarrassingly simple case with a single item and two agents with some value for it. Recently proposed relaxations of envy-freeness aim to serve as useful alternative fairness notions. In a line of research that emerged very recently, allocations are complemented with payments (or *subsidies*) to the agents [7,18]. Now, envy-freeness dictates that no agent prefers the allocation and payment of another agent to hers, and becomes a feasible goal. However, important questions arise related to the sparing use of money.

In this paper, we follow an optimization approach. We define and study the optimization problem SMEF (standing for Subsidy Minimization for Envy-Freeness). Given an allocation problem consisting of items and agents with valuations for the items, SMEF asks for an allocation that is envy-freeable using the minimum total amount of subsidies.

SMEF is NP-hard; this follows by the NP-hardness of deciding whether a given allocation problem has an envy-free allocation or not. Thus, we resort to *approximation algorithms* for SMEF. As multiplicative approximation guarantees are hopeless, our aim is to design algorithms that run in polynomial-time and compute an allocation that is envy-freeable with an amount of subsidies that does not exceed the minimum possible amount of subsidies (denoted χ) by much. In particular, we use the total valuation of all agents for all goods (denoted by sum v) as a benchmark and seek allocations that are envy-freeable with an amount of at most $\chi + \rho \cdot$ sum v as subsidies. The goal for the approximation guarantee ρ of an algorithm is to be as small as possible.

We initiate the study of SMEF and present two results. On the positive side, we design an algorithm that achieves an arbitrary low approximation guarantee of $\epsilon > 0$. When applied to allocation instances with a constant number of agents, the algorithm uses *dynamic programming* and runs in time that is polynomial in the number of items and $1/\epsilon$. On the negative side, we show that, in general, SMEF is not only hard to solve exactly, but also hard to approximate within a small constant. Unlike the folklore reduction[1] for proving hardness of envy-freeness, our proof uses a novel *approximation-preserving* reduction. Besides separating the general case from that with constantly many agents, our negative result indicates that achieving good approximation guarantees will be a challenging goal.

1.1 Related Work

The concept of envy-freeness was formally introduced by Foley [15] and Varian [29]. As envy-freeness may not be achievable when goods are indivisible, recent research has focused on defining approximations of envy-freeness. These include envy-freeness up to one good [8], envy-freeness up to any good [10], epistemic envy-freeness [4], and more. Still, achieving even them in polynomial time

---

[1] Notice that deciding whether an envy-free allocation exists for two agents with identical item valuations requires solving PARTITION, a well-known NP-hard problem [16].

can be challenging, and recent work has focused on approximation algorithms; see, e.g., [2,6,9,11,12,21,25].

The approach of mixing allocations with payments either from or to the agents has been extensively considered in the economics literature. A typical example is the rent division problem, where n items (rooms) and a fixed rent have to be divided among n agents in an envy-free manner [27,28]. Compensations to the agents were first considered by Maskin [22]. Subsequent papers consider unit-demand allocation problems, where each agent can get at most one item; see, e.g., [1]. Aragones [3] and Klijn [20] give polynomial-time algorithms that compute allocations and payments. More general models are studied by Haake et al. [17] and Meertens et al. [23].

In the AI literature, Chevaleyre et al. [13] consider allocation problems and monetary transfers between the agents. In a model that is the closest to ours, Halpern and Shah [18] aim to bound the amount of external subsidies assuming that all agent valuations for goods are in $[0, 1]$. Among several results, they conjectured that subsidies of $n-1$ suffice; an even stronger version of the conjecture was proved very recently by Brustle et al. [7].

1.2 Roadmap

The rest of the paper is structured as follows. We begin with preliminary definitions in Sect. 2. Our approximation algorithm is presented in Sect. 3 and our result on the hardness of approximation for SMEF is presented in Sect. 4. We conclude in Sect. 5.

2 Preliminaries

We consider allocation instances with a set M of m items and a set N of n agents. Each agent $i \in N$ has a valuation function $v_i : M \to \mathbb{R}_{\geq 0}$ over the items.[2] With some abuse of notation, we use $v_i(B)$ to denote the valuation of agent i for the set (or *bundle*) of items B. Valuations are additive, i.e., $v_i(B) = \sum_{g \in B} v_i(g)$. An allocation is simply a partition $X = (X_1, X_2, ..., X_n)$ of the items of M into n disjoint bundles, where agent $i \in N$ is supposed to get the bundle X_i. We use the abbreviations sum $v = \sum_{i \in N} v_i(M)$ and max $v = \max_{i \in N} v_i(M)$.

As usual, we define the *social welfare* of an allocation $X = (X_1, ..., X_n)$ to be $\mathrm{SW}(X, v) = \sum_{i \in N} v_i(X_i)$. An allocation $X = (X_1, X_2, ..., X_n)$ is *envy-free* if $v_i(X_i) \geq v_i(X_j)$ for every pair of agents i and j. Informally, envy-freeness requires that no agent envies the bundle allocated to any other agent compared to her own.

For an allocation $X = (X_1, ..., X_n)$ in an instance with agent valuations v, the *envy graph* $\mathrm{EG}(X, v)$, introduced by Lipton et al. [21], is an edge-weighted complete directed graph that has a node for each agent and the weight of the

---

[2] In our exposition, we assume that valuations are non-negative, even though our positive result can be extended to work without this assumption, in the model of [5] where items can be goods or chores.

directed edge (i, j) represents the "envy" of agent i for agent j. Using $G = \text{EG}(X, v)$ and $\text{wgt}_G(i, j)$ for the weight of the directed edge from node i to node j in the envy graph $\text{EG}(X, v)$, we define $\text{wgt}_G(i, j) = v_i(X_j) - v_i(X_i)$.

Following the modelling assumptions of [18], we also consider *payments* (or *subsidies*) to the agents, represented by a payment vector $\pi = \langle \pi_1, ..., \pi_n \rangle$ with non-negative entries, i.e., $\pi_i \geq 0$ for every agent $i \in N$. Below, we use the terms "payment" and "subsidy" interchangeably. Now, we say that the pair (X, π) of the allocation X and payment vector π is envy-free if $v_i(X_i) + \pi_i \geq v_i(X_j) + \pi_j$ for every pair of agents $i, j \in N$. Informally, this extended version of envy-freeness requires that no agent envies the bundle and the payment of any other agent compared to the bundle and payment she gets.

We say that allocation X is *envy-freeable* if there is a payment vector π so that the pair (X, π) is envy-free. Although the use of payments makes envy-freeness a feasible goal, not all allocations are envy-freeable. The following theorem, due to Halpern and Shah [18], gives sufficient and necessary conditions so that an allocation is envy-freeable.

Theorem 1 (Halpern and Shah [18]). *The following statements are equivalent:*

- *The allocation $X = (X_1, X_2, ..., X_n)$ is envy-freeable.*
- *The allocation X maximizes social welfare among all redistributions of its bundles to the agents.*
- *The envy graph $\text{EG}(X, v)$ contains no directed cycles of positive total weight.*

Detecting whether a given allocation X is envy-freeable can be done using the following linear program $\text{LP}(X, v)$:

$$\text{minimize} \quad \sum_{i \in N} \pi_i \tag{1}$$
$$\text{subject to:} \quad \pi_i - \pi_j \geq v_i(X_j) - v_i(X_i), \forall i, j \in N$$
$$\pi \geq 0$$

$\text{LP}(X, v)$ aims to find a payment vector π so that the envy-freeness constraints between pairs of agents are satisfied. In addition, it minimizes the total amount of payments. As it is observed by Halpern and Shah [18], the payment π_i of agent i obtained in this way is equal to the maximum total weight in any simple path that originates from node i in the envy graph $\text{EG}(X, v)$.

We study the optimization problem SMEF (standing for Subsidy Minimization for Envy-Freeness). Given an allocation instance, SMEF aims to compute an allocation that is envy-freeable with the minimum amount of subsidies. Since the problem of computing an envy-free allocation is NP-hard, SMEF is NP-hard as well.

We are interested in the design of approximation algorithms for SMEF. As algorithms with finite multiplicative approximation ratio are hopeless (since it is NP-hard to decide whether the minimum amount of subsidies is zero or not), we seek polynomial-time algorithms that compute an allocation that is envy-freeable with subsidies $\chi + \rho \cdot \text{sum } v$, with the approximation guarantee ρ being as low as possible.

As a warmup, consider the algorithm that allocates all items to the agent i^* who has maximum value for M and paying a subsidy of $v_{i^*}(M)$ to every other agent i. Clearly, this is a polynomial-time algorithm. The allocation obtained is envy-freeable since no redistribution of the bundles (i.e., giving all items to another agent) results in higher social welfare. And the particular payments are right: agent i^* is indifferent between the bundle M and the payment to any other agent, while the other agents are indifferent between the (equal) payments, and prefer their payment to getting the whole bundle M. It can be easily verified that the algorithm guarantees an amount of at most $\chi + (n-1) \max v \leq \chi + (n-1)\text{sum } v$ as subsidies; this is the best guarantee of this form for this algorithm in the worst-case.

3   An Approximation Algorithm

We now present an algorithm that does much better. The algorithm exploits ideas that have led to polynomial-time approximation schemes for combinatorial optimization problems like KNAPSACK; e.g., see [30]. It first discretizes all valuations to multiples of a discretization parameter. In this way, the different discretized valuations an agent can have for bundles of items in the new instance is small. This allows to classify all allocations into a relatively small number of classes, each defined by specific discretized valuation levels of each agent for all bundles. Dynamic programming is used to decide the classes that are non-empty and to select a representative allocation from each class. The final allocation is selected among all representative allocations, possibly after redistributing the bundles so that social welfare (with respect to the original valuations) is maximized (in order to get envy-freeability). This requires a call to linear program (1) to compute the minimum amount of subsidies for each representative allocation.

The classification of allocations guarantees that the algorithm will consider a representative allocation from the class that also contains the optimal one (i.e., the allocation that is envy-freeable with the minimum amount of subsidies overall). Our analysis shows that the amount of subsidies for making the representative allocation envy-free is close to optimal. Polynomial running time for the case of a constant number of agents follows by setting the discretization parameter appropriately.

We now present our algorithm in detail. It uses an accuracy parameter $\epsilon > 0$ and initially decides the value of the discretization parameter δ as follows:

$$\delta = \frac{\epsilon \max v}{4mn^2}.$$

First, the algorithm implicitly discretizes all agent valuations by defining new valuations $\tilde{v}$ as follows: for an agent i with valuation $v_i(g)$ for item g, the discretized valuation $\tilde{v}_i(g)$ is equal to $\lfloor v_i(g)/\delta \rfloor \delta$.

The algorithm uses an arbitrary ordering of the items in M; let $M = \{g_1, g_2, ..., g_m\}$, where the item indices are those in this ordering. The algorithm builds a table $\mathbf{T}$ which classifies all possible allocations of subsets of M. Consider an $(n^2 + 1)$-dimensional tuple $\tau = (t, P_{ij}, 1 \leq i, j \leq n)$, where t is an integer from 1 to m and P_{ij} is an integer from 0 to $\lfloor \max v/\delta \rfloor$, for every pair of agents i and j. The entry $\mathbf{T}(\tau)$ of the table indicates whether an allocation $A^t = (A_1^t, A_2^t, ..., A_n^t)$ of the first t items $g_1, ..., g_t$ of M to the n agents, satisfying $\tilde{v}_i(A_j^t) = P_{ij}\delta$ for every pair of agents i and j, exists ($\mathbf{T}(\tau) = 1$) or not ($\mathbf{T}(\tau) = 0$).

The entries of $\mathbf{T}$ are computed using the following recursive relation:

- For a tuple $\tau = (t, P_{ij}, 1 \leq i, j \leq n)$ with $t = 1$, the algorithm sets $\mathbf{T}(\tau) = 1$ if there exists $k \in [n]$ such that, for every $i \in [n]$, $\tilde{v}_i(g_1) = P_{ik}\delta$ and $P_{ij} = 0$ for every $j \neq k$. Otherwise, the algorithm sets $\mathbf{T}(\tau) = 0$.
- For a tuple $\tau = (t, P_{ij}, 1 \leq i, j \leq n)$ with $t > 1$, the algorithm sets $\mathbf{T}(\tau) = 1$ if there exists $k \in [n]$ and tuple $\tau' = (t - 1, P'_{ij}, 1 \leq i, j \leq n)$ such that, for every $i \in [n]$, $P_{ik} = P'_{ik} + \tilde{v}_i(g_t)/\delta$ and $P_{ij} = P'_{ij}$ for every $j \neq k$. Otherwise, the algorithm sets $\mathbf{T}(\tau) = 0$.

Essentially, each non-zero entry of $\mathbf{T}$ (i.e., $\mathbf{T}(\tau) = 1$) indicates a non-empty class $\mathcal{A}_\tau$ of (possibly partial, when the first argument of τ is an integer smaller than m) allocations. To compute a representative complete allocation $A_\tau \in \mathcal{A}_\tau$ among those implied by the non-zero entry corresponding to the tuple $(m, P_{ij}^m, 1 \leq i, j \leq n)$, the algorithm does the following for $t = m$ downto 2. Let $k \in [n]$ be such that $\mathbf{T}(\tau') = 1$ for a tuple $\tau' = (t - 1, P_{ij}^{t-1}, 1 \leq i, j \leq n)$ with $P_{ik}^{t-1} = P_{ik}^t - \tilde{v}_i(g_t)/\delta$ and $P_{ij}^{t-1} = P_{ij}^t$ for every pair of agents i and $j \neq k$. The algorithm assigns item g_t to agent k and proceeds to considering the next item. The first item g_1 is assigned to agent k such that $\mathbf{T}(\tau') = 1$ for a tuple $\tau' = (1, P_{ij}^1, 1 \leq i, j \leq n)$ with $P_{ik}^1 = \tilde{v}_i(g_1)/\delta$ and $P_{ij}^1 = 0$ for every pair of agents i and $j \neq k$.

Next, the algorithm redistributes the bundles of each allocation A_τ that represents a non-empty class $\mathcal{A}_\tau$ so that an allocation A'_τ of maximum social welfare (among those that distribute the particular bundles to the agents) is obtained (in terms of the original valuations). It solves LP(A'_τ, v) (for the original valuations) to compute the minimum amount of subsidies that make A'_τ envy-free. Among all allocations A'_τ, it outputs the one with the minimum amount of subsidies. The approximation guarantee of the algorithm is given by the next lemma.

Lemma 1. *Given an instance of SMEF that has an allocation that is envy-freeable with an amount of χ as total subsidies, the algorithm computes an allocation that is envy-freeable with total subsidies of at most $\chi + 4mn^2\delta$.*

Proof. Let τ be a full tuple such that $\mathcal{A}_\tau$ contains an allocation $O = (O_1, ..., O_n)$ that is envy-freeable with subsidies of χ. Since $\mathcal{A}_\tau$ is non-empty, it is $\mathbf{T}(\tau) = 1$. Let A be the allocation computed by the algorithm as representative of $\mathcal{A}_\tau$ and A' the allocation that is obtained after redistributing the bundles of A. By Theorem 1, A' is clearly envy-freeable; we will show that the corresponding subsidies are at most $\chi + 4mn^2\delta$. Clearly, the output of the algorithm will be envy-freeable with at most this amount of subsidies.

Let $\sigma \in \mathcal{L}(n)$ be the permutation over $[n]$ such that $A'_j = A_{\sigma(j)}$ for every $j \in [n]$. Let G and H be the envy graphs $EG(O, v)$ and $EG(A', v)$, respectively.

We now present the most crucial component of our analysis. It exploits the fact that both O and A belong to class $\mathcal{A}_\tau$ and uses the third statement of Theorem 1.

Lemma 2. *For every pair of agents i and j, there exists a (not necessarily simple) path $p(i, j)$ from node $\sigma(i)$ to node $\sigma(j)$ such that*

$$wgt_H(i,j) \leq \sum_{e \in p(i,j)} wgt_G(e) + 4m\delta.$$

Proof. In the proof, we will use the following simple lemma.

Lemma 3. *For every agent i and every two bundles B_1 and B_2 such that $\tilde{v}_i(B_1) = \tilde{v}_i(B_2)$, it holds that*

$$-|B_2|\delta \leq v_i(B_1) - v_i(B_2) \leq |B_1|\delta. \tag{2}$$

Proof. First observe that, by the definition of $\tilde{v}$ and its relation to v, for every agent i and item $g \in M$, it holds that $\tilde{v}_i(g) \leq v_i(g) \leq \tilde{v}_i(g) + \delta$. Hence, for every bundle B,

$$\tilde{v}_i(B) \leq v_i(B) \leq \tilde{v}_i(B) + |B|\delta.$$

The lemma follows by applying this inequality for bundles B_1 and B_2 and using the fact that $\tilde{v}_i(B_1) = \tilde{v}_i(B_2)$. □

We use the notation σ^{-1} to refer to the inverse permutation of σ, i.e., $\sigma^{-1}(k) = j$ when $k = \sigma(j)$. Consider the set C that contains edge $(k, \sigma^{-1}(k))$ for every agent k such that $k \neq \sigma^{-1}(k)$. C is either empty (if $k = \sigma^{-1}(k)$ for every agent k) or consists of disjoint directed cycles. For an agent i, if $\sigma^{-1}(i) \neq i$, we denote by C_i the set of nodes that are spanned by the cycle of C that includes node i. Otherwise, we define C_i to contain only node i.

Define the (not necessarily simple) path $p(i, j)$ from node $\sigma(i)$ to node $\sigma(j)$ to contain edge $(k, \sigma(k))$ for every node k in the set C_i besides node i and, if $i \neq \sigma(j)$, the directed edge $(i, \sigma(j))$.

For every pair of agents i and j, we have that the weight of the directed edge (i, j) in H is

$$wgt_H(i,j) \leq wgt_H(i,j) - \sum_{k \in C_i} wgt_H(k, \sigma^{-1}(k))$$

$$= v_i(A'_j) - v_i(A'_i) - \sum_{k \in C_i} \left(v_k(A'_{\sigma^{-1}(k)}) - v_k(A'_k) \right)$$

$$= v_i(A_{\sigma(j)}) - v_i(A_{\sigma(i)}) - \sum_{k \in C_i} \left(v_k(A_k) - v_k(A_{\sigma(k)}) \right)$$

$$\leq v_i(O_{\sigma(j)}) - v_i(O_{\sigma(i)}) - \sum_{k \in C_i} \left(v_k(O_k) - v_k(O_{\sigma(k)}) \right)$$

$$+ \left(|A_{\sigma(j)}| + |O_{\sigma(i)}| + \sum_{k \in C_i} |O_k| + \sum_{k \in C_i} |A_{\sigma(k)}| \right) \delta$$

$$\leq v_i(O_{\sigma(j)}) - v_i(O_{\sigma(i)}) - \sum_{k \in C_i} \left(v_k(O_k) - v_k(O_{\sigma(k)}) \right) + 4m\delta$$

$$= v_i(O_{\sigma(j)}) - v_i(O_i) + \sum_{k \in C_i \setminus \{i\}} \left(v_k(O_{\sigma(k)}) - v_k(O_k) \right) + 4m\delta$$

$$= \text{wgt}_G(i, \sigma(j)) + \sum_{k \in C_i \setminus \{i\}} \text{wgt}_G(k, \sigma(k)) + 4m\delta$$

$$= \sum_{e \in p(i,j)} \text{wgt}_G(e) + 4m\delta.$$

The first inequality follows since C_i consists of node i only (when $i = \sigma(i)$) or the edges $(k, \sigma^{-1}(k))$ for $k \in C_i$ form a directed cycle of non-positive total weight in H. The second inequality follows by applying Lemma 3 (recall that both allocations A and O belong to the class $\mathcal{A}_\tau$ and, hence, $\tilde{v}_\ell(A_q) = \tilde{v}_\ell(O_q)$ for every pair of agents ℓ and q). The third inequality follows since the bundles $A_{\sigma(k)}$ (respectively, O_k) for $k \in C_i$ are disjoint. The equalities are obvious or follow by the definition of the weights. $\qquad\square$

Now, let π' and π be the solutions of $\text{LP}(A', v)$ and $\text{LP}(O, v)$, respectively. Hence, $\chi = \text{Sub}(O, v) = \sum_{i=1}^{n} \pi_i$. We will use Lemma 2 to argue that

$$\pi'_i \leq \pi_{\sigma(i)} + 4mn\delta. \tag{3}$$

This will yield total subsidies of

$$\text{Sub}(A', v) = \sum_{i=1}^{n} \pi'_i \leq \sum_{i=1}^{n} \left(\pi_{\sigma(i)} + 4mn\delta \right) = \chi + 4mn^2\delta,$$

completing the proof.

Recall from Theorem 1 that the payment π'_ℓ (respectively, π_ℓ) is equal to the maximum path weight over all simple paths that originate from node ℓ in graph H (respectively, graph G). Let Q_ℓ be the corresponding simple path that is destined for some node s (and originates from node ℓ), i.e., $\pi'_\ell = \sum_{e \in Q_\ell} \text{wgt}_H(e)$. We construct the (not necessarily simple) path P_ℓ from node $\sigma(\ell)$ to node $\sigma(s)$ of G that consists of path $p(i, j)$ for every directed edge (i, j) in the path Q_ℓ. Using Lemma 2, we get

$$\pi'_\ell = \sum_{e \in Q_\ell} \text{wgt}_H(e) \leq \sum_{e \in Q_\ell} \left(\sum_{e' \in p(e)} \text{wgt}_G(e') + 4m\delta \right) \tag{4}$$

$$\leq \sum_{e \in Q_\ell} \sum_{e' \in p(e)} \text{wgt}_G(e') + 4mn\delta = \sum_{e \in P_\ell} \text{wgt}_G(e) + 4mn\delta.$$

The second inequality follows since path Q_ℓ is simple (and, hence, contains at most $n-1$ edges). Now, create the simple path P'_ℓ from node $\sigma(\ell)$ to node $\sigma(s)$ by removing the cycles in P_ℓ. Since graph G does not have any directed cycles of positive total weight (by Theorem 1), we have $\text{wgt}_G(P_\ell) \leq \text{wgt}_G(P'_\ell)$. Now, (4) yields

$$\pi'_\ell \leq \sum_{e\in P'_\ell} \text{wgt}_G(e) + 4mn\delta,$$

which implies (3) since P'_ℓ is a simple path that originates from node $\sigma(\ell)$.   □

The running time of the algorithm depends on the number of table entries, the number of steps required for computing each table entry using the recursive relation, the number of steps required to compute a representative allocation for a non-empty allocation class, the redistribution time, and the time required to solve the linear programs.

The dimensions of the table $\mathbf{T}$ are m for the first one that enumerates over all items, and at most $1+\lfloor \max v/\delta \rfloor = 1+\frac{4mn^2}{\epsilon}$ for each of the other dimensions. Overall, the size of the table is $\mathcal{O}\left(\left(\frac{m}{\epsilon}\right)^{n^2+1}\right)$. The computation of each table entry using the recursive relation needs the values in n^2 table entries that have been previously computed. In a representative allocation, the agent to which each of the m items is allocated requires time n^2 as well, i.e., time $\mathcal{O}(m)$ in total. The redistribution of the bundles can be implemented using a matching computation in a complete edge-weighted bipartite graph that has a node for each agent and for each bundle and the weight of an edge indicates the valuation of an agent for a bundle. As n is constant, this takes constant time. Also, the linear programs have constant size. In general, since n is a constant, it is ignored in the $\mathcal{O}$ notation unless it appears in the exponent. The above discussion is summarized in the next statement.

Theorem 2. *Let $\epsilon > 0$ be the accuracy parameter used by the algorithm. Given an instance of SMEF consisting of a constant number n of agents with valuations v over a set M of m items that has an envy-freeable allocation using an amount χ of subsidies, the algorithm runs in time $\mathcal{O}\left((m/\epsilon)^{n^2+2}\right)$ and computes an allocation that is envy-freeable using a total subsidy of at most $\chi + \epsilon \max v$.*

4   Hardness of Approximating SMEF

In this section, we show that approximation guarantees like the one in the statement of Theorem 2 are not possible when the number of agents is part of the input.

Theorem 3. *Approximating SMEF within an additive term of $3 \cdot 10^{-4} \text{sum } v$ is NP-hard.*

We prove Theorem 3 by presenting a reduction from Maximum 3-Dimensional Matching (MAX-3DM). An instance of MAX-3DM consists of three disjoint sets of elements $A = \{a_1, a_2, ..., a_n\}$, $B = \{b_1, b_2, ..., b_n\}$, and $C = \{c_1, c_2, ..., c_n\}$, each of size n, and a set T of m triplets of the form (a_i, b_j, c_k) with $a_i \in A$, $b_j \in B$, and $c_k \in C$. The objective is to compute a disjoint subset of T (or, simply, a 3D matching) of maximum size. The problem is well-known to be NP-hard not only to solve exactly [16] but also to approximate [19].

We will use the inapproximability result of Chlebík and Chlebíková [14], which applies to bounded instances of MAX-3DM in which each element appears in exactly two triplets (i.e., $m = 2n$); we will refer to this restriction of MAX-3DM as MAX-3DM-2. In particular, Chlebík and Chlebíková [14] show that it is NP-hard to distinguish between instances of MAX-3DM-2 with a 3D matching of size at least K and instances of MAX-3DM-2 in which any 3D matching has size at most $K - 0.01n$.[3]

4.1   The Reduction

We present our reduction and full proof for the case $\chi > 0$. We omit the case $\chi = 0$, which requires a minor modification of the reduction. On input an instance of MAX-3DM-2, our reduction constructs in polynomial time an instance of SMEF, in which the minimum amount of subsidies that can make some allocation envy-free is exactly $\chi(1 + \max\{K - L, 0\})$, where L is the size of the maximum 3D matching in the MAX-3DM-2 instance. Using the result of [14], we will get that it is NP-hard to distinguish between SMEF instances in which the minimum amount of subsidies is at most χ and instances in which it is at least $\chi(1 + 0.01n)$. Hence, SMEF will be proved to be NP-hard to approximate within $0.01n\chi$. Our construction will be such that sum $v < 30n\chi$. In this way, we will obtain a hardness of approximating SMEF within an additive term of (at least) $3 \cdot 10^{-4}$ sum v, as desired.

Our reduction is as follows. Given an instance of MAX-3DM-2 consisting of sets of elements A, B, and C, each of size n, and a set of $2n$ triplets T, the instance of SMEF has

- three agents 1, 2, and 3,
- three agents $J_1(t)$, $J_2(t)$, and $J_3(t)$ for every triplet $t \in T$,
- an item A_i for every element $a_i \in A$,
- an item B_i for every element $b_i \in B$,
- an item Γ_i for every element $c_i \in C$,
- three items Δ_t, Z_t, and Θ_t for every triplet $t \in T$, and
- an additional item Λ.

The agents $J_1(t)$, $J_2(t)$, and $J_3(t)$ that correspond to the triplet $t = (a_i, b_j, c_k)$ have valuations 0 for all items besides the items A_i, B_j, Γ_k, Δ_t, Z_t, and Θ_t.

---

[3] This statement is actually weaker than the one proved in [14]. However, it suffices for our purpose to prove hardness of approximation. Note that we have made no particular attempt to optimize our inapproximability threshold.

Agents 1, 2 have valuation 0 for all items besides item Λ and agent 3 has valuation zero for all items besides item Λ and items Θ_t for $t \in T$. Their remaining valuations are as follows:

	A_i	B_j	Γ_k	Δ_t	Z_t	Θ_t	Λ
1	0	0	0	0	0	0	χ
2	0	0	0	0	0	0	χK
3	0	0	0	0	0	χ	χK
$J_1(t)$	χ	χ	χ	3χ	3χ	0	0
$J_2(t)$	0	0	0	χ	χ	χ	0
$J_3(t)$	0	0	0	0	χ	0	0

Recall that each element belongs to exactly two triplets. Hence, two agents have positive value for item A_i (similarly for items B_j and Γ_k): agents $J_1(t_1)$ and $J_1(t_2)$ such that the triplets t_1 and t_2 contain element a_i (similarly for elements b_j and c_k). It is easy to see that either two or three agents have positive value for each item. For every triplet t, the agents $J_1(t)$, $J_2(t)$, and $J_3(t)$ have total valuation 9χ, 3χ, and χ, respectively. Taking into account that $K \leq n$, we obtain that sum $v < 30n\chi$.

4.2   Lower Bound on Subsidies

Consider an instance of SMEF constructed by our reduction and let X be an envy-freeable allocation in it. We will first lower-bound the minimum amount of subsidies that make X envy-free. First observe that X cannot give item Λ to agent 1; in that case, exchanging the bundles of agents 1 and 2 would result to an increase of the social welfare and, hence, X would not be envy-freeable. If X gives item Λ to agent 3, agents 1 and 2 would need subsidies of at least χ and χK, respectively, so that they do not envy agent 3. Hence, $\text{Sub}(A, v) \geq \chi(1 + K)$ in this case.

In the following, we will lower-bound the minimum total subsidies that make X envy-free assuming that item Λ is given to agent 2. Let θ be the number of items Θ_t for $t \in [2n]$ agent 3 gets. Then, agent 3 should be given a subsidy of at least $\chi \max\{K - \theta, 0\}$ so that she does not envy agent 2. Agent 1 needs a subsidy of $\chi \max\{K - \theta, 1\}$ so that she does not envy agents 1 and 2.

For a triplet $t = (a_i, b_j, c_k)$ in the original instance of MAX-3DM-2, we call it *full* if all items A_i, B_j, and Γ_k (which correspond to the elements of the triplet) have been allocated to the agents $J_1(t)$, $J_2(t)$, or $J_3(t)$. Otherwise, we call it *partial*. We call t *supported* if item Θ_t has been allocated to agent $J_2(t)$; otherwise, we call t *unsupported*.

In the next four lemmas, we lower-bound the total amount of subsidies the agents $J_1(t)$, $J_2(t)$, and $J_3(t)$ of a triplet t need, depending on the type of t.

Lemma 4. *The agents $J_1(t)$, $J_2(t)$, and $J_3(t)$ of a full and supported triplet t need subsidies of at least $\chi \max\{K - \theta - 2, 0\}$.*

Proof. Consider a full and supported triplet t. If agent $J_2(t)$ has value at most 2χ (i.e., getting Θ_t and at most one of the items Δ_t and Z_t), then she needs a subsidy of at least $\chi \max\{K - \theta - 2, 0\}$ so that she does not envy agent 3. If agent $J_2(t)$ has value 3χ by getting both items Δ_t and Z_t in addition to Θ_t, she needs a subsidy of at least $\chi \max\{K - \theta - 3, 0\}$, while then agents $J_1(t)$ and $J_3(t)$ need subsidies of at least $3\chi + \chi \max\{K - \theta - 3, 0\}$ and $\chi + \chi \max\{K - \theta - 3, 0\}$, respectively, so that they do not envy agent $J_2(t)$. In both cases, the total amount of subsidies of the agents $J_1(t)$, $J_2(t)$, and $J_3(t)$ is at least $\chi \max\{K - \theta - 2, 0\}$. $\square$

Lemma 5. *The agents $J_1(t)$, $J_2(t)$, and $J_3(t)$ of a full and unsupported triplet t need subsidies of at least $\chi \max\{K - \theta - 1, 0\}$.*

Proof. Consider a full and unsupported triplet t. If agent $J_2(t)$ has value at most χ (i.e., getting at most one of the items Δ_t and Z_t), then she needs a subsidy of at least $\chi \max\{K - \theta - 1, 0\}$ so that she does not envy agent 3. If agent $J_2(t)$ has value 2χ by getting both items Δ_t and Z_t, she needs a subsidy of at least $\chi \max\{K - \theta - 2, 0\}$, while then agents $J_1(t)$ and $J_3(t)$ need subsidies of at least $3\chi + \chi \max\{K - \theta - 2, 0\}$ and $\chi + \chi \max\{K - \theta - 2, 0\}$, respectively, so that they do not envy agent $J_2(t)$. In both cases, the total amount of subsidies of agents $J_1(t)$, $J_2(t)$, and $J_3(t)$ is at least $\chi \max\{K - \theta - 1, 0\}$. $\square$

Lemma 6. *The agents $J_1(t)$, $J_2(t)$, and $J_3(t)$ of a partial and supported triplet t need subsidies of at least $\chi \max\{K - \theta - 1, 0\}$.*

Proof. Let t be a partial and supported triplet. If agent $J_2(t)$ does not get items Δ_t and Z_t, then she gets only a value of χ from item Θ_t and needs a subsidy of at least $\chi \max\{K - \theta - 1, 0\}$ so that she does not envy agent 3.

If agent $J_2(t)$ gets item Δ_t but not item Z_t, she needs a subsidy of $\chi \max\{K - \theta - 2, 0\}$ so that she does not envy agent 3. Then, if agent $J_1(t)$ does not get item Z_t, her value is at most 2χ (from at most two of the items A_i, B_j, and Γ_k) and needs a subsidy of $\chi + \chi \max\{K - \theta - 2, 0\}$ so that she does not envy agent $J_2(t)$. If agent $J_3(t)$ does not get item Z_t, she needs a subsidy of at least $\chi + \chi \max\{K - \theta - 2, 0\}$ so that she does not envy agent $J_2(t)$.

If agent $J_2(t)$ gets item Z_t but not Δ_t, she needs a subsidy of $\chi \max\{K - \theta - 2, 0\}$ so that she does not envy agent 3 and agent $J_3(t)$ needs a subsidy of at least $\chi + \chi \max\{K - \theta - 2, 0\}$ so that she does not envy agent $J_2(t)$.

Finally, if agent $J_2(t)$ gets items Δ_t and Z_t, her value is 3χ and needs a subsidy of at least $\chi \max\{K - \theta - 3, 0\}$ so that she does not envy agent 3. Then, each of agents $J_1(t)$ and $J_3(t)$ need a subsidy of at least $\chi + \chi \max\{K - \theta - 3, 0\}$ so that they do not envy agent $J_2(t)$.

In all cases, the total amount of subsidies the agents $J_1(t)$, $J_2(t)$, and $J_3(t)$ need is at least $\chi \max\{K - \theta - 1, 0\}$. $\square$

Lemma 7. *The agents $J_1(t)$, $J_2(t)$, and $J_3(t)$ of a partial and unsupported triplet t need subsidies of at least $\chi \max\{K - \theta, 1\}$.*

Proof. Let t be a partial and unsupported triplet. If agent $J_2(t)$ gets both items Δ_t and Z_t, she needs a subsidy of $\chi \max\{K - \theta - 2, 0\}$ so that she does not envy agent 3, while agents $J_1(t)$ and $J_3(t)$ would then need subsidies of at least $4\chi + \chi \max\{K - \theta - 2, 0\}$ and $\chi + \chi \max\{K - \theta - 2, 0\}$, respectively, so that they do not envy agent $J_2(t)$.

If agent $J_2(t)$ gets only item Δ_t, she needs a subsidy of $\chi \max\{K - \theta - 1, 0\}$ so that she does not envy agent 3. Then, the agent who does not get item Z_t among $J_1(t)$ and $J_3(t)$ would need a subsidy of at least $\chi + \chi \max\{K - \theta - 1, 0\}$ so that she does not envy agent $J_2(t)$.

If agent $J_2(t)$ gets only item Z_t, she needs a subsidy of $\chi \max\{K - \theta - 1, 0\}$ so that she does not envy agent 3, while agent $J_3(t)$ needs a subsidy of at least $\chi + \chi \max\{K - \theta - 1, 0\}$ so that she does not envy agent $J_2(t)$.

Finally, if agent $J_2(t)$ gets no item (among Δ_t and Z_t), she needs a subsidy of at least χ so that she does not envy the agents who get items Δ_t and Z_t and a subsidy of at least $\chi \max\{K - \theta, 0\}$ so that she does not envy agent 3.

In all cases, the total amount of subsidies the agents $J_1(t)$, $J_2(t)$, and $J_3(t)$ need is at least $\chi \max\{K - \theta, 1\}$.   □

We now denote by L_1, L_2, P_1, and P_2, the number of full and supported, full and unsupported, partial and supported, and partial and unsupported triplets defined by X, respectively. Notice that the full triplets form a 3D matching. Denoting by L the maximum size over all 3D matchings of the MAX-3DM-2 instance, we have $L \geq L_1 + L_2$. Using Lemmas 4–7, and our observations for agents 1 and 3, we have that the total amount of subsidies X needs to become envy-free is

$$\text{Sub}(X, v) \geq \chi \, (L_1 \max\{K - \theta - 2, 0\} + L_2 \max\{K - \theta - 1, 0\} \qquad (5)$$
$$+ P_1 \max\{K - \theta - 1, 0\} + P_2 \max\{K - \theta, 1\}$$
$$+ \max\{K - \theta, 0\} + \max\{K - \theta, 1\}) \, .$$

We will distinguish between two cases for $K - \theta$. If $K - \theta \geq 2$, (5) yields

$$\text{Sub}(X, v) \geq \chi \, (L_2 + P_1 + 2P_2 + 4) = \chi \, (2n - L_1 + P_2 + 4)$$
$$\geq \chi(1 + \max\{K - L, 0\}).$$

Now, notice that θ, the number of items Θ_t agent 3 gets in X is upper-bounded by the number of unsupported triplets, i.e., $\theta \leq L_2 + P_2$. Thus, if $K - \theta \leq 1$, (5) yields

$$\text{Sub}(X, v) \geq \chi \, (P_2 + K - \theta + 1) \geq \chi \, (K - L_2 + 1) \geq \chi(1 + \max\{K - L, 0\}).$$

We conclude that the minimum amount of subsidies necessary to make X envy-free is at least $\chi(1 + \max\{K - L, 0\})$.

4.3   Upper Bound on Minimum Subsidies

We now present our upper bound on the minimum amount of subsidies for envy-freeness. Given a 3D matching $\mathcal{M}$ of maximum size L in the MAX-3DM-2 instance, we will construct an allocation for the SMEF instance and will show that it is envy-freeable with an amount of subsidies equal to $\chi(1 + \max\{K - L, 0\})$.

For defining the allocation, we partition $T \setminus \mathcal{M}$ in two disjoint sets of triplets T_1 and T_2 of size $2n - \max\{K, L\}$ and $\max\{K - L, 0\}$, respectively.

- For every triplet $t = (a_i, b_j, c_k) \in \mathcal{M}$, agent $J_1(t)$ gets items A_i, B_j, and Γ_k, agent $J_2(t)$ gets item Δ_t and agent $J_3(t)$ gets item Z_t.
- For every triplet $t = (a_i, b_j, c_k) \notin \mathcal{M}$, let $F(t)$ be the set of items that correspond to the elements of t that have not been included in triplets of $\mathcal{M}$. Note that, due to the maximality of $\mathcal{M}$, $F(t)$ has zero, one, or two elements among A_i, B_j, and Γ_k. For every triplet $t = (a_i, b_j, c_k) \in T_1$, agent $J_1(t)$ gets item Δ_t, agent $J_2(t)$ gets the items in $F(t)$, if any, and item Θ_t, and agent $J_3(t)$ gets item Z_t.
- For every triplet $t = (a_i, b_j, c_k) \in T_2$, agent $J_1(t)$ gets item Δ_t, agent $J_2(t)$ gets the items in $F(t)$, if any, and agent $J_3(t)$ gets item Z_t.
- Agent 3 gets item Θ_t for every triplet $t \in \mathcal{M} \cup T_2$.
- Agent 2 gets item Λ.
- Agent 1 gets no items.

We claim that the allocation above is envy-freeable by assigning a subsidy of χ to agent 1 and a subsidy of χ to agent $J_2(t)$ for every triplet $t \in T_2$ (if any).

Indeed, agent 1 has positive value only for item Λ, which is given to agent 2, who gets no subsidy. Also, no other agent gets a subsidy more than the subsidy χ that is given to agent 1. Hence, agent 1 is not envious. Agent 2 gets item Λ, which is the only item she values positively and much higher than the subsidy given to any other agent. Hence, agent 2 is not envious either. Agent 3 gets exactly $\max\{K, L\}$ items of total value $\chi \max\{K, L\}$. She does not envy agent 2 who gets item Λ (which agent 3 values for χK) since no subsidy is given to agent 2. Clearly, the value of agent 3 is much higher than the subsidy given to any other agent.

Consider a triplet $t = (a_i, b_j, c_k) \in \mathcal{M}$. Agent $J_1(t)$ has a value of 3χ for the items A_i, B_j, and Γ_k she gets. The remaining items for which she has positive valuation of 3χ have been given to agents $J_2(t)$ and $J_3(t)$, respectively. Since these agents do not get subsidies, agent $J_1(t)$ is not envious of them. Clearly, agent $J_1(t)$ is not envious of any other agent since she has zero value for all other items and no agent gets a subsidy more than χ. Agent $J_3(t)$ gets item Z_t, the only item for which she has positive value and does not envy any other agent since no one gets a subsidy higher than χ. Agent $J_2(t)$ gets a value of χ from item Δ_t and does not envy agent $J_3(t)$, who gets item Z_t, or agent 3, who gets item Θ_t, as these agents receive no subsidy. Clearly, agent $J_2(t)$ envies no other agent.

Now consider a triplet $t = (a_i, b_j, c_k) \notin \mathcal{M}$. Agent $J_1(t)$ has a value of 3χ for the item Δ_t she gets. The remaining items for which she has positive valuation have been allocated as follows. Item Z_t has been given to agent $J_3(t)$; clearly, agent $J_1(t)$ is not envious of $J_3(t)$ since the latter gets no subsidies. The items in $F(t)$ have been given to agent $J_2(t)$. Again, agent $J_1(t)$ is not envious of $J_2(t)$

since $F(t)$ contains at most two items (which agent $J_1(t)$ values for χ each) and agent $J_2(t)$ gets a subsidy of zero (if $t \in T_1$) or χ (if $i \in T_2$). Clearly, $J_1(t)$ does not envy any other agent. Agent $J_3(t)$ gets item Z_t, the only item for which she has positive value and does not envy any other agent since no one gets a subsidy higher than χ. Agent $J_2(t)$ gets a value of χ, either from item Θ_t (if $t \in T_1$) or as subsidy (if $t \in T_2$), and does not envy agent $J_1(t)$ who gets item Δ_t or agent 3 who gets item Θ_t only when $t \in T_2$; recall that these two agents never get subsidies. Again, agent $J_2(t)$ envies no other agent.

4.4 Adapting the Proof for the Case $\chi = 0$

The modification required in our reduction so that it covers the case $\chi = 0$ as well is to remove agent 1 and replace χ with 1 in the definition of valuations. In particular, the agents $J_1(t)$, $J_2(t)$, and $J_3(t)$ that correspond to the triplet $t = (a_i, b_j, c_k)$ have valuations 0 for all items besides the items A_i, B_j, Γ_k, Δ_t, Z_t, and Θ_t. Agent 2 has valuation 0 for all items besides item Λ and agent 3 has valuation zero for all items besides item Λ and items Θ_t for $t \in T$. The remaining valuations are now as follows:

	A_i	B_j	Γ_k	Δ_t	Z_t	Θ_t	Λ
2	0	0	0	0	0	0	K
3	0	0	0	0	0	1	K
$J_1(t)$	1	1	1	3	3	0	0
$J_2(t)$	0	0	0	1	1	1	0
$J_3(t)$	0	0	0	0	1	0	0

The same reasoning as in our proof for the case $\chi \neq 0$ gives a minimum amount of subsidies for the SMEF instance of exactly $\max\{K - L, 0\}$, where L is the maximum 3D matching size in the MAX-3DM-2 instance. In this way, we get that SMEF is NP-hard to approximate within $0.01n$ (i.e., it is NP-hard to distinguish between envy-free instances and instances that need subsidies of $0.01n$) and the construction satisfies sum $v < 30n$. This yields the desired inapproximability result in the statement of Theorem 3 for the case $\chi = 0$ as well.

5   Concluding Remarks

We have initiated the study of the optimization problem SMEF. The challenging open problem that deserves investigation is to close the gap between the trivial approximation guarantee of $n - 1$ in Sect. 2 and our negative result for super-constant numbers of agents in Sect. 4. Unfortunately, more sophisticated existing algorithms, such as the recent one by Brustle et al. [7], do not lead to better approximations.

We remark that $\max v$ could be used alternatively to sum in the definition of the approximation guarantees of SMEF. Actually, the guarantee for our dynamic programming algorithm is stated in terms of $\max v$. We can express the rest of our results using $\max v$ as well. First, the trivial algorithm presented at the end of Sect. 2 uses an amount of $\chi + (n-1)\max v$ as subsidies. Second, an adaptation of the current proof of the inapproximability result can easily give that approximating SMEF within an additive term of $c \cdot \max v$ for a constant c is NP-hard. The important observation is that $\max v < n\chi$ (or $\max v < n$ when $\chi = 0$) in our construction. Then, distinguishing between SMEF instances in which the minimum amount is at most χ and at least $\chi(1 + 0.01n)$ (or at least $0.01n$ when $\chi = 0$) requires to distinguish between SMEF instances in which the minimum amount is at most χ and at least $\chi + 0.01\max v$. So, the inapproximability constant is a bit higher in this case. The main advantage of adopting sum is that it makes the problem of computing the tight approximation factor more challenging.

Interestingly, an advantage of the trivial algorithm is that the particular payments incentivize the agents to report their valuations truthfully. What is the best possible approximation guarantee that can be obtained for SMEF by truthful algorithms? Unfortunately, a simple application of Myerson's characterization in single-item settings [24] indicates that no approximation guarantee better than $n-1$ is possible. Indeed, consider instances with a single item. By the characterization of envy-freeable allocations by Halpern and Shah [18] (i.e., the second statement in Theorem 1), we know that the agent with the highest valuation should get the item. Then, Myerson's characterization for truthful mechanisms in single parameters environments and our requirement for non-negative payments give us the specific form payments should have so that truthful reporting is a dominant strategy for all agents when this algorithm is used: if the agent i who gets the item receives payment of $p \geq 0$, agent t should get a payment of exactly $p + v_i - v_t$, where v_i and v_t are the payments of agents i and t. Now, consider specifically the instance in which one agent has value 1 for the item, and all other agents have value 0. Truthfulness requires (at least) a unit of subsidy to each agent that does not get the item (i.e., total subsidies of $n-1$ while sum $= 1$), even though there is clearly an allocation that is envy-free without any payments. This yields the claimed lower bound of $n-1$ in the approximation guarantee.

References

1. Alkan, A., Demange, G., Gale, D.: Fair allocation of indivisible goods and criteria of justice. Econometrica **59**(4), 1023–1039 (1991)
2. Amanatidis, G., Markakis, E., Ntokos, A.: Multiple birds with one stone: beating 1/2 for EFX and GMMS via envy cycle elimination. Theoret. Comput. Sci. **841**, 94–109 (2020)
3. Aragones, E.: A derivation of the money Rawlsian solution. Soc. Choice Welf. **12**(3), 267–276 (1995). https://doi.org/10.1007/BF00179981
4. Aziz, H., Bouveret, S., Caragiannis, I., Giagkousi, I., Lang, J.: Knowledge, fairness, and social constraints. In: Proceedings of the 32nd AAAI Conference on Artificial Intelligence (AAAI), pp. 4638–4645 (2018)

5. Aziz, H., Caragiannis, I., Igarashi, A., Walsh, T.: Fair allocation of indivisible goods and chores. In: Proceedings of the 28th International Joint Conference on Artificial Intelligence (IJCAI), pp. 53–59 (2019)
6. Barman, S., Krishnamurthy, S.K., Vaish, R.: Finding fair and efficient allocations. In: Proceedings of the 19th ACM Conference on Economics and Computation (EC), pp. 557–574 (2018)
7. Brustle, J., Dippel, J., Narayan, V.V., Suzuki, M., Vetta, A.: One dollar each eliminates envy. In: Proceedings of the 21st ACM Conference on Economics and Computation (EC), pp. 23–39 (2020)
8. Budish, E.: The combinatorial assignment problem: approximate competitive equilibrium from equal incomes. J. Polit. Econ. **119**(6), 1061–1103 (2011)
9. Caragiannis, I., Gravin, N., Huang, X.: Envy-freeness up to any item with high Nash welfare: the virtue of donating items. In: Proceedings of the 20th ACM Conference on Economics and Computation (EC), pp. 527–545 (2019)
10. Caragiannis, I., Kurokawa, D., Moulin, H., Procaccia, A.D., Shah, N., Wang, J.: The unreasonable fairness of maximum nash welfare. ACM Trans. Econ. Comput. **7**(3), 12:1–12:32 (2019)
11. Chaudhury, B.R., Kavitha, T., Mehlhorn, K., Sgouritsa, A.: A little charity guarantees almost envy-freeness. SIAM J. Comput. **50**(4), 1336–1358 (2021)
12. Chevaleyre, Y., Endriss, U., Estivie, S., Maudet, N.: Reaching envy-free states in distributed negotiation settings. In: Proceedings of the 20th International Joint Conference on Artificial Intelligence (IJCAI), pp. 1239–1244 (2007)
13. Chevaleyre, Y., Endriss, U., Maudet, N.: Distributed fair allocation of indivisible goods. Artif. Intell. **242**, 1–22 (2017)
14. Chlebík, M., Chlebíková, J.: Complexity of approximating bounded variants of optimization problems. Theoret. Comput. Sci. **354**(3), 320–338 (2006)
15. Foley, D.K.: Resource allocation and the public sector. Yale Econ. Essays **7**(1), 45–98 (1967)
16. Garey, M.R., Johnson, D.S.: Computers and Intractability: A Guide to the Theory of NP-Completeness. W. H. Freeman & Co., New York (1979)
17. Haake, C.J., Raith, M.G., Su, F.E.: Bidding for envy-freeness: a procedural approach to n-player fair-division problems. Soc. Choice Welf. **19**(4), 723–749 (2002)
18. Halpern, D., Shah, N.: Fair division with subsidy. In: Fotakis, D., Markakis, E. (eds.) SAGT 2019. LNCS, vol. 11801, pp. 374–389. Springer, Cham (2019). https://doi.org/10.1007/978-3-030-30473-7_25
19. Kann, V.: Maximum bounded 3-dimensional matching is MAXSNP-complete. Inf. Process. Lett. **37**(1), 27–35 (1991)
20. Klijn, F.: An algorithm for envy-free allocations in an economy with indivisible objects and money. Soc. Choice Welf. **17**(2), 201–215 (2000)
21. Lipton, R.J., Markakis, E., Mossel, E., Saberi, A.: On approximately fair allocations of indivisible goods. In: Proceedings of the 5th ACM Conference on Electronic Commerce (EC), pp. 125–131 (2004)
22. Maskin, E.: On the fair allocation of indivisible goods. In: Arrow and the Foundations of the Theory of Economic Policy (Essays in Honor of Kenneth Arrow), pp. 341–349. MacMillan (1987). https://doi.org/10.1007/978-1-349-07357-3_12
23. Meertens, M., Potters, J., Reijnierse, H.: Envy-free and pareto efficient allocations in economies with indivisible goods and money. Math. Soc. Sci. **44**(3), 223–233 (2002)
24. Myerson, R.B.: Optimal auction design. Math. Oper. Res. **6**(1), 58–73 (1981)
25. Plaut, B., Roughgarden, T.: Almost envy-freeness with general valuations. SIAM J. Discret. Math. **34**(2), 1039–1068 (2020)

26. Procaccia, A.D.: An answer to fair division's most enigmatic question: technical perspective. Commun. ACM **63**(4), 118 (2020)
27. Su, F.E.: Rental harmony: Sperner's lemma in fair division. Am. Math. Mon. **106**(10), 930–942 (1999)
28. Svensson, L.G.: Large indivisibles: an analysis with respect to price equilibrium and fairness. Econometrica **51**(4), 939–954 (1983)
29. Varian, H.R.: Equity, envy, and efficiency. J. Econ. Theory **9**(1), 63–91 (1974)
30. Vazirani, V.V.: Approximation Algorithms. Springer, Heidelberg (2001). https:// doi.org/10.1007/978-3-662-04565-7

Abstracts

Asymptotically Optimal Competitive Ratio for Online Allocation of Reusable Resources

Vineet Goyal[1], Garud Iyengar[1], and Rajan Udwani[2(✉)]

[1] Columbia University, New York, NY 10027, USA
{vgoyal,garud}@ieor.columbia.edu
[2] UC Berkeley, Berkeley, CA 94720, USA
rudwani@berkeley.edu

Abstract. We consider the problem of online allocation (matching, budgeted allocations, and assortments) of reusable resources where an adversarial sequence of resource requests is revealed over time and allocated resources are used/rented for a stochastic duration, drawn independently from known resource usage distributions. This problem is a fundamental generalization of well studied models in online resource allocation and assortment optimization. Previously, it was known that the greedy algorithm that simply makes the best decision for each arriving request is 0.5 competitive against clairvoyant benchmark that knows the entire sequence of requests in advance. We give a new algorithm that is $(1-1/e)$ competitive for arbitrary usage distributions and large resource capacities. This is the best possible guarantee for the problem.

Designing the optimal online policy for allocating reusable resources requires a reevaluation of the key trade off between conserving resources for future requests and being greedy. Resources that are currently in use may return "soon" but the time of return and types of future requests are both uncertain. At the heart of our algorithms is a new quantity that factors in the potential of reusability for each resource by (computationally) creating an asymmetry between identical units of the resource. We establish a performance guarantee for our algorithms by constructing a feasible solution to a novel LP *free* system of constraints. More generally, these ideas lead to a principled approach for integrating stochastic and combinatorial elements (such as reusability, customer choice, and budgeted allocations) in online resource allocation problems.

The full version of this paper is available at https://arxiv.org/pdf/2002.02430.pdf.

Keywords: Online resource allocation · Reusable resources · LP free analysis · Optimal competitive ratio

© Springer Nature Switzerland AG 2022
M. Feldman et al. (Eds.): WINE 2021, LNCS 13112, p. 543, 2022.
https://doi.org/10.1007/978-3-030-94676-0

Dynamic Bipartite Matching Market with Arrivals and Departures

Naonori Kakimura[1(✉)] and Donghao Zhu[2(✉)]

[1] Keio University, Yokohama, Japan
`kakimura@math.keio.ac.jp`
[2] Technical University of Munich, Munich, Germany
`donghao.zhu@in.tum.de`

Abstract. In this paper, we study a matching market model on a bipartite network where agents on each side arrive and depart stochastically by a Poisson process. For such a dynamic model, we design a mechanism that decides not only which agents to match, but also when to match them, to minimize the expected number of unmatched agents. The main contribution of this paper is to achieve theoretical bounds on the performance of local mechanisms with different timing properties. We show that an algorithm that waits to thicken the market, called the *Patient* algorithm, is exponentially better than the *Greedy* algorithm, i.e., an algorithm that matches agents greedily. This means that waiting has substantial benefits on maximizing a matching over a bipartite network. We remark that the Patient algorithm requires the planner to identify agents who are about to leave the market, and, under the requirement, the Patient algorithm is shown to be an optimal algorithm. We also show that, without the requirement, the Greedy algorithm is almost optimal. In addition, we consider the *1-sided algorithms* where only an agent on one side can attempt to match. This models a practical matching market such as a freight exchange market and a labor market where only agents on one side can make a decision. For this setting, we prove that the Greedy and Patient algorithms admit the same performance, that is, waiting to thicken the market is not valuable. This conclusion is in contrast to the case where agents on both sides can make a decision and the non-bipartite case by [Akbarpour et al., *Journal of Political Economy*, 2020].

Keywords: Bipartite matching · Markov chain · Online algorithm

The full version is available at https://arxiv.org/abs/2110.10824.

N. Kakimura—Supported by JSPS KAKENHI Grant Numbers JP17K00028, JP18H05291, 20H05795, and 21H03397.
D. Zhu—Supported by the Deutsche Forschungsgemeinschaft (DFG, German Research Foundation) - 277991500/GRK2201.

M. Feldman et al. (Eds.): WINE 2021, LNCS 13112, p. 544, 2022.
https://doi.org/10.1007/978-3-030-94676-0

Static Pricing for Multi-unit Prophet Inequalities (Extended Abstract)

Shuchi Chawla[1], Nikhil R. Devanur[2], and Thodoris Lykouris[3($\boxtimes$)]

[1] The University of Texas at Austin, Austin, USA
shuchi@cs.utexas.edu
[2] Amazon Research, Sunnyvale, USA
[3] Massachusetts Institute of Technology, Cambridge, USA
lykouris@mit.edu

The *prophet inequality* problem constitutes one of the cornerstones of online decision-making. A designer knows a set of n distributions from which random variables are sequentially realized in an arbitrary order. Once a random variable is realized, the designer decides whether to accept it or not; *at most one* realized random variable can be accepted. The objective is to maximize the value of the variable accepted, and the performance of the algorithm is evaluated against the *ex-post* maximum realized. In a beautiful result, Samuel-Cahn showed that a simple static threshold policy achieves the optimal competitive ratio for this problem. Samuel-Cahn's algorithm determines a threshold p such that the probability that there exists a realization exceeding the threshold is exactly $\frac{1}{2}$, and then accepts the first random variable that exceeds the threshold. This algorithm achieves a competitive ratio of $\frac{1}{2}$ against the ex-post optimum; no online algorithm, even one with adaptive thresholds, can obtain better performance.

Over the last few years, many extensions of the basic prophet inequality to more general feasibility constraints have been studied, and tight bounds on the competitive ratio have been established. However, one simple natural extension has largely remained open: where the designer is allowed to accept $k > 1$ random variables for some small value of k. This is called the *multi-unit* prophet inequality. When k is relatively large, then it is known that static threshold policies can achieve a competitive ratio of $1 - \Theta\left(\sqrt{\frac{\log(k)}{k}}\right)$ which goes to 1 as $k \to \infty$, and this ratio is asymptotically tight. However, (for example,) for $k = 2$ or 3, prior to our work, the best known competitive ratio of static thresholds remained $\frac{1}{2}$. Our work addresses this gap by answering the following question: *Can a static threshold policy achieve a better competitive ratio than $\frac{1}{2}$ for small $k = 2, 3, \ldots$?*

We develop an algorithm for finding a static threshold policy for the multi-unit prophet inequality that is sensitive to the supply k. Our algorithm is simple and practical. For any fixed price p, it estimates two statistics: (1) the fraction of items expected to be sold at that price, $\mu_k(p)$, and (2) the probability that not all units will sell out before all the customers have been served, $\delta_k(p)$. We then pick the static price p at which these two quantities are equal: $\mu_k(p) = \delta_k(p)$; this generalizes Samuel-Cahn's algorithm and proof via a min-max approach and shows that the worst-case competitive ratio is attained for Poisson distributions.

© Springer Nature Switzerland AG 2022
M. Feldman et al. (Eds.): WINE 2021, LNCS 13112, pp. 545–546, 2022.
https://doi.org/10.1007/978-3-030-94676-0

The competitive ratio of our policy for $k = 2, \cdots, 5$ is $0.585, 0.630, 0.660$, and 0.682 respectively, and scales as $1 - \Omega(\sqrt{\log k / k})$ for large k.

The full version can be found here: https://arxiv.org/abs/2007.07990.

Fairness Maximization Among Offline Agents in Online-Matching Markets

Will Ma[1], Pan Xu[2(✉)], and Yifan Xu[3]

[1] Columbia University, New York, NY 10027, USA
[2] New Jersey Institute of Technology, Newark, NJ 07102, USA
pxu@njit.edu
[3] Key Lab of CNII, MOE, Southeast University, Nanjing, China
xyf@seu.edu.cn

Abstract. Matching markets involve heterogeneous agents (typically from two parties) who are paired for mutual benefit. During the last decade, matching markets have emerged and grown rapidly through the medium of the Internet. They have evolved into a new format, called *Online Matching Markets* (OMMs), with examples ranging from crowd-sourcing to online recommendations to ridesharing. There are two features distinguishing OMMs from traditional matching markets. One is the dynamic arrival of one side of the market: we refer to these as *online agents* while the rest are *offline agents*. Examples of online and offline agents include keywords (online) and sponsors (offline) in Google Advertising; workers (online) and tasks (offline) in Amazon Mechanical Turk (AMT); riders (online) and drivers (offline when restricted to a short time window) in ridesharing. The second distinguishing feature of OMMs is the real-time decision-making element.

However, studies have shown that the algorithms making decisions in these OMMs leave disparities in the match rates of offline agents. For example, tasks in neighborhoods of low socioeconomic status rarely get matched to gig workers, and drivers of certain races/genders get discriminated against in matchmaking. In this paper, we propose online matching algorithms which optimize for either individual or group-level fairness among offline agents in OMMs. We present two linear-programming (LP) based sampling algorithms, which achieve online competitive ratios at least 0.725 for individual fairness maximization (IFM) and 0.719 for group fairness maximization (GFM), respectively. There are two key ideas helping us break the barrier of $1 - 1/e$. One is *boosting*, which is to adaptively re-distribute all sampling probabilities only among those offline available neighbors for every arriving online agent. The other is *attenuation*, which aims to balance the matching probabilities among offline agents with different mass allocated by the benchmark LP. We conduct extensive numerical experiments and results show that our boosted version of sampling algorithms are not only conceptually easy to implement but also highly effective in practical instances of fairness-maximization-related models.

Here is the arXiv link to the full version: https://arxiv.org/abs/2109.08934.

Keywords: Fairness maximization · Online-Matching Markets

The second author Pan Xu is partially supported by NSF CRII Award IIS-1948157.

© Springer Nature Switzerland AG 2022
M. Feldman et al. (Eds.): WINE 2021, LNCS 13112, p. 547, 2022.
https://doi.org/10.1007/978-3-030-94676-0

Funding Public Projects: A Case for the Nash Product Rule

Florian Brandl[1], Felix Brandt[2], Matthias Greger[2], Dominik Peters[3], Christian Stricker[2], and Warut Suksompong[4(✉)]

[1] Hausdorff Center for Mathematics, Universität Bonn, Bonn, Germany
florian.brandl@uni-bonn.de
[2] Institut für Informatik, Technische Universität München, Munich, Germany
{brandt,matthias.greger,christian.stricker}@tum.de
[3] Department of Computer Science, University of Toronto, Toronto, Canada
dominik@cs.toronto.edu
[4] School of Computing, National University of Singapore, Singapore, Singapore
warut@comp.nus.edu.sg

Abstract. We study a mechanism design problem where a community of agents wishes to fund public projects via voluntary monetary contributions by the community members. This serves as a model for public expenditure without an exogenously available budget, such as participatory budgeting or voluntary tax programs, as well as donor coordination when interpreting charities as public projects and donations as contributions. Our aim is to identify a mutually beneficial distribution of the individual contributions. In the preference aggregation problem that we study, agents with linear utility functions over projects report the amount of their contributions, and the mechanism determines a socially optimal distribution of the money. We identify a specific mechanism—the Nash product rule—which picks the distribution that maximizes the product of the agents' utilities weighted by their contributions. This rule arises naturally from a simple, dynamic procedure. The Nash product rule is Pareto efficient, and we prove that it satisfies attractive incentive properties: it spends each agent's contribution only on projects the agent finds acceptable, and agents are strongly incentivized to participate. We also derive impossibility theorems that show that strengthened versions of these two axioms are incompatible with Pareto efficiency.

Keywords: Public goods provision · Collective decision making · Participation incentives

This material is based on work supported by the Deutsche Forschungsgemeinschaft under grants BR 5969/1-1, BR 2312/11-1, and BR 2312/12-1, by the European Research Council (ERC) under grant number 639945, and by an NUS Start-up Grant. The full version of the paper is available at http://arxiv.org/pdf/2005.07997.pdf.

M. Feldman et al. (Eds.): WINE 2021, LNCS 13112, p. 548, 2022.
https://doi.org/10.1007/978-3-030-94676-0

Screening with Limited Information: The Minimax Theorem and a Geometric Approach

Zhi Chen[1], Zhenyu Hu[2], and Ruiqin Wang[2(✉)]

[1] City University of Hong Kong, Kowloon Tong, Kowloon, Hong Kong SAR
zhi.chen@cityu.edu.hk
[2] National University of Singapore, 21 Lower Kent Ridge Road, 119077 Singapore, Singapore
bizhuz@nus.edu.sg, ruiqin_wang@u.nus.edu

Abstract. A seller seeks a selling mechanism to maximize the worst-case revenue obtained from a buyer whose valuation distribution lies in a certain ambiguity set. For a generic convex ambiguity set, we show via the minimax theorem that strong duality holds between the problem of finding the optimal robust mechanism and a minimax pricing problem where the adversary first chooses a worst-case distribution and then the seller decides the best posted price mechanism. This observation connects prior literature that separately studies the primal (robust mechanism design) and problems related to the dual (*e.g.*, robust pricing, buyer-optimal pricing and personalized pricing). We provide a geometric approach to analytically solving the minimax pricing problem (and the robust pricing problem) for several important ambiguity sets such as the ones with mean and various dispersion measures, and with the Wasserstein metric. The solutions are then used to construct the optimal robust mechanism and to compare with the solutions to the robust pricing problem.

Keywords: Robust mechanism design · Moment condition · Mean-preserving contraction · Wasserstein metric

The full paper can be found at http://ssrn.com/abstract=3940212.

© Springer Nature Switzerland AG 2022
M. Feldman et al. (Eds.): WINE 2021, LNCS 13112, p. 549, 2022.
https://doi.org/10.1007/978-3-030-94676-0

Optimal DSIC Auctions for Correlated Private Values: Ex-Post Vs. Ex-Interim IR

Ido Feldman[1] and Ron Lavi[1,2(✉)]

[1] Technion – Israel Institute of Technology, Haifa, Israel
[2] University of Bath, Bath, UK

Abstract. We study Dominant-Strategy Incentive-Compatible (DSIC) revenue-maximizing auctions ("optimal" auctions) for a single-item and correlated private values. We give tight bounds on the ratio of the revenue of the optimal Ex-Post Individually Rational (EPIR) auction and the revenue of the optimal Ex-Interim Individually Rational (EIIR) auction. This bound is expressed as a non-decreasing function of the expected social welfare of the underlying distribution. In particular, we show a class of distributions on which this ratio cannot be lower bounded by any positive number. Thus, the restriction to EPIR auctions, which has been the de-facto standard in the computer science literature on auctions with correlated values, may significantly reduce the revenue that can be possibly extracted, as the revenue extracted by an EPIR auction might be an arbitrarily small fraction of the revenue extracted by an EIIR auction.

Keywords: Optimal auctions · Correlated private values · Full surplus extraction · The look-ahead auction

This research was partially supported by the ISF-NSFC joint research program (grant No. 2560/17). We thank Ronny Lempel for stimulating discussions.

M. Feldman et al. (Eds.): WINE 2021, LNCS 13112, p. 550, 2022.
https://doi.org/10.1007/978-3-030-94676-0

Throttling Equilibria in Auction Markets

Xi Chen, Christian Kroer, and Rachitesh Kumar[(✉)]

Columbia University, New York, NY 10027, USA
{xc2198,christian.kroer,rk3068}@columbia.edu

Abstract. Throttling is a popular method of budget management for online ad auctions in which the platform modulates the participation probability of an advertiser in order to smoothly spend her budget across many auctions. In this work, we investigate the setting in which all of the advertisers simultaneously employ throttling to manage their budgets, and we do so for both first-price and second-price auctions. We analyze the structural and computational properties of the resulting equilibria. For first-price auctions, we show that a unique equilibrium always exists, is well-behaved and can be computed efficiently via tâtonnement-style decentralized dynamics. In contrast, for second-price auctions, we prove that even though an equilibrium always exists, the problem of finding an equilibrium is PPAD-complete, there can be multiple equilibria, and it is NP-hard to find the revenue maximizing one. Finally, we compare the equilibrium outcomes of throttling to those of multiplicative pacing, which is the other most popular and well-studied method of budget management.

The full paper can be found at https://arxiv.org/abs/2107.10923.

Keywords: Auctions · Budget constraints · Computational advertising · Computational complexity · PPAD

X. Chen—Supported by NSF grants IIS-1838154 and CCF-1703925.

M. Feldman et al. (Eds.): WINE 2021, LNCS 13112, p. 551, 2022.
https://doi.org/10.1007/978-3-030-94676-0

Generalized Nash Equilibrium Problems with Mixed-Integer Variables

Tobias Harks⬤ and Julian Schwarz[✉]⬤

Augsburg University, Universitätsstraße 2, 86159 Augsburg, Germany
julian.schwarz@math.uni-augsburg.de

Abstract. We study generalized Nash equilibrium problems (GNEPs) with non-convex strategy spaces and non-convex cost functions. This general class of games includes the important case of games with mixed-integer variables for which only a few results are known in the literature. We present a new approach to characterize equilibria via a convexification technique using the Nikaido-Isoda function. To any given instance I of the GNEP, we derive a convexified instance I^{conv} and show that every feasible strategy profile for I is an equilibrium if and only if it is an equilibrium for I^{conv} and the convexified cost functions coincide with the initial ones. Based on this general result we identify important classes of GNEPs which allow us to reformulate the equilibrium problem via standard optimization problems.

1. We first define *quasi-linear* GNEPs in which for fixed strategies of the opponent players, the cost function of every player is linear and the convex hull of the respective strategy space is polyhedral. For this game class we reformulate the equilibrium problem for I^{conv} as a standard (non-linear) optimization problem.

2. We then study GNEPs with *joint constraint sets*. We introduce the new class of *projective-closed GNEPs* for which we show that I^{conv} falls into the class of jointly convex GNEPs. As an important application, we show that general GNEPs with shared binary sets $\{0, 1\}^k$ are projective-closed.

3. We demonstrate the applicability of our results by presenting a numerical study regarding the computation of equilibria for a class of quasi-linear and projective-closed GNEPs. It turns out that our characterization of a projective-closed GNEP via a jointly convex GNEP leads to an efficiently solvable reformulation of the original non-convex GNEP.

Keywords: Generalized Nash equilibrium problem · Mixed-integer variables · Nonconvex and discrete strategy space

---

This research was funded by the Deutsche Forschungsgemeinschaft (DFG, German Research Foundation) - HA 8041/4-1.

M. Feldman et al. (Eds.): WINE 2021, LNCS 13112, p. 552, 2022.
https://doi.org/10.1007/978-3-030-94676-0

In Which Matching Markets Do Costly Compatibility Inspections Lead to a Deadlock?

Nicole Immorlica[1], Yash Kanoria[2], and Jiaqi Lu[3(✉)]

[1] Microsoft Research, Redmond, USA
[2] Columbia Business School, New York, USA
[3] The Chinese University of Hong Kong, Shenzhen, China
ujiaqi@cuhk.edu.cn

With the aim of understanding congestion in matching markets, we study a matching market with N women and $M = \alpha N$ men who want to match with each other. An agent pair must perform a costly inspection to verify compatibility prior to matching with each other, and we assume they are willing to perform the inspection only if it is "mutually desirable", i.e., they mutually rank each other as their favorite potential partner who remains under consideration. The inspection and matching process progresses iteratively in the market as matches form (in the case of successful inspections) and incompatibilities are revealed. We ask which large random markets suffer from an information deadlock, i.e., in which markets will a constant fraction of agents get stuck waiting for a mutually desirable inspection to become available. We prove, by building on the machinery of message passing and density evolution from statistical physics, that the existence of an information deadlock is governed by the men-to-women ratio α, the average degree of women (or men) and the probability that an inspection is successful. We find a phase transition between the information deadlock regime and the deadlock-free regime (where a vanishingly small fraction of agents are stuck waiting) and study the dependence of deadlock and its size on market primitives. We find, e.g., that well connected markets suffer from deadlocks, and holding the degree of women fixed there is a deadlock for α below a certain threshold.

A complete version is available at https://papers.ssrn.com/sol3/papers.cfm?abstract_id=3697165.

M. Feldman et al. (Eds.): WINE 2021, LNCS 13112, p. 553, 2022.
https://doi.org/10.1007/978-3-030-94676-0

Contest Design with Threshold Objectives

Edith Elkind, Abheek Ghosh[✉], and Paul W. Goldberg

Department of Computer Science, University of Oxford, Oxford, UK
{edith.elkind,abheek.ghosh,paul.goldberg}@cs.ox.ac.uk

Abstract. We study contests where the designer's objective is an extension of the widely studied objective of maximizing the total output: The designer gets zero marginal utility from a player's output if the output of the player is very low or very high. We consider two variants of this setting, which correspond to two objective functions: *binary threshold*, where a player's contribution to the designer's utility is 1 if her output is above a certain threshold, and 0 otherwise; and *linear threshold*, where a player's contribution is linear in her output if the output is between a lower and an upper threshold, and becomes constant below the lower and above the upper threshold. For both of these objectives, we study (1) *rank-order allocation* contests, which assign prizes based on players' rankings only, and (2) general contests, which may use the numerical values of the players' outputs to assign prizes. We characterize the contests that maximize the designer's objective and indicate techniques to efficiently compute them. We also prove that for the linear threshold objective, a contest that distributes the prize equally among a fixed number of top-ranked players offers a factor-2 approximation to the optimal rank-order allocation contest.

Keywords: Contest theory · Mechanism design · All-pay auctions

Full version is available at https://arxiv.org/abs/2109.03179.

A. Ghosh—Supported by Clarendon Fund and SKP (Pathak) Scholarship.

M. Feldman et al. (Eds.): WINE 2021, LNCS 13112, p. 554, 2022.
https://doi.org/10.1007/978-3-030-94676-0

Confounding Equilibria for Platforms with Private Information on Promotion Value

Yonatan Gur[1], Gregory Macnamara[2], Ilan Morgenstern[1]([✉]),
and Daniela Saban[1]

[1] Stanford University Graduate School of Business, Stanford, CA 94305, USA
{ygur,ilanmor,dsaban}@stanford.edu
[2] Facebook Core Data Science, 1 Hacker Way, Menlo Park, CA 94025, USA

Keywords: Information design · Bayesian learning · Online marketplaces

Extended Abstract

Online marketplaces allow consumers to evaluate, compare, and purchase products while providing a channel for third-party sellers to reach a broad consumer base and increase demand for their products. As platforms seek to maintain a large consumer base, many platforms prioritize increasing consumer surplus by offering competitively priced products. At the same time, it is common practice in such marketplaces to let sellers determine their own price, but such flexibility may result in higher prices that reduce consumer surplus.

In this paper, we consider a platform facilitating trade between sellers and buyers with the objective of maximizing consumer surplus. Even though in many such marketplaces prices are set by revenue-maximizing sellers, platforms can influence prices through (i) price-dependent promotion policies that can increase demand for a product by featuring it in a prominent position on the webpage and (ii) the information revealed to sellers about the value of being promoted. Identifying effective joint information design and promotion policies is a challenging dynamic problem as sellers can sequentially learn the promotion value from sales observations and update prices accordingly. We introduce the notion of *confounding* promotion policies, which are designed to prevent a Bayesian seller from learning the promotion value (at the expense of the short-run loss of diverting consumers from the best product offering). Leveraging these policies, we characterize the maximum long-run average consumer surplus that is achievable through joint information design and promotion policies when the seller sets prices myopically. We then establish that these strategies are supported in a Bayesian Nash equilibrium, by showing that the seller's best response to the platform's optimal policy is to price myopically at every history. Moreover, the equilibrium we identify is platform-optimal within the class of horizon-maximin equilibria, in which strategies are not predicated on precise knowledge of the horizon length, and are designed to maximize payoff over the worst-case horizon. Our analysis allows one to identify practical long-run average optimal platform policies for a broad range of demand models.

© Springer Nature Switzerland AG 2022
M. Feldman et al. (Eds.): WINE 2021, LNCS 13112, p. 555, 2022.
https://doi.org/10.1007/978-3-030-94676-0

Author Index